Carbohydrate-Based Drugs Discovery

Carbohydrate-Based Drugs Discovery

Editors

Jian Yin
Jing Zeng
De-Cai Xiong

MDPI

Basel • Beijing • Wuhan • Barcelona • Belgrade • Novi Sad • Cluj • Manchester

Editors
Jian Yin
Key Laboratory of Carbohydrate Chemistry and Biotechnology, Ministry of Education, School of Biotechnology
Jiangnan University
Wuxi
China

Jing Zeng
Hubei Key Laboratory of Natural Medicinal Chemistry and Resource Evaluation, School of Pharmacy
Huazhong University of Science and Technology
Wuhan
China

De-Cai Xiong
The State Key Laboratory of Natural and Biomimetic Drugs, School of Pharmaceutical Sciences
Peking University
Beijing
China

Editorial Office
MDPI
St. Alban-Anlage 66
4052 Basel, Switzerland

This is a reprint of articles from the Special Issue published online in the open access journal *Molecules* (ISSN 1420-3049) (available at: https://www.mdpi.com/journal/molecules/special_issues/carbo_base_drug).

For citation purposes, cite each article independently as indicated on the article page online and as indicated below:

Lastname, A.A.; Lastname, B.B. Article Title. *Journal Name* **Year**, *Volume Number*, Page Range.

ISBN 978-3-7258-0432-0 (Hbk)
ISBN 978-3-7258-0431-3 (PDF)
doi.org/10.3390/books978-3-7258-0431-3

Contents

Preface

Carbohydrates play an essential role in the organism and have many bioactivities. Carbohydrate-based drug discovery is an up-and-coming area of research in medicinal chemistry. Bioactive carbohydrates open up a new source for drug development. More than 170 carbohydrate-based drugs have been successfully approved as anticoagulants, antitumor agents, antidiabetic agents, antibiotics, antiviral agents, and vaccines. However, most carbohydrates have low druggability. As such, new methods and strategies to improve carbohydrates' druggability are in high demand.

This Special Issue aimed to serve as an open forum where researchers can share their experiences and findings on carbohydrate-based drugs, with a focus on the discovery and development of new carbohydrate-based therapeutic agents. Contributions to this issue, in the form of both original research and review articles, cover the separation, identification, synthesis, and pharmacology of bioactive carbohydrates, including monosaccharides, oligosaccharides, polysaccharides, glycopeptides, glycoproteins, glycomimetics, etc.

Jian Yin, Jing Zeng, and De-Cai Xiong
Editors

molecules

MDPI

Article

Cancer Vaccines Based on Fluorine-Modified KH-1 Elicit Robust Immune Response

Yang Liu [1,†], Bohan Li [1,†], Xiujing Zheng [1], Decai Xiong [1,2,*] and Xinshan Ye [1,*]

1 State Key Laboratory of Natural and Biomimetic Drugs, School of Pharmaceutical Sciences, Peking University, Xue Yuan Rd. No. 38, Beijing 100191, China
2 The NMPA Key Laboratory for Quality Research and Evaluation of Carbohydrate-Based Medicine, Shandong University, 27 Shanda Nanlu, Jinan 250100, China
* Correspondence: decai@bjmu.edu.cn (D.X.); xinshan@bjmu.edu.cn (X.Y.)
† These authors contributed equally to this work.

Abstract: KH-1 is a tumor-associated carbohydrate antigen (TACA), which serves as a valuable target of antitumor vaccines for cancer immunotherapies. However, most TACAs are thymus-independent antigens (TD-Ag), and they tend to induce immunological tolerance, leading to their low immunogenicity. To overcome these problems, some fluorinated derivatives of the KH-1 antigen were designed, synthesized, and conjugated to the carrier protein CRM197 to form glycoconjugates, which were used for immunological studies with Freund's adjuvant. The results showed that fluorine-modified *N*-acyl KH-1 conjugates can induce higher titers of antibodies, especially IgG, which can recognize KH-1-positive cancer cells and can eliminate cancer cells through complement-dependent cytotoxicity (CDC). The trifluoro-modified KH-1-TF-CRM197 showed great potential as an anticancer vaccine candidate.

Keywords: tumor-associated carbohydrate antigens; KH-1; fluoro-modification; tumor vaccine

1. Introduction

Some aberrant glycans are overexpressed on the surface of tumor cells due to the abnormal expression of glycosidases and glycosyltransferases, and they are correlated with tumor cell adhesion and metastasis. These aberrant glycans are called tumor-associated carbohydrate antigens (TACAs). They are important immunotherapeutic targets against cancers [1–3]. Unfortunately, most TACAs are thymus-independent antigens, and they are also expressed by normal cells, leading to their low immunogenicity [4,5]. To enhance the immunogenicity of TACAs, they are usually conjugated with immunogenic carriers, and unnatural TACA analogues are also introduced through the modification of TACAs [6–11].

The KH-1 antigen is one of the TACAs found in human colon cancer cells. It is a branched nonasaccharide glycolipid containing two TACA epitopes (the tetrasaccharide of the Le^y antigen and the trisaccharide of the Le^x antigen) [12–14]. It is barely expressed by normal tissues but highly expressed by adenocarcinoma cells [15]. Compared with the low incidence (5–10%) of oncogenes and the deletion (<50%) of tumor-suppressing gene products, the incidence of the extended Le^y antigens expressed in colorectal cancer is up to about 90% [5]. Therefore, the KH-1 antigen is a potential immunotherapeutic target against adenocarcinoma. Tumor vaccines based on the KH-1 antigen have been proven to induce IgM and IgG antibodies against KH-1 and to show great safety due to low cross-reactivity with the Le^x antigen, which is also expressed by neutrophilic granulocytes [16,17]. However, as a self-antigen, the natural KH-1 antigen regularly leads to immunological tolerance, which is a major challenge in carbohydrate-based vaccine design.

The chemical synthesis of KH-1 is extremely valuable and challenging because it cannot be obtained in large quantities from natural resources and has a complex structure. At present, several research groups have completed the total synthesis of KH-1 through

check for updates

Citation: Liu, Y.; Li, B.; Zheng, X.; Xiong, D.; Ye, X. Cancer Vaccines Based on Fluorine-Modified KH-1 Elicit Robust Immune Response. *Molecules* **2023**, *28*, 1934. https://doi.org/10.3390/molecules28041934

Academic Editor: Iva Pashkuleva

Received: 12 January 2023
Revised: 12 February 2023
Accepted: 15 February 2023
Published: 17 February 2023

different strategies, in which Guo and our group have achieved relatively high synthetic yields due to the use of a one-pot glycosylation protocol [18–22]. Reliable and efficient synthesis provides support for the study of biological activities.

As fluorine has a similar atomic radius and lipophilicity to hydrogen and is a trace element for humans, fluoro-modifications of TACAs have been proven to increase the antitumor immunogenicity of TACAs. Previously, our group conducted a systematic study of fluorine-modified TACAs, such as STn and Globo H [23–26]. Compared with unmodified glycoconjugates, fluorine-modified glycoconjugates induced higher titers of IgG antibodies. Fluoro-modifications of TACAs can enhance the CDC effect of antisera on tumor cells. Based on our previous study, we hope to improve the immunogenicity of KH-1 through fluorinated modifications. Herein, we want to report the design and synthesis of these fluorine-modified KH-1 derivatives (Figure 1) and evaluate the immunogenicity of their corresponding glycoconjugates.

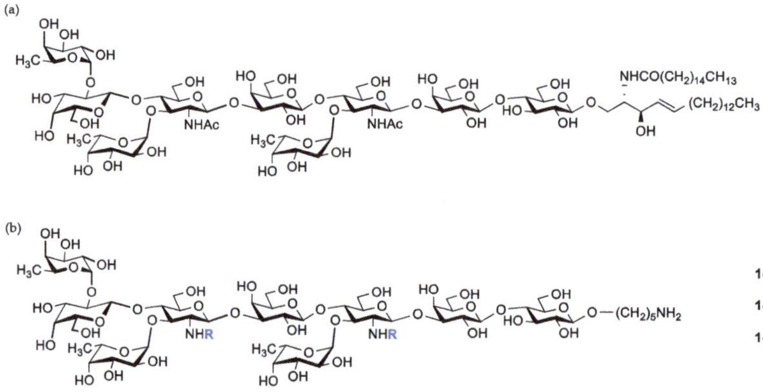

1-1 R = $COCH_3$

1-2 R = $COCF_3$

1-3 R = $COCHF_2$

Figure 1. (**a**) Native KH-1 antigen; (**b**) fluorine-modified and unmodified KH-1 derivatives in this work.

2. Results and Discussion

2.1. Synthesis of KH-1 Antigen and Its Fluorinated Derivatives

As described in our previous report [27], three disaccharide building blocks **4**, **5**, and **6** and one fucose building block **7** were prepared through dozens of steps. Next, the combination of photo-induced strategy [28,29] with a preactivation-based one-pot glycosylation protocol [30] was employed to obtain a linear hexasaccharide **3-1**, which was subjected to partial deprotection to yield hexasaccharide **3-2**. After screening the reaction conditions, the photo-induced fucosylation was finally realized, providing the KH-1 core nonasaccharide **2** in high yield (Scheme 1) [27].

The global deprotection of **2** in four successive steps was carried out to afford compound **1-1** smoothly. In order to generate the fluorinated derivatives, a series of conditions were attempted for the fluoroacetylation step. For the trifluoroacetylation reaction, the reaction conditions were assessed (Table S1), and the optimized conditions were as follows: methyl trifluoroacetate as the fluorinated reagent, 4-dimethylaminopyridine (DMAP) as the base, dry methanol as the solvent, and under room temperature for 48 h. Using similar fluoroacetylation conditions and deprotection steps, the fluorinated compounds **1-2** and **1-3** were prepared (Scheme 2). We also sought to obtain the monofluorinated derivative, but unfortunately, it was not successful due to the restrictions in reagent purchase and preparation.

2.2. Synthesis of Glycoconjugates

Diphtheria toxoid cross-reactive material 197 (CRM197) is a vaccine carrier approved by the FDA [31]. Globo H-CRM197 has been proven to induce higher titers of the IgG antibody than Globo H-KLH [32]. Moreover, it is effortless for keyhole limpet hemocyanin (KLH) to form polymers, which may disturb the quality control of vaccines [33]. Therefore, CRM197 was chosen as the carrier of the KH-1 antigen. The three nonasaccharide molecules mentioned above were covalently linked to the CRM197 protein and the BSA protein via a known linker [34], respectively (Scheme 3). Finally, six glycoconjugates were obtained. The carbohydrate loading levels were determined using the mass spectra analysis (Table 1 and Figures S1–S8).

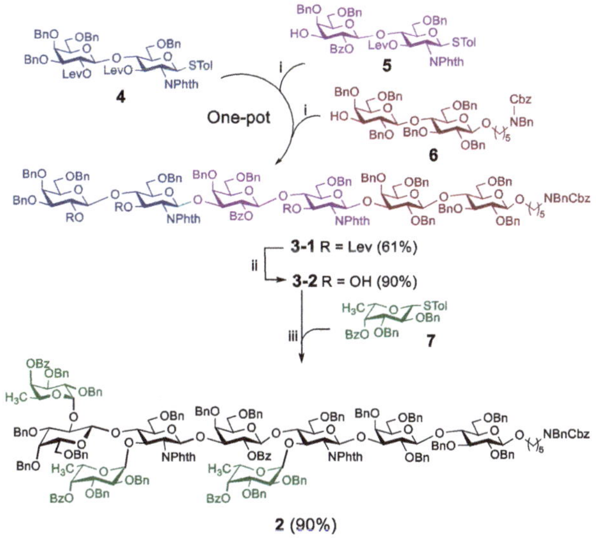

Scheme 1. Photo-induced synthesis of the fully protected KH-1 core nonasaccharide: (**i**) UV, 4 Å molecular sieves, Umemoto's reagent, $Cu(OTf)_2$, TTBP, −72 °C, 20 min; then acceptor, −72 °C to rt, 1.5 h; (**ii**) $H_2NNH_2{\cdot}HOAc$, CH_3OH/THF (1:10); (**iii**) UV, 4 Å molecular sieves, Umemoto's reagent, $Cu(OTf)_2$, TTBP, −50 °C, 4 min; then acceptor, −50 °C to rt, 2.5 h.

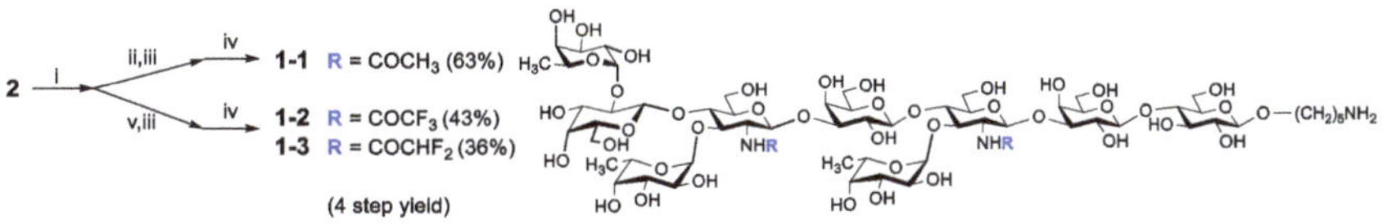

Scheme 2. Deprotection and synthesis of fluorinated derivatives: (**i**) $H_2NCH_2CH_2NH_2$/n-BuOH (1:3), 120 °C; (**ii**) Ac_2O/MeOH; (**iii**) LiOH, MeOH/THF (4:1), 80 °C; (**iv**) $Pd(OH)_2$, H_2, CH_2Cl_2/MeOH/H_2O (3:3:1); (**v**) DMAP, $ROCH_3$/MeOH.

2.3. Evaluation of the Antibodies

With the glycoconjugates in hand, the immunological properties of these conjugates were evaluated in mice. Groups of six female BALB/c mice were vaccinated four times at biweekly intervals with the CRM197 conjugates using Freund's adjuvant. After immunization, antisera were obtained and used for KH-1 antibody detection. Anti-KH-1 or anti-modified-KH-1 titers were firstly evaluated using ELISA (enzyme-linked immunosorbent assay) plates coated with KH-1-BSA or modified KH-1-BSA, using the pooled antisera of all immunized mice after the third or fourth vaccination (Table 2). We found that KH-1-DF-CRM197 and KH-1-TF-CRM197 induced a stronger KH-1-specific antibody response and elicited higher titers of anti-KH-1 IgG and IgM antibodies than KH-1-CRM197 after the third and fourth vaccination. The sera from preimmunized mice showed no titer of the IgG or IgM antibodies against the KH-1 antigen. The IgG/IgM ratio of KH-1-TF-CRM197 was particularly increased. For tumor immunotherapy, the IgG response is more functional than IgM on account of their properties, such as affinity maturation and immunological memory [35,36]. In addition, KH-1-DF-CRM197 and KH-1-TF-CRM197 elicited higher titers of anti-modified-KH-1 IgG antibodies than KH-1-CRM197, which again proved that fluoro-modifications of KH-1 can increase its antibody response.

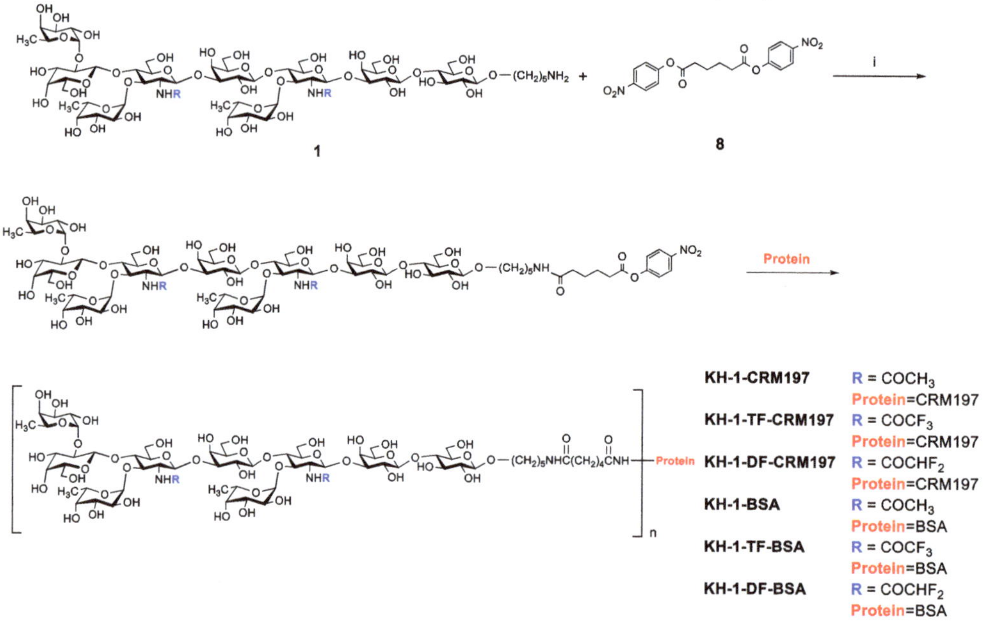

Scheme 3. Synthesis of glycoconjugates: (**i**) DMF, $(C_2H_5)_3N$.

Table 1. Carbohydrate loading levels of glycoconjugates.

Conjugate	Molecular Weight of Glycoconjugate	Total Molecular Weight of Carbohydrate on Conjugate	Molecular Weight of Single Nonasaccharide	Concentration of Carbohydrate (μg/mL)	Concentration of Protein (μg/mL)	Number of Nonasaccharide	Percentage of Carbohydrate Loading
KH-1-CRM197	67,285	8968	1596	213	1600	5.62	13.32%
KH-1-TF-CRM197	67,908	9591	1704	165	1167	5.63	14.12%
KH-1-DF-CRM197	70,134	11,817	1668	195	1145	7.08	16.85%
KH-1-BSA	80,787	14,350	1596	206	1156	8.99	17.76%
KH-1-TF-BSA	83,936	17,499	1704	221	1057	10.27	20.85%
KH-1-DF-BSA	79,742	13,305	1668	197	1079	7.98	16.69%

The anti-KH-1 titers of IgG and IgM for an individual mouse after the third and fourth vaccination were also detected using ELISA (Figure 2). The IgG and IgM titers of modified KH-1 conjugates, especially KH-1-TF-CRM197, were higher than those of KH-1-CRM197. This demonstrated that the modified KH-1 conjugates, especially KH-1-TF-CRM197, produced higher titers of anti-modified-KH-1 antibodies with cross-reactive efficiency for KH-1, resulting in an increase in the anti-KH-1 IgG and IgM antibodies.

The IgG subtypes in antisera after the third and fourth vaccinations were also detected with ELISA (Figure 3). Immunized with Freund's adjuvant, KH-1-TF-CRM197 elicited significantly more anti-KH-1 IgG1, IgG2a, IgG2b, and IgG3 antibodies than the unmodified KH-1-KLH. These results indicated that a mixed Th1/Th2 response was elicited by KH-1-TF-CRM197 because IgG1 antibodies are usually generated via the Th2 response, while activated Th1 cells may generate a prominent IgG2a response [37].

Table 2. Titers of the antibodies elicited by vaccines.

ELISA Titer Anti-KH-1 and Ratio of IgG/IgM after the Third and the Fourth Vaccination							ELISA Titer Anti-Modified-KH-1
Vaccine	After the Third Vaccination			After the Fourth Vaccination			After the Fourth Vaccination
	IgG	IgM	IgG/IgM	IgG	IgM	IgG/IgM	IgG*
KH-1-CRM197/FA	<100	242	-	45,268	8952	5.06	45,268
KH-1-DF-CRM197/FA	11,686	4480	2.61	101,447	20,437	4.96	216,853
KH-1-TF-CRM197/FA	848,249	6307	134.49	1,180,228	9670	122.05	2,259,394

IgG is the IgG titer of the mixed serum from 6 mice of the same group against KH-1-BSA, whereas IgG* is the titer against the corresponding modified KH-1-BSA. Mouse serum IgG obtained from the preimmune condition could not be detected when diluted 1:100.

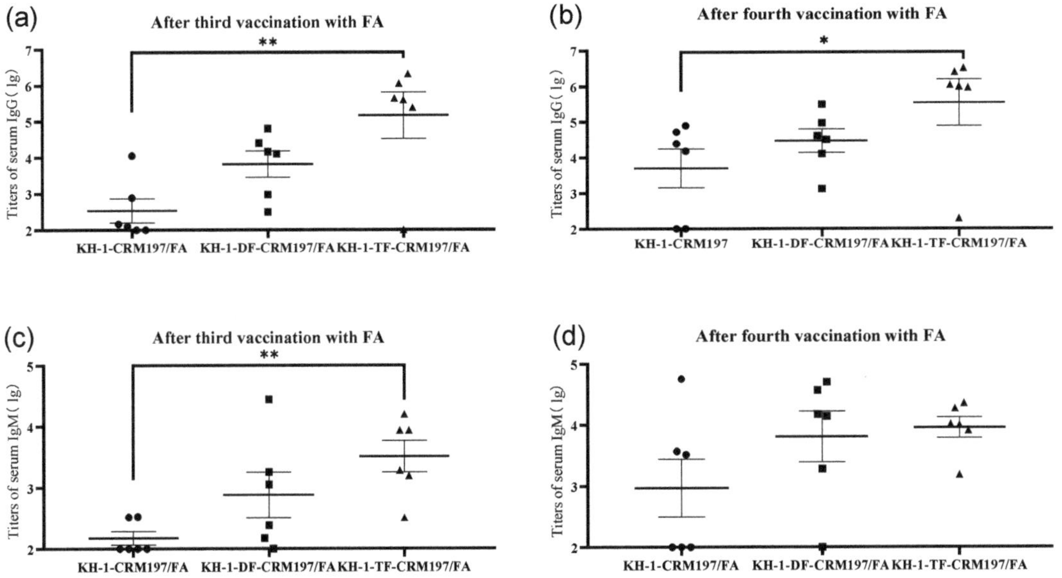

Figure 2. Serum antibody titers against KH-1-BSA in an individual mouse vaccinated with KH-1-CRM197, KH-1-DF-CRM197, or KH-1-TF-CRM197 after the 3rd (**a**,**c**) and 4th (**b**,**d**) vaccinations. Each data point represents the average titer of two detections in an individual mouse, and the black line in each series represents the mean titer of six mice in each group. The data of IgG titers were plotted in the form of log 10. * $p < 0.05$, ** $p < 0.01$.

2.4. Recognition of KH-1-Positive Tumor Cells with IgG Antibodies in Antisera

Flow cytometry was used to evaluate the capabilities of antisera induced by KH-1-CRM197, KH-1-DF-CRM197, or KH-1-TF-CRM197 to recognize the native KH-1 antigens of cancer cells (Figure 4). KH-1-positive MCF-7 human breast cancer cells were used as target cells, and the preimmune serum was used as a negative control [38]. The results showed that the IgG antibodies elicited by the three glycoconjugates could bind well with MCF-7 cells on which KH-1 antigens were overexpressed. This indicated a great cross-reactivity of antisera elicited by the fluorine-modified KH-1-CRM197 to the native KH-1 antigen on tumor cells, which is the foundation of the antitumor activities of fluorine-modified KH-1-CRM197. Furthermore, the IgG antibody induced by fluorine-modified glycoconjugates (KH-1-DF-CRM197 and KH-1-TF-CRM197) showed a stronger binding affinity to MCF-7 cells than the unmodified KH-1-CRM197, which was consistent with the trend of the IgG titer results.

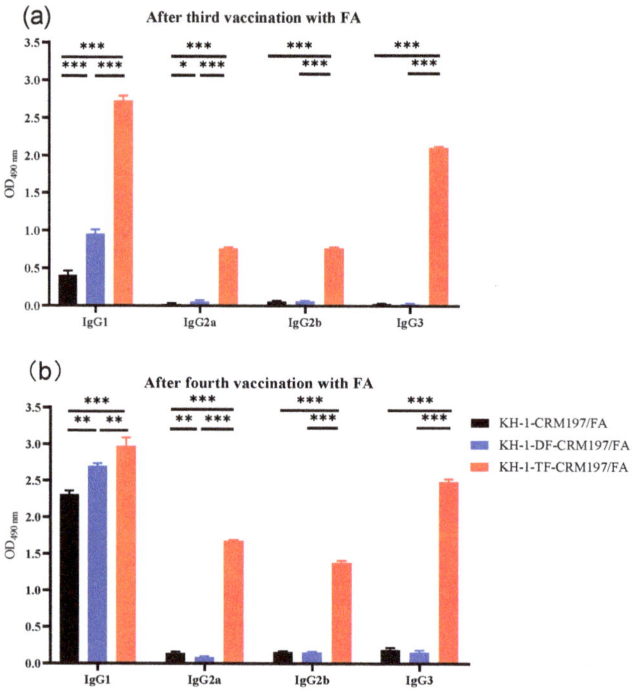

Figure 3. IgG subtypes after the third (**a**) and fourth (**b**) vaccinations with KH-1-KLH, KH-1-DF-KLH, and KH-1-TF-KLH with Freund's adjuvant using ELISA with a 1:1000 dilution of pooled sera. Results represent the mean ± SEM of three independent experiments. One-way ANOVA analysis was performed. * $p < 0.05$, ** $p < 0.01$, *** $p < 0.01$.

2.5. Complement-Dependent Cytotoxicity of Antisera Induced by Vaccines

According to previous reports [24,39,40], complement-dependent cytotoxicity (CDC) is a potent mechanism of cell killing. Therefore, CDC was studied to determine whether the immune response provoked by the fluorine-modified KH-1-CRM197 is useful for cancer immunotherapy. We found that the antisera elicited by the three glycoconjugates after the fourth vaccination were able to increase the lysis of KH-1-positive MCF-7 cells compared with that in the preimmune group (Figure 5). Fluoro-modifications on KH-1 can provoke the CDC of antisera against tumor cells. This might be associated with the fact that the fluorine-modified KH-1-CRM197 produced more IgG3 and IgM antibodies against the KH-1 antigen, which mainly acts via the complement [41].

Taken together, these results reveal that fluoro-modifications of the KH-1 antigen can improve its immunogenicity. This might be attributed to the increased enzymatic stability, enhanced affinity to T-cell receptor (TCR), and exogenous properties of fluoride antigens [11].

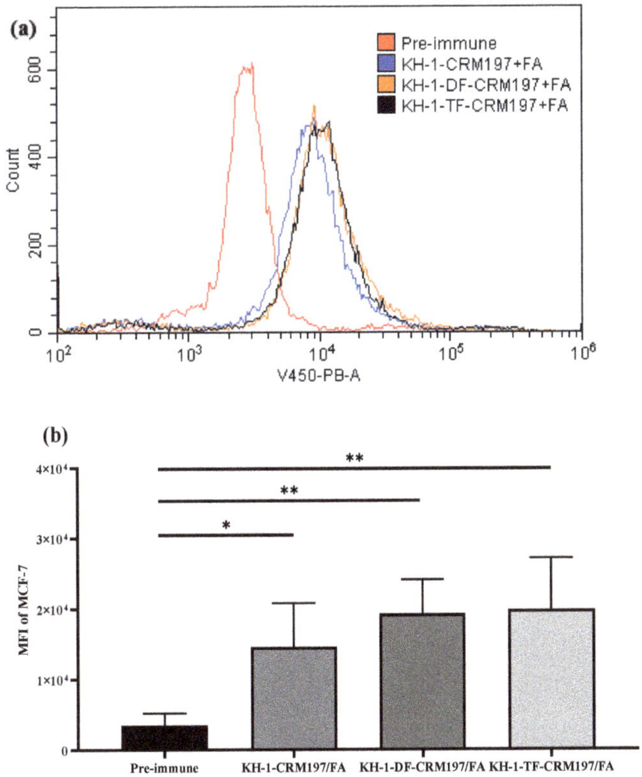

Figure 4. Recognition of human tumor cells by IgG antibodies in antisera: (**a**) serological IgG analysis results on MCF-7 human breast cancer cells after the fourth immunization via flow cytometry; (**b**) mean fluorescence intensity (MFI) of the IgG antibody's binding to MCF-7 cells. Results are representative of three independent experiments. One-way ANOVA analysis was performed. * $p < 0.05$, ** $p < 0.01$.

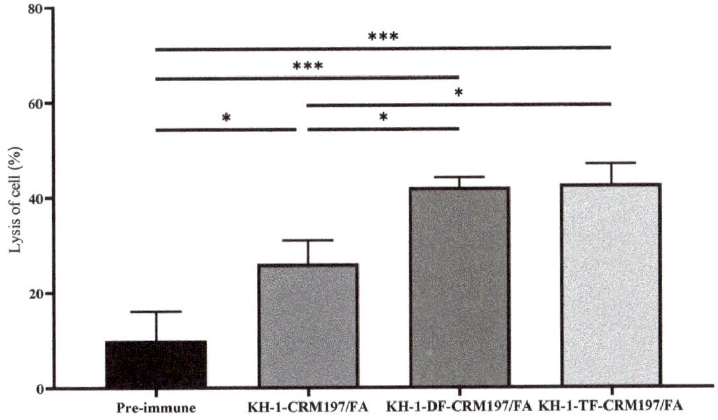

Figure 5. Antibodies elicited by glycoconjugates mediate complement-dependent cytotoxicity (CDC) to eliminate KH-1-positive MCF-7 cancer cells. Cell lysis of pooled sera (1:20 dilution) after the 4th vaccination. Data are the mean ± SEM of three independent experiments. Cytotoxicity is shown as percentage cell lysis determined by the lactate dehydrogenase (LDH) assay. Results are representative of three independent experiments. One-way ANOVA analysis was performed. * $p < 0.05$, *** $p < 0.01$.

3. Materials and Methods

3.1. *Chemical Synthesis*

The reactions were carried out in oven-dried glassware. All reactions were carried out under anhydrous conditions with freshly distilled solvents under a positive pressure of argon unless otherwise noted. Umemoto's reagent refers to *S*-(trifluoromethyl)dibenzothiophenium tetrafluoroborate. The reactions were monitored using thin-layer chromatography (TLC) on silica-gel-coated aluminum plates (60 F_{254}, E. Merck). Column chromatography was performed on silica gel (200–300 mesh). Gel filtration was performed on Sephadex LH-20 (Pharmacia). Optical rotations were obtained on a Hanon P850 Automatic Polarimeter. Experiments under ultraviolet (UV) irradiation were carried out using a safe and stable mercury lamp spotlight system purchased from Beijing Zhongjiao Jinyuan Technology Co., Ltd. (Item No. CEL-M500). ^{1}H NMR spectra were recorded at room temperature for solutions in $CDCl_3$ or D_2O with Avance III-400 or III-600 instruments (Bruker), and the chemical shifts were referenced to the peak for TMS (0 ppm, $CDCl_3$) or external CH_3OH (3.34 ppm, D_2O). ^{13}C NMR spectra were recorded using the same NMR spectrometers, and the chemical shifts were reported relative to the internal $CDCl_3$ (δ = 77.16 ppm) or external CH_3OH (49.70 ppm, D_2O). High-resolution mass spectrometry (HRMS) experiments were performed on a Waters Xevo G2 Q-TOF spectrometer or a Bruker APEX IV FTMS instrument.

3.1.1. Synthesis of Compound **1-1**

Compound **2** [27] (40.0 mg, 9.7 μmol) was dissolved in $H_2NCH_2CH_2NH_2$/*n*-BuOH (*v*/*v* = 1:3, 20 mL). The solution was heated at 120 °C for 6 h. The solution was concentrated under reduced pressure. The residue was dissolved in MeOH (5.0 mL) and then Ac_2O (1.0 mL) was added, and the reaction was performed at room temperature for 3 h. The solution was concentrated under reduced pressure. The residue was dissolved in THF/MeOH (*v*/*v* = 1:4, 5 mL), and then LiOH (50.0 mg, 1.2 mmol) was added. The mixture was heated at 80 °C overnight. H^+ resin was used for neutralization. The mixture was filtered, and the filtrate was concentrated under reduced pressure. The crude product was purified via column chromatography on silica gel (toluene: CH_3CN = 6:1, *v*/*v*) to afford white solids. The obtained product was dissolved in $CH_2Cl_2/MeOH/H_2O$ (*v*/*v*, 3:3:1, 7 mL), and then $Pd(OH)_2$ (10%, 15.0 mg) was added. The reaction was carried out under a hydrogen atmosphere of 40 psi for 30 h. The mixture was filtered, and the filtrate was concentrated under reduced pressure. The residue was purified using column chromatography on gel LH20 (H_2O: MeOH = 15:1, *v*/*v*) to afford **1-1** as white solids (10.0 mg, 63% yield).

^{1}H NMR (400 MHz, D_2O) δ 5.19 (d, *J* = 3.1 Hz, 1H), 5.03 (t, *J* = 3.6 Hz, 2H), 4.83–4.76 (m, 1H), 4.62 (d, *J* = 8.0 Hz, 2H), 4.45–4.31 (m, 4H), 4.19–4.11 (m, 1H), 4.07 (d, *J* = 3.3 Hz, 1H), 4.00 (d, *J* = 3.3 Hz, 1H), 3.95–3.30 (m, 46H), 3.24–3.16 (m, 1H), 2.92 (d, *J* = 7.5 Hz, 2H), 1.93 (s, 6H), 1.65–1.54 (m, 4H), 1.43–1.30 (m, 2H), 1.17 (d, *J* = 6.6 Hz, 3H), 1.15 (d, *J* = 6.7 Hz, 3H), 1.06 (d, *J* = 6.5 Hz, 3H). ^{13}C NMR (151 MHz, D_2O) δ 174.70, 174.66, 102.95, 102.50, 102.42, 101.99, 101.72, 100.20, 99.43, 98.69, 98.58, 82.08, 81.55, 78.40, 76.38, 75.32, 75.11, 74.86, 74.77, 74.45, 73.56, 73.04, 72.80, 71.94, 71.85, 71.70, 70.56, 70.08, 69.94, 69.73, 69.17, 68.75, 68.28, 68.21, 67.70, 67.65, 66.92, 66.79, 66.69, 61.47, 61.41, 61.27, 61.11, 60.95, 60.08, 59.78, 59.65, 48.86, 39.36, 28.14, 26.43, 22.24, 22.07, 15.45, 15.43, 15.26. HRMS (ESI): $[M + H]^+$ $C_{63}H_{110}N_3O_{43}{}^+$ *m*/*z* calcd. 1596.65, found 1596.64. The data are consistent with those reported previously [27].

3.1.2. Synthesis of Compound **1-2**

Compound **2** (40.0 mg, 9.7 μmol) was dissolved in $H_2NCH_2CH_2NH_2$/*n*-BuOH (*v*/*v* = 1:3, 20 mL). The solution was heated at 120 °C for 6 h. The solution was concentrated under reduced pressure. The residue was dissolved in THF/MeOH (*v*/*v* = 1:4, 5 mL), and then LiOH (50.0 mg, 1.2 mmol) was added. The solution was heated at 80 °C overnight. The solution was concentrated under reduced pressure. Then, the residue was dissolved in CH_2Cl_2 and washed with water, and then it was concentrated under reduced pressure and further dried under vacuum.

The residue was dissolved in MeOH (5.0 mL), and the reaction mixture was stirred for 10 min. 4-Dimethylaminopyridine (600 mg) was dissolved in MeOH (5.0 mL), stirred for 10 min, and then added to the reaction mixture. The reaction mixture was stirred for 20 min at room temperature and then cooled to 0 °C for 10 min. CF_3COOCH_3 (16 mL in total) was added 4 times every 5 min, and it was stirred at room temperature for 48 h. The mixture was concentrated under reduced pressure and further dried under vacuum.

The crude product was purified via column chromatography on silica gel (toluene: CH_3CN = 6:1, *v/v*) to afford white solids. The obtained product was dissolved in CH_2Cl_2/MeOH/H_2O (*v/v*, 3:3:1, 7 mL), and then $Pd(OH)_2$ (10%, 15.0 mg) was added. The reaction was carried out under a hydrogen atmosphere of 40 psi for 30 h. The mixture was filtered, and the filtrate was concentrated under reduced pressure. The residue was purified using column chromatography on gel LH20 (H_2O: MeOH = 15:1, *v/v*) to afford **1-2** as white solids (6.4 mg, 43% yield).

^{1}H NMR (400 MHz, D_2O) δ 5.16 (d, *J* = 3.1 Hz, 1H), 4.91 (t, *J* = 3.4 Hz, 2H), 4.43–4.28 (m, 4H), 4.16–4.13 (m, 1H), 4.05 (d, *J* = 2.8 Hz, 1H), 4.01–3.34 (m, 47H), 3.20–3.14 (m, 1H), 2.88 (t, *J* = 7.5 Hz, 2H), 1.56 (q, *J* = 7.8 Hz, 4H), 1.39–1.29 (m, 2H), 1.15 (d, *J* = 6.5 Hz, 3H), 1.12 (d, *J* = 6.6 Hz, 3H), 1.01 (d, *J* = 6.5 Hz, 3H). ^{13}C NMR (151 MHz, D_2O) δ 159.75, 159.63, 159.50, 159.38, 159.25, 159.12, 159.00, 158.88, 138.14, 118.63, 116.73, 114.84, 112.94, 106.73, 102.94, 101.97, 101.83, 101.80, 101.71, 100.17, 99.40, 99.25, 99.00, 98.85, 82.45, 81.73, 78.36, 76.28, 75.38, 75.14, 74.86, 74.84, 74.74, 74.51, 74.42, 73.53, 73.06, 72.83, 72.79, 72.62, 71.92, 71.84, 71.70, 71.56, 71.20, 70.38, 70.07, 69.78, 69.73, 69.57, 69.53, 69.24, 68.76, 68.26, 68.16, 68.13, 68.09, 67.51, 67.44, 66.94, 66.90, 66.82, 61.51, 61.44, 60.93, 60.52, 60.05, 59.67, 59.52, 56.62, 56.45, 54.99, 39.34, 39.27, 28.13, 26.39, 22.06, 20.03, 15.45, 15.40, 15.19. MALDI-TOF-MS: $[M + Na]^+$ $C_{63}H_{103}N_3O_{43}F_6Na^+$ *m/z* calcd. 1726.58, found 1726.67; $[M + K]^+$ $C_{63}H_{103}N_3O_{43}F_6K^+$ *m/z* calcd. 1742.55, found 1742.64.

3.1.3. Synthesis of Compound **1-3**

Compound **2** (40.0 mg, 9.7 μmol) was dissolved in $H_2NCH_2CH_2NH_2$/*n*-BuOH (*v/v* = 1:3, 20 mL). The solution was heated at 120 °C for 6 h. The solution was concentrated under reduced pressure. The residue was dissolved in THF/MeOH (*v/v* = 1:4, 5 mL), and then LiOH (50.0 mg, 1.2 mmol) was added. The mixture was heated at 80 °C overnight. The mixture was concentrated under reduced pressure. Then, the residue was dissolved in CH_2Cl_2 and washed with water, and then it was concentrated under reduced pressure and further dried under vacuum.

The residue was dissolved in MeOH (5.0 mL), and the reaction mixture was stirred for 10 min. 4-Dimethylaminopyridine (600 mg) was dissolved in MeOH (5.0 mL), stirred for 10 min, and then added to the reaction mixture. The reaction mixture was stirred for 20 min at room temperature and then cooled to 0 °C for 10 min. CF_2COOCH_3 (16 mL in total) was added 4 times every 5 min, and it was stirred at room temperature for 48 h. The mixture was concentrated under reduced pressure and further dried under vacuum.

The crude product was purified via column chromatography on silica gel (toluene: CH_3CN = 6:1, *v/v*) to afford white solids. The obtained product was dissolved in CH_2Cl_2/MeOH/H_2O (*v/v*, 3:3:1, 7 mL), and then $Pd(OH)_2$ (10%, 15.0 mg) was added. The reaction was carried out under a hydrogen atmosphere of 40 psi for 30 h. The mixture was filtered, and the filtrate was concentrated under reduced pressure. The residue was purified using column chromatography on gel LH20 (H_2O: MeOH = 15:1, *v/v*) to afford **1-3** as white solids (5.5 mg, 36% yield).

^{1}H NMR (400 MHz, D_2O) δ 6.05 (t, *J* = 53.6 Hz, 2H), 5.17 (d, *J* = 3.2 Hz, 1H), 4.95 (t, *J* = 3.5 Hz, 2H), 4.43–4.28 (m, 4H), 4.18–4.12 (m, 1H), 4.05 (d, *J* = 3.3 Hz, 1H), 4.00–3.35 (m, 47H), 3.19 (t, *J* = 8.6 Hz, 1H), 2.90 (t, *J* = 7.5 Hz, 2H), 1.65–1.51 (m, 4H), 1.44–1.29 (m, 2H), 1.16 (d, *J* = 6.6 Hz, 3H), 1.13 (d, *J* = 6.6 Hz, 3H), 1.02 (d, *J* = 6.6 Hz, 3H). ^{13}C NMR (151 MHz, D_2O) δ 166.38, 166.32, 166.21, 166.15, 166.04, 165.98, 158.31, 138.85, 110.76, 110.71,

109.12, 109.07, 107.50, 107.47, 107.42, 103.70, 102.78, 102.75, 102.68, 102.54, 100.94, 100.18, 99.61, 99.49, 83.07, 82.53, 79.14, 77.08, 76.14, 75.92, 75.64, 75.63, 75.52, 75.30, 75.24, 75.20, 74.31, 73.66, 73.56, 73.50, 72.70, 72.62, 72.47, 71.17, 70.84, 70.64, 70.50, 69.95, 69.53, 69.23, 69.04, 68.94, 68.35, 68.29, 67.70, 67.63, 67.53, 62.27, 62.20, 61.71, 60.83, 60.48, 60.33, 57.00, 56.84, 47.44, 43.02, 40.11, 40.06, 28.91, 27.16, 22.84, 20.80, 16.22, 16.18, 16.00, 12.54, 11.28, 8.98. MALDI-TOF-MS: $[M + Na]^+$ $C_{63}H_{105}N_3O_{43}F_4Na^+$ m/z calcd. 1690.60, found 1691.19; $[M + K]^+$ $C_{63}H_{105}N_3O_{43}F_4K^+$ m/z calcd. 1706.57, found 1707.16.

3.1.4. Synthesis of Compound **3-1**

4 Å MS (400 mg) was added to a 10 mL quartz two-necked reaction flask, baked at high temperature, and cooled under vacuum. Under Ar protection, compound **4** (20.0 mg, 17.6 μmol), Umemoto's reagent (16.0 mg, 44. 1 μmol), $Cu(OTf)_2$ (18.0 mg, 44.1 μmol), and anhydrous CH_2Cl_2 (4 mL) were added. The reaction mixture was stirred for 15 min and then cooled to −72 °C. The reaction flask was exposed to UV irradiation at −72 °C for 20 min. After the disappearance of **4** detected by TLC, the removal of UV irradiation was followed by the addition of a solution of compound **5** (15.4 mg, 14.7 μmol) and TTBP (6.6 mg, 26.4 μmol) in CH_2Cl_2 (0.5 mL) via syringe. The reaction mixture was stirred and slowly warmed to room temperature for 1.5 h. The reaction was cooled to −72 °C. Umemoto's reagent (16.0 mg, 44.1 μmol) and $Cu(OTf)_2$ (18.0 mg, 44.1 μmol) were added. The reaction flask was exposed to UV irradiation at −72 °C for 20 min. After the disappearance of the starting material detected by TLC, the removal of UV irradiation was followed by the addition of a solution of compound **6** (16.0 mg, 13.2 μmol) and TTBP (6.6 mg, 26.4 μmol) in CH_2Cl_2 (0.5 mL) via syringe. The reaction mixture was stirred and slowly warmed to room temperature for 1.5 h, and then quenched by Et_3N (0.1 mL). The mixture was filtered and the filtrate was concentrated under reduced pressure. The crude product was purified by column chromatography on silica gel (petroleum ether: ethyl acetate = 1:1, v/v) to afford **3-1** as a white solid (50.0 mg, 61% yield).

$[\alpha]_D^{25}$ -10.3 (c 0.24, $CHCl_3$); ^1H NMR (600 MHz, $CDCl_3$) δ 7.71 (d, J = 7.2 Hz, 1H), 7.62 (d, J = 6.9 Hz, 1H), 7.56-7.49 (m, 4H), 7.46 (t, J = 7.2 Hz, 1H), 7.35-7.08 (m, 71H), 7.03 (t, J = 7.5 Hz, 2H), 6.82 (s, 3H), 5.66 (dd, J = 9.8, 9.1 Hz, 1H), 5.59 (dd, J = 10.0, 9.3 Hz, 1H), 5.40 (d, J = 8.2 Hz, 1H), 5.36 (d, J = 8.4 Hz, 1H), 5.16-5.12 (m, 4H), 5.03 (d, J = 11.6 Hz, 1H), 4.97 (d, J = 11.4 Hz, 1H), 4.84 (d, J = 11.4 Hz, 2H), 4.78–4.74 (m, 1H), 4.65–4.60 (m, 3H), 4.53–4.35 (m, 16H), 4.28 (d, J = 11.8 Hz, 1H), 4.21–4.11 (m, 9H), 3.98 (d, J = 11.8 Hz, 1H), 3.93–3.86 (m, 5H), 3.82–3.70 (m, 5H), 3.54-3.46 (m, 5H), 3.43-3.19 (m, 16H), 3.11 (m, 1H), 2.88 (d, J = 9.6 Hz, 1H), 2.78–2.74 (m, 1H), 2.65–2.60 (m, 1H), 2.48–2.44 (m, 1H), 2.42–2.10 (m, 12H), 1.67–1.66 (m, 9H), 1.55–1.43 (m, 4H), 1.26–1.21 (m, 2H). ^{13}C NMR (151 MHz, $CDCl_3$) δ 206.57, 206.38, 206.18, 171.62, 171.53, 171.04, 167.41, 167.38, 167.33, 167.05, 164.05, 156.65, 156.09, 139.35, 138.99, 138.97, 138.67, 138.49, 138.37, 138.29, 138.26, 137.95, 137.91, 137.85, 137.66, 136.81, 136.73, 133.54, 133.34, 133.23, 132.63, 131.22, 131.16, 131.08, 130.90, 129.69, 129.35, 128.51, 128.46, 128.43, 128.39, 128.35, 128.20, 128.18, 128.16, 128.12, 128.11, 128.06, 128.02, 127.99, 127.92, 127.86, 127.83, 127.80, 127.77, 127.72, 127.65, 127.60, 127.41, 127.38, 127.32, 127.30, 127.09, 126.94, 126.64, 126.35, 123.30, 123.20, 123.02, 122.85, 103.41, 102.27, 100.41, 100.31, 99.46, 99.27, 82.84, 82.19, 81.53, 80.69, 80.32, 78.52, 76.53, 76.27, 75.88, 75.29, 75.07, 74.96, 74.88, 74.83, 74.59, 74.48, 74.42, 74.37, 73.95, 73.92, 73.62, 73.42, 73.38, 73.36, 73.17, 73.13, 72.97, 72.82, 72.80, 72.32, 71.98, 71.76, 71.48, 70.33, 70.24, 69.57, 69.47, 68.25, 68.18, 67.98, 67.62, 67.45, 67.08, 55.05, 54.81, 50.47, 50.15, 47.10, 46.14, 37.75, 37.63, 37.62, 29.89, 29.30, 29.24, 27.82, 27.77, 23.29, 27.46, 23.25. HRMS (ESI): $[M + 2NH_4]^{2+}$ $C_{185}H_{199}N_5O_{42}{}^{2+}$ m/z calcd. 1581.18, found 1581.19. The data are consistent with those reported previously [27].

3.1.5. Synthesis of Compound **3-2**

Compound **3-1** (50.0 mg, 16 μmol) was dissolved in a solution of THF/MeOH (v/v = 10: 1, 4.4 mL). $H_2NNH_2{\cdot}HOAc$ (22.0 mg, 240 μmol) was added. The mixture was stirred at room temperature overnight. The mixture was diluted with CH_2Cl_2, washed sequentially with water, and NaCl aq., and dried over anhydrous Na_2SO_4. The solution was filtered and the filtrate was concentrated under reduced pressure. The crude product was purified by column chromatography on silica gel (petroleum ether: ethyl acetate: CH_2Cl_2 = 1:1:0.25, $v/v/v$) to afford **3-2** as a white solid (41.5 mg, 90% yield).

$[\alpha]_D^{25}$ −0.78 (c 0.44, $CHCl_3$); ^{1}H NMR (600 MHz, $CDCl_3$) δ 7.76 (d, J = 7.6 Hz, 1H), 7.66–7.65 (m, 3H), 7.57–7.54 (m, 2H), 7.46 (t, J = 7.5 Hz, 1H), 7.38–7.08 (m, 74H), 7.03–7.00 (m, 4H), 6.91 (d, J = 7.6 Hz, 1H), 6.85–6.83 (m, 2H), 5.43 (dd, J = 9.0, 8.3 Hz, 1H), 5.31 (d, J = 8.4 Hz, 1H), 5.28 (d, J = 8.5 Hz, 1H), 5.13 (d, J = 9.5 Hz, 2H), 5.01 (d, J = 11.6 Hz, 2H), 4.97 (d, J = 11.4 Hz, 1H), 4.84 (d, J = 10.4 Hz, 1H), 4.81 (d, J = 11.4 Hz, 1H), 4.76 (t, J = 11.7 Hz, 1H), 4.68 (d, J = 12.1 Hz, 1H), 4.64–4.50 (m, 7H), 4.47–4.36 (m, 8H), 4.29–4.08 (m, 14H), 4.03–4.01 (m, 2H), 3.93–3.86 (m, 5H), 3.82–3.73 (m, 6H), 3.65–3.63 (m, 1H), 3.60–3.36 (m, 11H), 3.32–3.17 (m, 10H), 3.12–3.10 (m, 1H), 2.91–2.89 (m, 1H), 2.62–2.61 (m, 1H), 1.55–1.42 (m, 4H), 1.30-1.21 (m, 2H). ^{13}C NMR (151 MHz, $CDCl_3$) δ 167.69, 167.65, 167.60, 167.28, 164.24, 156.64, 156.07, 139.38, 139.00, 138.67, 138.45, 138.39, 138.33, 138.28, 138.17, 138.06, 137.84, 137.79, 137.31, 137.23, 136.79, 136.70, 133.44, 133.33, 132.83, 131.42, 131.35, 131.12, 129.74, 129.34, 128.69, 128.53, 128.44, 128.36, 128.26, 128.20, 128.18, 128.15, 128.14, 128.11, 128.09, 128.04, 128.01, 127.92, 127.84, 127.80, 127.74, 127.72, 127.64, 127.62, 127.59, 127.52, 127.49, 127.45, 127.38, 127.29, 127.26, 127.07, 126.97, 126.90, 126.64, 126.38, 123.30, 122.98, 122.95, 122.59, 104.16, 103.39, 102.25, 101.64, 99.74, 99.42, 83.31, 82.83, 81.99, 81.95, 81.89, 81.50, 80.31, 78.60, 76.63, 75.90, 75.44, 75.27, 74.87, 74.80, 74.59, 74.55, 74.43, 73.97, 73.81, 73.46, 73.44, 73.41, 73.32, 73.09, 72.96, 72.83, 72.63, 72.44, 72.34, 71.13, 70.95, 69.53, 69.46, 69.18, 68.99, 68.89, 68.40, 68.36, 68.28, 67.58, 67.06, 56.40, 56.01, 50.46, 50.14, 47.09, 46.13, 29.31, 29.27, 29.22, 27.85, 27.44, 23.29, 23.24. HRMS (ESI): $[M + 2NH_4]^{2+}$ $C_{170}H_{181}N_5O_{36}{}^{2+}$ m/z calcd. 1434.12, found 1434.13. The data are consistent with those reported previously [27].

3.1.6. Synthesis of Compound 2

4 Å MS (1.5 g) was baked at high temperature and cooled under vacuum, then was added to a 25 mL quartz three-necked reaction flask. Under Ar protection, donor **7** (30.0 mg, 54.1 μmol) was added. Umemoto's reagent (64.0 mg, 0.16 mmol) and $Cu(OTf)_2$ (59.0 mg, 0.16 mmol) were dissolved or suspended in anhydrous CH_2Cl_2 (7.0 mL) under ultrasound and then added via syringe. The reaction mixture was stirred for 5 min and then cooled to −50 °C. The reaction flask was exposed to UV irradiation for 4 min. The removal of UV irradiation was followed by the addition of a solution of the acceptor **3-2** (13.0 mg, 5.4 μmol) and TTBP (20.0 mg, 81.1 μmol) in CH_2Cl_2 (2.0 mL) via syringe. The reaction mixture was stirred and slowly warmed to room temperature in 2.5 h, and then quenched by Et_3N (0.1 mL). The mixture was filtered and the filtrate was concentrated under reduced pressure. The residue was purified by column chromatography on silica gel (petroleum ether: ethyl acetate = 2:1, v/v) to afford **2** as a white solid (36.0 mg, 90% yield).

$[\alpha]_D^{25}$ −5.21 (c 0.37, $CHCl_3$); ^{1}H NMR (600 MHz, $CDCl_3$) δ 8.05 (d, J = 7.2 Hz, 2H), 7.83 (d, J = 7.5 Hz, 2H), 7.72 (t, J = 7.8 Hz, 5H), 7.62–7.57 (m, 4H), 7.54–7.43 (m, 9H), 7.36–7.03 (m, 96H), 7.00–6.94 (m, 5H), 6.83–6.82 (m, 2H), 6.76–6.70 (m, 4H), 6.54 (d, J = 7.0 Hz, 2H), 6.45 (d, J = 7.5 Hz, 2H), 5.70 (d, J = 2.3 Hz, 1H), 5.67 (d, J = 3.8 Hz, 1H), 5.64–5.63 (m, 2H), 5.29 (dd, J = 8.7, 8.0 Hz, 1H), 5.19 (d, J = 8.5 Hz, 1H), 5.13 (d, J = 6.5 Hz, 2H), 5.07-5.00 (m, 4H), 4.95 (dd, J = 13.2, 6.5 Hz, 1H), 4.85–4.48 (m, 28H), 4.45–4.28 (m, 10H), 4.25–4.05 (m, 15H), 4.03–3.94 (m, 6H), 3.91–3.88 (m, 2H), 3.84–3.69 (m, 7H), 3.57–3.19 (m, 23H), 3.11 (d, J = 9.5 Hz, 2H), 2.86–2.84 (m, 1H), 1.54–1.44 (m, 4H), 1.27 (d, J = 6.5 Hz, 3H), 1.11 (d, J = 6.5 Hz, 3H), 0.98 (d, J = 6.5 Hz, 3H). ^{13}C NMR (151 MHz, $CDCl_3$) δ 168.33, 168.31,166.73, 166.69, 166.52, 165.91, 163.92, 156.66, 156.10, 139.36, 138.99, 138.69, 138.44, 138.41, 138.32, 138.23, 138.15, 138.09, 138.05, 137.99, 137.91, 137.88, 137.83, 137.61, 137.54, 137.33, 136.84, 136.75, 133.56, 133.07, 132.88, 132.44, 132.32, 131.68, 131.04, 130.53, 130.25, 130.20, 129.96, 129.94, 129.86, 129.74, 129.72, 129.48, 129.39, 128.55, 128.53, 128.47, 128.43, 128.38, 128.35, 128.33, 128.21, 128.20, 128.15, 128.12, 128.06, 128.02, 127.99, 127.94, 127.87, 127.86, 127.82, 127.78, 127.77, 127.74, 127.69, 127.67, 127.64, 127.60, 127.54, 127.50, 127.46, 127.43, 127.37, 127.32, 127.29, 127.27, 127.24, 127.18, 127.13, 127.00, 126.99, 126.92, 126.90, 126.86, 126.70, 126.22, 126.14, 123.26, 103.45, 102.59, 100.12, 99.93, 99.91, 99.59, 98.41, 98.20, 97.92, 83.90, 82.86, 81.81, 81.54, 80.05,

78.30, 77.42, 76.68, 76.06, 75.95, 75.81, 75.37, 74.98, 74.85, 74.62, 74.21, 74.11, 73.92, 73.72, 73.65, 73.52, 73.32, 73.23, 73.15, 73.12, 72.91, 72.87, 72.84, 72.76, 72.58, 72.50, 72.47, 71.98, 71.75, 71.71, 71.60, 71.56, 71.48, 71.23, 71.17, 71.08, 71.01, 69.63, 69.49, 68.21, 68.02, 67.74, 67.61, 67.59, 67.09, 65.27, 65.23, 65.22, 56.93, 56.76, 50.50, 50.19, 47.12, 46.16, 29.33, 27.88, 27.48, 23.31, 15.94, 15.85, 15.43. HRMS (ESI): $[M + 2NH_4]^{2+}$ $C_{251}H_{259}N_5O_{51}{}^{2+}$ m/z calcd. 2079.39, found 2079.39. The data are consistent with those reported previously [27].

3.1.7. Synthesis of CRM197 Glycoconjugates

Compound **8** (20 mg) was dissolved in DMF (150 µL) in a micro-centrifuge tube, and then triethylamine (8–10 µL) was added. The color of the solution turned yellow. Nonasaccharide (3 mg) was dissolved in DMF (150 µL) in a micro-centrifuge tube, and then it was slowly added to the solution of compound **8**. The micro-centrifuge tube was washed for dissolving the nonasaccharide with DMF (100 µL), and then it was also added to the reaction mixture and mixed well. The reaction mixture was stirred for 4 h at room temperature. The reaction solution was transferred to a micro-thick-walled reaction vial, and then it was concentrated under reduced pressure and further dried under vacuum. After that, the residue was dissolved in water, and it was washed with CH_2Cl_2. The aqueous phase was evaporated at room temperature. The residue was dissolved in 0.01 M PBS buffer (600 µL, pH 7.6), and a 0.01 M PBS solution of protein CRM197 (400 µL, 10 mg/mL) was added. Then, the solution was rotated at room temperature without mechanical stirring for 24 h. After that, the solution was centrifuged (4 °C, 12,000 *g*) with an ultrafiltration centrifuge tube (Millipore®, 0.5 mL, 30 kD) for 5 min and washed three times with ultrapure water. The remaining samples in the ultrafiltration centrifuge tube were diluted and taken out to obtain an aqueous solution containing CRM197 glycoconjugates.

3.1.8. Synthesis of BSA Glycoconjugates

Compound **8** (10 mg) was dissolved in DMF (100 µL) in a micro-centrifuge tube, and then triethylamine (8–10 µL) was added. The color of the solution turned yellow. Nonasaccharide (1.5 mg) was dissolved in DMF (60 µL) in a micro-centrifuge tube, and then it was slowly added to the solution of compound **8**. The micro-centrifuge tube was washed for dissolving the nonasaccharide with DMF (60 µL), and then it was also added to the reaction mixture and mixed well. The reaction mixture was stirred for 4 h at room temperature. The reaction solution was transferred to a micro-thick-walled reaction vial, and then it was concentrated under reduced pressure and further dried under vacuum. After that, the residue was dissolved in water, and it was washed with CH_2Cl_2. The aqueous phase was evaporated at room temperature. The residue was dissolved in 0.01 M PBS buffer (300 µL, pH 7.6), and a 0.01 M PBS solution of protein BSA (200 µL, 10 mg/mL) was added. Then, the solution was rotated at room temperature without mechanical stirring for 24 h. After that, the solution was centrifuged (4 °C, 12,000 *g*) with an ultrafiltration centrifuge tube (Millipore®, 0.5 mL, 30 kD) for 5 min, and it was washed three times with ultrapure water. The remaining samples in the ultrafiltration centrifuge tube were diluted and taken out to obtain an aqueous solution containing BSA glycoconjugates.

3.2. *Immunization of Mice*

Groups of six mice (female pathogen-free BALB/c, age 6–8 weeks, Number, SCXKjing 2016-0010, from the Department of Laboratory Animal Science, Peking University Health Science Center) were immunized four times at a two-week interval with unmodified KH-1-CRM197 or modified glycoconjugates (KH-1-DF-CRM197 and KH-1-TF-CRM197) (each containing 2 µg of carbohydrate in PBS, which was the diluent of the mother liquor mentioned in Table 1). Each vaccination was conducted through a subcutaneous (s.c.) injection with a mixture of the adjuvant (first with Freund's complete adjuvant and then Freund's incomplete adjuvant, Sigma). The mice were bled via the tail vein 1 day before the first vaccination, 13 days after the third vaccination, and 14 days after the fourth vaccination. Blood was clotted to obtain sera, which were stored at −80 °C. On day 56, the mice were euthanized for a series of analyses described below. The animals used in this study were well cared for, and the experiments were approved by the Peking University Health Science Center (LA2017096).

3.3. Serological Assays

The antigen-specific antibody titers of the sera were assessed using ELISA. An ELISA plate was coated with 100 μL of KH-1-BSA (including 0.02 μg of carbohydrate) overnight at 4 °C (0.1 M bicarbonate buffer, pH = 9.6). After being washed three times with PBST (0.05% Tween-20 in PBS), micro-wells were blocked with 3% BSA (200 μL/well). The original serum was serially diluted with 1% BSA. After the plate was washed, the serially diluted sera were added to the micro-wells (100 μL/well) and incubated for 1 h at 37 °C. For the IgG subclass assay, diluted sera (1:1000, from 13 days after the third vaccination) were added instead. Then, the plate was washed and incubated with a 1:5000 dilution of horseradish-peroxidase-conjugated goat anti-mouse IgG, IgG1, IgG2a, IgG2b, IgG3, or IgM (Southern Biotechnology Associates, Inc., Birmingham, AL, USA) for 1 h at 37 °C. Finally, the *o*-phenylenediamine (OPD) substrate was added to the plate in the dark for 15 min; the reaction was terminated by 1 M H_2SO_4 and then read at 490 nm. The antibody titer was defined as the highest dilution showing an absorbance of 0.1, after subtracting background. Meanwhile, the anti-modified-KH-1 antibody titers (sera from 14 days after the fourth vaccination) were determined using ELISA, with the plate coated by the corresponding modified KH-1-BSA conjugates instead.

3.4. Flow Cytometry

The binding of antisera to MCF-7 human breast cancer cells that highly expressed native KH-1 was tested through flow cytometry. Firstly, MCF-7 cells (5×10^5 cells/well) were washed in PBS with a 3% fetal bovine serum and incubated with a test serum (25 μL, finally diluted 1/20, from preimmunization and 14 days after the fourth vaccination) for 30 min on ice. Then, a goat anti-mouse IgG antibody labeled with AF405 (100 μL, diluted 1/500) was added and incubated at room temperature for 30 min. The MFI of the stained cells was analyzed using a FACScan (Becton Dickinson).

3.5. Complement-Dependent Cytotoxicity (CDC) Assay

CDC was assayed on MCF-7 cells that highly expressed the KH-1 antigen with a non-radioactive cytotoxicity assay kit (Promega). MCF-7 cells (25 μL, 1×10^4 cells/well) were seeded into 96-well plates (Corning). Diluted sera (from preimmunization and 14 days after the fourth vaccination) were added and incubated at 37 °C for 2 h (the final concentration of sera was 1:20). After washing the cells twice, a rabbit complement (1:20) was added to the cells and incubated at 37 °C for another 4 h. After centrifugation, the cell supernatants (50 μL/well) were isolated and transferred to another 96-well plate. LDH assay reagents (50 μL/well) were added and incubated at room temperature protected from light for 30 min, and then a stop solution (1 M H_2SO_4, 50 μL/well) was added to each well of the plate. The absorptions of these plates were read at 490 nm wavelength using a microplate reader. The assays were performed in triplicate. The percentage of cell lysis was calculated according to the following formula:

$$Cell\ lysis(\%) = \frac{Experimental - Target\ Spontaneous}{Target\ Maxium - Target\ Spontaneous} \times 100\% \quad (1)$$

4. Conclusions

In conclusion, a native KH-1 antigen and two fluorine-modified KH-1 antigens were designed and synthesized, and the antigens were conjugated with the CRM197 protein for vaccination. The results showed that fluoro-modifications on the KH-1 antigen, especially KH-1-TF, significantly enhanced the immunogenicity of the KH-1 antigen. KH-1-TF-CRM197 elicited higher titers of anti-KH-1 IgG and IgM antibodies than the unmodified

KH-1-CRM197. The IgG antibody elicited by KH-1-TF-CRM197 showed a stronger CDC effect than that of KH-1-CRM197 to KH-1-positive tumor cells. Moreover, KH-1-TF-CRM197 elicited a mixed Th1/Th2 response, which is helpful for eliminating tumor cells. These results are encouraging because they lay a foundation for improving the efficacy of KH-1 conjugate and prove the feasibility of the fluorine modification strategy for the development of carbohydrate-based antitumor vaccines, thus holding potential for effective cancer immunotherapy.

Supplementary Materials: The following supporting information can be downloaded at: https://www.mdpi.com/article/10.3390/molecules28041934/s1. Table S1: Screening of reaction conditions for trifluoroacetylation; Figure S1. The MALDI-TOF-MS result of CRM197; Figure S2. The MALDI-TOF-MS result of KH-1-CRM197; Figure S3. The MALDI-TOF-MS result of KH-1-DF-CRM197; Figure S4. The MALDI-TOF-MS result of KH-1-TF-CRM197; Figure S5. The MALDI-TOF-MS result of BSA; Figure S6. The MALDI-TOF-MS result of KH-1-BSA; Figure S7. The MALDI-TOF-MS result of KH-1-DF-BSA; Figure S8. The MALDI-TOF-MS result of KH-1-TF-BSA; Figure S9. 1H NMR of compound 1-2(400 MHz, D2O); Figure S10. 13C NMR of compound 1-2(151 MHz, D2O); Figure S11. 1H NMR of compound 1-3(400 MHz, D2O); Figure S12. 13C NMR of compound 1-3(151 MHz, D2O).

Author Contributions: B.L. conducted all the synthesis and compound characterization. Y.L. conducted all biological work. X.Z. helped with some biological work. D.X. and X.Y. designed and supervised the project. All authors have read and agreed to the published version of the manuscript.

Funding: This research was funded by grants from the National Key Research and Development Program of China (Grant No. 2022YFC3400800), the National Natural Science Foundation of China (Grant Nos. 22177003, 22237001), the Fundamental Research Funds for the Central Universities and the Open Projects Fund of NMPA Key Laboratory for Quality Research and Evaluation of Carbohydrate-based Medicine (Grant No. 2021QRECM01).

Institutional Review Board Statement: Not applicable.

Informed Consent Statement: Not applicable.

Data Availability Statement: The data presented in this study are available in this article and Supplementary Material, or on request from the corresponding author.

Conflicts of Interest: The authors declare no conflict of interest.

Sample Availability: Not applicable.

References

1. Pinho, S.S.; Reis, C.A. Glycosylation in cancer: Mechanisms and clinical implications. *Nat. Rev. Cancer* **2015**, *15*, 540–555. [CrossRef]
2. Sanders, D.S.; Kerr, M.A. Lewis blood group and CEA related antigens; coexpressed cell-cell adhesion molecules with roles in the biological progression and dissemination of tumors. *Mol. Pathol.* **1999**, *52*, 174–178. [CrossRef]
3. Thurin, M. Tumor-associated glycans as targets for immunotherapy: The Wistar institute experience/legacy. *Monoclon. Antib. Immunodiagn. Immunother.* **2021**, *40*, 89–100. [CrossRef]
4. Heimburg-Molinaro, J.; Lum, M.; Vijay, G.; Jain, M.; Almogren, A.; Rittenhouse-Olson, K. Cancer vaccines and carbohydrate epitopes. *Vaccine* **2011**, *29*, 8802–8826. [CrossRef]
5. Hakomori, S. Tumor-associated carbohydrate antigens defining tumor malignancy: Basis for development of anti-cancer vaccines. *Adv. Exp. Med. Biol.* **2001**, *491*, 369–402. [CrossRef]
6. Livingston, P.O. Augmenting the immunogenicity of carbohydrate tumor antigens. *Semin. Cancer Biol.* **1995**, *6*, 357–366. [CrossRef]
7. Krug, L.M.; Ragupathi, G.; Ng, K.K.; Hood, C.; Jennings, H.J.; Guo, Z.; Kris, M.G.; Miller, V.; Pizzo, B.; Tyson, L.; et al. Vaccination of small cell lung cancer patients with polysialic acid or *N*-propionylated polysialic acid conjugated to keyhole limpet hemocyanin. *Clin Cancer Res.* **2004**, *10*, 916–923. [CrossRef]
8. Sadraei, S.I.; Reynolds, M.R.; Trant, J.F. The synthesis and biological characterization of acetal-free mimics of the tumor-associated carbohydrate antigens. *Adv. Carbohydr. Chem. Biochem.* **2017**, *74*, 137–237. [CrossRef]
9. Weiwer, M.; Huang, F.; Chen, C.-C.; Yuan, X.; Tokuzaki, K.; Tomiyama, H.; Linhardt, R.J. Synthesis and evaluation of anticancer vaccine candidates: *C*-glycoside analogues of STn and PSA. *ACS Symp. Ser.* **2008**, *990*, 216–238.
10. Bundle, D.R.; Rich, J.R.; Jacques, S.; Yu, H.N.; Nitz, M.; Ling, C.C. Thiooligosaccharide conjugate vaccines evoke antibodies specific for native antigens. *Angew. Chem. Int. Ed.* **2005**, *44*, 7725–7729. [CrossRef]

11. Hoffmann-Roder, A.; Kaiser, A.; Wagner, S.; Gaidzik, N.; Kowalczyk, D.; Westerlind, U.; Gerlitzki, B.; Schmitt, E.; Kunz, H. Synthetic antitumor vaccines from tetanus toxoid conjugates of MUC1 glycopeptides with the Thomsen-Friedenreich antigen and a fluorine-substituted analogue. *Angew. Chem. Int. Ed.* **2010**, *49*, 8498–8503. [CrossRef]
12. Nudelman, E.; Levery, S.B.; Kaizu, T.; Hakomori, S. Novel fucolipids of human adenocarcinoma: Characterization of the major Le^y antigen of human adenocarcinoma as trifucosylnonaosyl Le^y glycolipid (III3FucV3FucVI2FucnLc6). *J. Biol. Chem.* **1986**, *261*, 11247–11253. [CrossRef]
13. Kaizu, T.; Levery, S.B.; Nudelman, E.; Stenkamp, R.E.; Hakomori, S. Novel fucolipids of human adenocarcinoma: Monoclonal antibody specific for trifucosyl Ley (III3FucV3FucVI2FucnLc6) and a possible three-dimensional epitope structure. *J. Biol. Chem.* **1986**, *261*, 11254–11258. [CrossRef]
14. Hellström, I.; Garrigues, H.J.; Garrigues, U.; Hellström, K.E. Highly tumor-reactive, internalizing, mouse monoclonal antibodies to Le(y)-related cell surface antigens. *Cancer Res.* **1990**, *50*, 2183–2190. [CrossRef]
15. Kim, Y.S.; Yuan, M.; Itzkowitz, S.H.; Sun, Q.B.; Kaizu, T.; Palekar, A.; Trump, B.F.; Hakomori, S. Expression of Ley and extended Ley blood group-related antigens in human malignant, premalignant, and nonmalignant colonic tissues. *Cancer Res.* **1986**, *46*, 5985–5992. [CrossRef]
16. Stocks, S.C.; Albrechtsen, M.; Kerr, M.A. Expression of the CD15 differentiation antigen (3-fucosyl-N-acetyl-lactosamine, LeX) on putative neutrophil adhesion molecules CR3 and NCA-160. *Biochem. J.* **1990**, *268*, 275–280. [CrossRef]
17. Buskas, T.; Li, Y.; Boons, G.J. Synthesis of a dimeric Lewis antigen and the evaluation of the epitope specificity of antibodies elicited in mice. *Chem. Eur. J.* **2005**, *11*, 5457–5467. [CrossRef]
18. Deshpande, P.P.; Danishefsky, S.J. Total synthesis of the potential anticancer vaccine KH-1 adenocarcinoma antigen. *Nature* **1997**, *387*, 164–166. [CrossRef]
19. Spassova, M.K.; Bornmann, W.G.; Ragupathi, G.; Sukenick, G.; Livingston, P.O.; Danishefsky, S.J. Synthesis of selected Le^Y and KH-1 analogues: A medicinal chemistry approach to vaccine optimization. *J. Org. Chem.* **2005**, *70*, 3383–3395. [CrossRef]
20. Hummel, G.; Schmidt, R.R. A versatile synthesis of the lactoneo-series antigens—Synthesis of sialyl dimer Lewis X and of dimer Lewis Y. *Tetrahedron Lett.* **1997**, *38*, 1173–1176. [CrossRef]
21. Routenberg Love, K.; Seeberger, P.H. Automated solid-phase synthesis of protected tumor-associated antigen and blood group determinant oligosaccharides. *Angew. Chem. Int. Ed.* **2004**, *116*, 612–615. [CrossRef]
22. Li, Q.; Guo, Z. Synthesis of the cancer-associated KH-1 antigen by block assembly of its backbone structure followed by one-step grafting of three fucose residues. *Org. Lett.* **2017**, *19*, 6558–6561. [CrossRef] [PubMed]
23. Yang, F.; Zheng, X.J.; Huo, C.X.; Wang, Y.; Zhang, Y.; Ye, X.-S. Enhancement of the immunogenicity of synthetic carbohydrate vaccines by chemical modifications of STn antigen. *ACS Chem. Biol.* **2011**, *6*, 252–259. [CrossRef] [PubMed]
24. Zhai, C.; Zheng, X.J.; Song, C.; Ye, X.-S. Synthesis and immunological evaluation of *N*-acyl modified Globo H derivatives as anticancer vaccine candidates. *RSC Med. Chem.* **2021**, *12*, 1239–1243. [CrossRef]
25. Song, C.; Zheng, X.J.; Guo, H.; Cao, Y.; Zhang, F.; Li, Q.; Ye, X.-S.; Zhou, Y. Fluorine-modified sialyl-Tn-CRM197 vaccine elicits a robust immune response. *Glycoconj. J.* **2019**, *36*, 399–408. [CrossRef]
26. Song, C.; Zheng, X.J.; Liu, C.C.; Zhou, Y.; Ye, X.-S. A cancer vaccine based on fluorine-modified sialyl-Tn induces robust immune responses in a murine model. *Oncotarget* **2017**, *8*, 47330–47343. [CrossRef]
27. Li, B.-H.; Yao, W.; Yang, H.; Wu, C.; Xiong, D.-C.; Yin, Y.; Ye, X.-S. Total synthesis of tumor-associated KH-1 antigen core nonasaccharide via photo-induced glycosylation. *Org. Chem. Front.* **2020**, *7*, 1255–1259. [CrossRef]
28. Huang, X.; Huang, L.; Wang, H.; Ye, X.-S. Iterative one-pot synthesis of oligosaccharides. *Angew. Chem. Int. Ed.* **2004**, *43*, 5221–5224. [CrossRef]
29. Mao, R.Z.; Guo, F.; Xiong, D.-C.; Li, Q.; Duan, J.; Ye, X.-S. Photoinduced C-S bond cleavage of thioglycosides and glycosylation. *Org. Lett.* **2015**, *17*, 5606–5609. [CrossRef]
30. Mao, R.-Z.; Xiong, D.-C.; Guo, F.; Li, Q.; Duan, J.; Ye, X.-S. Light-driven highly efficient glycosylation reactions. *Org. Chem. Front.* **2016**, *3*, 737–743. [CrossRef]
31. Badahdah, A.M.; Rashid, H.; Khatami, A. Update on the use of meningococcal serogroup C CRM_{197}-conjugate vaccine (Meningitec) against meningitis. *Expert. Rev. Vaccines* **2016**, *15*, 9–29. [CrossRef] [PubMed]
32. Huang, Y.L.; Hung, J.T.; Cheung, S.K.; Lee, H.Y.; Chu, K.C.; Li, S.T.; Lin, Y.C.; Ren, C.T.; Cheng, T.J.; Hsu, T.L.; et al. Carbohydrate-based vaccines with a glycolipid adjuvant for breast cancer. *Proc. Natl. Acad. Sci. USA* **2013**, *110*, 2517–2522. [CrossRef] [PubMed]
33. Harris, J.R.; Markl, J. Keyhole limpet hemocyanin (KLH): A biomedical review. *Micron* **1999**, *30*, 597–623. [CrossRef] [PubMed]
34. Guo, K.; Chu, C.C.; Chkhaidze, E.; Katsarava, R. Synthesis and characterization of novel biodegradable unsaturated poly(ester amide)s. *J. Polym. Sci. A Polym. Chem.* **2005**, *43*, 1463–1477. [CrossRef]
35. Guttormsen, H.K.; Paoletti, L.C.; Mansfield, K.G.; Jachymek, W.; Jennings, H.J.; Kasper, D.L. Rational chemical design of the carbohydrate in a glycoconjugate vaccine enhances IgM-to-IgG switching. *Proc. Natl. Acad. Sci. USA* **2008**, *105*, 5903–5908. [CrossRef] [PubMed]
36. Wei, M.-M.; Wang, Y.-S.; Ye, X.-S. Carbohydrate-based vaccines for oncotherapy. *Med. Res. Rev.* **2018**, *38*, 1003–1026. [CrossRef] [PubMed]

37. Finkelman, F.D.; Holmes, J.; Katona, I.M.; Urban, J.F.; Beckmann, M.P.; Park, L.S.; Schooley, K.A.; Coffman, R.L.; Mosmann, T.R.; Paul, W.E. Lymphokine control of in vivo immunoglobulin isotype selection. *Annu. Rev. Immunol.* **1990**, *8*, 303–333. [CrossRef] [PubMed]
38. Sakamoto, J.; Furukawa, K.; Cordon-Cardo, C.; Yin, B.W.; Rettig, W.J.; Oettgen, H.F.; Old, L.J.; Lloyd, K.O. Expression of Lewis a, Lewis b, X, and Y blood group antigens in human colonic tumors and normal tissue and in human tumor-derived cell lines. *Cancer Res.* **1986**, *46*, 1553–1561.
39. Guo, J.; Jiang, W.; Li, Q.; Jaiswal, M.; Guo, Z. Comparative immunological studies of tumor-associated Lewis X, Lewis Y, and KH-1 antigens. *Carbohydr. Res.* **2020**, *492*, 107999. [CrossRef]
40. Bordron, A.; Bagacean, C.; Tempescul, A.; Berthou, C.; Bettacchioli, E.; Hillion, S.; Renaudineau, Y. Complement system: A neglected pathway in immunotherapy. *Clin. Rev. Allergy. Immunol.* **2020**, *58*, 155–171. [CrossRef]
41. Sörman, A.; Zhang, L.; Ding, Z.; Heyman, B. How antibodies use complement to regulate antibody responses. *Mol. Immunol.* **2014**, *61*, 79–88. [CrossRef] [PubMed]

molecules

MDPI

Article

Kukhtin–Ramirez-Reaction-Inspired Deprotection of Sulfamidates for the Synthesis of Amino Sugars

Ting Li [1,†], Bingbing Xu [1,†], Dengxian Fu [1], Qian Wan [1] and Jing Zeng [1,2,*]

1 Hubei Key Laboratory of Natural Medicinal Chemistry and Resource Evaluation, School of Pharmacy, Huazhong University of Science and Technology, Wuhan 430030, China
2 Shandong Provincial Key Laboratory of Carbohydrate Chemistry and Glycobiology, Shandong University, Qingdao 266237, China
* Correspondence: zengjing0052@hust.edu.cn
† These authors contributed equally to this work.

Abstract: Herein, we present a mild strategy for deprotecting cyclic sulfamidates via the Kukhtin–Ramirez reaction to access amino sugars. The method features the removal of the sulfonic group of cyclic sulfamidates, which occurs through an N-H insertion reaction that implicates the Kukhtin–Ramirez adducts, followed by a base-promoted reductive N-S bond cleavage. The mild reaction conditions of the protocol enable the formation of amino alcohols including analogs that bear multiple functional groups.

Keywords: amino alcohols; 3-aminosugars; sulfamidates; deprotection; Kukhtin–Ramirez reaction

1. Introduction

Amino alcohols are important skeletons which are widely distributed in pharmaceuticals and biologically active natural products (Figure 1a) [1–3]. Amino alcohols also play important roles in organic synthesis as synthons, ligands, auxiliaries, and chiral catalysts [4–8]. This significance has inspired tremendous efforts to devise elegant synthetic methods for the construction of amino alcohols [9–15]. Among them, the utility of sulfamate esters [16–21] as precursors of amino groups to form cyclic sulfamidate [22–25] via substitution [26,27], condensation [28–30], C-H amination [31–34], C-H aziridination [35], etc. [36–38], has been well established. This has emerged as one of the most prominent methods to produce amino alcohols, due to the ready availability of the materials, the high efficiency of transformations, as well as the well-controlled regioselectivity and stereoselectivity (Figure 1b).

3-Amino deoxy sugars represent a special type of amino alcohols found in many carbohydrate-based antibiotics [39,40]. This strategy has also been incorporated into our study to prepare various 3-amino deoxy sugars (Figure 1c) [41–43]. However, the subsequent removal of the SO_2 group of the cyclic sulfamidate to deliver free amino alcohols presented a notable challenge. Conventional deprotection methods employ strong reducing reagents such as $LiAlH_4$, AlH_3, and so on [30,44]. Apparently, the functional group tolerance is largely hampered by these conditions, wherein esters, ketones, aldehydes, and so on must be avoided altogether. Another common deprotection method is hydrolysis under acidic or alkaline conditions, but epimerization is always encountered for secondary alcohols [26,45]. To address these limitations and, more importantly, to gain expedite access to diversified 3-amino sugars, we have developed a new deprotection method for the SO_2 group of sulfamidates under mild conditions.

Citation: Li, T.; Xu, B.; Fu, D.; Wan, Q.; Zeng, J. Kukhtin–Ramirez-Reaction-Inspired Deprotection of Sulfamidates for the Synthesis of Amino Sugars. *Molecules* **2023**, *28*, 182. https://doi.org/10.3390/molecules28010182

Academic Editor: Zaher Judeh

Received: 30 November 2022
Revised: 14 December 2022
Accepted: 21 December 2022
Published: 25 December 2022

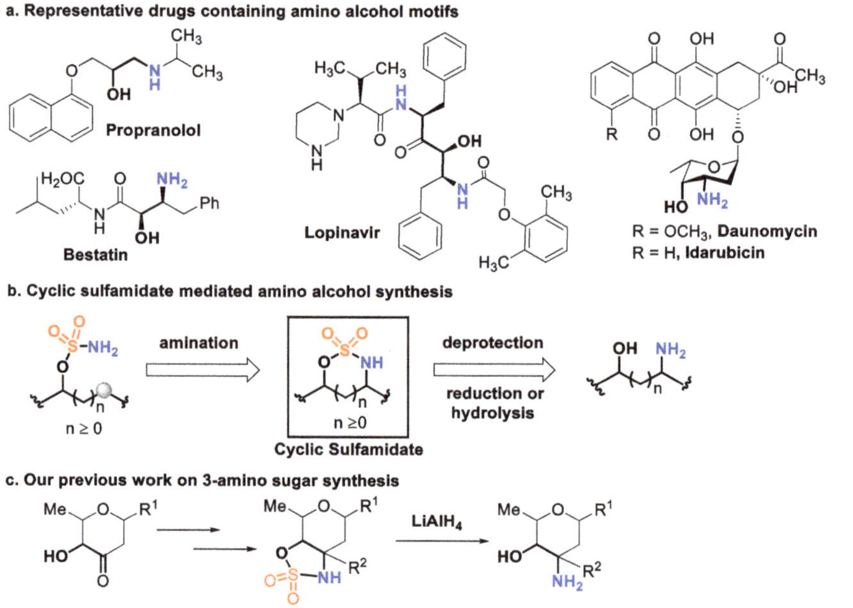

Figure 1. Representative drugs containing amino alcohol motifs and established approaches to synthesizing amino alcohols.

2. Results and Discussion

Our reaction design was inspired by a three-decade-old reaction known as the Kukhtin–Ramirez reaction, which was independently discovered by Kukhtin [46] and Ramirez [47,48]. In this reaction, the redox condensation of a 1,2-dicarbonyl compound with a trivalent phosphorus derivative produces a pentacoordinate dioxaphospholene **Ia**, which exists in equilibrium with a tetracoordinate oxyphosphonium enolate **Ib** (Scheme 1a). Due to their unique properties, these species that are known as the Kukhtin–Ramirez adducts have been well explored in X-H insertion [49–52], reductive addition [53], cycloaddition [54–56], etc. [57–60]. Very recently, Fier et al. described an ingenious solution to degrade secondary sulfonamides into the corresponding sulfinates by virtue of the Kukhtin–Ramirez adducts [61]. This chemistry involves the addition of sulfonamides onto the Kukhtin–Ramirez adducts to form an N-H insertion intermediate **II**, which undergoes further degradation through a base-promoted reductive cleavage of the N-S bond (Scheme 1b). This unprecedented example of the cleavage of a strong sulfonamide S-N bond led us to envision a similar protocol that might be amenable to cleave the sulfamidate S-N bond to deliver an intermediate (**V**) containing both sulfinate and imine functionalities. The corresponding amino alcohol would be revealed upon hydrolysis (Scheme 1c).

With this idea in mind, our investigation commenced with the deprotection of disaccharide **1a** as the model reaction. The requisite cyclic sulfamidate **1a** used in this study was prepared with the application of the corresponding glycals as starting materials [43]. Initially, disaccharide **1a** was subjected to the Kukhtin–Ramirez intermediate formed from ethyl benzoylformate and tris(dimethylamino)phosphorus (Table 1, entry 1). To our delight, the N-H insertion reaction proceeded smoothly to generate N-sulfonyl phenylglycine ester **3a** in 95% yield, which set the stage for the deprotection reaction. Subsequently, BTMG was added into the system as a base to facilitate the S-N bond cleavage (Table 1, entry 2). As expected, the S-N bond was efficiently cleaved with the removal of the SO_2 group, but imino ester **4a** was obtained in 79% yield instead of the target free amino alcohols **2a**. This indicated the occurrence of intramolecular esterification prior to the hydrolysis. To avoid this competing reaction, the second step was carried out in an aqueous solution of THF

(THF:H_2O, 1:1). Following this modification, the desired free amino sugar **2a** was obtained in 88% yield. Other than BTMG, DBU and KOH could also yield the target product in 86% and 80% yields, respectively. However, weak bases such as K_2CO_3 and Et_3N lead to dramatic decreases in yield. Interestingly, the basic anion exchange resin Ambersep® 900 (OH) could cleave the S-N bond effectively. This could simplify the product isolation, although the yield of **2a** would be slightly compromised. Considering the product isolation convenience and the price of the reagent (especially in large-scale preparations), DBU was selected as the base. In principle, the 1,2-dicarbonyl entity could be fully recovered, but the ethyl benzoylformate used in this reaction was hydrolyzed under the strong basic conditions. Several other 1,2-dicarbonyl reagents [62–64] were subsequently examined to circumvent this process. Unfortunately, none of the examined reagents could promote the preceding N-H insertion reaction (see the SI for the screening of 1,2-dicarbonyl compounds).

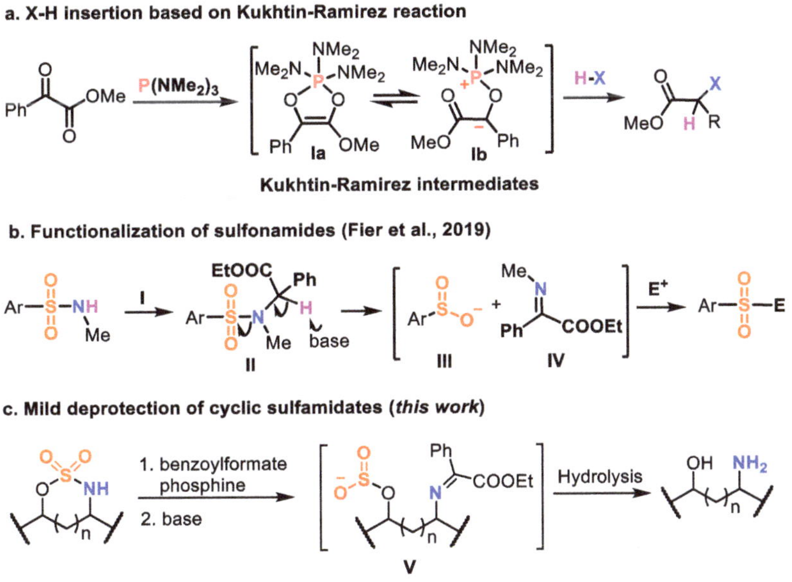

Scheme 1. Application of Kukhtin–Ramirez reaction [61].

Table 1. Reaction development.

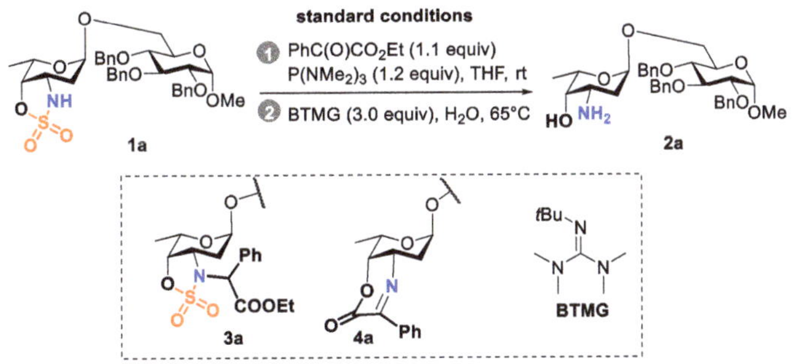

Table 1. *Cont.*

Entry	Variation from Standard Conditions	Yield [a]
1	Step 1 only	**3a**, 95%
2	Without H_2O	**4a**, 79%
3	Standard conditions	**2a**, 88%
4	DBU instead of BTMG	**2a**, 86%
5	KOH instead of BTMG	**2a**, 80%
6	K_2CO_3 instead of BTMG	**2a**, 16%
7	Et_3N instead of BTMG	**2a**, trace
8	Ambersep® 900 (OH) instead of BTMG	**2a**, 80%

[a] Yield of isolated product.

With the optimized conditions in place, we then surveyed the scope and limitations of the method, especially in attaining our ultimate goal to prepare 3-aminosugars (Scheme 2). After examining a series of 3-aminosugars, it could be concluded that: (1) The glycosidic bonds (including both O- and C-analogs) were left intact and the optical purities of the α- and β-glycosidic bonds were not eroded at all. (2) All D- and L-sugars of the 3,5-cis or 3,5-trans configuration could undertake the transformation smoothly to produce the cis amino sugars in good yields. (3) Acid-labile groups such as benzylidine acetals (**2b**, **2i**), isopropylidene ketals (**2c**, **2d**), and other ketals (**2f**) as well as alkenes (**2e**) were well tolerated. (4) Functional groups that are generally sensitive to reductive conditions such as esters (**2e**), ketones (**2m**), and iodine (**2i**) endured the established conditions. However, it is worth mentioning that the ester group was hydrolyzed under the strong basic conditions, with the exception of an α,β-unsaturated ester that could furnish **2e** in good yield. (5) The latent glycosyl donors SPTB ((*S*-2-(2-propylthio)benzyl, **2b**) and OPTB (*O*-2-(2-propylthio)benzyl, **2c**) featured in the interrupted Pummer reaction mediated (IPRm) glycosylations were well compatible [65–69], and could be transformed into the corresponding active SPSB/OPSB glycosyl donors via oxidation, indicating the potential for the further elongation of the sugar chain.

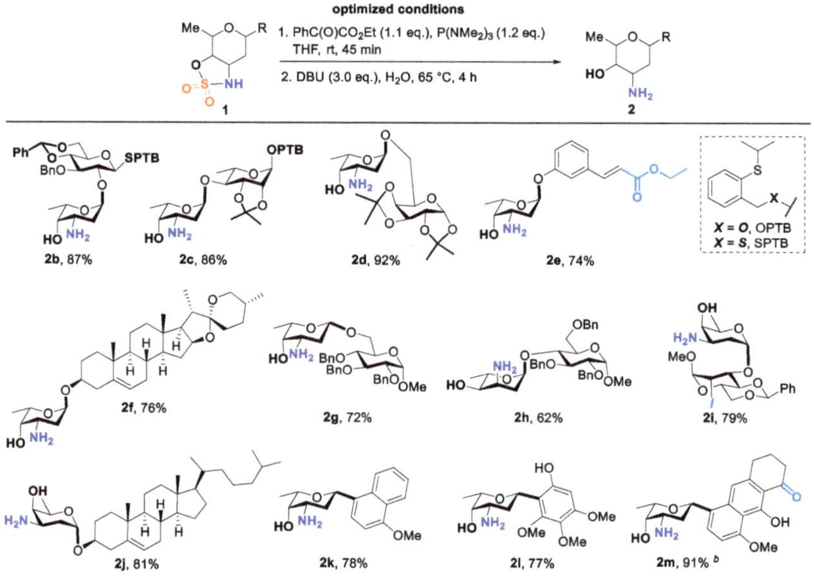

Scheme 2. Substrate scope [a]. [a] Yield of isolated product. [b] c = 0.05 M.

To verify the general synthetic utility of this protocol, we sought to merge the well-defined alcohol-induced amination reactions [31–34] with this deprotection protocol to modify naturally occurring or biologically important alcohols by introducing amino groups at nearby positions (Scheme 3). As an example, cholesterol was subjected to in-situ-generated sulfamoyl chloride [16,18], followed by Rh-catalyzed C-H amination conditions [32]. This sequence produced the five-membered cyclic sulfamidate **6a** in 38% yield. Subsequently, the application of the optimized deprotection conditions gave rise to the β-amino alcohol analog of cholesterol **7a** in 70% yield. The implementation of a similar protocol for the synthetic modification of indole-3-propanol also successfully introduced the amino group at the γ-position. To further showcase the applicability, a 1 mmol scale reaction of **6b** was performed under the optimized conditions. The desired amino alcohol **7b** was obtained in a comparable yield (82%).

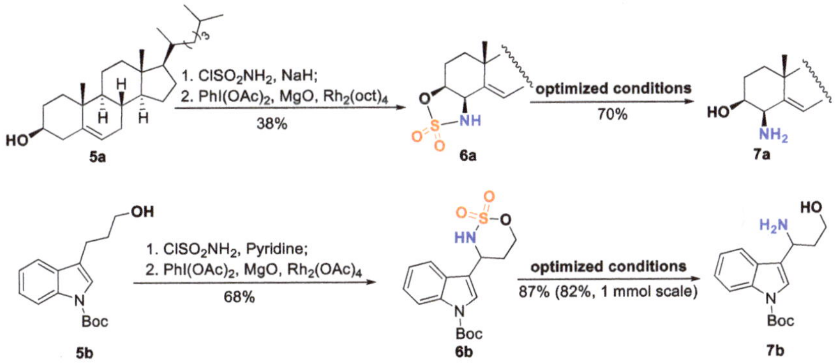

Scheme 3. Amino alcohol synthesis through cyclic sulfamidate.

3. Experimental Section

3.1. General

All the commercially available chemicals were purchased from Alfa, Innochem, and Adamas and used without further purification. The solvents for the reactions were dried on an Innovative Technologies Pure Solv400 solvent purifier. All the reactions were monitored using thin-layer chromatography over silica-gel-coated TLC plates (Yantai Chemical Industry Research Institute). The spots on the TLC were visualized by warming 10% H_2SO_4-(10% H_2SO_4 in ethanol) or 10% phosphomolybdic-acid (10% phosphomolybdic acid in ethanol) -sprayed plates on a hot plate. Column chromatography was performed using silica gel (Qingdao Marine Chemical Inc., Qingdao, China). NMR spectra were recorded with a Bruker AM-400 spectrometer (400 MHz) or Bruker Ascend TM-600 spectrometer (600 MHz). The ^{1}H and ^{13}C NMR chemical shifts were referenced against the solvent or solvent impurity peaks for $CDCl_3$ at δ_H 7.24 and δ_C 77.23, for CD_2Cl_2 at δ_H 5.32 and δ_C 53.80, and for DMSO-d_6 at δ_H 2.50 and δ_C 39.52 ppm, respectively. Optical rotations were measured at 25 °C with a Rudolph Autopol IV automatic polarimeter using a quartz cell with a 2 mL capacity and a 1 dm path length. Concentrations (c) are given in g/100 mL. High-resolution mass spectra were recorded with a Bruker micrOTOF II spectrometer using electrospray ionization (ESI). The copies of ^{1}H and ^{13}C NMR spectra of the new compounds are provided in the Supplementary Material.

3.2. Procedures for Compound **3a** and **4a**

3.2.1. Procedures for Compound **3a**

Ethyl 2-((3aS,4*S*,6*R*,7a*S*)-4-methyl-2,2-dioxido-6-(((2*R*,3*R*,4*S*,5*R*,6*S*)-3,4,5-tris(benzyloxy)-6-methoxytetrahydro-2*H*-pyran-2-yl)methoxy)tetrahydropyrano [4,3-*d*][1,2,3]oxathiazol-1(4*H*)-yl)-2-phenylacsetate (**3a**).

To a solution of **1a** (20.0 mg, 0.031 mmol) in THF (0.31 mL, C = 0.1 M), Ph(CO)CO_2Et (5.3 μL, 0.034 mmol) and P(NMe_2)$_3$ (6.7 μL, 0.037 mmol) were added sequentially. After stirring for 45 min at room temperature, the mixture was concentrated and purified using silica gel chromatography to obtain **3a** (23.8 mg, 95%) as a colorless syrup. The major isomer: R_f = 0.71 (petroleum ether-EtOAc 2:1). $[\alpha]_D^{25}$ −19.6 (c, 1.34 in $CHCl_3$). The readings for ^{1}H NMR spectra (400 MHz, $CDCl_3$) were δ 7.37–7.25 (m, 18H, Ar-H), 7.14 (dd, *J* = 7.6, 2.8 Hz, 2H, Ar-H), 5.28 (s, 1H, CH), 4.98–4.95 (m, 2H, $PhCH_2$, H-4′), 4.78 (d, *J* = 11.2 Hz, 3H, $PhCH_2$), 4.66 (d, *J* = 12.0 Hz, 1H, $PhCH_2$), 4.53 (d, *J* = 3.6 Hz, 2H, H-1, H-1′), 4.36–4.21 (m, 4H, CH_2, $PhCH_2$, H-3′), 4.02 (dq, *J* = 6.4, 1.6 Hz, 1H, H-5′), 3.93 (t, *J* = 9.2 Hz, 1H, H-3), 3.66 (d, *J* = 11.6 Hz, 1H, H-6a), 3.61 (dd, *J* = 10.4, 5.2 Hz, 1H, H-5), 3.46 (dd, *J* = 9.6, 3.6 Hz, 1H, H-2), 3.30 (dd, *J* = 10.8, 5.6 Hz, 1H, H-6b), 3.26–3.21 (m, 4H, H-4, OMe), 1.77–1.69 (m, 1H, H-2′a), 1.29–1.25 (m, 6H, H-6′, Me), and 1.05 (dd, *J* = 13.2, 6.4 Hz, 1H, H-2′b). The readings for the ^{13}C NMR spectra (100 MHz, $CDCl_3$) were δ 170.8, 138.9, 138.3, 133.6, 129.4, 129.3, 128.7, 128.6, 128.3, 128.2, 128.2, 128.0, 127.9, 127.9, 98.0, 96.8, 82.3, 81.5, 80.3, 78.1, 76.0, 75.2, 73.6, 70.1, 66.8, 62.7, 62.2, 61.2, 55.1, 53.0, 32.4, 16.8, and 14.2. The HRMS calculation for $C_{44}H_{51}NO_{12}S$ was [M + Na]$^+$: 840.3024, found: 840.3041. The minor isomer: R_f = 0.58 (petroleum ether-EtOAc 2:1). $[\alpha]_D^{25}$ −43.8 (c, 1.17 in $CHCl_3$). The readings for the ^{1}H NMR spectra (400 MHz, $CDCl_3$) were δ 7.44–7.42 (m, 2H, Ar-H), 7.36–7.26 (m, 16H, Ar-H), 7.21–7.18 (m, 2H, Ar-H), 5.15 (s, 1H, CH), 4.97 (d, *J* = 10.8 Hz, 1H, $PhCH_2$), 4.82 (d, *J* = 10.8 Hz, 1H, $PhCH_2$), 4.80–4.75 (m, 3H, $PhCH_2$, H-1′), 4.65 (d, *J* = 12.4 Hz, 1H, $PhCH_2$), 4.51 (d, *J* = 3.6 Hz, 1H, H-1), 4.41 (d, *J* = 11.2 Hz, $PhCH_2$), 4.26 (dd, *J* = 4.0, 1.2 Hz, 1H, H-4′), 4.23–4.12 (m, 2H, CH_2), 3.92 (t, *J* = 9.2 Hz, 1H, H-3), 3.86 (qd, *J* = 6.4, 1.6 Hz, 1H, H-5′), 3.76 (ddd, *J* = 11.2, 6.4, 4.4 Hz, 1H, H-3′), 3.68 (dd, *J* = 10.4, 0.8 Hz, 1H, H-6a), 3.61 (dd, *J* = 10.4, 5.6 Hz, 1H, H-5), 3.41 (dd, *J* = 9.6, 3.2 Hz, 1H, H-2), 3.38 (dd, *J* = 10.8, 6.0 Hz, 1H, H-6b), 3.26 (t, *J* = 9.2 Hz, 1H, H-4), 3.23 (s, 3H, OMe), 2.23–2.16 (m, 1H, H-2′a), 1.98 (dd, *J* = 14.0, 6.4 Hz, 1H, H-2′b), and 1.22–1.18 (m, 6H, H-6′, Me). The readings for the ^{13}C NMR spectra (100 MHz, $CDCl_3$) were δ 168.9, 138.9, 138.3, 138.2, 133.3, 129.9, 129.3, 129.1, 128.7, 128.7, 128.6, 128.3, 128.2, 128.1, 128.0, 127.9, 98.0, 96.8, 82.3, 80.3, 80.2, 78.0, 76.0, 75.3, 73.5, 70.1, 66.6, 64.0, 62.6, 62.3, 55.2, 54.3, 30.8, 16.8, and 14.1. The HRMS calculation for $C_{44}H_{51}NO_{12}S$ was [M + Na]$^+$: 840.3024, found: 840.3051.

3.2.2. Procedures for Compound **4a**

(4a*S*,5*S*,7*R*,8a*S*)-5-methyl-2-phenyl-7-(((2*R*,3*R*,4*S*,5*R*,6*S*)-3,4,5-tris(benzyloxy)-6-methoxytetrahydro-2*H*-pyran-2-yl)methoxy)-4a,7,8,8a-tetrahydropyrano [3,4-*b*][1,4]oxazin-3(5*H*)-one (**4a**).

To a solution of **1a** (20.0 mg, 0.031 mmol) in THF (0.31 mL, C = 0.1 M), Ph(CO)CO_2Et (5.3 μL, 0.034 mmol) and P(NMe_2)$_3$ (6.7 μL, 0.037 mmol) were added sequentially. After stirring for 45 min at room temperature, BTMG was added, and the mixture was stirred for 4 h at 65 °C. The mixture was concentrated and purified using silica gel chromatography to obtain **4a** (17.0 mg, 79%) as a white solid. R_f = 0.61 (petroleum ether-EtOAc 2:1), m.p. 159–160 °C. $[\alpha]_D^{25}$ −128.3 (c, 1.36 in $CHCl_3$). The readings for the ^{1}H NMR spectra (400 MHz, $CDCl_3$) were δ 7.95–7.82 (m, 2H, Ar-H), 7.47–7.26 (m, 18H, Ar-H), 5.00 (d, *J* = 11.2 Hz, 1H, $PhCH_2$), 4.90 (d, *J* = 11.2 Hz, 1H, $PhCH_2$), 4.84–4.77 (m, 3H, $PhCH_2$, H-1), 4.66 (d, *J* = 12.0 Hz, 1H, $PhCH_2$), 4.58 (d, *J* = 3.6 Hz, 1H, H-1′), 4.56 (d, *J* = 11.2 Hz, $PhCH_2$), 4.52 (ddd, *J* = 10.8, 4.8, 3.2 Hz, 1H, H-3′), 4.29 (d, *J* = 1.6 Hz, 1H, H-4′), 4.07 (q, *J* = 6.4 Hz, 1H, H-5′), 4.01 (t, *J* = 9.2 Hz, 1H, H-3), 3.85 (dd, *J* = 10.8, 1.6 Hz, 1H, H-6a), 3.76 (ddd, *J* = 10.0, 4.8, 1.2 Hz, 1H, H-5), 3.55 (dd, *J* = 10.8, 5.2 Hz, 1H, H-6b), 3.52–3.46 (m, 2H, H-2, H-4), 3.38 (s, 3H, OMe), 2.15 (dd, *J* = 13.2, 4.8 Hz, 1H, H-2′a), 1.42 (td, *J* = 12.8, 3.2 Hz, 1H, H-2′b), and 1.33 (d, *J* = 6.4 Hz, 3H, H-6′). The readings for the ^{13}C NMR spectra (100 MHz, $CDCl_3$) were δ 158.8, 155.9, 138.9, 138.4, 138.4, 134.3, 131.4, 129.0, 128.7, 128.6, 128.5, 128.3, 128.2, 128.1, 127.8, 98.2, 96.7, 82.4, 80.3, 78.0, 75.9, 75.3, 75.0, 73.6, 70.2, 66.6, 63.6, 55.3, 51.5, 30.7, and 15.9. The HRMS calculation for $C_{42}H_{45}NO_9$ [M + Na]$^+$: 730.2987, found: 730.2989.

3.3. General Procedure for Deprotection and Characterization of the Products

To a solution of the substrate (1.0 equiv) in THF (C = 0.1 M), Ph(CO)CO_2Et(1.1 equiv) and P(NMe_2)$_3$ (1.2 equiv) were added sequentially. After stirring for 45 min at room temperature, DBU (3.0 equiv relative to the starting substrate) and H_2O (the amount of water was equal to that of the THF) were added sequentially, and the mixture was stirred for 4 h at 65 °C. The mixture was extracted with CH_2Cl_2 after removing the THF by concentration. The organic layer was washed with saturated $NaHCO_3$ and brine, dried over Na_2SO_4, and concentrated in vacuo. The residue was purified with column chromatography on silica gel (dichloromethane-methanol gradient elution, with 0.5% or 1% $NH_3 \cdot H_2O$) to obtain the desired product.

(2*S*,3*S*,4*S*,6*R*)-4-amino-2-methyl-6-(((2*R*,3*R*,4*S*,5*R*,6*S*)-3,4,5-tris(benzyloxy)-6-methoxytetrahydro-2*H*-pyran-2-yl)methoxy)tetrahydro-2*H*-pyran-3-ol (**2a**).

According to the General Procedure, **1a** (20.0 mg, 0.031 mmol) was used to obtain **2a** as a white solid in 86% yield. R_f = 0.43 (CH_2Cl_2-MeOH 10:1), m.p. 181–182 °C. $[\alpha]_D^{25}$ −29.6 (c, 0.45 in $CHCl_3$).The readings for the ^{1}H NMR spectra (400 MHz, $CDCl_3$) were δ 7.27–7.24 (m, 4H, Ar-H), 7.23–7.14 (m, 11H, Ar-H), 4.89 (d, *J* = 10.8 Hz, 1H, PhCH_2), 4.78 (d, *J* = 11.2 Hz, 1H, PhCH_2), 4.71 (d, *J* = 10.8 Hz, 1H, PhCH_2), 4.69 (d, *J* = 12.0 Hz, 1H, PhCH_2), 4.64 (br s, 1H, H-1′), 4.56 (d, *J* = 12.0 Hz, PhCH_2), 4.48 (d, *J* = 3.6 Hz, 1H, H-1), 4.44 (d, *J* = 11.2 Hz, 1H, PhCH_2), 3.89 (t, *J* = 9.2 Hz, 1H, H-3), 3.77 (q, *J* = 6.4 Hz, 1H, H-5′), 3.72 (dd, *J* = 10.8, 1.6 Hz, 1H, H-6a), 3.65 (ddd, *J* = 10.0, 4.8, 1.2 Hz, 1H, H-5), 3.42–3.37 (m, 3H, H-2, H-4, H-6b), 3.28–3.25 (m, 4H, H-4′, OMe), 3.15–3.08 (m, 1H, H-3′), 1.52 (dd, *J* = 9.2, 2.4 Hz, 2H, H-2′a, H-2′b), and 1.11 (d, *J* = 6.8 Hz, 3H, H-6′). The readings for the ^{13}C NMR spectra (100 MHz, $CDCl_3$) were δ 138.9, 138.4, 138.4, 128.7, 128.6, 128.3, 128.2, 128.1, 128.0, 127.9, 98.1, 97.6, 82.3, 80.2, 78.1, 76.0, 75.2, 73.5, 71.1, 70.2, 66.3, 66.0, 55.3, 46.4, 32.8, and 17.1. The HRMS calculation for $C_{36}H_{45}NO_9$ was $[M + Na]^+$: 658.2987, found: 658.3006.

(2*S*,3*S*,4*S*,6*S*)-4-amino-6-(((2*R*,4a*R*,6*S*,7*R*,8*S*,8a*R*)-8-(benzyloxy)-6-((2-(isopropylthio) benzyl)thio)-2-phenylhexahydropyrano [3,2-*d*][1,3]dioxin-7-yl)oxy)-2-methyltetrahydro-2*H*-pyran-3-ol (**2b**).

According to the General Procedure, **1b** (20.0 mg, 0.027 mmol) was used to obtain **2b** as a white solid in 87% yield. R_f = 0.13 (CH_2Cl_2-MeOH 15:1), m.p. 57–58 °C. $[\alpha]_D^{25}$ −121.5 (c, 0.61 in $CHCl_3$). The readings for the ^{1}H NMR spectra (400 MHz, $CDCl_3$) were δ 7.46–7.41 (m, 3H, Ar-H), 7.36–7.34 (m, 3H, Ar-H), 7.30–7.25 (m, 6H, Ar-H), 7.21–7.14 (m, 2H, Ar-H), 5.54 (s, 1H, PhCHO_2), 5.35 (d, *J* = 3.2 Hz, 1H, H-1′), 4.93 (d, *J* = 11.2 Hz, 1H, PhCH_2), 4.60 (d, *J* = 11.2 Hz, 1H, PhCH_2), 4.36–4.31 (m, 3H, H-1), 4.14 (d, *J* = 13.2 Hz, 1H, PhCH_2S), 4.01 (d, *J* = 13.2 Hz, PhCH_2S), 3.78–3.66 (m, 4H), 3.40–3.33 (m, 3H), 3.03 (d, *J* = 10.8 Hz, 1H), 1.61 (dd, *J* = 13.2, 4.0 Hz, 1H, H-2′a), 1.46–1.40 (m, 1H, H-2′b), 1.28–1.23 (m, 6H, $(CH_3)_2$CH), and 1.18 (d, *J* = 6.8 Hz, 3H, H-6′). The readings for the ^{13}C NMR spectra (100 MHz, $CDCl_3$) were δ 139.7, 138.1, 137.3, 135.5, 132.8, 130.0, 129.0, 128.5, 128.3, 128.0, 127.8, 127.7, 126.8, 126.0, 101.2, 98.4, 84.1, 84.0, 81.8, 75.1, 71.2, 70.0, 68.8, 66.9, 46.3, 38.8, 33.1, 29.8, 23.3, 23.1, and 16.8. The HRMS calculation for $C_{36}H_{45}NO_7S_2$ was $[M + H]^+$: 668.2710, found: 668.2688.

(2*S*,3*S*,4*S*,6*S*)-4-amino-6-(((3a*R*,4*R*,6*S*,7*S*,7a*R*)-4-((2-(isopropylthio)benzyl)oxy)-2,2,6-trimethyltetrahydro-4*H*-[1,3]dioxolo [4,5-*c*]pyran-7-yl)oxy)-2-methyltetrahydro-2*H*-pyran-3-ol (**2c**).

According to the General Procedure, **1c** (20.0 mg, 0.036 mmol) was used to obtain **2c** as a colorless syrup in 86% yield. R_f = 0.30 (CH_2Cl_2-MeOH 10:1). $[\alpha]_D^{25}$ −93.9 (c, 1.20 in $CHCl_3$).The readings for the ^{1}H NMR spectra (400 MHz, $CDCl_3$) were δ 7.43 (dd, *J* = 8.0, 1.6 Hz, 1H, Ar-H), 7.39 (dd, *J* = 7.2, 1.6 Hz, 1H, Ar-H), 7.27–7.24 (m, 1H, Ar-H), 7.23–7.19 (m, 1H, Ar-H), 5.45 (d, *J* = 2.4 Hz, 1H, H-1′), 5.07 (s, 1H, H-1), 4.85 (d, *J* = 12.0 Hz, 1H, PhCH_2), 4.62 (d, *J* = 11.6 Hz, 1H, PhCH_2), 4.19–4.15 (m, 1H), 4.12 (d, *J* = 5.2 Hz, 1H), 3.88 (q, *J* = 6.8 Hz, H-5′), 3.78–3.71 (m, 1H), 3.52 (dd, *J* = 10.0, 7.2 Hz, 1H, H-4), 3.40 (br s, 1H), 3.38–3.31 (m, 1H), 3.16 (br s, 1H), 1.73–1.61 (m, 4H, H-2′a, H-2′b, NH_2), 1.52 (s, 3H, Me), 1.31 (s, 3H, Me), and 1.28–1.19 (m, 12H, H-6, H-6′, $(CH_3)_2$CH),. The readings for the ^{13}C NMR spectra (100 MHz, $CDCl_3$) were δ 138.8, 135.4, 132.8, 129.5, 128.5, 127.1, 109.5, 96.8,

95.6, 79.2, 76.5, 76.4, 71.4, 67.8, 66.5, 64.7, 46.5, 38.9, 33.1, 28.1, 26.7, 23.4, 23.3, 18.7, 18.2, and 17.2. The HRMS calculation for $C_{25}H_{39}NO_7S$ was $[M + H]^+$: 498.2520, found: 498.2541.

(2*S*,3*S*,4*S*,6*R*)-4-amino-2-methyl-6-(((3a*R*,5*R*,5a*S*,8a*S*,8b*R*)-2,2,7,7-tetramethyltetrahydro-5*H*-bis([1,3]dioxolo)[4,5-*b*:4′,5′-*d*]pyran-5-yl)methoxy)tetrahydro-2*H*-pyran-3-ol (**2d**).

According to the General Procedure, **1d** (15.5 mg, 0.034 mmol) was used to obtain **2d** as a colorless syrup in 92% yield. R_f = 0.33 (CH_2Cl_2-MeOH 10:1). $[\alpha]_D^{25}$ −99.5 (c, 1.17 in $CHCl_3$). The readings for the ^1H NMR spectra (400 MHz, $CDCl_3$) were δ 5.51 (d, *J* = 5.2 Hz, 1H, H-1), 4.88 (d, *J* = 2.8 Hz, 1H, H-1′), 4.58 (dd, *J* = 7.6, 2.4 Hz, 1H), 4.29 (dd, *J* = 5.2, 2.4 Hz, 1H), 4.21 (dd, *J* = 8.0, 1.6 Hz, 1H), 3.97–3.90 (m, 2H), 3.79 (dd, *J* = 10.0, 6.0 Hz, 1H), 3.54 (dd, *J* = 10.0, 6.8 Hz, 1H), 3.38 (d, *J* = 2.4 Hz, 1H), 3.21 (d, *J* = 8.8 Hz, 1H), 1.72–1.59 (m, 4H, H-2′a, H-2′b, NH_2), 1.51 (s, 3H, Me), 1.42 (s, 3H, Me), 1.31 (s, 3H, Me), 1.30 (s, 3H, Me), and 1.23 (d, *J* = 6.4 Hz, 3H, H-6′). The readings for the ^{13}C NMR spectra (100 MHz, $CDCl_3$) were δ 109.5, 108.7, 97.3, 96.5, 71.5, 71.4, 70.9, 70.8, 67.0, 66.1, 65.6, 46.6, 33.0, 26.3, 26.2, 25.2, 24.8, and 17.2. The HRMS calculation for $C_{18}H_{31}NO_8$ was $[M + H]^+$: 390.2122, found: 390.2138.

Ethyl (E)-3-(3-(((2*S*,4*S*,5*S*,6*S*)-4-amino-5-hydroxy-6-methyltetrahydro-2*H*-pyran-2-yl)oxy)phenyl)acrylate (**2e**).

According to the General Procedure, **1e** (12.0 mg, 0.031 mmol) was used to obtain **2e** as a colorless syrup in 74% yield. R_f= 0.23 (CH_2Cl_2-MeOH 10:1). $[\alpha]_D^{25}$ −110.9 (c, 0.60 in $CHCl_3$).The readings for the ^1H NMR spectra (400 MHz, $CDCl_3$) were δ 7.62 (d, *J* = 16.0 Hz, 1H, CH = CH), 7.27 (t, *J* = 7.6 Hz, 1H, Ar-H), 7.20 (br s, 1H, Ar-H), 7.14 (d, *J* = 7.6 Hz, 1H, Ar-H), 7.07 (dd, *J* = 8.4, 1.6 Hz, 1H, Ar-H), 6.39 (d, *J* = 16.0 Hz, 1H, CH = CH), 5.60 (br s, 1H, H-1), 4.24 (q, *J* = 7.2 Hz, 2H, CH_2), 3.96 (q, *J* = 6.4 Hz, 1H, H-5), 3.48–3.44 (m, 2H, H-3, H-4), 1.86 (dd, *J* = 8.8, 2.4 Hz, 2H, H-2a, H-2b), 1.32 (t, *J* = 7.2 Hz, 3H, Me), and 1.22 (d, *J* = 6.8 Hz, 3H, H-6). The readings for the ^{13}C NMR spectra (100 MHz, $CDCl_3$) were δ 167.2, 157.5, 144.6, 136.1, 130.1, 121.9, 118.9, 118.4, 115.8, 96.1, 71.0, 67.2, 60.7, 46.4, 32.8, 17.2, and 14.5. The HRMS calculation for $C_{17}H_{23}NO_5$ was $[M + H]^+$: 322.1649, found: 322.1676.

(2*S*,3*S*,4*S*,6*R*)-4-amino-2-methyl-6-(((4*S*,5′*R*,6a*R*,6b*S*,8a*S*,8b*R*,9*S*,10*R*,11a*S*,12a*S*,12b*S*) - 5′,6a,8a,9-tetramethyl-1,3,3′,4,4′,5,5′,6,6a,6b,6′,7,8,8a,8b,9,11a,12,12a,12b-icosahydrospiro [naphtho[2′,1′:4,5]indeno [2,1-*b*]furan-10,2′-pyran]-4-yl)oxy)tetrahydro-2*H*-pyran-3-ol (**2f**).

According to the General Procedure, **1f** (20.0 mg, 0.033 mmol) was used to obtain **2f** as a white solid in 76% yield. R_f= 0.32 (CH_2Cl_2-MeOH 10:1), m.p. 243–244 °C. $[\alpha]_D^{25}$ −154.8 (c, 1.09 in $CHCl_3$). The readings for the ^1H NMR spectra (400 MHz, $CDCl_3$) were δ 5.31 (d, *J* = 5.2 Hz, 1H, C = CH), 5.00 (d, *J* = 3.2 Hz, 1H, H-1), 4.38 (q, *J* = 7.2 Hz, 1H), 3.94 (q, *J* = 6.4 Hz, 1H, H-5), 3.47–3.32 (m, 4H), 3.24 (d, *J* = 10.4 Hz, 1H), 2.32 (ddd, *J* = 13.2, 4.4, 1.6 Hz, 1H), 2.15 (td, *J* = 12.4, 2.4 Hz, 1H), 2.02–1.92 (m, 2H), 1.85–1.48 (m, 18H), 1.46–1.38 (m, 2H), 1.30–1.25 (m, 1H), 1.22 (d, *J* = 6.8 Hz, 3H, H-6), 1.17 (dd, *J* = 12.4, 4.4 Hz, 1H), 1.13–1.05 (m, 2H), 1.00 (s, 3H, Me), 0.96–0.89 (m, 4H), and 0.78–0.75 (m, 6H, Me). The readings for the ^{13}C NMR spectra (100 MHz, $CDCl_3$) were δ 140.9, 121.7, 109.5, 95.4, 81.0, 76.3, 71.5, 67.1, 66.1, 62.4, 56.7, 50.3, 46.6, 41.8, 40.5, 40.0, 38.9, 37.6, 37.1, 33.6, 32.3, 32.1, 31.7, 31.6, 30.5, 29.7, 29.0, 21.1, 19.6, 17.3, 17.2, 16.5, and 14.7. The HRMS calculation for $C_{33}H_{53}NO_5$ was $[M + H]^+$: 544.3997, found: 544.4011.

(2*S*,3*S*,4*S*,6*S*)-4-amino-2-methyl-6-(((2*R*,3*R*,5*R*,6*S*)-3,4,5-tris(benzyloxy)-6-methoxytetrahydro-2*H*-pyran-2-yl)methoxy)tetrahydro-2*H*-pyran-3-ol (**2g**).

According to the General Procedure, **1g** (20.0 mg, 0.031 mmol) was used to obtain **2g** as a white solid in 72% yield. R_f = 0.13 (CH_2Cl_2-MeOH 15:1). The readings for the ^1H NMR spectra (600 MHz, $CDCl_3$) were δ 7.31–7.18 (m, 15H, Ar-H), 4.91 (d, *J* = 11.4 Hz, 1H, $PhCH_2$), 4.79 (d, *J* =10.8 Hz, 1H, $PhCH_2$), 4.78 (d, *J* = 10.8 Hz, 1H, $PhCH_2$), 4.73 (d, *J* = 12.0 Hz, 1H, $PhCH_2$), 4.66 (d, *J* = 10.8 Hz, 1H, $PhCH_2$), 4.60 (d, *J* = 12.6 Hz, 1H, $PhCH_2$), 4.56 (d, *J* = 3.6 Hz, 1H, H-1), 4.40 (dd, *J* = 9.6, 1.8 Hz, 1H, H-1′), 4.10 (dd, *J* = 11.4, 3.6 Hz, 1H, H-6a), 3.92 (t, *J* = 9.6 Hz, 1H, H-3), 3.68–3.61 (m, 2H, H-5, H-6b), 3.54 (t, *J* = 9.6 Hz, 1H, H-4), 3.47 (dd, *J* = 9.6, 3.6 Hz, 1H, H-2), 3.42 (q, *J* = 6.6 Hz, 1H, H-5′), 3.33 (d, *J* = 1.8 Hz, 1H, H-4′), 3.30 (s, 3H, OMe), 2.97 (brs, 3H, NH_2, OH), 2.92–2.87 (m, 1H, H-3′), 1.85–1.78 (m, 1H, H-2′a), 1.51–1.43 (m, 1H, H-2′b), and 1.19 (d, *J* = 6.6 Hz, 3H, H-6′). The readings for the ^{13}C NMR spectra (150 MHz, $CDCl_3$) were δ 139.0, 138.6, 138.4, 128.6, 128.6, 128.5, 128.3, 128.2,

128.1, 128.0, 127.9, 127.7, 100.4, 98.3, 82.2, 80.1, 77.8, 75.9, 75.2, 73.6, 71.9, 70.1, 70.1, 66.9, 55.3, 50.6, 34.4, and 17.1.

(2*S*,3*R*,4*R*,6*S*)-4-amino-6-(((2*R*,3*R*,4*S*,5*R*,6*S*)-4,5-bis(benzyloxy)-2-((benzyloxy)methyl)-6-methoxytetrahydro-2*H*-pran-3-yl)oxy)-2-methyltetrahydro-2H-pyran-3-ol (**2h**).

According to the General Procedure, **1h** (21.0 mg, 0.032 mmol) was used to obtain **2h** as a colorless syrup in 62% yield. R_f = 0.58 (CH_2Cl_2-MeOH 10:1). $[\alpha]_D^{25}$ −15.0 (c, 0.56 in $CHCl_3$). The readings for the ^{1}H NMR spectra (400 MHz, $CDCl_3$) were δ 7.35–7.25 (m, 15H, Ar-H), 5.25 (d, J = 3.2 Hz, 1H, H-1′), 4.77 (d, J = 11.6 Hz, 2H, $PhCH_2$), 4.69 (m, 2H, $PhCH_2$, H-1), 4.65 (d, J = 11.6 Hz, 1H, $PhCH_2$), 4.56 (d, J = 11.6 Hz, 1H, $PhCH_2$), 4.65 (d, J = 11.6 Hz, 1H, $PhCH_2$), 4.06 (d, J = 1.2 Hz, 1H, H-4), 3.92 (t, J = 10.0 Hz, 1H), 3.84 (qd, J = 10.0, 2.8 Hz, 2H), 3.61–3.56 (m, 3H), 3.38 (s, 3H, OMe), 3.06 (dd, J = 9.2, 4.4 Hz, 1H), 3.00 (m, 1H), 2.06–2.03 (m, 1H, H-2′a), 1.79 (dt, J = 14.4, 4.4 Hz, 1H, H-2′b), and 1.18 (d, J = 6.0 Hz, 3H, H-6′). The readings for the ^{13}C NMR spectra (100 MHz, $CDCl_3$) were δ 138.6, 138.5, 138.1, 128.7, 128.6, 128.6, 128.4, 128.0, 128.0, 127.8, 127.7, 99.0, 98.7, 78.9, 75.8, 74.8, 73.8, 73.5, 73.4, 71.0, 70.1, 69.6, 65.2, 55.6, 47.5, 36.4, and 18.2. The HRMS calculation for $C_{34}H_{43}NO_8$ was $[M + H]^+$: 594.3061, found: 594.3078.

(2*R*,3*R*,4*R*,6*R*)-4-amino-6-(((2*S*,6*S*,7*R*,8*R*,8a*S*)-7-iodo-6-methoxy-2-phenylhexahydropyrano[3,2-*d*][1,3]dioxin-8-yl)oxy)-2-methyltetrahydro-2-pyran-3-ol (**2i**).

According to the General Procedure, **1i** (19.0 mg, 0.033 mmol) was used to obtain **2i** as a white solid in 79% yield. R_f = 0.36 (CH_2Cl_2-MeOH 10:1), m.p. 89–90 °C. $[\alpha]_D^{25}$ +22.8 (c, 1.48 in $CHCl_3$). The readings for the ^{1}H NMR spectra (400 MHz, $CDCl_3$) were δ 7.43–7.39 (m, 2H, Ar-H), 7.37–7.30 (m, 3H, Ar-H), 5.57 (s, 1H, $PhCHO_2$), 5.18 (d, J = 3.2 Hz, 1H, H-1′), 5.05 (s, 1H, H-1), 4.36 (d, J = 4.4 Hz, 1H, H-2), 4.25 (dd, J = 9.6, 4.0 Hz, 1H, H-6a), 4.02–3.95 (m, 2H, H-4, H-5′), 3.90 (td, J = 10.0, 4.0 Hz, 1H, H-5), 3.84 (t, J = 10.0 Hz, 1H, H-6b), 3.48 (dd, J = 9.6, 4.4 Hz, 1H, H-3), 3.45 (br s, 1H, H-4′), 3.36 (s, 3H, OMe), 3.30 (ddd, J = 12.0, 4.8, 2.8 Hz, 1H, H-3′), 1.74 (dd, J = 13.2, 4.8 Hz, 1H, H-2′a), 1.65 (td, J = 12.4, 3.6 Hz, 1H, H-2′b), and 1.25 (d, J = 6.8 Hz, 3H, H-6′). The readings for the ^{13}C NMR spectra (100 MHz, $CDCl_3$) were δ 137.5, 129.2, 128.4, 126.2, 104.0, 101.8, 99.3, 80.8, 71.6, 70.9, 68.9, 67.4, 64.8, 55.3, 46.5, 35.2, 32.6, and 17.2. The HRMS calculation for $C_{20}H_{28}INO_7$ was $[M + H]^+$: 522.0983, found: 522.0996.

(2*R*,3*R*,4*R*,6*S*)-4-amino-6-(((3*S*,8*S*,9*S*,10*R*,13*R*,14*S*,17*R*)-10,13-dimethyl-17-((*R*)-6-methylheptan-2-yl)-2,3,4,7,8,9,10,11,12,13,14,15,16,17-tetradecahydro-1*H*-cyclopenta[a]phenanthren-3-yl)oxy)-2-methyltetrahydro-2*H*-pyran-3-ol (**2j**).

According to the General Procedure, **1j** (24.0 mg, 0.042 mmol) was used to obtain **2j** as a white solid in 81% yield. R_f = 0.26 (CH_2Cl_2-MeOH 10:1), m.p. 185–186 °C. $[\alpha]_D^{25}$ +54.0 (c, 1.62 in $CHCl_3$). The readings for the ^{1}H NMR spectra (400 MHz, $CDCl_3$) were δ 5.32 (d, J = 4.8 Hz, 1H, C = CH), 4.99 (d, J = 3.2 Hz, 1H, H-1), 3.96 (q, J = 6.4 Hz, 1H, H-5), 3.44 (m, 1H,), 3.39 (d, J = 2.4 Hz, 1H, H-4), 3.25 (ddd, J = 11.6, 4.4, 2.8 Hz, 1H, H-3), 2.32–2.21 (m, 2H), 2.00–1.92 (m, 2H), 1.84–1.77 (m, 4H), 1.71–1.63 (m, 2H), 1.61–1.37 (m, 4H), 1.51–1.36 (m, 5H), 1.36–1.28 (m, 3H), 1.23 (d, J = 6.4 Hz, 4H, H-6), 1.20–0.99 (m, 8H), 0.98–0.92 (m, 5H), 0.89 (d, J = 6.4 Hz, 3H, Me), 0.85 (d, J = 1.6 Hz, 3H, Me), 0.83 (d, J =1.6 Hz, 3H, Me), and 0.65 (s, 3H, Me). The readings for the ^{13}C NMR spectra (100 MHz, $CDCl_3$) were δ 141.2, 121.9, 95.2, 76.3, 71.4, 66.1, 57.0, 56.4, 50.4, 46.6, 42.6, 40.4, 40.0, 39.7, 37.3, 37.0, 36.4, 36.0, 33.6, 32.2, 32.1, 28.4, 28.2, 28.1, 24.5, 24.1, 23.0, 22.8, 21.3, 19.6, 18.9, 17.2, and 12.1. The HRMS calculation for $C_{33}H_{57}NO_3$ was $[M + H]^+$: 516.4411, found: 516.4422.

(2*S*,3*S*,4*S*,6*S*)-4-amino-6-(4-methoxynaphthalen-1-yl)-2-methyltetrahydro-2*H*-pyran-3-ol (**2k**).

According to the General Procedure, **1k** (20.0 mg, 0.057 mmol) was used to obtain **2k** as a white solid in 78% yield. R_f = 0.29 (CH_2Cl_2-MeOH 10:1), m.p. 112–113 °C. $[\alpha]_D^{25}$ −86.6 (c, 0.56 in $CHCl_3$). The readings for the ^{1}H NMR spectra (400 MHz, $CDCl_3$) were δ 8.28 (d, J = 8.0 Hz, 1H, Ar-H), 7.98 (d, J = 8.4 Hz, 1H, Ar-H), 7.53–7.43 (m, 3H, Ar-H), 6.76 (d, J = 8.0 Hz, 1H, Ar-H), 5.00 (d, J = 10.4 Hz, 1H, H-1), 3.97 (s, 3H, OMe), 3.82 (q, J = 6.4 Hz, 1H, H-5), 3.52 (br s, 1H, H-4), 3.18 (d, J = 10.4 Hz, 1H, H-3), 1.96–1.84 (m, 2H, H-2a, H-2b), and 1.39 (d, J = 6.4 Hz, 3H, H-6). The readings for the ^{13}C NMR spectra

(100 MHz, $CDCl_3$) were δ 155.5, 131.8, 129.3, 126.8, 126.0, 125.1, 123.6, 123.3, 122.8, 103.3, 75.7, 75.6, 71.5, 55.7, 52.2, 35.7, and 17.9. The HRMS calculation for $C_{17}H_{21}NO_3$ was [M + H]$^+$: 288.1594, found: 288.1597.

(2*S*,3*S*,4*S*,6*S*)-4-amino-6-(6-hydroxy-2,3,4-trimethoxyphenyl)-2-methyltetrahydro-2*H*-pyran-3-ol (**2l**).

According to the General Procedure, **1l** (20.0 mg, 0.053 mmol) was used to obtain **2l** as a white solid in 77% yield. R_f = 0.24 (CH_2Cl_2-MeOH 10:1), m.p. 178–179 °C. $[\alpha]_D^{25}$ −71.0 (c, 0.67 in $CHCl_3$). The readings for the ^{1}H NMR spectra (400 MHz, $CDCl_3$) were δ 6.23 (s, 1H, Ar-H), 4.90 (dd, *J* = 11.6, 2.8 Hz, 1H, H-1), 3.86 (s, 3H, OMe), 3.78 (s, 3H, OMe), 3.74 (s, 3H, OMe), 3.65 (q, *J* = 6.4 Hz, 1H, H-5), 3.45 (d, *J* = 2.4 Hz, 1H, H-4), 3.16 (d, *J* = 10.4 Hz, 1H, H-3), 1.89 (q, *J* = 12.0 Hz, 1H, H-2a), 1.61 (dq, *J* = 13.2, 3.2 Hz, 1H, H-2b), and 1.36 (d, *J* = 6.4 Hz, 3H, H-6). The readings for the ^{13}C NMR spectra (100 MHz, $CDCl_3$) were δ 153.7, 152.4, 150.1, 135.1, 111.5, 97.3, 75.3, 73.6, 69.8, 61.5, 61.1, 56.1, 50.7, 33.9, and 17.8. The HRMS calculation for $C_{15}H_{23}NO_6$ was [M + H]$^+$: 314.1598, found: 314.1625.

5-((2*S*,4*S*,5*S*,6*S*)-4-amino-5-hydroxy-6-methyltetrahydro-2*H*-pyran-2-yl)-9-hydroxy-8-methoxy-3,4-dihydroanthracen-1(2*H*)-one (**2m**).

According to the General Procedure, **1m** (15.0 mg, 0.035 mmol) was used to obtain **2m** as a yellow solid in 91% yield. R_f = 0.27 (CH_2Cl_2-MeOH 10:1), m.p. 166–167 °C. $[\alpha]_D^{25}$ −154.0 (c, 0.82 in $CHCl_3$). The readings for the ^{1}H NMR spectra (400 MHz, $CDCl_3$) were δ 7.66 (d, *J* = 8.4 Hz, 1H, Ar-H), 7.11 (s, 1H, Ar-H), 6.77 (d, *J* = 8.4 Hz, 1H, Ar-H), 4.87 (d, *J* = 10.4 Hz, 1H, H-1), 3.98 (s, 3H, OMe), 3.80 (q, *J* = 6.4 Hz, 1H, H-5), 3.52 (d, *J* = 2.0 Hz, 1H, H-4), 3.18 (dt, *J* = 11.6, 3.6 Hz, 1H, H-3), 2.99 (t, *J* = 6.0 Hz, 2H, CH_2), 2.74 (t, *J* = 6.4 Hz, 2H, CH_2), 2.13–2.06 (m, 2H, CH_2), 1.91 (dd, *J* = 13.2, 2.4 Hz, 1H, H-2a), 1.81–1.72 (m, 3H, H-2b, NH_2), and 1.39 (d, *J* = 6.8 Hz, 3H, H-6). The readings for the ^{13}C NMR spectra (100 MHz, $CDCl_3$) were δ 204.8, 166.5, 159.9, 139.5, 137.3, 129.1, 128.6, 115.5, 112.0, 111.8, 105.3, 75.7, 75.4, 71.4, 56.4, 52.1, 39.2, 35.6, 31.0, 23.0, and 17.9. The HRMS calculation for $C_{21}H_{25}NO_5$ was [M + H]$^+$: 372.1805, found: 372.1790.

(3*S*,4*R*,8*S*,9*S*,10*R*,13*R*,14*S*,17*R*)-4-amino-10,13-dimethyl-17-((*R*)-6-methylheptan-2-yl)-2,3,4,7,8,9,10,11,12,13,14,15,16,17-tetradecahydro-1*H*-cyclopenta[*a*]phenanthren-3-ol (**7a**).

To a solution of **6a** (23.0 mg, 0.050 mmol, 1.0 equiv) in THF (0.5 mL), Ph(CO)CO_2Et (0.055 mmol, 1.1 equiv) and $P(NMe_2)_3$ (0.060 mmol, 1.2 equiv) were added sequentially. After stirring for 45 min at room temperature, DBU (0.200 mmol, 4.0 equiv) and H_2O (0.5 mL) were added sequentially, and the mixture was stirred for 4 h at 65 °C. The mixture was extracted with CH_2Cl_2 after removing the THF by concentration. The organic layer was washed with saturated $NaHCO_3$ and brine, dried over Na_2SO_4, and concentrated in vacuo. The crude residue was purified using column chromatography on silica gel (dichloromethane-methanol gradient elution, with 1% $NH_3{\cdot}H_2O$) to obtain **7a** as a white solid in 70% yield. R_f = 0.40 (CH_2Cl_2-MeOH 10:1), m.p. 158–159°C. $[\alpha]_D^{25}$ −41.3 (c, 0.76 in $CHCl_3$). The readings for the ^{1}H NMR spectra (400 MHz, $CDCl_3$) were δ 5.54 (s, 1H, C = CH), 3.52–3.46 (m, 2H), 2.42 (br s, 3H, OH, NH2), 2.06–1.96 (m, 2H), 1.84–1.72 (m, 3H), 1.65–1.50 (m, 4H), 1.42–1.23 (m, 8H), 1.13–1.03 (m, 12H), 0.89 (d, *J* = 6.4 Hz, 3H, Me), 0.85 (d, *J* = 1.6 Hz, 3H, Me), 0.83 (d, *J* = 1.6 Hz, 3H, Me), and 0.65 (s, 3H, Me). The readings for the ^{13}C NMR spectra (100 MHz, $CDCl_3$) were δ 127.2, 71.5, 57.3, 56.3, 50.5, 42.5, 39.9, 39.7, 36.8, 36.4, 36.3, 36.0, 32.4, 32.1, 28.4, 28.2, 25.9, 24.5, 24.1, 23.0, 22.8, 21.9, 20.7, 18.9, and 12.1. The HRMS calculation for $C_{27}H_{47}NO$ was [M + H]$^+$: 402.3730, found: 402.3717.

tert-butyl 3-(1-amino-3-hydroxypropyl)-1*H*-indole-1-carboxylate (**7b**).

To a solution of **6b** (20.0 mg, 0.079 mmol, 1.0 equiv) in THF (0.79 mL), Ph(CO)CO_2Et (0.158 mmol, 2.0 equiv) and $P(NMe_2)_3$ (0.166 mmol, 2.1 equiv) were added sequentially. After stirring for 45 min at room temperature, DBU (0.474 mmol, 6.0 equiv) and H_2O (0.79 mL) were added sequentially, and the mixture was stirred for 4 h at 65 °C. The mixture was extracted with CH_2Cl_2 after removing the THF by concentration. The organic layer was washed with saturated $NaHCO_3$ and brine, dried over Na_2SO_4, and concentrated in vacuo. The crude residue was purified using column chromatography on silica gel (dichloromethane-methanol gradient elution, with 1% $NH_3{\cdot}H_2O$) to obtain **7b** as a yellow oil in 87% yield. According to the above procedure, **6b** (352.4 mg, 1.0 mmol) was used to obtain **7b** as a yellow oil in 82% yield. R_f = 0.25 (CH_2Cl_2-MeOH 10:1). The readings for the ^{1}H NMR spectra (400 MHz, $CDCl_3$) were δ 8.12 (d, J = 7.2 Hz, 1H, Ar-H), 7.54 (d, J = 8.4 Hz, 2H, Ar-H, C = CH), 7.30 (t, J = 7.2 Hz, 1H, Ar-H), 7.21 (t, J = 7.2 Hz, 1H, Ar-H), 4.47 (dd, J = 7.6, 4.0 Hz, 1H), 3.84 (t, J = 5.2 Hz, 2H), 3.18 (s, 3H, OH, NH_2), 2.10–1.98 (m, 2H), and 1.64 (s, 9H, $(CH_3)_3C$). The readings for the ^{13}C NMR spectra (100 MHz, $CDCl_3$) were δ 149.9, 136.0, 128.7, 124.9, 122.8, 122.0, 119.2, 115.8, 84.1, 62.2, 48.9, 37.8, and 28.4. The HRMS calculation for $C_{16}H_{22}N_2O_3$ was $[M + H]^+$: 291.1703, found: 291.1691.

4. Conclusions

In conclusion, the investigation described above has led to the development of a practical method to smoothly convert cyclic sulfamidates into amino alcohols under mild conditions. This highly efficient deprotection method is initiated with the Kukhtin–Ramirez reaction. It exhibited operational simplicity, which provided a solution to the deprotection problem encountered in our synthesis of rare amino sugar. In addition, this approach allows the construction of valuable building blocks and structurally complex compounds containing amino alcohol motifs.

Supplementary Materials: The following supporting information can be downloaded at https://www.mdpi.com/article/10.3390/molecules28010182/s1: synthesis of compounds **1a–1m**, **6a–6b**, and 1,2-dicarbonyl reagents (**S3–S5**); Table S1: screening of 1,2-dicarbonyl compounds; ^{1}H NMR spectra for **6a–6b** and **S3–S5** [70,71]; ^{1}H and ^{13}C NMR spectra for compounds **2a–2m**, **3a**, and **4a**; 2D HSQC and COSY NMR spectra for compounds **2a**, **3a**, and **4a**.

Author Contributions: Conceptualization, J.Z.; methodology, J.Z. and Q.W.; investigation, T.L. and B.X.; writing—original draft preparation, T.L.; writing—review and editing, J.Z., Q.W. and D.F.; supervision and funding acquisition, J.Z. and Q.W. All authors have read and agreed to the published version of the manuscript.

Funding: This research was funded by the National Natural Science Foundation of China (22077039, 22277033, 21772050, 21761132014), the National Science and Technology Innovation 2030—Major program of "Brain Science and Brain-Like Research" (2022ZD0211800), and the Open Projects Fund of the Shandong Key Laboratory of Carbohydrate Chemistry and Glycobiology, Shandong University (2021CCG02).

Institutional Review Board Statement: Not applicable.

Informed Consent Statement: Not applicable.

Data Availability Statement: Data are included within the manuscript or the Supplementary Data File.

Conflicts of Interest: The authors declare no conflict of interest.

References

1. Heravi, M.M.; Lashaki, T.B.; Fattahi, B.; Zadsirjan, V. Application of asymmetric Sharpless aminohydroxylation in total synthesis of natural products and some synthetic complex bio-active molecules. *RSC Adv.* **2018**, *8*, 6634–6659. [CrossRef] [PubMed]
2. Dimarco, A.; Gaetani, M.; Orezzi, P.; Scarpinato, B.M.; Silvestrini, R.; Soldati, M.; Dasdia, T.; Valentini, L. 'Daunomycin', a new antibiotic of the rhodomycin group. *Nature* **1964**, *201*, 706–707. [CrossRef] [PubMed]
3. Bergmeier, S.C.; Stanchina, D.M. Acylnitrene route to vicinal amino alcohols. Application to the synthesis of (-)-Bestatin and analogues. *J. Org. Chem.* **1999**, *64*, 2852–2859. [CrossRef] [PubMed]
4. Ager, D.J.; Prakash, I.; Schaad, D.R. 1,2-Amino alcohols and their heterocyclic derivatives as chiral auxiliaries in asymmetric synthesis. *Chem. Rev.* **1996**, *96*, 835–876. [CrossRef] [PubMed]
5. Fache, F.; Schulz, E.; Tommasino, M.L.; Lemaire, M. Nitrogen-containing ligands for asymmetric homogeneous and heterogeneous catalysis. *Chem. Rev.* **2000**, *100*, 2159–2231. [CrossRef] [PubMed]
6. Shrestha, B.; Rose, B.T.; Olen, C.L.; Roth, A.; Kwong, A.C.; Wang, Y.; Denmark, S.E. A unified strategy for the asymmetric synthesis of highly substituted 1,2-amino alcohols leading to highly substituted bisoxazoline ligands. *J. Org. Chem.* **2021**, *86*, 3490–3534. [CrossRef] [PubMed]

7. Malkov, A.V.; Kabeshov, M.A.; Bella, M.; Kysilka, O.; Malyshev, D.A.; Pluháčková, K.; Kočovský, P. Vicinal amino alcohols as organocatalysts in asymmetric cross-aldol reaction of ketones: Application in the synthesis of convolutamydine A. *Org. Lett.* **2007**, *9*, 5473–5476. [CrossRef]
8. Heravi, M.M.; Zadsirjan, V.; Farajpour, B. Applications of oxazolidinones as chiral auxiliaries in the asymmetric alkylation reaction applied to total synthesis. *RSC Adv.* **2016**, *6*, 30498–30551. [CrossRef]
9. Lait, S.M.; Rankic, D.A.; Keay, B.A. 1,3-Aminoalcohols and their derivatives in asymmetric organic synthesis. *Chem. Rev.* **2007**, *107*, 767–796. [CrossRef]
10. Donohoe, T.J.; Callens, C.K.; Flores, A.; Lacy, A.R.; Rathi, A.H. Recent developments in methodology for the direct oxyamination of olefins. *Chemistry* **2011**, *17*, 58–76. [CrossRef]
11. Palomo, C.; Oiarbide, M.; Laso, A. Recent advances in the catalytic asymmetric nitroaldol (Henry) reaction. *Eur. J. Org. Chem.* **2007**, *2007*, 2561–2574. [CrossRef]
12. Gupta, P.; Mahajan, N. Biocatalytic approaches towards the stereoselective synthesis of vicinal amino alcohols. *New J. Chem.* **2018**, *42*, 12296–12327. [CrossRef]
13. Karjalainen, O.K.; Koskinen, A.M. Diastereoselective synthesis of vicinal amino alcohols. *Org. Biomol. Chem.* **2012**, *10*, 4311–4326. [CrossRef] [PubMed]
14. Bergmeier, S.C. The synthesis of vicinal amino alcohols. *Tetrahedron* **2000**, *56*, 2561–2576. [CrossRef]
15. Burchak, O.N.; Py, S. Reductive cross-coupling reactions (RCCR) between C=N and C=O for β-amino alcohol synthesis. *Tetrahedron* **2009**, *65*, 7333–7356. [CrossRef]
16. Appel, R.; Berger, G. Hydrazinsulfonsäure-amide, I. Über das Hydrazodisulfamid. *Chem. Ber.* **1958**, *91*, 1339–1341. [CrossRef]
17. Armitage, I.; Berne, A.M.; Elliott, E.L.; Fu, M.; Hicks, F.; McCubbin, Q.; Zhu, L. N-(tert-butoxycarbonyl)-N-[(triethylenediammonium) sulfonyl] azanide: A convenient sulfamoylation reagent for alcohols. *Org. Lett.* **2012**, *14*, 2626–2629. [CrossRef]
18. Okada, M.; Iwashita, S.; Koizumi, N. Efficient general method for sulfamoylation of a hydroxyl group. *Tetrahedron Lett.* **2000**, *41*, 7047–7051. [CrossRef]
19. Rapp, P.B.; Murai, K.; Ichiishi, N.; Leahy, D.K.; Miller, S.J. Catalytic sulfamoylation of alcohols with activated aryl sulfamates. *Org. Lett.* **2020**, *22*, 168–174. [CrossRef]
20. Sguazzin, M.A.; Johnson, J.W.; Magolan, J. Hexafluoroisopropyl sulfamate: A useful reagent for the synthesis of sulfamates and sulfamides. *Org. Lett.* **2021**, *23*, 3373–3378. [CrossRef]
21. Wang, H.M.; Xiong, C.D.; Chen, X.Q.; Hu, C.; Wang, D.Y. Preparation of sulfamates and sulfamides using a selective sulfamoylation agent. *Org. Lett.* **2021**, *23*, 2595–2599. [CrossRef] [PubMed]
22. Collet, F.; Dodd, R.H.; Dauban, P. Catalytic C-H amination: Recent progress and future directions. *Chem. Commun.* **2009**, 5061–5074. [CrossRef] [PubMed]
23. Bower, J.F.; Rujirawanich, J.; Gallagher, T. N-heterocycle construction via cyclic sulfamidates. Applications in synthesis. *Org. Biomol. Chem.* **2010**, *8*, 1505–1519. [CrossRef] [PubMed]
24. Pham, Q.H.; Hyland, C.J.T.; Pyne, S.G. Five-membered cyclic sulfamidate imines: Versatile scaffolds for organic synthesis. *Org. Biomol. Chem.* **2020**, *18*, 7467–7484. [CrossRef]
25. Meléndez, R.E.; Lubell, W.D. Synthesis and reactivity of cyclic sulfamidites and sulfamidates. *Tetrahedron* **2003**, *59*, 2581–2616. [CrossRef]
26. Nicolaou, K.C.; Huang, X.; Snyder, S.A.; Bheema Rao, P.; Bella, M.; Reddy, M.V. A novel regio- and stereoselective synthesis of sulfamidates from 1,2-diols using Burgess and related reagents: A facile entry into β-amino alcohols. *Angew. Chem. Int. Ed.* **2002**, *41*, 834–838. [CrossRef]
27. Nicolaou, K.C.; Snyder, S.A.; Longbottom, D.A.; Nalbandian, A.Z.; Huang, X. New uses for the Burgess reagent in chemical synthesis: Methods for the facile and stereoselective formation of sulfamidates, glycosylamines, and sulfamides. *Chemistry* **2004**, *10*, 5581–5606. [CrossRef]
28. Lee, H.K.; Kang, S.; Choi, E.B. Stereoselective synthesis of norephedrine and norpseudoephedrine by using asymmetric transfer hydrogenation accompanied by dynamic kinetic resolution. *J. Org. Chem.* **2012**, *77*, 5454–5460. [CrossRef]
29. Kang, S.; Han, J.; Lee, E.S.; Choi, E.B.; Lee, H.K. Enantioselective synthesis of cyclic sulfamidates by using chiral Rhodium-catalyzed asymmetric transfer hydrogenation. *Org. Lett.* **2010**, *12*, 4184–4187. [CrossRef]
30. Wang, Y.Q.; Yu, C.B.; Wang, D.W.; Wang, X.B.; Zhou, Y.G. Enantioselective synthesis of cyclic sulfamidates via Pd-catalyzed hydrogenation. *Org. Lett.* **2008**, *10*, 2071–2074. [CrossRef]
31. Alderson, J.M.; Phelps, A.M.; Scamp, R.J.; Dolan, N.S.; Schomaker, J.M. Ligand-controlled, tunable silver-catalyzed C-H amination. *J. Am. Chem. Soc.* **2014**, *136*, 16720–16723. [CrossRef] [PubMed]
32. Espino, C.G.; Wehn, P.M.; Chow, J.; Du Bois, J. Synthesis of 1,3-difunctionalized amine derivatives through selective C−H bond oxidation. *J. Am. Chem. Soc.* **2001**, *123*, 6935–6936. [CrossRef]
33. Paradine, S.M.; Griffin, J.R.; Zhao, J.; Petronico, A.L.; Miller, S.M.; Christina White, M. A manganese catalyst for highly reactive yet chemoselective intramolecular C(sp^3)-H amination. *Nat Chem.* **2015**, *7*, 987–994. [CrossRef] [PubMed]
34. Hazelard, D.; Nocquet, P.A.; Compain, P. Catalytic C–H amination at its limits: Challenges and solutions. *Org. Chem. Front.* **2017**, *4*, 2500–2521. [CrossRef]
35. Ruppel, J.V.; Kamble, R.M.; Zhang, X.P. Cobalt-catalyzed intramolecular C−H amination with arylsulfonyl azides. *Org. Lett.* **2007**, *9*, 4889–4892. [CrossRef] [PubMed]

36. Jiang, C.; Lu, Y.; Hayashi, T. High performance of a palladium phosphinooxazoline catalyst in the asymmetric arylation of cyclic N-sulfonyl ketimines. *Angew. Chem. Int.Ed.* **2014**, *53*, 9936–9939. [CrossRef]
37. Higginbotham, M.C.; Bebbington, M.W. Gold(I)-catalysed synthesis of cyclic sulfamidates by intramolecular allene hydroamination. *Chem. Commun.* **2012**, *48*, 7565–7567. [CrossRef]
38. Chen, Y.J.; Chen, Y.H.; Feng, C.G.; Lin, G.Q. Enantioselective Rhodium-catalyzed arylation of cyclic N-sulfamidate alkylketimines: A new access to chiral β-alkyl-β-aryl amino alcohols. *Org. Lett.* **2014**, *16*, 3400–3403. [CrossRef]
39. Weymouth-Wilson, A.C. The role of carbohydrates in biologically active natural products. *Nat. Prod. Rep.* **1997**, *14*, 99–110. [CrossRef]
40. William Lown, J. Anthracycline and anthraquinone anticancer agents: Current status and recent developments. *Pharmacol. Ther.* **1993**, *60*, 185–214. [CrossRef]
41. Zeng, J.; Sun, G.; Yao, W.; Zhu, Y.; Wang, R.; Cai, L.; Liu, K.; Zhang, Q.; Liu, X.W.; Wan, Q. 3-Aminodeoxypyranoses in glycosylation: Diversity-oriented synthesis and assembly in oligosaccharides. *Angew. Chem. Int.Ed.* **2017**, *56*, 5227–5231. [CrossRef] [PubMed]
42. Zeng, J.; Wang, R.; Yao, W.; Zhang, S.; Sun, G.; Liao, Z.; Meng, L.; Wan, Q. Diversified synthesis and α-selective glycosylation of 3-amino-2,3,6-trideoxy sugars. *Org. Chem. Front.* **2018**, *5*, 3391–3395. [CrossRef]
43. Fu, D.; Zhang, S.; Xu, B.; Peng, P.; Wan, Q.; Zeng, J. Selective reduction leading to 3,5-*cis*-3-aminosugars: Synthesis and stereoselective glycosylation. *J. Org. Chem.* **2022**. [CrossRef] [PubMed]
44. Yamashita, S.; Himuro, M.; Hayashi, Y.; Hirama, M. Remote C–H bond functionalization of androstane C-ring: C12-amination. *Tetrahedron Lett.* **2013**, *54*, 1307–1308. [CrossRef]
45. Liu, M.-Q.; Jiang, T.; Chen, W.-W.; Xu, M.-H. Highly enantioselective Rh/chiral sulfur-olefin-catalyzed arylation of alkyl-substituted non-benzofused cyclic N-sulfonyl ketimines. *Org. Chem. Front.* **2017**, *4*, 2159–2162. [CrossRef]
46. Kukhtin, V.A. Some new types of Arbuzov's regrouping. *Dokl. Akad. Nauk SSSR* **1958**, *121*, 466–469.
47. Ramirez, F. Condensations of carbonyl compounds with phosphite esters. *Pure Appl. Chem.* **1964**, *9*, 337–370. [CrossRef]
48. Ramirez, F. Oxyphosphoranes. *Acc. Chem. Res.* **1968**, *1*, 168–174. [CrossRef]
49. Miller, E.J.; Zhao, W.; Herr, J.D.; Radosevich, A.T. A nonmetal approach to α-heterofunctionalized carbonyl derivatives by formal reductive X-H insertion. *Angew. Chem. Int. Ed.* **2012**, *51*, 10605–10609. [CrossRef]
50. Harpp, D.N.; Mathiaparanam, P. 1,3,2-Dioxaphospholene-sulfenyl chloride condensation. Scope and mechanism. *J. Org. Chem.* **1972**, *37*, 1367–1374. [CrossRef]
51. Zhao, W.; Yan, P.K.; Radosevich, A.T. A phosphetane catalyzes deoxygenative condensation of α-keto esters and carboxylic acids via PIII/PV=O redox cycling. *J. Am. Chem. Soc.* **2015**, *137*, 616–619. [CrossRef] [PubMed]
52. Zhao, W.; Fink, D.M.; Labutta, C.A.; Radosevich, A.T. A Csp^3-Csp^3 bond forming reductive condensation of α-keto esters and enolizable carbon pronucleophiles. *Org. Lett.* **2013**, *15*, 3090–3093. [CrossRef] [PubMed]
53. Haugen, K.C.; Rodriguez, K.X.; Chavannavar, A.P.; Oliver, A.G.; Ashfeld, B.L. Phosphine-mediated addition of 1,2-dicarbonyls to diazenes: An umpolung approach toward N-acyl hydrazone synthesis. *Tetrahedron Lett.* **2015**, *56*, 3527–3530. [CrossRef]
54. Zhang, L.; Lu, H.; Xu, G.Q.; Wang, Z.Y.; Xu, P.F. PPh_3 mediated reductive annulation reaction between isatins and electron deficient dienes to construct spirooxindole compounds. *J. Org. Chem.* **2017**, *82*, 5782–5789. [CrossRef] [PubMed]
55. Zhou, R.; Yang, C.; Liu, Y.; Li, R.; He, Z. Diastereoselective synthesis of functionalized spirocyclopropyl oxindoles via $P(NMe_2)_3$-mediated reductive cyclopropanation. *J. Org. Chem.* **2014**, *79*, 10709–10715. [CrossRef]
56. Rodriguez, K.X.; Vail, J.D.; Ashfeld, B.L. Phosphorus(III)-mediated stereoconvergent formal [4+1]-cycloannulation of 1,2-dicarbonyls and *o*-quinone methides: A multicomponent assembly of 2,3-dihydrobenzofurans. *Org. Lett.* **2016**, *18*, 4514–4517. [CrossRef]
57. Corre, E.; Foucaud, A. Synthesis of 2-acetyl-3-aryl-1,1-dicyano-2-methyl-cyclopropanes. *J. Chem. Soc. D* **1971**, 570a. [CrossRef]
58. Choi, G.; Kim, H.E.; Hwang, S.; Jang, H.; Chung, W.J. Phosphorus(III)-mediated, tandem deoxygenative geminal chlorofluorination of 1,2-diketones. *Org. Lett.* **2020**, *22*, 4190–4195. [CrossRef]
59. Wang, S.R.; Radosevich, A.T. $P(NMe_2)_3$-mediated umpolung alkylation and nonylidic olefination of α-keto esters. *Org. Lett.* **2015**, *17*, 3810–3813. [CrossRef]
60. Calcatelli, A.; Denton, R.M.; Ball, L.T. Modular synthesis of α,α-diaryl α-amino esters via Bi(V)-mediated arylation/S_N2-displacement of Kukhtin–Ramirez intermediates. *Org. Lett.* **2022**, *24*, 8002–8007. [CrossRef]
61. Fier, P.S.; Kim, S.; Maloney, K.M. Reductive cleavage of secondary sulfonamides: Converting terminal functional groups into versatile synthetic handles. *J. Am. Chem. Soc.* **2019**, *141*, 18416–18420. [CrossRef] [PubMed]
62. Zhang, J.-R.; Liao, Y.-Y.; Deng, J.-C.; Tang, Z.-L.; Xu, Y.-L.; Xu, L.; Tang, R.-Y. DABCO-promoted decarboxylative acylation: Synthesis of α-keto and α,β-unsaturated amides or esters. *Asian J. Org. Chem.* **2017**, *6*, 305–312. [CrossRef]
63. Yu, T.T.; Nizalapur, S.; Ho, K.K.K.; Yee, E.; Berry, T.; Cranfield, C.G.; Willcox, M.; Black, D.S.; Kumar, N. Design, Synthesis and biological evaluation of N-sulfonylphenyl glyoxamide-based antimicrobial peptide mimics as novel antimicrobial agents. *ChemistrySelect* **2017**, *2*, 3452–3461. [CrossRef]
64. Ryu, H.; Seo, J.; Ko, H.M. Synthesis of spiro[oxindole-3,2′-pyrrolidine] derivatives from benzynes and azomethine ylides through 1,3-dipolar cycloaddition reactions. *J. Org. Chem.* **2018**, *83*, 14102–14109. [CrossRef] [PubMed]

65. Shu, P.; Xiao, X.; Zhao, Y.; Xu, Y.; Yao, W.; Tao, J.; Wang, H.; Yao, G.; Lu, Z.; Zeng, J.; et al. Interrupted Pummerer reaction in latent-active glycosylation: Glycosyl donors with a recyclable and regenerative leaving group. *Angew. Chem. Int. Ed.* **2015**, *54*, 14432–14436. [CrossRef]
66. Xiao, X.; Zhao, Y.; Shu, P.; Zhao, X.; Liu, Y.; Sun, J.; Zhang, Q.; Zeng, J.; Wan, Q. Remote activation of disarmed thioglycosides in latent-active glycosylation via interrupted Pummerer reaction. *J. Am. Chem. Soc.* **2016**, *138*, 13402–13407. [CrossRef]
67. Zhao, X.; Zeng, J.; Meng, L.; Wan, Q. Application of interrupted Pummerer reaction mediated (IPRm) glycosylation in natural product synthesis. *Chem. Rec.* **2020**, *20*, 743–751. [CrossRef]
68. Xiao, X.; Zeng, J.; Fang, J.; Sun, J.; Li, T.; Song, Z.; Cai, L.; Wan, Q. One-pot relay glycosylation. *J. Am. Chem. Soc.* **2020**, *142*, 5498–5503. [CrossRef]
69. Cai, L.; Chen, Q.; Liang, Z.; Fu, D.; Meng, L.; Zeng, J.; Wan, Q. Recyclable fluorous-tag assisted two-directional oligosaccharide synthesis enabled by interrupted Pummerer reaction mediated glycosylation. *Chem. Sci.* **2022**, *13*, 8759–8765. [CrossRef]
70. Paradine, S.M.; White, M.C. Iron-catalyzed intramolecular allylic C-H amination. *J. Am. Chem.Soc.* **2012**, *134*, 2036–2039. [CrossRef]
71. Zhang, X.; Wang, L. TBHP/I2-promoted oxidative coupling of acetophenones with amines at room temperature under metal-free and solvent-free conditions for the synthesis of α-ketoamides. *Green Chem.* **2012**, *14*, 2141–2145. [CrossRef]

molecules

MDPI

Review

Impact of N-Linked Glycosylation on Therapeutic Proteins

Baoquan Chen [1], Wenqiang Liu [1], Yaohao Li [1], Bo Ma [1], Shiying Shang [2,*] and Zhongping Tan [1,*]

1 State Key Laboratory of Bioactive Substance and Function of Natural Medicines, Institute of Materia Medica, Chinese Academy of Medical Sciences and Peking Union Medical College, Beijing 100050, China
2 Center of Pharmaceutical Technology, School of Pharmaceutical Sciences, Tsinghua University, Beijing 100084, China
* Correspondence: shangshiying@tsinghua.edu.cn (S.S.); zhongping.tan@imm.pumc.edu.cn (Z.T.)

Abstract: Therapeutic proteins have unique advantages over small-molecule drugs in the treatment of various diseases, such as higher target specificity, stronger pharmacological efficacy and relatively low side effects. These advantages make them increasingly valued in drug development and clinical practice. However, although highly valued, the intrinsic limitations in their physical, chemical and pharmacological properties often restrict their wider applications. As one of the most important post-translational modifications, glycosylation has been shown to exert positive effects on many properties of proteins, including molecular stability, and pharmacodynamic and pharmacokinetic characteristics. Glycoengineering, which involves changing the glycosylation patterns of proteins, is therefore expected to be an effective means of overcoming the problems of therapeutic proteins. In this review, we summarize recent efforts and advances in the glycoengineering of erythropoietin and IgG monoclonal antibodies, with the goals of illustrating the importance of this strategy in improving the performance of therapeutic proteins and providing a brief overview of how glycoengineering is applied to protein-based drugs.

Keywords: glycosylation; proteins; erythropoietin; monoclonal antibodies

check for updates

Citation: Chen, B.; Liu, W.; Li, Y.; Ma, B.; Shang, S.; Tan, Z. Impact of N-Linked Glycosylation on Therapeutic Proteins. *Molecules* **2022**, 27, 8859. https://doi.org/10.3390/molecules27248859

Academic Editors: Jian Yin, Jing Zeng and De-Cai Xiong

Received: 30 October 2022
Accepted: 30 November 2022
Published: 13 December 2022

1. Introduction

According to the central dogma, the genetic information carried on DNA is transcribed to RNA and then translated to protein, with protein generally considered the functional end product in the process. However, it has been demonstrated that this process is actually far more complex than what is initially described by the central dogma. The number of expressed proteins could be orders of magnitude greater than the number of protein-coding genes, due to alternative splicings, variable promoter usage, post-translational modifications (PTMs), and other regulatory mechanisms. Among these mechanisms, PTMs are important contributors to the vast diversity of proteomes and can lead to an exponential increase in the complexity of the proteome, relative to that of the transcriptome or genome. A wide range of PTMs, including phosphorylation, glycosylation, ubiquitination, acetylation, and methylation, have been identified [1] (Figure 1). Of these, glycosylation is the most common and complex PTM on secreted proteins. It is estimated that about 85% of secreted proteins are glycosylated [2].

Protein glycosylation occurs mainly in the endoplasmic reticulum (ER) and the Golgi apparatus, where glycosyl donors are covalently linked to target glycosyl acceptors (such as proteins and lipids) through enzyme-catalyzed processes involving approximately 200 glycosyltransferases [3]. There are two main types of protein glycosylation: N-linked glycosylation (N-glycosylation) and O-linked glycosylation (O-glycosylation). In N-glycosylation, the glycans are covalently attached to the side-chain nitrogen (N) atoms of the Asn residues in the N-X-S/T sequons, where X is any amino acid except proline. In O-glycosylation, the side-chain oxygen (O) atoms of the Ser/Thr residues are used as the connection points for the glycans.

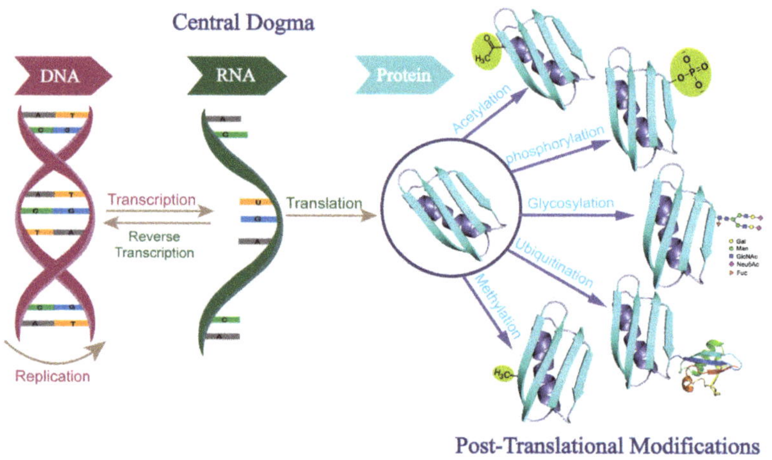

Figure 1. Central dogma of molecular biology and different forms of post-translational modifications. Abbreviations: Gal, galactose; Man, mannose; GlcNAc, N-acetylglucosamine; Neu5Ac, sialic acid; Fuc, fucose.

Unlike the synthesis of DNA, RNA and protein, glycosylation is not a template-driven process. It is regulated by many different factors, such as the relative accessibility of potential glycosylation sites and the availability of activated glycosyl donors and glycosyltransferases [4]. Due to the lack of tight control of the process of glycan biosynthesis, glycoproteins secreted from cells usually exist as complex mixtures of up to a hundred different glycosylated protein isoforms (glycoforms), which differ in their glycosylation sites and/or glycan structures [5,6]. The compositions of glycoform mixtures can vary significantly depending on the cell types and expression conditions [5,6]. Different glycoforms have different biological properties and functions [7,8]. It thus appears quite possible to develop new therapeutic proteins, or to improve the efficacy of existing protein-based drugs, by changing the glycosylation patterns of proteins (glycoengineering).

Therapeutic proteins as macromolecules have favorable characteristics, such as higher specificity, better efficacy and lower side effects, compared to the small-molecule drugs that have been used clinically for centuries. This is why they are now widely accepted and administered to patients with cancer or other life-threatening diseases [9]. However, due to their complex structures and large sizes, therapeutic proteins also have unfavorable characteristics, such as limited solubility, stability [10] and biological properties, which could lead to less desirable outcomes in clinical use, and ineffective or even harmful treatments. Substantial efforts have been devoted to minimizing these problems [11]. With a continuously increasing number of tools available for manipulating glycosylation sites and glycan structures (glycosylation patterns), glycoengineering has become an attractive strategy for achieving such a goal [12,13].

Previous glycoengineering efforts have demonstrated the feasibility of this strategy [14]. However, the number of successful applications has so far been limited, due to insufficient understanding of the structure–function relationship of protein glycosylation, and a lack of reliable scientific theories to guide the glycoengineering design process. To date, the most well-known examples of glycoengineering are erythropoietin (EPO) and immunoglobulin G (IgG) antibodies. In this review, we will summarize and discuss current knowledge about the glycoengineering of these two types of representative therapeutic proteins, with the goals of providing a brief overview of the studies undertaken and the current status of this research area, and of facilitating the future application of glycoengineering to develop more successful protein-based drugs. In addition to these two representative examples, there are many coagulation factors, cytokines, and hormone-based therapeutic proteins whose properties have been reported to be affected by glycosylation. A detailed description of these reported findings is beyond the scope of this mini review, and the interested reader is referred to the excellent recent review articles for more comprehensive information [14–18].

2. Erythropoietin

Human erythropoietin (HuEPO) is a cytokine. It is mainly secreted by renal interstitial cells, but a small amount can also be synthesized by hepatocytes (Figure 2A). Its expression is regulated by the blood oxygen level and controlled by the hypoxia-inducible transcription factor-1 (HIF-1) [19,20]. HuEPO was first isolated and purified by Goldwasser et al. in 1977 from the urine of patients with aplastic anemia [21]. Subsequent studies found that the HuEPO gene encodes a protein precursor of 193 amino acids. Cleavage of a 27-amino-acid signal peptide from the N-terminus of this precursor yields a protein of 166 amino acids [22] (Figure 2B). The C-terminal arginine residue is proteolytically removed prior to secretion, resulting in a mature protein of 165 amino acids.

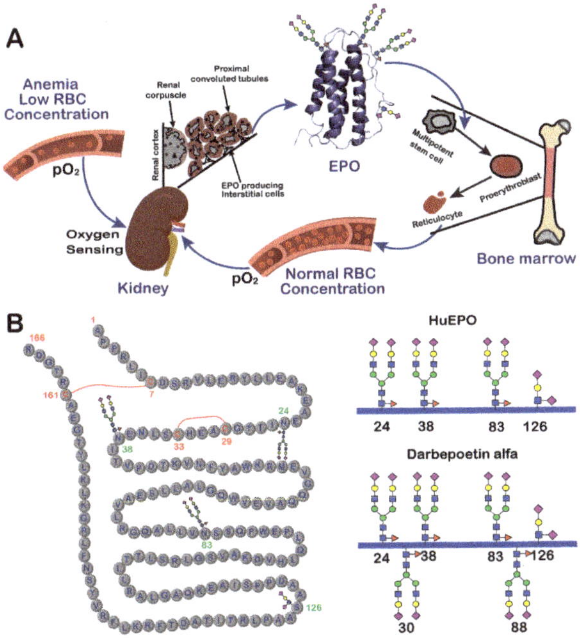

Figure 2. The production, function and structure of EPO. (**A**) A schematic view of the feedback loop mechanism of EPO production and function. (**B**) The amino acid sequence and glycan structures of different EPO variants.

HuEPO consists of four α-helices and contains two disulfide bonds, one between Cys7 and Cys161, and the other between Cys29 and Cys33. Natural HuEPO is a heavily glycosylated protein. It has a much higher molecular weight (about 30 kDa) than the unglycosylated one (about 18 kDa), with the glycan moiety comprising approximately 40% of its total molecular weight. HuEPO has three N-glycosylation sites at Asn24, Asn38 and Asn83, and one O-glycosylation site at Ser126 [23]. Characterization of the glycosylation of HuEPO revealed that the three N-glycans on this protein typically have highly sialylated bi-, tri- or tetra-antennary structures, and the O-glycan has a mucin-core-1-type structure. The same as most other glycoproteins, HuEPO isolated from human urine always exists as heterogeneous mixtures of glycoforms. Different glycoforms have been demonstrated to have different properties and functions. For example, it was found that HuEPO glycoforms treated with sialidase, which catalyzes the removal of the terminal sialic acid residues from glycoproteins, had no in vivo erythropoietic activity [24].

It was not possible to obtain sufficient amounts of HuEPO from human urine to meet the clinical needs of patients. In order to address this gap, in 1985, Jacobs and Lin successfully cloned and expressed the gene encoding HuEPO [25,26]. This achievement made it possible to produce recombinant human erythropoietin (rHuEPO) in Chinese hamster ovary cells on a manufacturing scale. In 1989, rHuEPO was approved by the US Food and Drug Administration (FDA) for clinical use under the trade names Epogen®/Procrit® and Eprex®. Since then, rHuEPO has become one of the most successful glycoprotein drugs. It is now widely employed for the treatment of anemia of various causes, such as renal anemia and tumor-related anemia. This application has changed the way of treating patients with end-stage renal disease on chronic hemodialysis, where blood transfusion treatment used to be the only means of survival. The administration of rHuEPO not only improves hemoglobin levels and anemia symptoms, but also strongly stimulates bone marrow erythroid progenitor cells to increase the number of mature red blood cells [27] (Figure 2A).

As a typical therapeutic protein, rHuEPO shares the same disadvantages as other protein-based drugs. For example, it must be administered by injection and, because of a relatively short half-life, multiple injections are required to maintain an effective therapeutic level, which frequently leads to low quality of life for patients [28]. One of the main parameters responsible for the short half-life of EPO is believed to be the rate of body clearance. At present, the exact pathway via which EPO is removed from the circulation has not been fully elucidated. It is generally speculated that this mainly occurs in the liver and kidneys, and is mediated by receptors, such as the EPO receptor (EPOR), on the cell surface. Previous findings suggest that the interaction between EPO and EPOR promotes the cellular uptake and degradation of rHuEPO through endocytosis, and that the disappearance rate of rHuEPO is directly related to the number of EPORs: if the total number of EPORs in chemotherapy patients is low, the EPO clearance rate is also low [29]. Studies have also shown that EPO glycoforms lacking sialic acid could be recognized and rapidly cleared by asialoglycoprotein receptors (ASGPR) on the surface of hepatocytes [30]. However, more than 90% of glycans in rHuEPO are fully sialylated, so the ASGPR-mediated process may be the major mechanism for the clearance of rHuEPO, only after sialic acid is removed from the serum by sialidase [31].

To improve the compliance of anemia patients, novel EPO derivatives with extended in vivo half-lives have long been the focus of research in the field of medicine. Inspired by the observation that the molecular size of glycans, and the number of sialic acids, have a significant effect on the clearance rate of proteins, a glycoengineered long-lasting EPO derivative, darbepoetin alfa (trade name Aranesp®), was developed and launched by Amgen in 2001. Compared to natural human EPO, darbepoetin alfa has two additional N-linked glycans at positions 30 and 88 (Figure 2B) [32]. This change increases its half-life in the circulation by a factor of 3, and reduces the frequency of its administration to once every 1 or 2 weeks [33]. The improvement in the properties of darbepoetin alfa makes it a very successful therapeutic agent. In 2021, the global sales of Aranesp® reached USD 1.5 billion [34].

The enhanced and prolonged biological effects of darbepoetin alfa are apparently due to a greater resistance to degradation and not due to a higher binding affinity for EPOR. Its binding affinity to EPOR is actually lower than that of rHuEPO, which is likely to be the result of the increased glycan density and sialic acid content. The large size of the glycans, together with their dynamic properties, may have the capability of sterically hindering the interactions between darbepoetin alfa and EPOR. At the same time, the increased charges carried by the additional sialic acid residues may also have a negative effect on the binding to EPOR [35]. Overall, the combined effects of these two factors may act to partially reduce EPOR-binding-mediated endocytosis, and thus increase the half-life of darbepoetin alfa.

The additional two sialylated N-linked glycans at Asn30 and Asn88 increase the size of darbepoetin alfa. This increase is believed to contribute to the improvement in the pharmacokinetics of the drug, which is likely to be related to the renal clearance of proteins (which occurs primarily through glomerular filtration). The glomerular filtration rate decreases as the size of the protein increases, and the molecular-weight threshold limiting the glomerular filtration is approximately 40 Å [36]. When the size of a protein is small, it can readily pass through the glomerular filtration barrier under normal conditions, exhibiting a rapid clearance from the circulation. When the size approaches 40 kDa, the glomerular filtration rate drops significantly. The introduction of two additional N-glycans changes the molecular weight of darbepoetin alfa by approximately 10 kDa, to about 40 kDa, which is a 22% increase compared to rHuEPO. At the same time, the N-glycans can occupy a large space, further increasing the overall size of darbepoetin alfa. Its larger size is apparently effective in reducing the glomerular filtration rate. The circulation time of darbepoetin alfa is extended from 8 h to about 25 h [37].

Glycoengineering as a strategy to increase protein size with a view to reducing the glomerular filtration rate also has some limitations. First of all, if the protein is too small, the addition of a large amount of glycans is required to make it possible to extend the size of the protein to the level approaching the glomerular filtration threshold. Such modification could significantly affect the interaction between the protein and its targets, by mechanisms such as the steric blocking of binding sites. Second, the effect of the added glycans on the glomerular filtration rate depends on many factors, including the glycosylation site and the glycan orientation. In order to achieve an optimal effect, the properties and functions of a series of glycoforms, with many different glycosylation patterns, should be analyzed and compared. The preparation of these glycoforms could be time-consuming and costly. Third, if the protein size exceeds 40 kDa, further increasing its size by glycoengineering is unlikely to contribute much to the prolongation of the circulation time.

Protein size change alone is not sufficient to fully explain the significantly prolonged clearance time of darbepoetin alfa. Studies also suggest that the clearance time may be closely related to the level of sialylation [38]. The terminal sialic acid residues of circulating glycoproteins can protect them against clearance by ASGPR, thus leading to longer serum half-lives. Darbepoetin alfa contains 5 N-linked glycans and up to 22 sialic acids. In addition to increasing the size of this therapeutic protein, the highly sialylated glycans also contribute to suppressing the binding of ASGPR, thereby inhibiting the endocytosis mediated by ASGPR and subsequent degradation by lysosomal proteases. Another type of receptor that is involved in the elimination of glycoproteins is the mannose receptor (ManR) [39]. It can recognize glycans with mannose as their terminal residues. Again, terminal sialylation can minimize the binding of ManR and prolong the action of proteins.

3. Monoclonal Antibodies

rhuEPO is a relatively small protein, with a molecular weight of about 30 kDa. Its biological activity can be increased by introducing additional N-linked glycosylation sites onto the protein surface to prolong its circulation in the blood. However, such a glycoengineering strategy is not equally useful for large therapeutic proteins, such as monoclonal antibodies (mAbs), which have an average molecular weight of approximately 150 kDa. Therefore, the glycoengineering studies in the area of mAbs are not focused on extending the circulation time by introducing new glycans onto their surface, but rather on fine-tuning the structures of glycans that are naturally found on mAbs. Accordingly, the primary task in the glycoengineering of therapeutic antibodies is to gain a better understanding of the correlation between glycan structure and function.

It is well known that the immune system consists of a variety of cells, organs, and pro- and anti-inflammatory mediators throughout the body. These components form complex networks that interact with and modulate each other through cascades and positive and negative feedback mechanisms to maintain normal inflammation and immunity. Exogenous or endogenous stresses may disrupt this delicate balance, leading to the development of various immunological diseases. Traditionally, these diseases are treated by the administration of non-specific immunosuppressive and immunomodulatory agents, such as glucocorticoids, to regulate immune response. Although effective, such treatment may induce side effects due to the non-specific nature of the agents. In order to prevent the side effects and reach the desired treatment results, in recent years, more specific immunomodulatory therapies, such as mAbs, have been developed [40].

Antibodies play a central role in the function of the human immune system. They can bind to a variety of soluble antigens and block the antigens from binding to receptors on human cells (Figure 3A). They are also able to induce malignant or infected cell death through complement-dependent cytotoxicity (CDC), antibody-dependent cellular cytotoxicity (ADCC), and phagocytosis [41,42]. There are five different classes of antibodies that have been identified in humans: immunoglobulins G (IgG), IgM, IgA, IgE, and IgD. They share the same basic four-chain structure, but have different heavy chains (Figure 3B). IgG has the functions of recognizing, neutralizing and eliminating threats, and is the most abundant immunoglobulin in human serum. It accounts for about 75% of the total human immunoglobulin, and most therapeutic antibodies are of the IgG class. Adalimumab (trade name Humira®), the world's first fully human therapeutic mAb, is based on the IgG1 isotype. It was approved by the FDA in 2002 for the treatment of rheumatoid arthritis (RA), and its sales reached USD 22 billion in 2021 [43].

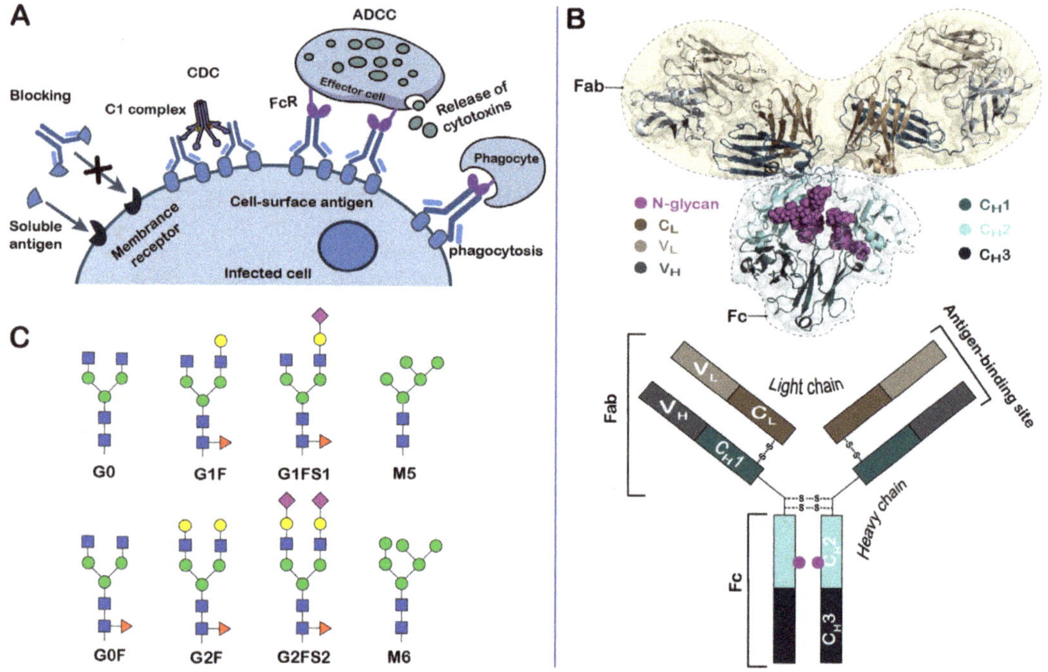

Figure 3. Structure, function and glycosylation of antibodies. (**A**) Schematic representation of the biological functions of antibodies; (**B**) a representative three-dimensional structure of IgG (PDB ID: 1IGY); (**C**) the N-glycan structures found on the IgG antibodies.

IgG antibodies are composed of two light chains and two heavy chains, which are arranged into two Fab (fragment antigen binding) regions and one Fc (fragment crystallizable) region (Figure 3B). The Fc region is mainly responsible for interacting with various receptors and complement proteins [44]. It is composed of the second and third constant domains of the heavy chains (C_H2 and C_H3). The Fc region of IgG bears a highly conserved

N-glycosylation site at Asn297, within the C_H2 domain, that is essential for Fc-receptor-mediated activity [45]. The same as is observed in most glycoproteins; glycosylation at Asn297 is also highly heterogeneous, with more than 30 different glycan structures detected in the serum IgG. The glycans that are covalently linked to the Asn297 residue contain a common heptaglycan biantennary core structure (G0, four GlcNAc and three mannose residues). The core structure can be further extended differently by fucosylation (G0F), galactosylation of one or two arms (G1, G2), and addition of terminal sialic acids in the presence of galactose (G1S1, G2S2) (Figure 3C). The extended structures can differ greatly in their percentages. For example, in the consistency evaluation of 381 batches of recombinant adalimumab manufactured by AbbVie from 2000 to 2013, it was found that the terminally ungalactosylated N-glycans (G0F + G0F-GlcNAc), terminally galactosylated N-glycans (G1F + G2F), and terminally mannosylated N-glycans (M5 + M6) accounted for 74.28% $\pm$ 1.75, 18.45% $\pm$ 1.80 and 7.29% $\pm$ 0.76 of the total glycans attached to the Asn297 residue, respectively [46].

It is generally believed that the diverse N-glycan structures could confer different biological effects to therapeutic antibodies. These effects may be beneficial for the treatment of diseases by improving the therapeutic properties, or may adversely affect the biological functions [47]. For example, it was found that altered IgG glycosylation patterns in mice and humans were often accompanied by autoimmune diseases, such as rheumatoid arthritis, especially when the structures of the glycans lack the terminal sialic acid and galactose residues (G0). At the same time, it is also known that intravenous immunoglobulin (IVIG), a purified IgG fraction obtained from healthy donors, has anti-inflammatory properties, and that high-dose IVIG can be used for the treatment of autoimmune neutropenia in childhood and autoimmune hemolytic anemia [48]. In 2006, Ravetch and coworkers showed that the distinct properties (pro-inflammatory versus anti-inflammatory properties) observed for IgG antibodies are likely to be the result of the differential sialylation of the N-linked glycan at Asn297 in the Fc domain [49]. Conformational studies revealed that glycosylation may be essential for the binding of IgG Fc to Fcγ receptors by stabilizing the conformation of the heavy chains [50,51]. In addition to altering the pro- and anti-inflammatory activities of antibodies, the glycosylation of IgG Fc at Asn297 also has a profound influence on ADCC, which could be triggered by the binding of the Fc domain to the receptor FcγRIIIa. In a study assessing the effect of fucosylation on the properties of the Rituximab biosimilar Truxima, the results showed that the binding affinity of Fc to FcγRIIIa, and the ADCC activity, tend to negatively correlate with the level of core fucosylation [52].

IgG can also be glycosylated in two Fab regions. The Fab is composed of two constant domains (C_H1 and C_L), as well as variable heavy (V_H) and light (V_L) domains. About 15–25% of the IgG antibodies in human serum are N-glycosylated in the variable domains [53]. Similar to the structures of glycans present on the Fc region, the majority of the N-linked glycans found on the Fab regions are also of the complex biantennary type. The most striking difference between the glycosylation of the Fc and Fab domains lies in the extension of the core heptaglycan. The percentages of bisecting GlcNAc and terminal galactose and sialic acid are higher in Fab glycans, while the percentages of core fucose are lower. Overall, the Fab glycans are more complex and heterogeneous than the Fc glycans.

Results from previous studies have provided initial evidence that N-glycans can influence many properties of Fab. For example, it was found that the N-linked glycosylation, introduced by somatic hypermutation (SHM) in the V_H/V_L regions of the autoantibodies isolated from patients with rheumatoid arthritis, could modulate the binding of Fab to the antigen citrullinated histone (cit-H2B) [54]. In another example, the antigen binding was tested using several anti-adalimumab and anti-infliximab antibody mutants, in which the naturally occurring Fab glycans were removed. The results showed that although some Fab N-glycans have no measurable effect on antigen binding, the presence of some Fab glycans, especially those in anti-adalimumab antibodies, could lead to higher binding affinity to their antigens [55]. Different mechanisms may account for the effect of Fab glycosylation on their binding affinity to different antigens, including the steric hindrance effect caused by the bulky glycans and the charge–charge interaction caused by the terminal sialic acids.

Fab N-glycans can also play a role in increasing the stability of antibodies. In a study comparing the differences between the thermostability of anti-adalimumab and anti-infliximab antibodies with naturally acquired Fab glycans and mutants without Fab glycans, three out of five tested mutants showed lower melting temperatures [56]. Studies have also suggested Fab N-glycosylation may affect the aggregation propensity, solubility and in vivo half-life of mAbs [57,58]. However, these conclusions are presently based on a limited body of evidence and further research is needed to define the effects of these glycans.

4. Conclusions

As one of the most widely occurring and complex post-translational modifications, glycosylation has recently attracted great attention, especially in the field of therapeutic proteins, because of its capability to simultaneously improve multiple properties [59,60]. Theoretically, it is possible to maximize the performance of therapeutic proteins by optimizing their glycosylation patterns (glycosylation sites and glycan structures) through glycoengineering. The validity of this hypothesis has been demonstrated by the development of Darbepoetin alfa and IgG. However, similar successful examples are very rare, especially in the area of therapeutic antibody discovery and development. This situation is mainly due to the lack of deep knowledge of the structure function of antibody glycosylation, and the lack of reliable and simple glycoengineering tools. In order to address these problems, more research efforts should be devoted to gaining a better understanding of antibody glycosylation and to continuing to develop protein glycoengineering technology.

Author Contributions: B.C., W.L., Y.L., B.M., S.S. and Z.T. wrote the paper. All authors have read and agreed to the published version of the manuscript.

Funding: The National Key R & D Program of China (Grant No. 2018YFE0111400), the CAMS Innovation Fund for Medical Sciences (CIFMS, 2021-I2M-1-026), the Training Program of the Major Research Plan of National Natural Science Foundation of China (Grant No. 91853120), the National Major Scientific and Technological Special Project of China (Grant Nos. 2018ZX09711001-005 and 2018ZX09711001-013), and the NIH Research Project Grant Program (R01 EB025892).

Institutional Review Board Statement: Not applicable.

Informed Consent Statement: Not applicable.

Data Availability Statement: Not applicable.

Acknowledgments: We would like to thank the Ministry of Science of Technology of China, the Chinese Academy of Medical Sciences and Peking Union Medical College, the National Natural Science Foundation of China, the State Key Laboratory of Bioactive Substance and Function of Natural Medicines, Institute of Materia Medica, and the National Institute of Health of the United States for funding.

Conflicts of Interest: The funders had no role in the design of the study; in the collection, analyses, or interpretation of data; in the writing of the manuscript, or in the decision to publish the results.

Sample Availability: Not applicable.

References

1. Ramazi, S.; Zahiri, J. Posttranslational modifications in proteins: Resources, tools and prediction methods. *Database* **2021**, *2021*, baab012. [CrossRef] [PubMed]
2. Spiro, R.G. Protein glycosylation: Nature, distribution, enzymatic formation, and disease implications of glycopeptide bonds. *Glycobiology* **2002**, *12*, 43R–56R. [CrossRef] [PubMed]
3. Aebi, M. N-linked protein glycosylation in the ER. *Biochim. Biophys. Acta* **2013**, *1833*, 2430–2437. [CrossRef] [PubMed]
4. Saito, T.; Yagi, H.; Kuo, C.W.; Khoo, K.H.; Kato, K. An embeddable molecular code for Lewis X modification through interaction with fucosyltransferase 9. *Commun. Biol.* **2022**, *5*, 676. [CrossRef] [PubMed]

5. Rudd, P.M.; Dwek, R.A. Glycosylation: Heterogeneity and the 3D structure of proteins. *Crit. Rev. Biochem. Mol. Biol.* **1997**, *32*, 1–100. [CrossRef] [PubMed]
6. Goh, J.B.; Ng, S.K. Impact of host cell line choice on glycan profile. *Crit. Rev. Biotechnol.* **2018**, *38*, 851–867. [CrossRef]
7. Guan, X.; Chaffey, P.K.; Zeng, C.; Greene, E.R.; Chen, L.; Drake, M.R.; Chen, C.; Groobman, A.; Resch, M.G.; Himmel, M.E.; et al. Molecular-scale features that govern the effects of O-glycosylation on a carbohydrate-binding module. *Chem. Sci.* **2015**, *6*, 7185–7189. [CrossRef]
8. Wada, R.; Matsui, M.; Kawasaki, N. Influence of N-glycosylation on effector functions and thermal stability of glycoengineered IgG1 monoclonal antibody with homogeneous glycoforms. *MAbs* **2019**, *11*, 350–372. [CrossRef]
9. Owczarek, B.; Gerszberg, A.; Hnatuszko-Konka, K. A brief reminder of systems of production and chromatography-based recovery of recombinant protein biopharmaceuticals. *Biomed. Res. Int.* **2019**, *2019*, 4216060. [CrossRef]
10. Schuster, J.; Koulov, A.; Mahler, H.C.; Detampel, P.; Huwyler, J.; Singh, S.; Mathaes, R. In vivo stability of therapeutic proteins. *Pharm. Res.* **2020**, *37*, 23. [CrossRef]
11. Marshall, S.A.; Lazar, G.A.; Chirino, A.J.; Desjarlais, J.R. Rational design and engineering of therapeutic proteins. *Drug Discov. Today* **2003**, *8*, 212–221. [CrossRef] [PubMed]
12. Sola, R.J.; Griebenow, K. Glycosylation of therapeutic proteins: An effective strategy to optimize efficacy. *BioDrugs* **2010**, *24*, 9–21. [CrossRef] [PubMed]
13. Sinclair, A.M.; Elliott, S. Glycoengineering: The effect of glycosylation on the properties of therapeutic proteins. *J. Pharm. Sci.* **2005**, *94*, 1626–1635. [CrossRef]
14. Dammen-Brower, K.; Epler, P.; Zhu, S.; Bernstein, Z.J.; Stabach, P.R.; Braddock, D.T.; Spangler, J.B.; Yarema, K.J. Strategies for glycoengineering therapeutic proteins. *Front. Chem.* **2022**, *10*, 863118. [CrossRef] [PubMed]
15. Majewska, N.I.; Tejada, M.L.; Betenbaugh, M.J.; Agarwal, N. N-Glycosylation of IgG and IgG-like recombinant therapeutic proteins: Why is it important and how can we control it? *Annu. Rev. Chem. Biomol. Eng.* **2020**, *11*, 311–338. [CrossRef]
16. Thompson, N.; Wakarchuk, W. O-glycosylation and its role in therapeutic proteins. *Biosci. Rep.* **2022**, *42*, BSR20220094. [CrossRef]
17. Delobel, A. Glycosylation of therapeutic proteins: A critical quality attribute. *Methods Mol. Biol.* **2021**, *2271*, 1–21. [CrossRef]
18. Zhong, X.; D'Antona, A.M.; Scarcelli, J.J.; Rouse, J.C. New opportunities in glycan engineering for therapeutic proteins. *Antibodies* **2022**, *11*, 5. [CrossRef]
19. Semenza, G.L. Regulation of mammalian O2 homeostasis by hypoxia-inducible factor 1. *Annu. Rev. Cell Dev. Biol.* **1999**, *15*, 551–578. [CrossRef]
20. Lee, J.; Vernet, A.; Gruber, N.G.; Kready, K.M.; Burrill, D.R.; Way, J.C.; Silver, P.A. Rational engineering of an erythropoietin fusion protein to treat hypoxia. *Protein Eng. Des. Sel.* **2021**, *34*, gzab025. [CrossRef]
21. Miyake, T.; Kung, C.K.; Goldwasser, E. Purification of human erythropoietin. *J. Biol. Chem.* **1977**, *252*, 5558–5564. [CrossRef] [PubMed]
22. Recny, M.A.; Scoble, H.A.; Kim, Y. Structural characterization of natural human urinary and recombinant DNA-derived erythropoietin. Identification of des-arginine 166 erythropoietin. *J. Biol. Chem.* **1987**, *262*, 17156–17163. [CrossRef] [PubMed]
23. Lai, P.H.; Everett, R.; Wang, F.F.; Arakawa, T.; Goldwasser, E. Structural characterization of human erythropoietin. *J. Biol. Chem.* **1986**, *261*, 3116–3121. [CrossRef] [PubMed]
24. Lowy, P.H.; Keighley, G.; Borsook, H. Inactivation of erythropoietin by neuraminidase and by mild substitution reactions. *Nature* **1960**, *185*, 102–103. [CrossRef] [PubMed]
25. Jacobs, K.; Shoemaker, C.; Rudersdorf, R.; Neill, S.D.; Kaufman, R.J.; Mufson, A.; Seehra, J.; Jones, S.S.; Hewick, R.; Fritsch, E.F.; et al. Isolation and characterization of genomic and cDNA clones of human erythropoietin. *Nature* **1985**, *313*, 806–810. [CrossRef]
26. Lin, F.K.; Suggs, S.; Lin, C.H.; Browne, J.K.; Smalling, R.; Egrie, J.C.; Chen, K.K.; Fox, G.M.; Martin, F.; Stabinsky, Z.; et al. Cloning and expression of the human erythropoietin gene. *Proc. Natl. Acad. Sci. USA* **1985**, *82*, 7580–7584. [CrossRef]
27. Lombardero, M.; Kovacs, K.; Scheithauer, B.W. Erythropoietin: A hormone with multiple functions. *Pathobiology* **2011**, *78*, 41–53. [CrossRef]
28. Leader, B.; Baca, Q.J.; Golan, D.E. Protein therapeutics: A summary and pharmacological classification. *Nat. Rev. Drug Discov.* **2008**, *7*, 21–39. [CrossRef]
29. Nalbant, D.; Saleh, M.; Goldman, F.D.; Widness, J.A.; Veng-Pedersen, P. Evidence of receptor-mediated elimination of erythropoietin by analysis of erythropoietin receptor mRNA expression in bone marrow and erythropoietin clearance during anemia. *J. Pharmacol. Exp. Ther.* **2010**, *333*, 528–532. [CrossRef]
30. D'Souza, A.A.; Devarajan, P.V. Asialoglycoprotein receptor mediated hepatocyte targeting-strategies and applications. *J. Control. Release* **2015**, *203*, 126–139. [CrossRef]
31. Glanz, V.Y.; Kashirskikh, D.A.; Grechko, A.V.; Yet, S.F.; Sobenin, I.A.; Orekhov, A.N. Sialidase activity in human blood serum has a distinct seasonal pattern: A pilot study. *Biology* **2020**, *9*, 184. [CrossRef] [PubMed]
32. Macdougall, I.C.; Padhi, D.; Jang, G. Pharmacology of darbepoetin alfa. *Nephrol. Dial. Transplant.* **2007**, *22* (Suppl. S4), iv2–iv9. [CrossRef] [PubMed]
33. Egrie, J.C.; Dwyer, E.; Browne, J.K.; Hitz, A.; Lykos, M.A. Darbepoetin alfa has a longer circulating half-life and greater in vivo potency than recombinant human erythropoietin. *Exp. Hematol.* **2003**, *31*, 290–299. [CrossRef] [PubMed]
34. Chairman and CEO Letter and Amgen Inc. 2021 Annual Report. Available online: https://investors.amgen.com/static-files/1789c7c6-0a07-49d4-bd81-e1afc7cb1f6d (accessed on 29 March 2022).

35. Darling, R.J.; Kuchibhotla, U.; Glaesner, W.; Micanovic, R.; Witcher, D.R.; Beals, J.M. Glycosylation of erythropoietin affects receptor binding kinetics: Role of electrostatic interactions. *Biochemistry* **2002**, *41*, 14524–14531. [CrossRef]
36. Koeppen, B.M.; Stanton, B.A. Glomerular filtration and renal blood flow. In *Renal Physiology*; Elsevier: Amsterdam, The Netherlands, 2013; pp. 27–43.
37. Macdougall, I.C.; Gray, S.J.; Elston, O.; Breen, C.; Jenkins, B.; Browne, J.; Egrie, J. Pharmacokinetics of novel erythropoiesis stimulating protein compared with epoetin alfa in dialysis patients. *J. Am. Soc. Nephrol.* **1999**, *10*, 2392–2395. [CrossRef]
38. Kwak, C.Y.; Park, S.Y.; Lee, C.G.; Okino, N.; Ito, M.; Kim, J.H. Enhancing the sialylation of recombinant EPO produced in CHO cells via the inhibition of glycosphingolipid biosynthesis. *Sci. Rep.* **2017**, *7*, 13059. [CrossRef]
39. Lee, S.J.; Evers, S.; Roeder, D.; Parlow, A.F.; Risteli, J.; Risteli, L.; Lee, Y.C.; Feizi, T.; Langen, H.; Nussenzweig, M.C. Mannose receptor-mediated regulation of serum glycoprotein homeostasis. *Science* **2002**, *295*, 1898–1901. [CrossRef]
40. Mastrangeli, R.; Palinsky, W.; Bierau, H. Glycoengineered antibodies: Towards the next-generation of immunotherapeutics. *Glycobiology* **2019**, *29*, 199–210. [CrossRef]
41. Bournazos, S.; Wang, T.T.; Dahan, R.; Maamary, J.; Ravetch, J.V. Signaling by antibodies: Recent progress. *Annu. Rev. Immunol.* **2017**, *35*, 285–311. [CrossRef]
42. Wang, X.; Mathieu, M.; Brezski, R.J. IgG Fc engineering to modulate antibody effector functions. *Protein Cell* **2018**, *9*, 63–73. [CrossRef]
43. AbbVie Reports Full-Year and Fourth-Quarter 2021 Financial Results. Available online: https://news.abbvie.com/news/press-releases/abbvie-reports-full-year-and-fourth-quarter-2021-financial-results.htm (accessed on 2 February 2022).
44. Reusch, D.; Tejada, M.L. Fc glycans of therapeutic antibodies as critical quality attributes. *Glycobiology* **2015**, *25*, 1325–1334. [CrossRef] [PubMed]
45. Beck, A.; Wagner-Rousset, E.; Bussat, M.C.; Lokteff, M.; Klinguer-Hamour, C.; Haeuw, J.F.; Goetsch, L.; Wurch, T.; Van Dorsselaer, A.; Corvaia, N. Trends in glycosylation, glycoanalysis and glycoengineering of therapeutic antibodies and Fc-fusion proteins. *Curr. Pharm. Biotechnol.* **2008**, *9*, 482–501. [CrossRef]
46. Tebbey, P.W.; Varga, A.; Naill, M.; Clewell, J.; Venema, J. Consistency of quality attributes for the glycosylated monoclonal antibody Humira(R) (adalimumab). *MAbs* **2015**, *7*, 805–811. [CrossRef] [PubMed]
47. Ravetch, J.V.; Nimmerjahn, F. *Fc Mediated Activity of Antibodies Structural and Functional Diversity Preface*; Thermo Fisher: Waltham, MA, USA, 2019; Volume 423, pp. V–VI.
48. Anthony, R.M.; Nimmerjahn, F.; Ashline, D.J.; Reinhold, V.N.; Paulson, J.C.; Ravetch, J.V. Recapitulation of IVIG anti-inflammatory activity with a recombinant IgG Fc. *Science* **2008**, *320*, 373–376. [CrossRef] [PubMed]
49. Kaneko, Y.; Nimmerjahn, F.; Ravetch, J.V. Anti-inflammatory activity of immunoglobulin G resulting from Fc sialylation. *Science* **2006**, *313*, 670–673. [CrossRef]
50. Barb, A.W.; Prestegard, J.H. NMR analysis demonstrates immunoglobulin G N-glycans are accessible and dynamic. *Nat. Chem. Biol.* **2011**, *7*, 147–153. [CrossRef]
51. Subedi, G.P.; Hanson, Q.M.; Barb, A.W. Restricted motion of the conserved immunoglobulin G1 N-glycan is essential for efficient FcgammaRIIIa binding. *Structure* **2014**, *22*, 1478–1488. [CrossRef]
52. Lee, K.H.; Lee, J.; Bae, J.S.; Kim, Y.J.; Kang, H.A.; Kim, S.H.; Lee, S.J.; Lim, K.J.; Lee, J.W.; Jung, S.K.; et al. Analytical similarity assessment of rituximab biosimilar CT-P10 to reference medicinal product. *MAbs* **2018**, *10*, 380–396. [CrossRef]
53. Van de Bovenkamp, F.S.; Hafkenscheid, L.; Rispens, T.; Rombouts, Y. The Emerging Importance of IgG Fab Glycosylation in Immunity. *J. Immunol.* **2016**, *196*, 1435–1441. [CrossRef]
54. Corsiero, E.; Carlotti, E.; Jagemann, L.; Perretti, M.; Pitzalis, C.; Bombardieri, M. H and L Chain Affinity Maturation and/or Fab N-Glycosylation Influence Immunoreactivity toward Neutrophil Extracellular Trap Antigens in Rheumatoid Arthritis Synovial B Cell Clones. *J. Immunol.* **2020**, *204*, 2374–2379. [CrossRef]
55. Van de Bovenkamp, F.S.; Derksen, N.I.L.; Ooijevaar-de Heer, P.; van Schie, K.A.; Kruithof, S.; Berkowska, M.A.; van der Schoot, C.E.; H, I.J.; van der Burg, M.; Gils, A.; et al. Adaptive antibody diversification through N-linked glycosylation of the immunoglobulin variable region. *Proc. Natl. Acad. Sci. USA* **2018**, *115*, 1901–1906. [CrossRef] [PubMed]
56. Van de Bovenkamp, F.S.; Derksen, N.I.L.; van Breemen, M.J.; de Taeye, S.W.; Ooijevaar-de Heer, P.; Sanders, R.W.; Rispens, T. Variable Domain N-Linked Glycans Acquired During Antigen-Specific Immune Responses Can Contribute to Immunoglobulin G Antibody Stability. *Front. Immunol.* **2018**, *9*, 740. [CrossRef] [PubMed]
57. Nakamura, H.; Kiyoshi, M.; Anraku, M.; Hashii, N.; Oda-Ueda, N.; Ueda, T.; Ohkuri, T. Glycosylation decreases aggregation and immunogenicity of adalimumab Fab secreted from Pichia pastoris. *J. Biochem.* **2021**, *169*, 435–443. [CrossRef] [PubMed]
58. Reslan, M.; Sifniotis, V.; Cruz, E.; Sumer-Bayraktar, Z.; Cordwell, S.P.; Kayser, V. Enhancing the stability of adalimumab by engineering additional glycosylation motifs. *Int. J. Biol. Macromol.* **2020**, *158*, 189–196. [CrossRef]
59. Mereiter, S.; Balmana, M.; Campos, D.; Gomes, J.; Reis, C.A. Glycosylation in the Era of Cancer-Targeted Therapy: Where Are We Heading? *Cancer Cell* **2019**, *36*, 6–16. [CrossRef]
60. Higel, F.; Seidl, A.; Sorgel, F.; Friess, W. N-glycosylation heterogeneity and the influence on structure, function and pharmacokinetics of monoclonal antibodies and Fc fusion proteins. *Eur. J. Pharm. Biopharm.* **2016**, *100*, 94–100. [CrossRef]

molecules

MDPI

Review

An Overview of Antitumour Activity of Polysaccharides

Hongzhen Jin [1], Maohua Li [1], Feng Tian [1], Fan Yu [2,*] and Wei Zhao [1,3,*]

1 College of Pharmacy, Nankai University, 38 Tongyan Road, Jinnan District, Tianjin 300350, China
2 College of Life Sciences, Nankai University, Weijin Road, Nankai District, Tianjin 300350, China
3 State Key Laboratory of Medicinal Chemical Biology, Nankai University, 38 Tongyan Road, Jinnan District, Tianjin 300350, China
* Correspondence: fanyu@nankai.edu.cn (F.Y.); wzhao@nankai.edu.cn (W.Z.)

Abstract: Cancer incidence and mortality are rapidly increasing worldwide; therefore, effective therapies are required in the current scenario of increasing cancer cases. Polysaccharides are a family of natural polymers that hold unique physicochemical and biological properties, and they have become the focus of current antitumour drug research owing to their significant antitumour effects. In addition to the direct antitumour activity of some natural polysaccharides, their structures offer versatility in synthesizing multifunctional nanocomposites, which could be chemically modified to achieve high stability and bioavailability for delivering therapeutics into tumor tissues. This review aims to highlight recent advances in natural polysaccharides and polysaccharide-based nanomedicines for cancer therapy.

Keywords: anticancer; polysaccharides; drug delivery systems; nanomedicines

check for updates

Citation: Jin, H.; Li, M.; Tian, F.; Yu, F.; Zhao, W. An Overview of Antitumour Activity of Polysaccharides. *Molecules* **2022**, *27*, 8083. https://doi.org/10.3390/molecules27228083

Academic Editors: Jian Yin, Jing Zeng and De-Cai Xiong

Received: 28 October 2022
Accepted: 17 November 2022
Published: 21 November 2022

1. Introduction

In the coming years, cancer is expected to become the main cause of death and the most important obstacle to extending life expectancy in the world. Lung cancer is the most common cancer and the leading cause of cancer death (18.4% of total cancer deaths), closely followed by colorectal cancer (9.2%), stomach cancer (8.2%), and liver cancer (8.2%) [1]. There are three common cancer therapeutics, including surgery, radiation therapy, and chemotherapy, as well as other emerging therapies, such as molecular targeted therapy. However, the serious side effects and drug resistance of chemotherapy and other treatments are becoming major obstacles in current cancer research. Hence, it is very important to develop a new type of anticancer agent with ideal antitumour activity and extremely low toxicity.

Polysaccharides are carbohydrates that participate in almost all aspects of organisms and play various important biological functions [2]. Polysaccharides consist of 10 or more monosaccharides linked together by glycosidic bonds, which can be linear or contain branched chains. Importantly, monosaccharide composition, molecular weight (MW), and polysaccharide attachment affect its structure, and its structure further affects its properties and functional mechanisms [3]. According to their source, polysaccharides can be classified into natural polysaccharides and semisynthetic polysaccharides. Natural polysaccharides are distributed in many organisms. Then, the natural polysaccharide is further chemically or enzymatically modified to obtain semisynthetic polysaccharides. So far, researchers have found that polysaccharides have a wide range of biological effects, including anticancer, antibiotic, antioxidant, anticoagulant, and immuno-stimulation activities.

The antitumor effect of polysaccharides was first discovered by Nauts et al. in 1946, which can effectively relieve the symptoms of cancer patients [4]. Ample evidence indicated that polysaccharides can inhibit tumors through direct anticancer activity, such as inducing apoptosis of tumor cells and inhibiting migration (Table 1). In addition, the structure of polysaccharides provides versatility for the synthesis of multi-functional nanocomposites, which can achieve high stability and bioavailability through chemical modification, thus delivering therapeutic drugs to tumor tissues [5]. This review used keywords (anticancer/polysaccharides/drug delivery systems/nanomedicines) to search in PubMed and Web of Science databases, and selected qualified high-level papers for systematic sorting and summary. In this paper, we aim to systematically summarize the research findings in the past decade, and the different structures of anticancer polysaccharides from different sources and polysaccharide-based nanomedicines for cancer treatment are reviewed, which provides theoretical support for the design and development of polysaccharide preparations.

Table 1. Performance and structural features of natural anticancer polysaccharides.

Natural Polysaccharides	Performances	Structural Features
Polysaccharides from plants	Target Twist/AKR1C2/NF-1 pathway	acidic protein–polysaccharide
Polysaccharides from animals	Antiangiogenic properties	GlcN-GlcA or GlcN-IdoA
Polysaccharides from fungi	Inhibiting JAK2/STAT3 signaling pathway	β-(1→3) glucose linkages

2. Polysaccharides from Plants

2.1. Panax ginseng C. A. Meyer *Polysaccharides*

Panax ginseng C. A. Meyer (*P. ginseng*) is a precious medicine that has been used for thousands of years, also known as ginseng [6]. Ginseng is composed of multiple active components, including ginsenosides and polysaccharides. Studies have proven that polysaccharides are one of the most important components in *P. ginseng* and participate in immunomodulation, antitumour, and antidiabetic activities [7].

P. ginseng polysaccharide contains starch-like glucans and pectin [8]. Pectin is a plant-derived neutral polysaccharide with abundant resources for its amounts and categories. Many types of pectin polysaccharides are associated with anticancer activity. Pectin, with Panax ginseng C. A. Meyer Polysaccharidesvery complex structure, typically contains galacturonic acid (GalA), galactose (Gal), arabinose (Ara), and rhamnose (Rha) residues [9]. Pectin could be divided into five types: homogalacturonan (HG), type I rhamnogalacturonans (RG-I), type II rhamnogalacturonans (RG-II), xylagalgalacturonan (XGA), and Apio galgalacturonan (AGA), based on the different structural characteristics [10]. HG is characterized by α-(1→4)-D-GalA repeat units as the backbone [11], whereas RG-I is composed of Ara, galactans, and L-fucose (L-fuc) in the sidechains [12]. RG-II and XGA are both derivatives of HG [10]. The components of P. ginseng pectin include HG and RG-I, as well as GalA, Gal, Ara, and Rha [13].

To date, many kinds of pectin have been isolated and identified from ginseng, and some of them have been identified as having antitumour activity, as described in Table 2.

Table 2. Ginseng polysaccharides with antitumour activity.

Compound	Structure Features	MW	Antitumor Mechanism	Ref.
PGPW1	97.4% carbohydrate and 1.2% uronic acid	~3.5 × 105 Da	Not been elucidated	[14,15]
PGP2a	Acidic protein–polysaccharide	~3.2 × 104 Da	Target Twist/AKR1C2/NF-1 pathway	[16]
RG-I	RG-I and side chains AG-I	~6 × 104 Da	Bound to galectin-3	[17]
MCGP-1	The ratio of Rha/GalA is 0.82	1.649 × 105 Da	Might be related to the Ara residues linked to the surface of the polysaccharide	[18]
MCGP-2	Mainly composed of GalA, Ara, Gal, Rha, and Glc	1.644 × 105 Da	The same mechanism as MCGP-1	[18]
MCGP-3	The characteristic compositions of RG-I pectin	1.572 × 105 Da	The same mechanism as MCGP-1 and contains disaccharide [-(1, 4)-α-D-GalAp-(1, 2. -α-L-Rhap-]	[18]
MCGP-4	The characteristic compositions of RG-I pectin	1.673 × 105 Da	The same mechanism as MCGP-1	[18]
MCGP-5	The ratio of Rha/GalA is 0.24	1.600 × 105 Da	The same mechanism as MCGP-1	[18]
MCGP-6	Mainly composed of GalA, Ara, Gal, Rha, and Glc	1.592 × 105 Da	The same mechanism as MCGP-1	[18]
MCGP-7	Mainly composed of GalA, Ara, Gal, Rha, and Glc	1.520 × 105 Da	The same mechanism as MCGP-1	[18]

2.2. Portulaca oleracea *L. Polysaccharides*

P. oleracea L., a traditional Chinese herbal medicine, is known as MaChiXian in Chinese and purslane in English. It exhibits a range of biological activities, such as anti-inflammatory, antioxidant, and antiaging [29–32]. *P. oleracea* L. polysaccharides (POL-P) are major bioactive components of purslane with antitumour activity. Zhou et al. purified a homogeneous POL-P, which contains Gal, Ara, Man, and Glc. Then, they evaluated an animal model transplanted with sacroma 180 and found that it had pronounced antitumour effects [33]. Another POL-P, named POL-P3b, inhibits cancer cell growth, and the mechanism involves triggering DNA damage and inducing apoptosis [34]. Further research also showed that POL-P3b inhibits the proliferation of HeLa cells, and the possible antitumor mechanism is through downregulating the TLR4 downstream signaling pathway and inducing cell apoptosis [35]. In addition, POL-P3b could also decrease the growth of cervical carcinoma, suggesting the antitumour mechanism via stimulating the TLR4/PI3K/AKTNF-κB signaling pathway [36].

In addition to direct antitumour effects, Lee et al. go deeply into the immune-enhancing characteristics of POL-P. The preliminary results showed that POL-P increased the viability of CY-treated splenocytes because of CY-induced immunosuppression [37]. POL-P also enhances the immune efficiency of the breast cancer dendritic cell vaccine [38]. Ding et al. found that POL-P can improve lipopolysaccharide-induced inflammation and barrier dysfunction of the porcine intestinal epithelium monolayer [39].

Ginseng polysaccharide could also significantly inhibit the growth of Lewis lung carcinoma tumor [19]. In addition, one selenium-modified polysaccharide, sGP, has been reported. The experimental results indicate that sGP enhances apoptosis in HL-60 cells, demonstrating that chemical modification methods to obtain high contents of selenium polysaccharides could be developed as a novel antitumour therapy [20].

2.3. Angelica Sinensis *(Oliv.) Diels Polysaccharides*

The root of *A. sinensis*, known as Danggui, is a celebrated Chinese medicinal herb [21]. *A. sinensis* possesses a wide range of pharmacological activities, including hematopoiesis, immunomodulation, antioxidant, and anticancer activities [22–25]. Polysaccharides are the most important active constituents in Danggui, and numerous *A. sinensis* polysaccharides (ASPs) have been identified. The majority of ASPs contain GalA, Gal, Ara, Rha, mannose (Man), and glucose (Glc) with various molar ratios. Wei et al. also proved that APSs could induce apoptosis in cancer cells via regulation of the JAK/STAT of the transcription pathway [26]. Key kinases in the JAK/STAT and PI3K/AKT pathways were also downregulated by ASPs' stimulation in another study [27]. ASPs have also been utilized in drug delivery systems. Wang et al. prepared doxorubicin (DOX)-loaded nanoparticles and proved that it can inhibit the growth of HepG2 multicellular spheres [28].

2.4. Lycium barbarum *L. Polysaccharides*

L. barbarum, known as wolfberry in China, is a herbal medicine [40]. Polysaccharides are one of the most investigated, as they are considered to be mainly responsible for different biological effects among all *L. barbarum* components [41]. Zhao et al. extracted polysaccharides from Chinese wolfberry fruits and proved that it could induce MCF-7 cell apoptosis. Cao et al. isolated and characterized another polysaccharide, named CF1, with an MW of 1540.10 $\pm$ 48.78 kDa. Their results showed that CF1 also exhibited effective cell growth inhibition in vitro [42]. Then, Cao et al. conducted further research and exploration. Eventually, they found that the antitumour mechanism of CF1 was associated with the PI3K/AKT pathway [43].

2.5. Ginkgo biloba *Polysaccharides*

G. biloba, known as yinxing in China, is a traditional Chinese herb. Polysaccharides are bioactive compounds isolated from *G. biloba*, with a wide variety of physiological functions such as antitumor activity. Kong et al. reported a selenium (Se)-containing polysaccharide purified from the leaves of *G. biloba*, and proved that it induced human bladder cancer T24 cell apoptosis through a mitochondria-dependent pathway [44].

2.6. Seeds' Polysaccharides

Seeds are one of the important sources of plant polysaccharides and accumulated evidence has demonstrated that these polysaccharides show superior anticancer activity, as described in Table 3.

Table 3. Seeds' polysaccharides with anticancer activity.

Plants Species	Types of Carcinoma Cell Lines	Ref.
Peony seeds	Pc-3/HCT-116/MCF-7/Hela	[45]
Chenopodium quinoa seeds	SMMC 7721/MCF-7	[46]
Psidium guajava L. seeds	MCF-7	[47]

2.7. Citrus Polysaccharides

Citrus pectin is a neutral polysaccharide isolated from the pulp and peel of citrus fruits, which consists of HG and RG-I [48]. Modified citrus pectin (MCP) is a nonbranched polysaccharide and is high in Gal extracted from citrus pectin by enzymatic hydrolysis, high temperature, and high pH [49]. The shorter and nonbranched MCP could recognize and bind tightly with galectin-3 [50], whose overexpression was related to a variety of malignant tumors [51]. The combination mechanism of MCP and galectin-3 is that the former can recognize galectin-3 on the surface of cancer cells and then inhibit tumor metastasis [49,50]. However, citrus pectin from a neutral resource is unable to interact with galectin-3 owing to its limited solubility in water.

It has been reported that MCP inhibits myeloma/prostate cancer/bladder tumor [52]/ gastrointestinal cancer [53] via interaction with galectin-3. Conti et al. found that MCP is a potential sensitizer targeting galectin-3 for prostate cancer radiotherapy [54]. Fabi et al. demonstrated that MCP fractions with different molecular sizes can have different effects on the development of malignant tumors [55]. In addition, pectic from Aegle marmelos L. could potentially inhibit skin cancer [56]. Additionally, pectin polysaccharides extracted from tomato, papaya, or olive have been reported to possess the activity of inhibiting galactose lectin-3. The pectin polysaccharide fraction from papaya pulp and olive showed inhibitory effects on colon cancer [57] and bladder cancer [58], respectively, through interactions with galectin-3.

2.8. Marine Algae Polysaccharides

Marine algae are one of the richest resources in the ocean, and contain a variety of active components, such as peptides and polysaccharides [59]. According to the thallus color, marine algae are usually divided into red seaweed, brown seaweed, and green seaweed. Marine algal polysaccharide (MAP) is a unique polysaccharide, which is different from land plant polysaccharides in composition, substitution, and linkage [60]. The major MAP contains carrageenan of red algae, fucoidan and laminarans of brown algae, and ulvan of green algae, comprising monosaccharide subunits such as Gal, Ara, Glc, Man, fucose, xylose, glucuronic acid (GlcA), mannuronic acid (ManA), and iduronic acid (IdoA) [61,62] (Figure 1).

According to a previous study, polysaccharides fractionated from brown seaweed Sargassum (S.) show superior anticancer activity. For example, a study showed that sulfated polysaccharides could inhibit proliferation in A549 cells via induced mitochondria-mediated intrinsic apoptosis and cell cycle arrest [63]. Rajendran et al. obtained polysaccharide fractions (SWP1) from *S.* wightii and found that it showed a dose-dependent manner inhibition of proliferation and migration of cancer cells. Further research reveals that the mechanism of SWP1 inducing apoptosis in cancer cells is via cutting the mitochondrial membrane and damaging the nucleus, as well as increasing caspase 3/9 activity [64].

Fucoidan, a sulfated polysaccharide rich in fucose, has antitumour activities [65]. The experimental results of Kang et al. also prove that fucoidan possesses anti-proliferation of B16 melanoma cell [66]. Alginate oligosaccharide was prepared from alginate sodium using alginate lyase and can reduce tumor size by improving the antioxidant and anti-inflammatory capacities of patients [67]. The red seaweed sulfated polysaccharide from Acanthophora spicifera (Vahl) Borgeson exhibited apoptotic effects in lung cancer cells [68]. In addition, polysaccharides isolated from two microalgae sources showed certain ant-hepatoma activity in vitro mainly through the induction of apoptosis [69,70].

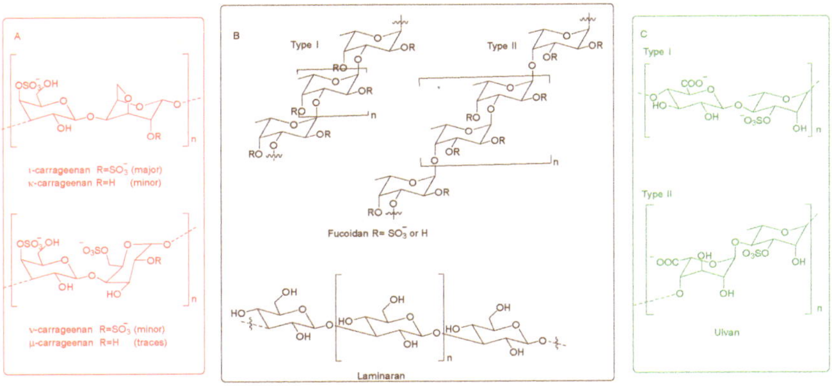

Figure 1. The major MAP in red seaweed (**A**), brown seaweed (**B**), and green seaweed (**C**).

2.9. *Other Plant Sources of Polysaccharides*

2.9.1. Polysaccharides with Anti-Lung Cancer Activity

Ni et al. successfully separated HRWP-A, a natural pectin, from Hippophae rhamnoides berries. HRWP-A effectively inhibits the growth of lung cancer in vivo and promotes NK cell activity and CTL mechanism by enhancing lymphocyte proliferation and macrophage activity [71]. HCA4S1 was separated from Houttuynia cordata, and bioactivity tests suggested that it exerts anticancer action via inducing cell cycle arrest and apoptosis on lung cancer cells [72]. Additionally, Glehnia littoralis polysaccharide effectively inhibits the proliferation and migration of A549 cell lines and induces cell apoptosis [73]. Lee et al. showed that the bioactive polysaccharides from Achyranthes bidentata exhibit potential anti-metastasis effects with the mechanisms of blocking the epithelial-to-mesenchymal transition process [74].

2.9.2. Polysaccharides with Anti-Pancreatic Cancer Activity

Lonicera japonica and *Lycium ruthenicum* pectin have certain inhibitory effects on pancreatic cancer in vitro. LJ-02–1 is an RG-I polysaccharide, and bioactivity tests suggested that it might inhibit BxPC-3 and PANC-1 cell growth [75]. LRP3-S1 could also inhibit the growth of pancreatic cancer cells via downregulating the protein expression of p-FAK and p-p38 MAP kinase [76].

2.9.3. Polysaccharides with Anticancer Activity

In addition to lung cancer and pancreatic cancer, polysaccharides from other species of plants have also been reported for the use of other malignant tumors, as shown in Table 4.

Table 4. Polysaccharides from other species of plants with antitumour activity.

Plants Species	Structure Features	Types of Carcinoma Cell Lines	Ref.
Broccoli	Comprised of Ara, Gal, and Rha with a molar ratio of 5.3:0.8:1.0	HepG2, Siha cervical, MDA-MB-231	[77]
Gleoestereum incarnatum	Composed of Gal, Glc, xylose, and Man at molar ratios of 1:4.25:1.14:1.85	HepG2	[78]
Zizyphus jujuba cv.Muzao	Presence of RG-I domains and typical pectic polysaccharides, with homogalacturonan (methyl and acetyl esterified)	HepG2	[79]
Taxus chinensis var.mairei fruits		S180	[80]
Huperzia serrata	Composed of Gal, Glc, Ara, Rha, Man, GalA, and so on	Skov3 and A2780	[81]
Dandelion	α-type polysaccharides, consisted of Glc, Gal, Ara, arabinose rhamnose, and GlcA	HepG2	[82,83]
Dendrobium nobile Lindl	Composed of Gal, Glc, Ara, Rha, Man, and so on	Sarcoma 180	[84]

3. Polysaccharides from Animals

3.1. *Polysaccharides from Mammals*

Glycosaminoglycans (GAGs) are natural linear polydisperse heteropolysaccharides distributed in both vertebrates and invertebrates, with molecular weights up to several million Dalton [85]. Evidence obtained from glycobiology studies suggests that GAGs can recognize and interact with numerous proteins, and thus possess extensive biological functions [86]. GAGs are one class of glycostructures of the extracellular matrix (ECM). There are four classes of GAGs, each according to the constitution of the repeating disaccharide units, which consist of heparin (HP)/heparan sulfate (HS), hyaluronan (HA), chondroitin sulfate (CS)/dermatan sulfate (DS), and keratan sulfate (KS) (Figure 2) [85,87]. Except for HA, other compounds contain O-sulfonation, N-acetylation, and N-sulfonation modifications, and this polyanionic character allows GAGs to bind to positively charged moieties, including plasma proteins, growth factors, and so on [87]. These molecules are a kind of ubiquitous molecule with extensive biological functions and, of course, they are also widely used as therapeutics, for example, HP is an anticoagulant, while CS is generally used to treat osteoarthritis [88]. In addition, further understanding of GAG's structure–function relationships has also led to the discovery of novel pharmaceuticals for the possible treatment of serious diseases, such as antitumor agents. In light of GAGs related to tumorigenesis, its application in drug development has been the focus of two main directions: (I) using GAGs as the target of therapeutic strategies and (II) utilizing the specificity and excellent physical and chemical properties of GAGs to deliver targeted cancer drugs [89].

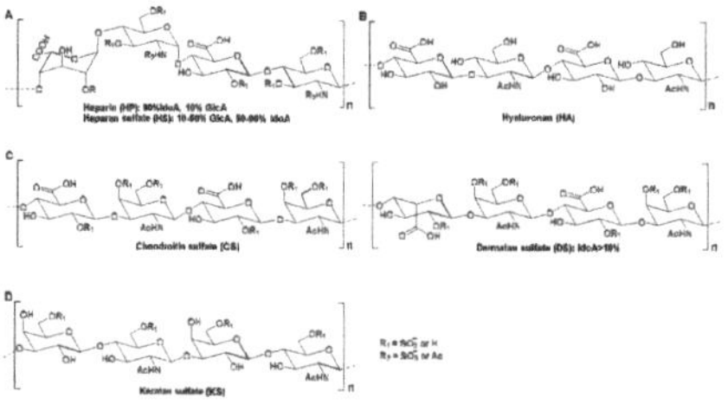

Figure 2. Four classes of mammalian GAGs and their potential sulfation sites. (**A**) (HP)/heparan sulfate (HS), (**B**) Hyaluronan (HA), (**C**) Chondroitin sulfate (CS)/Dermatan sulfate (DS), and (**D**) Keratan sulfate (KS).

3.1.1. Heparin/Heparan Sulfate

HP has been used as an anticoagulant for more than 80 years, and it is a true biologic and can be purified from bovine lung or porcine mucosa. The anticoagulant activity of HP is mostly owing to the action of a precise pentasaccharide sequence that acts in accordance with antithrombin-III (AT-III), a serine protease inhibitor [90]. As an important member of the linear GAG family, HP and HS are composed of sulfated disaccharide repeating units of either GlcA- or IdoA-linked glucosamine (GlcN) residues (Figure 2A). HP is, on the whole, more highly sulfated than HS. Depending on the sources and molecular weight differences, HP is classified into the following three classes: (I) unfractionated heparin (UFH), extracted from many animal sources, with an MW of approximately 14,000 Da; (II) low molecular weight heparin (LMWH), prepared from UFH, with a MW of approximately 3500~6000 Da; and (III) ultralow molecular weight (ULMWH), generally referring to the chemically synthesized pentasaccharide fondaparinux sodium, with the trade name Arixtra.

HP, including UFH and LMWH, is used in the treatment of cancer-associated venous thromboembolism (VTE), and LMWH is recommended as the nursing standard for the treatment of established VTE [91–93]. Preclinical data support that coagulation inhibition greatly limits tumor metastasis in some experimental models, and it has been demonstrated that LMWH can effectively inhibit metastasis of solid malignant tumors [94]. In addition to anticoagulant activity, HP may possess direct anticancer benefits because of its antiangiogenic properties [95]. The antiangiogenesis mechanism is that HP binds to vascular endothelial growth factor (VEGF) and then inhibits the phosphorylation of VEGF receptor (VEGFR) [96]. Furthermore, HP is an inhibitor of heparanase, which is overexpressed in tumors, and heparin can bind with P-selectin to significantly inhibit tumor cell adhesion [97,98]. As natural resourced polysaccharides, HP are often described as nonimmunogenic and nontoxic, driving the desire to employ them in nanoformulations for cancer management. Because of the above factors, HP plays an important role in cancer treatment, as shown in Table 5.

Table 5. Application of HP in antitumour therapy.

Compound	HP Combination Types	Anticancer Mechanisms	Types of Cancer	Ref.
LHT	HP–drug conjugate	Antiangiogenic properties	Pancreatic cancer cells-bearing mice	[99]
Oral LMWH conjugate (LHTD4)	HP–drug conjugate	Antiangiogenic properties	A549 lung cancer cells	[100]
Tinzaparin, a LMWH	HP fragments	Reverses the cisplatin resistance in A2780cis cells	A2780cis cells	[101]
Deoxycholic acid conjugatedHP fragments (HFD)	HP–drug conjugate	Inhibiting VEGF165	SCC7 cells	[102]
LMWH-Suramin	HP–drug conjugate	Inhibiting VEGF165	SCC7-bearing mouse model	[103]
HP-suramin/PEGylated protamine	HP–drug conjugate	Antiangiogenic properties	SCC7-bearing mouse model	[104]
HP-functionalized Pluronic nanoparticles	Polymeric nanoparticles	Antiangiogenic properties and drug combination	Gastric cancers	[105]
Heparin/polyethyleneglycol (PEG) hydrogel	Nanogels	Antiangiogenic properties and drug combination	Breast cancer	[106]
LMWH-poloxamer	Nanogels	Enhancing the efficacies, minimizing the side effects ofdalteparin, and exhibiting a good thermosensitivity	Xenograft S180 sarcoma tumor	[107]
HP-containing cryogel microcarriers	Polyelectrolyte complex nanoparticles	Reversible strong electrostatic interaction	Metastatic breast cancer	[108]
HP-Folate-Tat-Taxol	Polyelectrolyte complex nanoparticles	Negatively charged nanoparticles may cause lower toxic effect	Breastcancer cells	[109]
LMWH–quercetin conjugate	HP–drug conjugate	Antiangiogenic properties	MCF-7 tumor cells	[110]
HP-Poloxamer	HP-coated inorganic nanoparticles	Antiangiogenic properties and drug combination	HeLa cells	[111]
Heparosan-cystamine-vitamin E succinate	Nanogels	Increase tumor selectivity and improve the therapeutic effect	MGC80-3 tumor cells	[112]
LMWH-TOS	Polyelectrolyte complex nanoparticles	Antiangiogenic properties and drug combination	4T1 solid tumor model	[113]
HP–folate–retinoic acid bioconjugates	Polyelectrolyte complex nanoparticles	Drug combination	HeLa cells	[114]
HP-reduced graphene oxide nanocomposites	Polyelectrolyte complex nanoparticles	Combinational chemotherapy and photothermal therapy	MCF-7 and A549cells	[115]
PEGylated HP-based nanomedicines	Polyelectrolyte complex nanoparticles	Photodynamic therapy	4T1 cells	[116]

3.1.2. Hyaluronan

HA normally exists in the form of long-chain nonsulfated polysaccharides, which are the main component of the ECM in cells [117]. The repeated disaccharide unit of HA is composed of GlcA β (1→3) GlcNAc, and each disaccharide unit passes through a β (1→4) glycosidic bond (Figure 2B). Native HA, extracted from many animal sources, is present as a linear polymer with an average molecular weight of approximately 106~107 Da [118]. Likewise, HA with strong hydrophilicity could form a very viscous gel that helps to maintain tissue integrity [119]. In addition to being a structural part of tissues, HA is the ligand of the cluster of differentiation (CD) protein CD44 receptor [118]. CD44 is a complex transmembrane receptor protein that is overexpressed by many tumor types [117]. Hence, specific ligation with HA-CD44 enables HA-based drug delivery (containing HA–drug conjugates, nanogels, polymeric nanoparticles, and HA-coated organic and inorganic nanoparticles) to target diseased cells that express these receptors (Figure 3). In addition, HA combined with drugs or drug carriers could solve some solubility problems [118].

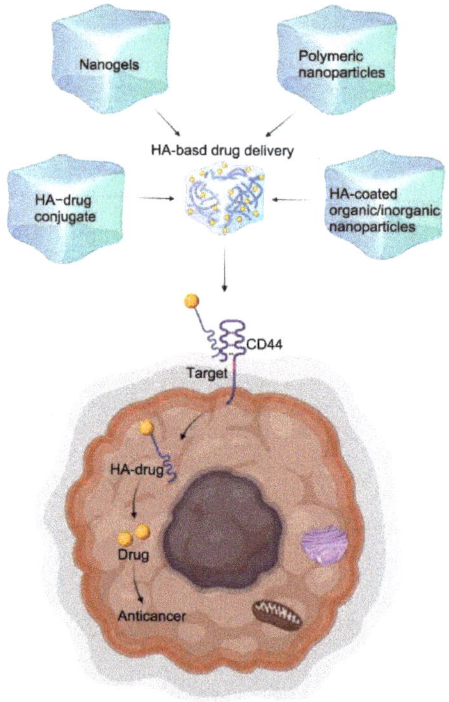

Figure 3. Mechanism of action of HA-based drug delivery targeting CD44.

To date, HA has been widely used in anticancer drug delivery, either associating HA with drugs to form conjugates or producing hydrogels, for the local delivery of various drugs, including antitumoral agents, owing to its biocompatibility, biodegradability, nontoxicity, nonimmunogenicity, and as a ligand of CD44. The application of these nanoparticles in various cancer therapies is shown in Table 6.

Table 6. Application of HA in antitumour therapy.

Compound	HA Combination Types	Anticancer Mechanisms	Types of Cancer	Ref.
Carbon nanotubes-Chitosan (CHI)-HA-DOX	Polymeric nanoparticles	CD44-targeted, hydrophilic	HeLa cells	[120]
HA-DOX-afatinib-CaP	Polymeric nanoparticles	CD44-targeted, high-densitycarboxyl groups	A549 lung cancer cells	[121]
HA-Curcumin (Cur)	Nanogels	CD44-targeted	A549 lung cancer cells	[122]
HA-Sinulariolide	Polymeric nanoparticles	CD44-targeted	A549 lung cancer cells	[123]
HA-Cur-prodrug-CaP	Polymeric nanoparticles	CD44-targeted	MB-MDA-231 mouse model	[124]
HA-cystamin-pyrenyl-Ir(III)	Polymeric nanoparticles	CD44-targeted, hydrophilic	A549 tumor-bearing mice	[125]
HA-DOX-cisplatin	Nanogels	CD44-targeted	A2780 cell lines	[126]
HA-keratin-DOX	Nanogels	CD44-targeted, negative charge and good hydrophilicity	4T1 and B16 cells	[127]
HA-Pemetrexed	HA–drug conjugate	CD44-targeted, as a prognostic marker in malignant pleural mesothelioma	Malignant pleuralmesothelioma model	[128]
HA-fluvastatin-encapsulating liposomes	Polymeric nanoparticles	CD44-targeted, hydrophilic barrier	Breast cancer stem cellxenografted mouse model	[129]
HA-coated silica/hydroxyapatite-DOX	HA-coated inorganic nanoparticles	CD44-targeted	4T1 tumor-bearing mice	[130]
HA-sclareol/poly-lactic-co-glycolic acid	HA-coated inorganic nanoparticles	CD44-targeted, hydrophilic	MCF-7 and MDA-MB468 cell lines	[131]
HA-coated camptothecin	HA-coated inorganic nanoparticles	CD44-targeted	MDA-MB-231 cells	[132]
HA and poly-(N-ε-carbobenzyloxy-L-lysine)	Polymeric nanoparticles	CD44-targeted	HepG2 tumor-bearing mice	[133]
Ursolic acid-loadedin a poly-L-lysine coat and HA	HA-coated organic nanoparticles	CD44-targeted	SCC-7 xenograft tumor model	[134]
folic acid- and dopamine-decorated HA	HA-coated organic nanoparticles	CD44-targeted	B16 melanoma model	[135]
HA-$Cu_{2-x}S$	HA-coated organic nanoparticles	CD44-targeted, biocompatibility	CT26.WT cells-bearing mice	[136]
HA Conjugated ZincProtoporphyrin	HA conjugated cincprotoporphyrin	CD44-targeted	C26 colon cancer cells	[137]
Irinotecan-loaded self-agglomerating HA	Polymeric nanoparticles	CD44-targeted	H23 non-small-cell lung cancer cells	[138]
HA-SuperparamagneticIron Oxide	Polyelectrolyte complex nanoparticles	CD44-targeted	U87MG cells	[139]

3.1.3. Chondroitin Sulfate/Dermatan Sulfate

The repeated disaccharide unit of CS is comprised of GlcA β (1→3) GlcNAc, and each disaccharide unit passes through a β (1→4) glycosidic bond (Figure 2C). CS can be divided into five types according to their different modification types and sulfonation forms, as shown in Table 7 [87,139]. After rare C5 isomerization of CS GlcA into IdoA, a special type CS-B of CS, DS, is produced (Figure 2C). As with other GAGs, CS is a special anionic acid polysaccharide with high biocompatibility and specificity, and is a promising drug carrier for cancer treatment.

Table 7. Types of CS.

CS Types	Major Disaccharide Unit	Other Disaccharide Unit
CS-A	GlcA-GalNAc4S	GlcA-GalNAc/GlcA2S-GalNAc
CS-B(DS)	IdoA-GalNAc4S	IdoA2S-GalNAc4S/GlcA3S-GalNAc
CS-C	GlcA-GalNAc6S	IdoA-GalNAc4S6S/GlcA3S-GalNAc4S
CS-D	GlcA2S-GalNAc6S	IdoA2S-GalNAc4S6S/GlcA3S-GalNAc4S6S
CS-E	GlcA-GalNAc4S6S	IdoA2S-GalNAc/GlcA3S-GalNAc6S

Curcumin-loaded CS/chitosan nanoparticles inhibited the apoptosis of lung cancer cells, whereas loading CS/chitosan hydrogel with curcumin exhibited cytotoxicity-inducing effects in HeLa, HT29, and PC3 cancer cells [140,141]. Curcumin-loaded zein and CS self-assembled nanoparticles also exhibited anti neoplastic activity on HepG2, MCF-7, and HeLa cells [142]. In colorectal cancer cells, folate-targeted nanostructured chitosan/CS complex carriers, CS–chitosan nanoparticle carriers encapsulating black rice anthocyanins, and CS-based smart hydrogels could heighten the delivery of antitumor drugs to tumor cells [143–146].

Similar to HA, CS has a great targeting ability for the cluster CD44, which is over-expressed in particular cancer cells [147]. Therefore, the surface functionalization of CS-endowed nanoparticles has been successfully used for the treatment of colon cancer [148]. Moreover, a codelivery vector including CS loaded with small interfering RNA and paclitaxel has been proven to have a mighty targeting effect towards CD44-overexpressing cancer cells [149]. CS-based multi-walled carbon nanotubes can precisely target CD44 receptors overexpressed on triple-negative breast cancer specific cells [150]. In addition, combined application of CS with doxorubicin or quercetin (chemicalsensitizer) can enhance chemical photodynamic therapy and overcome multidrug resistance [151,152].

3.1.4. Keratan Sulfate

KS is localized in the ECM of different tissues, has a relatively small molecular weight, and ranges from 5 to 30 repeating disaccharide subunits. KS is composed of Galβ(1→4)GlcNAc, and each disaccharide unit passes through a β (1→3) glycosidic bond. It is different from other GAGs because its uronic acid moiety is partially replaced by neutral Gal units (Figure 2D) [87]. As a class of GAGs, the potential of KS in the delivery of anticancer drugs needs to be further developed.

3.2. *Polysaccharides Derived from Marine Animals*

3.2.1. Chondroitin Sulfate from Sturgeon and Cartilage

As mentioned earlier, CS is a natural polymer and is widely distributed in the cartilage and bone of animals. Herein, a sturgeon (*Acipenser*)-derived CS significantly inhibits tumor progression of HCT-116 mice model by inhibiting proliferation and inducing apoptosis [153]. Moreover, a novel CS-E exhibits dose-dependent antimetastatic activity [154].

3.2.2. Sulfated Polysaccharides from Sea Cucumber

Sulfated polysaccharides are one of the main components of sea cucumber, which have a wide range of biological activities. Ermakova et al. isolated sulfated fucans and proved that it exhibits anticancer activity against the cancer cell lines [155].

3.2.3. Polysaccharides from Common Cockles

Research by Pye et al. shows that the sulfated polysaccharide has antiproliferative activity on chronic myeloid leukemia and relapsing acute lymphoblastic leukemia cell lines. They identified that sulfated polysaccharides are a unique marine-derived HP/HS-like polysaccharide [156].

4. Polysaccharides from Fungi

4.1. Lentinan

Lentinan (LNT), a neutral polysaccharide extracted from Lentinus edodes, has been widely used in Asia. LNT is a kind of β-(1→3)-D-glucan and its repeating unit is shown in Figure 4. The primary structure of LNT consists of two lateral β-(1→6) glucose branches on five β-(1→3) glucose linkages [157].

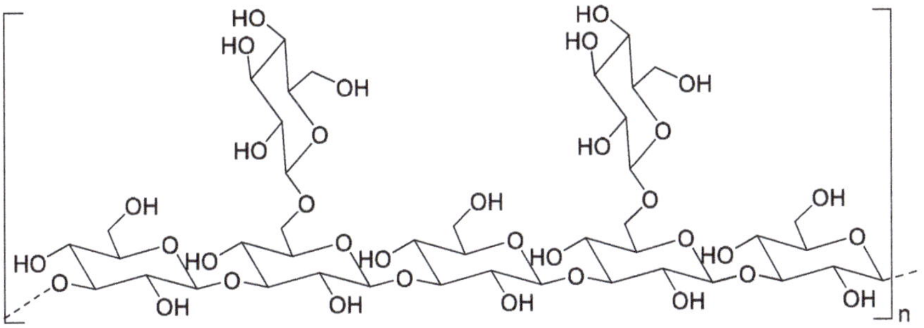

Figure 4. The repeating unit of the LNT structure.

The antitumour activity of LNT and its synergistic effect with various chemicals or other therapies have been extensively studied. Wu et al. reported that LNT can effectively delay the development of lung adenocarcinoma by upregulating miR-216a-5p and inhibiting the JAK2/STAT3 signaling pathway (Figure 5) [158]. LNT as an adjuvant has been prepared into lentinan calcium carbonate (LNT-$CaCO_3$) microspheres and has potential use as a vaccine delivery system [159]. Chen et al. used LNT as a modifier to synthesize stable and efficient selenium nanoparticles (SeNPs), which can effectively inhibit the growth of solid tumors [160]. Additionally, LNT-coated selenium nanoparticles (SeNPs@LNT) could restore the dysfunctional immune cells in the malignant pleural perfusion microenvironment [161].

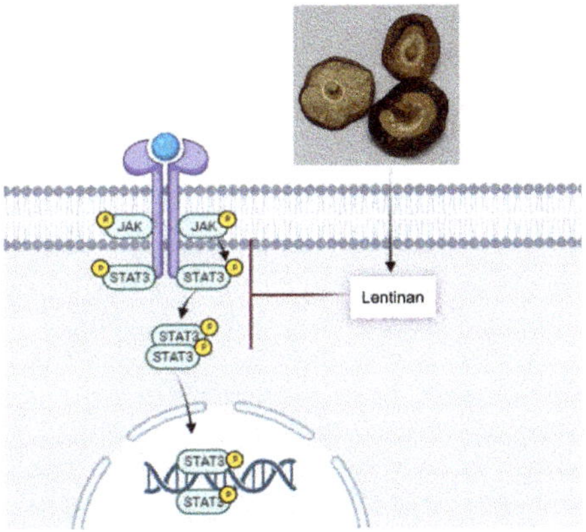

Figure 5. The mechanism action of the Lentinan.

4.2. Ganoderma lucidum *Polysaccharide*

Ganoderma lucidum (*G. lucidum*) is one of the most famous folk medicines in China [162]. Most *G. lucidum* polysaccharides (GLPs) are β-glucans with an MW distribution of 103–106 Da. Ding et al. reported a neutral polysaccharide, GLSA50-1B, with a (1→6)(1→4)-β-D-glucan (Figure 6A) [163]. Fang et al. identified a branched β-D-(1→3)-glucan, named PSGL-I-1A (Figure 6B) [164]. WGLP, a water-soluble polysaccharide, was obtained from spores of *Ganoderma lucidum* (Fr.) Karst and its repeating unit is shown in Figure 6C [165].

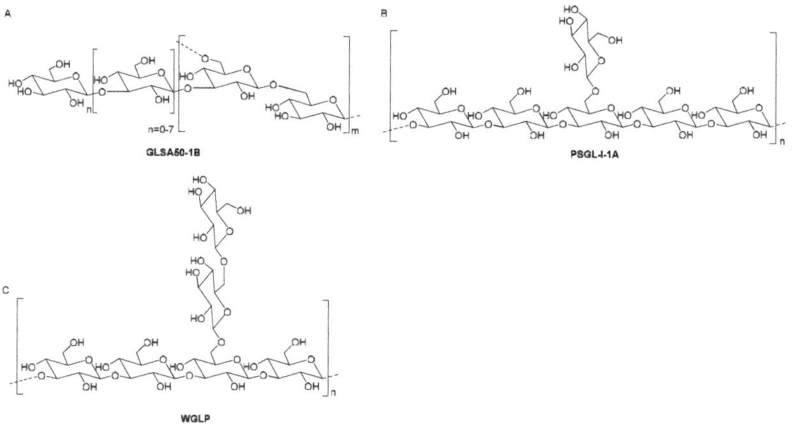

Figure 6. The structures of GLPs. (**A**) GLSA50-1B, (**B**) PSGL-I-1A, (**C**) WGLP.

Crude polysaccharides from *G. lucidum* work with dacarbazine to inhibit the growth of melanoma tumors [166]. A fucoxylomannan from *G. lucidum* showed effective antiproliferative effects [167]. Ding et al reported that WGLP can significantly inhibit the growth of tumor in vivo at a certain concentration without drug-related toxicity [165]. The water-soluble polysaccharide WSG is effective against lung cancer and tongue cancer [168,169]. In addition, Lin et al. found that the combination of WSG and cisplatin can inhibit cell activity and induce apoptosis [169]. The application of GLPs on gold nanocomposites can be activated effectively for dendritic cells and T lymphocytes in breast cancer-bearing mouse models and inhibit the growth and metastasis of tumors [170]. In addition, GLP-conjugated bismuth sulfide nanoparticles can effectively assist tumor radiotherapy via radiosensitization and dendritic cell activation [171].

5. Conclusions

It is predicted that the global number of cancer patients will reach 34 million in 2070, with a doubling of the incidence of all cancers combined relative to 2020 [172]. More and more evidence shows that polysaccharides have great anticancer potential. Polysaccharides are a class of biological macromolecules produced by plants, animals, and fungi, which have received extensive attention in recent years owning to their high therapeutic efficacy and low toxicity. Some polysaccharides isolated from the leaves, seeds, roots, and bark of plants show a certain direct anticancer effect, with mechanisms involved in regulating multiple proteins or signal transduction pathways. Besides, the unique structure diversities and physiochemical properties of polysaccharides lay the foundation for developing various nanocarriers. Drug delivery methods based on polysaccharides nanomaterials help to achieve targeted delivery of immunotherapeutic agents to immune cell subtypes and effectively improve the therapeutic effect of drug carriers. In addition, the degradation products of polysaccharides are normal monosaccharides in vivo and can be recycled by cells without accumulation in the tissue.

In a word, this article reviews the latest progress of polysaccharides and polysaccharide-based nanomaterials and their applications in cancer immunotherapy. The anticancer properties of polysaccharides are mainly mediated through two ways: (I) direct cytotoxicity and (II) as a targeted nano carrier platform, which carries traditional anticancer drugs. Although there are still many unsolved problems in this field, the clinical value and broad application prospects of anticancer polysaccharides make them an important direction of new drug development.

Author Contributions: Writing original draft preparation, H.J.; writing review and editing, H.J., M.L. and F.T.; visualization and supervision, W.Z. and F.Y. All authors have read and agreed to the published version of the manuscript.

Funding: This work was supported by the National Key R&D Program of China (2018YFA0507204) and the National Natural Science Foundation of China (22077068) and the Fundamental Research Funds for the Central Universities.

Institutional Review Board Statement: Not applicable.

Informed Consent Statement: Not applicable.

Data Availability Statement: Not applicable.

Conflicts of Interest: The authors declare no conflict of interest.

References

1. Bray, F.; Ferlay, J.; Soerjomataram, I.; Siegel, R.L.; Torre, L.A.; Jemal, A. Global cancer statistics 2018: GLOBOCAN estimates of incidence and mortality worldwide for 36 cancers in 185 countries. *CA A Cancer J. Clin.* **2018**, *68*, 394–424. [CrossRef]
2. Schjoldager, K.T.; Narimatsu, Y.; Joshi, H.J.; Clausen, H. Global view of human protein glycosylation pathways and functions. *Nat. Rev. Mol. Cell Biol.* **2020**, *21*, 729–749. [CrossRef]
3. Li, N.; Wang, C.; Georgiev, M.I.; Bajpai, V.K.; Tundis, R.; Gandara, J.S.; Lu, X.; Xiao, J.; Tang, X.; Qiao, T. Advances in dietary polysaccharides as anticancer agents: Structure-activity relationship. *Trends Food Sci. Technol.* **2021**, *111*, 360–377. [CrossRef]
4. Zong, A.; Cao, H.; Wang, F. Anticancer polysaccharides from natural resources: A review of recent research. *Carbohydr. Polym.* **2012**, *90*, 1395–1410. [CrossRef]
5. Zeng, Y.; Xiang, Y.; Sheng, R.; Tomás, H.; Rodrigues, J.; Gu, Z.; Zhang, H.; Gong, Q.; Luo, K. Polysaccharide-based nanomedicines for cancer immunotherapy: A review. *Bioact. Mater.* **2021**, *6*, 3358–3382. [CrossRef] [PubMed]
6. Liu, L.; Xu, F.-R.; Wang, Y.Z. Traditional uses, chemical diversity and biological activities of *Panax*, L. (Araliaceae): A review. *J. Ethnopharmacol.* **2020**, *263*, 112792. [CrossRef] [PubMed]
7. Zhao, B.; Lv, C.; Lu, J. Natural occurring polysaccharides from Panax ginseng C. A. Meyer: A review of isolation, structures, and bioactivities. *Int. J. Biol. Macromol.* **2019**, *133*, 324–336. [CrossRef]
8. Sun, L.; Wu, D.; Ning, X.; Yang, G.; Lin, Z.; Tian, M.; Zhou, Y. α-Amylase-assisted extraction of polysaccharides from Panax ginseng. *Int. J. Biol. Macromol.* **2015**, *75*, 152–157. [CrossRef]
9. Sun, L.; Ropartz, D.; Cui, L.; Shi, H.; Ralet, M.C.; Zhou, Y. Structural characterization of rhamnogalacturonan domains from Panax ginseng C. A. Meyer. *Carbohydr Polym.* **2019**, *203*, 119–127. [CrossRef]
10. Yue, F.; Xu, J.; Zhang, S.; Hu, X.; Wang, X.; Lü, X. Structural features and anticancer mechanisms of pectic polysaccharides: A review. *Int. J. Biol. Macromol.* **2022**, *209*, 825–839. [CrossRef]
11. Maxwell, E.G.; Belshaw, N.J.; Waldron, K.W.; Morris, V.J. Pectinan emerging newbioactive food polysaccharide. *Trends Food Sci. Technol.* **2012**, *24*, 64–73. [CrossRef]
12. Shakhmatov, E.G.; Makarova, E.N.; Belyy, V.A. Structural studies of biologically active pectin-containing polysaccharides of pomegranate Punica granatum. *Int. J. Biol. Macromol.* **2019**, *122*, 29–36. [CrossRef] [PubMed]
13. Zhang, X.; Yu, L.; Bi, H.; Li, X.; Ni, W.; Han, H.; Li, N.; Wang, B.; Zhou, Y.; Tai, G. Total fractionation and characterization of the water-soluble polysaccharides isolated from *Panax* ginseng C.A. Meyer. *Carbohydr. Polym.* **2009**, *77*, 544–552. [CrossRef]
14. Li, C.; Cai, J.; Geng, J.; Li, Y.; Wang, Z.; Li, R. Purification, characterization and an-ticancer activity of a polysaccharide from *Panax* ginseng. *Int. J. Biol. Macromol.* **2012**, *51*, 968–973. [CrossRef]
15. Cai, J.-P.; Wu, Y.-J.; Li, C.; Feng, M.Y.; Shi, Q.T.; Li, R.; Wang, Z.Y.; Geng, J.S. Panax ginseng polysaccharide suppresses metastasis by modulating Twist expression in gastric cancer. *Int. J. Biol. Macromol.* **2013**, *57*, 22–25. [CrossRef]
16. Li, C.; Tian, Z.-N.; Cai, J.P.; Chen, K.X.; Zhang, B.; Feng, M.Y.; Shi, Q.T.; Li, R.; Qin, Y.; Geng, J.S. *Panax* ginseng polysaccharide induces apoptosis by targeting Twist/AKR1C2/NF-1 pathway in human gastric cancer. *Carbohydr. Polym.* **2014**, *102*, 103–109. [CrossRef]
17. Gao, X.; Zhi, Y.; Sun, L.; Peng, X.; Zhang, T.; Xue, H.; Tai, G.; Zhou, Y. The inhibitory effects of a rhamnogalacturonan I (RG-I) domain from ginseng pectin on galectin-3 and its structure-activity relationship. *J. Biol. Chem.* **2013**, *288*, 33953–33965. [CrossRef]

18. Jia, H.; Zhao, B.; Zhang, F.; Santhanam, R.K.; Wang, X.; Lu, J. Extraction, structural characterization, and anti-hepatocellular carcinoma activity of polysaccharides from Panax ginseng meyer. *Front. Oncol.* **2021**, *11*, 4905. [CrossRef]
19. Zhou, X.; Shi, H.; Jiang, G.; Zhou, Y.; Xu, J. Antitumour activities of ginseng polysaccharide in C57BL/6 mice with Lewis lung carcinoma. *Tumor Biol.* **2014**, *35*, 12561–12566. [CrossRef]
20. Liao, K.; Bian, Z.; Xie, D.; Peng, Q. A selenium-modified ginseng polysaccharide promotes the apoptosis in human promyelocytic leukemia (HL-60) cells via a mitochondrial-mediated pathway. *Biol. Trace Elem. Res.* **2017**, *177*, 64–71. [CrossRef]
21. Chen, X.-P.; Li, W.; Xiao, X.F.; Zhang, L.L.; Liu, C.X. Phytochemical and pharmacological studies on Radix Angelica sinensis. *Chin. J. Nat. Med.* **2013**, *11*, 577–587. [CrossRef] [PubMed]
22. Younas, F.; Aslam, B.; Muhammad, F.; Mohsin, M.; Raza, A.; Faisal, M.N.; Hassan, S.; Majeed, W. Haematopoietic effects of Angelica sinensis root cap polysaccharides against lisinopril-induced anaemia in albino rats. *Pharm. Biol.* **2017**, *55*, 108–113. [CrossRef] [PubMed]
23. Pan, S.; Jiang, L.; Wu, S. Stimulating effects of polysaccharide from Angelica sinensis on the nonspecific immunity of white shrimps (*Litopenaeus vannamei*). *Fish Shellfish Immun.* **2018**, *74*, 170–174. [CrossRef]
24. Wang, Y.; Li, X.; Chen, X.; Zhao, P.; Qu, Z.; Ma, D.; Zhao, C.; Gao, W. Effect of stir-frying time during Angelica Sinensis Radix processing with wine on physicochemical, structure properties and bioactivities of polysaccharides. *Process Biochem.* **2019**, *81*, 188–196. [CrossRef]
25. Zhou, W.-J.; Wang, S.; Hu, Z.; Zhou, Z.Y.; Song, C.J. Angelica sinensis polysaccharides promotes apoptosis in human breast cancer cells via CREB-regulated caspase-3 activation. *Biochem. Biophys. Res. Commun.* **2015**, *467*, 562–569. [CrossRef]
26. Fu, Z.; Li, Y.; Yang, S.; Ma, C.; Zhao, R.; Guo, H.; Wei, H. Angelica sinensis polysaccharide promotes apoptosis by inhibiting JAK/STAT pathway in breast cancer cells. *Trop. J. Pharm. Res.* **2019**, *18*, 2247–2253.
27. Yang, J.; Shao, X.; Jiang, J.; Sun, Y.; Wang, L.; Sun, L. Angelica sinensis polysaccharide inhibits proliferation, migration, and invasion by downregulating microRNA-675 in human neuroblastoma cell line SH-SY5Y. *Cell Biol. Int.* **2018**, *42*, 867–876. [CrossRef]
28. Zhang, Y.; Cui, Z.; Mei, H.; Xu, J.; Zhou, T.; Cheng, F.; Wang, K. Angelica sinensis polysaccharide nanoparticles as a targeted drug delivery system for enhanced therapy of liver cancer. *Carbohydr. Polym.* **2019**, *219*, 143–154. [CrossRef]
29. Tleubayeva, M.I.; Datkhayev, U.M.; Alimzhanova, M.; Ishmuratova, M.Y.; Korotetskaya, N.V.; Abdullabekova, R.M.; Flisyuk, E.V.; Gemejiyeva, N.G. Component composition and antimicrobial activity of CO2 extract of portulaca oleracea, growing in the territory of kazakhstan. *Sci. World J.* **2021**, *2021*, 1–10. [CrossRef]
30. Kim, K.-H.; Park, E.-J.; Jang, H.J.; Lee, S.J.; Park, C.S.; Yun, B.S.; Lee, S.W.; Rho, M.C. 1-Carbomethoxy-β-Carboline, derived from portulaca oleracea L., ameliorates LPS-mediated inflammatory response associated with MAPK signaling and nuclear translocation of NF-κB. *Molecules* **2019**, *24*, 4042. [CrossRef]
31. Tian, X.; Ding, Y.; Kong, Y.; Wang, G.; Wang, S.; Cheng, D. Purslane (*Portulacae oleracea* L.) attenuates cadmium-induced hepatorenal and colonic damage in mice: Role of chelation, antioxidant and intestinal microecological regulation. *Phytomedicine* **2021**, *92*, 153716. [CrossRef]
32. Zhang, W.; Zheng, B.; Deng, N.; Wang, H.; Li, T.; Liu, R.H. Effects of ethyl acetate fractional extract from Portulaca oleracea L. (PO-EA) on lifespan and healthspan in Caenorhabditis elegans. *J. Food Sci.* **2020**, *85*, 4367–4376. [CrossRef]
33. Shena, H.; Tang, G.; Zeng, G.; Yang, Y.; Cai, X.; Li, D.; Liu, H.; Zhou, N. Purification and characterization of an antitumour polysaccharide from *Portulaca oleracea* L. *Carbohydr. Polym.* **2013**, *93*, 395–400. [CrossRef] [PubMed]
34. Zhao, R.; Gao, X.; Cai, Y.; Shao, X.; Jia, G.; Huang, Y.; Qin, X.; Wang, J.; Zheng, X. Antitumour activity of *Portulaca oleracea* L. polysaccharides against cervical carcinoma in vitro and in vivo. *Carbohydr. Polym.* **2013**, *96*, 376–383. [CrossRef] [PubMed]
35. Zhao, R.; Zhang, T.; Ma, B.; Li, X. Antitumour activity of portulaca oleracea L. polysaccharide on heLa cells through inducing TLR4/NF-kB signaling. *Nutr. Cancer* **2017**, *69*, 131–139. [CrossRef] [PubMed]
36. Zhao, R.; Shao, X.; Jia, G.; Huang, Y.; Liu, Z.; Song, B.; Hou, J. Anti-cervical carcinoma effect of *Portulaca oleracea* L. polysaccharides by oral administration on intestinal dendritic cells. *BMC Complement. Altern. Med.* **2019**, *19*, 1–10. [CrossRef]
37. Park, Y.M.; Lee, H.Y.; Kang, Y.G.; Park, S.H.; Lee, B.G.; Park, Y.J.; Oh, H.G.; Moon, D.I.; Kim, Y.P.; Park, D.S.; et al. Immune-enhancing effects of *Portulaca oleracea* L.-based complex extract in cyclophosphamide-induced splenocytes and immunosuppressed rats. *Food Agric. Immunol.* **2018**, *30*, 13–24. [CrossRef]
38. Jia, G.; Shao, X.; Zhao, R.; Zhang, T.; Zhou, X.; Yang, Y.; Li, T.; Chen, Z.; Liu, Y. *Portulaca oleracea* L. polysaccharides enhance the immune efficacy of dendritic cell vaccine for breast cancer. *Food Function* **2021**, *12*, 4046–4059. [CrossRef]
39. Zhuang, S.; Ming, K.; Ma, N.; Sun, J.; Wang, D.; Ding, M.; Ding, Y. *Portulaca oleracea* L. polysaccharide ameliorates lipopolysaccharide-induced inflammatory responses and barrier dysfunction in porcine intestinal epithelial monolayers. *J. Funct. Foods* **2022**, *91*, 104997. [CrossRef]
40. Masci, A.; Carradori, S.; Casadei, M.A.; Paolicelli, P.; Petralito, S.; Ragno, R.; Cesa, S. *Lycium barbarum* polysaccharides: Extraction, purification, structura characterization and evidence about hypoglycaemic and hypolipidaemic effects. A review. *Food Chem.* **2018**, *254*, 377–389. [CrossRef]
41. Amagase, H.; Farnsworth, N.R. A review of botanical characteristics, phytochemistry, clinical relevance in efficacy and safety of Lycium barbarum fruit (GOJI). *Food Res. Int.* **2011**, *44*, 1702–1717. [CrossRef]
42. Chen, F.; Ran, L.; Mi, J.; Yan, Y.; Lu, L.; Jin, B.; Li, X.; Cao, Y. Isolation, characterization and antitumour effect on DU145 cells of a main polysaccharide in pollen of chinese wolfberry. *Molecules* **2018**, *23*, 2430. [CrossRef] [PubMed]

43. Ran, L.; Chen, F.; Zhang, J.; Mi, J.; Lu, L.; Yan, Y.; Cao, Y. Antitumour effects of pollen polysaccharides from Chinese wolfberry on DU145 cells via the PI3K/AKT pathway in vitro and in vivo. *Int. J. Biol. Macromol.* **2020**, *152*, 1164–1173. [CrossRef] [PubMed]
44. Chen, D.; Sun, S.; Cai, D.; Kong, G. Induction of mitochondrial-dependent apoptosis in T24 cells by a selenium (Se)-containing polysaccharide from Ginkgo biloba L. leaves. *Int. J. Biol. Macromol.* **2017**, *101*, 126–130. [CrossRef] [PubMed]
45. Zhang, F.; Shi, J.-J.; Thakur, K.; Hu, F.; Zhang, J.G.; Wei, Z.J. Anti-cancerous potential of polysaccharide fractions extracted from peony seed dreg on various human cancer cell lines via cell cycle arrest and apoptosis. *Front. Pharmacol.* **2017**, *8*, 102. [CrossRef]
46. Hua, Y.; Zhang, J.; Zou, L.; Fu, C.; Li, P.; Zhao, G. Chemical characterization, antioxidant, immune-regulatin and anticancer activities of a novel bioactive polysaccharide from Chenopodium quinoa seeds. *Int. J. Biol. Macromol.* **2017**, *99*, 622–629. [CrossRef] [PubMed]
47. Lin, H.-C.; Lin, J.-Y. GSF3, a polysaccharide from guava (*Psidium guajava* L.) seeds, inhibits MCF-7 breast cancer cell growth via increasing Bax/Bcl-2 ratio or Fas mRNA expression levels. *Int. J. Biol. Macromol.* **2020**, *161*, 1261–1271. [CrossRef]
48. Kaya, M.; Sousa, A.G.; Crépeau, M.J.; Sørensen, S.O.; Ralet, M.C. Characterization of citrus pectin samples extracted under different conditions: Influence of acid type and pH of extraction. *Ann. Bot.* **2014**, *114*, 1319–1326. [CrossRef]
49. Glinsky, V.V.; Raz, A. Modified citrus pectin anti-metastatic properties: One bullet, multiple targets. *Carbohydr. Res.* **2009**, *344*, 1788–1791. [CrossRef]
50. Nangia-Makker, P.; Hogan, V.; Honjo, Y.; Baccarini, S.; Tait, L.; Bresalier, R.; Raz, A. Inhibition of human cancer cell growth and metastasis in nude mice by oral intake of modified citrus pectin. *J. Natl. Cancer Inst.* **2002**, *94*, 1854–1862. [CrossRef]
51. Ahmed, H.; Alsadek, D.M.M. Galectin-3 as a potential target to prevent cancer metastasis. *Clin. Med. Insights: Oncol.* **2015**, *9*, 113–121. [CrossRef] [PubMed]
52. Fang, T.; Liu, D.-D.; Ning, H.; Liu, D.; Sun, J.; Huang, X.; Dong, Y.; Geng, M.; Yun, S.; Yan, J.; et al. Modified citrus pectin inhibited bladder tumor growth through downregulation of galectin-3. *Acta Pharmacol. Sin.* **2018**, *39*, 1885–1893. [CrossRef] [PubMed]
53. Wang, S.; Li, P.; Lu, S.M.; Ling, Z.Q. Chemoprevention of low-molecular-weight citrus pectin (LCP) in gastrointestinal cancer cells. *Int. J. Biol. Sci.* **2016**, *12*, 746–756. [CrossRef] [PubMed]
54. Conti, S.; Vexler, A.; Hagoel, L.; Kalich-Philosoph, L.; Corn, B.W.; Honig, N.; Shtraus, N.; Meir, Y.; Ron, I.; Eliaz, I.; et al. Modified citrus pectin as a potential sensitizer for radiotherapy in prostate cancer. *Integr. Cancer Ther.* **2018**, *17*, 1225–1234. [CrossRef]
55. do Prado, S.B.R.; Shiga, T.M.; Harazono, Y.; Hogan, V.A.; Raz, A.; Carpita, N.C.; Fabi, J.P. Migration and proliferation of cancer cells in culture are differentially affected by molecular size of modified citrus pectin. *Carbohydr. Polym.* **2019**, *211*, 141–151. [CrossRef]
56. Pynam, H.; Dharmesh, S.M. A xylorhamnoarabinogalactan I from Bael (*Aegle marmelos* L.) modulates UV/DMBA induced skin cancer via galectin-3 & gut microbiota. *J. Funct. Foods* **2019**, *60*, 103425.
57. do Prado, S.B.R.; Mourão, P.A.S.; Fabi, J.P. Chelate-soluble pectin fraction from papaya pulp interacts with galectin-3 and inhibits colon cancer cell proliferation. *Int. J. Biol. Macromol.* **2019**, *126*, 170–178. [CrossRef]
58. Bermudez-Oria, A.; Rodriguez-Gutierrez, G.; Fátima, R.S.; Marta, S.C.; Juan, F.B. Antiproliferative activity of olive extract rich in polyphenols and modified pectin on bladder cancer cells. *J. Med. Food* **2020**, *23*, 719–727. [CrossRef]
59. Tanna, B.; Mishra, A. Nutraceutical potential of seaweed polysaccharides: Structure, bioactivity, safety, and toxicity. *Compr. Rev. Food Sci. Food Saf.* **2019**, *18*, 817–831. [CrossRef]
60. Zheng, L.-X.; Chen, X.-Q. Current trends in marine algae polysaccharides: The digestive tract, microbial catabolism, and prebiotic potential. *Int. J. Biol. Macromol.* **2020**, *151*, 344–354. [CrossRef] [PubMed]
61. Jiao, G.; Yu, G.; Zhang, J.; Ewart, H.S. Chemical structures and bioactivities of sulfated polysaccharides from marine algae. *Mar. Drugs* **2011**, *9*, 196–223. [CrossRef] [PubMed]
62. Tziveleka, L.-A.; Ioannou, E.; Roussis, V. Ulvan, a bioactive marine sulfated polysaccharide as a key constituent of hybrid biomaterials: A review. *Carbohydr. Polym.* **2019**, *218*, 355–370. [CrossRef] [PubMed]
63. Liu, G.; Kuang, S.; Wu, S.; Jin, W.; Sun, C. A novel polysaccharide from *Sargassum integerrimum* induces apoptosis in A549 cells and prevents angiogensis in vitro and in vivo. *Sci. Rep.* **2016**, *6*, 1–12. [CrossRef] [PubMed]
64. Vaikundamoorthy, R.; Krishnamoorthy, V.; Vilwanathan, R.; Rajendran, R. Structural characterization and anticancer activity (MCF7 and MDA-MB-231) of polysaccharides fractionated from brown seaweed *Sargassum wightii*. *Int. J. Biol. Macromol.* **2018**, *111*, 1229–1237. [CrossRef]
65. Senthilkumar, K.; Manivasagan, P.; Venkatesan, J.; Kim, S.K. Brown seaweed fucoidan: Biological activity and apoptosis, growth signaling mechanism in cancer. *Int. J. Biol. Macromol.* **2013**, *60*, 366–374. [CrossRef]
66. Wang, Z.-J.; Xu, W.; Liang, J.W.; Wang, C.S.; Kang, Y. Effect of fucoidan on B16 murine melanoma cell melanin formation and apoptosis. *Afr. J. Tradit. Complement. Altern. Med.* **2017**, *14*, 149–155. [CrossRef]
67. Chen, J.; Hu, Y.; Zhang, L.; Wang, Y.; Wang, S.; Zhang, Y.; Guo, H.; Ji, D.; Wang, Y. Alginate oligosaccharide DP5 exhibits antitumor effects in osteosarcoma patients following surgery. *Front. Pharmacol.* **2017**, *8*, 623. [CrossRef]
68. Anand, J.; Sathuvan, M.; Babu, G.V.; Sakthivel, M.; Palani, P.; Nagaraj, S. Bioactive potential and composition analysis of sulfated polysaccharide from *Acanthophora spicifera* (Vahl) Borgeson. *Int. J. Biol. Macromol.* **2018**, *111*, 1238–1244. [CrossRef]
69. Chen, X.; Song, L.; Wang, H.; Liu, S.; Yu, H.; Wang, X.; Li, R.; Liu, T.; Li, P. Partial characterization, the immune modulation and anticancer activities of sulfated polysaccharides from filamentous *microalgae tribonema* sp. *Molecules* **2019**, *24*, 322. [CrossRef]
70. Yang, S.; Wan, H.; Wang, R.; Hao, D. Sulfated polysaccharides from *Phaeodactylum tricornutum*: Isolation, structural characteristics, and inhibiting HepG2 growth activity in vitro. *PeerJ* **2019**, *7*, e6409. [CrossRef]

71. Wanga, H.; Gao, T.; Du, Y.; Yang, H.; Wei, L.; Bi, H.; Ni, W. Anticancer and immunostimulating activities of a novel homogalacturonan from *Hippophae rhamnoides* L. berry. *Carbohydr. Polym.* **2015**, *131*, 288–296. [CrossRef] [PubMed]
72. Han, K.; Jin, C.; Chen, H.; Wang, P.; Yu, M.; Ding, K. Structural characterization and anti-A549 lung cancer cells bioactivity of a polysaccharide from Houttuynia cordata. *Int. J. Biol. Macromol.* **2018**, *120*, 288–296. [CrossRef] [PubMed]
73. Wu, J.; Gao, W.; Song, Z.; Xiong, Q.; Xu, Y.; Han, Y.; Yuan, J.; Zhang, R.; Cheng, Y.; Fang, J.; et al. Anticancer activity of polysaccharide from Glehnia littoralis on human lung cancer cell line A549. *Int. J. Biol. Macromol.* **2018**, *106*, 464–472. [CrossRef] [PubMed]
74. Zhong, C.; Yang, J.; Lu, Y.; Xie, H.; Zhai, S.; Zhang, C.; Luo, Z.; Chen, X.; Fang, X.; Jia, L. *Achyranthes bidentata* polysaccharide can safely prevent NSCLC metastasis by targeting EGFR and EMT. *Signal Transduct. Target. Ther.* **2020**, *5*, 178. [CrossRef] [PubMed]
75. Lina, L.; Wang, P.; Du, Z.; Wang, W.; Cong, Q.; Zheng, C.; Jin, C.; Ding, K.; Shao, C. Structural elucidation of a pectin from flowers of *Lonicera japonica* and its antipancreatic cancer activity. *Int. J. Biol. Macromol.* **2016**, *88*, 130–137. [CrossRef]
76. Zhanga, S.; He, F.; Chen, X.; Ding, K. Isolation and structural characterization of a pectin from Lycium ruthenicum Murr and its anti-pancreatic ductal adenocarcinoma cell activity. *Carbohydr. Polym.* **2019**, *223*, 115104. [CrossRef]
77. Xu, L.; Cao, J.; Chen, W. Structural characterization of a broccoli polysaccharide and evaluation of anticancer cell proliferation effects. *Carbohydr. Polym.* **2015**, *126*, 179–184. [CrossRef]
78. Zhang, Z.F.; Lv, G.Y.; Jiang, X.; Cheng, J.H.; Fan, L.F. Extraction optimization and biological properties of a polysaccharide isolated from *Gleoestereum incarnatum*. *Carbohydr. Polym.* **2015**, *117*, 185–191. [CrossRef]
79. Wang, Y.; Liu, X.; Zhang, J.; Liu, G.; Liu, Y.; Wang, K.; Yang, M.; Cheng, H.; Zhao, Z. Structural characterization and in vitro antitumour activity of polysaccharides from *Zizyphus jujuba* cv. Muzao. *RSC Adv.* **2015**, *5*, 7860–7867. [CrossRef]
80. Zhao, C.; Li, Z.; Li, C.; Yang, L.; Yao, L.; Fu, Y.; He, X.; Shi, K.; Lu, Z. Optimized extraction of polysaccharides from *Taxus chinensis* var. *mairei* fruits and its antitumour activity. *Int. J. Biol. Macromol.* **2015**, *75*, 192–198. [CrossRef]
81. Feng, Y.-N.; Zhang, X.-F. Polysaccharide extracted from *Huperzia serrata* using response surface methodology and its biological activity. *Int. J. Biol. Macromol.* **2020**, *157*, 267–275. [CrossRef] [PubMed]
82. Ren, F.; Li, J.; Yuan, X.; Wang, Y.; Wu, K.; Kang, L.; Luo, Y.; Zhang, H.; Yuan, Z. Dandelion polysaccharides exert anticancer effect on Hepatocellular carcinoma by inhibiting PI3K/AKT/mTOR pathway and enhancing immune response. *J. Funct. Foods* **2019**, *55*, 263–274. [CrossRef]
83. Ren, F.; Wu, K.; Yang, Y.; Yang, Y.; Wang, Y.; Li, J. Dandelion polysaccharide exerts anti-angiogenesis effect on hepatocellular carcinoma by regulating VEGF/HIF-1a expression. *Front. Pharmacol.* **2020**, *11*, 460. [CrossRef] [PubMed]
84. Wang, J.-H.; Luo, J.-P. Comparison of antitumor activities of different polysaccharide fractions from the stems of Dendrobium nobile Lindl. *Carbohydr. Polym.* **2010**, *79*, 114–118. [CrossRef]
85. Yeung, B.K.S.; Chong, P.Y.C.; Petillo, P.A. Synthesis of Glycosaminoglycans. *J. Carbohydr. Chem.* **2002**, *21*, 799–865. [CrossRef]
86. Pomin, V.H.; Mulloy, B. Glycosaminoglycans and proteoglycans. *Pharmaceuticals* **2018**, *11*, 27. [CrossRef]
87. Mende, M.; Bednarek, C.; Wawryszyn, M.; Sauter, P.; Biskup, M.B.; Schepers, U.; Bräse, S. Chemical synthesis of glycosaminoglycans. *Chem. Rev.* **2016**, *116*, 8193–8255. [CrossRef]
88. Volpi, N. Therapeutic applications of glycosaminoglycans. *Curr. Med. Chem.* **2006**, *13*, 1799–1810. [CrossRef]
89. Berdiaki, A.; Neagu, M.; Giatagana, E.M.; Kuskov, A.; Tsatsakis, A.M.; Tzanakakis, G.N.; Nikitovic, D. Glycosaminoglycans: Carriers and Targets for Tailored Anti-Cancer Therapy. *Biomolecules* **2021**, *11*, 395. [CrossRef]
90. Li, W.; Johnson, D.J.D.; Esmon, C.T.; Huntington, J.A. Structure of the antithrombin-thrombin-heparin ternary complex reveals the antithrombotic mechanism of heparin. *Nat. Struct. Mol. Biol.* **2004**, *11*, 857–862. [CrossRef]
91. Khorana, A.A.; Streiff, M.B.; Farge, D.; Mandala, M.; Debourdeau, P.; Cajfinger, F.; Marty, M.; Falanga, A.; Lyman, G.H. Venous thromboembolism prophylaxis and treatment in cancer: A consensus statement of major guidelines panels and call to action. *J. Clin. Oncol.* **2009**, *27*, 4919–4926. [CrossRef] [PubMed]
92. Farge, D.; Bounameaux, H.; Brenner, B.; Cajfinger, F.; Debourdeau, P.; Khorana, A.A.; Pabinger, I.; Solymoss, S.; Douketis, J.; Kakkar, A. International clinical practice guidelines including guidance for direct oral anticoagulants in the treatment and prophylaxis of venous thromboembolism in patients with cancer. *Lancet Oncol.* **2016**, *17*, e452–e466. [CrossRef]
93. Frere, C.; Benzidia, I.; Marjanovic, Z.; Farge, D. Recent advances in the management of cancer-associated thrombosis: New Hopes but New Challenges. *Cancers* **2019**, *11*, 71. [CrossRef] [PubMed]
94. Gil-Bernabé, A.M.; Lucotti, S.; Muschel, R.J. Coagulation and metastasis: What does the experimental literature tell us? *Br. J. Haematol.* **2013**, *162*, 433–441. [CrossRef]
95. Walenga, J.M.; Lyman, G.H. Evolution of heparin anticoagulants to ultralow-molecular-weight heparins: A review of pharmacologic and clinical differences and applications in patients with cancer. *Crit. Rev. Oncol. Hematol.* **2013**, *88*, 1–18. [CrossRef]
96. Zhang, W.; Swanson, R.; Izaguirre, G.; Xiong, Y.; Lau, L.F.; Olson, S.T. The heparin-binding site of antithrombin is crucial for antiangiogenic activity. *Blood* **2005**, *106*, 1621–1628. [CrossRef]
97. Rohloff, J.; Zinke, J.; Schoppmeyer, K.; Tannapfel, A.; Witzigmann, H.; Mössner, J.; Wittekind, C.; Caca, K. Heparanase expression is a prognostic indicator for postoperative survival in pancreatic adenocarcinoma. *Br. J. Cancer* **2002**, *86*, 1270–1275. [CrossRef]
98. Ludwig, R.J.; Boehme, B.; Podda, M.; Henschler, R.; Jager, E.; Tandi, C.; Boehncke, W.H.; Zollner, T.M.; Kaufmann, R.; Gille, J. Endothelial P-selectin as a target of heparin action in experimental melanoma lung metastasis. *Cancer Res.* **2004**, *64*, 2743–2750. [CrossRef]

99. Hwang, H.H.; Jeong, H.J.; Yun, S.; Byun, Y.; Okano, T.; Kim, S.W.; Lee, D.Y. Anticancer effect of heparin–taurocholate conjugate on orthotopically induced exocrine and endocrine pancreatic cancer. *Cancers* **2021**, *13*, 5775. [CrossRef]
100. Kim, J.-y.; Al-Hilal, T.A.; Chung, S.W.; Kim, S.Y.; Ryu, G.H.; Son, W.C.; Byun, Y. Antiangiogenic and anticancer effect of an orally active low molecular weight heparin conjugates and its application to lung cancer chemoprevention. *J. Control. Release* **2015**, *199*, 122–131. [CrossRef]
101. Pfankuchena, D.B.; Stölting, D.P.; Schlesinger, M.; Royer, H.D.; Bendas, G. Low molecular weight heparin tinzaparin antagonizes cisplatin resistance of ovarian cancer cells. *Biochem. Pharmacol.* **2015**, *97*, 147–157. [CrossRef] [PubMed]
102. Park, J.; Jeong, J.-H.; Al-Hilal, T.A.; Kim, J.; Byun, Y. Size controlled heparin fragment—deoxycholic acid conjugate showed anticancer property by inhibiting VEGF165. *Bioconjugate Chem.* **2015**, *26*, 932–940. [CrossRef] [PubMed]
103. Park, J.; Kim, J.-y.; Hwang, S.R.; Mahmud, F.; Byun, Y. Chemical conjugate of low molecular weight heparin and suramin fragment inhibits tumor growth possibly by blocking VEGF165. *Mol. Pharm.* **2015**, *12*, 3935–3942. [CrossRef]
104. Parka, J.; Hwang, S.R.; Choi, J.U.; Alam, F.; Byun, Y. Self-assembled nanocomplex of PEGylated protamine and heparin–suramin conjugate for accumulation at the tumor site. *Int. J. Pharm.* **2018**, *535*, 38–46. [CrossRef]
105. Yang, Y.-C.; Cai, J.; Yin, J.; Zhang, J.; Wang, K.L.; Zhang, Z.T. Heparin-functionalized Pluronic nanoparticles to enhance the antitumour efficacy of sorafenib in gastric cancers. *Carbohydr. Polym.* **2016**, *136*, 782–790. [CrossRef]
106. Seib, F.P.; Tsurkan, M.; Freudenberg, U.; Kaplan, D.L.; Werner, C. Heparin-modified polyethylene glycol microparticle aggregates for focal cancer chemotherapy. *ACS Biomater. Sci. Eng.* **2016**, *2*, 2287–2293. [CrossRef]
107. Li, J.; Pan, H.; Qiao, S.; Li, Y.; Wang, J.; Liu, W.; Pan, W. The utilization of lowmolecular weight heparin-poloxamer associated Laponite nanoplatform for safe and efficient tumor therapy. *Int. J. Biol. Macromol.* **2019**, *134*, 63–72. [CrossRef]
108. Newlanda, B.; Varricchio, C.; Körner, Y.; Hoppe, F.; Taplan, C.; Newland, H.; Eigel, D.; Tornillo, G.; Pette, D.; Brancale, A.; et al. Focal drug administration via heparin-containing cryogel microcarriers reduces cancer growth and metastasis. *Carbohydr. Polym.* **2020**, *245*, 116504. [CrossRef] [PubMed]
109. Wang, D.; Luo, W.; Wen, G.; Yang, L.; Hong, S.; Zhang, S.; Diao, J.; Wang, J.; Wei, H.; Li, Y.; et al. Synergistic effects of negatively charged nanoparticles assisted by ultrasound on the reversal multidrug resistance phenotype in breast cancer cells. *Ultrason. Sonochem.* **2017**, *34*, 448–457. [CrossRef] [PubMed]
110. Tian, F.; Dahmani, F.Z.; Qiao, J.; Ni, J.; Xiong, H.; Liu, T.; Zhou, J.; Yao, J. A targeted nanoplatform codelivering chemotherapeutic and antiangiogenic drugs as a tool to reverse multidrug resistance in breast cancer. *Acta Biomater.* **2018**, *75*, 398–412. [CrossRef]
111. Thi, T.T.H.; Tran, D.-H.N.; Bach, L.G.; Vu-Quang, H.; Nguyen, D.C.; Park, K.D.; Nguyen, D.H. Functional magnetic core-shell system-based iron oxide nanoparticle coated with biocompatible copolymer for anticancer drug delivery. *Pharmaceutics* **2019**, *11*, 120.
112. Qiu, L.; Ge, L.; Long, M.; Mao, J.; Ahmed, K.S.; Shan, X.; Zhang, H.; Qin, L.; Lv, G.; Chen, J. Redox-responsive biocompatible nanocarriers based on novel heparosan polysaccharides for intracellular anticancer drug delivery. *Asian J. Pharm. Sci.* **2020**, *15*, 83–94. [CrossRef] [PubMed]
113. Guo, R.; Long, Y.; Lu, Z.; Deng, M.; He, P.; Li, M.; He, Q. Enhanced stability and efficacy of GEM-TOS prodrug by coassembly with antimetastatic shell LMWH-TOS. *Acta Pharm. Sin. B* **2020**, *10*, 1977–1988. [CrossRef] [PubMed]
114. Trana, T.H.; Bae, B.-c.; Lee, Y.; Na, K.; Huh, K.M. Heparin-folate-retinoic acid bioconjugates for targeted delivery of hydrophobic photosensitizers. *Carbohydr. Polym.* **2013**, *92*, 1615–1624. [CrossRef] [PubMed]
115. Shi, X.; Wang, Y.; Sun, H.; Chen, Y.; Zhang, X.; Xu, J.; Zhai, G. Heparin-reduced graphene oxide nanocomposites for curcumin delivery: In vitro, in vivo and molecular dynamics simulation study. *Biomater. Sci.* **2019**, *7*, 1011. [CrossRef] [PubMed]
116. Wu, Y.; Li, F.; Zhang, X.; Li, Z.; Zhang, Q.; Wang, W.; Pan, D.; Zheng, X.; Gu, Z.; Zhang, H.; et al. Tumor microenvironment-responsive PEGylated heparin-pyropheophorbide-a nanoconjugates for photodynamic therapy. *Carbohydr. Polym.* **2021**, *255*, 117490. [CrossRef]
117. Chaudhry, G.-e.S.; Akim, A.; Zafar, M.N.; Safdar, N.; Sung, Y.Y.; Muhammad, T.S.T. Understanding hyaluronan receptor (CD44) interaction, HA-CD44 activated potential targets in cancer therapeutics. *Adv. Pharm. Bull.* **2021**, *11*, 426–438. [CrossRef]
118. Dosio, F.; Arpicco, S.; Stella, B.; Fattal, E. Hyaluronic acid for anticancer drug and nucleic acid delivery. *Adv. Drug Deliv. Rev.* **2016**, *97*, 204–236. [CrossRef]
119. Espejo-Román, J.M.; Rubio-Ruiz, B.; Cano-Cortés, V.; Cruz-López, O.; Gonzalez-Resines, S.; Domene, C.; Conejo-García, A.; Sánchez-Martín, R.M. Selective anticancer therapy based on a HA-CD44 interaction inhibitor loaded on polymeric nanoparticles. *Pharmaceutics* **2022**, *14*, 788. [CrossRef]
120. Mo, Y.; Wang, H.; Liu, J.; Lan, Y.; Guo, R.; Zhang, Y.; Xue, W.; Zhang, Y. Controlled release and targeted delivery to cancer cells of doxorubicin from polysaccharide-functionalised single-walled carbon nanotubes. *J. Mater. Chem. B* **2015**, *3*, 1846–1855. [CrossRef]
121. Chen, W.; Wang, F.; Zhang, X.; Hu, J.; Wang, X.; Yang, K.; Huang, L.; Xu, M.; Li, Q.; Fu, L. Overcoming ABCG2-mediated multidrug resistance by a mineralized hyaluronan-drug nanocomplex. *J. Mater. Chem. B* **2016**, *4*, 6652–6661. [CrossRef] [PubMed]
122. Teong, B.; Lin, C.-Y.; Chang, S.J.; Niu, G.C.C.; Yao, C.H.; Chen, I.F.; Kuo, S.M. Enhanced anticancer activity by curcumin-loaded hydrogel nanoparticle derived aggregates on A549 lung adenocarcinoma cells. *J. Mater. Sci. Mater. Med.* **2015**, *26*, 1–15. [CrossRef] [PubMed]
123. Hsiao, K.Y.; Wu, Y.-J.; Liu, Z.; Chuang, C.; Huang, H.; Kuo, S. Anticancer effects of sinulariolide-conjugated hyaluronan nanoparticles on lung adenocarcinoma cells. *Molecules* **2016**, *21*, 297. [CrossRef] [PubMed]

124. Chen, D.; Dong, X.; Qi, M.; Song, X.; Sun, J. Dual pH/redox responsive and CD44 receptor targeting hybrid nanochrysalis based on new oligosaccharides of hyaluronan conjugates. *Carbohydr. Polym.* **2017**, *157*, 1272–1280. [CrossRef]
125. Cai, Z.; Zhang, H.; Wei, Y.; Wei, Y.; Xie, Y.; Cong, F. Reduction- and pH-sensitive hyaluronan nanoparticles for delivery of iridium(III) anticancer drugs. *Biomacromolecules* **2017**, *18*, 2102–2117. [CrossRef]
126. Zhang, W.; Tung, C.-H. Redox-responsive cisplatin nanogels for anticancer drug Delivery. *Chem. Commun.* **2018**, *54*, 8367–8370. [CrossRef] [PubMed]
127. Sun, Z.; Yi, Z.; Cui, X.; Chen, X.; Su, W.; Ren, X.; Li, X. Tumor-targeted and nitric oxide-generated nanogels of keratin and hyaluronan for enhanced cancer therapy. *Nanoscale* **2018**, *10*, 12109–12122. [CrossRef]
128. Amano, Y.; Ohta, S.; Sakura, K.L.; Ito, T. Pemetrexed-conjugated hyaluronan for the treatment of malignant pleural mesothelioma. *Eur. J. Pharm. Sci.* **2019**, *138*, 105008. [CrossRef]
129. Yu, J.S.; Shin, D.H.; Kim, J.S. Repurposing of fluvastatin as an anticancer agent against breast cancer stem cells via encapsulation in a hyaluronan-conjugated liposome. *Pharmaceutics* **2020**, *12*, 1133. [CrossRef]
130. Kang, Y.; Sun, W.; Li, S.; Li, M.; Fan, J.; Du, J.; Liang, X.J.; Peng, X. Oligo Hyaluronan-Coated Silica/Hydroxyapatite Degradable Nanoparticles for Targeted Cancer Treatment. *Adv. Sci.* **2019**, *6*, 1–11. [CrossRef]
131. Cosco, D.; Mare, R.; Paolino, D.; Salvatici, M.C.; Cilurzo, F.; Fresta, M. Sclareol-loaded hyaluronan-coated PLGA nanoparticles: Physico-chemical properties and in vitro anticancer features. *Int. J. Biol. Macromol.* **2019**, *132*, 550–557. [CrossRef] [PubMed]
132. Wang, J.; Muhammad, N.; Li, T.; Wang, H.; Liu, Y.; Liu, B.; Zhan, H. Hyaluronic acid-coated camptothecin nanocrystals for targeted drug delivery to enhance anticancer efficacy. *Mol. Pharm.* **2020**, *17*, 2411–2425. [CrossRef] [PubMed]
133. Yang, H.; Miao, Y.; Chen, L.; Li, Z.; Yang, R.; Xu, X.; Liu, Z.; Zhang, L.M.; Jiang, X. Redox-responsive nanoparticles from disulfide bond-linked poly-(N-ε-carbobenzyloxy-L-lysine)-grafted hyaluronan copolymers as theranostic nanoparticles for tumor-targeted MRI and chemotherapy. *Int. J. Biol. Macromol.* **2020**, *148*, 483–492. [CrossRef] [PubMed]
134. Poudel, K.; Gautam, M.; Maharjan, S.; Jeong, J.H.; Choi, H.G.; Khan, G.M.; Yong, C.S.; Kim, J.O. Dual stimuli-responsive ursolic acid-embedded nanophytoliposome for targeted antitumour therapy. *Int. J. Pharm.* **2020**, *582*, 119330. [CrossRef]
135. Cong, Z.; Zhang, L.; Ma, S.Q.; Lam, K.S.; Yang, F.F.; Liao, Y.H. Size-transformable hyaluronan stacked self-assembling peptide nanoparticles for improved transcellular tumor penetration and photo−chemo combination therapy. *ACS Nano* **2020**, *14*, 1958–1970. [CrossRef]
136. Gao, X.; Wei, M.; Ma, D.; Yang, X.; Zhang, Y.; Zhou, X.; Li, L.; Deng, Y.; Yang, W. Engineering of a hollow-structured Cu_2-XS nano-homojunction platform for near infrared-triggered infected wound healing and cancer therapy. *Adv. Funct. Mater.* **2021**, *31*, 2106700. [CrossRef]
137. Gao, S.; Islam, R.; Fang, J. Tumor environment-responsive hyaluronan conjugated zinc protoporphyrin for targeted anticancer photodynamic therapy. *J. Pers. Med.* **2021**, *11*, 136. [CrossRef]
138. Kim, J.-E.; Park, Y.-J. Hyaluronan self-agglomerating nanoparticles for non-small cell lung cancer targeting. *Cancer Nanotechnol.* **2022**, *13*, 1–24. [CrossRef]
139. Chang, Y.-L.; Liao, P.B.; Wu, P.H.; Chang, W.J.; Lee, S.Y.; Huang, H.M. Cancer cytotoxicity of a hybrid hyaluronan-superparamagnetic iron oxide nanoparticle material: An in-vitro evaluation. *Nanomaterials* **2022**, *12*, 496. [CrossRef]
140. Jacquinet, J.C.; Lopin-Bon, C.; Vibert, A. From polymer to size-defined oligomers: A highly divergent and stereocontrolled construction of chondroitin sulfate A, C, D, E, K, L, and M oligomers from a single precursor: Part 2. *Chem. Eur.* **2009**, *15*, 9579–9595. [CrossRef]
141. Jardim, K.V.; Joanitti, G.A.; Azevedo, R.B.; Parize, A.L. Physico-chemical characterization and cytotoxicity evaluation of curcumin loaded in chitosan/chondroitin sulfate nanoparticles. *Mater. Sci. Eng. C* **2015**, *56*, 294–304. [CrossRef] [PubMed]
142. Bárbara, S.; Cátia, C.; Nunes, S.; Panice, M.R.; Scariot, D.B.; Nakamura, C.V.; Muniz, E.C. Manufacturing micro/nano chitosan/chondroitin sulfate curcumin-loaded hydrogel in ionic liquid: A new biomaterial effective against cancer cells. *Int. J. Biol. Macromol.* **2021**, *180*, 88–96.
143. Yuan, Y.; Ma, M.; Zhang, S.; Liu, C.; Chen, P.; Li, H.; Wang, D.; Xu, Y. Effect of sophorolipid on the curcumin-loaded ternary composite nanoparticles self-assembled from zein and chondroitin sulfate. *Food Hydrocoll.* **2021**, *113*, 106493. [CrossRef]
144. Soe, Z.C.; Poudel, B.K.; Nguyen, H.T.; Thapa, R.K.; Ou, W.; Gautam, M.; Poudel, K.; Jin, S.G.; Jeong, J.H.; Ku, S.K.; et al. Folate-targeted nanostructured chitosan/chondroitin sulfate complex carriers for enhanced delivery of bortezomib to colorectal cancer cells. *Asian J. Pharm. Sci.* **2019**, *14*, 40–51. [CrossRef] [PubMed]
145. Liang, T.; Zhang, Z.; Jing, P. Black rice anthocyanins embedded in self-assembled chitosan/chondroitin sulfate nanoparticles enhance apoptosis in HCT-116 cells. *Food Chem.* **2019**, *301*, 125280. [CrossRef] [PubMed]
146. Barkat, K.; Ahmad, M.; Minhas, M.U.; Khalid, I.; Malik, N.S. Chondroitin sulfate-based smart hydrogels for targeted delivery of oxaliplatin in colorectal cancer: Preparation, characterization and toxicity evaluation. *Polym. Bull.* **2020**, *77*, 6271–6297. [CrossRef]
147. Li, M.; Sun, J.; Zhang, W.; Zhao, Y.; Zhang, S.; Zhang, S. Drug delivery systems based on CD44-targeted glycosaminoglycans for cancer therapy. *Carbohydr. Polym.* **2021**, *251*, 117103. [CrossRef]
148. Zu, M.; Ma, L.; Zhang, X.; Xie, D.; Kang, Y.; Xiao, B. Chondroitin sulfate-functionalized polymeric nanoparticles for colon cancer-targeted chemotherapy. *Colloids Surf. B: Biointerfaces* **2019**, *177*, 399–406. [CrossRef]
149. Chen, Y.; Li, B.; Chen, X.; Wu, M.; Ji, Y.; Tang, G.; Ping, Y. A supramolecular codelivery strategy for combined breast cancer treatment and metastasis prevention. *Chin. Chem. Lett.* **2020**, *31*, 1153–1158. [CrossRef]

150. Singhai, N.J.; Maheshwari, R.; Jain, N.K.; Ramteke, S. Chondroitin sulfate and α-tocopheryl succinate tethered multiwalled carbon nanotubes for dual-action therapy of triple-negative breast cancer. *J. Drug Deliv. Sci. Technol.* **2020**, *60*, 102080. [CrossRef]
151. Zhang, Z.; Ma, L.; Luo, J. Chondroitin sulfate-modified liposomes for targeted co-delivery of doxorubicin and retinoic acid to suppress breast cancer lung metastasis. *Pharmaceutics* **2021**, *13*, 406. [CrossRef] [PubMed]
152. Shi, X.; Yang, X.; Liu, M.; Wang, R.; Qiu, N.; Liu, Y.; Yang, H.; Ji, J.; Zhai, G. Chondroitin sulfate-based nanoparticles for enhanced chemo-photodynamic therapy overcoming multidrug resistance and lung metastasis of breast cancer. *Carbohydr. Polym.* **2021**, *254*, 117459. [CrossRef] [PubMed]
153. Wu, R.; Shang, N.; Gui, M.; Yin, J.; Li, P. Sturgeon (*acipenser*)-derived chondroitin sulfate suppresses human colon cancer HCT-116 both *in vitro* and *in vivo* by inhibiting proliferation and inducing apoptosis. *Nutrients* **2020**, *12*, 1130. [CrossRef] [PubMed]
154. Peng, C.; Wang, Q.; Jiao, R.; Xu, Y.; Han, N.; Wang, W.; Zhu, C.; Li, F. A novel chondroitin sulfate E from *Dosidicus gigas* cartilage and its antitumour metastatic activity. *Carbohydr. Polym.* **2021**, *262*, 117971. [CrossRef]
155. Thinh, P.D.; Ly, B.M.; Usoltseva, R.V.; Shevchenko, N.M.; Rasin, A.B.; Anastyuk, S.D.; Malyarenko, O.S.; Zvyagintseva, T.N.; San, P.T.; Ermakova, S.P. A novel sulfated fucan from Vietnamese sea cucumber Stichopus variegatus: Isolation, structure and anticancer activity in vitro. *Int. J. Biol. Macromol.* **2018**, *117*, 1101–1109. [CrossRef]
156. Aldairi, A.F.; Ogundipe, O.D.; Pye, D.A. Antiproliferative Activity of Glycosaminoglycan-Like Polysaccharides Derived from Marine Molluscs. *Mar. Drugs* **2018**, *16*, 63. [CrossRef]
157. Sasaki, T.; Takasuka, N. Further Study of the Structure of Lentinan, an Antitumour Polysaccharide from Lentinus Edodes. *Carbohydr. Res.* **1976**, *47*, 99–104. [CrossRef]
158. Chen, Q.; Zheng, Y.; Chen, X.; Ge, P.; Wang, P.; Wu, B. Upregulation of miR-216a-5p by lentinan targeted inhibition of JAK2/STAT3 signaling pathway to reduce lung adenocarcinoma cell stemness, promote apoptosis, and slow down the lung adenocarcinoma, mechanisms. *Front. Oncol.* **2021**, *11*, 778096. [CrossRef]
159. Liu, Z.; Yu, L.; Gu, P.; Bo, R. Preparation of lentinan-calcium carbonate microspheres and their application as vaccine adjuvants. *Carbohydr. Polym.* **2020**, *245*, 116520. [CrossRef]
160. Yang, F.; Huang, J.; Liu, H.; Lin, W.; Li, X.; Zhu, X.; Chen, T. Lentinan-functionalized selenium nanosystems with high permeability infiltrate solid tumors by enhancing transcellular transport. *Nanoscale* **2020**, *12*, 14494–14503. [CrossRef]
161. Song, Z.; Luo, W.; Zheng, H.; Zeng, Y.; Wang, J.; Chen, T. Translational nanotherapeutics reprograms immune microenvironment in malignant pleural effusion of lung adenocarcinoma. *Adv. Healthc. Mater.* **2021**, *10*, 2100149. [CrossRef]
162. Lu, J.; He, R.; Sun, P.; Zhang, F.; Linhardt, R.J. Molecular mechanisms of bioactive polysaccharides from *Ganoderma lucidum* (Lingzhi). a review. *Int. J. Biol. Macromol.* **2020**, *150*, 765–774. [CrossRef] [PubMed]
163. Dong, Q.; Wang, Y.; Shi, L.; Yao, J.; Li, J.; Ma, F.; Ding, K. A novel water-soluble β-D-glucan isolated from the spores of Ganoderma lucidum. *Carbohydr. Res.* **2012**, *353*, 100–105. [CrossRef] [PubMed]
164. Bao, X.-F.; Zhen, Y.; Ruan, L.; Fang, J.N. Purification, characterization, and modification of T lymphocyte-stimulating polysaccharide from spores of Ganoderma lucidum. *Chem. Pharm. Bull.* **2002**, *50*, 623–629. [CrossRef] [PubMed]
165. Fu, Y.; Shi, L.; Ding, K. Structure elucidation and antitumour activity in vivo of a polysaccharide from spores of *Ganoderma lucidum* (Fr.) Karst. *Int. J. Biol. Macromol.* **2019**, *141*, 693–699. [CrossRef]
166. Liu, H.; Amakye, W.K.; Ren, J. Codonopsis pilosula polysaccharide in synergy with dacarbazine inhibits mouse melanoma by repolarizing M2-like tumor-associated macrophages into M1-like tumor-associated macrophages. *Biomed. Pharmacother.* **2021**, *142*, 112016. [CrossRef]
167. da Silva Milhorini, S.; de Lima Bellan, D.; Zavadinack, M.; Simas, F.F.; Smiderle, F.R.; de Santana-Filho, A.P.; Sassaki, G.L.; Iacomini, M. Antimelanoma effect of a fucoxylomannan isolated from Ganoderma lucidum fruiting bodies. *Carbohydr. Polym.* **2022**, *294*, 119823. [CrossRef] [PubMed]
168. Hsu, W.-H.; Qiu, W.-L.; Tsao, S.M.; Tseng, A.J.; Lu, M.K.; Hua, W.J.; Cheng, H.C.; Hsu, H.Y.; Lin, T.Y. Effects ofWSG, a polysaccharide from Ganoderma lucidum, on suppressing cell growth and mobility of lung cancer. *Int. J. Biol. Macromol.* **2020**, *165*, 1604–1613. [CrossRef]
169. Hsu, W.-H.; Hua, W.-J.; Qiu, W.L.; Tseng, A.J.; Cheng, H.C.; Lin, T.Y. WSG, a glucose-enriched polysaccharide from Ganoderma lucidum, suppresses tongue cancer cells via inhibition of EGFR-mediated signaling and potentiates cisplatin-induced apoptosis. *Int. J. Biol. Macromol.* **2021**, *193*, 1201–1208. [CrossRef]
170. Zhang, S.; Pang, G.; Chen, C.; Qin, J.; Yu, H.; Liu, Y.; Zhang, X.; Song, Z.; Zhao, J.; Wang, F.; et al. Effective cancer immunotherapy by Ganoderma lucidum polysaccharide-gold nanocomposites through dendritic cell activation and memory T-cell response. *Carbohydr. Polym.* **2019**, *205*, 192–202. [CrossRef]
171. Yu, H.; Yang, Y.; Jiang, T.; Zhang, X.; Zhao, Y.; Pang, G.; Feng, Y.; Zhang, S.; Wang, F.; Wang, Y.; et al. Effective Radiotherapy in Tumor Assisted by Ganoderma lucidum Polysaccharide-Conjugated Bismuth Sulfide Nanoparticles through Radiosensitization and Dendritic Cell Activation. *ACS Appl. Mater. Interfaces* **2019**, *11*, 27536–27547. [CrossRef] [PubMed]
172. Soerjomataram, I.; Bray, F. Planning for tomorrow: Global cancer incidence and the role of prevention 2020–2070. *Nat. Rev. Clin. Oncol.* **2021**, *18*, 663–672. [CrossRef] [PubMed]

molecules

Communication

Stereoselective Synthesis of 2-Deoxythiosugars from Glycals

Xueying You [1], Yifei Cai [1], Chenyu Xiao [1], Lijuan Ma [1], Yong Wei [2], Tianpeng Xie [2], Lei Chen [2,*] and Hui Yao [1,*]

1 Hubei Key Laboratory of Natural Products Research and Development, Key Laboratory of Functional Yeast (China National Light Industry), College of Biological and Pharmaceutical Sciences, China Three Gorges University, Yichang 443002, China
2 Yichang Humanwell Pharmaceutical Co., Ltd., Yichang 443000, China
* Correspondence: chenleiyf@renfu.com.cn (L.C.); yaohui@ctgu.edu.cn (H.Y.)

Abstract: 2-deoxythiosugars are more stable than 2-deoxysugars occurring broadly in bioactive natural products and pharmaceutical agents. An effective and direct methodology to stereoselectively synthesize α-2-deoxythioglycosides catalyzed by AgOTf has been developed. Various alkyl thiols and thiophenols were explored and the desired products were formed in good yields with excellent α-selectivity. This method was further applied to the syntheses of *S*-linked disaccharides and late-stage 2-deoxyglycosylation of estrogen, L-menthol, and zingerone thiols successfully.

Keywords: 2-deoxythioglycosides; glycosylation; silver triflate; stereoselective synthesis; glycals

Citation: You, X.; Cai, Y.; Xiao, C.; Ma, L.; Wei, Y.; Xie, T.; Chen, L.; Yao, H. Stereoselective Synthesis of 2-Deoxythiosugars from Glycals. *Molecules* **2022**, *27*, 7979. https://doi.org/10.3390/molecules27227979

Academic Editors: Jian Yin, Jing Zeng and De-Cai Xiong

Received: 31 October 2022
Accepted: 15 November 2022
Published: 17 November 2022

1. Introduction

2-deoxysugars occur broadly in bioactive natural products and pharmaceutical agents [1–3]. They have been applied as clinical drugs in treating various diseases, including heart failure, cancers, and bacterial and viral infections [4–7]. However, the glycosidic bond of 2-deoxysugar is easily hydrolyzed by enzymes or acids, resulting in a short half-life in vivo, which limits its application in drug development [8,9]. As sulfur is in the same group as oxygen, it is often used as a bioisostere to replace oxygen atoms in medicinal chemistry, indicating a longer half-life and better biological activity [10–13]. Stereoselective synthesis of 2-deoxythiosugar is quite challenging because there is no C2-group to direct the anomeric selectivity through the neighboring participation effect. Many efforts have been made to develop the strategies of 2-deoxythioglycosylation. Conventionally, 2-deoxysugar synthesis studies could be performed with saturated glycosyl donors, the high stereoselectivity of which relied on a specific structure of a well-assembled glycosyl donor [14,15]. On the other hand, unsaturated glycosyl donors (glycals) could be applied directly to construct 2-deoxysugars stereoselectively [16,17]. Toste and coworkers developed an effective synthesis of 2-deoxyglycosides from glycal donors mediated by a catalytic Re(V)-oxo complex [18]. Recently, Wan's group successfully achieved access to α-1,1′-2-deoxy thioglycosides stereoselectively catalyzed by $ReOCl_3(SMe_2)(OPPh_3)$ as shown in Scheme 1a [19]. With the development of organocatalysis, Kancharla's group utilized a bulky pyridinium salt 2,4,6-tri-*tert*-butylpyridine-hydrochloric acid (TTBPy·HCl) to catalyze the reaction between glycals and thiols giving 2-deoxy-β-galactosides with an α/β ratio from 3.8:1 to β only (Scheme 1b) [20]. However, rhenium catalysts are quite expensive and the synthesis of organocatalysts is tedious. Therefore, the stereoselective synthesis of 2-deoxythioglycosides is still highly challenging. In view of our long-standing interest in exploring 3,4-*O*-carbonate-glycal donors [21–27], herein, we report an effective and stereoselective 2-deoxythioglycosylation catalyzed by the commercially available catalyst silver trifluoromethanesulfonate (AgOTf) as shown in Scheme 1c. The target α-2-deoxythiosugars were able to be obtained in good yields with high stereoselectivity under mild conditions.

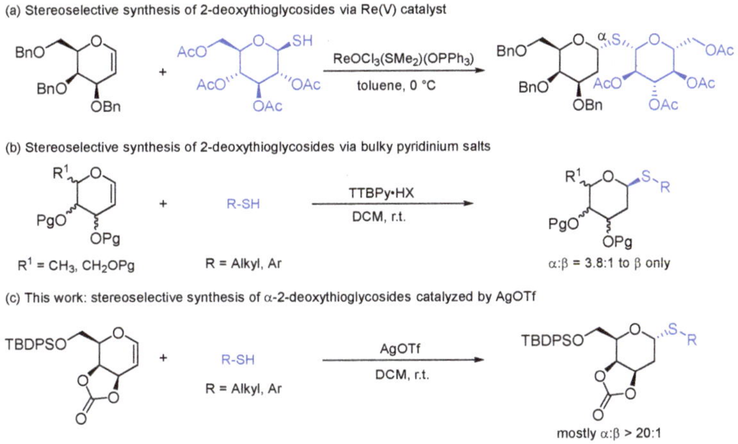

Scheme 1. Synthesis of 2-deoxythioglycosides from glycal donors.

2. Results

First, 3,4-*O*-carbonate galactal **1a** was adopted as a glycosyl donor and *p*-methylthiophenol **2a** was adopted as a glycosyl acceptor to optimize the conditions for 2-deoxythioglycosylation (Table 1). Initially, $Cu(OTf)_2$ was examined as the catalyst, giving 2-deoxysugar **3a** in a yield of 32% with high α-selectivity (α/β > 20:1) in dichloromethane (entry 1). Then, various Lewis acids were screened (entries 2–8). For example, $Hg(OTf)_2$ could increase the yield to 59% (entry 2), while $Yb(OTf)_3$ decreased the yield to 30% (entry 3). $Fe(OTf)_3$ was also able to catalyze this reaction to give the desired product in a 64% yield (entry 4). There was no reaction when $Ni(OTf)_2$ and $Dy(OTf)_3$ were examined (entries 5, 7). AgOTf could increase the yield to 80% (entry 8), which was the best catalyst for this reaction. Then various solvents were examined, including toluene, THF, MeCN, dimethyl carbonate (DMC), and diethyl carbonate (DEC) (entries 9–15). The yield was lower (43%) in toluene (entry 9), while no reaction was observed in THF (entry 10), acetonitrile (entry 11), DMF (entry 14), and DMSO (entry 15). The green solvent (DMC, entry 12) was applied, and the yield decreased to 57%, while DEC (entry 13) was also attempted to proceed with the reaction, but only a 33% yield was observed. Therefore, the condition was finalized as Ag(OTf) as the catalyst in DCM at room temperature.

With the optimized condition in hand, the substrate scope was first explored using thiophenols (Scheme 2). The aryl 2-deoxythioglycosides (**3a-3c**) were obtained using *p*-, *m*- and *o*-thiocresols as acceptors in yields of 80–82% and with an α/β ratio from 15:1 to > 20:1 determined by ^1H NMR. Fortunately, the crystal of **3c** was obtained and the configuration was confirmed as α-2-deoxy-D-galactoside by single-crystal X-ray diffraction. 3,4-Dimethylthiophenol also worked well giving **3d** in an 82% yield and with a ratio of α/β = 20:1. Besides methyl substitution thiophenols, 4-methoxythiophenol reacted with 3,4-*O*-carbonate-D-galactal **1a** to form **3e** in a 73% yield with a ratio of α/β > 20:1. The bulkier substitute group on the phenyl ring also tolerated this method well. For example, 4-*tert*-butylthiophenol was applied to generate **3f** in an 81% yield and with an α/β ratio of 18:1. Thiophenol 2-deoxysugar **3g** was obtained stereoselectively (α/β > 20:1) in an 80% yield. When thiophenols were substituted by electron-withdrawing groups, the desired products (**3h-3j**) could still be formed in good yields but stereoselectivity decreased slightly. The 2-deoxythiosugar **3h** was formed in a yield of 75% with a ratio of α:β = 11:1 using 4-bromothiophenol as the acceptor. 2-Bromothiophenol and 4-fluorothiophenol were examined and gave target products in a 71–74% yield at the ratios of α:β = 15:1 and α:β = 12:1, respectively.

Table 1. Condition optimization for the 2-deoxythioglycosylation [a].

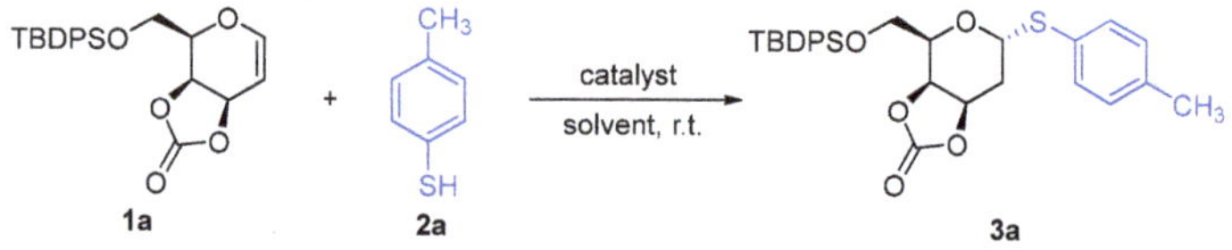

Entry	Catalyst	Solvent	Yield (%) [b]
1	$Cu(OTf)_2$	DCM	32
2	$Hg(OTf)_2$	DCM	59
3	$Yb(OTf)_3$	DCM	30
4	$Fe(OTf)_3$	DCM	64
5	$Ni(OTf)_2$	DCM	N.R.
6	$Sc(OTf)_3$	DCM	50
7	$Dy(OTf)_3$	DCM	N.R.
8	AgOTf	DCM	80
9	AgOTf	toluene	43
10	AgOTf	THF	N.R.
11	AgOTf	MeCN	N.R.
12	AgOTf	DMC	57
13	AgOTf	DEC	33
14	AgOTf	DMF	N.R.
15	AgOTf	DMSO	N.R.

[a] Unless otherwise specified, all reactions were carried out with 0.1 mmol of **1a**, 0.11 mmol of **2a**, and 10 mol% catalyst in 2 mL solvent, at room temperature for 24 h under N_2 atmosphere. [b] Isolated yield, $\alpha/\beta > 20{:}1$ by ^1H NMR. N.R. = No Reaction.

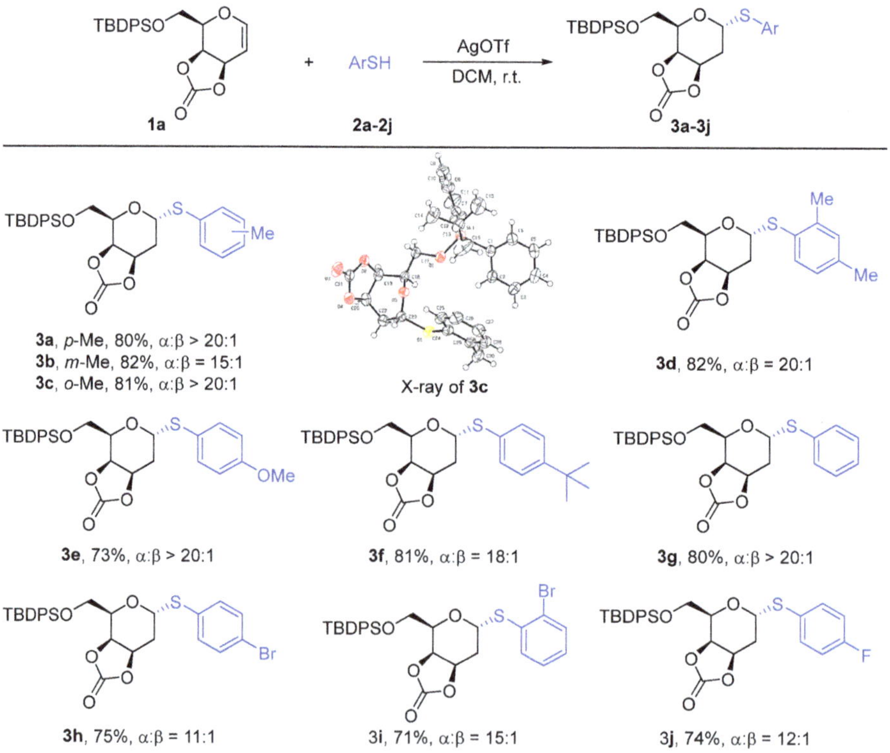

Scheme 2. Substrate scope of aryl 2-deoxythioglycosides.

2-deoxythioglycosides synthesis and purified by flash column chromatography giving colorless oil 43.5 mg, yield 71%; ^{1}H NMR (400 MHz, $CDCl_3$) δ 7.64–7.69 (m, 4H, Ar-H), 7.55 (ddd, J = 7.9, 2.5, 1.5 Hz, 2H, Ar-H), 7.50–7.35 (m, 6H, Ar-H), 7.17–7.20 (m,1H, Ar-H), 7.06–7.11 (m, 1H, Ar-H), 5.71 (dd, J = 9.8, 6.5 Hz, 1H, H-1), 5.12 (ddd, J = 8.6, 4.0, 2.2 Hz, 1H, H-3), 4.95 (dd, J = 8.6, 1.6 Hz, 1H, H-4), 4.21 (ddd, J = 7.8, 6.3, 1.7 Hz, 1H, H-5), 3.85 (dd, J = 10.2, 7.5 Hz, 1H, H-6), 3.79 (dd, J = 10.2, 6.3 Hz, 1H, H-6′), 2.74 (ddd, J = 15.9, 6.6, 3.5 Hz, 1H, H-2), 1.98 (ddd, J = 15.8, 9.8, 2.9 Hz, 1H, H-2′), 1.04 (s, 9H, Si-^{t}Bu).^{13}C NMR (100 MHz, $CDCl_3$) δ 154.2, 135.7, 135.6, 135.2, 133.1, 132.8, 132.0, 130.1, 130.0, 128.6, 128.20, 128.0, 125.2, 79.3, 76.8, 72.3, 68.9, 61.3, 28.6, 26.9, 19.3. HRMS (ESI) m/z: calcd. for $C_{29}H_{31}BrO_5SSiNa^+$ $(M + Na)^+$ 621.0737, found 621.0735; $[\alpha]_D^{25}$ = +82.4 (c = 1.0, $CHCl_3$).

4-Fluorophenyl-1-thio-6-O-(tert-butyldiphenylsilyl)-3,4-O-carbonate-2-deoxy-α-D-galactopyranoside (**3j**). The title compound was prepared according to the general procedure of 2-deoxythioglycosides synthesis and purified by flash column chromatography giving colorless oil 39.8 mg, yield 74%; ^{1}H NMR (400 MHz, $CDCl_3$) δ 7.79–7.67 (m, 4H, Ar-H), 7.56–7.40 (m, 8H, Ar-H), 6.90–6.93 (m, 2H, Ar-H), 5.46 (dd, J = 9.9, 6.5 Hz, 1H, H-1), 5.07 (ddd, J = 8.6, 4.2, 2.0 Hz, 1H, H-3), 4.90 (dd, J = 8.7, 1.6 Hz, 1H, H-4), 4.16–4.20 (m, 1H, H-5), 3.87 (dd, J = 10.4, 7.2 Hz, 1H, H-6), 3.80 (dd, J = 10.3, 6.3 Hz, 1H, H-6′), 2.68 (ddd, J = 15.9, 6.5, 3.4 Hz, 1H, H-2), 1.89 (ddd, J = 15.8, 9.8, 2.9 Hz, 1H, H-2′), 1.09 (s, 9H, Si-^{t}Bu).^{13}C NMR (100 MHz, $CDCl_3$) δ 163.0 (d, J = 246.7 Hz), 154.1, 135.7, 135.6, 135.1 (d, J = 8.2 Hz), 133.1, 132.8, 130.1, 130.0, 128.0, 116.2 (d, J = 21.6 Hz), 81.0, 73.4, 72.4, 68.7, 61.7, 28.8, 26.9, 19.4. HRMS (ESI) m/z: calcd. for $C_{29}H_{31}FO_5SSiNa^+$ $(M + Na)^+$ 561.1538, found 561.1541; $[\alpha]_D^{25}$ = +86.3 (c = 1.0, $CHCl_3$).

Ethyl-1-thio-6-O-(tert-butyldiphenylsilyl)-3,4-O-carbonate-2-deoxy-α-D-galactopyranoside (**5a**). The title compound was prepared according to the general procedure of 2-deoxythioglycosides synthesis and purified by flash column chromatography giving colorless oil 40.6 mg, yield 86%; ^{1}H NMR (400 MHz, $CDCl_3$) δ 7.67–7.71 (m, 4H, Ar-H), 7.60–7.36 (m, 6H, Ar-H), 5.39 (dd, J = 9.3, 6.5 Hz, 1H, H-1), 5.02 (ddd, J = 8.4, 4.8, 2.4 Hz, 1H, H-3), 4.86 (dd, J = 8.5, 1.7 Hz, 1H, H-4), 4.23–4.01 (m, 1H, H-5), 3.85 (dd, J = 10.3, 7.3 Hz, 1H, H-6), 3.78 (dd, J = 10.3, 6.4 Hz, 1H, H-6′), 2.67–2.75 (m, 1H, H-2), 2.63–2.49 (m, 2H, CH_2CH_3), 1.79 (ddd, J = 15.8, 9.3, 3.2 Hz, 1H, H-2′), 1.24 (dd, J = 13.4, 6.7 Hz, 3H, CH_2CH_3), 1.08 (s, 9H, Si-^{t}Bu). ^{13}C NMR (100 MHz, $CDCl_3$) δ 154.4, 135.7, 135.6, 133.2, 132.9, 130.1, 130.0, 128.0, 127.9, 76.7, 73.7, 72.5, 68.0, 61.7, 29.2, 26.9, 24.7, 19.3, 15.0. HRMS (ESI) m/z: calcd. For $C_{25}H_{32}O_5SsiNa^+$ $(M + Na)^+$ 495.1632, found 495.1634; $[\alpha]_D^{25}$ = +54.2 (c = 1.0, $CHCl_3$).

Octyl-1-thio-6-O-(tert-butyldiphenylsilyl)-3,4-O-carbonate-2-deoxy-α-D-galactopyranoside (**5b**). The title compound was prepared according to the general procedure of 2-deoxythioglycosides synthesis and purified by flash column chromatography giving colorless oil 43.3 mg, yield 78%; ^{1}H NMR (400 MHz, $CDCl_3$) δ 7.68–7.72 (m, 4H, Ar-H), 7.40–7.50 (m, 6H, Ar-H), 5.35 (dd, J = 9.3, 6.5 Hz, 1H, H-1), 5.02 (ddd, J = 8.4, 4.5, 2.3 Hz, 1H, H-3), 4.87 (dd, J = 8.5, 1.7 Hz, 1H, H-4), 4.13 (ddd, J = 7.4, 6.9, 1.6 Hz, 1H, H-5), 3.85 (dd, J = 10.2, 7.3 Hz, 1H, H-6), 3.78 (dd, J = 10.2, 6.4 Hz, 1H, H-6′), 2.68 (ddd, J = 12.9, 8.1, 6.6 Hz, 1H, H-2), 2.57–2.41 (m, 2H), 1.79 (ddd, J = 15.8, 9.3, 3.1 Hz, 1H, H-2′), 1.59–1.51 (m, 2H, CH_2), 1.41–1.12 (m, 10H, CH_2), 1.08 (s, 9H, Si-^{t}Bu), 0.89 (dd, J = 7.6, 5.7 Hz, 3H, CH_3). ^{13}C NMR (100 MHz, $CDCl_3$) δ 154.4, 135.7, 135.6, 133.2, 132.9, 130.1, 130.0, 128.0, 127.9, 77.1, 73.7, 72.5, 67.9, 61.7, 31.9, 30.7, 29.9, 29.3, 29.2, 29.2, 29.0, 26.9, 22.8, 19.3, 14.2. HRMS (ESI) m/z: calcd. for $C_{31}H_{44}O_5SSiNa^+$ $(M + Na)^+$ 579.2571, found 579.2556; $[\alpha]_D^{25}$ = +51.4 (c = 1.0, $CHCl_3$).

n-Buty-1-thio-6-O-(tert-butyldiphenylsilyl)-3,4-O-carbonate-2-deoxy-α-D-galactopyranoside (**5c**). The title compound was prepared according to the general procedure of 2-deoxythioglycosides synthesis and purified by flash column chromatography giving colorless oil 41.5 mg, yield 83%; ^{1}H NMR (400 MHz, $CDCl_3$) δ 7.48–7.71 (m, 4H, Ar-H), 7.55–7.36 (m, 6H, Ar-H), 5.35 (dd, J = 9.3, 6.5 Hz, 1H, H-1), 5.02 (ddd, J = 8.5, 4.4, 2.4 Hz, 1H, H-3), 4.87 (dd, J = 8.5, 1.7 Hz, 1H, H-4), 4.13 (ddd, J = 8.3, 6.4, 1.6 Hz, 1H, H-5), 3.85 (dd, J = 10.2, 7.4 Hz, 1H, H-6), 3.77 (dd, J = 10.2, 6.3 Hz, 1H, H-6′), 2.69 (ddd, J = 14.7, 8.1, 6.7 Hz, 1H, H-2), 2.60–2.46 (m, 2H, CH_2), 1.78 (ddd, J = 15.7, 9.3, 3.1 Hz, 1H, H-2′), 1.60–1.49 (m, 2H, CH_2), 1.41–1.32 (m, 2H, CH_2), 1.08 (s, 9H, Si-^{t}Bu), 0.87 (dd, J = 14.6, 6.9 Hz, 3H, CH_3). ^{13}C NMR

(100 MHz, $CDCl_3$) δ 154.4, 135.7, 135.6, 133.2, 132.9, 130.1, 130.0, 128.0, 127.9, 77.1, 73.7, 72.5, 67.9, 61.6, 31.9, 30.4, 29.2, 26.9, 22.0, 19.4, 13.7. HRMS (ESI) m/z: calcd. for $C_{27}H_{36}O_5SSiNa^+$ $(M + Na)^+$ 523.1945, found 523.1945; $[\alpha]_D^{25}$ = +36.4 (c = 1.0, $CHCl_3$).

Isobutyl-1-thio-6-O-(tert-butyldiphenylsilyl)-3,4-O-carbonate-2-deoxy-α-D-galactopyranoside (**5d**). The title compound was prepared according to the general procedure of 2-deoxythioglycosides synthesis and purified by flash column chromatography giving colorless oil 41.5 mg, yield 83%; ^1H NMR (400 MHz, $CDCl_3$) δ 7.67–7.71(m, 4H, Ar-H), 7.54–7.34 (m, 6H, Ar-H), 5.31 (dd, *J* = 9.3, 6.5 Hz, 1H, H-1), 5.02 (ddd, 8.6, 4.6, 2.4, 1H, H-3), 4.88 (dd, *J* = 8.5, 1.6 Hz, 1H, H-4), 4.19–4.04 (m, 1H, H-5), 3.85 (dd, *J* = 10.2, 7.5 Hz, 1H, H-6), 3.77 (dd, *J* = 10.2, 6.3 Hz, 1H, H-6′), 2.79–2.50 (m, 2H, H-2, C*H*HCH(CH_3)$_2$), 2.42 (dd, *J* = 12.9, 7.2 Hz, 1H, CH*H*CH(CH_3)$_2$), 1.84–1.74 (m, 2H, H-2′, CH_2C*H*(CH_3)$_2$), 1.08 (s, 9H, Si-^{t}Bu), 0.94 (dd, *J* = 6.6, 2.7 Hz, 6H, C*H*$_3$CHC*H*$_3$). ^{13}C NMR (100 MHz, $CDCl_3$) δ 154.4, 135.7, 135.6, 135.6, 133.2, 133.0, 130.1, 130.0, 128.0, 127.9, 77.7, 73.7, 72.5, 67.9, 61.7, 39.7, 29.4, 28.8, 26.9, 22.1, 22.0, 19.4. HRMS (ESI) m/z: calcd. for $C_{27}H_{36}O_5SSiNa^+$ $(M + Na)^+$ 523.1945, found 523.1942; $[\alpha]_D^{25}$ = +63.3 (c = 1.0, $CHCl_3$).

sec-Butyl-1-thio-6-O-(tert-butyldiphenylsilyl)-3,4-O-carbonate-2-deoxy-α-D-galactopyranoside (**5e**). The title compound was prepared according to the general procedure of 2-deoxythioglycosides synthesis and purified by flash column chromatography giving colorless oil 41.0 mg, yield 82%; ^1H NMR (400 MHz, $CDCl_3$) δ 7.67–7.71 (m, 4H, Ar-H), 7.53–7.37 (m, 6H, Ar-H), 5.43 (dd, *J* = 9.3, 6.5 Hz, 1H, H-1), 5.00–5.05 (m, 1H, H-3), 4.89 (ddd, *J* = 8.3, 6.1, 1.7 Hz, 1H, H-4), 4.13–4.17 (m, 1H, H-5), 3.92–3.81 (m, 1H, H-6), 3.75 (ddd, *J* = 10.1, 6.1, 4.0 Hz, 1H, H-6′), 2.86–2.95 (m, 1H, C*H*CH_2CH_3), 2.53 (dddd, *J* = 15.7, 6.2, 3.8, 2.1 Hz, 1H, H-2), 1.81 (dddd, *J* = 15.8, 9.3, 4.5, 3.2 Hz, 1H, H-2′), 1.57–1.41 (m, 2H, CHC*H*$_2$$CH_3$), 1.29 (d, *J* = 6.9 Hz, 2H, CHC*H*$_2$H), 1.20 (d, *J* = 7.0 Hz, 1H, $CHCH_2$*H*), 1.08 (s, 9H, Si-^{t}Bu), 0.91 (dt, *J* = 23.3, 7.4 Hz, 3H, $CHCH_2$C*H*$_3$). ^{13}C NMR (100 MHz, $CDCl_3$) δ 154.4, 135.7, 135.7, 135.6, 135.6, 133.2, 130.1, 130.0, 128.0, 127.9, 76.6, 73.6, 72.5, 67.8, 61.6, 41.8, 40.0, 30.2, 26.9, 21.3, 19.4, 11.4. ^{13}C NMR (100 MHz, $CDCl_3$) δ.154.4, 135.7, 135.7, 135.6, 135.6, 133.2, 130.1, 130.0, 128.0, 127.9, 75.8, 73.6, 72.5, 67.8, 61.6, 41.2, 39.3, 29.4, 26.9, 21.3, 19.4, 11.3. HRMS (ESI) m/z: calcd. for $C_{27}H_{36}O_5SSiNa^+$ $(M + Na)^+$ 523.1945, found 523.1946; $[\alpha]_D^{25}$ = +35.6 (c = 1.0, $CHCl_3$).

tert-Butyl-1-thio-6-O-(tert-butyldiphenylsilyl)-3,4-O-carbonate-2-deoxy-α-D-galactopyranoside (**5f**). The title compound was prepared according to the general procedure of 2-deoxythioglycosides synthesis and purified by flash column chromatography giving colorless oil 40.0 mg, yield 80%; ^1H NMR (400 MHz, $CDCl_3$) δ 7.69–7.72 (m, 4H, Ar-H), 7.56–7.37 (m, 6H, Ar-H), 5.56 (dd, *J* = 9.4, 6.5 Hz, 1H, H-1), 5.03 (ddd, *J* = 8.4, 4.8, 2.4 Hz, 1H, H-3), 4.91 (dd, *J* = 8.5, 1.6 Hz, 1H, H-4), 4.13 (ddd, *J* = 7.9, 5.7, 1.6 Hz, 1H, H-5), 3.85 (dd, *J* = 10.0, 8.3 Hz, 1H, H-6), 3.72 (dd, *J* = 10.0, 5.8 Hz, 1H, H-6′), 2.49 (ddd, *J* = 15.7, 6.6, 3.8 Hz, 1H, H-2), 1.81 (ddd, *J* = 15.8, 9.4, 3.1 Hz, 1H, H-2′), 1.31 (s, 9H, C-^{t}Bu), 1.07 (s, 9H, Si-^{t}Bu).^{13}C NMR (100 MHz, $CDCl_3$) δ 154.6, 135.7, 135.6, 133.2, 132.8, 130.1, 130.0, 128.0, 127.9, 75.4, 73.5, 72.6, 67.6, 61.2, 44.5, 31.8, 29.4, 26.9, 19.4. HRMS (ESI) m/z: calcd. for $C_{27}H_{36}O_5SSiNa^+$ $(M + Na)^+$ 523.1945, found 523.1949; $[\alpha]_D^{25}$ = +40.2 (c = 1.0, $CHCl_3$).

Benzyl-1-thio-6-O-(tert-butyldiphenylsilyl)-3,4-O-carbonate-2-deoxy-α-D-galactopyranoside (**5g**). The title compound was prepared according to the general procedure of 2-deoxythioglycosides synthesis and purified by flash column chromatography giving colorless oil 43.8 mg, yield 82%; ^1H NMR (400 MHz, $CDCl_3$) δ 7.70–7.73 (m, 4H, Ar-H), 7.55–7.33 (m, 6H, Ar-H), 7.30–7.10 (m, 5H, Ar-H), 5.21 (dd, *J* = 9.2, 6.6 Hz, 1H, H-1), 5.00 (ddd, *J* = 8.4, 4.6, 2,4 Hz, 1H, H-3), 4.86 (dd, *J* = 8.5, 1.7 Hz, 1H, H-4), 4.14–4.18(m, 1H, H-5), 3.92–3.83 (m, 2H, H-6, PhC*H*H), 3.76 (dd, *J* = 10.3, 6.4 Hz, 1H, H-6′), 3.66 (d, *J* = 13.6 Hz, 1H, PhCH*H*), 2.46 (ddd, *J* = 15.7, 6.6, 3.8 Hz, 1H, H-2), 1.77 (ddd, *J* = 15.8, 9.2, 3.2 Hz, 1H, H-2′), 1.11 (s, 9H, Si-^{t}Bu).^{13}C NMR (100 MHz, $CDCl_3$) δ 154.3, 137.9, 135.7, 135.6, 133.2, 132.9, 130.1, 130.0, 129.0, 128.7, 128.0, 127.3, 76.0, 73.6, 72.5, 68.2, 61.7, 34.5, 28.8, 27.0, 19.4. HRMS (ESI) m/z: calcd. for $C_{30}H_{34}O_5SSiNa^+$ $(M + Na)^+$ 557.1788, found 557.1777; $[\alpha]_D^{25}$ = +140.3 (c = 1.0, $CHCl_3$).

Besides the substrate study of the thiophenol acceptors, the alkyl thiols were also employed to synthesize 2-deoxythioglycosides, including *S*-linked disaccharides. As shown in Scheme 3, two primary thiols ethyl and *n*-octyl mercaptan were utilized to react with glycal **1a** to form 2-deoxythiosugars **5a** and **5b** in high yields (78–86%) with excellent stereoselectivity (α/β > 20:1). When *n*-butyl, *iso*-butyl, and *sec*-butyl mercaptan were employed, the corresponding α-2-deoxythioglycosides **5c**-**5e** were all able to be generated in high yields (>80%). It is worth noting that the bulky *tert*-butyl mercaptan could also be well compatible with this method and gave **5f** in an 80% yield with exclusive α-selectivity, indicating that the strategy was less affected by steric hindrance. Benzyl mercaptan was applied to form the 2-deoxythiogalactoside **5g** in a yield of 82% with a ratio of α/β > 20:1. Methyl thioglycolate was also applied as a glycosyl acceptor and the desired product **5h** was obtained in an 80% yield. Encouraged by these observations, we continued to study the synthesis of *S*-linked α-disaccharides. For example, **5i** was achieved successfully from 3,4-*O*-carbonate-galactal **1** and 1,2:5,6-di-*O*-isopropylidene-α-D-allofuranose in a yield of 67% with exclusive α-selectivity. The α,β-1,1′-2-deoxythioglycosides **5j** was obtained from glycal **1** and 1-thio-β-D-glucose tetraacetate in a yield of 69% with excellent stereoselectivity (α/β > 20:1).

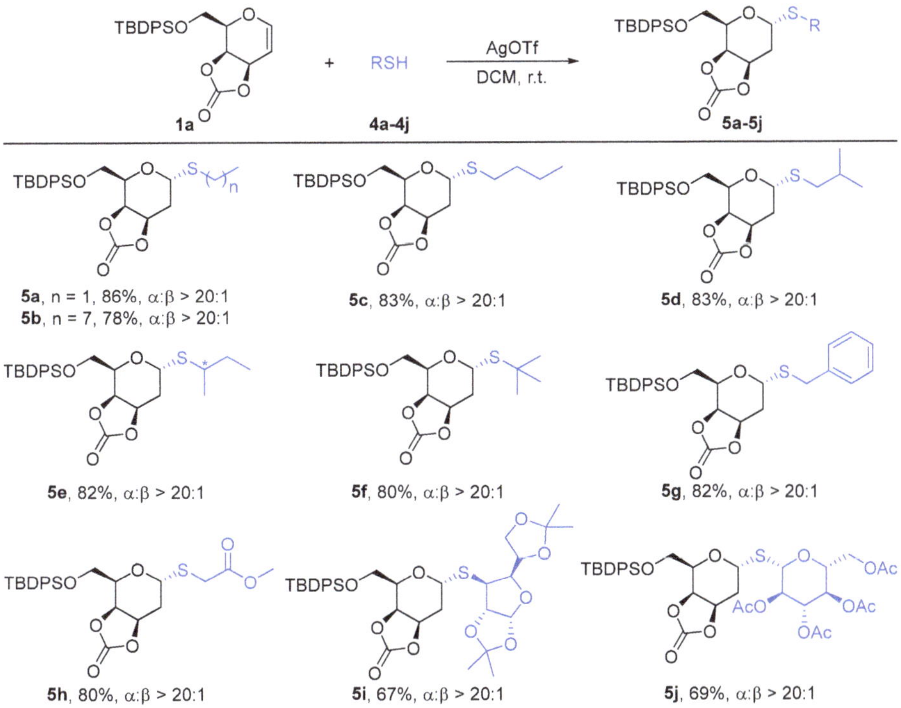

Scheme 3. Substrate scope of alkyl 2-deoxythioglycosides.

These successes stimulated us to further apply this 2-deoxythioglycosylation methodology to the modification of bioactive natural products (In the Supplementary Materials). Estrogen is a female sex hormone, one of the three main endogenous estrogens, which plays a vital role in human life and is also used as a medicine in clinical treatment [28–30]. S-Linked estrone 2-deoxygalactoside **7a** was successfully achieved by AgOTf at room temperature with a 78% yield and excellent α-selectivity (Scheme 4). L-Menthol is the main component of peppermint and also demonstrates analgesic, antibacterial, and anti-inflammatory effects [31]. L-Menthol-2-deoxythioglycoside **7b** was generated in a 60%

yield with high α-selectivity. Zingerone extracted from ginger indicates antioxidant, anti-inflammatory and anti-cancer, and antibacterial bioactivity [32–34], and it was also converted to be a thiol to react with glycal **1a** giving 2-deoxythiosugar **7c** in a 72% yield with a ratio of α/β > 20:1.

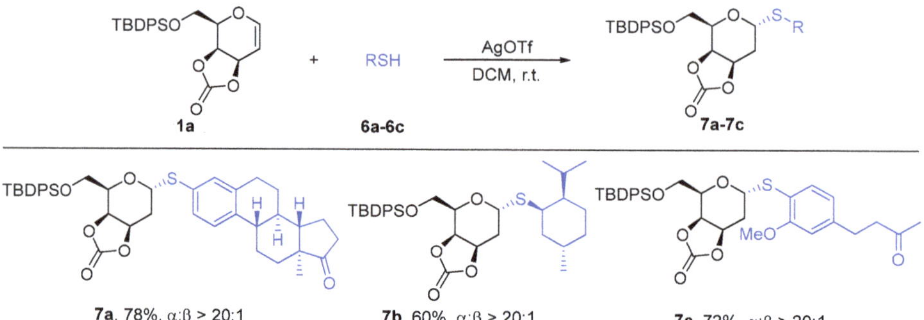

Scheme 4. Late-stage 2-deoxythioglycosylation of natural product thiols.

Based on the results and literature research, a possible mechanism of AgOTf-catalyzed 2-deoxythioglycosylation was proposed as shown in Scheme 5. Silver triflate coordinated with the double bond of glycal from the bottom face to form intermediate **A** because of the steric effect [35,36]. After the pronation of the reactive olefin, the oxocarbenium ion **B** would be generated [1,4,37]. Alkylthio anion (RS⁻) would attack the anomeric position from the bottom face to yield α-2-deoxythioglycosides because the 3,4-*O*-carbonate ring would block the upper face.

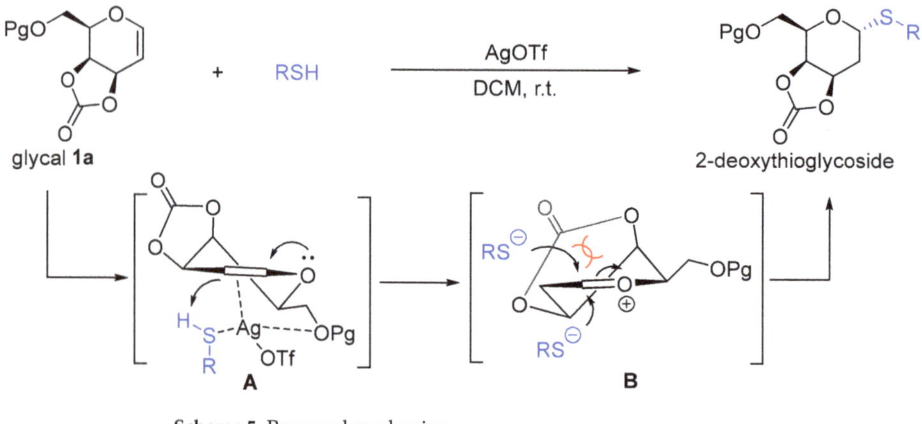

Scheme 5. Proposed mechanism.

3. Conclusions

In conclusion, we have developed an effective strategy to synthesize 2-deoxythioglycosides using 3,4-*O*-carbonate glycal donors catalyzed by silver triflate in mild conditions. This reaction could tolerate various alkyl thiols and thiophenols, all the target products were obtained in moderate to good yields with excellent α-selectivity. *S*-Linked disaccharides and late-stage functionalization of natural product thiols were achieved successfully as well. The results of this study suggest that this method may be a promising alternative way to access the 2-deoxythioglycosides applied in natural product synthesis and drug development.

4. Materials and Methods

General Procedure. The 3,4-*O*-carbonate glycal donor (0.100 mmol) and thiol reagent (0.110 mmol) were added to anhydrous dichloromethane (2.00 mL) in a Schlenk tube, followed by adding silver triflate (0.01 mmol) under N_2 atmosphere. The reaction mixture was stirred at room temperature and monitored by TLC. Then, aqueous sodium bicarbonate was added to quench the reaction, extracted with dichloromethane, washed by aqueous sodium bicarbonate and dried by sodium sulfate. The organic layer was collected and removed under reduced pressure to afford a crude product which was purified by silica gel flash chromatography with a gradient solvent system (petroleum ether/ethyl acetate as eluent) to yield 2-deoxythioglycosides.

4-Methylphenyl-1-thio-6-O-(tert-butyldiphenylsilyl)-3,4-O-carbonate-2-deoxy-α-D-galactopyranoside (**3a**). The title compound was prepared according to the general procedure of 2-deoxythioglycosides synthesis and purified by flash column chromatography giving colorless oil 42.7 mg, yield 80%; ^{1}H NMR (400 MHz, $CDCl_3$) δ 7.73–7.63 (m, 4H, Ar-H), 7.47–7.39 (m, 6H, Ar-H), 7.33–7.27 (m, 2H, Ar-H), 7.01–7.03 (m, 2H, Ar-H), 5.45 (dd, *J* = 9.8, 6.5 Hz, 1H, H-1), 5.05 (ddd, *J* = 8.8, 4.4, 2.0 Hz, 1H, H-3), 4.90 (dd, *J* = 8.6, 1.6 Hz, 1H, H-4), 4.16 (ddd, *J* = 7.7, 6.0, 1.6 Hz, 1H, H-5), 3.84 (dd, *J* = 10.2, 7.6 Hz, 1H, H-6), 3.77 (dd, *J* = 10.2, 6.2 Hz, 1H, H-6′), 2.64 (ddd, *J* = 15.8, 6.5, 3.5 Hz, 1H, H-2), 2.29 (s, 3H, $PhCH_3$), 1.88 (ddd, *J* = 15.9, 9.8, 3.0 Hz, 1H, H-2′), 1.06 (s, 9H, Si-^{t}Bu). ^{13}C NMR (100 MHz, $CDCl_3$) δ 154.2, 138.4, 135.7, 135.6, 133.2, 133.1, 132.8, 130.1, 130.0, 129.9, 129.4, 128.0, 80.8, 73.5, 72.4, 68.5, 61.6, 28.9, 26.9, 21.3, 19.4. HRMS (ESI) m/z: calcd. for $C_{30}H_{34}O_5SSiNa^+$ $(M + Na)^+$ 557.1788, found 557.1786; $[\alpha]_D^{25}$ = +45.7 (c = 1.0, $CHCl_3$).

3-Methylphenyl-1-thio-6-O-(tert-butyldiphenylsilyl)-3,4-O-carbonate-2-deoxy-α-D-galactopyranoside (**3b**). The title compound was prepared according to the general procedure of 2-deoxythioglycosides synthesis and purified by flash column chromatography giving colorless oil 43.8 mg, yield 82%; ^{1}H NMR (400 MHz, $CDCl_3$) δ 7.69 (ddd, *J* = 8.0, 5.2, 1.6 Hz, 4H, Ar-H), 7.58–7.34 (m, 6H, Ar-H), 7.27–7.20 (m, 2H, Ar-H), 7.19–7.03 (m, 2H, Ar-H), 5.56 (dd, *J* = 9.8, 6.5 Hz, 1H, H-1), 5.08 (ddd, *J* = 8.4, 4.0, 2.0 Hz, 1H, H-3), 4.93 (dd, *J* = 8.6, 1.6 Hz, 1H, H-4), 4.19 (ddd, *J* = 7.7, 6.0, 1.6 Hz, 1H, H-5), 3.87 (dd, *J* = 10.2, 7.7 Hz, 1H, H-6), 3.80 (dd, *J* = 10.2, 6.1 Hz, 1H, H-6′), 2.68 (ddd, *J* = 15.8, 6.5, 3.5 Hz, 1H, H-2), 2.27 (s, 3H, $PhCH_3$), 1.92 (ddd, *J* = 15.9, 9.8, 3.0 Hz, 1H, H-2′), 1.08 (s, 9H, Si-^{t}Bu). ^{13}C NMR (100 MHz, $CDCl_3$) δ 154.2, 138.9, 135.7, 135.6, 133.2, 133.1, 132.9, 132.8, 130.1, 130.0, 129.3, 128.9, 128.9, 128.0, 80.5, 73.4, 72.4, 68.6, 61.5, 29.0, 26.9, 21.4, 19.4. HRMS (ESI) m/z: calcd. for $C_{30}H_{34}O_5SSiNa^+$ $(M + Na)^+$ 557.1788, found 557.1790; $[\alpha]_D^{25}$ = +62.2 (c = 1.0, $CHCl_3$).

2-Methylphenyl-1-thio-6-O-(tert-butyldiphenylsilyl)-3,4-O-carbonate-2-deoxy-α-D-galactopyranoside (**3c**). The title compound was prepared according to the general procedure of 2-deoxythioglycosides synthesis and purified by flash column chromatography giving white solid 43.2 mg, yield 81%; m. p.: 155.7–157.1 °C. ^{1}H NMR (400 MHz, $CDCl_3$) δ 7.64–7.67 (m, 4H, Ar-H), 7.73–7.45 (m, 7H, Ar-H), 7.20–7.09 (m, 2H, Ar-H), 7.03–7.07 (m, 1H, Ar-H), 5.53 (dd, *J* = 9.8, 6.4 Hz, 1H, H-1), 5.07 (ddd, *J* = 8.6, 4.2, 2.0 Hz, 1H, H-3), 4.93 (dd, *J* = 8.7, 1.6 Hz, 1H, H-4), 4.17 (ddd, *J* = 8.0, 6.0, 1.7 Hz, 1H, H-5), 3.83 (dd, *J* = 10.1, 8.0 Hz, 1H, H-6), 3.75 (dd, *J* = 10.1, 6.0 Hz, 1H, H-6′), 2.67 (ddd, *J* = 15.8, 6.5, 3.5 Hz, 1H, H-2), 2.33(s, 3H, $PhCH_3$), 1.96 (ddd, *J* = 15.9, 9.8, 2.9 Hz, 1H, H-2′), 1.04 (s, 9H, Si-^{t}Bu). ^{13}C NMR (100 MHz, $CDCl_3$) δ 154.2, 139.4, 135.7, 135.6, 133.2, 132.8, 132.8, 130.3, 130.1, 130.0, 128.0, 128.0, 126.8, 79.8, 73.4, 72.4, 68.6, 61.4, 29.2, 26.9, 20.9, 19.4. HRMS (ESI) m/z: calcd. for $C_{30}H_{34}O_5SSiNa^+$ $(M + Na)^+$ 557.1788, found 557.1793; $[\alpha]_D^{25}$ = +43.2 (c = 1.0, $CHCl_3$)

2,4-Methoxylphenyl-1-thio-6-O-(tert-butyldiphenylsilyl)-3,4-O-carbonate-2-deoxy-D-galactopyranoside (**3d**). The title compound was prepared according to the general procedure of 2-deoxythioglycosides synthesis and purified by flash column chromatography giving colorless oil 44.9 mg, yield 82%; ^{1}H NMR (400 MHz, $CDCl_3$) δ 7.77–7.64 (m, 4H, Ar-H), 7.52–7.40 (m, 6H, Ar-H), 7.31–7.33 (m, 1H, Ar-H), 7.03–6.99 (m, 1H, Ar-H), 6.85–6.88 (m, 1H, Ar-H), 5.45 (dd, *J* = 9.8, 6.4 Hz, 1H, H-1), 5.13–5.03 (m, 1H, H-3), 4.96 (dd, *J* = 8.6, 1.6 Hz, 1H, H-4), 4.18 (ddd, *J* = 7.8, 5.8, 1.6 Hz, 1H, H-5), 3.86 (dd, *J* = 10.1, 8.2 Hz, 1H, H-6), 3.76 (dd, *J* = 10.0, 5.9 Hz, 1H, H-6′), 2.68 (ddd, *J* = 15.8, 6.4, 3.5 Hz, 1H, H-2), 2.33 (s, 3H,

PhCH_3), 2.28 (s, 3H, PhCH_3), 1.96 (ddd, J = 15.9, 9.9, 2.9 Hz, 1H, H-2'), 1.08 (s, 9H, Si-^{t}Bu). ^{13}C NMR (100 MHz, $CDCl_3$) δ 154.3, 140.0, 138.5, 135.7, 135.6, 134.0, 133.2, 132.8, 131.3, 130.1, 130.0, 128.7, 128.0, 127.5, 80.2, 73.4, 72.4, 68.4, 61.4, 29.1, 26.9, 21.2, 21.0, 19.4. HRMS (ESI) m/z: calcd. for $C_{31}H_{36}O_5SSiNa^+$ (M + Na)$^+$ 571.1945, found 571.1941; $[\alpha]_D^{25}$ = +68.4 (c = 1.0, $CHCl_3$).

4-Methoxylphenyl-1-thio-6-O-(tert-butyldiphenylsilyl)-3,4-O-carbonate-2-deoxy-α-D-galactopyranside (**3e**). The title compound was prepared according to the general procedure of 2-deoxythioglycosides synthesis and purified by flash column chromatography giving colorless oil 40.1 mg, yield 73%; ^{1}H NMR (400 MHz, $CDCl_3$) δ 7.69–7.72 (m, 4H, Ar-H), 7.41–7.48 (m, 6H, Ar-H), 7.35–7.38 (m, 2H, Ar-H), 6.74–6.76 (m, 2H, Ar-H), 5.38 (dd, J = 9.8, 6.4 Hz, 1H, H-1), 5.06 (ddd J = 8.6, 4.4, 2.2Hz, 1H, H-3), 4.91 (dd, J = 8.6, 1.6 Hz, 1H, H-4), 4.18 (ddd, J = 7.7, 6.1, 1.6 Hz, 1H, H-5), 3.87 (dd, J = 10.3, 7.5 Hz, 1H, H-6), 3.80 (dd, J = 9.4, 5.3 Hz, 1H, H-6'), 3.77 (s, 3H, PhOCH_3), 2.65 (ddd, J = 15.9, 6.5, 3.5 Hz, 1H, H-2), 1.89 (ddd, J = 15.8, 9.8, 3.0 Hz, 1H, H-2'), 1.10 (s, 9H, Si-^{t}Bu).^{13}C NMR (100 MHz, $CDCl_3$) δ 160.2, 154.2, 135.7, 135.7, 135.6, 133.2, 132.8, 130.1, 130.0, 128.0, 123.3, 114.7, 81.2, 73.5, 72.5, 68.5, 61.7, 55.4, 28.9, 27.0, 19.4. HRMS (ESI) m/z: calcd. for $C_{30}H_{34}O_6SSiNa^+$ (M + Na)$^+$ 573.1738, found 573.1739; $[\alpha]_D^{25}$ = +90.6 (c = 1.0, $CHCl_3$).

4-tert-Butylphenyl-1-thio-6-O-(tert-butyldiphenylsilyl)-3,4-O-carbonate-2-deoxy-α-D-galactopyranoside (**3f**). The title compound was prepared according to the general procedure of 2-deoxythioglycosides synthesis and purified by flash column chromatography giving colorless oil 46.7 mg, yield 81%; ^{1}H NMR (400 MHz, $CDCl_3$) δ 7.69–7.72 (m, 4H, Ar-H), 7.52–7.38 (m, 6H, Ar-H), 7.41–7.34 (m, 2H, Ar-H), 7.30–7.22 (m, 2H, Ar-H), 5.51 (dd, J = 9.7, 6.5 Hz, 1H, H-1), 5.07 (ddd, J = 8.6, 4.2, 2.2 Hz, 1H, H-3), 4.92 (dd, J = 8.7, 1.7 Hz, 1H, H-4), 4.22 (ddd, J = 7.6, 6.2, 1.6 Hz, 1H, H-5), 3.88 (dd, J = 10.2, 7.5 Hz, 1H, H-6), 3.82 (dd, J = 10.2, 6.2 Hz, 1H, H-6'), 2.67 (ddd, J = 15.9, 6.5, 3.5 Hz, 1H, H-2), 1.91 (ddd, J = 15.9, 9.8, 3.0 Hz, 1H, H-2'), 1.29 (s, 9H, C-^{t}Bu), 1.09 (s, 9H, Si-^{t}Bu).^{13}C NMR (100 MHz, $CDCl_3$) δ 154.2, 151.4, 135.7, 135.6, 133.2, 132.8, 132.6, 130.1, 130.0, 129.6, 128.0, 126.2, 80.7, 73.5, 72.4, 68.5, 61.6, 34.7, 31.3, 29.0, 26.9, 19.4. HRMS (ESI) m/z: calcd. for $C_{33}H_{40}O_5SSiNa^+$ (M + Na)$^+$ 599.2258, found 599.2259; $[\alpha]_D^{25}$ = +104.8 (c = 1.0, $CHCl_3$).

Phenyl-1-thio-6-O-(tert-butyldiphenylsilyl)-3,4-O-carbonate-2-deoxy-α-D-galactopyranoside (**3g**). The title compound was prepared according to the general procedure of 2-deoxythioglycosides synthesis and purified by flash column chromatography giving colorless oil 41.6 mg, yield 80%; ^{1}H NMR (400 MHz, $CDCl_3$) δ 7.70 (ddd, J = 8.0, 4.0, 1.7 Hz, 4H, Ar-H), 7.53–7.34 (m, 8H, Ar-H), 7.27–7.21 (m, 3H, Ar-H), 5.57 (dd, J = 9.7, 6.5 Hz, 1H, H-1), 5.08 (ddd, J = 8.4, 4.0, 2.4 Hz, 1H, H-3), 4.92 (dd, J = 8.6, 1.6 Hz, 1H, H-4), 4.20 (ddd, J = 7.6, 6.2, 1.7 Hz, 1H, H-5), 3.87 (dd, J = 10.2, 7.5 Hz, 1H, H-6), 3.81 (dd, J = 10.3, 6.2 Hz, 1H, H-6'), 2.69 (ddd, J = 15.9, 6.5, 3.5 Hz, 1H, H-2), 1.92 (ddd, J = 15.8, 9.8, 3.0 Hz, 1H, H-2'), 1.08 (s, 9H, Si-^{t}Bu).^{13}C NMR (100 MHz, $CDCl_3$) δ 154.2, 135.7, 135.6, 133.4, 133.2, 132.8, 132.3, 130.1, 130.0, 129.1, 128.0, 80.5, 73.4, 72.4, 68.6, 61.6, 29.0, 26.9, 19.4. HRMS (ESI) m/z: calcd. for $C_{29}H_{32}O_5SSiNa^+$ (M + Na)$^+$ 543.1632, found 543.1642; $[\alpha]_D^{25}$ = +67.9 (c = 1.0, $CHCl_3$).

4-Bromophenyl-1-thio-6-O-(tert-butyldiphenylsilyl)-3,4-O-carbonate-2-deoxy-α-D-galactopyranoside (**3h**). The title compound was prepared according to the general procedure of 2-deoxythioglycosides synthesis and purified by flash column chromatography giving colorless oil 46.0 mg, yield 75%; ^{1}H NMR (400 MHz, $CDCl_3$) δ 7.66–7.70 (m, 4H, Ar-H), 7.51–7.39 (m, 6H, Ar-H), 7.36–7.20 (m, 3H, Ar-H), 7.29 (d, J = 2.4 Hz, 1H, Ar-H), 5.54 (dd, J = 9.8, 6.5 Hz, 1H, H-1), 5.07 (ddd, J = 8.6, 4.4, 2.2 Hz, 1H, H-3), 4.90 (dd, J = 8.6, 1.7 Hz, 1H, H-4), 4.15–4.19 (m, 1H, H-5), 3.87 (dd, J = 10.4, 7.1 Hz, 1H, H-6), 3.79 (dd, J = 10.3, 6.3 Hz, 1H, H-6'), 2.69 (ddd, J = 15.9, 6.5, 3.5 Hz, 1H, H-2), 1.89 (ddd, J = 15.9, 9.9, 3.0 Hz, 1H, H-2'), 1.07 (s, 9H, Si-^{t}Bu).^{13}C NMR (100 MHz, $CDCl_3$) δ 154.1, 135.7, 135.6, 133.7, 133.6, 133.1, 132.8, 132.6, 132.3, 132.2, 130.1, 130.0, 128.0, 122.3, 80.5, 73.4, 72.3, 68.8, 61.6, 28.8, 26.9, 19.4. HRMS (ESI) m/z: calcd. for $C_{29}H_{31}BrO_5SSiNa^+$ (M + Na)$^+$ 621.0737, found 621.0734; $[\alpha]_D^{25}$ = +45.6 (c = 1.0, $CHCl_3$).

2-Bromophenyl-1-thio-6-O-(tert-butyldiphenylsilyl)-3,4-O-carbonate-2-deoxy-α-D-galactopyranoside (**3i**). The title compound was prepared according to the general procedure of

2-deoxythioglycosides synthesis and purified by flash column chromatography giving colorless oil 43.5 mg, yield 71%; ^{1}H NMR (400 MHz, $CDCl_3$) δ 7.64–7.69 (m, 4H, Ar-H), 7.55 (ddd, *J* = 7.9, 2.5, 1.5 Hz, 2H, Ar-H), 7.50–7.35 (m, 6H, Ar-H), 7.17–7.20 (m,1H, Ar-H), 7.06–7.11 (m, 1H, Ar-H), 5.71 (dd, *J* = 9.8, 6.5 Hz, 1H, H-1), 5.12 (ddd, *J* = 8.6, 4.0, 2.2 Hz, 1H, H-3), 4.95 (dd, *J* = 8.6, 1.6 Hz, 1H, H-4), 4.21 (ddd, *J* = 7.8, 6.3, 1.7 Hz, 1H, H-5), 3.85 (dd, *J* = 10.2, 7.5 Hz, 1H, H-6), 3.79 (dd, *J* = 10.2, 6.3 Hz, 1H, H-6′), 2.74 (ddd, *J* = 15.9, 6.6, 3.5 Hz, 1H, H-2), 1.98 (ddd, *J* = 15.8, 9.8, 2.9 Hz, 1H, H-2′), 1.04 (s, 9H, Si-^{t}Bu).^{13}C NMR (100 MHz, $CDCl_3$) δ 154.2, 135.7, 135.6, 135.2, 133.1, 132.8, 132.0, 130.1, 130.0, 128.6, 128.20, 128.0, 125.2, 79.3, 76.8, 72.3, 68.9, 61.3, 28.6, 26.9, 19.3. HRMS (ESI) m/z: calcd. for $C_{29}H_{31}BrO_5SSiNa^+$ $(M + Na)^+$ 621.0737, found 621.0735; $[\alpha]_D^{25}$ = +82.4 (c = 1.0, $CHCl_3$).

4-Fluorophenyl-1-thio-6-O-(tert-butyldiphenylsilyl)-3,4-O-carbonate-2-deoxy-α-D-galactopyranoside (**3j**). The title compound was prepared according to the general procedure of 2-deoxythioglycosides synthesis and purified by flash column chromatography giving colorless oil 39.8 mg, yield 74%; ^{1}H NMR (400 MHz, $CDCl_3$) δ 7.79–7.67 (m, 4H, Ar-H), 7.56–7.40 (m, 8H, Ar-H), 6.90–6.93 (m, 2H, Ar-H), 5.46 (dd, *J* = 9.9, 6.5 Hz, 1H, H-1), 5.07 (ddd, *J* = 8.6, 4.2, 2.0 Hz, 1H, H-3), 4.90 (dd, *J* = 8.7, 1.6 Hz, 1H, H-4), 4.16–4.20 (m, 1H, H-5), 3.87 (dd, *J* = 10.4, 7.2 Hz, 1H, H-6), 3.80 (dd, *J* = 10.3, 6.3 Hz, 1H, H-6′), 2.68 (ddd, *J* = 15.9, 6.5, 3.4 Hz, 1H, H-2), 1.89 (ddd, *J* = 15.8, 9.8, 2.9 Hz, 1H, H-2′), 1.09 (s, 9H, Si-^{t}Bu).^{13}C NMR (100 MHz, $CDCl_3$) δ 163.0 (d, *J* = 246.7 Hz), 154.1, 135.7, 135.6, 135.1 (d, *J* = 8.2 Hz), 133.1, 132.8, 130.1, 130.0, 128.0, 116.2 (d, *J* = 21.6 Hz), 81.0, 73.4, 72.4, 68.7, 61.7, 28.8, 26.9, 19.4. HRMS (ESI) m/z: calcd. for $C_{29}H_{31}FO_5SSiNa^+$ $(M + Na)^+$ 561.1538, found 561.1541; $[\alpha]_D^{25}$ = +86.3 (c = 1.0, $CHCl_3$).

Ethyl-1-thio-6-O-(tert-butyldiphenylsilyl)-3,4-O-carbonate-2-deoxy-α-D-galactopyranoside (**5a**). The title compound was prepared according to the general procedure of 2-deoxythioglycosides synthesis and purified by flash column chromatography giving colorless oil 40.6 mg, yield 86%; ^{1}H NMR (400 MHz, $CDCl_3$) δ 7.67–7.71 (m, 4H, Ar-H), 7.60–7.36 (m, 6H, Ar-H), 5.39 (dd, *J* = 9.3, 6.5 Hz, 1H, H-1), 5.02 (ddd, *J* = 8.4, 4.8, 2.4 Hz, 1H, H-3), 4.86 (dd, *J* = 8.5, 1.7 Hz, 1H, H-4), 4.23–4.01 (m, 1H, H-5), 3.85 (dd, *J* = 10.3, 7.3 Hz, 1H, H-6), 3.78 (dd, *J* = 10.3, 6.4 Hz, 1H, H-6′), 2.67–2.75 (m, 1H, H-2), 2.63–2.49 (m, 2H, CH_2CH_3), 1.79 (ddd, *J* = 15.8, 9.3, 3.2 Hz, 1H, H-2′), 1.24 (dd, *J* = 13.4, 6.7 Hz, 3H, CH_2CH_3), 1.08 (s, 9H, Si-^{t}Bu). ^{13}C NMR (100 MHz, $CDCl_3$) δ 154.4, 135.7, 135.6, 133.2, 132.9, 130.1, 130.0, 128.0, 127.9, 76.7, 73.7, 72.5, 68.0, 61.7, 29.2, 26.9, 24.7, 19.3, 15.0. HRMS (ESI) m/z: calcd. For $C_{25}H_{32}O_5SsiNa^+$ $(M + Na)^+$ 495.1632, found 495.1634; $[\alpha]_D^{25}$ = +54.2 (c = 1.0, $CHCl_3$).

Octyl-1-thio-6-O-(tert-butyldiphenylsilyl)-3,4-O-carbonate-2-deoxy-α-D-galactopyranoside (**5b**). The title compound was prepared according to the general procedure of 2-deoxythioglycosides synthesis and purified by flash column chromatography giving colorless oil 43.3 mg, yield 78%; ^{1}H NMR (400 MHz, $CDCl_3$) δ 7.68–7.72 (m, 4H, Ar-H), 7.40–7.50 (m, 6H, Ar-H), 5.35 (dd, *J* = 9.3, 6.5 Hz, 1H, H-1), 5.02 (ddd, *J* = 8.4, 4.5, 2.3 Hz, 1H, H-3), 4.87 (dd, *J* = 8.5, 1.7 Hz, 1H, H-4), 4.13 (ddd, *J* = 7.4, 6.9, 1.6 Hz, 1H, H-5), 3.85 (dd, *J* = 10.2, 7.3 Hz, 1H, H-6), 3.78 (dd, *J* = 10.2, 6.4 Hz, 1H, H-6′), 2.68 (ddd, *J* = 12.9, 8.1, 6.6 Hz, 1H, H-2), 2.57–2.41 (m, 2H), 1.79 (ddd, *J* = 15.8, 9.3, 3.1 Hz, 1H, H-2′), 1.59–1.51 (m, 2H, CH_2), 1.41–1.12 (m, 10H, CH_2), 1.08 (s, 9H, Si-^{t}Bu), 0.89 (dd, *J* = 7.6, 5.7 Hz, 3H, CH_3). ^{13}C NMR (100 MHz, $CDCl_3$) δ 154.4, 135.7, 135.6, 133.2, 132.9, 130.1, 130.0, 128.0, 127.9, 77.1, 73.7, 72.5, 67.9, 61.7, 31.9, 30.7, 29.9, 29.3, 29.2, 29.2, 29.0, 26.9, 22.8, 19.3, 14.2. HRMS (ESI) m/z: calcd. for $C_{31}H_{44}O_5SSiNa^+$ $(M + Na)^+$ 579.2571, found 579.2556; $[\alpha]_D^{25}$ = +51.4 (c = 1.0, $CHCl_3$).

n-Buty-1-thio-6-O-(tert-butyldiphenylsilyl)-3,4-O-carbonate-2-deoxy-α-D-galactopyranoside (**5c**). The title compound was prepared according to the general procedure of 2-deoxythioglycosides synthesis and purified by flash column chromatography giving colorless oil 41.5 mg, yield 83%; ^{1}H NMR (400 MHz, $CDCl_3$) δ 7.48–7.71 (m, 4H, Ar-H), 7.55–7.36 (m, 6H, Ar-H), 5.35 (dd, *J* = 9.3, 6.5 Hz, 1H, H-1), 5.02 (ddd, *J* = 8.5, 4.4, 2.4 Hz, 1H, H-3), 4.87 (dd, *J* = 8.5, 1.7 Hz, 1H, H-4), 4.13 (ddd, *J* = 8.3, 6.4, 1.6 Hz, 1H, H-5), 3.85 (dd, *J* = 10.2, 7.4 Hz, 1H, H-6), 3.77 (dd, *J* = 10.2, 6.3 Hz, 1H, H-6′), 2.69 (ddd, *J* = 14.7, 8.1, 6.7 Hz, 1H, H-2), 2.60–2.46 (m, 2H, CH_2), 1.78 (ddd, *J* = 15.7, 9.3, 3.1 Hz, 1H, H-2′), 1.60–1.49 (m, 2H, CH_2), 1.41–1.32 (m, 2H, CH_2), 1.08 (s, 9H, Si-^{t}Bu), 0.87 (dd, *J* = 14.6, 6.9 Hz, 3H, CH_3). ^{13}C NMR

(100 MHz, $CDCl_3$) δ 154.4, 135.7, 135.6, 133.2, 132.9, 130.1, 130.0, 128.0, 127.9, 77.1, 73.7, 72.5, 67.9, 61.6, 31.9, 30.4, 29.2, 26.9, 22.0, 19.4, 13.7. HRMS (ESI) m/z: calcd. for $C_{27}H_{36}O_5SSiNa^+$ $(M + Na)^+$ 523.1945, found 523.1945; $[\alpha]_D^{25}$ = +36.4 (c = 1.0, $CHCl_3$).

Isobutyl-1-thio-6-O-(tert-butyldiphenylsilyl)-3,4-O-carbonate-2-deoxy-α-D-galactopyranoside (**5d**). The title compound was prepared according to the general procedure of 2-deoxythioglycosides synthesis and purified by flash column chromatography giving colorless oil 41.5 mg, yield 83%; ^{1}H NMR (400 MHz, $CDCl_3$) δ 7.67–7.71(m, 4H, Ar-H), 7.54–7.34 (m, 6H, Ar-H), 5.31 (dd, *J* = 9.3, 6.5 Hz, 1H, H-1), 5.02 (ddd, 8.6, 4.6, 2.4, 1H, H-3), 4.88 (dd, *J* = 8.5, 1.6 Hz, 1H, H-4), 4.19–4.04 (m, 1H, H-5), 3.85 (dd, *J* = 10.2, 7.5 Hz, 1H, H-6), 3.77 (dd, *J* = 10.2, 6.3 Hz, 1H, H-6′), 2.79–2.50 (m, 2H, H-2, C*H*HCH(CH_3)$_2$), 2.42 (dd, *J* = 12.9, 7.2 Hz, 1H, CH*H*CH(CH_3)$_2$), 1.84–1.74 (m, 2H, H-2′, CH_2*CH*(CH_3)$_2$), 1.08 (s, 9H, Si-^{t}Bu), 0.94 (dd, *J* = 6.6, 2.7 Hz, 6H, *CH*$_3$CH*CH*$_3$). ^{13}C NMR (100 MHz, $CDCl_3$) δ 154.4, 135.7, 135.6, 135.6, 133.2, 133.0, 130.1, 130.0, 128.0, 127.9, 77.7, 73.7, 72.5, 67.9, 61.7, 39.7, 29.4, 28.8, 26.9, 22.1, 22.0, 19.4. HRMS (ESI) m/z: calcd. for $C_{27}H_{36}O_5SSiNa^+$ $(M + Na)^+$ 523.1945, found 523.1942; $[\alpha]_D^{25}$ = +63.3 (c = 1.0, $CHCl_3$).

sec-Butyl-1-thio-6-O-(tert-butyldiphenylsilyl)-3,4-O-carbonate-2-deoxy-α-D-galactopyranoside (**5e**). The title compound was prepared according to the general procedure of 2-deoxythioglycosides synthesis and purified by flash column chromatography giving colorless oil 41.0 mg, yield 82%; ^{1}H NMR (400 MHz, $CDCl_3$) δ 7.67–7.71 (m, 4H, Ar-H), 7.53–7.37 (m, 6H, Ar-H), 5.43 (dd, *J* = 9.3, 6.5 Hz, 1H, H-1), 5.00–5.05 (m, 1H, H-3), 4.89 (ddd, *J* = 8.3, 6.1, 1.7 Hz, 1H, H-4), 4.13–4.17 (m, 1H, H-5), 3.92–3.81 (m, 1H, H-6), 3.75 (ddd, *J* = 10.1, 6.1, 4.0 Hz, 1H, H-6′), 2.86–2.95 (m, 1H, C*H*CH_2CH_3), 2.53 (dddd, *J* = 15.7, 6.2, 3.8, 2.1 Hz, 1H, H-2), 1.81 (dddd, *J* = 15.8, 9.3, 4.5, 3.2 Hz, 1H, H-2′), 1.57–1.41 (m, 2H, CH*CH*$_2$$CH_3$), 1.29 (d, *J* = 6.9 Hz, 2H, CH*CH*$_2$H), 1.20 (d, *J* = 7.0 Hz, 1H, $CHCH_2$*H*), 1.08 (s, 9H, Si-^{t}Bu), 0.91 (dt, *J* = 23.3, 7.4 Hz, 3H, $CHCH_2$*CH*$_3$). ^{13}C NMR (100 MHz, $CDCl_3$) δ 154.4, 135.7, 135.7, 135.6, 135.6, 133.2, 130.1, 130.0, 128.0, 127.9, 76.6, 73.6, 72.5, 67.8, 61.6, 41.8, 40.0, 30.2, 26.9, 21.3, 19.4, 11.4. ^{13}C NMR (100 MHz, $CDCl_3$) δ.154.4, 135.7, 135.7, 135.6, 135.6, 133.2, 130.1, 130.0, 128.0, 127.9, 75.8, 73.6, 72.5, 67.8, 61.6, 41.2, 39.3, 29.4, 26.9, 21.3, 19.4, 11.3. HRMS (ESI) m/z: calcd. for $C_{27}H_{36}O_5SSiNa^+$ $(M + Na)^+$ 523.1945, found 523.1946; $[\alpha]_D^{25}$ = +35.6 (c = 1.0, $CHCl_3$).

tert-Butyl-1-thio-6-O-(tert-butyldiphenylsilyl)-3,4-O-carbonate-2-deoxy-α-D-galactopyranoside (**5f**). The title compound was prepared according to the general procedure of 2-deoxythioglycosides synthesis and purified by flash column chromatography giving colorless oil 40.0 mg, yield 80%; ^{1}H NMR (400 MHz, $CDCl_3$) δ 7.69–7.72 (m, 4H, Ar-H), 7.56–7.37 (m, 6H, Ar-H), 5.56 (dd, *J* = 9.4, 6.5 Hz, 1H, H-1), 5.03 (ddd, *J* = 8.4, 4.8, 2.4 Hz, 1H, H-3), 4.91 (dd, *J* = 8.5, 1.6 Hz, 1H, H-4), 4.13 (ddd, *J* = 7.9, 5.7, 1.6 Hz, 1H, H-5), 3.85 (dd, *J* = 10.0, 8.3 Hz, 1H, H-6), 3.72 (dd, *J* = 10.0, 5.8 Hz, 1H, H-6′), 2.49 (ddd, *J* = 15.7, 6.6, 3.8 Hz, 1H, H-2), 1.81 (ddd, *J* = 15.8, 9.4, 3.1 Hz, 1H, H-2′), 1.31 (s, 9H, C-^{t}Bu), 1.07 (s, 9H, Si-^{t}Bu).^{13}C NMR (100 MHz, $CDCl_3$) δ 154.6, 135.7, 135.6, 133.2, 132.8, 130.1, 130.0, 128.0, 127.9, 75.4, 73.5, 72.6, 67.6, 61.2, 44.5, 31.8, 29.4, 26.9, 19.4. HRMS (ESI) m/z: calcd. for $C_{27}H_{36}O_5SSiNa^+$ $(M + Na)^+$ 523.1945, found 523.1949; $[\alpha]_D^{25}$ = +40.2 (c = 1.0, $CHCl_3$).

Benzyl-1-thio-6-O-(tert-butyldiphenylsilyl)-3,4-O-carbonate-2-deoxy-α-D-galactopyranoside (**5g**). The title compound was prepared according to the general procedure of 2-deoxythioglycosides synthesis and purified by flash column chromatography giving colorless oil 43.8 mg, yield 82%; ^{1}H NMR (400 MHz, $CDCl_3$) δ 7.70–7.73 (m, 4H, Ar-H), 7.55–7.33 (m, 6H, Ar-H), 7.30–7.10 (m, 5H, Ar-H), 5.21 (dd, *J* = 9.2, 6.6 Hz, 1H, H-1), 5.00 (ddd, *J* = 8.4, 4.6, 2,4 Hz, 1H, H-3), 4.86 (dd, *J* = 8.5, 1.7 Hz, 1H, H-4), 4.14–4.18(m, 1H, H-5), 3.92–3.83 (m, 2H, H-6, PhC*H*H), 3.76 (dd, *J* = 10.3, 6.4 Hz, 1H, H-6′), 3.66 (d, *J* = 13.6 Hz, 1H, PhCH*H*), 2.46 (ddd, *J* = 15.7, 6.6, 3.8 Hz, 1H, H-2), 1.77 (ddd, *J* = 15.8, 9.2, 3.2 Hz, 1H, H-2′), 1.11 (s, 9H, Si-^{t}Bu).^{13}C NMR (100 MHz, $CDCl_3$) δ 154.3, 137.9, 135.7, 135.6, 133.2, 132.9, 130.1, 130.0, 129.0, 128.7, 128.0, 127.3, 76.0, 73.6, 72.5, 68.2, 61.7, 34.5, 28.8, 27.0, 19.4. HRMS (ESI) m/z: calcd. for $C_{30}H_{34}O_5SSiNa^+$ $(M + Na)^+$ 557.1788, found 557.1777; $[\alpha]_D^{25}$ = +140.3 (c = 1.0, $CHCl_3$).

2-Methoxy-2-oxoethyl-1-thio-6-O-(tert-butyldiphenylsilyl)-3,4-O-carbonate-2-deoxy-α-D-galactopyranooside (**5h**). The title compound was prepared according to the general procedure of 2-deoxythioglycosides synthesis and purified by flash column chromatography giving colorless oil 41.2 mg, yield 80%; ^{1}H NMR (400 MHz, $CDCl_3$) δ 7.67–7.70 (m, 4H, Ar-H), 7.54–7.37 (m, 6H, Ar-H), 5.50 (dd, *J* = 9.5, 6.6 Hz, 1H, H-1), 5.05 (ddd, *J* = 8.6, 4.0, 2.2 Hz, 1H, H-3), 4.88 (dd, *J* = 8.6, 1.7 Hz, 1H, H-4), 4.14–4.03 (m, 1H, H-5), 3.84 (dd, *J* = 10.2, 7.2 Hz, 1H, H-6), 3.77 (dd, *J* = 10.2, 6.4 Hz, 1H, H-6′), 3.67 (s, 3H, CH_2COOCH_3), 3.52 (d, *J* = 15.5 Hz, 1H, *CH*HCOOCH$_3$), 3.17 (d, *J* = 15.5 Hz, 1H, CH*H*COOCH$_3$), 2.61 (ddd, *J* = 15.8, 6.5, 3.5 Hz, 1H, H-2), 1.76 (ddd, *J* = 15.8, 9.5, 3.1 Hz, 1H, H-2′), 1.08 (s, 9H, Si-^{t}Bu).^{13}C NMR (100 MHz, $CDCl_3$) δ 170.7, 154.2, 135.7, 135.6, 133.1, 132.8, 130.1, 130.0, 128.0, 77.4, 73.5, 72.3, 68.4, 61.5, 52.7, 31.5, 28.6, 26.9, 19.4. HRMS (ESI) m/z: calcd. for $C_{26}H_{32}O_7SSiNa^+$ $(M + Na)^+$ 539.1530, found 539.1519; $[\alpha]_D^{25}$ = +28.0 (c = 1.0, $CHCl_3$).

3-S-(6-O-(tert-Butyldiphenylsilyl)-3,4-O-carbonate-2-deoxy-a-D-galactopyranosyl)-1,2,3,4-di-O-isopropylidene-β-D-glucofuranoside (**5i**). The title compound was prepared according to the general procedure of 2-deoxythioglycosides synthesis and purified by flash column giving colorless oil 45.9 mg, yield 67%; ^{1}H NMR (400 MHz, $CDCl_3$) δ 7.66 (ddd, J = 7.8, 3.6, 1.7 Hz, 4H, Ar-H), 7.57–7.35 (m, 6H, Ar-H), 5.93 (d, J = 3.6 Hz, 1H, H-1b), 5.05 (dd, J = 9.6, 6.0 Hz, 1H, H-1a), 4.97–5.00 (m, 1H, H-3a), 4.85 (dd, J = 8.3, 1.7 Hz, 1H, H-4a), 4.45 (dd, J = 3.5, 1.8 Hz, 1H, H-2b), 4.30–4.33 (m, 1H, H-5b), 4.04–4.08 (m, 1H, H-5a), 3.92–3.65 (m, 5H, H-6a, 6′a, H-4b, H-7b, 7′b), 3.60 (dd, J = 4.5, 2.0 Hz, 1H, H-3b), 2.41 (ddd, J = 15.7, 5.6, 4.2 Hz, 1H, H-2a), 1.89 (ddd, J = 15.7, 6.4, 3.7 Hz, 1H, H-2′a), 1.57 (s, 3H, CH_3), 1.49 (s, 3H, CH_3), 1.45 (s, 3H, CH_3), 1.31 (s, 3H, CH_3), 1.06 (s, 9H, Si-^{t}Bu).^{13}C NMR (100 MHz, $CDCl_3$) δ 154.4, 135.7, 135.6, 133.2, 132.9, 130.1, 130.0, 128.0, 112.3, 106.4, 95.21, 83.6, 83.2, 77.9, 73.8, 73.5, 72.1, 67.9, 67.7, 61.7, 47.5, 31.6, 29.7, 27.5, 27.1, 27.0, 26.6, 19.4. HRMS (ESI) m/z: calcd. for $C_{35}H_{47}O_{10}SSi^+$ $(M + H)^+$ 687.2654, found 687.2670; $[\alpha]_D^{25}$ = +27.2 (c = 1.0, $CHCl_3$).

6-S-(6-O-(tert-butyldiphenylsilyl)-3,4-O-carbonate-2-deoxy-a-D-galactopyranosyl)-,2,3,4,6-tetra-O-acetyl-β-D-glucopyranoside (**5j**). The title compound was prepared according to the general procedure of 2-deoxythioglycosides synthesis and purified by flash column chromatography giving white solid 53.0 mg, yield 69%; m.p.: 144.6–146.3 °C. ^{1}H NMR (400 MHz, $CDCl_3$) δ 7.84–7.56 (m, 4H, Ar-H), 7.65–7.35 (m, 6H, Ar-H), 5.56 (dd, *J* = 9.8, 6.4 Hz, 1H, H-1a), 5.12–5.17(m, 1H, H-3b), 5.09–5.02 (m, 2H, H-3a, H-2b), 5.02–4.97 (m, 2H, H-4a, H-4b), 4.61 (d, *J* = 10.1 Hz, 1H, H-1b), 4.11 (ddd, *J* = 8.0, 6.8, 1.5 Hz, 1H, H-5), 4.00 (dd, *J* = 12.5, 4.8 HZ, 1H, H-6a), 3.90 (dd, *J* = 12.5, 2.2 Hz, 1H, H-6′a), 3.90–3.79 (m, 2H, H-6, 6′b), 3.60 (ddd, *J* = 10.1, 4.7, 2.2 Hz, 1H, H-5b), 2.56 (ddd, *J* = 15.8, 6.4, 3.3 Hz, 1H, H-2a), 2.20–1.97 (m, 9H, OAc), 1.89–1.82 (m, 4H, H-2′a, OAc), 1.07 (s, 9H, Si-^{t}Bu). ^{13}C NMR (100 MHz, $CDCl_3$) δ 170.6, 170.3, 169.5, 169.2, 154.1, 135.7, 135.6, 133.1, 132.7, 130.1, 130.0, 128.0, 83.3, 77.6, 76.4, 73.9, 73.1, 72.02, 71.1, 68.2, 67.9, 61.8, 60.7, 29.0, 26.9, 20.8, 20.7, 19.4. HRMS (ESI) m/z: calcd. for $C_{37}H_{46}O_{14}SSiNa^+$ $(M + Na)^+$ 797.2270, found 797.2265; $[\alpha]_D^{25}$ = +21.1 (c = 1.0, $CHCl_3$).

Estronyl-1-thio-6-O-(tert-butyldiphenylsilyl)-3,4-O-carbonate-2-deoxy-α-D-galactopyranoside (**7a**). The title compound was prepared according to the general procedure of 2-deoxythioglycosides synthesis and purified by flash column chromatography giving colorless oil 54.3 mg, yield 78%; ^{1}H NMR (400 MHz, $CDCl_3$) δ 7.80–7.64 (m, 4H, Ar-H), 7.52–7.36 (m, 6H, Ar-H), 7.26–7.10 (m, 3H, Ar-H), 5.51 (dd, *J* = 9.8, 6.5 Hz, 1H, H-1), 5.08 (ddd, *J* = 8.6, 4.1, 2.4 Hz, 1H, H-3), 4.94 (dd, *J* = 8.6, 1.6 Hz, 1H, H-4), 4.20 (ddd, *J* = 7.6, 5.9, 1.6 Hz, 1H, H-5), 3.88 (dd, *J* = 10.2, 7.8 Hz, 1H, H-6), 3.79 (dd, *J* = 10.2, 6.0 Hz, 1H, H-6′), 2.83–2.74 (m, 2H, Ph*CH$_2$*), 2.68 (ddd, *J* = 15.9, 6.5, 3.4 Hz, 1H, H-2), 2.52 (dd, *J* = 18.9, 8.6 Hz, 1H, PhC*H*), 2.34–2.38 (m, 1H, COC*H*H), 2.22–2.23 (m, 1H, COCH*H*), 2.23–2.04 (m, 3H, CH_2, C*H*H), 2.04–1.86 (m, 3H, H-2′, CH_2), 1.71–1.53 (m, 3H, CH_2, CH*H*), 1.50–1.37 (m, 2H, *C*H*C*H), 1.09 (s, 9H, Si-^{t}Bu), 0.91 (s, 3H, CH_3). ^{13}C NMR (100 MHz, $CDCl_3$) δ 220.9, 154.2, 140.1, 137.5, 135.7, 135.6, 133.3, 133.2, 132.8, 130.3, 130.1, 130.0, 129.9, 128.0, 126.2, 80.7, 73.4, 72.4, 68.5, 61.5, 50.6, 48.1, 44.4, 38.0, 36.0, 31.7, 29.3, 28.9, 26.9, 26.4, 25.7, 21.7, 19.4, 13.9. HRMS (ESI) m/z: calcd. For $C_{41}H_{48}O_5SsiNa^+$ $(M + Na)^+$ 719.2833, found 719.2830; $[\alpha]_D^{25}$ = +114.6 (c = 1.0, $CHCl_3$).

L-Menthyl-1-thio-6-O-(tert-butyldiphenylsilyl)-3,4-O-carbonate-2-deoxy-α-D-galactopyranoside **(7b)**. The title compound was prepared according to the general procedure of 2-deoxythioglycosides synthesis and purified by flash column chromatography giving colorless oil 34.9 mg, yield 60%; ^{1}H NMR (400 MHz, $CDCl_3$) δ 7.83–7.55 (m, 4H, Ar-H), 7.55–7.34 (m, 6H, Ar-H), 5.32 (dd, *J* = 9.7, 6.4 Hz, 1H, H-1), 5.06 (ddd, *J* = 8.7, 3.1 Hz, 1H, H-3), 5.01–5.08 (m, 1H, H-4), 4.14 (ddd, *J* = 9.5, 5.5, 1.5 Hz, 1H, H-5), 3.84–3.88 (m, 1H, H-6), 3.74 (dd, *J* = 9.6, 5.5 Hz, 1H, H-6′), 3.36–3,37 (m, 1H, SC*H*), 2.55 (ddd, *J* = 15.8, 6.4, 3.5 Hz, 1H, H-2), 1.88–1.95 (m, 2H, CH, CH), 1.83 (ddd, *J* = 15.8, 9.7, 2.8 Hz, 1H, H-2′), 1.77–1.69 (m, 2H, CH_2), 1.52–1.43 (m, 1H, C*H*H), 1.25–1.13 (m, 1H, CH*H*), 1.08 (s, 9H, Si-^{t}Bu), 1.05–0.99 (m, 2H, CH, C*H*H), 0.88–0.92 (m, 4H, CH_3, CH*H*), 0.83 (d, *J* = 6.6 Hz, 3H, CH_3), 0.67 (d, *J* = 6.5 Hz, 3H, CH_3). ^{13}C NMR (100 MHz, $CDCl_3$) δ 154.5, 135.6, 135.5, 133.2, 132.8, 130.1, 130.0, 128.0, 127.9, 75.4, 73.5, 72.6, 67.5, 61.0, 48.3, 44.5, 40.3, 35.4, 29.9, 29.5, 26.9, 26.6, 26.4, 22.2, 21.0, 20.4, 19.4. HRMS (ESI) m/z: calcd. for $C_{33}H_{46}O_5SSiNa^+$ (M + Na)$^+$ 605.2727, found 605.2728; $[\alpha]_D^{25}$ = +56.7 (c = 1.0, $CHCl_3$).

Zingerone-1-thio-6-O-(tert-butyldiphenylsilyl)-3,4-O-carbonate-2-deoxy-α-D-galactopyranoside **(7c)**. The title compound was prepared according to the general procedure of 2-deoxythioglycosides synthesis and purified by flash column chromatography giving colorless oil 41.2 mg, yield 72%; ^{1}H NMR (400 MHz, $CDCl_3$) δ 7.81–7.63 (m, 4H), 7.52–7.34 (m, 6H, Ar-H), 7.33–7.14 (m, 1H, Ar-H), 6.69 (d, *J* = 1.6 Hz, 1H, Ar-H), 6.60 (dd, *J* = 7.8, 1.7 Hz, 1H, Ar-H), 5.61 (dd, *J* = 9.6, 6.5 Hz, 1H, H-1), 5.08–5.10 (m, 1H, H-3), 4.95 (dd, *J* = 8.7, 1.6 Hz, 1H, H-4), 4.22 (ddd, *J* = 7.9, 5.9, 1.6 Hz, 1H, H-5), 3.94–3.78 (m, 4H, H-6, OMe), 3.70 (dd, *J* = 10.0, 6.0 Hz, 1H, H-6′), 2.82–2.86 (m, 2H, C*H*$_2$CH$_2$), 2.77–2.63 (m, 3H, H-2, CH$_2$C*H*$_2$), 2.14 (s, 3H, CH$_2$COC*H*$_3$), 1.92 (ddd, *J* = 15.8, 9.6, 3.0 Hz, 1H, H-2′), 1.07 (s, 9H, Si-^{t}Bu). ^{13}C NMR (100 MHz, $CDCl_3$) δ 207.7, 158.4, 154.3, 143.3, 135.7, 135.6, 134.5, 133.2, 132.8, 130.0, 129.9, 128.0, 121.0, 118.1, 111.3, 78.6, 73.5, 72.5, 68.23, 61.4, 55.9, 45.0, 30.2, 29.8, 28.9, 26.9, 19.3. HRMS (ESI) m/z: calcd. for $C_{34}H_{40}O_7SSiNa^+$ (M + Na)$^+$ 643.2165, found 643.2154; $[\alpha]_D^{25}$ = +99.8 (c = 1.0, $CHCl_3$).

Supplementary Materials: The following supporting information can be downloaded at: https://www.mdpi.com/article/10.3390/molecules27227979/s1, general information, NMR spectra and X-ray crystal structure and data, Figure S1: Synthesis of estrone thiol, Figure S2: Synthesis of L-menthol thiol, Figure S3: Synthesis of zingerone thiol, Figure S4: ORTEP drawing of compound 3c showing thermal ellipsoids at the 50% probability level (CCDC: 2212881), Table S1: Crystal data and structure refinement for **3c**. The data for known compounds were checked in comparison with literature for consistency [38–41].

Author Contributions: Conceptualization, H.Y. and L.C.; methodology, X.Y.; validation, Y.C., C.X. and L.M.; formal analysis, Y.W.; investigation, T.X.; data curation, X.Y.; writing—original draft preparation, X.Y and H.Y.; writing—review and editing, H.Y.; supervision, H.Y.; project administration, H.Y and L.C.; funding acquisition, H.Y. All authors have read and agreed to the published version of the manuscript.

Funding: This research was funded by the 111 Project (D20015) and the Educational Commission of Hubei Province of China (D20221204).

Institutional Review Board Statement: Not applicable.

Informed Consent Statement: Not applicable.

Data Availability Statement: Data are included within the manuscript or the supplementary data file.

Acknowledgments: The authors acknowledged the NMR analysis support of Nianyu Huang from Hubei Key Laboratory of Natural Products Research and Development, China Three Gorges University.

Conflicts of Interest: The authors declare no conflict of interest.

Sample Availability: Samples of the compounds are available from the authors.

References

1. Bennett, C.S.; Galan, M.C. Methods for 2-deoxyglycoside synthesis. *Chem. Rev.* **2018**, *118*, 7931–7985. [CrossRef] [PubMed]
2. Liu, M.; Liu, K.M.; Xiong, D.C.; Zhang, H.; Li, T.; Li, B.; Qin, X.; Bai, J.; Ye, X.-S. Stereoselective electro-2-deoxyglycosylation from glycals. *Angew. Chem. Int. Ed.* **2020**, *59*, 15204–15208. [CrossRef] [PubMed]
3. Cao, X.; Du, X.; Jiao, H.; An, Q.; Chen, R.; Fang, P.; Wang, J.; Yu, B. Carbohydrate-based drugs launched during 2000−2021. *Acta Pharm. Sin. B* **2022**, *12*, 3783–3821. [CrossRef] [PubMed]
4. Zeng, J.; Xu, Y.; Wang, H.; Meng, L.; Wan, Q. Recent progress on the synthesis of 2-deoxy glycosides. *Sci. China Chem.* **2017**, *60*, 1162–1179. [CrossRef]
5. Wan, L.Q.; Zhang, X.; Zou, Y.; Shi, R.; Cao, J.G.; Xu, S.Y.; Deng, L.F.; Zhou, L.; Gong, Y.; Shu, X.; et al. Nonenzymatic stereoselective *S*-glycosylation of polypeptides and proteins. *J. Am. Chem. Soc.* **2021**, *143*, 11919–11926. [CrossRef] [PubMed]
6. Li, G.; Dao, Y.; Mo, J.; Dong, S.; Shoda, S.-I.; Ye, X.-S. Protection-free site-directed peptide or protein *S*-glycosylation and its application in the glycosylation of glucagon-like peptide 1. *CCS Chem.* **2021**, *4*, 2316–2323. [CrossRef]
7. Pal, K.B.; Guo, A.; Das, M.; Lee, J.; Báti, G.; Yip, B.R.P.; Loh, T.-P.; Liu, X.-W. Iridium-promoted deoxyglycoside synthesis: Stereoselectivity and mechanistic insight. *Chem. Sci.* **2021**, *12*, 2209–2216. [CrossRef]
8. Adhikari, S.; Baryal, K.N.; Zhu, D.; Li, X.; Zhu, J. Gold-catalyzed synthesis of 2-deoxy glycosides using *S*-but-3-ynyl thioglycoside donors. *ACS Catal.* **2013**, *3*, 57–60. [CrossRef]
9. Zhu, J.; Baryal, K. Stereoselective synthesis of *S*-linked 2-deoxy sugars. *Synlett* **2013**, *25*, 308–312. [CrossRef]
10. Zhu, F.; Miller, E.; Zhang, S.Q.; Yi, D.; O'Neill, S.; Hong, X.; Walczak, M.A. Stereoretentive C(sp^3)-S cross-coupling. *J. Am. Chem. Soc.* **2018**, *140*, 18140–18150. [CrossRef]
11. Tamburrini, A.; Colombo, C.; Bernardi, A. Design and synthesis of glycomimetics: Recent advances. *Med. Res. Rev.* **2020**, *40*, 495–531. [CrossRef] [PubMed]
12. Liu, Y.; Jiao, Y.; Luo, H.; Huang, N.; Lai, M.; Zou, K.; Yao, H. Catalyst-controlled regiodivergent synthesis of 1- and 3-thiosugars with high stereoselectivity and chemoselectivity. *ACS Catal.* **2021**, *11*, 5287–5293. [CrossRef]
13. Hua, M.; Sun, Y.; Zhang, X.; Yao, H.; Huang, N. Convenient cobalt-catalyzed stereoselective synthesis of β-D-thioglucosides. *Chin. J. Org. Chem* **2022**, *42*, 2140–2154. [CrossRef]
14. Hoang, K.M.; Lees, N.R.; Herzon, S.B. Programmable synthesis of 2-deoxyglycosides. *J. Am. Chem. Soc.* **2019**, *141*, 8098–8103. [CrossRef]
15. Zhu, D.Y.; Baryal, K.N.; Adhikari, S.; Zhu, J.L. Direct synthesis of 2-deoxy-β-glycosides via anomeric *O*-alkylation with secondary electrophiles. *J. Am. Chem. Soc.* **2014**, *136*, 3172–3175. [CrossRef] [PubMed]
16. Paul, S.; Jayaraman, N. Catalytic ceric ammonium nitrate mediated synthesis of 2-deoxy-1-thioglycosides. *Carbohydr. Res.* **2004**, *339*, 2197–2204. [CrossRef]
17. Meng, S.; Li, X.; Zhu, J. Recent advances in direct synthesis of 2-deoxy glycosides and thioglycosides. *Tetrahedron* **2021**, *88*, 132140. [CrossRef]
18. Sherry, B.D.; Loy, R.N.; Toste, F.D. Rhenium(V)-catalyzed synthesis of 2-deoxy-α-glycosides. *J. Am. Chem. Soc.* **2004**, *126*, 4510–4511. [CrossRef]
19. Zhao, X.; Wu, B.; Shu, P.; Meng, L.; Zeng, J.; Wan, Q. Rhenium(V)-catalyzed synthesis of 1,1′-2-deoxy thioglycosides. *Carbohydr. Res.* **2021**, *508*, 108415. [CrossRef]
20. Mukherji, A.; Addanki, R.B.; Halder, S.; Kancharla, P.K. Sterically strained Brønsted pair catalysis by bulky pyridinium salts: Direct stereoselective synthesis of 2-deoxy and 2,6-dideoxy-*β*-thioglycosides from glycals. *J. Org. Chem.* **2021**, *86*, 17226–17243. [CrossRef]
21. Yao, H.; Zhang, S.; Leng, W.-L.; Leow, M.-L.; Xiang, S.; He, J.-X.; Liao, H.; Le Mai Hoang, K.; Liu, X.-W. Catalyst-controlled stereoselective *O*-glycosylation: Pd(0) vs Pd(II). *ACS Catal.* **2017**, *7*, 5456–5460. [CrossRef]
22. Lai, M.; Othman, K.A.; Yao, H.; Wang, Q.; Feng, Y.; Huang, N.; Liu, M.; Zou, K. Open-air stereoselective construction of *C*-aryl glycosides. *Org. Lett.* **2020**, *22*, 1144–1148. [CrossRef] [PubMed]
23. Wang, Y.; Yao, H.; Hua, M.; Jiao, Y.; He, H.; Liu, M.; Huang, N.; Zou, K. Direct *N*-glycosylation of amides/amines with glycal donors. *J. Org. Chem.* **2020**, *85*, 7485–7493. [CrossRef] [PubMed]
24. Abdulmajeed Othman, K.; Cai, J.; Xie, R.; Zhou, X.; Yao, H.; Huang, N. Synthesis and absolute configuration of (2*R*,3*S*,4*Z*,6*Z*)-1,3-bis(benzyloxy)-8-chloro-7-((*E*)-(2-(2,4-dinitro-phenyl)hydrazono)methyl)octa-4,6-dien-2-ol. *Chin. J. Struct. Chem.* **2020**, *39*, 1781–1787.
25. Feng, Y.; Su, H.; Mukula Otukol, B.J.; Zhang, X.; Yao, H.; Huang, N. Synthesis and crystal sructure of *tert*-butyl(((2*R*,3*R*,6*R*)-3-hydroxy-6-(nitromethyl)-3,6-dihydro-2*H*-pyran-2-yl)methyl)carbonate. *Chin. J. Struct. Chem.* **2021**, *40*, 1205–1212.
26. Wang, Q.; Lai, M.; Luo, H.; Ren, K.; Wang, J.; Huang, N.; Deng, Z.; Zou, K.; Yao, H. Stereoselective *O*-glycosylation of glycals with arylboronic acids using air as the oxygen source. *Org. Lett.* **2022**, *24*, 1587–1592. [CrossRef]
27. Ding, W.-Y.; Liu, H.-H.; Cheng, J.K.; Yao, H.; Xiang, S.-H.; Tan, B. Palladium catalyzed decarboxylative β-*C*-glycosylation of glycals with oxazol-5-(4*H*)-ones as acceptors. *Org. Chem. Front.* **2022**, *9*, 6149–6155. [CrossRef]
28. Chernikova, N.; Frantsiyants, E.; Molseyenko, T.; Komarova, E.; Adamyan, M.; Nikitin, I. Content of estrone and estrogen metabolites in tissues of hysterocarcinoma, adenomyosis, and hysteromyoma tissue. *J. Clin. Oncol.* **2014**, *32*, e22214. [CrossRef]
29. Escandon, P.; Nicholas, S.E.; Cunningham, R.L.; Murphy, D.A.; Riaz, K.M.; Karamichos, D. The role of estriol and estrone in keratoconic stromal sex hormone receptors. *Int. J. Mol. Sci.* **2022**, *23*, 916. [CrossRef]
30. Canario, C.; Matias, M.; Brito, V.; Santos, A.O.; Falcao, A.; Silvestre, S.; Alves, G. New estrone oxime derivatives: Synthesis, cytotoxic evaluation and docking studies. *Molecules* **2021**, *26*, 2687. [CrossRef]
31. Zhao, R.L.; He, Y.M. Network pharmacology analysis of the anti-cancer pharmacological mechanisms of *Ganoderma lucidum* extract with experimental support using Hepa1-6-bearing C57 BL/6 mice. *J. Ethnopharmacol.* **2018**, *210*, 287–295. [CrossRef]

32. Kung, M.L.; Lin, P.Y.; Huang, S.T.; Tai, M.H.; Hsieh, S.L.; Wu, C.C.; Yeh, B.W.; Wu, W.J.; Hsieh, S. Zingerone nanotetramer strengthened the polypharmacological efficacy of zingerone on human hepatoma cell lines. *ACS Appl. Mater. Interfaces* **2019**, *11*, 137–150. [CrossRef] [PubMed]
33. Wali, A.F.; Rehman, M.U.; Raish, M.; Kazi, M.; Rao, P.G.M.; Alnemer, O.; Ahmad, P.; Ahmad, A. Zingerone [4-(3-methoxy-4-hydroxyphenyl)-butan-2] attenuates lipopolysaccharide-induced inflammation and protects rats from sepsis associated multi organ damage. *Molecules* **2020**, *25*, 5127. [CrossRef]
34. Heo, K.T.; Park, K.W.; Won, J.; Lee, B.; Jang, J.H.; Ahn, J.O.; Hwang, B.Y.; Hong, Y.S. Construction of an artificial biosynthetic pathway for Zingerone production in *Escherichia coli* using benzalacetone synthase from *Piper methysticum*. *J. Agric. Food. Chem.* **2021**, *69*, 14620–14629. [CrossRef] [PubMed]
35. Sau, A.; Williams, R.; Palo-Nieto, C.; Franconetti, A.; Medina, S.; Galan, M.C. Palladium-catalyzed direct stereoselective synthesis of deoxyglycosides from glycals. *Angew. Chem. Int. Ed.* **2017**, *56*, 3640–3644. [CrossRef] [PubMed]
36. Kumar, M.; Reddy, T.R.; Gurawa, A.; Kashyap, S. Copper(II)-catalyzed stereoselective 1,2-addition *vs.* Ferrier glycosylation of "armed" and "disarmed" glycal donors. *Org. Biomol. Chem.* **2020**, *18*, 4848–4862. [CrossRef]
37. Kumar, M.; Gurawa, A.; Kumar, N.; Kashyap, S. Bismuth-catalyzed stereoselective 2-deoxyglycosylation of disarmed/armed glycal donors. *Org. Lett.* **2022**, *24*, 575–580. [CrossRef]
38. Jin, M.; Ren, W.; Qian, D.W.; Yang, S.D. Direct allylic C(sp^3)-H alkylation with 2-naphthols via cooperative palladium and copper catalysis: Construction of cyclohexadienones with quaternary carbon centers. *Org. Lett.* **2018**, *20*, 7015–7019. [CrossRef]
39. van den Hoogenband, A.; Lange, J.H.M.; Bronger, R.P.J.; Stoit, A.R.; Terpstra, J.W. A simple, base-free preparation of S-aryl thioacetates as surrogates for aryl thiols. *Tetrahedron Lett.* **2010**, *51*, 6877–6881. [CrossRef]
40. Li, P.-K.; Pillai, R.; Young, B.L.; Bender, W.H.; Martino, D.M.; Lint, F.-T. Synthesis and biochemical studies of estrone sulfatase inhibitors. *Steroids* **1993**, *58*, 106–111.
41. Hsu, J.-L.; Fang, J.-M. Stereoselective synthesis of δ-lactones from 5-oxoalkanals via one-pot sequential acetalization, tishchenko reaction, and lactonization by cooperative catalysis of samarium ion and mercaptan. *J. Org. Chem.* **2001**, *66*, 8573–8584. [CrossRef] [PubMed]

molecules

Review

Recent Research and Application Prospect of Functional Oligosaccharides on Intestinal Disease Treatment

Tong Xu [1,2,†], Ruijie Sun [1,3,†], Yuchen Zhang [1], Chen Zhang [1], Yujing Wang [1], Zhuo A. Wang [1,*] and Yuguang Du [1,*]

1 State Key Laboratory of Biochemical Engineering, Institute of Process Engineering, Chinese Academy of Sciences, Beijing 100190, China
2 School of Chemical Engineering, University of Chinese Academy of Sciences, Beijing 100049, China
3 Key Laboratory of Carbohydrate Chemistry and Biotechnology, Ministry of Education, School of Biotechnology, Jiangnan University, Wuxi 214000, China
* Correspondence: wangzhuo@ipe.ac.cn (Z.A.W.); ygdu@ipe.ac.cn (Y.D.); Tel.: +86-10-8254-5070 (Z.A.W. & Y.D.)
† These authors contributed equally to this work.

Abstract: The intestinal tract is an essential digestive organ of the human body, and damage to the intestinal barrier will lead to various diseases. Functional oligosaccharides are carbohydrates with a low degree of polymerization and exhibit beneficial effects on human intestinal health. Laboratory experiments and clinical studies indicate that functional oligosaccharides repair the damaged intestinal tract and maintain intestinal homeostasis by regulating intestinal barrier function, immune response, and intestinal microbial composition. Functional oligosaccharides treat intestinal disease such as inflammatory bowel disease (IBD) and colorectal cancer (CRC) and have excellent prospects for therapeutic application. Here, we present an overview of the recent research into the effects of functional oligosaccharides on intestinal health.

Keywords: functional oligosaccharides; gut microbiota; intestinal barriers; intestinal diseases

check for updates

Citation: Xu, T.; Sun, R.; Zhang, Y.; Zhang, C.; Wang, Y.; Wang, Z.A.; Du, Y. Recent Research and Application Prospect of Functional Oligosaccharides on Intestinal Disease Treatment. *Molecules* **2022**, *27*, 7622. https://doi.org/10.3390/molecules27217622

Academic Editors: Jian Yin, Jing Zeng and De-Cai Xiong

Received: 29 September 2022
Accepted: 21 October 2022
Published: 7 November 2022

1. Introduction

With the continuous improvement of people's quality of life in recent years, the diet structure has changed from simple to complex. Today, the global nutrition situation is complicated. On the one hand, hunger and malnutrition are the dominant concerns in low- and middle-income countries. On the other hand, millions of people are at increased risk of developing diet-related chronic diseases, for example, intestinal disease, heart disease, and diabetes. Taking inflammatory bowel disease (IBD) as an example, the incidence of IBD has increased year by year worldwide over the past decade [1,2], with the highest incidence of IBD in developed countries [3]. More than 2 million people in North America and 3.2 million people in Europe are afflicted with IBD [4]. With the development trend of globalization, IBD is becoming more and more common in developing countries such as Brazil and China [5]. In Brazil, Crohn's disease (CD) and ulcerative colitis (UC) increased by 11.1% and 14.9%, respectively, from 1988 to 2012 [4]. In China, there were 350,000 IBD patients in 2014, which is expected to increase by 4.2 times by 2025 and an approximate 70% increase in UC and 30% in CD (data from CCDC). IBD is also a risk factor for colitis-associated colorectal cancer (CA-CRC), which causes death in about 15% of patients with IBD [6]. In 2020, the global number of CRC cases was close to 2,000,000 and accounted for 9.7% of the global cancer population (data from IARC). There are 1.5 million CRC patients in the United States. In recent years, the incidence and mortality of CRC have decreased, but there are still about 150,000 new patients each year [7]. In contrast, the number of CRC cases in China has been progressively higher than in the United States in recent years, with 2.6 times the patients of the United States. Chinese CRC patients account for 31% of the patients worldwide. From 1990 to 2019, the number of CRC cases in China increased by 700% (data from World Bank IHME-GBD). In addition, irritable bowel syndrome (IBS) is one of the most common intestinal disorders in clinical practice. The prevalence of IBS in Western countries is 10% to 20%, and the prevalence of IBS in China is 5.7%. In 2016, there were approximately 754 million people with IBS worldwide, and it is expected to reach 830 million by 2025 (data from Data monitor). In summary, with the increasing number of intestinal diseases represented by IBD, IBS, and CRC, their prevention and treatment is gradually becoming an important issue for domestic and international research.

The intestine is a vital digestive organ responsible for the digestion and absorption of nutrients, and the intestinal barrier prevents pathogenic bacteria, toxins, and other harmful substances from entering the intestine's circulatory system [8–10]. The intestinal barrier is comprised of the epithelial and mucus barrier, immune barrier, and biological barrier, which together maintain the health and homeostasis of the intestinal tract. The intestinal epithelial and mucus barrier is mainly composed of single-layer cells connecting proteins and chemical substances in the intestinal epithelium. A variety of transmembrane proteins further constitute a complex protein network between adjacent cells. The integrity of the intestinal epithelial barrier depends on the link complexes in the protein network, including the tight junction, adhesion junction, and bridge and gap junction [11]. The chemical substances are composed of mucus, digestive fluid, antibacterial components, and other compounds secreted by the intestinal mucosa and microorganisms. The epithelial and mucus barrier prevents the penetration of harmful bacteria and toxins [12]. The intestinal tract is also the largest immune organ in the human body. The intestinal immune barrier includes intestinal-related lymphoid tissue (GALT), diffuse immune cells, and immune factors [13]. Microorganisms colonized in the intestine are considered intestinal biological barriers. Many laboratory and clinical studies have confirmed that the damaged intestinal barrier may lead to overactive immune responses in the intestinal microenvironment or the uncontrolled growth of microbial flora, leading to various diseases [14]. The effect of functional oligosaccharides on intestinal barrier function and health is illustrated in Figure 1.

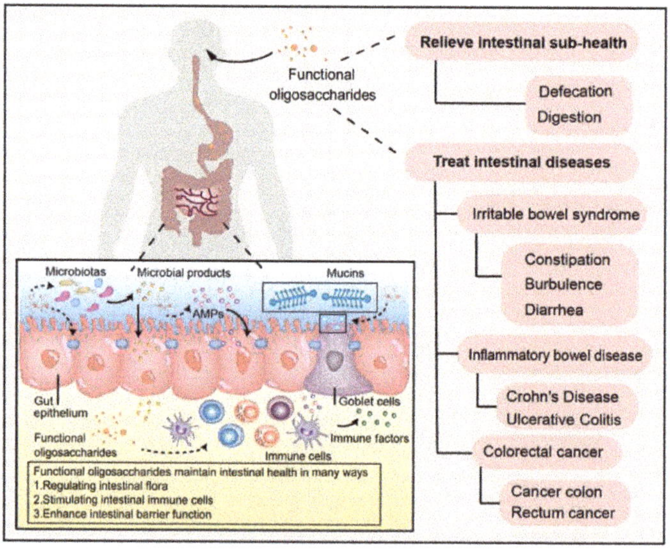

Figure 1. Schematic overview of the effects of functional oligosaccharides on intestinal barrier function and health. AMPs: antimicrobial peptides.

Current treatment strategies for intestinal diseases include micro-ecological regulation therapy [15], surgical treatment [16,17], and drug therapy [17,18]. In recent years, micro-ecological agents including probiotics, prebiotics, and diet fibers have drawn more and more attention to treating intestinal diseases. The International Association for Probiotics and Prebiotics (ISAPP) redefined prebiotics in 2016: the host microorganisms selectively use them to make them healthy substrates [19,20]. Prebiotics are diverse and are divided into carbohydrate sources and non-carbohydrate sources, with functional oligosaccharides as the principal source. Functional oligosaccharides are carbohydrate oligomers with branched or straight chains of 2–20 monosaccharide molecules linked through glycosidic bonds. Here we mainly introduce the representative functional oligosaccharides: isomaltooligosaccharide (IMO), fructooligosaccharides (FOS), xylooligosaccharides (XOS), galactooligosaccharides (GOS), chitosan oligosaccharides (COS), and human milk oligosaccharides (HMOs). The structure of the functional oligosaccharides is shown in Figure 2.

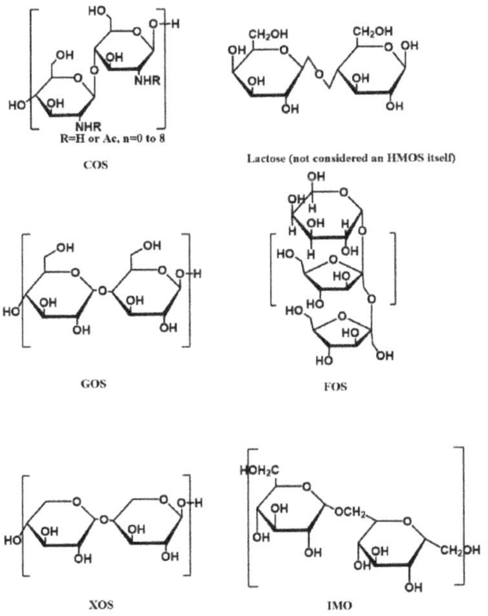

Figure 2. Schematic diagram of the structure of common functional oligosaccharides.

Various studies have shown that functional oligosaccharides can ease intestinal injury and treat intestinal diseases by maintaining and repairing intestinal barriers [21–23]. As of April 2022, there were 112 registered clinical trials (data from ClinicalTrials.gov, accessed on 1 June 2022) related to assessing beneficial effects of functional oligosaccharides on human health, including 28 IBS-related studies and 33 IBD-related studies (Some details of the studies are shown in Table 1).

It is well accepted that functional oligosaccharides such as raffinose oligosaccharide (ROS) [24], FOS, and GOS [25] can affect specific groups of the microbial community in vitro and in vivo to promote their growth and metabolic activity, thereby maintaining host gut health benefits [26]. In addition, functional oligosaccharides are also considered to interact directly with the host and exert local positive effects on inflammation and barrier function by regulating immunity and intestinal epithelial cell signal transduction [27]. Functional oligosaccharides have different effects on the host intestine due to their different monosaccharide composition, degree of polymerization, and linkage types [28,29]. A number of studies have been carried out regarding the activity of functional oligosaccharides affecting intestinal barrier function. This review focuses on the latest research on functional oligosaccharides and their effects on intestinal health, especially their interaction with intestinal flora, immunity, and disease treatment.

Table 1. Clinical study of common functional oligosaccharides in intestine-related diseases.

Functional Oligosaccharides	Study Title	Year	Conditions	Interventions	Actual Enrollment
FOS	Dietary Treatment of Crohn's Disease	2006–2021	Crohn's disease, inflammatory bowel disease	drug: active fructo-oligosaccharide drug: placebo fructo-oligosaccharide	73
scFOS	Effects of scFOS on Stool Frequency in People With Functionnal Constipation	2013–2018	functional constipation	dietary supplement: short-chain fructo-oligosaccharides dietary supplement: maltodextrin	120
IMO	Prebiotic Effects of Isomalto-oligosaccharide	2015–2017	intestinal microbiota,	dietary supplement: isomalto-oligosaccharide	54
GOS	GOS to Reduce Symptom Severity in IBS (EGIS)	2021–	irritable bowel syndrome, irritable bowel syndrome—constipation, irritable bowel syndrome—diarrhoea, irritable bowel syndrome—mixed	dietary supplement: galactooligosaccharides (GOS) dietary supplement: maltodextrine	210
HMO	Human Milk Oligosaccharides (HMOs) for Irritable Bowel Syndrome (IBS) (HIBS)	2022–	irritable bowel syndrome, IBS—irritable bowel syndrome	dietary supplement: human milk oligosaccharide mix other: placebo	500

2. Effects of Functional Oligosaccharides on the Intestinal Barrier

2.1. Biological Barriers

Functional oligosaccharides can be selectively fermented into short-chain fatty acids (SCFA) in the gut [30] to maintain intestinal function and health by controlling the growth of pathogenic microorganisms, reducing pH, preventing peptide degradation, and the formation of toxic compounds [31,32]. Functional oligosaccharides can be used directly by the microbiota as a carbon source. Furthermore, some studies have found that functional oligosaccharides such as inulin-derived FOS can also increase the colonization sites of probiotics in the intestinal tract [33]. Our previous studies also found that COS promoted the growth, metabolic activity, and metabolite concentration changes of probiotics represented by *Akkermansia muciniphila* by affecting specific populations in microbial groups; reduced the adhesion, invasion, and colonization of intestinal pathogens represented by *Escherichia coli*; and inhibited the occurrence and development of enteritis, thereby maintaining intestinal health [34,35].

There is a correspondence between functional oligosaccharides and probiotics. Functional oligosaccharides exhibit a complex degree of polymerization and glycosidic bonding [36,37], and probiotics utilize functional oligosaccharides with a diversity of transporter proteins and glycosidic hydrolases [38,39]. Therefore, the growth promotion effects of functional oligosaccharides on probiotics are species-specific. For example, butyrate-producing strains showed different growth curves in the presence of FOS, GOS, and XOS [40,41]. The same kind of FOS, due to their different sources, have different effects on the growth of probiotics. Studies have shown that inulin FOS have more noticeable effects on the growth of *Bifidobacterium* than sucrose FOS. The molecular mechanism of the metabolism of FOS, GOS, and milk-derived oligosaccharides by probiotics has also been studied, and the unique intake mechanism for functional oligosaccharides plays an active role. A brief summary of the coincidence relationship between common functional oligosaccharides and probiotics is shown in Table 2.

Table 2. Sources of common functional oligosaccharides and their mode of action with probiotics.

Functional Oligosaccharide	Source	Composition	Advantage Probiotics	Transport Pathway	References
GOS	human milk, cow's milk	monosaccharide and number: glucose 1, galactose 2–5; connection mode: β-1,4, β-1,6	*B. adolescentis, B. bifidum, B. longum, B. infantis, B. breve, B. animalis, B. catenulatum; L. reuteri, L. plantarum, L. paracasei, L. agili, L. fermentium, L. acidophilus, L. salivarius, L. casei, L. rhamnosus, L. bulgaricus, L. delbrueckii, Lactobacillus johnsonii, Lactobacillus gasseri; S. thermophilus*	LacEF, LacA, LacS, ABC, GPH, LacL, LacM	[36,42–46]
FOS	fruits, vegetables, honey, Jerusalem artichoke, cicory	monosaccharide and number: glucose 1, fructose 2–4; connection mode: α-1,2, β-1,2	*B. adolescentis, B. longum, B. breve, B. animalis, B. infantis, B. pseudolongum; L. reuteri, L. acidophilus, L. salivarius, L. plantarum, L. fermentium, L. casei, L. bulgaricus; Clostridium, Streptococcus, Coprococcus, Enterococcus*	PTS, ABC, MFS, LacS	[43–49]
IMO	corn steep liquor, honey, sugar cane juice	monosaccharide and number: glucose 2–5; connection mode: at least 1 α-1,6	*B. animalis, B. adolescentis, B. bifidum, B. longum, B. infantis, B. breve; L. plantarum, L. rhamnosus, L. paracasei, L. agilis, L. acidophilus, L. reuteri, L. lactic, L. delbrueckii, L. casei; S. lactic, S. thermophilus*	ABC, MalEFG-MsmK, PTS, MFS, MIP	[50–52]
XOS	birch, corncob, straw, bamboo	monosaccharide and number: xylose 2–7; connection mode: β-1,4	*B. adolescentis, B. longum, B. breve, B. animalis, B. catenulatum, B. pseudocatenulatum, B. thermophilum; L. plantarum, L. brevis, L. rhamnosus, L. fermentium, L. acidophilus, L. salivarius, L. casei, L. crispatus, L. lactis, L. mucosae, L. sakei, L. zeae, L. reuteri; Enterococcus faecalis and Enterococcus faecium*	ABC, MFS	[53–55]
COS	shrimp and crab shell, fungal cell wall	monosaccharide and number: N-acetyl-D-glucosamine 2–20; connection mode: β-1,4	*B. bifidium; L. brevis, L. casei, L. acidophilus; Akkermansia, S. thermophilus*	CsnEFG, SBP, PTS, ABC	[34,56–59]

Table 2. *Cont.*

Functional Oligosaccharide	Source	Composition	Advantage Probiotics	Transport Pathway	References
HMO	breast milk, amniotic fluid	monosaccharide and number: glucose, N-acetyl-D-glucosamine, galactose, fucose, N-acetylneuraminic acid; connection mode: α-1,2, α-1,3, α-1,4, α-2,3, α-2,6	*B. infantis*, *B. longum*, *B. breve*, *B. bifidum*; *L. acidophilus*; *Bacteroides fragilis*, *Bacteroides vulgatus*, *Bacteroides thetaiotaomicron*	ABC	[60–63]

Based on previous and our own research on functional oligosaccharides affecting proliferation and colonization of probiotics and considering the specificity and complexity of the interaction between intestinal flora and functional oligosaccharides, it is critical to further study the effects of functional oligosaccharides with different structures on the changes in intestinal metabolites, bacterial gene expression, and potential molecular mechanisms in maintaining intestinal barrier function.

2.2. Immune Barrier

Functional oligosaccharide plays a positive role in the intestinal immune barrier. Indirectly, functional oligosaccharides can be fermented by probiotic to produce SCFA, which regulate the activity of T cells, B cells, and dendritic cells [14,64]. For example, oral administration of FOS increased the level of SCFA, including butyrate, which increased the level of regulatory T cells in the mesenteric lymph nodes of mice [65,66]. In addition, some functional oligosaccharides have also been found to directly act on intestinal-associated immune cells and immune factors, providing beneficial effects on intestinal diseases such as allergies or IBD [67,68]. Specifically, functional oligosaccharides can stimulate Toll-like receptors and induce the differentiation of immune cells represented by T and B cells to regulate intestinal immunity [67,68]. Functional oligosaccharides also regulate the secretion of inflammatory factors represented by IFN-γ, IL-5, and IL-6 in the intestine and increase the content of immunoglobulin represented by IgA, IgM, and IgG. For example, some studies have found that FOS and GOS act as TLR4 agonists in intestinal epithelial cells; activating the TLR4-NF-κB pathway; and reducing pro-inflammatory factors such as IL-12p35, IL-8, and TNF-α [69]. FOS and arabinogalactan oligosaccharides regulate the immune-related parameters in GALT, secondary lymphoid tissue, and peripheral circulation [70]. We summarize the regulatory effects of different functional oligosaccharides on the intestinal immune barrier in Figure 3 and Table 3.

Table 3. The mode of action of common functional oligosaccharides on the intestinal immune barrier.

Functional Oligosaccharides	Immune Cells	Immune Factors	References
GOS	NK cells, T cells, phagocytes	increase IgA, IgM, IL-8, IL-10, IFN-γ; decrease IL-6, IL-18, IL-13, IL-33	[71]
FOS	B cells, T cells, macrophages, leukocytes	increase IgG, IgE, IFN-γ, IL-10; decrease IL-5, IL-6	[72–75]
IMO	T cells, phagocytes	increase lysozyme, IgE, IgG, IgA, IgM, IL-2, IFN-γ; decrease IL-5, IL-6, IL-13	[76–78]
XOS	B cells, T cells, NK cells, macrophages	increase IgG, IgA, IgM; decrease TLR2	[79]

Table 3. *Cont.*

Functional Oligosaccharides	Immune Cells	Immune Factors	References
COS	macrophages	increase CCL20, IgA, MHCII, TGF-131, pIgR; decrease CCL15, CCL25, ICL25, IL-1β, IL-4, IL-6, IL-8, IL-13, TNF-α	[80–83]
HMO	macrophages, T cell	increase INF-γ, IL-10 decrease IL-4, IL-6, IL-8, TNF-α, IL-1β, GM-CSF2, IL-17C, PF4, CXCL1, CCL20	[84–89]

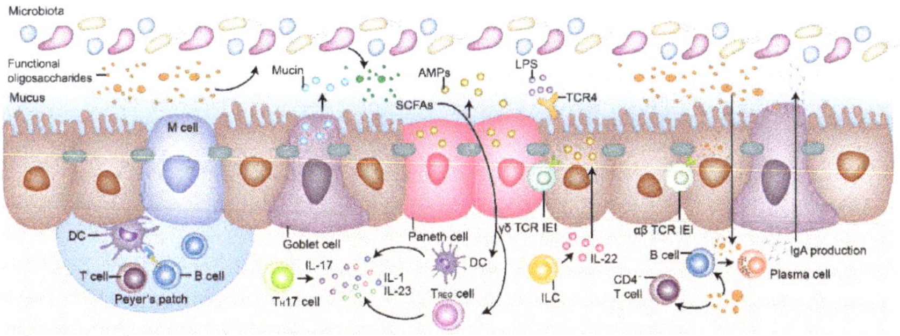

Figure 3. Schematic diagram of the modulating effect of functional oligosaccharides on the intestinal immune barrier. SCFAs: short chain fatty acids; AMPs: antimicrobial peptides; LPS: lipopolysaccharide; M cell: membranous/microfold cell; DC: dendritic cell; T cell: T-lymphocyte cell; B cell: B-lymphocyte cell; Th17 cell: T helper cell 17; IL-1/17/22/23: interleukin 1/17/22/23; Treg cell: regulatory T cell; TCR: T cell receptor; ILC: innate lymphoid cells; CD4+ T cell: cluster of differentiation 4 T cell; IgA: immunoglobulin A.

2.3. Epithelial and Mucus Barrier

The human gastrointestinal tract has no relevant enzyme system to hydrolyze functional oligosaccharides [90,91]. However, functional oligosaccharides exhibit excellent benefits in the composition and maintenance of intestinal epithelium, either directly or indirectly. It is well accepted that functional oligosaccharides are utilized by gut microbes [92,93] to produce metabolites such as short-chain fatty acids (SCFA), which regulate host cell growth, differentiation, apoptosis, and physiological functions in the intestine [94,95]. Moreover, recent evidence suggests that functional oligosaccharides such as COS, GOS, and cello-oligosaccharides could directly affect the permeability and integrity of intestinal epithelial cells by improving colonic epithelial cell transmembrane resistance and reducing intestinal epithelial cell permeability to fluorescein isothiocyanate-glucan [96,97]. Studies show that functional oligosaccharides could upregulate the expression of specific tight junction proteins of epithelial cells [98]. The mechanism of functional oligosaccharides regulating intestinal epithelial cell homeostasis has not been fully explored.

Functional oligosaccharides can also affect the production of mucin and antimicrobial peptides by host cells [97,99]. For example, feeding 1 g/d of GOS to rats with severe pancreatitis can significantly improve their mucus defects [100]. This improvement effect is related to the structure of functional oligosaccharides and the dose of the functional oligosaccharides used. However, only a few studies considered exploring the protective effect of a functional oligosaccharide dose on intestinal mucus barrier function. The study noted that berberine promoted the proliferation of *Akkermansia* in a dose- and time-dependent manner in mice, with 300 mg/kg of berberine showing a two-fold higher proliferation rate than 200 mg/kg. The investigators also demonstrated that this proliferation works by promoting the secretion of mucins, especially mucin-2 [101]. The structure-function relationship and action mechanism of functional oligosaccharides need to be further analyzed and evaluated.

3. Application of Functional Oligosaccharides in Intestinal Diseases

3.1. Colorectal Cancer

The colon environment, including imbalanced intestinal microflora and mutations in the Wnt signaling pathway are the leading causes of CRC [102,103]. The current treatments for CRC include chemotherapy, radiotherapy, and surgery, but most of them are accompanied by high-risk complications, and the success rate is limited. Therefore, new early treatment strategies are needed [104]. The use of functional oligosaccharides in preventing CRC may be promising. Studies show FOS and GOS can reduce the severity of colon cancer in rats and mice induced by 1,2-dimethylhydrazine by reducing the number of colon ACF [105–107]. Researchers have found that low-degree FOS are more effective in treating early colon cancer in mice induced by DMH [108] and significantly reducing the risk of colon cancer in animal models [109]. There are two aspects regarding the inhibitory effect of functional oligosaccharides on colorectal cancer. First, functional oligosaccharides affect the homeostasis of intestinal microflora by promoting the growth and colonization of intestinal probiotics and upregulating production of metabolites such as SCFA, which inhibit the proliferation and differentiation of colon tumor cells [104,110] and regulates exogenous metabolic enzymes that stimulate the activation and metabolism of carcinogens [111,112]. Furthermore, functional oligosaccharides directly regulate the functions of intestinal GALT and other immune cells, influence gene expression levels of cancer cells, and promote cancer cell apoptosis [109].

The clinical data also show that functional oligosaccharides have a positive effect on the immunological indexes of colon cancer and microbial flora abundance [113]. However, some clinical data point out that functional oligosaccharides do not significantly reduce the mortality of colorectal cancer in women after menopause [102,113]. There is no clear explanation for the structure-activity relationship, dosage, and individual differences of functional oligosaccharides, which may also be the main reason for restricting the clinical trials of functional oligosaccharides in the treatment of colorectal cancer. Therefore, the clinical treatment of CRC with functional oligosaccharides remains unconfirmed. Consequently, research on new technologies such as combining probiotics and functional oligosaccharides as targeted therapeutic agents for colon cancer based on host–guest chemistry is also an aspect worth exploring [114].

3.2. Inflammatory Bowel Disease

IBD is a chronic nonspecific gastrointestinal inflammatory disease that destroys the intestinal mucosal structure and floral balance, leading to abnormal systemic biochemical indexes [115]. The etiology of inflammatory bowel disease is not clear, while comprehensive factors such as intestinal flora, immunity, environment, and gene susceptibility might be involved.

The DSS-induced mouse colitis model is one of the widely recognized models for studying the pathogenesis of IBD and evaluating potential therapeutic methods [116]. Growing evidence supports the potential of functional oligosaccharides to treat inflammatory diseases, including colitis. FOS and GOS in vitro affect immunity by binding to TLR on monocytes, macrophages, and intestinal epithelial cells and regulating cytokine production and immune cell maturation [69,117–120]. In addition, animal models and clinical studies have shown that functional oligosaccharides reduce the intestinal inflammatory response and IBD symptoms [121,122]. A clinical study focused on enteritis after abdominal radiotherapy (RT) found that FOS supplementation in patients' daily diet can stimulate the proliferation of *Lactobacilus* and *Bifidobacterium*, thereby repairing intestinal mucosal damage during RT and preventing the occurrence and development of IBD [123]. Our previous studies have found that COS treatment upregulates the expression of occludin in the proximal colon of diabetic mice [34], alleviates DSS-induced mucosal defects in IBD, and protects the intestinal mucosal barrier function of ulcerative colitis mice [97].

Future studies need to understand how functional oligosaccharides regulate the disease-related signaling pathways, drive different cellular processes and regulate intestinal functions, and conduct the mechanism of functional oligosaccharides as a drug adjuvant or substitute in the treatment of IBD.

3.3. Irritable Bowel Syndrome

IBS is a chronic disease affected by stress and eating habits. It is characterized by abdominal pain, mucosal and immune functions, and changes in the intestinal microbial structure. Dietary patterns, the intestinal microbial structure, inflammatory response, and other factors can aggravate the symptoms of irritable bowel syndrome. Dietary interventions are recommended to control the disease due to the efficacy and tolerance of common drug treatments.

Evidence shows that the ecological imbalance of intestinal and mucosal colon microflora in IBS is usually characterized by the reduction of the *Bifidobacterium* species [124–127]. Some studies have found that supplementing probiotics to regulate intestinal microflora are effective in treating IBS [128,129]. Some clinical studies have also found that low-dose functional oligosaccharides, such as FOS, can alleviate the symptoms of IBS patients through increasing the concentration of SCFA [130]. In contrast, a low FODMAP diet has gradually become the standard method for the treatment of IBS worldwide. This method can alleviate the clinical symptoms of IBS patients by limiting the daily intake of short-chain fermentable carbohydrates (low fermentable oligosaccharides, disaccharides, monosaccharides, and polyols (FODMAP)). Studies have consistently proven the clinical efficacy of a low FODMAP diet in patients with IBS [130]. In fact, the low FODMAP diet has clinical efficacy, but it reduces the abundance of intestinal *Bifidobacterium*, which is not conducive to the thorough treatment of IBS patients. In view of the pathogenic factors and pathogenesis of IBS and the complexity and diversity of individual microbial communities, we need to consider these two interventions for further research and consider individualized diagnoses according to clinical symptoms.

4. Application Prospect of Functional Oligosaccharides in the Intestinal Tract

Glycans generally have complex monosaccharide composition, glycosidic bond type and degree of polymerization, and their structural complexity is much higher than that of proteins and nucleic acids. In the past decade, glycoscience, with the support of governments, has made a lot of progress, and it has revealed the role of glycans in inflammatory responses and immune system regulation, cardiovascular diseases, intestinal diseases, and cancers. A variety of functional oligosaccharides have shown kinds of activities in intestinal barrier protection and repair and have demonstrated the great promise of glycans in intestinal disease treatment. However, the structure-activity relationship and molecular mechanism have not been fully elucidated. Glycan-based products used in related research are often a mixture of glycans with slightly different structural characters and subject to variations in different preparation methods and raw material sources. Recent studies have shown that small changes in the structure of glycans have significant effects on their activities, so the accurate analysis and preparation of glycan products is the key to clarify their structure-activity relationships and develop functional glycan products, for example, structural analysis and pharmacokinetic study of glycans by liquid chromatography-tandem mass spectrometry (LC-MS/MS) [131]. On the other hand, based on the different activities of different sugar chains, the combination of several different glycans have received more attention and applications [64,132,133]. In the face of the complex microbial and host environment in the intestinal tract, products containing multiple different glycans will play a greater role. However, there is a lack of in-depth research on the compounding mechanism and synergistic effect of multiple glycan recipes.

Author Contributions: Introduction, T.X. and R.S.; literature collection, T.X., R.S., Y.Z., C.Z. and Y.W.; writing—original draft preparation, T.X. and R.S.; writing—review and editing, T.X. and R.S.; supervision, Z.A.W. and Y.D. All authors have read and agreed to the published version of the manuscript.

Funding: National Key R&D Program of China grant number 2019YFD0902000.

Institutional Review Board Statement: Not applicable.

Informed Consent Statement: Not applicable.

Conflicts of Interest: The authors declare no conflict of interest.

References

1. Chow, D.K.L.; Leong, R.W.L.; Tsoi, K.K.F.; Ng, S.S.M.; Leung, W.K.; Wu, J.C.Y.; Wong, V.W.S.; Chan, F.K.L.; Sung, J.J.Y. Long-term Follow-up of Ulcerative Colitis in the Chinese Population. *Am. J. Gastroenterol.* **2009**, *104*, 647–654. [PubMed]
2. Kaplan, G.G. The global burden of IBD: From 2015 to 2025. *Nat. Rev. Gastroenterol. Hepatol.* **2015**, *12*, 720–727. [CrossRef] [PubMed]
3. Ng, S.C.; Shi, H.Y.; Hamidi, N.; Underwood, F.E.; Tang, W.; Benchimol, E.I.; Panaccione, R.; Ghosh, S.; Wu, J.C.Y.; Chan, F.K.L.; et al. Worldwide incidence and prevalence of inflammatory bowel disease in the 21st century: A systematic review of population-based studies. *Lancet* **2017**, *390*, 2769–2778. [CrossRef]
4. Ananthakrishnan, A.N.; Kaplan, G.G.; Ng, S.C. Changing Global Epidemiology of Inflammatory Bowel Diseases: Sustaining Health Care Delivery Into the 21st Century. *Clin. Gastroenterol. Hepatol.* **2020**, *18*, 1252–1260. [CrossRef] [PubMed]
5. Victoria, C.R.; Sassak, L.Y.; de Carvalho Nunes, H.R. Incidence and prevalence rates of inflammatory bowel diseases, in midwestern of Sao Paulo State, Brazil. *Arq. Gastroenterol.* **2009**, *46*, 20–25. [CrossRef] [PubMed]
6. Marynczak, K.; Wlodarczyk, J.; Sabatowska, Z.; Dziki, A.; Dziki, L.; Wlodarczyk, M. Colitis-Associated Colorectal Cancer in Patients with Inflammatory Bowel Diseases in a Tertiary Referral Center: A Propensity Score Matching Analysis. *J. Clin. Med.* **2022**, *11*, 866. [CrossRef]
7. Siegel, R.L.; Miller, K.D.; Goding Sauer, A.; Fedewa, S.A.; Butterly, L.F.; Anderson, J.C.; Cercek, A.; Smith, R.A.; Jemal, A. Colorectal cancer statistics, 2020. *CA Cancer J. Clin.* **2020**, *70*, 145–164. [CrossRef]
8. Zhang, Y.; Li, J.X.; Zhang, Y.; Wang, Y.L. Intestinal microbiota participates in nonalcoholic fatty liver disease progression by affecting intestinal homeostasis. *World J. Clin. Cases* **2021**, *9*, 6654–6662. [CrossRef]
9. Lai, Y.J.; Masatoshi, H.; Ma, Y.B.; Guo, Y.M.; Zhang, B.K. Role of Vitamin K in Intestinal Health. *Front. Immunol.* **2022**, *12*, 791565. [CrossRef]
10. Zhou, B.L.; Yuan, Y.T.; Zhang, S.S.; Guo, C.; Li, X.L.; Li, G.Y.; Xiong, W.; Zeng, Z.Y. Intestinal Flora and Disease Mutually Shape the Regional Immune System in the Intestinal Tract. *Front. Immunol.* **2020**, *11*, 575. [CrossRef]
11. Turner, J.R. Intestinal mucosal barrier function in health and disease. *Nat. Rev. Immunol.* **2009**, *9*, 799–809. [CrossRef] [PubMed]
12. Di Tommaso, N.; Gasbarrini, A.; Ponziani, F.R. Intestinal Barrier in Human Health and Disease. *Int. J. Environ. Res. Public Health* **2021**, *18*, 12836. [CrossRef] [PubMed]
13. Hao, W.; Hao, C.; Wu, C.; Xu, Y.; Jin, C. Aluminum induced intestinal dysfunction via mechanical, immune, chemical and biological barriers. *Chemosphere* **2022**, *288*, 132556. [CrossRef] [PubMed]
14. Chelakkot, C.; Ghim, J.; Ryu, S.H. Mechanisms regulating intestinal barrier integrity and its pathological implications. *Exp. Mol. Med.* **2018**, *50*, 1–9. [CrossRef] [PubMed]
15. Magen, R.; Shaoul, R. Alternative & complementary treatment for pediatric inflammatory bowel disease. *Transl. Pediatr.* **2019**, *8*, 428–435. [PubMed]
16. Delaney, C.P.; Fazio, V.W. Crohn's Disease of the Small Bowel. *Surg. Clin. N. Am.* **2001**, *81*, 137–158. [CrossRef]
17. Na, S.Y.; Moon, W. Perspectives on Current and Novel Treatments for Inflammatory Bowel Disease. *Gut Liver* **2019**, *13*, 604–616. [CrossRef]
18. Collij, V.; Festen, E.A.M.; Alberts, R.; Weersma, R.K. Drug Repositioning in Inflammatory Bowel Disease Based on Genetic Information. *Inflamm. Bowel Dis.* **2016**, *22*, 2562–2570. [CrossRef]
19. Salminen, S.; Collado, M.C.; Endo, A.; Hill, C.; Lebeer, S.; Quigley, E.M.M.; Sanders, M.E.; Shamir, R.; Swann, J.R.; Szajewska, H.; et al. The International Scientific Association of Probiotics and Prebiotics (ISAPP) consensus statement on the definition and scope of postbiotics. *Nat. Rev. Gastroenterol. Hepatol.* **2021**, *18*, 671.
20. Bindels, L.B.; Delzenne, N.M.; Cani, P.D.; Walter, J. Towards a more comprehensive concept for prebiotics. *Nat. Rev. Gastroenterol. Hepatol.* **2015**, *12*, 303–310. [CrossRef]
21. Langen, M.; Dieleman, L.A. Prebiotics in Chronic Intestinal Inflammation. *Inflamm. Bowel Dis.* **2009**, *15*, 454–462. [CrossRef] [PubMed]
22. Szilagyi, A. Use of prebiotics for inflammatory bowel disease. *Can. J. Gastroenterol. Hepatol.* **2005**, *19*, 505–510. [CrossRef]
23. Li, P.H.; Lu, W.C.; Chan, Y.J.; Zhao, Y.P.; Nie, X.B.; Jiang, C.X.; Ji, Y.X. Feasibility of Using Seaweed (*Gracilaria coronopifolia*) Synbiotic as a Bioactive Material for Intestinal Health. *Foods* **2019**, *8*, 623. [CrossRef] [PubMed]

24. Liang, Y.; Wang, Y.; Wen, P.; Chen, Y.; Ouyang, D.; Wang, D.; Zhang, B.; Deng, J.; Chen, Y.; Sun, Y.; et al. The Anti-Constipation Effects of Raffino-Oligosaccharide on Gut Function in Mice Using Neurotransmitter Analyses, 16S rRNA Sequencing and Targeted Screening. *Molecules* **2022**, *27*, 2235. [CrossRef] [PubMed]
25. Raouani, N.E.H.; Claverie, E.; Randoux, B.; Chaveriat, L.; Yaseen, Y.; Yada, B.; Martin, P.; Cabrera, J.C.; Jacques, P.; Reignault, P.; et al. Bio-Inspired Rhamnolipids, Cyclic Lipopeptides and a Chito-Oligosaccharide Confer Protection against Wheat Powdery Mildew and Inhibit Conidia Germination. *Molecules* **2022**, *27*, 6672. [CrossRef]
26. Bamigbade, G.B.; Subhash, A.J.; Kamal-Eldin, A.; Nyström, L.; Ayyash, M. An Updated Review on Prebiotics: Insights on Potentials of Food Seeds Waste as Source of Potential Prebiotics. *Molecules* **2022**, *27*, 5947. [CrossRef]
27. Cunningham, M.; Azcarate-Peril, M.A.; Barnard, A.; Benoit, V.; Grimaldi, R.; Guyonnet, D.; Holscher, H.D.; Hunter, K.; Manurung, S.; Obis, D.; et al. Shaping the Future of Probiotics and Prebiotics. *Trends Microbiol.* **2021**, *29*, 667–685. [CrossRef]
28. Akbari, P.; Fink-Gremmels, J.; Willems, R.; Difilippo, E.; Schols, H.A.; Schoterman, M.H.C.; Garssen, J.; Braber, S. Characterizing microbiota-independent effects of oligosaccharides on intestinal epithelial cells: Insight into the role of structure and size. *Eur. J. Nutr.* **2017**, *56*, 1919–1930. [CrossRef]
29. Li, W.; Wang, K.Q.; Sun, Y.; Ye, H.; Hu, B.; Zeng, X.X. Influences of structures of galactooligosaccharides and fructooligosaccharides on the fermentation in vitro by human intestinal microbiota. *J. Funct. Foods* **2015**, *13*, 158–168. [CrossRef]
30. Van Loo, J. The specificity of the interaction with intestinal bacterial fermentation by prebiotics determines their physiological efficacy. *Nutr. Res. Rev.* **2004**, *17*, 89–98. [CrossRef]
31. Cummings, J.H.; Macfarlane, G.T. The Control and Consequences of Bacterial Fermentation in the Human Colon. *J. Appl. Bacteriol.* **1991**, *70*, 443–459. [CrossRef] [PubMed]
32. Jarrett, S.; Ashworth, C.J. The role of dietary fibre in pig production, with a particular emphasis on reproduction. *J. Anim. Sci. Biotechnol.* **2018**, *9*, 783–793. [CrossRef] [PubMed]
33. Edogawa, S.; Peters, S.A.; Jenkins, G.D.; Gurunathan, S.V.; Sundt, W.J.; Johnson, S.; Lennon, R.J.; Dyer, R.B.; Camilleri, M.; Kashyap, P.C.; et al. Sex differences in NSAID-induced perturbation of human intestinal barrier function and microbiota. *FASEB J.* **2018**, *32*, 6615–6625. [CrossRef] [PubMed]
34. Zhang, C.; Jiao, S.M.; Wang, Z.A.; Du, Y.G. Exploring Effects of Chitosan Oligosaccharides on Mice Gut Microbiota in in vitro Fermentation and Animal Model. *Front. Microbiol.* **2018**, *9*, 2388. [CrossRef] [PubMed]
35. Jing, B.; Xia, K.; Zhang, C.; Jiao, S.; Zhu, L.; Wei, J.; Wang, Z.A.; Chen, N.; Tu, P.; Li, J.; et al. Chitosan Oligosaccharides Regulate the Occurrence and Development of Enteritis in a Human Gut-On-a-Chip. *Front. Cell Dev. Biol.* **2022**, *10*, 877892. [CrossRef]
36. Thongaram, T.; Hoeflinger, J.L.; Chow, J.; Miller, M.J.; Chemistry, F. Prebiotic Galactooligosaccharide Metabolism by Probiotic Lactobacilli and Bifidobacteria. *J. Agric. Food Chem.* **2017**, *65*, 4184–4192. [CrossRef]
37. Wang, S.; Pan, J.H.; Zhang, Z.S.; Yan, X.B. Investigation of dietary fructooligosaccharides from different production methods: Interpreting the impact of compositions on probiotic metabolism and growth. *J. Funct. Foods* **2020**, *69*, 103955. [CrossRef]
38. Murtini, D.; Aryantini, N.P.; Sujaya, I.N.; Urashima, T.; Fukuda, K. Effects of prebiotic oligosaccharides consumption on the growth and expression profile of cell surface-associated proteins of a potential probiotic Lactobacillus rhamnosus FSMM15. *Biosci. Microbiota. Food Health* **2016**, *35*, 41–49. [CrossRef]
39. Theilmann, M.C.; Fredslund, F.; Svensson, B.; Lo Leggio, L.; Abou Hachem, M. Substrate preference of an ABC importer corresponds to selective growth on β-(1,6)-galactosides in Bifidobacterium animalis subsp. lactis. *J. Biol. Chem.* **2019**, *294*, 11701–11711. [CrossRef]
40. Rawi, M.H.; Zaman, S.A.; Pa'ee, K.F.; Leong, S.S.; Sarbini, S.R. Prebiotics metabolism by gut-isolated probiotics. *J. Food Sci. Technol.* **2020**, *57*, 2786–2799. [CrossRef]
41. Scott, K.P.; Grimaldi, R.; Cunningham, M.; Sarbini, S.R.; Wijeyesekera, A.; Tang, M.L.K.; Lee, J.C.Y.; Yau, Y.F.; Ansell, J.; Theis, S.; et al. Developments in understanding and applying prebiotics in research and practice—An ISAPP conference paper. *J. Appl. Microbiol.* **2020**, *128*, 934–949. [CrossRef] [PubMed]
42. Kittibunchakul, S.; Maisc Hb Erger, T.; Domig, K.; Kneifel, W.; Nguyen, H.M.; Haltrich, D.; Nguyen, T.H.J.M. Fermentability of a Novel Galacto-Oligosaccharide Mixture by *Lactobacillus* spp. and *Bifidobacterium* spp. *Molecules* **2018**, *23*, 3352. [CrossRef] [PubMed]
43. Hernandez-Hernandez, O.; Muthaiyan, A.; Moreno, F.J.; Montilla, A.; Sanz, M.L.; Ricke, S.C. Effect of prebiotic carbohydrates on the growth and tolerance of Lactobacillus. *Food Microbiol.* **2012**, *30*, 355–361. [CrossRef]
44. Maathuis, A.; Van, D.; Schoterman, M.; Venema, K. Galacto-oligosaccharides have prebiotic activity in a dynamic in vitro colon model using a ^{13}C-labeling technique. *J. Nutr.* **2012**, *142*, 1205–1212. [CrossRef] [PubMed]
45. Fischer, C.; Kleinschmidt, T. Valorisation of sweet whey by fermentation with mixed yoghurt starter cultures with focus on galactooligosaccharide synthesis. *Int. Dairy J.* **2021**, *119*, 105068. [CrossRef]
46. Richards, P.J.; Lafontaine, G.; Connerton, P.L.; Liang, L.; Connerton, I.F. Galacto-Oligosaccharides Modulate the Juvenile Gut Microbiome and Innate Immunity To Improve Broiler Chicken Performance. *mSystems* **2020**, *5*, e00827-19. [CrossRef]
47. Kaplan, H.; Hutkins, R.W. Fermentation of Fructooligosaccharides by Lactic Acid Bacteria and Bifidobacteria. *Appl. Environ. Microbiol.* **2000**, *66*, 2682–2684. [CrossRef]
48. Mandadzhieva, T.; Ignatova-Ivanova, T.; Kambarev, S.; Iliev, I.; Ivanova, I. Utilization of different prebiotics by *Lactobacillus* spp. and *Lactococcus* spp. *Biotechnol. Biotechnol. Equip.* **2011**, *25*, 117–120. [CrossRef]

49. Parhi, P.; Song, K.P.; Choo, W.S. Growth and survival of Bifidobacterium breve and Bifidobacterium longum in various sugar systems with fructooligosaccharide supplementation. *J. Food Sci. Technol.* **2022**, *59*, 3775–3786. [CrossRef]
50. Hu, Y.; Ketabi, A.; Buchko, A.; Ganzle, M.G. Metabolism of isomalto-oligosaccharides by Lactobacillus reuteri and bifidobacteria. *Lett. Appl. Microbiol.* **2013**, *57*, 108–114. [CrossRef]
51. Lu, K.; Zhao, X.H. Yields of three acids during simulated fermentation of inulin and xylo-oligosaccharides enhanced by six exogenous strains. *J. Food Meas. Charact.* **2016**, *11*, 696–703.
52. Chockchaisawasdee, S.; Stathopoulos, C.E.; Research, N. Viability of Streptococcus thermophilus, Lactobacillus delbrueckii ssp bulgaricus, Lactobacillus acidophilus and Lactobacillus casei in fermented milk supplemented with isomalto-oligosaccharides derived from banana flour. *J. Food Nutr. Res.* **2011**, *50*, 125–132.
53. Moura, P.; Barata, R.; Carvalheiro, F.; Gírio, F.; Loureiro-Dias, M.C.; Esteves, M.P. In vitro fermentation of xylo-oligosaccharides from corn cobs autohydrolysis by Bifidobacterium and Lactobacillus strains. *Food Sci. Technol.* **2007**, *40*, 963–972. [CrossRef]
54. Maria, A.; Margarita, T.; Iilia, I.; Iskra, I.J.B.; Equipment, B. Gene expression of enzymes involved in utilization of xylooligosaccharides by Lactobacillus strains. *Biotechnol. Biotechnol. Equip.* **2014**, *28*, 941–948. [CrossRef] [PubMed]
55. Mao, B.; Gu, J.; Li, D.; Cui, S.; Zhao, J.; Zhang, H.; Chen, W. Effects of Different Doses of Fructooligosaccharides (FOS) on the Composition of Mice Fecal Microbiota, Especially the Bifidobacterium Composition. *Nutrients* **2018**, *10*, 1105. [CrossRef]
56. Lee, H.W.; Park, Y.S.; Jung, J.S.; Shin, W.S. Chitosan oligosaccharides, dp 2–8, have prebiotic effect on the *Bifidobacterium bifidium* and *Lactobacillus* sp. *Anaerobe* **2002**, *8*, 319–324. [CrossRef]
57. Fukamizo, T.; Kitaoku, Y.; Suginta, W.J.I. Periplasmic solute-binding proteins: Structure classification and chitooligosaccharide recognition. *Int. J. Biol. Macromol.* **2019**, *128*, 985–993. [CrossRef]
58. Berg, T.; Schild, S.; Reidl, J. Regulation of the chitobiose-phosphotransferase system in Vibrio cholerae. *Arch. Microbiol.* **2007**, *187*, 433. [CrossRef]
59. Zheng, X.; Zhang, X.; Xiong, C. Effects of chitosan oligosaccharide-nisin conjugates formed by Maillard reaction on the intestinal microbiota of high-fat diet-induced obesity mice model. *Food Qual. Saf.* **2019**, *3*, 169–177. [CrossRef]
60. Bai, Y.; Tao, J.; Zhou, J.; Fan, Q.; Liu, M.; Hu, Y.; Xu, Y.; Zhang, L.; Yuan, J.; Li, W.; et al. Fucosylated Human Milk Oligosaccharides and N-Glycans in the Milk of Chinese Mothers Regulate the Gut Microbiome of Their Breast-Fed Infants during Different Lactation Stages. *mSystems* **2018**, *3*, e00206-18. [CrossRef]
61. Thongaram, T.; Hoeflinger, J.L.; Chow, J.M.; Miller, M.J. Human milk oligosaccharide consumption by probiotic and human-associated bifidobacteria and lactobacilli. *J. Dairy Sci.* **2017**, *100*, 7825–7833. [CrossRef] [PubMed]
62. Wang, J.; Chen, C.; Yu, Z.; He, Y.; Yong, Q.; Newburg, D.S. Relative fermentation of oligosaccharides from human milk and plants by gut microbes. *Eur. Food Res. Technol.* **2017**, *243*, 133–146. [CrossRef]
63. Salli, K.; Hirvonen, J.; Siitonen, J.; Ahonen, I.; Maukonen, J.J.; Chemistry, F. Selective Utilization of the Human Milk Oligosaccharides 2′-Fucosyllactose, 3-Fucosyllactose, and Difucosyllactose by Various Probiotic and Pathogenic Bacteria. *J. Agric. Food Chem.* **2020**, *69*, 170–182. [CrossRef] [PubMed]
64. Frei, R.; Akdis, M.; O'Mahony, L. Prebiotics, probiotics, synbiotics, and the immune system: Experimental data and clinical evidence. *Curr. Opin. Gastroenterol.* **2015**, *31*, 153–158. [CrossRef] [PubMed]
65. Vonk, M.M.; Diks, M.A.P.; Wagenaar, L.; Smit, J.J.; Pieters, R.H.H.; Garssen, J.; van Esch, B.; Knippels, L.M.J. Improved Efficacy of Oral Immunotherapy Using Non-Digestible Oligosaccharides in a Murine Cow's Milk Allergy Model: A Potential Role for Foxp3+ Regulatory T Cells. *Front. Immunol.* **2017**, *8*, 1230. [CrossRef]
66. Vonk, M.M.; Blokhuis, B.R.J.; Diks, M.A.P.; Wagenaar, L.; Smit, J.J.; Pieters, R.H.H.; Garssen, J.; Knippels, L.M.J.; van Esch, B.C.A.M. Butyrate Enhances Desensitization Induced by Oral Immunotherapy in Cow's Milk Allergic Mice. *Mediat. Inflamm.* **2019**, *2019*, 9062537. [CrossRef]
67. Pujari, R.; Banerjee, G. Impact of prebiotics on immune response: From the bench to the clinic. *Immunol. Cell Biol.* **2021**, *99*, 255–273. [CrossRef]
68. Del Fabbro, S.; Calder, P.C.; Childs, C.E. Microbiota-independent immunological effects of non-digestible oligosaccharides in the context of inflammatory bowel diseases. *Proc. Nutr. Soc.* **2020**, *79*, 468–478. [CrossRef]
69. Ortega-Gonzalez, M.; Ocon, B.; Romero-Calvo, I.; Anzola, A.; Guadix, E.; Zarzuelo, A.; Suarez, M.D.; de Medina, F.S.; Martinez-Augustin, O. Nondigestible oligosaccharides exert nonprebiotic effects on intestinal epithelial cells enhancing the immune response via activation of TLR4-NF kappa B. *Mol. Nutr. Food Res.* **2014**, *58*, 384–393. [CrossRef]
70. Bodera, P. Influence of prebiotics on the human immune system (GALT). *Recent Pat. Inflamm.* **2008**, *2*, 149–153. [CrossRef]
71. Ayechu-Muruzabal, V.; van de Kaa, M.; Mukherjee, R.; Garssen, J.; Stahl, B.; Pieters, R.J.; van't Land, B.; Kraneveld, A.D.; Willemsen, L.E.M. Modulation of the Epithelial-Immune Cell Crosstalk and Related Galectin Secretion by DP3-5 Galacto-Oligosaccharides and β-3′Galactosyllactose. *Biomolecules* **2022**, *12*, 384. [CrossRef]
72. Pandey, K.R.; Naik, S.R.; Vakil, B.V. Probiotics, prebiotics and synbiotics—A review. *J. Food Sci. Technol.* **2015**, *52*, 7577–7587. [CrossRef] [PubMed]
73. Hosono, A.; Ozawa, A.; Kato, R.; Ohnishi, Y.; Nakanishi, Y.; Kimura, T.; Nakamura, R. Dietary Fructooligosaccharides Induce Immunoregulation of Intestinal IgA Secretion by Murine Peyer's Patch Cells. *Biosci. Biotechnol. Biochem.* **2014**, *67*, 758–764. [CrossRef] [PubMed]
74. Csernus, B.; Czeglédi, L. Physiological, antimicrobial, intestine morphological, and immunological effects of fructooligosaccharides in pigs. *Arch. Anim. Breed.* **2020**, *63*, 325–335. [CrossRef] [PubMed]

75. Nawaz, A.; Bakhsh Javaid, A.; Irshad, S.; Hoseinifar, S.H.; Xiong, H.J.F.; Immunology, S. The functionality of prebiotics as immunostimulant: Evidences from trials on terrestrial and aquatic animals. *Fish Shellfish Immunol.* **2018**, *76*, 272–278. [CrossRef]
76. Zhu, F.G.; Kandimalla, E.R.; Yu, D.; Agrawal, S.J.; Immunology, C. Oral administration of a synthetic agonist of Toll-like receptor 9 potently modulates peanut-induced allergy in mice. *J. Allergy Clin. Immunol.* **2007**, *120*, 631–637. [CrossRef]
77. Wang, X.X.; Song, X.P.; Wu, X.J.; Zhong, Y.L. Effects of Graded Levels of Isomaltooligosaccharides on the Performance, Immune Function and Intestinal Status of Weaned Pigs. *Asian-Australas. J. Anim. Sci.* **2016**, *29*, 250–256. [CrossRef]
78. Mizubuchi, H.; Yajima, T.; Aoi, N.; Tomita, T.; Yoshikai, Y. Isomalto-oligosaccharides polarize Th1-like responses in intestinal and systemic immunity in mice. *J. Nutr.* **2005**, *135*, 2857–2861. [CrossRef]
79. Luo, D.; Li, J.; Xing, T.; Zhang, L.; Gao, F. Effects of xylo-oligosaccharides and coated sodium butyrate on intestinal development, intestinal mucosal immunity function and cecal microbial composition of broilers. *J. Nanjing Agric. Univ.* **2022**, *45*, 131–140.
80. Yousef, M.; Pichyangkura, R.; Soodvilai, S.; Chatsudthipong, V.; Muanprasat, C.J. Chitosan oligosaccharide as potential therapy of inflammatory bowel disease: Therapeutic efficacy and possible mechanisms of action. *Pharmacol. Res.* **2012**, *66*, 66–79. [CrossRef]
81. Bahar, B.; O'Doherty, J.V.; Maher, S.; Mcmorrow, J.; Sweeney, T. Chitooligosaccharide elicits acute inflammatory cytokine response through AP-1 pathway in human intestinal epithelial-like (Caco-2) cells. *Mol. Immunol.* **2012**, *51*, 283–291. [CrossRef] [PubMed]
82. Wen, J.; Niu, X.; Chen, S.; Chen, Z.; Wu, S.; Wang, X.; Yong, Y.; Liu, X.; Yu, Z.; Ma, X.; et al. Chitosan oligosaccharide improves the mucosal immunity of small intestine through activating SIgA production in mice: Proteomic analysis. *Int. Immunopharmacol.* **2022**, *109*, 108826. [CrossRef] [PubMed]
83. Vo, T.S.; Kong, C.S.; Kim, S.K. Inhibitory effects of chitooligosaccharides on degranulation and cytokine generation in rat basophilic leukemia RBL-2H3 cells. *Carbohydr. Polym.* **2011**, *84*, 649–655. [CrossRef]
84. Cheng, L.; Kong, C.; Wang, W.; Groeneveld, A.; Nauta, A.; Groves, M.R.; Kiewiet, M.B.G.; de Vos, P. The Human Milk Oligosaccharides 3-FL, Lacto-N-Neotetraose, and LDFT Attenuate Tumor Necrosis Factor-α Induced Inflammation in Fetal Intestinal Epithelial Cells In Vitro through Shedding or Interacting with Tumor Necrosis Factor Receptor 1. *Mol. Nutr. Food Res.* **2021**, *65*, e2000425. [CrossRef]
85. Li, M.; Monaco, M.H.; Wang, M.; Comstock, S.S.; Kuhlenschmidt, T.B.; Fahey, G.C., Jr.; Miller, M.J.; Kuhlenschmidt, M.S.; Donovan, S.M. Human milk oligosaccharides shorten rotavirus-induced diarrhea and modulate piglet mucosal immunity and colonic microbiota. *ISME J.* **2014**, *8*, 1609–1620. [CrossRef]
86. Lane, J.A.; O'Callaghan, J.; Carrington, S.D.; Hickey, R.M. Transcriptional response of HT-29 intestinal epithelial cells to human and bovine milk oligosaccharides. *Br. J. Nutr.* **2013**, *110*, 2127–2137.
87. Ehrlich, A.M.; Pacheco, A.R.; Henrick, B.M.; Taft, D.; Xu, G.; Huda, M.N.; Mishchuk, D.; Goodson, M.L.; Slupsky, C.; Barile, D.; et al. Indole-3-lactic acid associated with Bifidobacterium-dominated microbiota significantly decreases inflammation in intestinal epithelial cells. *BMC Microbiol.* **2020**, *20*, 357. [CrossRef]
88. Li, A.; Li, Y.; Zhang, X.; Zhang, C.; Li, T.; Zhang, J.; Li, C. The human milk oligosaccharide 2′-fucosyllactose attenuates β-lactoglobulin-induced food allergy through the miR-146a-mediated toll-like receptor 4/nuclear factor-κB signaling pathway. *J. Dairy Sci.* **2021**, *104*, 10473–10484. [CrossRef]
89. Azagra-Boronat, I.; Massot-Cladera, M.; Mayneris-Perxachs, J.; Knipping, K.; van't Land, B.; Tims, S.; Stahl, B.; Garssen, J.; Franch, À.; Castell, M.; et al. Immunomodulatory and Prebiotic Effects of 2′-Fucosyllactose in Suckling Rats. *Front. Immunol.* **2019**, *10*, 1773. [CrossRef]
90. El Kaoutari, A.; Armougom, F.; Gordon, J.I.; Raoult, D.; Henrissat, B. The abundance and variety of carbohydrate-active enzymes in the human gut microbiota. *Nat. Rev. Microbiol.* **2013**, *11*, 497–504. [CrossRef]
91. Cantarel, B.L.; Lombard, V.; Henrissat, B. Complex Carbohydrate Utilization by the Healthy Human Microbiome. *PLoS ONE* **2012**, *7*, e28742. [CrossRef] [PubMed]
92. Jang, K.B.; Kim, S.W. Role of milk carbohydrates in intestinal health of nursery pigs: A review. *J. Anim. Sci. Biotechnol.* **2022**, *13*, 6. [CrossRef] [PubMed]
93. Li, B.; Schroyen, M.; Leblois, J.; Wavreille, J.; Soyeurt, H.; Bindelle, J.; Everaert, N. Effects of inulin supplementation to piglets in the suckling period on growth performance, postileal microbial and immunological traits in the suckling period and three weeks after weaning. *Arch. Anim. Nutr.* **2018**, *72*, 425–442. [CrossRef] [PubMed]
94. Lindberg, J.E. Fiber effects in nutrition and gut health in pigs. *J. Anim. Sci. Biotechnol.* **2014**, *5*, 15. [CrossRef] [PubMed]
95. van der Aar, P.J.; Molist, F.; van der Klis, J.D. The central role of intestinal health on the effect of feed additives on feed intake in swine and poultry. *Anim. Feed Sci. Technol.* **2017**, *233*, 64–75. [CrossRef]
96. Sun, X.; Wan, J.; Cao, J.; Si, Y.; Wang, Q. Progress in lytic polysaccharide monooxygenase. *Chin. J. Biotechnol.* **2018**, *34*, 177–187.
97. Wang, Y.J.; Wen, R.; Liu, D.D.; Zhang, C.; Wang, Z.A.; Du, Y.G. Exploring Effects of Chitosan Oligosaccharides on the DSS-Induced Intestinal Barrier Impairment In Vitro and In Vivo. *Molecules* **2021**, *26*, 2199. [CrossRef]
98. Collado, M.C.; Grzeskowiak, L.; Salminen, S. Probiotic strains and their combination inhibit in vitro adhesion of pathogens to pig intestinal mucosa. *Curr. Microbiol.* **2007**, *55*, 260–265. [CrossRef]
99. Johnson-Henry, K.C.; Pinnell, L.J.; Waskow, A.M.; Irrazabal, T.; Martin, A.; Hausner, M.; Sherman, P.M. Short-Chain Fructo-oligosaccharide and Inulin Modulate Inflammatory Responses and Microbial Communities in Caco2-bbe Cells and in a Mouse Model of Intestinal Injury. *J. Nutr.* **2014**, *144*, 1725–1733. [CrossRef]
100. Zhong, Y.; Cai, D.L.; Cai, W.; Geng, S.S.; Chen, L.Y.; Han, T. Protective effect of galactooligosaccharide-supplemented enteral nutrition on intestinal barrier function in rats with severe acute pancreatitis. *Clin. Nutr.* **2009**, *28*, 575–580. [CrossRef]

101. Dong, C.; Yu, J.; Yang, Y.; Zhang, F.; Su, W.; Fan, Q.; Wu, C.; Wu, S. Berberine, a potential prebiotic to indirectly promote Akkermansia growth through stimulating gut mucin secretion. *Biomed. Pharmacother.* **2021**, *139*, 111595. [CrossRef]
102. Mahdavi, M.; Laforest-Lapointe, I.; Microorganisms, E.M.J. Preventing Colorectal Cancer through Prebiotics. *Microorganisms* **2021**, *9*, 1325. [CrossRef] [PubMed]
103. Seidel, D.V.; Azcárate-Peril, M.A.; Chapkin, R.S.; Turner, N.D. Shaping functional gut microbiota using dietary bioactives to reduce colon cancer risk. *Semin. Cancer Biol.* **2017**, *46*, 191–204. [CrossRef] [PubMed]
104. Liong, M.T. Roles of Probiotics and Prebiotics in Colon Cancer Prevention: Postulated Mechanisms and In-Vivo Evidence. *Int. J. Mol. Sci.* **2008**, *9*, 854–863. [CrossRef] [PubMed]
105. Hughes, R.; Rowland, I.R. Stimulation of apoptosis by two prebiotic chicory fructans in the rat colon. *Carcinogenesis* **2001**, *22*, 43–47. [CrossRef] [PubMed]
106. Buddington, K.K.; Donahoo, J.B.; Buddington, R.K. Dietary oligofructose and inulin protect mice from enteric and systemic pathogens and tumor inducers. *J. Nutr.* **2002**, *132*, 472. [CrossRef]
107. Qamar, T.R.; Iqbal, S.; Syed, F.; Nasir, M.; Rehman, H.; Iqbal, M.A.; Liu, R.H. Impact of Novel Prebiotic Galacto-Oligosaccharides on Various Biomarkers of Colorectal Cancer in Wister Rats. *Int. J. Mol. Sci.* **2017**, *18*, 1785. [CrossRef]
108. Verma, A.; Shukla, G.J. Administration of prebiotic inulin suppresses 1,2 dimethylhydrazine dihydrochloride induced procarcinogenic biomarkers fecal enzymes and preneoplastic lesions in early colon carcinogenesis in Sprague Dawley rats. *J. Funct. Foods* **2013**, *5*, 991–996. [CrossRef]
109. Bornet, F.R.J.; Brouns, F. Immune-stimulating and Gut Health-promoting Properties of Short-chain Fructo-oligosaccharides. *Nutr. Rev.* **2002**, *60*, 326–334.
110. Fernández, J.; Redondo-Blanco, S.; Gutiérrez-Del-Río, I.; Miguélez, E.; Villar, C.J.; Lombo, F. Colon microbiota fermentation of dietary prebiotics towards short-chain fatty acids and their roles as anti-inflammatory and antitumour agents: A review. *J. Funct. Foods* **2016**, *25*, 511–522. [CrossRef]
111. Ambalam, P.; Raman, M.; Purama, R.K.; Doble, M. Probiotics, prebiotics and colorectal cancer prevention. *Best Pract. Res. Clin. Gastroenterol.* **2016**, *30*, 119–131. [CrossRef] [PubMed]
112. Raman, M.; Ambalam, P.; Kondepudi, K.K.; Pithva, S.; Kothari, C.; Patel, A.T.; Purama, R.K.; Dave, J.; Vyas, B. Potential of probiotics, prebiotics and synbiotics for management of colorectal cancer. *Gut Microbes* **2013**, *4*, 181–192. [CrossRef] [PubMed]
113. Geier, M.S.; Butler, R.N.; Howarth, G.S. Therapy, Probiotics, prebiotics and synbiotics: A role in chemoprevention for colorectal cancer? *Cancer Biol. Ther.* **2006**, *5*, 1265–1269. [CrossRef] [PubMed]
114. Yan, W.A.; Peng, W.A.; Tgha, B.; Rjl, C.; Mhz, A.; Hong, W.A.; Zyc, B.J. Encapsulation of phycocyanin by prebiotics and polysaccharides-based electrospun fibers and improved colon cancer prevention effects. *Int. J. Biol. Macromol.* **2020**, *149*, 672–681.
115. Mulder, D.J.; Noble, A.J.; Justinich, C.J.; Duffin, J.M. A tale of two diseases: The history of inflammatory bowel disease. *J. Crohns Colitis* **2014**, *8*, 341–348. [CrossRef]
116. Wijmenga, C. Expressing the differences between Crohn disease and ulcerative colitis. *PLoS Med.* **2005**, *2*, 719–720. [CrossRef]
117. Newburg, D.S.; Ko, J.S.; Leone, S.; Nanthakumar, N.N. Human Milk Oligosaccharides and Synthetic Galactosyloligosaccharides Contain 3′-, 4-, and 6′-Galactosyllactose and Attenuate Inflammation in Human T84, NCM-460, and H4 Cells and Intestinal Tissue Ex Vivo. *J. Nutr.* **2016**, *146*, 358–367. [CrossRef]
118. Vogt, L.; Ramasamy, U.; Meyer, D.; Pullens, G.; Venema, K.; Faas, M.M.; Schols, H.A.; de Vos, P. Immune Modulation by Different Types of beta 2- > 1-Fructans Is Toll-Like Receptor Dependent. *PLoS ONE* **2013**, *8*, e68367. [CrossRef]
119. Capitan-Canadas, F.; Ortega-Gonzalez, M.; Guadix, E.; Zarzuelo, A.; Suarez, M.D.; de Medina, F.S.; Martinez-Augustin, O. Prebiotic oligosaccharides directly modulate proinflammatory cytokine production in monocytes via activation of TLR4. *Mol. Nutr. Food Res.* **2014**, *58*, 1098–1110. [CrossRef]
120. Lehmann, S.; Hiller, J.; van Bergenhenegouwen, J.; Knippels, L.M.J.; Garssen, J.; Traidl-Hoffmann, C. In Vitro Evidence for Immune-Modulatory Properties of Non-Digestible Oligosaccharides: Direct Effect on Human Monocyte Derived Dendritic Cells. *PLoS ONE* **2015**, *10*, e0132304. [CrossRef]
121. Casellas, F.; Borruel, N.; Torrejon, A.; Varela, E.; Antolin, M.; Guarner, F.; Malagelada, J.R. Oral oligofructose-enriched inulin supplementation in acute ulcerative colitis is well tolerated and associated with lowered faecal calprotectin. *Aliment. Pharmacol. Ther.* **2007**, *25*, 1061–1067. [CrossRef] [PubMed]
122. Lindsay, J.O.; Whelan, K.; Stagg, A.J.; Gobin, P.; Al-Hassi, H.O.; Rayment, N.; Kamm, M.A.; Knight, S.C.; Forbes, A. Clinical, microbiological, and immunological effects of fructo-oligosaccharide in patients with Crohn's disease. *Gut* **2006**, *55*, 348–355. [CrossRef] [PubMed]
123. Garcia-Peris, P.; Velasco, C.; Hernandez, M.; Lozano, M.A.; Paron, L.; De, L.; Breton, I.; Camblor, M.; Guarner, F. Effect of inulin and fructo-oligosaccharide on the prevention of acute radiation enteritis in patients with gynecological cancer and impact on quality-of-life: A randomized, double-blind, placebo-controlled trial. *Eur. J. Clin. Nutr.* **2016**, *70*, 170–174. [CrossRef] [PubMed]
124. Crouzet, L.; Gaultier, E.; Del'Homme, C.; Cartier, C.; Delmas, E.; Dapoigny, M.; Fioramonti, J.; Bernalier-Donadille, A. The hypersensitivity to colonic distension of IBS patients can be transferred to rats through their fecal microbiota. *Neurogastroenterol. Motil.* **2013**, *25*, e272–e282. [CrossRef] [PubMed]
125. Collins, S.M. A role for the gut microbiota in IBS. *Nat. Rev. Gastroenterol. Hepatol.* **2014**, *11*, 497–505. [CrossRef]

126. Kerckhoffs, A.P.M.; Samsom, M.; van der Rest, M.E.; de Vogel, J.; Knol, J.; Ben-Amor, K.; Akkermans, L.M.A. Lower Bifidobacteria counts in both duodenal mucosa-associated and fecal microbiota in irritable bowel syndrome patients. *World J. Gastroenterol.* **2009**, *15*, 2887–2892. [CrossRef]
127. Parkes, G.C.; Rayment, N.B.; Hudspith, B.N.; Petrovska, L.; Lomer, M.C.; Brostoff, J.; Whelan, K.; Sanderson, J.D. Distinct microbial populations exist in the mucosa-associated microbiota of sub-groups of irritable bowel syndrome. *Neurogastroenterol. Motil.* **2012**, *24*, 31–39. [CrossRef]
128. Hungin, A.P.S.; Mulligan, C.; Pot, B.; Whorwell, P.; Agreus, L.; Fracasso, P.; Lionis, C.; Mendive, J.; de Foy, J.M.P.; Rubin, G.; et al. Systematic review: Probiotics in the management of lower gastrointestinal symptoms in clinical practice—An evidence-based international guide. *Aliment. Pharmacol. Ther.* **2013**, *38*, 864–886. [CrossRef]
129. McKenzie, Y.A.; Alder, A.; Anderson, W.; Wills, A.; Goddard, L.; Gulia, P.; Jankovich, E.; Mutch, P.; Reeves, L.B.; Singer, A.; et al. British Dietetic Association evidence-based guidelines for the dietary management of irritable bowel syndrome in adults. *J. Hum. Nutr. Diet.* **2012**, *25*, 260–274. [CrossRef]
130. Staudacher, H.M.; Whelan, K. The low FODMAP diet: Recent advances in understanding its mechanisms and efficacy in IBS. *Gut* **2017**, *66*, 1517–1527. [CrossRef]
131. Jang, S.I.; Eom, H.Y.; Hwang, J.H.; Kim, L.; Lee, J.H. Simultaneous Quantification of 3′- and 6′-Sialyllactose in Rat Plasma Using Liquid Chromatography-Tandem Mass Spectrometry and Its Application to a Pharmacokinetic Study. *Molecules* **2021**, *26*, 1177. [CrossRef] [PubMed]
132. Zheng, D.W.; Li, R.Q.; An, J.X.; Xie, T.Q.; Han, Z.Y.; Xu, R.; Fang, Y.; Zhang, X.Z. Prebiotics-Encapsulated Probiotic Spores Regulate Gut Microbiota and Suppress Colon Cancer. *Adv. Mater.* **2020**, *32*, e2004529. [CrossRef] [PubMed]
133. Ouwehand, A.C.; Tlkk, S.; Salminen, S. The Effect of Digestive Enzymes on the Adhesion of Probiotic Bacteria In Vitro. *J. Food Sci.* **2010**, *66*, 856–859. [CrossRef]

molecules

MDPI

Article

Design, Synthesis, and Bioassay of 2′-Modified Kanamycin A

Ribai Yan [1,*], Xiaonan Li [2], Yuheng Liu [3] and Xinshan Ye [2,*]

1 National Demonstration Center for Experimental Pharmacy Education, School of Pharmaceutical Sciences, Peking University, Beijing 100191, China
2 State Key Laboratory of Natural and Biomimetic Drugs, School of Pharmaceutical Sciences, Peking University, Beijing 100191, China
3 Department of Medical Chemistry, College of Pharmaceutical Science, Hebei Medical University, Shijiazhuang 050017, China
* Correspondence: yanribai@bjmu.edu.cn (R.Y.); xinshan@bjmu.edu.cn (X.Y.)

Abstract: Chemical modification of old drugs is an important way to obtain new ones, and it has been widely used in developing new aminoglycoside antibiotics. However, many of the previous modifying strategies seem arbitrary for their lack of support from structural biological detail. In this paper, based on the structural information of aminoglycoside and its drug target, we firstly analyzed the reason that some 2′-*N*-acetylated products of aminoglycosides caused by aminoglycoside-modifying enzyme AAC(2′) can partially retain activity, and then we designed, synthesized, and evaluated a series of 2′-modified kanamycin A derivatives. Bioassay results showed our modifying strategy was feasible. Our study provided valuable structure–activity relationship information, which would help researchers to develop new aminoglycoside antibiotics more effectively.

Keywords: antibiotic; aminoglycoside; structural modification; structure–activity relationship

1. Introduction

In order to fight against drug-resistant bacteria, researchers need to constantly develop new antibiotics. A feasible way to acquire new antibiotics is to modify the old ones. Structural modification on old drugs not only offers new chemical entities, but also affords valuable information about the relationship between molecular structure and its activity. As a kind of clinically important antibiotic, aminoglycosides are also facing the problem of resistance. The most common mechanism for their resistance is that the bacteria acquire the capability to produce aminoglycoside-modifying enzymes [1–3]. After enzymic modification, the resulting products decrease or lose their affinity to the drug target and then decrease or lose their antibacterial activity. To tackle this problem, proactive structural modification on these drugs have been extensively investigated [4–7]. Through chemical modification, some sensitive groups on the drug molecules are eliminated or masked. Thus, the resulting products no longer are the proper substrates for the modifying enzymes, and then the activity remains.

To modify aminoglycosides, various chemical strategies have been developed, such as modifying the aminoglycoside core structure, developing aminoglycoside-heteroconjugates, and introducing various alkyl/aryl substituents and acyl substituents at different positions on aminoglycoside scaffolds [8]. However, in the view of the mutual interaction between drug and drug target, some of them seem arbitrary. Ribosomal 16S RNA A-site of bacteria is the pivot for codon recognition during protein synthesis. It is established that most aminoglycosides bind to this domain in a specific mode and then exhibit antibacterial activity [9–12]. Obviously, this specific drug–target interaction mode should be deemed as a critical reference when considering the modification strategy. Guided by the information of structural biology, some advances in the chemical modification of aminoglycosides have been achieved [13–17]. Herein, we report a modifying strategy for kanamycin based on the structural information.

check for updates

Citation: Yan, R.; Li, X.; Liu, Y.; Ye, X. Design, Synthesis, and Bioassay of 2′-Modified Kanamycin A. *Molecules* **2022**, *27*, 7482. https://doi.org/10.3390/molecules27217482

Academic Editor: José Luis de Paz

Received: 14 October 2022
Accepted: 28 October 2022
Published: 2 November 2022

2. Result and Discussion

2.1. Design

It is well known that aminoglycoside-modifying enzymes include three categories: aminoglycoside phosphotransferases (APHs), aminoglycoside acetyltransferases (AACs), and aminoglycoside nucleotidyltransferase (ANTs). Among them, AACs can add acetyl groups to the corresponding free amino groups on aminoglycosides in the presence of acetyl coenzyme A. The acylated products usually lose antibacterial activity. However, in a research paper on AAC (2′), it was noted that some 2′-*N*-acetylated products definitely remained active; although they were much weaker than their parent compounds [18]. For example, 2′-*N*-acetyl arbekacin (**2**) showed a considerably low MIC (minimum inhibitory concentration) range of 1.56–3.13 μg/mL against a variety of bacteria, while its parent compound arbekacin (**1**) showed a range of 0.20–0.78 μg/mL (Figure 1). For another example, the 2′-*N*-acetyl neomycin also showed distinct activity in the same research. These facts suggest that modification on 2′-amino does not necessarily mean completely losing activity. Since arbekacin and neomycin share the same core structure, namely neamine (**3**), a reasonable inference is that such a modification does not block the binding of the drugs to their target profoundly. In recent years, some 3D structures of A Site in complex with aminoglycosides have been disclosed, which affords a brand-new perspective to explain such a phenomenon. The common aminoglycosides usually contain a pseudodisaccharide unit named neamine (or its variants), and the extra substituent attaches on the 5-position or 6-position of 2-deoxystreptamine (2-DOS, ring II) (Figure 2). Despite the distinct structural difference, the 3D complexes show that the neamine moiety still binds to A site in a very similar way, whether in terms of the binding position or the binding conformation (Figure 3a) [19,20]. Another important feature is that the binding site looks quite spacious, and the 2′-NH_2 (or 2′-OH) points to a vacant space of the major groove of the RNA helix. Under this condition, an additional acetyl could be well accommodated, and then 2′-*N*-acetylated aminoglycosides could still bind to A Site and retain a part of activity. Since the 2′-position is closely adjacent to the acidic backbone of the nucleic acid, acylation of the amino group on this site may significantly reduce the electrostatic force between them, which may account for the partial loss of activity. In studies with AAC(2′)s, some chemical modification on C-2′ had been carried out [21]. Neomycin, paromomycin, and ribostamycin had been used as targets for modification, and several small groups, such as methyl, ethyl, propyl, or glycyl were installed on 2′-NH_2. Some modified products showed good activity against AAC(2′)-producing bacteria.

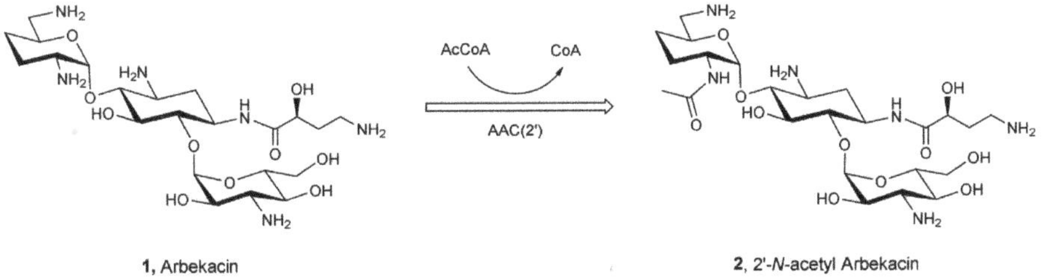

Figure 1. Schematic illustration for the action of AAC(2′) on arbekacin.

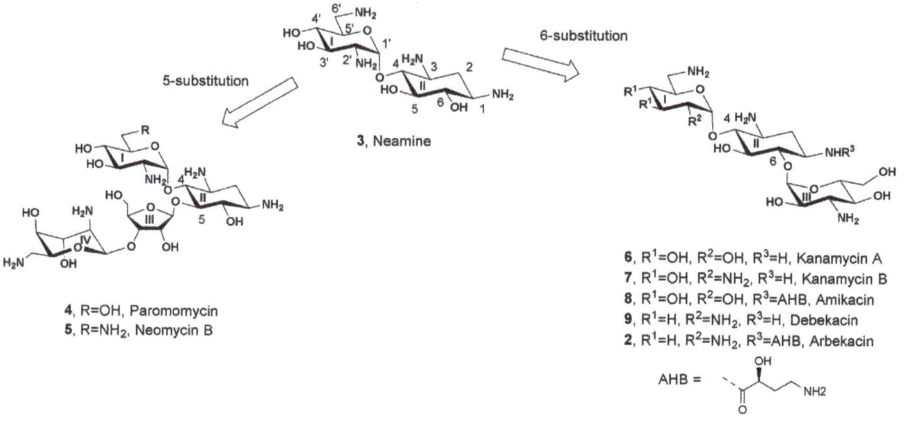

Figure 2. Structures of neamine and several representative aminoglycosides. The neamine core is composed of ring I and ring II.

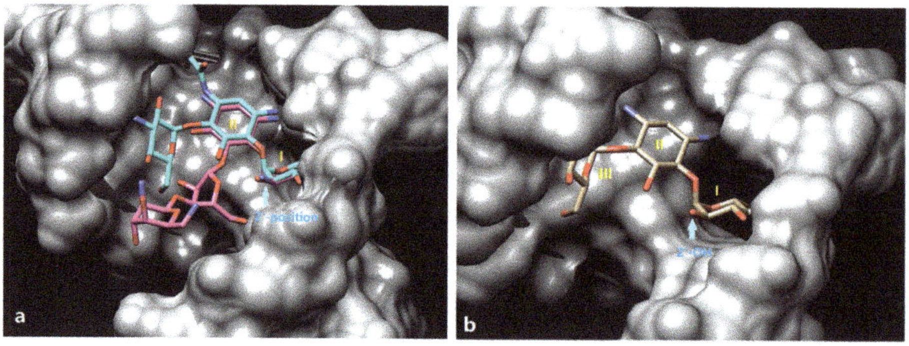

Figure 3. Crystal structures of neamine-containing aminoglycosides binding to A site. (**a**) The complexes of amikacin (cyanic, PDB ID: 4P20) and neomycin B (pink, PDB ID: 2ET4) with the decoding site are superimposed (based on nucleic acid chains). The neamine moiety is marked with Roman numerals I and II. The nucleic acid chains are colored in grey; (**b**) the complex of kanamycin A binding to A site (PDB ID: 2ESI). In both figures, 2′-position is indicated.

Usually, AAC(2′)s do not modify 2′-OH, so the antibiotics with a hydroxyl group on C-2′, such as kanamycin A, can remain active on the bacteria expressing AAC(2′)s [18]. However, 2′-OH could be a potential target for modifying enzyme APHs and/or ANTs, which can attach a phosphoryl or a nucleotidyl to a hydroxyl group, and then lead to resistance, so a proactive chemical modification seems meaningful. On the other hand, since the room that 2′-substituents point to is so vacant, a variety of groups with different features should also be able to be integrated into this position (Figure 3b). Based on these considerations, we used kanamycin A as parent compound, designed, synthesized a series of 2′-modified derivatives, and evaluated, and the relationship between activity and the features of introduced groups was also assayed.

2.2. Synthesis

Our synthesis started with commercially available sulfate of kanamycin A (**6**) (Scheme 1). According to the literature, we firstly synthesized 1,3,6′,3″-tetraazido-4″,6″-*O*-benzylidene-3′,4′,2″-tri-*O*-benzylkanamycin A (**13**) through seven steps [22]. Following this, after a silver oxide-promoted regioselective allylation, we successfully obtained key 2′-allylated kanamycin A

(**14**), which was confirmed by a serial of 2D NMR spectroscopy. Benzylation of 5-OH of **14** gave fully protected intermediate **15**, and then a potassium osmate catalyzed dihydroxylation in the presence of *N*-methylmorpholine-*N*-oxide was carried out; **15** smoothly transformed into diol **16** in high yield. Following this, treatment of the resulting diol with sodium periodate afforded key intermediate aldehyde **17**. Reduction of **17** with sodium borohydride gave alcohol **18** (Scheme 2). After this, tosylation and then substitution with sodium azide achieved **19**. Compound **18** and **19** underwent a Staudinger reduction and then a palladium-catalyzed hydrogenation, and finally transformed into products **20** and **21**, respectively. On the other hand, **17** coupled with different amines via reductive amination to afford **22a–22e**. All these intermediates transformed into the final products **23a–23e** with the same procedure as for **18** and **19**.

2.3. Bioactivity Assay

In order to determine the activity of newly synthesized compounds, we chose ten typical bacteria as test objects, both Gram-positive and Gram-negative bacteria were included. It contained two strains of *E. coli* (ATCC 25922 and ATCC 35218, G^-), three strains of *S. aureus* (ATCC 29213, ATCC 25923, and ATCC 33591, G^+), two strains of *K. pneumoniae* (ATCC 70063 and ATCC 13883, G^-), one strain of *E. faecalis* (ATCC 29212, G^+), and two strains of *P. aeruginosa* (ATCC 27853 and PAO1, G^-). All compounds were tested against these microorganisms by the microdilution assay, and the minimum inhibitory concentrations (MIC) in μg/mL were recorded. Kanamycin A served as the control.

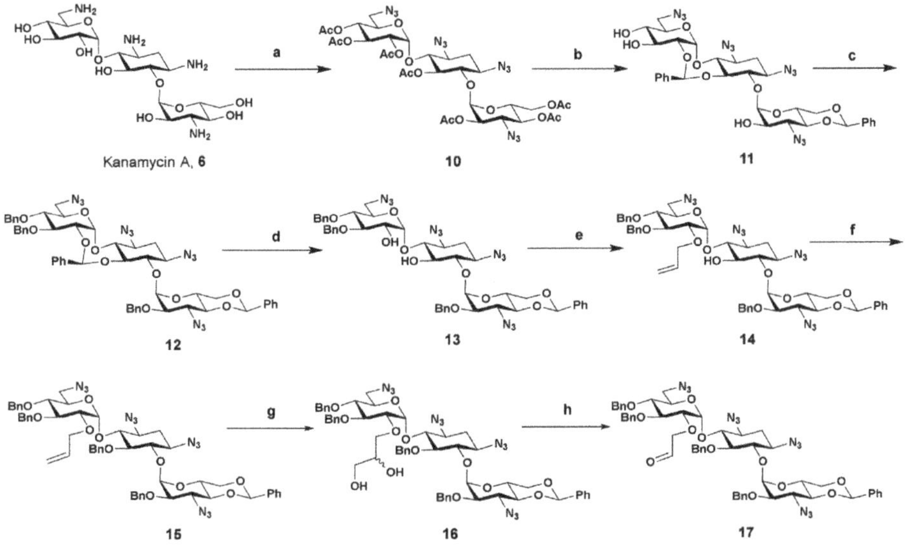

Scheme 1. Conditions: (**a**) (**i**) TfN_3, $CuSO_4$, CH_3CN, H_2O, NEt_3, ice bath; (**ii**) Ac_2O, pyridine, rt, 86% over two steps; (**b**) (**i**) MeOH, NaOMe, rt, 99%; (**ii**) $PhCH(OCH_3)_2$, DMF, p-TsOH, 68 °C, 46%; (**c**) BnBr, NaH, DMF, 93%; (**d**) (**i**)THF, MeOH, HCl (aq), 50 °C, 93%; (**ii**) 1.2 equiv of $PhCH(OCH_3)_2$, CH_3CN, p-TsOH, rt, 75%; (**e**) Allyl bromide, Ag_2O, TBAI, Toluene, 90%; (**f**) BnBr, NaH, DMF, 86%; (**g**) $K_2OsO_4{\bullet}2H_2O$, NMO, H_2O, acetone, 57%; (**h**) $NaIO_4$, MeOH, CH_2Cl_2, 100%.

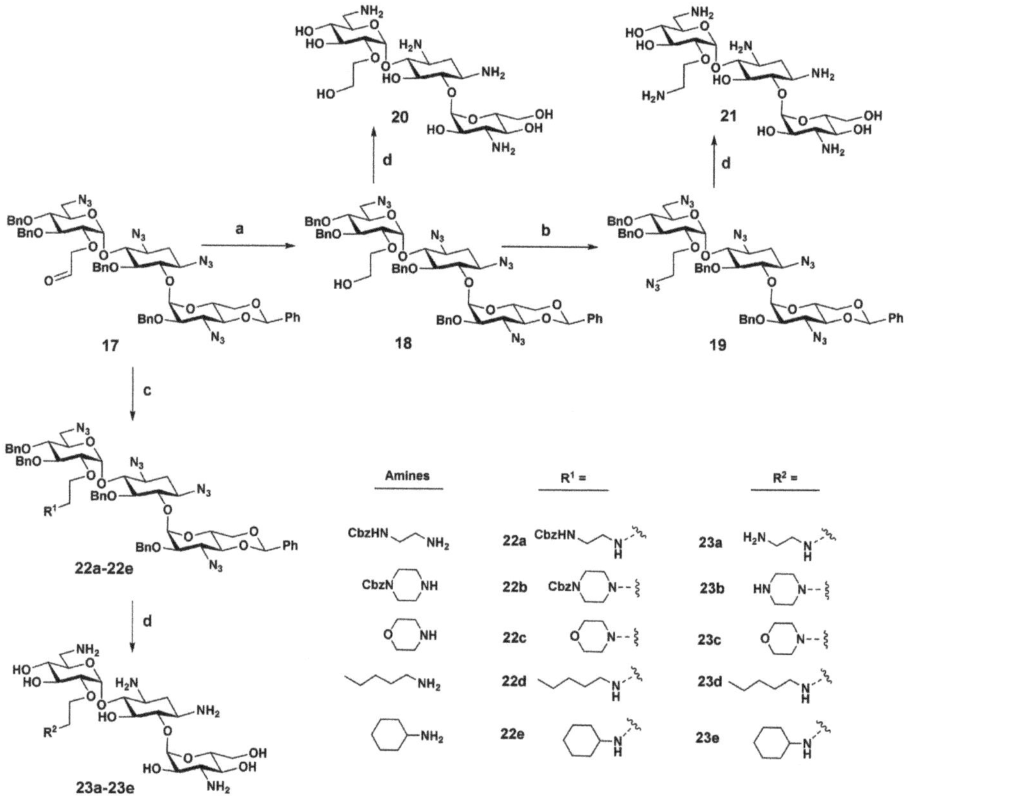

Scheme 2. Conditions: (**a**) $NaBH_4$, MeOH, CH_2Cl_2, ice bath, 92%; (**b**) (**i**) TsCl, Pyridine, rt; (**ii**) NaN_3, DMF, 84% over two steps; (**c**) Amines, $NaBH(OAc)_3$, 1,2-dichloroethane, rt, 52–67%; (**d**) (**i**) PMe_3, THF, H_2O, NaOH; (**ii**) H_2, Pd/C, MeOH, THF, 55–78% over two steps.

As shown in Table 1, most compounds exhibited distinct activity, and the activity was close to their parent compound kanamycin A generally. Among them, compound **21** showed somewhat better potency than kanamycin A. Except in the case of ATCC 25922, the MIC values of compound **21** were lower than or equal to that of kanamycin A in the case of the other nine bacteria. When it comes to compound **20**, in comparison with kanamycin A, the activity did not increase. The major structural difference between kanamycin A, compound **20**, and **21** is that compound **21** has one more amino group than the other two. It looks like that additional amino groups may benefit the activity. However, compound **23a**, which is the product obtained by adding a second amino group at the end of the 2′-position of compound **20**, did not exhibit further improvement in activity. On the contrary, the activity was slightly reduced in several cases in comparison with compound **21**. In the case of compound **23b**, in which the flexible 2-amino-ethylamino of **23a** was replaced by a more rigid piperazinyl, the activity had also not improved. When the piperazinyl in compound **23b** was replaced by a similarly hydrophilic morpholine ring, the resulting product compound, **23c**, gave almost the same activity result as **23b**. After studying the impact of the terminal hydrophilic groups on activity, we also checked whether the hydrophobic groups on the end of the 2′-position affected the activity. Compound **23d** and **23e** bear a pentyl and a cyclohexyl, respectively, at the terminal of 2′-substituent. To our surprise, these two compounds displayed comparable activity to that of other new derivatives. Compound **23e**, which bears a bulky and rigid cyclohexyl, even showed a very similar result to that of compound **21**. Based on these results, we can conclude that hydrophobicity and basicity of the terminal of 2′-substituents do not affect the activity remarkably, and the size of the groups also does not. Since the space that 2′-postion faces is so spacious and empty, it is no surprise that a good tolerance for the diversity of the introduced groups can be observed during chemical modifying.

Table 1. The results of antibacterial activities of synthesized compounds.

	MIC (μg/mL)									
	E. coli		*S. aureus*			*K. pneumoniae*		*E. faecalis*	*P. aeruginosa*	
Compd	**ATCC 35218**	**ATCC 25922**	**ATCC 29213**	**ATCC 25923**	**ATCC 33591**	**ATCC 700603**	**ATCC 13883**	**ATCC 29212**	**ATCC 27853**	**PAO1**
Kan	2	4	2	4	64	16	1	64	>128	8
20	4	32	1	4	64	8	2	64	>128	16
21	2	16	0.5	0.5	8	2	1	64	128	2
23a	4	>128	0.5	4	8	4	2	64	128	16
23b	4	64	1	4	128	8	4	128	>128	>128
23c	8	32	0.5	4	128	8	4	64	>128	128
23d	8	64	1	8	32	16	4	64	>128	128
23e	2	16	1	4	8	4	1	64	>128	16

Among the tested bacteria, some are antibiotic-resistant strains. *K. pneumoniae* K6 (ATCC 700603) is a clinical isolate and is reported to be resistant to some aminoglycosides (caused by ANT(2″)). Our antimicrobial susceptibility test showed that it was resistant to kanamycin A (MIC = 16 μg/mL). The bioassay results showed our modifying strategy brought some advantage, and all new compounds gave lower or equal MIC values to kanamycin A. Of note, **21**, **23a**, and **23f** displayed a four- to eight-fold enhancement in activity. ATCC 33591 is a methicillin-resistant *S. aureus* (MRSA). Many MRSA strains contain genes encoded for APH(3′), ANT(4′), and AAC(6′)/APH(2″), which render the bacteria resistant to many aminoglycosides. In our research, the MIC for kanamycin A against ATCC 33591 is 64 μg/mL. Three compounds, **21**, **23a**, and **23e** displayed eight-times enhancement in activity. In the case of PAO1, a prominent pathogen for its intrinsic resistance to many antibiotics, among newly synthesized derivatives, **21** also showed four-fold more potency than kanamycin. However, in the cases of the drug-resistant ATCC29212 (expressing efflux pump) and ATCC 27853 (expressing APH(3′)), none of new compounds exhibited a better outcome. In general, modification on the 2′-position may bring some benefits in treating drug-resistant bacteria, but this kind of improvement is mild and hard to predict.

3. Materials and Methods

3.1. Chemistry

General: All commercial reagents were used without further purification unless otherwise stated. Anhydrous *N,N*-dimethylformamide (DMF) was obtained by distilling commercial product over P_2O_5 under reduced pressure. Routine ^1H and ^{13}C nuclear magnetic resonance spectra were recorded on the Bruker AVANCE III 400 (Bruker Scientific Co. Ltd., Zurich, Switzerland) spectrometer or Bruker AVANCE III 600 spectrometer (Bruker Scientific Co. Ltd., Zurich, Switzerland). Samples were dissolved in deuterated chloroform ($CDCl_3$) or deuterium oxide (D_2O), and tetramethylsilane (TMS) was used as reference. High resolution mass spectra were recorded on the Thermo Scientific Orbitrap Fusion Lumos Mass Spectrometer (Thermo Fisher Scientific Inc., San Jose, CA, USA). Analytical thin-layer chromatography (TLC) was performed on Merck Silica Gel 60 F254. Compounds were visualized by UV light (254 nm) and/or by staining with a yellow solution containing $Ce(NH_4)_2(NO_3)_6$ (0.5 g) and $(NH_4)_6Mo_7O_{24}\bullet 4H_2O$ (24.0 g) in 6% H_2SO_4 (500 mL) or ninhydrin solution in ethyl acetate (5%) followed by heating. NMR spectra for new compounds can be found in Supplementary Materials section.

2′-O-Allyl-1,3,6′,3″-tetraazido-4″,6″-O-benzylidene-3′,4′,2″-tri-O-benzylkanamycin A (**14**): To a flask was added **13** (121 mg, 0.13 mmol), toluene (2 mL), freshly prepared silver (I) oxide (80 mg, 0.34 mmol), tetrabutylammonium iodide (10 mg, 0.03 mmol) and the allyl bromide (35mg, 0.29 mmol). The mixture was stirred overnight at room temperature and TLC showed the starting material disappeared. The reaction mixture was filtered and the filtrate was concentrated to residue. The gross product was purified by silica gel chromatography with the mixed solvent (petroleum/ethyl acetate from 15:1 to 6:1) as eluent to give the titled compound (112 mg, 0.12 mmol, 90% yield) as a colorless semisolid. ^{1}H NMR (600 MHz, $CDCl_3$) δ 7.50–7.49 (m, 2H), 7.43–7.42 (m, 2H), 7.39–7.26 (m, 16H), 5.91–5.85 (m, 1H), 5.52 (s, 1H), 5.25–5.22 (m, 2H), 5.18–5.16 (m, 1H), 5.09 (d, J = 3.6 Hz, 1H), 4.91–4.83 (m, 4H), 4.77 (d, J = 11.8 Hz, 1H), 4.70 (d, J = 1.7 Hz, 1H), 4.62 (d, J = 11.0 Hz, 1H), 4.48 (td, J_1 = 5.0 Hz, J_2 = 10.0 Hz, 1H), 4.34–4.31 (m, 1H), 4.22 (dd, J_1 = 5.0 Hz, J_2 = 10.0 Hz, 1H), 4.18–4.15 (m, 1H), 4.07–4.02 (m, 2H), 3.96 (t, J = 9.4 Hz, 1H), 3.65–3.54 (m, 5H), 3.50 (d, J_1 = 3.6 Hz, $J2$ = 9.8 Hz, 1H), 3.45–3.63 (m, 4H), 3.30–3.21 (m, 2H), 2.43 (ddd, J_1 = J_2 = 4.5 Hz, J_3 = 13.2 Hz, 1H), 1.60 (ddd, J_1 = J_2 = J_3 = 12.6 Hz, 1H). ^{13}C NMR (150 MHz, $CDCl_3$) δ 138.17, 137.77, 137.18, 137.09, 133.14, 128.95, 128.54, 128.48, 128.22, 128.21, 128.13, 128.01, 127.90, 127.80, 127.78, 126.04, 119.43, 101.32, 101.20, 97.64, 86.51, 81.81, 80.20, 80.09, 79.79, 78.10, 77.54, 75.60, 75.31, 74.13, 74.08, 73.10, 71.09, 68.78, 62.78, 61.94, 60.38, 58.76, 51.19, 32.21. HRMS (ESI/APCI) calculated for ($C_{49}H_{54}N_{12}O_{11}Na$) [M + Na^+]: 1009.3927, found: 1009.3962.

2′-O-Allyl-1,3,6′,3″-tetraazido-4″,6″-O-benzylidene-5,3′,4′,2″-tetra-O-benzylkanamycin A (**15**): To a stirred solution of **14** (226 mg, 0.23 mmol) in anhydrous DMF (3 mL) was added sodium hydride (26 mg, 60% in mineral oil, 0.65 mmol) and the resulting mixture was stirred for 30 min at room temperature, then benzyl bromide (59 mg, 0.35 mmol) was added in one portion. After stirring for 4 h, the reaction mixture was poured into water (50 mL) and the aqueous layer was extracted with ethyl acetate (3 × 15 mL). The combined organic phase was washed with brine, dried over anhydrous sodium sulfate, and concentrated under a vacuum. The residue was purified by silica gel column chromatography with the mixed solvent (petroleum ether/ethyl acetate 15:1 to 9:1) as eluent to give product **15** (217 mg, 0.20 mmol, 86% yield) as a white solid. ^{1}H NMR (400 MHz, $CDCl_3$) δ 7.44–7.24 (m, 20H), 7.19–7.11 (m, 4H), 7.05 (t, J = 7.4 Hz, 1H), 5.55 (d, J = 3.7 Hz, 1H), 5.52 (d, J = 4.0 Hz, 1H), 5.51–5.43 (m, 1H), 5.28 (s, 1H), 5.10 (d, J = 12.4 Hz, 1H), 4.98–4.74 (m, 8H), 4.58 (d, J = 11.3 Hz, 1H), 4.31–4.27 (m, 1H), 4.02–3.92 (m, 3H), 3.86–3.80 (m, 2H), 3.75 (d, J = 9.4 Hz, 1H), 3.67–3.50 (m, 6H), 3.46–3.36 (m, 4H), 3.28 (dd, J_1 = 3.8 Hz, $J2$ = 9.9 Hz, 1H), 3.22 (t, J = 9.8 Hz, 1H), 2.40 (ddd, J_1 = J_2 = 4.5 Hz, $J3$ = 13.2 Hz, 1H), 1.54 (ddd, J_1 = J_2 = J_3 = 12.9 Hz, 1H). ^{13}C NMR (100 MHz, CDCl3) δ 138.45, 138.04, 137.58, 137.27, 136.78, 134.25, 128.84, 128.59, 128.46, 128.44, 128.37, 128.19, 128.17, 127.92, 127.83, 127.74, 127.62, 127.10, 126.32, 125.19, 117.48, 101.39, 97.33, 96.32, 82.83, 81.81, 79.59, 78.70, 78.28, 78.11, 77.92, 77.15, 75.52, 74.96, 74.51, 73.44, 73.31, 70.76, 68.58, 62.89, 61.47, 60.17, 59.19, 51.37, 32.16. HRMS (ESI/APCI) calculated for ($C_{56}H_{60}N_{12}O_{11}Na$) [M + Na^+]: 1099.4397, found: 1099.4412.

2′-O-(2,3-Dihydroxypropyl)-1,3,6′,3″-tetraazido-4″,6″-O-benzylidene-5,3′,4′,2″-tetra-O-benzylkanamycin A (**16**): To a flask was added **15** (213 mg, 0.20 mmol), potassium osmate (VI) dihydrate (4 mg, 0.01 mmol), *N*-methylmorpholine-*N*-oxide solution in water (94 mg, 0.40 mmol, 50% *w*/*w*), acetone (5 mL), and water (0.1 mL) in sequence. After stirring overnight at room temperature, TLC showed the reaction was completed. The reaction mixture was poured into aqueous solution of sodium thiosulfate pentahydrate (50 mL, 1% *w*/*w*) and stirred for 30 min, then the resulting mixture was extracted with ethyl acetate (3 × 10 mL). The organic layer was combined, dried over anhydrous sodium sulfate, and concentrated under a vacuum. The gross product was purified by silica gel chromatography with the mixed solvent (petroleum/ethyl acetate from 8:1 to 4:1) as the solvent to give the titled compound **16** (125 mg, 0.11 mmol, 57% yield) as a colorless semisolid. ^{1}H NMR (400 MHz, $CDCl_3$) δ 7.44–7.14 (m, 24H), 7.10 (t, J = 7.2 Hz, 1H), 5.54 (t, J = 4.0 Hz, 1H), 5.48 (t, J = 4.1 Hz, 1H), 5.34 (s, 1H), 5.03–4.94 (m, 2H), 4.86 (d, J = 11.2 Hz, 1H), 4.82–4.68 (m, 4H), 4.59 (d, J = 11.2 Hz, 1H), 4.21–4.08 (m, 2H), 3.94–3.87 (m, 2H), 3.79–3.42 (m, 9H), 3.41–3.15 (m, 8H), 3.08 (br, 1H), 2.43–2.35 (m, 1H), 1.68–1.58 (m, 1H). HRMS (ESI/APCI) calculated for ($C_{56}H_{62}N_{12}NaO_{13}$) [M + Na^+]: 1133.4452, found: 1133.4487.

2′-O-(2-Oxoethyl)-1,3,6′,3″-tetraazido-4″,6″-O-benzylidene-5,3′,4′,2″-tetra-O-benzylkanamycin A (**17**): To a flask was added **16** (125 mg, 0.11 mmol), methanol (5 mL), and sodium periodate (47 mg, 0.22 mmol) and the resulting mixture was stirred at room temperature. After 4 h, TLC showed all starting material was consumed. The reaction mixture was poured into aqueous solution of sodium thiosulfate pentahydrate (50 mL, 0.4% *w*/*w*) and stirred for 30 min, then the resulting mixture was extracted with ethyl acetate (3 × 10 mL). The organic layer was combined, dried over anhydrous sodium sulfate, and concentrated under a vacuum. The resulting product **17** (122 mg, 0.11 mmol, 100%) was used in the following reactions directly without further purification. ^{1}H NMR (400 MHz, $CDCl_3$) δ ^{1}H NMR (400 MHz, $CDCl_3$) δ 9.09 (s, 1H), 7.44–7.16 (m, 22H), 7.10–7.04 (m, 3H), 5.56 (d, *J* = 3.8 Hz, 1H), 5.51 (d, *J* = 3.8 Hz, 1H), 5.28 (s, 1H), 5.07 (d, *J* = 12.5 Hz, 2H), 4.96 (d, *J* = 12.4 Hz, 1H), 4.86-4.76 (m, 4H), 4.68 (d, *J* = 11.2 Hz, 1H), 4.59 (d, *J* = 11.2 Hz, 1H), 4.29–4.25 (m, 1H), 4.03–3.96 (m, 3H), 3.86–3.79 (m, 2H), 3.74–3.61 (m, 3H), 3.59–3.49 (m, 3H), 3.46–3.38 (m, 5H), 3.25–3.19 (m, 2H), 2.41 (ddd, J_1 = J_2 = 4.7 Hz, J_3 = 13.0 Hz, 1H), 1.68 (ddd, J_1 = J_2 = J_3 = 12.6 Hz, 1H). HRMS (ESI/APCI) calculated for ($C_{55}H_{58}N_{12}NaO_{12}$) [M + Na^+]: 1101.4189, found: 1101.4218.

2′-O-(2-Hydroxyethyl)-1,3,6′,3″-tetraazido-4″,6″-O-benzylidene-5,3′,4′,2″-tetra-O-benzylkanamycin A (**18**): The product **17** (110 mg, 0.10 mmol) was dissolved in the mixture of methanol (3 mL) and dichloromethane (2 mL), and to the solution was added sodium borohydride (15 mg, 0.40 mmol) in several portions while stirring at ice bath temperature. After 30 min, TLC showed the reaction was completed. The solvent was removed, and the gross product was purified with silica gel chromatography by using the mixed solvent (petroleum ether/ethyl acetate 20:1 to 10:1) as eluent to give product **18** (101 mg, 0.09 mmol, 92% yield) as a white semisolid. ^{1}H NMR (400 MHz, $CDCl_3$) δ 7.43–7.24 (m, 20H), 7.20–7.15 (m, 4H), 7.08 (t, *J* = 7.2 Hz, 1H), 5.53 (d, *J* = 4.2 Hz, 1H), 5.52 (d, *J* = 4.3 Hz, 1H), 5.31 (s, 1H), 5.05 (d, *J* = 12.3 Hz, 1H), 4.97 (d, *J* = 12.2 Hz, 1H), 4.88 (d, *J* = 11.2 Hz, 1H), 4.82–4.75 (m, 4H), 4.59 (d, *J* = 11.3 Hz, 1H), 4.25-4.21 (m, 1H), 4.06 (dd, J_1 = 5.0 Hz, J_2 = 10.2 Hz, 1H), 3.96-3.85 (m, 2H), 3.66–3.29 (m, 14 H), 3.24 (t, *J* = 9.8 Hz, 1H), 2.65 (t, *J* = 6.1 Hz, 1H), 2.39 (dt, J_1 = 4.3 Hz, J_2 = 13.2, Hz, 1H), 1.63 (ddd, J_1 = J_2 = J_3 = 12.7 Hz, 1H). 13C NMR (100 MHz, $CDCl_3$) δ 137.97, 137.83, 137.26, 137.23, 136.81, 128.87, 128.63, 128.58, 128.50, 128.47, 128.18, 128.16, 127.97, 127.93, 127.83, 127.76, 127.47, 126.27, 125.72, 101.41, 97.21, 96.43, 82.29, 81.35, 80.54, 79.60, 78.55, 78.43, 78.03, 77.76, 75.80, 75.03, 74.95, 73.45, 73.29, 71.01, 68.60, 62.91, 62.31, 61.40, 60.18, 59.92, 51.20, 32.35. HRMS (ESI/APCI) calculated for ($C_{55}H_{60}N_{12}NaO_{12}$) [M + Na^+]: 1103.4346, found: 1103.4369.

2′-O-(2-Azidoethyl)-1,3,6′,3″-tetraazido-4″,6″-O-benzylidene-5,3′,4′,2″-tetra-O-benzylkanamycin A (**19**): To the solution of **18** (65 mg, 0.06 mmol) in pyridine (3 mL) was added p-toluene sulfonyl chloride (46 mg, 0.18 mmol) in one portion, and the resulting mixture was stirred overnight. Then, the solvent was removed, and the residue was purified by column chromatography on silica gel using petroleum ether/ethyl acetate (10:1 to 5:1) as eluent to afford the tosylate intermediate, which was mixed with sodium azide (12 mg, 0.18 mmol) and DMF (3 mL). After stirring at 80 °C for 4 h, the reaction mixture was poured into 50 mL of water and the aqueous layer was extracted with ethyl acetate (3 × 10 mL). The combined organic phase was dried over Na_2SO_4 and concentrated under a vacuum. The crude was purified by column chromatography on silica gel by using the mixture of petroleum ether and ethyl acetate (10:1 to 5:1) as eluent to afford **19** (56 mg, 0.05 mmol, 84% yield). ^{1}H NMR (600 MHz, $CDCl_3$) δ 7.43–7.38 (m, 4H), 7.35–7.24 (m, 16H), 7.18 (t, *J* = 7.7 Hz, 2H), 7.12–7.06 (m, 3H), 5.56–5.55 (m, 2H), 5.29 (s, 1H), 5.09 (d, *J* = 12.3 Hz, 1H), 4.99 (d, *J* = 12.3 Hz, 1H), 4.87–4.76 (m, 5H), 4.58 (d, *J* = 11.3 Hz, 1H), 4.29–4.26 (m, 1H), 4.01 (dd, J_1 = 4.9 Hz, J_2 = 10.1 Hz, 1H), 3.95 (t, *J* = 9.4 Hz, 1H), 3.87–3.82 (m, 2H), 3.75 (t, *J* = 9.4 Hz, 1H), 3.66 (t, *J* = 9.4 Hz, 1H), 3.61–3.51 (m, 5H), 3.46–3.36 (m, 4H), 3.25–3.22 (m, 2H), 3.15–3.11 (m, 1H), 3.04–2.96 (m, 2H), 2.40 (ddd, J_1 = J_2 = 4.6 Hz, J_3 = 13.3, Hz, 1H), 1.64 (ddd, J_1 = J_2 = J_3 = 12.7 Hz, 1H). ^{13}C NMR (150 MHz, $CDCl_3$) δ 138.39, 137.90, 137.54, 137.21, 136.74, 128.85, 128.58, 128.45, 128.41, 128.19, 127.92, 127.87, 127.82, 127.63, 127.60, 127.22, 126.29, 125.14, 101.39, 96.89, 96.32, 82.78, 81.58, 80.20, 79.55, 78.35, 78.01, 77.76, 75.41, 74.94, 74.46, 73.30, 70.76,

70.75, 68.57, 62.87, 61.46, 60.15, 59.25, 51.25, 50.81, 32.14. HRMS (ESI/APCI) calculated for ($C_{55}H_{59}N_{15}NaO_{11}$) [M + Na^+]: 1128.4411, found: 1128.4442.

2′-O-[[2-[2-(benzyloxycarbonylamino)ethyl]amino]-ethyl]-1,3,6′,3″-tetraazido-4″,6″-O-benzylidene-5,3′,4′,2″-tetra-O-benzylkanamycin A (**22a**): To a flask was added **17** (53 mg, 0.05 mmol), 1,2-dichloroethane (3 mL), and 1-(benzyloxycarbonylamino)-2-aminoethane (48 mg, 0.25 mmol) in one portion. The resulting mixture was stirred for 30 min, then sodium triacetoxyborohyride (53 mg, 0.25 mmol) was added in portions. After stirring overnight at room temperature, TLC showed a minor amount of **17** remained. Then, another portion of sodium triacetoxyborohyride (32 mg, 0.15 mmol) was added. After stirring for another 8 h, the reaction was complete. The solvent was removed, and the residue was purified by silica gel chromatography (petroleum/ethyl acetate from 8:1 to 4:1) to give the titled compound **22a** (38 mg, 0.03 mmol, 67% yield) as a colorless semisolid. ^1H NMR (400 MHz, $CDCl_3$) δ 7.45–7.09 (m, 29H), 7.06 (t, J = 7.2 Hz, 1H), 5.57 (d, J = 3.6 Hz, 1H), 5.53 (d, J = 3.6 Hz, 1H), 5.33 (s, 1H), 5.06 (s, 2H), 5.01–4.93 (m, 2H), 4.88–4.74 (m, 5H), 4.58 (d, J = 11.2 Hz, 1H), 4.26–4.20 (m, 1H), 4.02 (dd, J_1 = 10.1, J_2 = 4.8 Hz, 1H), 3.96–3.80 (m, 3H), 3.75 (t, J = 9.4 Hz, 1H), 3.68 (t, J = 9.4 Hz, 1H), 3.62–3.21 (m, 13H), 3.01 (br, 2H), 2.53–2.28 (m, 5H), 1.65 (ddd, J_1 = J_2 = J_3 = 12.6 Hz, 1H). ^{13}C NMR (100 MHz, $CDCl_3$) δ 156.46, 138.30, 137.83, 137.28, 136.78, 136.64, 128.88, 128.58, 128.49, 128.18, 128.15, 128.10, 128.06, 127.98, 127.93, 127.89, 127.76, 127.53, 127.41, 126.23, 125.96, 101.35, 96.77, 96.24, 82.64, 81.23, 80.15, 79.66, 78.46, 78.15, 78.05, 77.70, 75.38, 74.96, 74.90, 73.30, 70.93, 68.61, 66.60, 62.94, 61.47, 60.09, 59.71, 51.29, 48.60, 48.34, 39.80, 32.19. HRMS (ESI/APCI) calculated for ($C_{65}H_{73}N_{14}O_{13}$) [M + H^+]: 1257.5476, found: 1257.5442.

2′-O-[2-(4-benzyloxycarbonylpiperizyl)-ethyl]-1,3,6′,3″-tetraazido-4″,6″-O-benzylidene-5,3′,4′,2″-tetra-O-benzylkanamycin A (**22b**): This compound was synthesized through reductive amination by coupling **17** with 1-benzyloxycarbonylpiperize with the same procedure as described in the preparation of **22a**. 55% yield, colorless semisolid. ^1H NMR (400 MHz, $CDCl_3$) δ 7.43–7.22 (m, 25H), 7.16–7.10 (m, 4H), 7.00 (t, J = 7.5 Hz, 1H), 5.56 (d, J = 3.6 Hz, 1H), 5.54 (d, J = 3.7 Hz, 1H), 5.28 (s, 1H), 5.10–5.06 (m, 3H), 4.93 (d, J = 12.1 Hz, 1H), 4.86–4.75 (m, 5H), 4.58 (d, J = 11.2 Hz, 1H), 4.31–4.27 (m, 1H), 3.97 (dd, J_1 = 5.0 Hz, J_2 = 10.3 Hz, 1H), 3.91 (t, J = 9.4 Hz, 1H), 3.84–3.72 (m, 3H), 3.67–3.36 (m, 10H), 3.30–3.20 (m, 7H), 2.40 (ddd, J_1 = J_2 = 4.4 Hz, $J3$ = 13.1Hz, 1H), 2.29–2.22 (m, 1H), 2.12–1.94 (m, 5H), 1.64 (ddd, J_1 = J_2 = J_3 = 12.7 Hz, 1H). ^{13}C NMR (100 MHz, $CDCl_3$) δ 155.09, 138.45, 137.84, 137.46, 137.21, 136.73, 128.85, 128.59, 128.52, 128.47, 128.39, 128.18, 128.09, 127.92, 127.56, 127.29, 127.16, 126.25, 125.24, 101.25, 97.18, 96.28, 82.67, 81.52, 80.11, 79.55, 78.3, 77.94, 77.71, 75.16, 74.95, 73.3, 70.8, 68.52, 67.12, 62.83, 61.43, 60.12, 59.22, 57.30, 52.88, 51.25, 43.46, 32.16, 29.70. HRMS (ESI/APCI) calculated for ($C_{67}H_{75}N_{14}O_{13}$) [M + H^+]: 1283.5633, found: 1283.5602.

2′-O-[2-(4-Morpholinyl)-ethyl]-1,3,6′,3″-tetraazido-4″,6″-O-benzylidene-5,3′,4′,2″-tetra-O-benz ylkanamycin A (**22c**): This compound was synthesized through reductive amination by coupling **17** with morpholine with the same procedure as described in the preparation of **22a**. 68% yield, colorless semisolid. ^1H NMR (400 MHz, $CDCl_3$) δ 7.44–7.21 (m, 20H), 7.17–7.10 (m, 4H), 7.01 (t, J = 7.4 Hz, 1H), 5.57 (d, J = 3.7 Hz, 1H), 5.54 (d, J = 3.7 Hz, 1H), 5.28 (s, 1H), 5.08 (d, J = 12.2 Hz, 1H), 4.93 (d, J = 12.2 Hz, 1H), 4.89–4.75 (m, 5H), 4.58 (d, J = 11.3 Hz, 1H), 4.32–4.27 (m, 1H), 3.99–3.90 (m, 2H), 3.85–3.71 (m, 3H), 3.67–3.36 (m, 14H), 3.31–3.20 (m, 3H), 2.40 (ddd, J_1 = J_2 = 4.5 Hz, J_3 = 13.1, Hz, 1H), 2.30–2.23 (m, 1H), 2.12–2.02 (m, 3H), 1.64 (ddd, J_1 = J_2 = J_3 = 13.0 Hz, 1H). ^{13}C NMR (100 MHz, $CDCl_3$) δ138.51, 137.84, 137.46, 137.24, 136.76, 128.83, 128.61, 128.58, 128.45, 128.36, 128.18, 128.15, 127.91, 127.89, 127.50, 127.29, 127.13, 126.24, 125.37, 101.26, 97.08, 96.28, 82.61, 81.40, 80.12, 79.59, 78.34, 77.97, 77.70, 77.23, 77.02, 76.81, 75.09, 74.93, 74.49, 73.3, 70.85, 69.59, 68.54, 66.65, 62.85, 61.46, 60.12, 59.3, 57.7, 53.62, 51.28, 32.17, 29.69. HRMS (ESI/APCI) calculated for ($C_{59}H_{68}N_{13}O_{12}$) [M + H^+]: 1150.5105, found: 1150.5063.

2′-O-(2-n-Pentylamino-ethyl)-1,3,6′,3″-tetraazido-4″,6″-O-benzylidene-5,3′,4′,2″-tetra-O-benzy lkanamycin A (**22d**): This compound was synthesized through reductive amination by coupling **17** with n-pentylamine with the same procedure as described in the preparation of **22a**. 64% yield, colorless semisolid. ^{1}H NMR (400 MHz, $CDCl_3$) δ 7.45–7.21 (m, 20H), 7.18–7.14 (m, 4H), 7.06 (t, J = 7.3 Hz, 1H), 5.56–5.54 (m, 2H), 5.30 (s, 1H), 5.05 (d, J = 12.1 Hz, 1H), 4.95 (d, J = 12.1 Hz, 1H), 4.86–4.75 (m, 5H), 4.57 (d, J = 11.2 Hz, 1H), 4.29–4.23 (m, 1H), 4.01 (dd, J_1 = 4.8, J_2 = 10.1 Hz, 1H), 3.92 (t, J = 9.3 Hz, 1H), 3.89–3.81 (m, 2H), 3.77 (t, J = 9.3 Hz, 1H), 3.70–3.20 (m, 13H), 2.54–2.51 (m, 1H), 2.46–2.22 (m, 4H), 1.65 (ddd, J_1 = J_2 = J_3 = 12.6 Hz, 1H), 1.33–1.06 (m, 6H), 0.84 (t, J = 7.2 Hz, 3H). ^{13}C NMR (100 MHz, $CDCl_3$) δ 138.34, 137.83, 137.35, 137.17, 136.71, 128.84, 128.55, 128.43, 128.38, 128.14, 127.92, 127.85, 127.60, 127.43, 127.32, 126.24, 125.49, 101.33, 96.80, 96.28, 82.58, 81.18, 80.19, 79.54, 78.30, 78.00, 77.83, 77.53, 75.33, 74.90, 74.64, 73.27, 71.25, 70.83, 68.54, 62.87, 61.39, 60.11, 59.46, 51.28, 49.61, 49.14, 32.21, 29.37, 29.16, 22.49, 14.04. HRMS (ESI/APCI) calculated for ($C_{60}H_{72}N_{13}O_{11}$) [M + H^+]: 1150.5469, found: 1150.5430.

2′-O-(2-Cyclohexylamino-ethyl)-1,3,6′,3″-tetraazido-4″,6″-O-benzylidene-5,3′,4′,2″-tetra-O-benzylkanamycin A (**22e**): This compound was synthesized through reductive amination by coupling **17** with cyclohexylamine with the same procedure as described in the preparation of **22a**. 52% yield, colorless semisolid. ^{1}H NMR (400 MHz, $CDCl_3$) δ 7.45–7.22 (m, 20H), 7.19–7.15 (m, 4H), 7.08 (t, J = 7.3 Hz, 1H), 5.55–5.53 (m, 2H), 5.32 (s, 1H), 5.02 (d, J = 12.1 Hz, 1H), 4.94 (d, J = 12.0 Hz, 1H), 4.86–4.73 (m, 5H), 4.56 (d, J = 11.3 Hz, 1H), 4.26–4.20 (m, 1H), 4.01 (dd, J_1 = 4.8 Hz, J_2 = 10.1Hz, 1H), 3.94–3.83 (m, 3H), 3.78 (t, J = 9.3 Hz, 1H), 3.69–3.31 (m, 12H), 3.25 (t, J = 9.8 Hz, 1H), 2.64–2.50 (m, 2H), 2.38 (ddd, J_1 = J_2 = 4.4 Hz, J_3 = 13.0 Hz, 1H), 2.19–2.12 (m, 1H), 1.71–1.51 (m, 4H), 1.12–0.80 (m, 6H). ^{13}C NMR (100 MHz, $CDCl_3$) δ 138.36, 137.83, 137.44, 137.26, 136.80, 128.95, 128.66, 128.52, 128.24, 128.02, 127.97, 127.74, 127.49, 126.32, 125.74, 101.44, 96.74, 96.36, 82.55, 81.03, 80.11, 79.63, 78.43, 78.12, 77.94, 77.83, 75.26, 74.89, 74.68, 73.37, 71.05, 68.63, 62.97, 61.47, 60.16, 59.60, 56.70, 51.37, 45.57, 32.26, 31.88, 25.64, 24.86. HRMS (ESI/APCI) calculated for ($C_{61}H_{72}N_{13}O_{11}$) [M + H^+]: 1162.5469, found: 1162.5486.

General procedure for the deprotection of compounds **18**, **19**, and **22a**–**22e**: The starting compound (0.03–0.08 mmol) was dissolved in the mixture of tetrahydrofuran (2 mL) and water (1 mL). Then, 50 mg of sodium hydroxide was added. The mixture was stirred for 10 min at room temperature, then 1 mL of trimethylphosphine solution in tetrahydrofuran (1 M) was added. After TLC showed the reaction was completed, the mixture was concentrated and the residue was passed through a short column (silica gel) with eluents as the following: methanol (30 mL), and methanol/ammonia solution in methanol (7 M) (50 mL/5 mL). The proper fractions were collected, combined, and concentrated. The resulting amine was then dissolved in the mixture of methanol (5 mL) and tetrahydrofuran (1 mL), and the pH value of the resulting solution was adjusted to 3–4 with hydrochloric acid (1 M). Then, Pd/C (10%, 50 mg) was added. The mixture was subjected to hydrogenolysis for 2–5 days. After TLC showed that only one spot appeared, the mixture was filtered through a pad of celite. To the filtrate was added 0.1 mL of triethylamine, and then a small volume of silica gel was added. The solvent was removed, and the resulting mixture was transferred to a silica gel column. After eluting the column with methanol (50 mL), methanol/concentrated aqueous ammonia (100 mL/10 mL), and methanol/concentrated aqueous ammonia (50 mL/10 mL), the fractions with the desired product were collected and combined. After removal of solvent, the gross product was redissolved in acetic acid solution in water (0.05 mol/L, 5 mL) and the resulting solution was freeze-dried. Thus, we obtained the final products **20**, **21**, and **23a**–**23e**.

2′-O-(2-Hydroxyethyl)-kanamycin A (**20**): 78% yield, white amorphous powder. ^{1}H NMR (600 MHz, D_2O) δ 5.70 (d, J = 3.8 Hz, 1H), 5.10 (d, J = 3.6 Hz, 1H), 3.99–3.95 (m, 1H), 3.93–3.80 (m, 8H), 3.77–3.70 (m, 4H), 3.67 (t, J = 10.1 Hz, 1H), 3.58–3.45 (m, 4H), 3.43–3.36 (m, 2H), 3.15 (dd, J_1 = 8.1 Hz, J_2 = 13.4 Hz, 1H), 2.50 (ddd, J_1 = J_2 = 4.2 Hz, J_3 = 12.5 Hz, 1H), 1.94 (s, 12H), 1.90 (ddd, J_1 = J_2 = J_3 = 12.5 Hz, 1H). ^{13}C NMR (150 MHz, D_2O) δ 180.83, 101.36, 95.18, 84.83, 79.81, 78.73, 73.94, 73.51, 73.34, 72.56, 71.64, 69.27, 68.88, 66.16, 61.30, 60.67, 55.65, 50.51, 48.36, 41.04, 28.46, 23.19. HRMS (ESI/APCI) calculated for ($C_{20}H_{41}N_4O_{12}$) [M + H]$^+$ requires m/z 529.2715, found m/z 529.2734.

2′-O-(2-Aminoethyl)-kanamycin A (**21**): 58% yield, white amorphous powder. ^{1}H NMR (600 MHz, D_2O) δ 5.83 (d, J = 3.8 Hz, 1H), 5.11 (d, J = 3.7 Hz, 1H), 4.09–4.06 (m, 1H), 3.95–3.87

(m, 7H), 3.83 (dd, J_1 = 2.2 Hz, J_2 = 12.3 Hz, 1H), 3.77–3.73 (m, 2H), 3.67 (t, J = 10.1 Hz, 1H), 3.57–3.40 (m, 6H), 3.30–3.26 (m, 1H), 3.21–3.18 (m, 2H), 2.49 (ddd, J_1 = J_2 = 4.1 Hz, J_3 = 12.6 Hz, 1H), 1.93 (s, 15H), 1.89 (ddd, J_1 = J_2 = J_3 = 12.7 Hz, 1H). ^{13}C NMR (150 MHz, D_2O) δ 181.37, 101.31, 95.57, 84.95, 79.72, 78.29, 74.25, 73.65, 71.76, 71.69, 69.30, 68.84, 67.98, 66.27, 60.78, 55.67, 50.44, 48.92, 40.98, 39.94, 28.89, 23.53. HRMS (ESI/APCI) calculated for ($C_{20}H_{42}N_5O_{11}$) [M + H]$^+$ requires *m/z* 528.2875, found *m/z* 528.2892.

2′-O-[2-(4-Mopholinyl)-ethyl]-kanamycin A (**23c**): 62% yield, white amorphous powder. ^{1}H NMR (600 MHz, D_2O) δ 5.89 (d, J = 3.7 Hz, 1H), 5.11 (d, J = 3.6 Hz, 1H), 4.17 (ddd, J_1 = J_2 = 4.4 Hz, J_3 = 12.2 Hz, 1H), 4.02–3.73 (m, 14H), 3.67 (t, J = 10.1 Hz, 1H), 3.58–3.39 (m, 12H), 3.19 (dd, J_1 = 7.7 Hz, J_2 = 13.4 Hz, 1H), 2.50 (ddd, J_1 = J_2 = 4.2 Hz, J_3 = 12.5 Hz, 1H), 1.97–1.91 (m, 16H). ^{13}C NMR (150 MHz, D_2O) δ 180.26, 100.56, 94.57, 84.09, 78.86, 76.79, 73.67, 72.88, 71.06, 70.71, 68.53, 68.06, 65.43, 63.94, 63.70, 59.93, 56.19, 54.88, 51.76, 49.59, 48.24, 40.22, 27.90, 22.56. HRMS (ESI/APCI) calculated for ($C_{24}H_{48}N_5O_{12}$) [M + H]$^+$ requires *m/z* 598.3294, found *m/z* 598.3315.

2′-O-[2-[(2-Aminoethyl)amino]-ethyl]-kanamycin A (**23a**): 55% yield, white amorphous powder. ^{1}H NMR (600 MHz, D_2O) δ 5.84 (d, J = 3.7 Hz, 1H), 5.13 (d, J = 3.6 Hz, 1H), 4.09–4.05(m, 1H), 3.99–3.92 (m, 3H), 3.91–3.66 (m, 8H), 3.55–3.15 (m, 13H), 2.40 (ddd, J_1 = J_2 = 4.2 Hz, J_3 = 12.7 Hz, 1H), 1.92 (s, 17H), 1.80 (ddd, J_1 = J_2 = J_3 = 12.6 Hz, 1H). ^{13}C NMR (150 MHz, D_2O) δ 182.09, 101.16, 95.41, 85.46, 79.81, 79.43, 74.52, 73.57, 71.86, 71.78, 69.16, 68.97, 67.93, 66.37, 60.74, 55.70, 50.68, 49.06, 48.30, 45.48, 41.09, 37.26, 30.25, 23.95. HRMS (ESI/APCI) calculated for ($C_{22}H_{47}N_6O_{11}$) [M + H]$^+$ requires *m/z* 571.3297, found *m/z* 571.3318

2′-O-[2-[(2-Piperizinylethyl)amino]-ethyl]-kanamycin A (**23b**): 60% yield, white amorphous powder. ^{1}H NMR (600 MHz, D_2O) δ 5.78 (d, J = 3.7 Hz, 1H), 5.10 (d, J = 3.6 Hz, 1H), 3.97-3.91 (m, 4H), 3.86-3.81 (m, 4H), 3.78–3.74 (m, 2H), 3.70–3.64 (m, 2H), 3.50–3.33 (m, 6H), 3.27 (t, J = 4.9 Hz, 4H), 3.17 (dd, J_1 = 8.0 Hz, J_2 = 13.4 Hz, 1H), 2.87–2.80 (m, 4H), 2.76–2.72 (m, 2H), 2.39 (ddd, J_1 = J_2 = 4.1 Hz, J_3 = 12.7 Hz, 1H), 1.90 (s, 15H), 1.77 (ddd, J_1 = J_2 = J_3 = 12.5 Hz, 1H). ^{13}C NMR (150 MHz, D_2O) δ 181.98, 101.17, 95.39, 85.46, 79.95, 79.54, 74.40, 73.54, 72.00, 71.76, 69.10, 68.93, 67.92, 66.28, 60.67, 57.12, 55.71, 50.68, 49.89, 48.94, 43.57, 41.06, 30.12, 23.87. HRMS (ESI/APCI) calculated for ($C_{24}H_{49}N_6O_{11}$) [M + H]$^+$ requires *m/z* 597.3454, found *m/z* 597.3473.

2′-O-[2-(n-Pentylamino)-ethyl]-kanamycin A (**23d**): 68% yield, white amorphous powder. ^{1}H NMR (600 MHz, D_2O) δ 5.88 (d, J = 3.5 Hz, 1H), 5.14 (d, J = 3.4 Hz, 1H), 4.14–4.11 (m, 1H), 4.00–3.84 (m, 8H), 3.78–3.75 (m, 2H), 3.69 (t, J = 10.1 Hz, 1H), 3.57–3.41 (m, 6H), 3.36–3.33 (m, 1H), 3.29–3.25 (m, 1H), 3.22–3.18 (m, 1H), 3.09 (t, J = 7.7 Hz, 2H), 2.46 (ddd, J_1 = J_2 = 4.5 Hz, J_3 = 12.5 Hz, 1H), 1.93 (s, 17H), 1.90 (ddd, J_1 = J_2 = J_3 = 12.5 Hz, 1H), 1.73–1.68 (m, 2H), 1.39–1.32 (m, 4H), 0.90 (t, J = 6.9 Hz, 3H). ^{13}C NMR (150 MHz, D_2O) δ 181.71, 101.27, 95.52, 85.16, 79.68, 78.52, 74.64, 73.65, 71.87, 71.62, 69.24, 68.90, 66.66, 66.30, 60.77, 55.68, 50.57, 49.12, 48.34, 47.56, 41.09, 29.51, 28.57, 25.87, 23.71, 22.21, 13.77. HRMS (ESI/APCI) calculated for ($C_{25}H_{52}N_5O_{11}$) [M + H]$^+$ requires *m/z* 598.3658, found *m/z* 598.3678.

2′-O-[2-(Cyclohexylamino)-ethyl]-kanamycin A (**23e**): 64% yield, white amorphous powder. ^{1}H NMR (600 MHz, D_2O) δ 5.84 (d, J = 3.7 Hz, 1H), 5.11 (d, J = 3.7 Hz, 1H), 4.12–4.09 (m, 1H), 3.96–3.80 (m, 8H), 3.77–3.65 (m, 3H), 3.55–3.34 (m, 7H), 3.29–3.25 (m, 1H), 3.21–3.12 (m, 2H), 2.42 (ddd, J_1 = J_2 = 4.5 Hz, J_3 = 12.4 Hz, 1H), 2.08 (br, 2H), 1.92 (s, 12H), 1.85–1.79 (m, 3H), 1.69–1.64 (m, 1H), 1.40–1.27 (m, 4H), 1.21–1.14 (m, 1H). ^{13}C NMR (150 MHz, D_2O) δ 181.90, 101.25, 95.59, 85.29, 79.65, 79.08, 74.59, 73.63, 71.79, 71.63, 69.21, 68.89, 66.87, 66.28, 60.74, 58.02, 55.69, 50.59, 49.09, 44.59, 41.01, 29.81, 29.70, 29.50, 25.19, 24.68, 24.65, 23.88. HRMS (ESI/APCI) calculated for ($C_{26}H_{52}N_5O_{11}$) [M + H]$^+$ requires *m/z* 610.3658, found *m/z* 610.3680.

3.2. Bioassay

Ten bacterial strains were selected to evaluate the minimal inhibitory concentration (MIC) of compounds. All newly synthesized compounds were tested in the form of acetate. Corning 96-well plates were utilized for this test. Briefly, tested strains were seeded into 200 μL Mueller–Hinton (MH) broth per well with a concentration of 10^5 CFU/mL. Subsequently, an aliquot of sample stock was added, with a series of final concentrations of 1–128 μg/mL. All of the mixtures were incubated at 37 °C for 24 h. The MICs were determined by measuring the optical density at 600 nm. The sterilized water (0 μg/mL) was used as the control; all of the tests were performed in triplicate.

4. Conclusions

According to the fact that some modified products of aminoglycosides by AAC(2′) remain active, the possible reason was analyzed by means of some structural biology data. It was deduced that the 2′-position of neamine-containing aminoglycosides is a proper position for modification. Based on this hypothesis, we designed, synthesized, and evaluated a series of 2′-modified derivatives of kanamycin A. As expected, all derivatives exhibited moderate to good antibacterial activity. The structure–activity relationship showed that the feature of the introduced groups on the 2′-position, including the number of amino groups, rigidity, hydrophobicity, and bulk, had a mild impact on activity. All of these results were believed to be attributed to the fact that 2′-substituents point to a vacant space. On the other hand, proactive chemical modification on the 2′-position may bring some benefits to fight against drug-resistant bacteria, but cannot achieve a strong and broad-spectrum effect.

Supplementary Materials: The following supporting information can be downloaded at: https://www.mdpi.com/article/10.3390/molecules27217482/s1, NMR Spectra.

Author Contributions: Conceiving, R.Y.; investigation, R.Y., X.L. and Y.L.; writing, R.Y.; supervision, X.Y.; project administration, X.Y.; funding acquisition, R.Y. and X.Y. All authors have read and agreed to the published version of the manuscript.

Funding: This research was funded by the National Natural Science Foundation of China (Grant No. 21877006).

Institutional Review Board Statement: Not applicable.

Informed Consent Statement: Not applicable.

Data Availability Statement: The data presented in this study are available on request from the corresponding author.

Acknowledgments: We thank Qin Li, Jun Li and Yuan Wang for data collection.

Conflicts of Interest: The authors declare no conflict of interest.

References

1. Magnet, S.; Blanchard, J.S. Molecular Insights into Aminoglycoside Action and Resistance. *Chem. Rev.* **2005**, *105*, 477–497. [CrossRef] [PubMed]
2. Ramirez, M.S.; Tolmasky, M.E. Aminoglycoside Modifying Enzymes. *Drug Resist. Updat.* **2010**, *13*, 151–171. [CrossRef] [PubMed]
3. Garneau-Tsodikova, S.; Labby, K.J. Mechanisms of Resistance to Aminoglycoside Antibiotics: Overview and Perspectives. *MedChemComm* **2016**, *7*, 11–27. [CrossRef]
4. Ye, X.S.; Zhang, L.H. Aminoglycoside Mimetics as Small-Molecule Drugs Targeting RNA. *Curr. Med. Chem.* **2002**, *9*, 929–939. [CrossRef]
5. Zhou, J.; Wang, G.; Zhang, L.-H.; Ye, X.-S. Modifications of Aminoglycoside Antibiotics Targeting RNA. *Med. Res. Rev.* **2007**, *27*, 279–316. [CrossRef]
6. Houghton, J.L.; Green, K.D.; Chen, W.; Garneau-Tsodikova, S. The Future of Aminoglycosides: The End or Renaissance? *ChemBioChem* **2010**, *11*, 880–902. [CrossRef]
7. Obszynski, J.; Loidon, H.; Blanc, A.; Weibel, J.-M.; Pale, P. Targeted modifications of neomycin and paromomycin: Towards resistance-free antibiotics? *Bioorg. Chem.* **2022**, *126*, 105824. [CrossRef]
8. Chandrika, N.T.; Garneau-Tsodikova, S. Comprehensive review of chemical strategies for the preparation of new aminoglycosides and their biological activities. *Chem. Soc. Rev.* **2018**, *47*, 1189–1249. [CrossRef]
9. Moazed, D.; Noller, H.F. Interaction of Antibiotics with Functional Sites in 16S Ribosomal RNA. *Nature* **1987**, *327*, 389–394. [CrossRef]
10. Woodcock, J.; Moazed, D.; Cannon, M.; Davies, J.; Noller, H.F. Interaction of Antibiotics with A- and P-site-specific Bases in 16S Ribosomal RNA. *EMBO J.* **1991**, *10*, 3099–3103. [CrossRef]

11. Fourmy, D.; Recht, M.I.; Blanchard, S.C.; Puglisi, J.D. Structure of The A-site of Escherichia Coli 16S Ribosomal RNA Complexed with An Aminoglycoside Antibiotic. *Science* **1996**, *274*, 1367–1371. [CrossRef] [PubMed]
12. Carter, A.P.; Clemons, W.M.; Brodersen, D.E.; Morgan-Warren, R.J.; Wimberly, B.T.; Ramakrishnan, V. Functional Insights from The Structure of The 30S Ribosomal Subunit and Its Interactions with Antibiotics. *Nature* **2000**, *407*, 340–348. [CrossRef] [PubMed]
13. Bastida, A.; Hidalgo, A.; Chiara, J.L.; Torrado, M.; Corzana, F.; Pérez-Cañadillas, J.M.; Groves, P.; Garcia-Junceda, E.; Gonzalez, C.; Jimenez-Barbero, J.; et al. Exploring the Use of Conformationally Locked Aminoglycosides as a New Strategy to Overcome Bacterial Resistance. *J. Am. Chem. Soc.* **2006**, *128*, 100–116. [CrossRef] [PubMed]
14. Hanessian, S.; Szychowski, J.; Adhikari, S.S.; Vasquez, G.; Kandasamy, P.; Swayze, E.E.; Migawa, M.T.; Ranken, R.; François, B.; Wirmer-Bartoschek, J.; et al. Structure-Based Design, Synthesis, and A-Site rRNA Cocrystal Complexes of Functionally Novel Aminoglycoside Antibiotics: C2″ Ether Analogues of Paromomycin. *J. Med. Chem.* **2007**, *50*, 2352–2369. [CrossRef]
15. Kondo, J.; Pachamuthu, K.; François, B.; Szychowski, J.; Hanessian, S.; Westhof, E. Crystal Structure of the Bacterial Ribosomal Decoding Site Complexed with a Synthetic Doubly Functionalized Paromomycin Derivative: A New Specific Binding Mode to an A-Minor Motif Enhances in vitro Antibacterial Activity. *ChemMedChem* **2007**, *2*, 1631–1638. [CrossRef]
16. Yan, R.-B.; Yuan, M.; Wu, Y.; You, X.; Ye, X.-S. Rational design and synthesis of potent aminoglycoside antibiotics against resistant bacterial strains. *Bioorg. Med. Chem.* **2011**, *19*, 30–40. [CrossRef]
17. Kanazawa, H.; Saavedra, O.M.; Maianti, J.P.; Young, S.A.; Izquierdo, L.; Smith, T.K.; Hanessian, S.; Kondo, J. Structure-Based Design of a Eukaryote-Selective Antiprotozoal Fluorinated Aminoglycoside. *ChemMedChem* **2018**, *13*, 1541–1548. [CrossRef]
18. Hotta, K.; Zhu, C.-B.; Ogata, T.; Sunada, A.; Ishikawa, J.; Mizuno, S.; Ikeda, Y.; Kondo, S. Enzymatic 2′-N-acetylation of arbekacin and antibiotic activity of its product. *J. Antibiot.* **1996**, *49*, 458–464. [CrossRef]
19. Vicens, Q.; Westhof, E. Molecular recognition of aminoglycoside antibiotics by ribosomal RNA and resistance enzymes: An analysis of x-ray crystal structures. *Biopolymers* **2003**, *70*, 42–57. [CrossRef]
20. François, B.; Russell, R.J.M.; Murray, J.B.; Aboul-Ela, F.; Masquida, B.; Vicens, Q.; Westhof, E. Crystal structures of complexes between aminoglycosides and decoding A site oligonucleotides: Role of the number of rings and positive charges in the specific binding leading to miscoding. *Nucleic Acids Res.* **2005**, *33*, 5677–5690. [CrossRef]
21. Sati, G.C.; Sarpe, V.A.; Furukawa, T.; Mondal, S.; Mantovani, M.; Hobbie, S.N.; Vasella, A.; Böttger, E.C.; Crich, D. Modification at the 2′-Position of the 4,5-Series of 2-Deoxystreptamine Aminoglycoside Antibiotics to Resist Aminoglycoside Modifying Enzymes and Increase Ribosomal Target Selectivity. *ACS Infect. Dis.* **2019**, *5*, 1718–1730. [CrossRef] [PubMed]
22. Zhang, W.X.; Chen, Y.; Liang, Q.Z.; Li, H.; Jin, H.; Zhang, L.; Meng, X.; Li, Z. Design, Synthesis, and Antibacterial Activities of Conformationally Constrained Kanamycin A Derivatives. *J. Org. Chem.* **2013**, *78*, 400–409. [CrossRef] [PubMed]

molecules

MDPI

Review

Recent Advances on Natural Aryl-*C*-glycoside Scaffolds: Structure, Bioactivities, and Synthesis—A Comprehensive Review

Chen-Fu Liu

School of Pharmaceutical Sciences, Gannan Medical University, Ganzhou 341000, China; chenfu@gmu.edu.cn

Abstract: Aryl-C-glycosides, of both synthetic and natural origin, are of great significance in medicinal chemistry owing to their unique structures and stability towards enzymatic and chemical hydrolysis as compared to *O*-glycosides. They are well-known antibiotics and potent enzyme inhibitors and possess a wide range of biological activities such as anticancer, antioxidant, antiviral, hypoglycemic effects, and so on. Currently, a number of aryl-C-glycoside drugs are on sale for the treatment of diabetes and related complications. This review summarizes the findings on aryl-C-glycoside scaffolds over the past 20 years, concerning new structures (over 200 molecules), their bioactivities—including anticancer, anti-inflammatory, antioxidant, antivirus, glycation inhibitory activities and other pharmacological effects—as well as their synthesis.

Keywords: aryl-C-glycoside; novel structures; bioactivity; synthesis

check for updates

Citation: Liu, C.-F. Recent Advances on Natural Aryl-C-glycoside Scaffolds: Structure, Bioactivities, and Synthesis—A Comprehensive Review. *Molecules* **2022**, *27*, 7439. https://doi.org/10.3390/molecules27217439

Academic Editors: Jian Yin, Jing Zeng and De-Cai Xiong

Received: 17 October 2022
Accepted: 31 October 2022
Published: 1 November 2022

1. Introduction

Aryl-C-glycosides (ACGs), natural secondary metabolites, are glycosides in which the anomeric center is covalently linked to the carbon atom of arenes or heterocycles (C_{sp3}-C_{sp2}) [1]. They tend to have high oral bioavailability and reach high plasma levels without needing to be converted to a prodrug. The stable linkage between the sugar and the arene or heterocycle moieties is resistant to enzymatic hydrolysis, allowing these compounds to interfere with DNA and RNA synthases more efficiently [2]. Both natural and synthetic aryl-C-glycosides are of marked pharmaceutical interest, and many of them have proven to be efficient antibiotics, antitumor agents, and antidiabetics. Therefore, aryl-C-glycosides have received considerable attention [3].

Beginning with the structural elucidation of the first aryl-C-glycoside in 1970 [4], the study of aryl-C-glycosides has continued to flourish through the constant isolation/characterization of new aryl-C-glycoside natural products, the characterization of the corresponding biosynthetic pathways/enzymes, and the development of aryl-C-glycoside synthetic methods. As their name suggests, the core structure of aryl-C-glycosides is often abundantly decorated with substituents such as aromatic acids (e.g., gallic acid, caffeic acid, vanillic acid, benzoic acid, and ferulic acid) and various saccharides (e.g., L-rhamnose, D-xylose, D-glucose, D-galactose, and L-arabinose) through ester or glycosidic linkages, respectively. The outstanding activity of aryl-C-glycosides against diverse diseases proves their importance in medicinal chemistry research. Several reviews on aryl C-glycosides regarding their isolation and purification, structure elucidation, synthesis and biosynthesis, and pharmacological activities have been published [5–7]. Recently, interest in aryl-C-glycoside has been growing, as shown by a significantly increasing volume of literature describing novel structures, diverse bioactivities, general synthesis, and evident roles of these molecules in the prevention and treatment of various human diseases [8]. Such rich information prompted me to review papers on novel aryl-C-glycoside structures, pharmacological activities, and chemical synthesis published in the last two decades. This review will highlight the new structures (over 200 new aryl-C-glycoside molecules), bioactivities, and synthetic approaches to the preparation of aryl-C-glycosides, which were published in prestigious journals such as Journal of the American Chemical Society, Angewandte Chemie International Edition, Journal of Natural Products, Organic Letters, Phytochemistry and in related peer-reviewed natural product research journals, from 2002 to 2022. A summary of the structures, bioactivities, and origins of recently discovered aryl-C-glycoside molecules is provided in Table 1.

Table 1. Structures, Bioactivities, Sources of Aryl-C-glycoside Molecules.

Entry	Name	Carbohydrate	Bioactivities	Source	Ref.
1	Apigenosylide B (**16**)	β-D-glucose	α-glucosidase inhibitory activity	*Machilus japonica*	[9]
2	Speciflavoside A (**32**)	β-D-glucose	antiviral activity	*Lilium speciosum* var. *gloriosoides Baker*	[10]
3	Compounds **34–36**	β-D-glucose	cytotoxicity activity	*Lemna japonica*	[11]
4	Nelumboside B (**40**)	β-D-glucose	antioxidant activity	*Nelumbo nucifera*	[12]
5	Compound **42**	β-D-glucose	antioxidant activity	*Gentiana piasezkii*	[13]
6	Compounds **45–46**	β-D-glucose	anti-complementary activity	*Trollius chinensis*	[14]
7	Compound **68**	β-D-glucose	macrophage respiratory burst inhibitory activity	*Cyperus rotundus*	[15]
8	Diandraflavone (**81**)	β-D-glucose	selective inhibition on superoxide anion	*Drymaria diandra*	[16]
9	Nervilifordin J (**94**)	β-D-galactose	anti-inflammatory activity	*Nervilia fordii*	[17]
10	Chafurosides A–B (**107–108**)	β-D-mannose	anti-inflammation	oolong tea	[18]
11	Compound **109**	β-L-fucose	glycation inhibitory activity	the style of *Zea mays* L.	[19]
12	Compound **112**	β-L-boivinose	glycation inhibitory activity	the Style of *Zea mays*	[20]
13	Glomexanthones A–C (**131, 129, 132**)	β-D-glucose	neuroprotective effects	*Polygala glomerata*	[21]
14	Shamimoside (**135**)	β-D-glucose	antioxidant activity	*Bombax ceiba*	[22]
15	Compound **136**	β-D-glucose	neuroprotective activity	*Swertia punicea*	[23]
16	Calophymembranside C (**141**)	β-D-glucose	transcriptional inhibitory activity of RXRα	*Calophyllum membranaceum*	[24]
17	Arenicolin A (**152**)	β-D-glucose	anticancer activity	*Penicillium arenicola*	[25]
18	Compound **155**	β-D-glucose	antipyretic activity	*Melicope pteleifolia*	[26]
19	Ardimerin digallate (**169**)	β-D-glucose	inhibitory activity on HIV-1 and HIV-2 ribonuclease H	*Ardisia japonica*	[27]
20	Kunzeachromones A–F (**172–177**)	β-D-glucose	antivirus activity	*Kunzea ambigua.*	[28]
21	Neopetrosins A–D (**180–183**)	α-D-mannose	hepatoprotective activity	*Neopetrosia chaliniformis*	[29]
22	Konamycins A–B (**188–189**)	β-D-amicetose	radical scavenging activity	*Streptomyces hyaluromycini* MB-PO13^{T}	[30]
23	Monacyclinones I, J (**195, 192**)	α-L-amino sugar	anticancer activity	*Streptomyces* sp. HDN15129	[31]
24	Marmycins A and B (**198–199**)	α-L-amino sugar	anticancer activity	actinomycete related to the genus *Streptomyces*	[32]
25	Aciculatin (**200**)	β-D-digitoxopyranose	cytotoxic, anti-inflammatory, anti-arthritis activity	*Chrysopogon aciculatus*	[33]

Table 1. *Cont.*

Entry	Name	Carbohydrate	Bioactivities	Source	Ref.
26	Compound **201**	*β*-D-digitoxopyranose	anticancer activity	*Chrysopogon aciculatis*	[33]
27	Compound **202**	*β*-D-boivinopyranose	anticancer activity	*Chrysopogon aciculatis*	[33]
28	Aciculatinone (**204**)	3-keto-*β*-D-digitoxopyranose	anticancer activity	*Chrysopogon aciculatis*	[33]
29	Isocassiaoccidentalin B (**206**)	*β*-D-6-Deoxy-ribo-hexos-3-ulose	free-radical scavenging activity	*Cassia nomame*	[34]
30	Grincamycins B–D (**207–209**)	*β*-D-olivose	anticancer activity	*Streptomyces lusitanus* SCSIO LR32	[35]
31	Marangucyclines A–B (**210–211**)	*β*-D-olivose	anticancer activity	*Streptomyces* sp. SCSIO 11594	[36]

2. Structures

2.1. Flavonoid C-Glycosides

Apigenin and luteolin represent the major parent nuclei of flavonoid C-glycosides; the most common glycan moieties linked to C-glycosides are D-glucose, D-galactose, D-xylose, D-mannose, D-ribose, L-fucose, L-arabinose, and L-rhamnose.

2.1.1. Flavonoid C-Glucosides

To the best of my knowledge, Flavonoid *C*-*α*-D-glucopyranosides, Flavonoid *C*-*α*-L-glucopyranosides, and Flavonoid *C*-*β*-L-glucopyranosides have not been isolated and identified from natural source. Flavonoid *C*-*β*-D-glucopyranosides are the largest group of isolated aryl-*C*-glycosides. Since 2002, more than 80 new compounds have been isolated and identified. It was found that the D-glucose unit is directly attached to the flavonoid or flavonoid derivatives in *β* configuration. Compared with the known aryl-C-glycosides previously reported, some of the new ones differ in their core structure, while others differ in the number and/or position of their substituents. The sites of glycosylation in the flavonoids are usually C6 and/or C8. Very few examples are known of C-glycosylation occurring at position C4′.

Figure 1 illustrates the new aryl-C-glycosides with varied core structures or special substituents. These structures differ in the aglycon and or glucoside portions. The core structures are often abundantly decorated with substituents such as aromatic acids (e.g., benzoic acid, 2-methyl butyric acid, gallic acid, veratric acid, sinapic acid, ferulic acid, and acetic acid) and various saccharides (e.g., L-rhamnose, D-xylose, D-glucose, D-galactose, and L-arabinose) through ester or glycosidic linkages, respectively. These isolated structures are often associated with other known compounds, such as flavonoids, lignan, sesquiterpene, steroid, alkaloids, *O*-glycosides, etc.

Besides the well-known flavonoid C-glycosides vitexin, isovitexin, orientin, isoorientin, schaftoside, isoshaftoside, neoshaftoside, and their *O*-glycosylated and or *O*-acylated derivatives, more and more flavonoid C-glycosides were disclosed. These 6- or 8-*C*-*β*-D-glucopyranosides can be divided into five groups: (1) mono-C-glycosylated flavonoids **1–3** [12,37], **39–42** [12,13,37]; (2) 6, 8-di-C-glycosylated flavonoids **4** [38], **43** [39]; (3) flavonoid 6- or 8-*C*-*β*-D-glucopyranoside-7-*O*-*β*-D-glucopyranosides **5–10** [40–43], **44** [44]; (4) 2″- or 6″-*O*-glycosylated flavonoid 6- or 8-*C*-*β*-D-glucopyranosides **11–31** [9,38,45–50], **45–55** [14,51–57]; (5) 2″- or 6″-*O*-acylated flavonoid 6- or 8-*C*-*β*-D-glucopyranosides **32–38** [10,11,55,56], **56–80** [57–61]. Some of the above compounds are glycosylated with *β*-D-glucose, a disaccharide consisting of *β*-D-glucose glycosylated with a monosaccharide at position C2 or C6, or *β*-D-glucose acylated at C2 or C6. Such glycosylation with various sugars and/or acylation with various acids is a common strategy used by nature to introduce structural diversity and different biological activities in natural products.

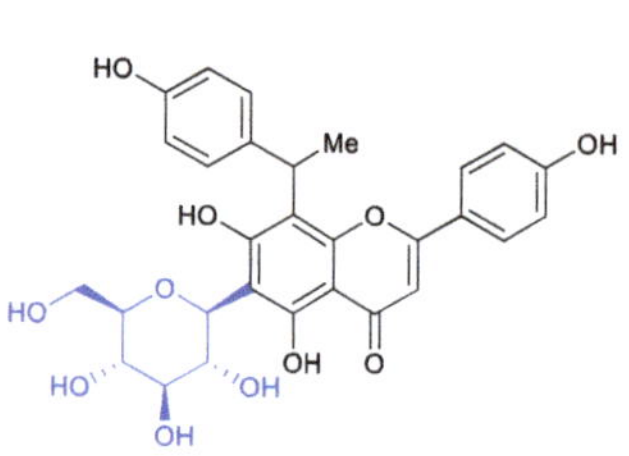

1, cucumerin B

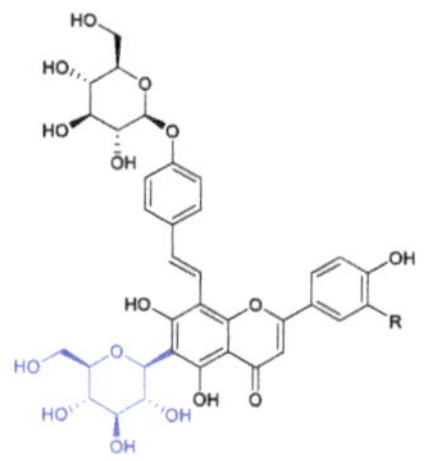

2, nelumboside A, R = OH
3, nelumboside C, R = H

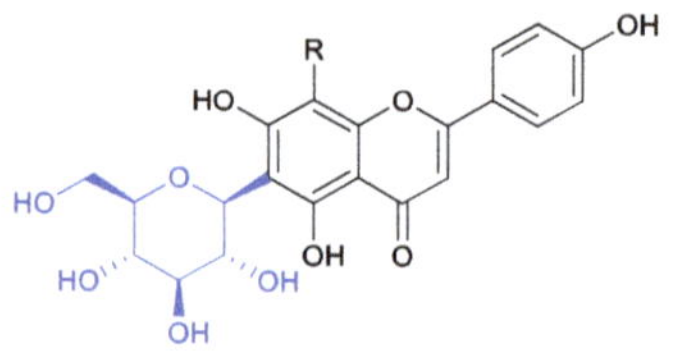

4, vicenin-3, R = *β*-D-xylopyranosyl

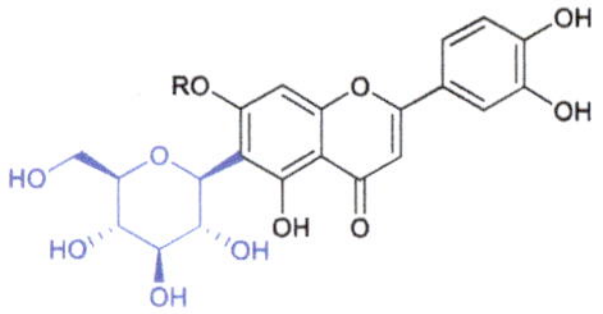

5, R = *α*-L-arabinopyranosyl

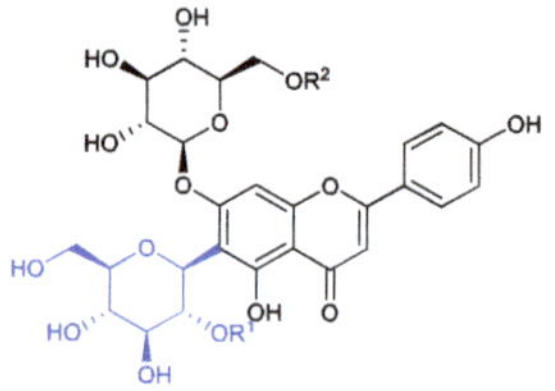

6, divarioside A, R[1] = *α*-L-rhamnopyranosyl, R[2] = isoferuloyl; **7**, divarioside B, R[1] = *β*-D-glucopyranosyl, R[2] = isoferuloyl

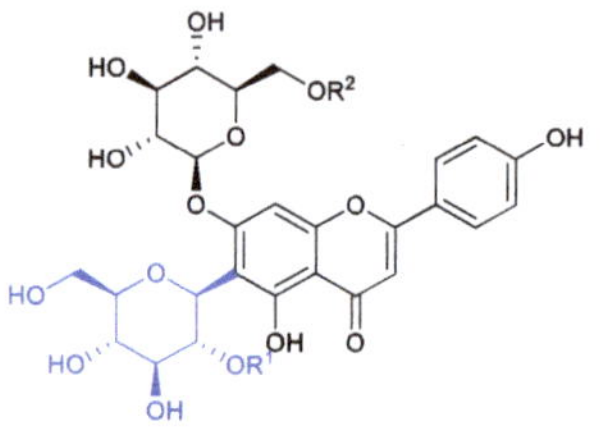

8, sileneside D, R[1] = *β*-D-glucopyranosyl, R[2] = H; **9**, sileneside G, R[1] = *β*-D-xylopyranosyl, R[2] = isoferuloyl

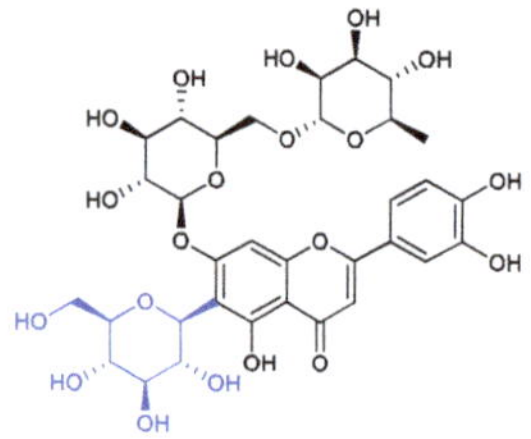

10

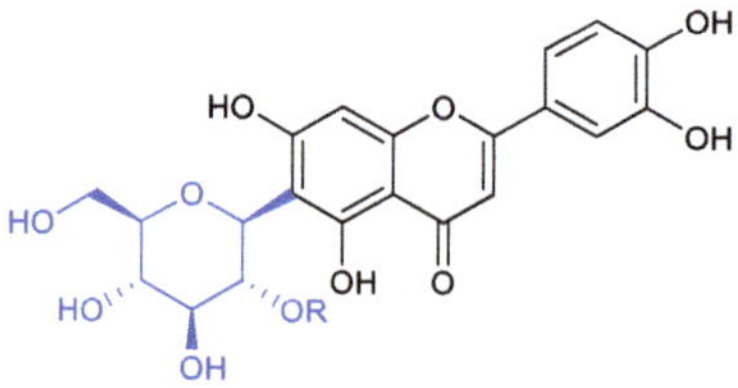

11, R = *α*-L-rhamnopyranosyl

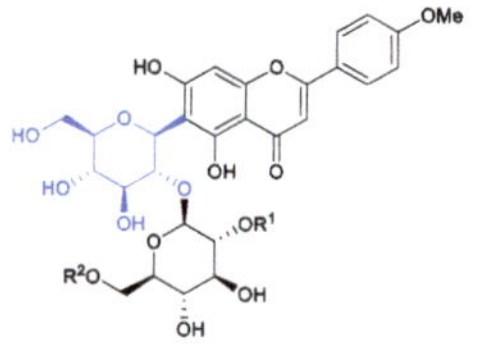

12, paraquinin A, R[1] = R[2] = H; **13**, paraquinin B, R[1] = *α*-L-rhamnopyranosyl, R[2] = H; **14**, paraquinin C, R[1] = *α*-L-rhamnopyranosyl, R[2] = E-feruloyl

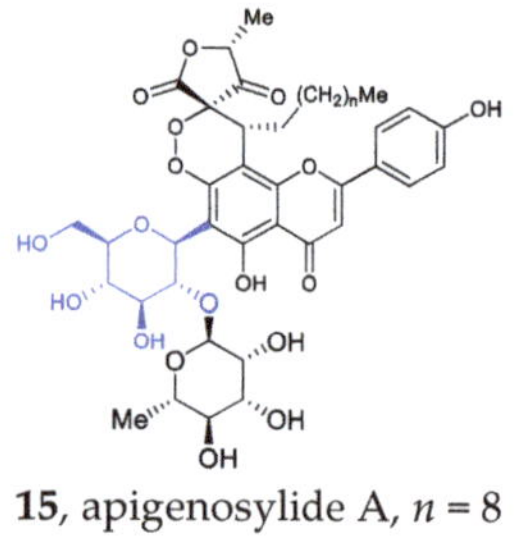

15, apigenosylide A, *n* = 8
16, apigenosylide B, *n* = 6

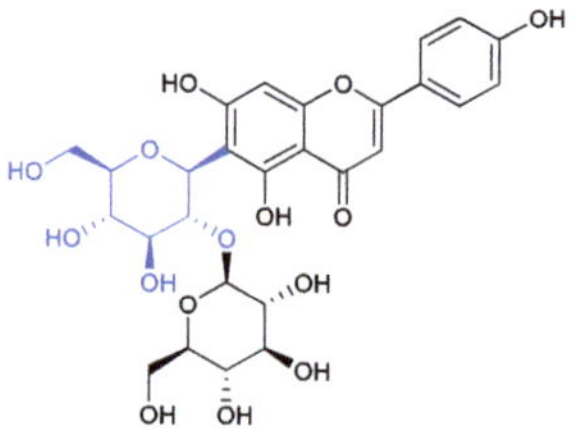

17

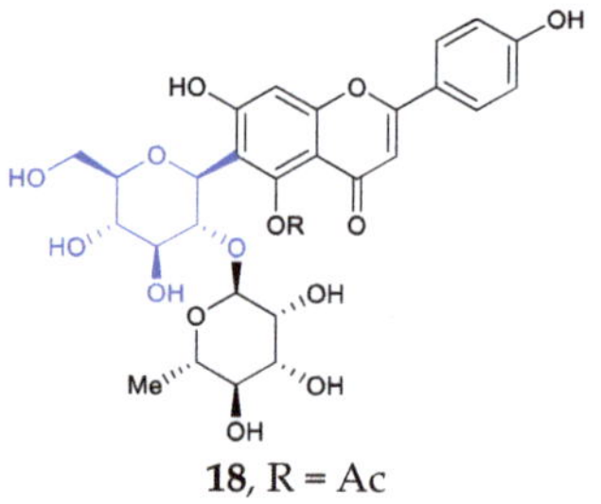

18, R = Ac
19, R = H

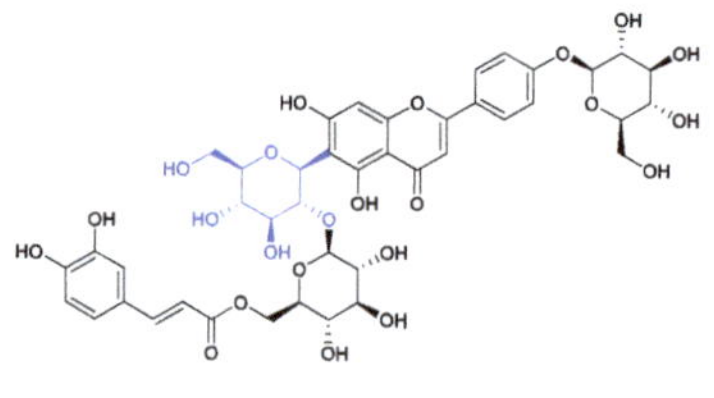

Figure 1. *Cont.*

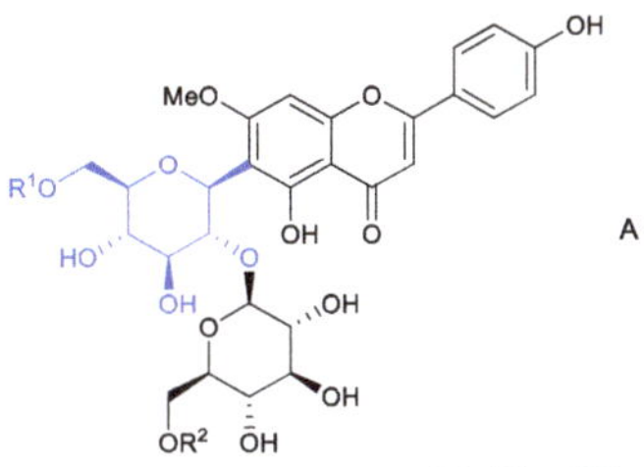

A = 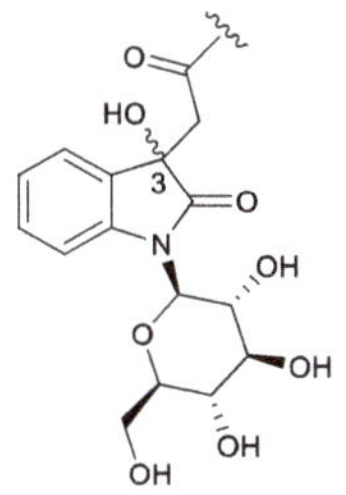

20, R¹ = R² = H
21, R¹ = H, R² = A(3S)
22, R¹ = H, R² = A(3R)
23, R¹ = A(3S), R² = H
24, R¹ = A(3R), R² = H
25, R¹ = A(3S), R² = feruloyl
26, R¹ = A(3R), R² = feruloyl
27, R¹ = H, R² = feruloyl

28, gentiflavone A

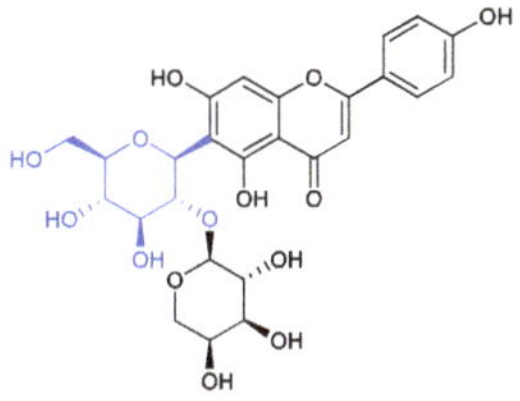

29

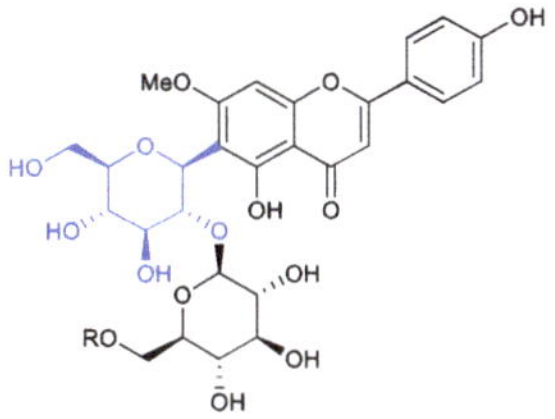

30, spinosin, R = H
31, R = 6‴-feruloyl

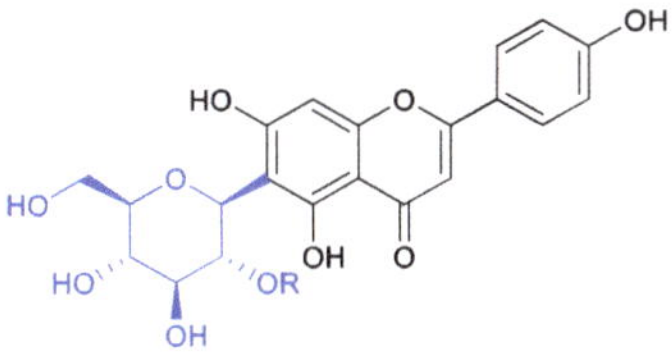

32, speciflavoside A, R = 2-methyl-butyryl
33, R = galloyl

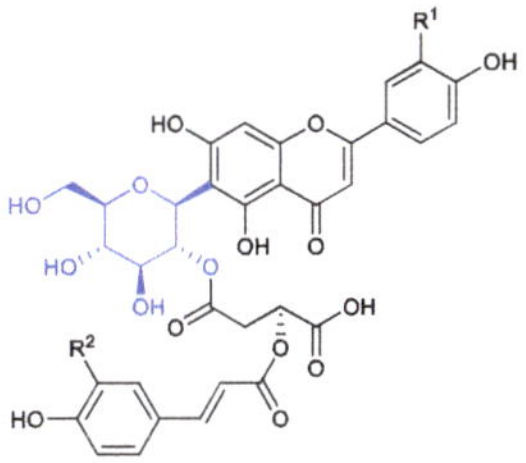

34, R¹ = OH, R² = OH
35, R¹ = H, R² = OH
36, R¹ = OH, R² =H

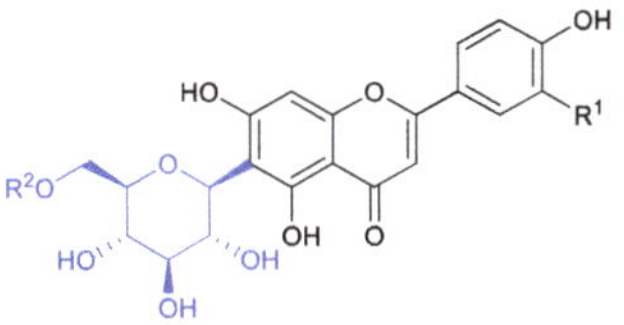

37, R¹ = H, R² = 3-hydroxy-3-methylglutaroyl
38, R¹ = OH, R² = Ac

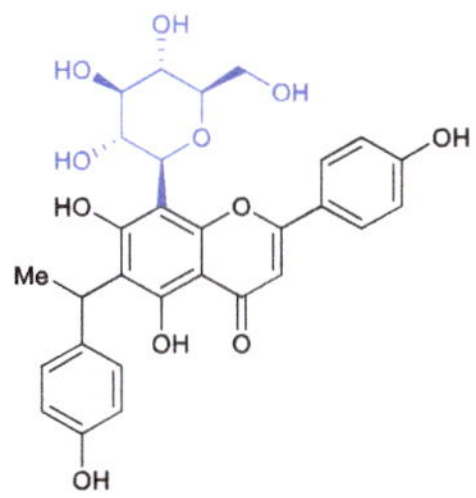

39, cucumerin A

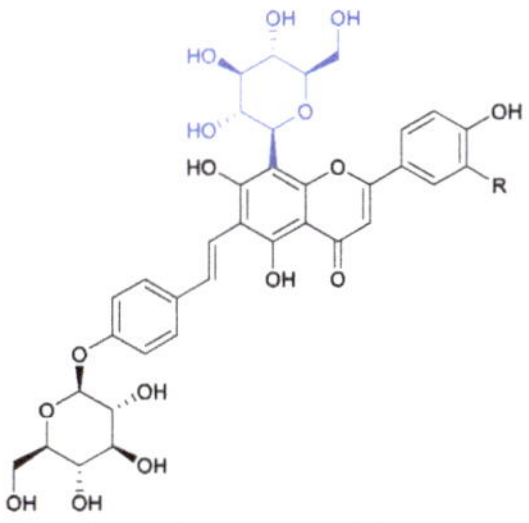

40, nelumboside B, R = OH; **41**, nelumboside D, R = H

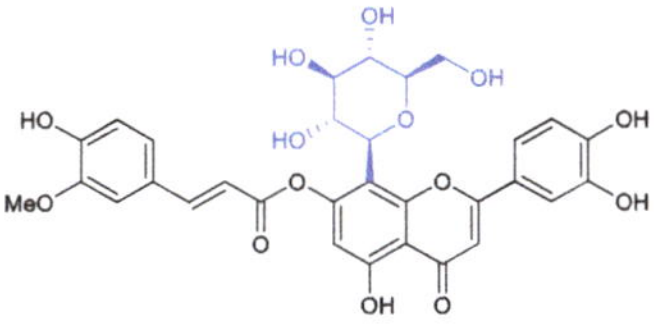

42

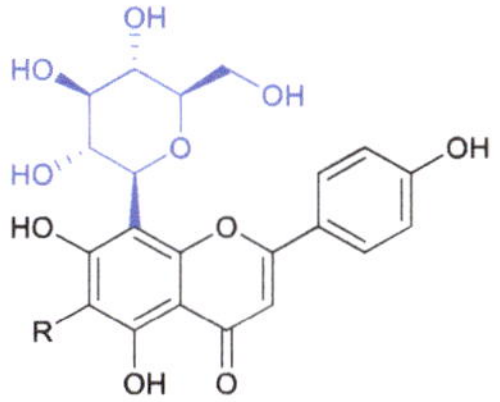

43, vicenin 1, R = β-D-xylopyranosyl

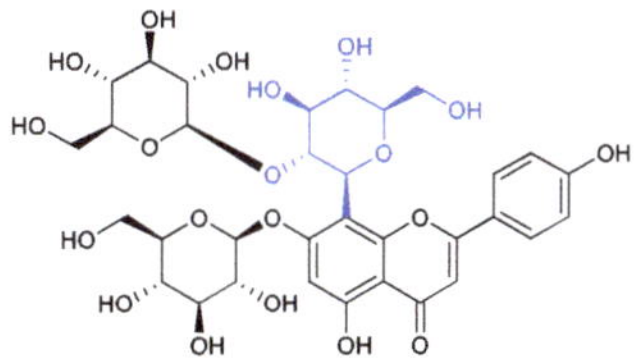

44

Figure 1. *Cont.*

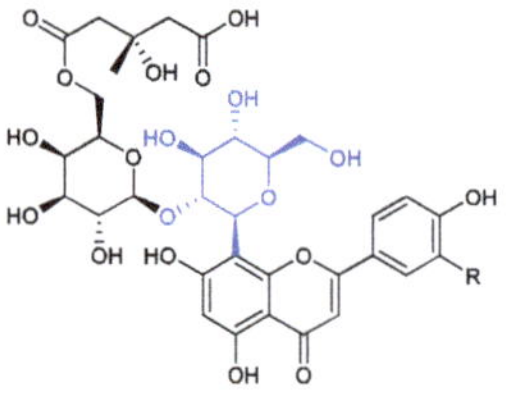

45, R = H
46, R = OH

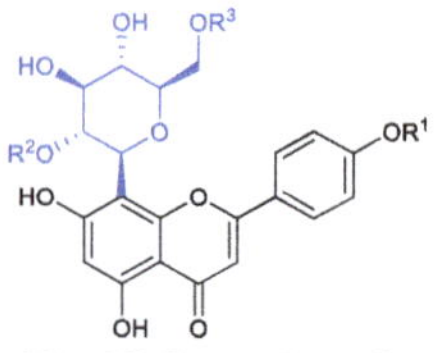

47, R¹ = R³ = H, R² = *β*-L-arabinopyranosyl; **48**, R¹ = R³ = H, R² = *α*-D-xylopyranosyl

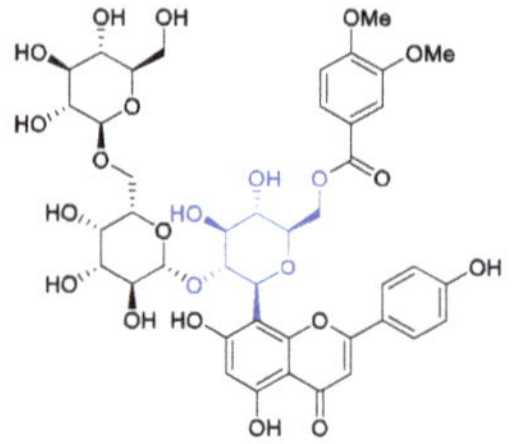

49, trollisin A

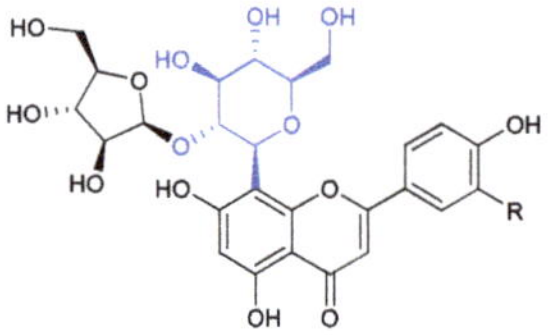

50, R = H
51, R = OH

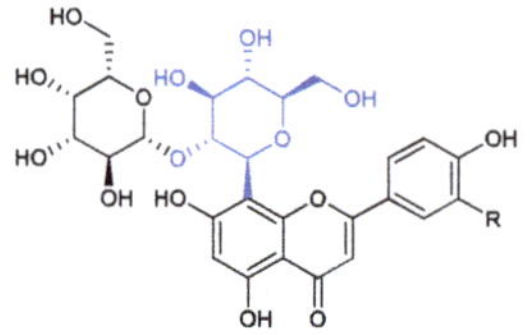

52, R = H
53, R = OH

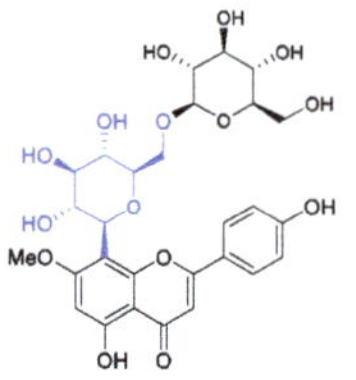

54, trollisin B

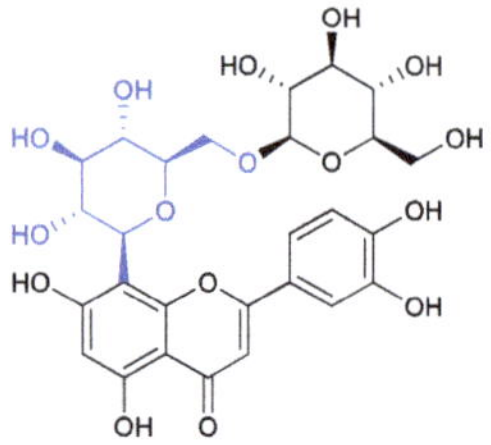

55

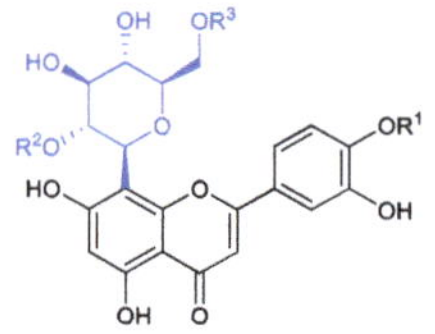

56, R¹ = R² = H, R³ = Ac; **57**, R¹ = R³ = H, R² = 2-methylbutyryl; **58**, R¹ = R³ = H, R² = veratroyl; **59**, R¹ = Me, R² = 2-methylbutyryl, R³ = H

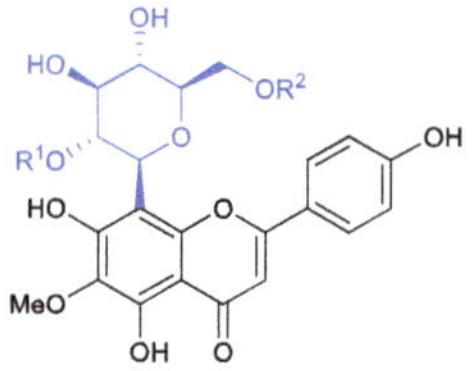

60, R¹ = syringoyl, R² = H
61, R¹ = H, R² = Ac
62, R¹ = syringoyl, R² = Ac

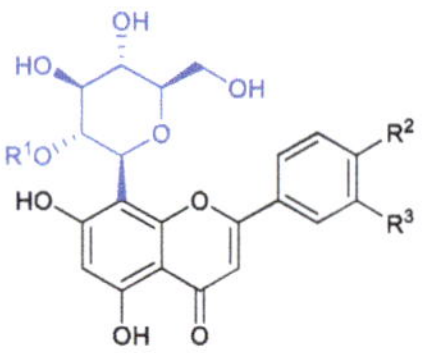

63, R¹ = vanilloyl, R² = OH, R³ = H; **64**, R¹ = veratroyl, R² = OH, R³ = H; **65**, R¹ = veratroyl, R² = OH, R³ = OH; **66**, R¹ = E-feruloyl, R² = R³ = OH; **67**, R¹ = galloyl, R² = OH, R³ = H

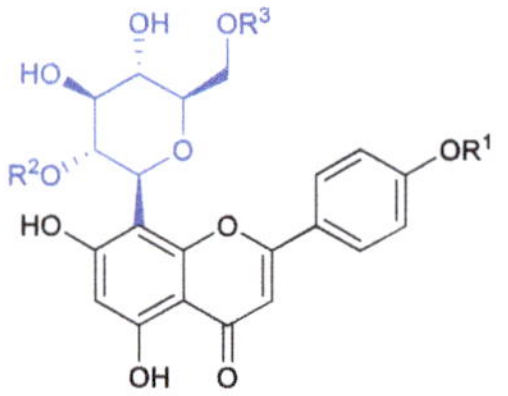

68, R¹ = Me, R² = 2-methylbutyryl, R³ = H; **69**, R¹ = R² = H, R³ = Ac; **70**, R¹ = R³ = H, R² = 2-methylbutyryl; **71**, R¹ = R³ = H, R² = 3-methoxylcaffeoyl

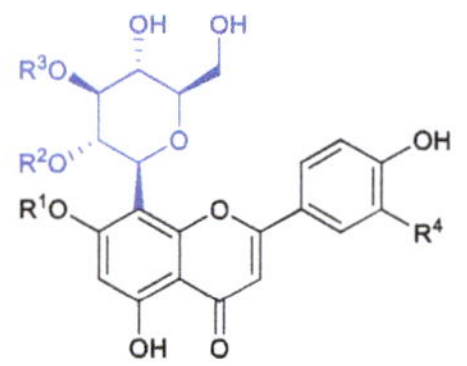

72, R¹ = Me, R² = 2-methylbutyryl, R³ = R⁴ = H; **73**, R¹ = Me, R² = R⁴ = H, R³ = 2-methylbutyryl; **74**, R¹ = R² = R⁴ = H, R³ = 2-methylbutyryl; **75**, R¹ = R³ = H, R² = 2-methylbutyryl, R⁴ = OH

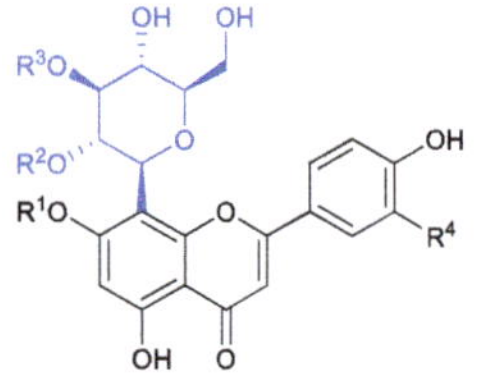

76, R¹ = R³ = R⁴ = H, R² = 2-methylbutyryl; **77**, R¹ = H, R² = veratroyl, R³ = R⁴ = H; **78**, R¹ = H, R² = veratroyl, R³ = H, R⁴ = OH

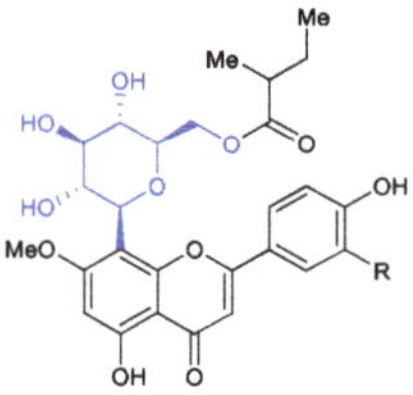

79, R = H
80, R = OH

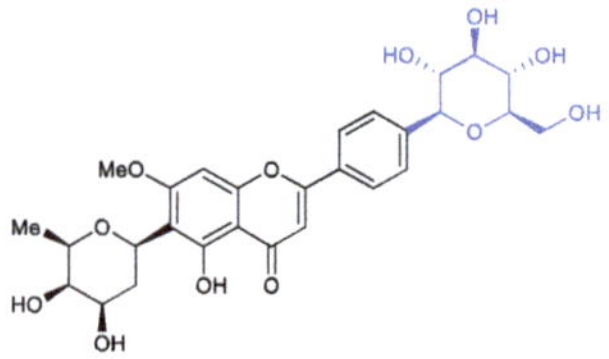

81, diandraflavone

Figure 1. Chemical structures of flavonoid 6- or 8-*C*-*β*-D-glucopyranosides (**1-80**) and flavonoid 4′-*C*-*β*-D-glucopyranoside (**81**).

It is noteworthy that flavonoid di-C-glycoside **81** features the β-D-glucopyranosyl and β-D-oliopyranosyl groups attached to C-4′ and C-6, respectively [16]. This kind of compound is very rare in nature.

2.1.2. Flavonoid C-Galactosides, C-Arabinosides, C-Xylosides, C-Mannosides, C-Fucosides, C-Boivinosides, and C-Riboside

Compared with flavonoid C-glucosides, flavonoid C-galactosides are less common in nature. Compounds **82–94** have diversified structures which can be classified into two groups: (1) flavonoid 6-C-β-D-galactopyranosides, **82–85** [62–64]; (2) flavonoid 8-C-β-D-galactopyranosides, **86–94** [64–66]. To date, no flavonoids galactosylated at other sites than C6 and C8 have been discovered. The hydroxyl groups of D-galactose or other sugars are often acylated or glycosylated, which leads to multiple complex structures with unique functions.

Figure 2 illustrates the recently discovered structures of flavonoid C-arabinosides **95–101** [67–69]. Di-C-glycosylation in flavonoids **96–101** usually takes place at positions C6 and C8. Both C-α-L-arabinosides and C-β-L-arabinosides are associated with β-D-glucose, β-D-galactose, and β-D-Xylose, with glycosylation sites shifting between C6 and C8.

The sites of C-glycosylation of flavonoid C-xylosides are usually C6 and or C8. Compounds **102–106** can be divided into two groups: (1) flavonoid mono-C-xylopyranosides **102–103** and (2) di-C-glycosylflavonoids **4, 43, 82, 91, 104–106** [38,39,62,69,70]. All discovered flavonoid C-xylosides are in β configuration.

Other flavonoid C-glycosides, including flavonoid C-mannosides **107–108**, flavonoid C-fucosides **109–110**, flavonoid C-boivinosides **111–112**, and flavonoid C-riboside **113**, are rarely found in the nature [71]. The site of C-glycosylation is normally C6.

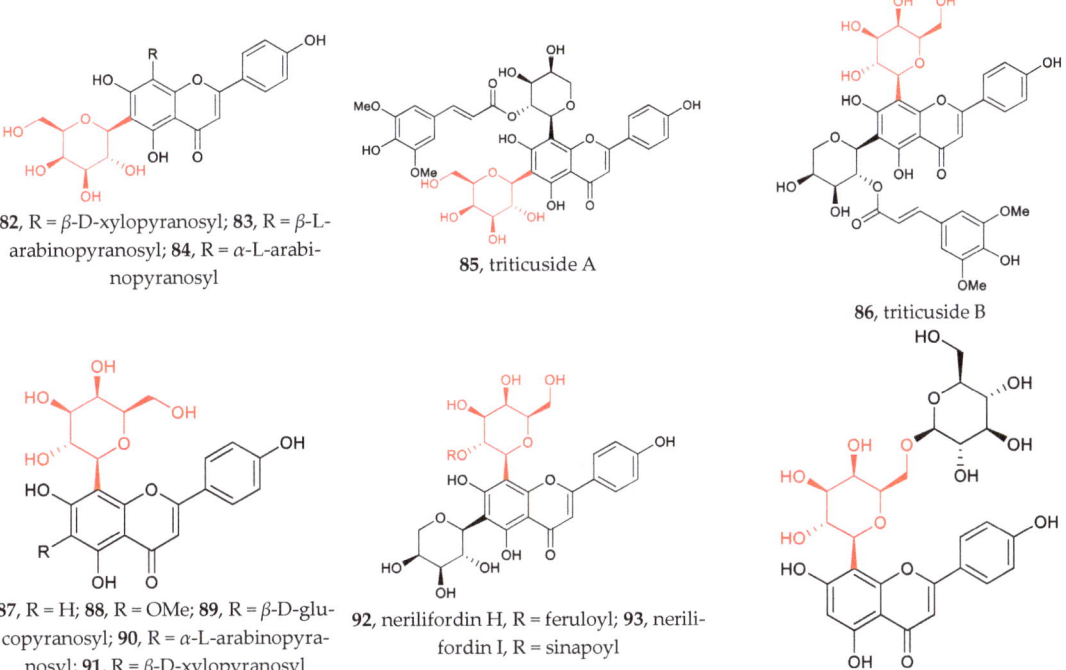

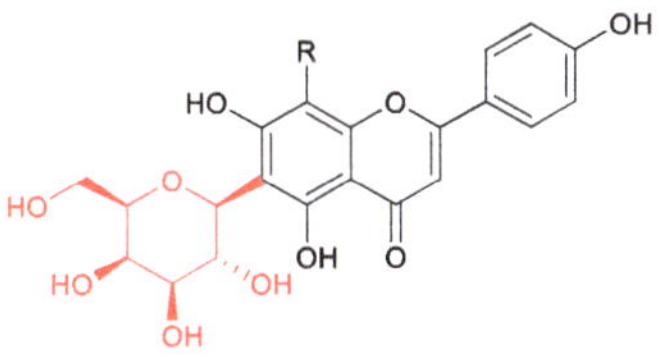

82, R = β-D-xylopyranosyl; **83**, R = β-L-arabinopyranosyl; **84**, R = α-L-arabinopyranosyl

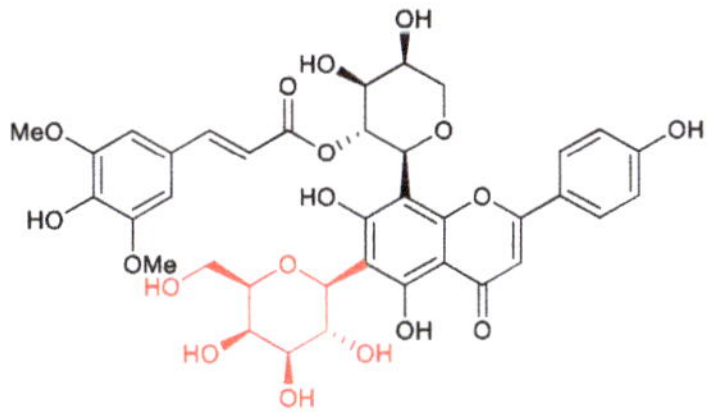

85, triticuside A

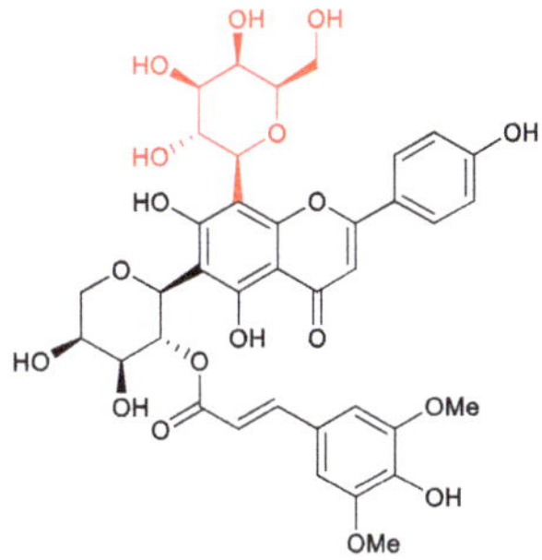

86, triticuside B

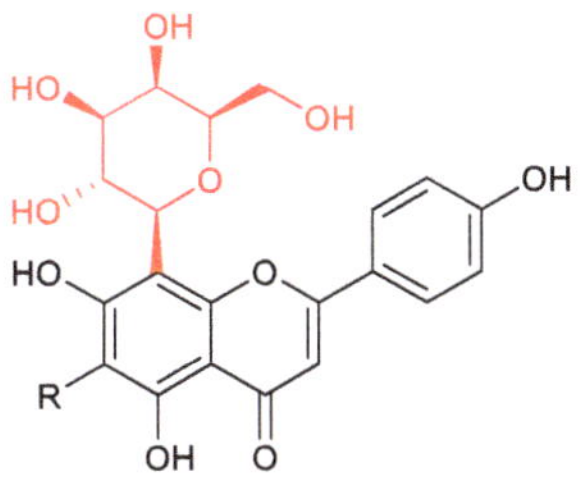

87, R = H; **88**, R = OMe; **89**, R = β-D-glucopyranosyl; **90**, R = α-L-arabinopyranosyl; **91**, R = β-D-xylopyranosyl

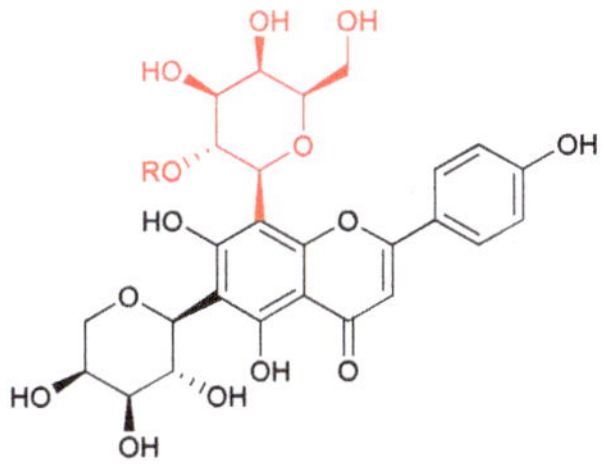

92, nerilifordin H, R = feruloyl; **93**, nerilifordin I, R = sinapoyl

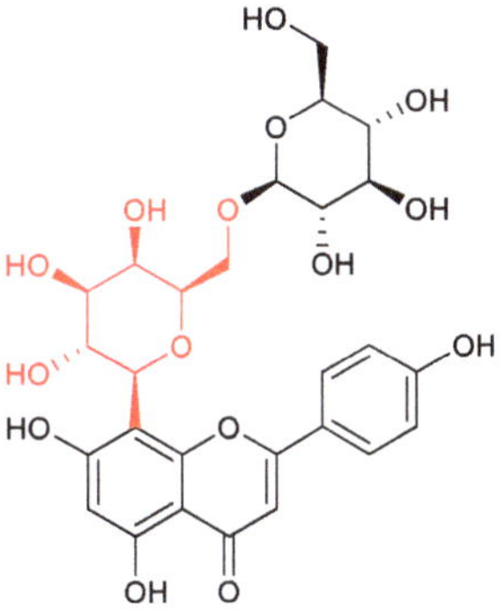

94, nervilifordin J

Figure 2. *Cont.*

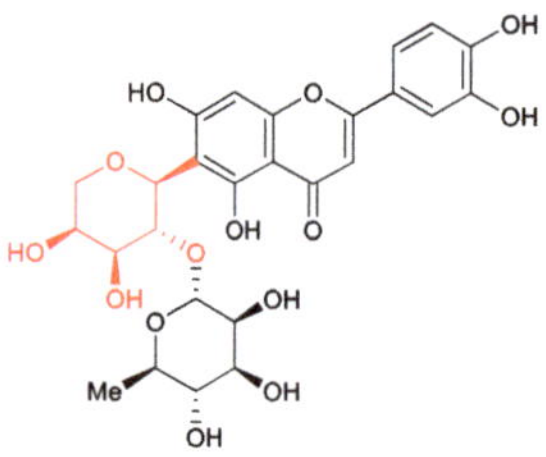

95

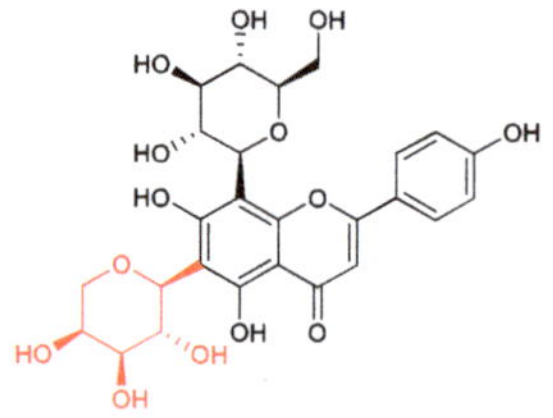

96

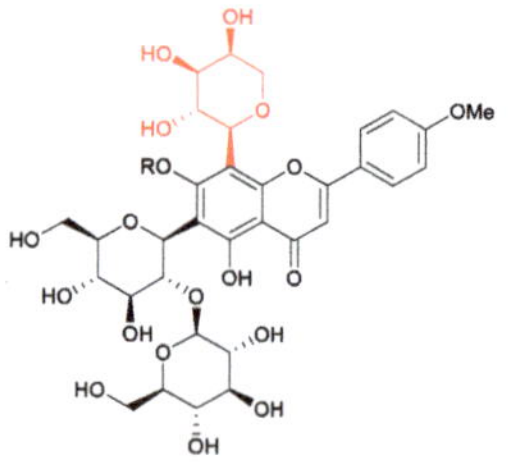

97, sileneside E, R = H
98, sileneside F, R = Me

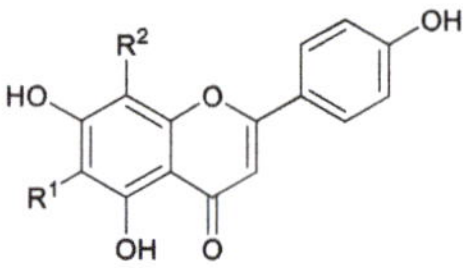

99, R¹ = R² = *α*-L-arabinopyranosyl; **100**, R¹ = *α*-L-arabinopyranosyl, R² = *β*-L-arabinopyranosyl; **101**, R¹ = *β*-L-arabinopyranosyl, R² = *α*-L-arabinopyranosyl

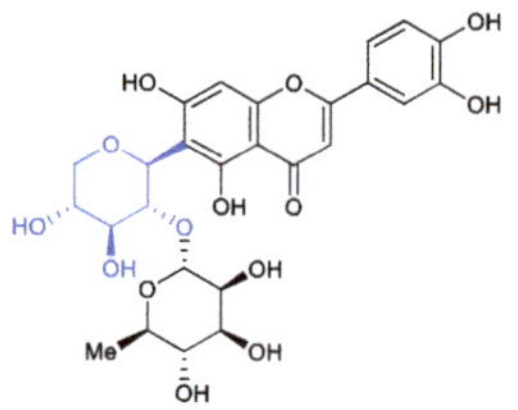

102

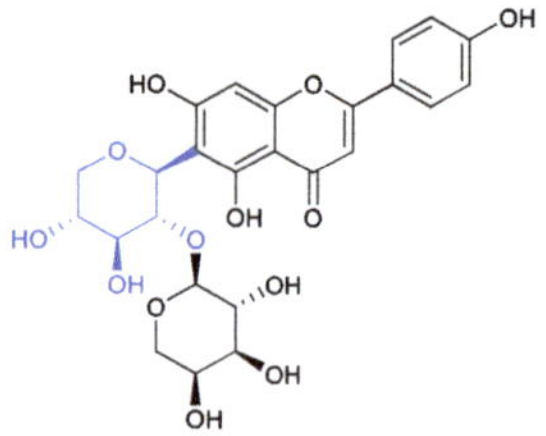

103

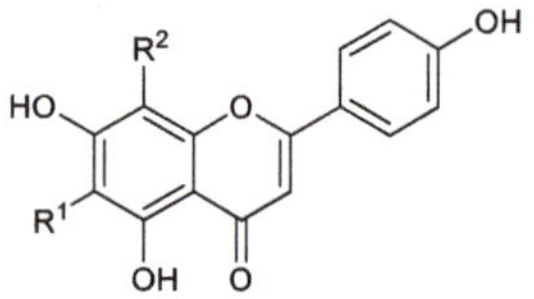

104, R¹ = *β*-D-xylopyranosyl, R² = *α*-L-arabinopyranosyl; **105**, R¹ = *α*-L-arabinopyranosyl, R² = *β*-D- xylopyranosyl

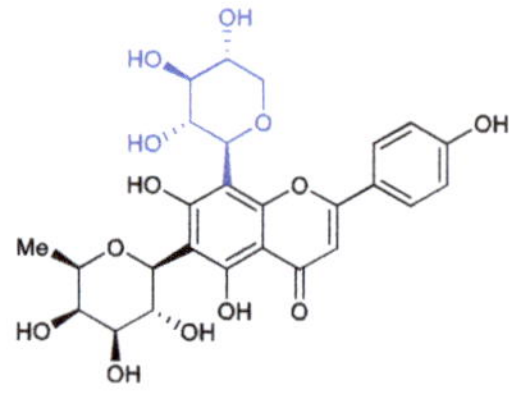

106, malloflavoside

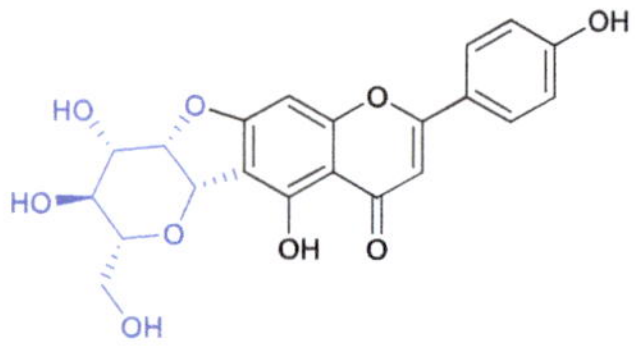

107, chafuroside A

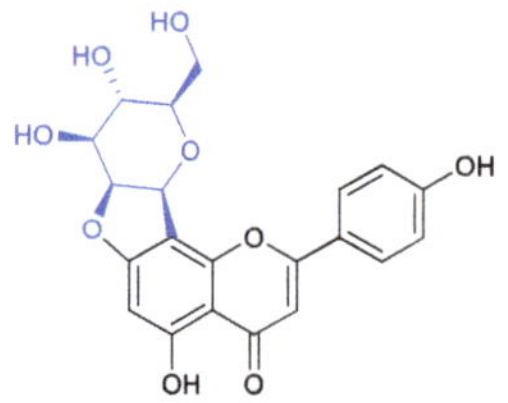

108, chafuroside B

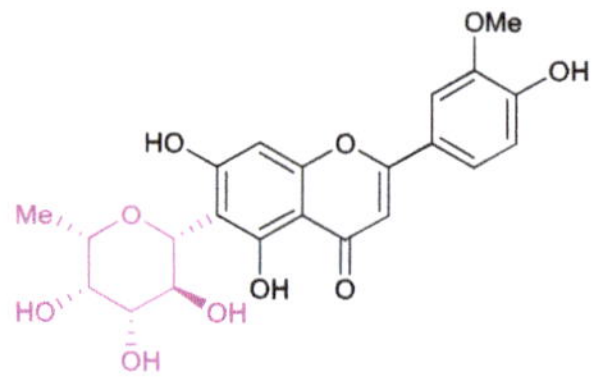

109

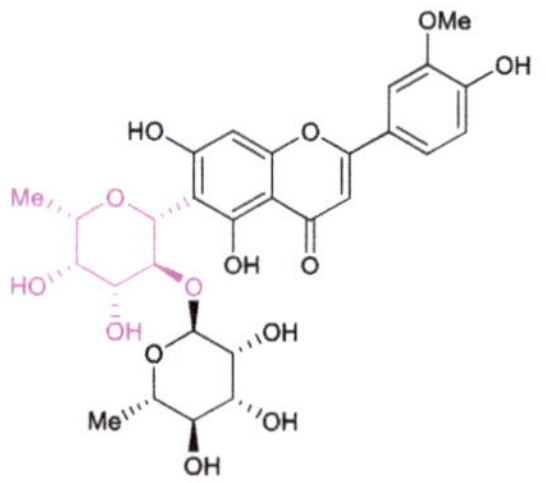

110

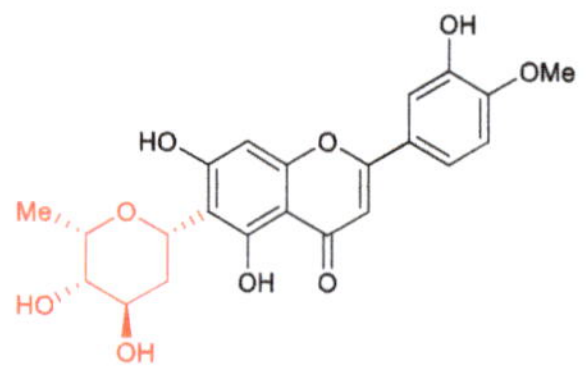

111

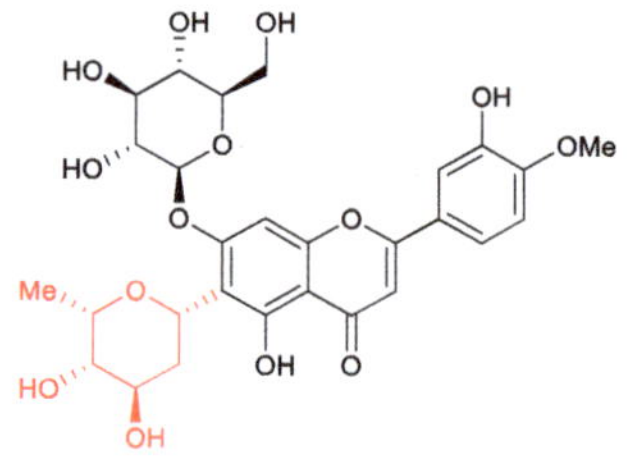

112

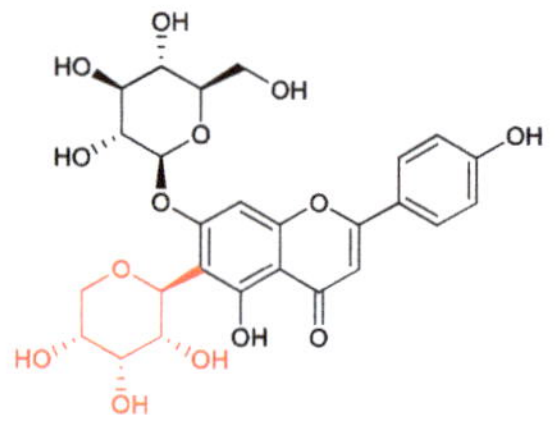

113

Figure 2. Chemical structures of flavonoid *C*-galactosides (**82–94**), *C*-arabinosides (**95–101**), *C*-xylosides (**102–106**), *C*-mannosides (**107–108**), *C*-fucosides (**109–110**), *C*-boivinosides (**111–112**), and *C*-riboside (**113**).

2.2. Other Flavonoid C-Glycosides

Other flavonoid C-glycosides include isoflavone [72,73], flavanone [74,75], dihydrochalcone [74], flavanonol [76,77], and flavonol C-glycosides [76]. The newly isolated structures are illustrated in Figure 3. Among them, Compound **120** possesses a unique scaffold featuring a C-β-D-glucose core linked to the flavanone at position C5′. This type of aryl-C-glycoside is a rare example of a natural aryl C-glycoside. It is very clear that all compounds are C-β-D-glucopyranosides (**114–127**), some of which are glycosylated with a monosaccharide at position C2 or C6 (Figure 3).

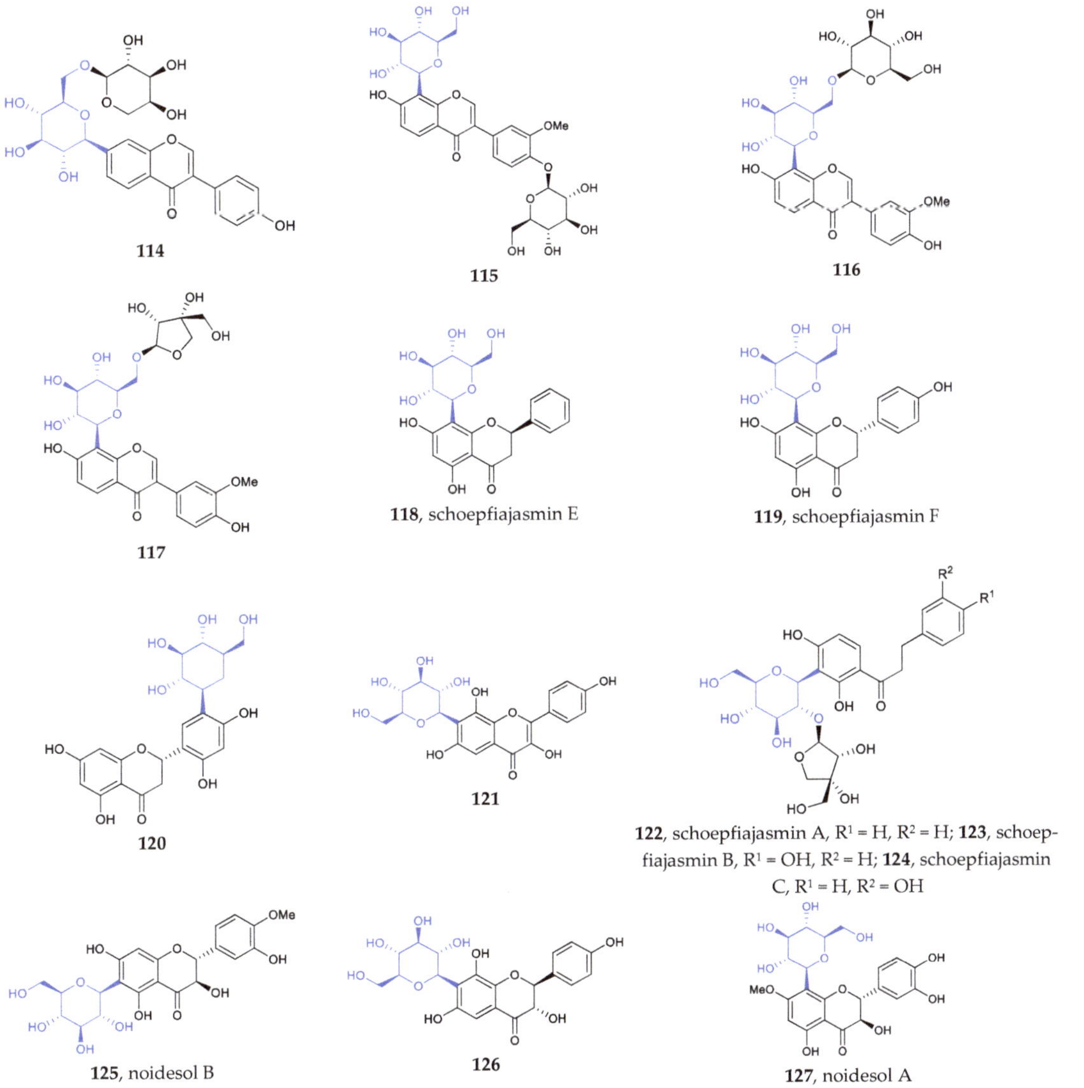

Figure 3. Chemical structures of isoflavone C-β-D-glycosides (**114–117**), flavanone C-glycosides (**118–120**), flavonol C-glycoside **121**, dihydrochalcone C-glycosides (**122–124**), and flavanonol C-glycosides (**125–127**).

2.3. *Xanthone C-Glycosides*

The sites where the sugar is linked to the xanthone are usually C2 (**128–133, 136–139**) and C4 (**134–135**) [21–23,78,79]. It seems that only *C-β*-D-glucosides have been isolated from nature. This kind of aryl *C*-glycosides include mono-xanthone *C*-glycosides, di-xanthone *C*-glycosides, and tri-xanthone *C*-glycosides (Figure 4).

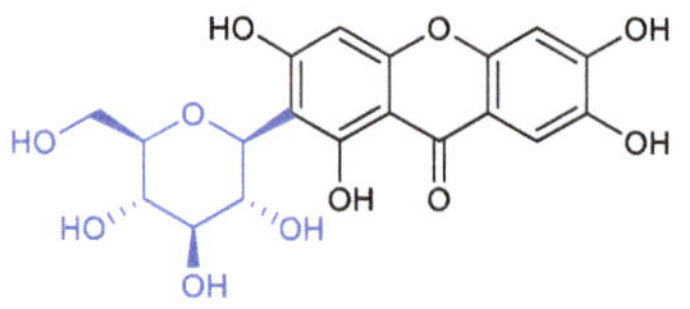

128

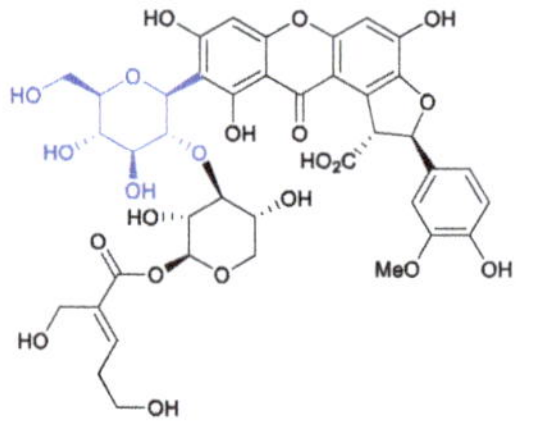

129, glomexanthone B

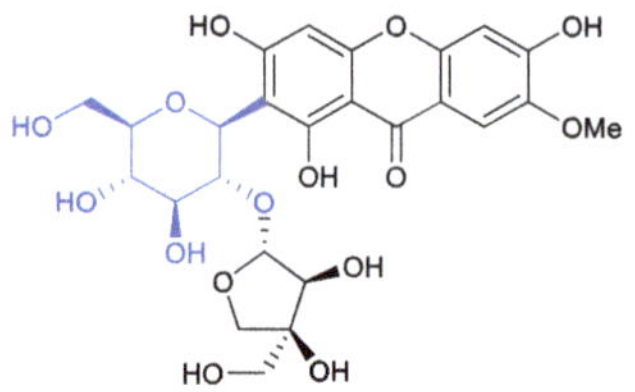

130, polygalaxanthone XI

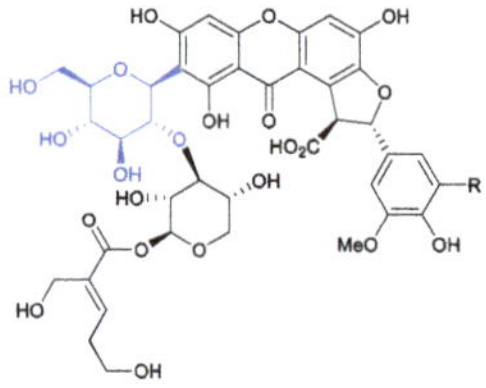

131, glomexanthone A, R = H; **132**, glomexanthone C, R = OMe

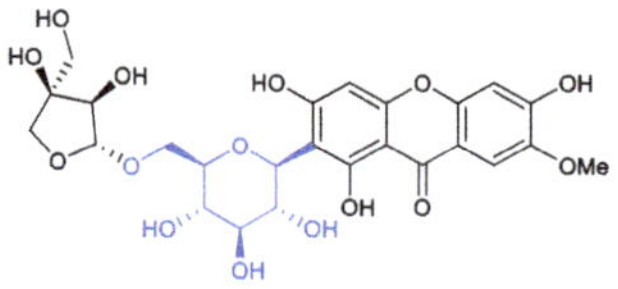

133, polygalaxanthone III

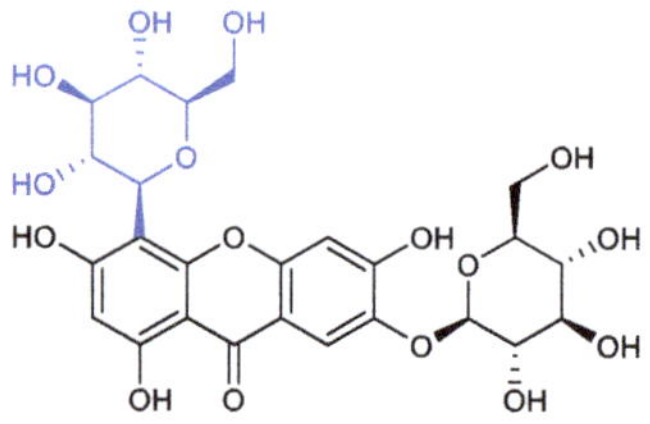

134

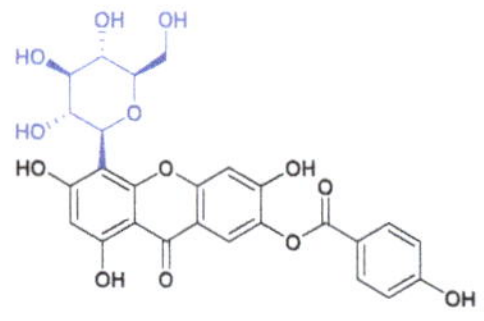

135, shamimoside

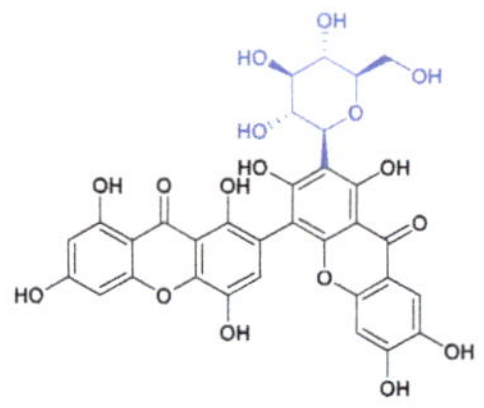

136

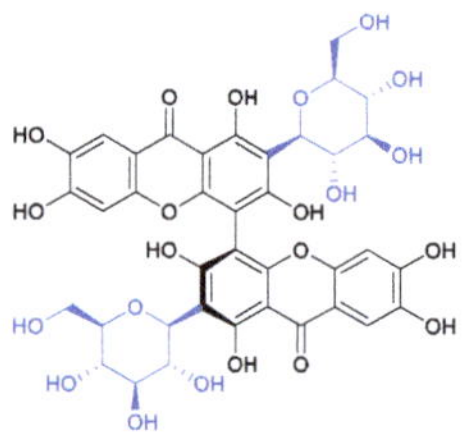

137

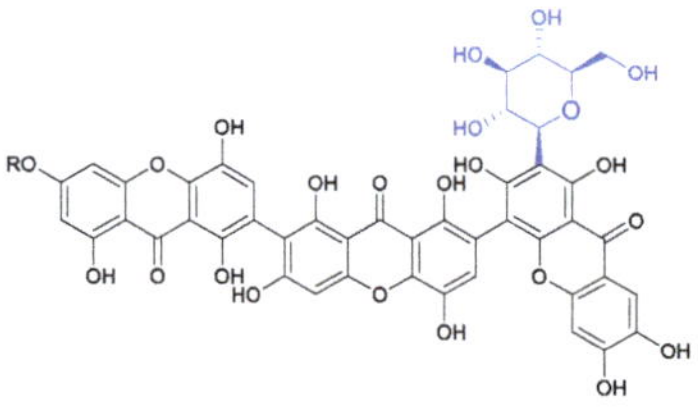

138, puniceaside D R = H
139, puniceaside E R = Me

Figure 4. Chemical structures (**128–139**) of xanthone *C*-glycosides.

2.4. *Phenyl C-Glycosides*

Phenyl C-glycosides were considered the simplest aryl-C-glycosides. The disclosed structures **140–169** feature *C-β*-D-glucopyranosides, with variations in the substitution on the benzene ring (Figure 5) [24–27,80–84]. The site of *C*-glycosylation is usually the ortho position of the phenolic hydroxyl group. The substitution groups of benzene include acyl, alkenyl, hydroxyl, alkyl, and alkoxyl moieties, providing aryl ketone, aryl vinyl, phenol, alkylbenzene, and aryl ether, respectively. The 2′-hydroxylation and 6′-hydroxylation of D-glucose are often accompanied by esterification by gallic acid (compounds **154, 169**), *trans-p*-coumaric acid (compounds **156**, **159–160**), ferulic acid (compounds **157**, **160–161**), and benzoic acid (compound **158**).

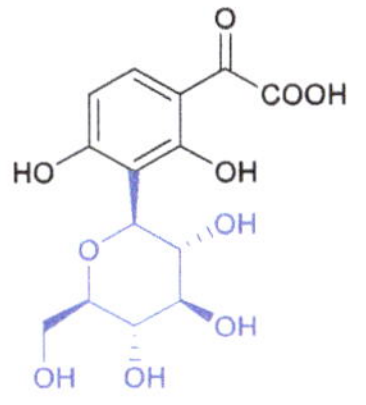

140

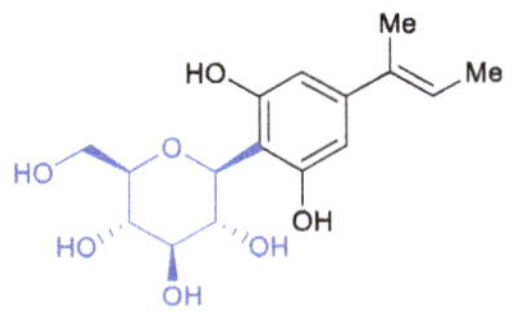

141, calophymembranside C

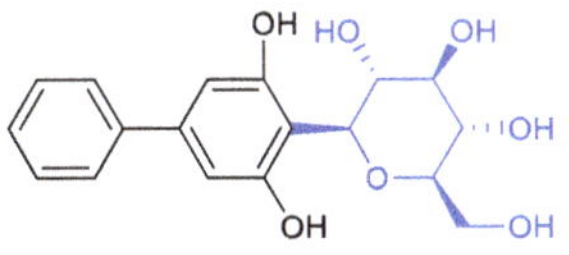

142, biphenyl *C*-glycoside

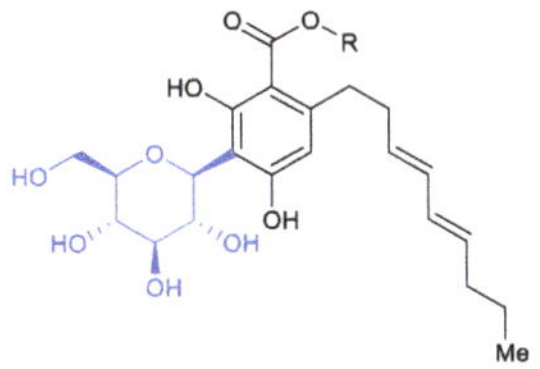

143, carnemycin B, R = Me
144, carnemycin A, R = Et

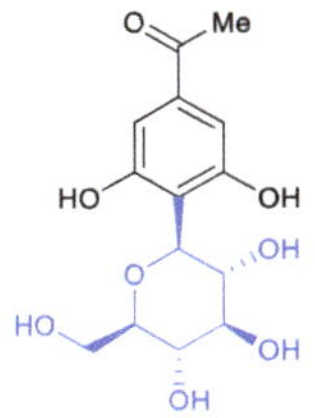

145, calophymembranside D

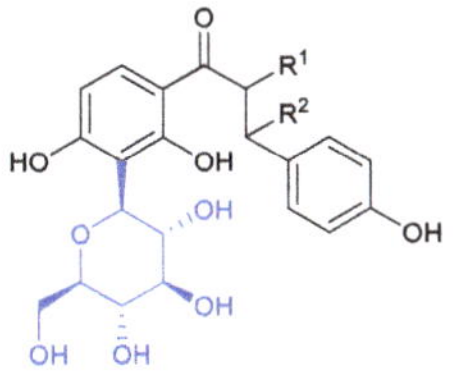

146, R[1] = (*S*)-OH, R[2] = H
147, R[1] = I-OH, R[2] = H
148, R[1] = H, R[2] = OH

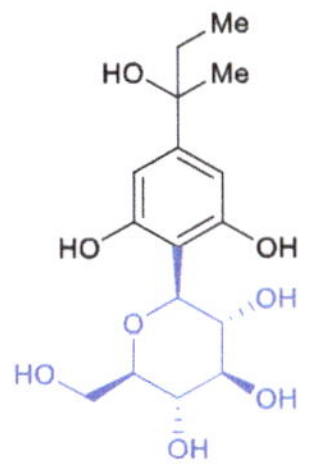

149, calophymembranside E

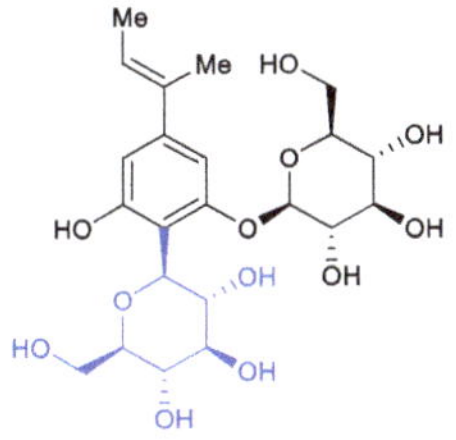

150, calophymembranside F

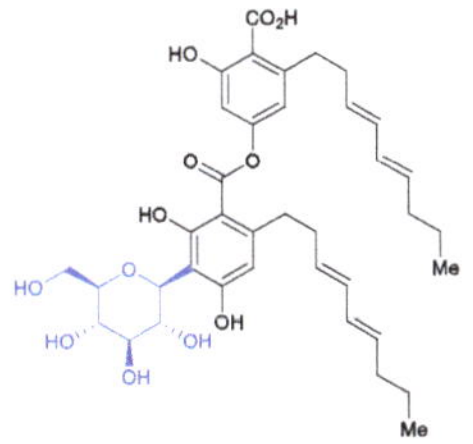

151, stromemycin

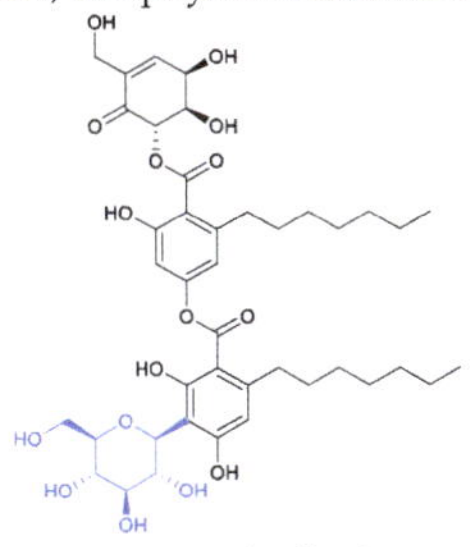

152, arenicolin A

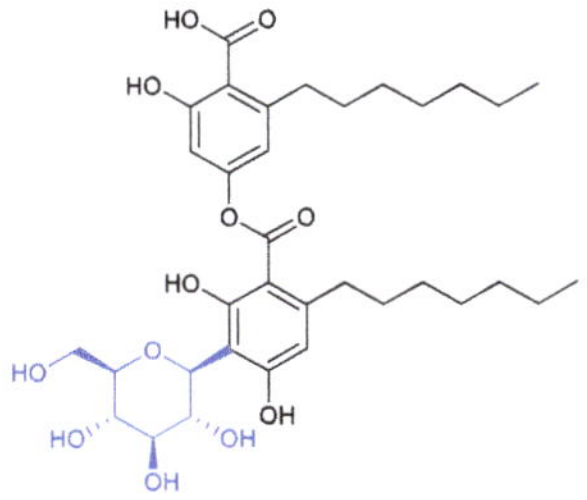

153, arenicolin B

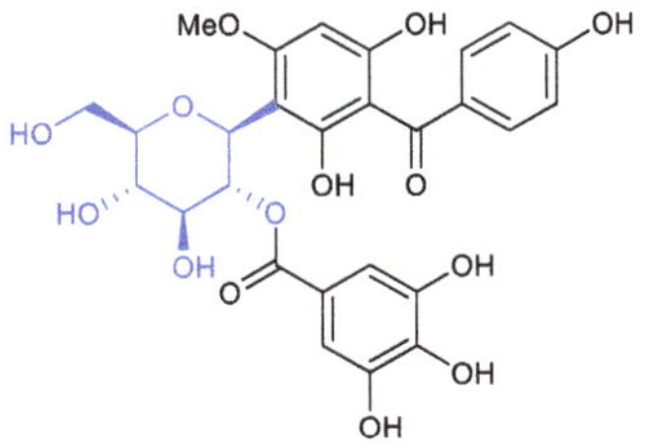

154

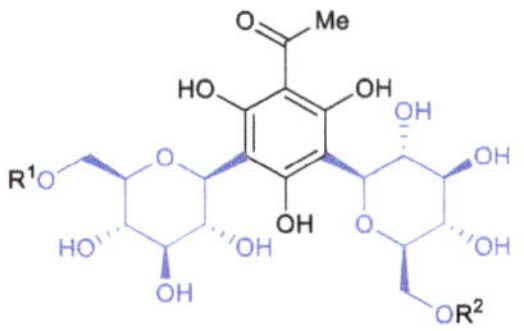

155, R[1] = R[2] = H; **156**, R[1] = H, R[2] = *trans-p*-coumaroyl; **157**, R[1] = H, R[2] = *trans*-feruloyl; **158**, R[1] = H, R[2] = benzoyl; **159**, R[1] = R[2] = *trans-p*-coumaroyl; **160**, R[1] = *trans-p*-coumaroyl, R[2]

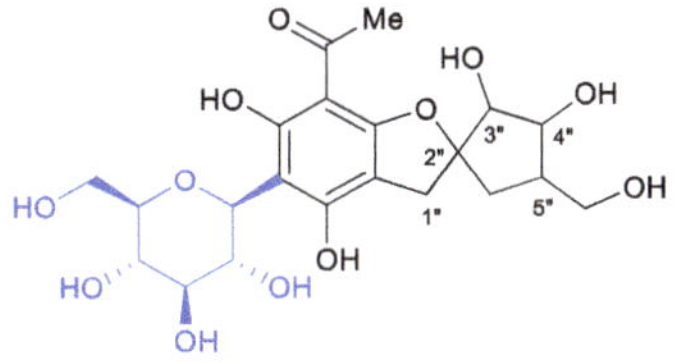

162, (2″ R, 3″ S, 4″ S, 5″ R)
163, (2″ R, 3″ S, 4″ R, 5″ R)
164, (2″ S, 3″ R, 4″ S, 5″ R)
165, (2″ R, 3″ R, 4″ R, 5″ S)
166, (2″ S, 3″ S, 4″ S, 5″ R)

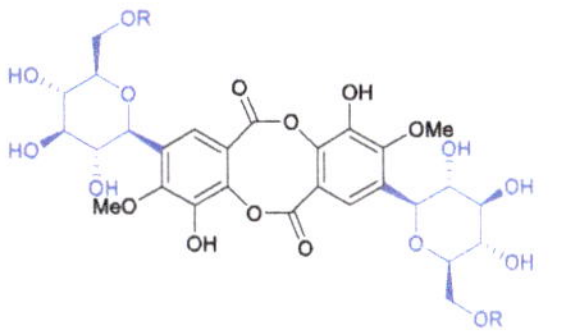

168, ardimerin, R = H
169, ardimerin digallate, R = galloyl

Figure 5. *Cont.*

= *trans*- feruloyl; **161**, R¹ = R² = *trans*-feruloyl

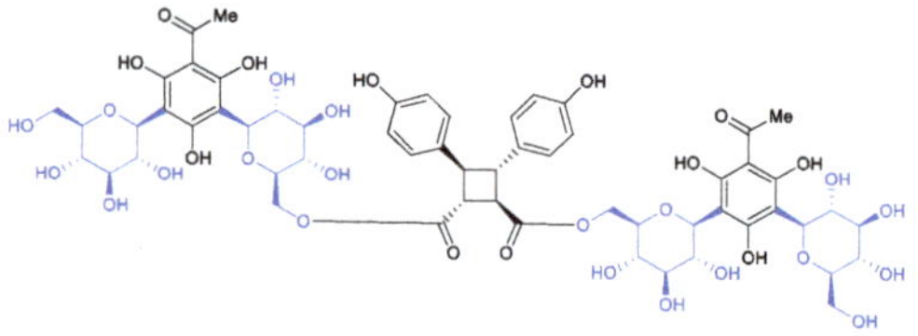

167

Figure 5. Chemical structures (**140–169**) of phenyl *C*-glycosides.

2.5. Heteroaryl C-Glycosides

The recently discovered heteroaryl *C*-glycosides include dihydrobenzofuran *C*-glycosides (**170–171**), chromone *C*-glycosides (**172–177**), indole *C*-glycosides (**178–184**), and other heteroaryl *C*-glycosides (**185–187**) [28,29,85,86]. Except for the common *C*-*β*-glucosides, *C*-*α*-D-mannopyranosides and those containing erythrose are rare examples of aryl-*C*-glycosides (Figure 6).

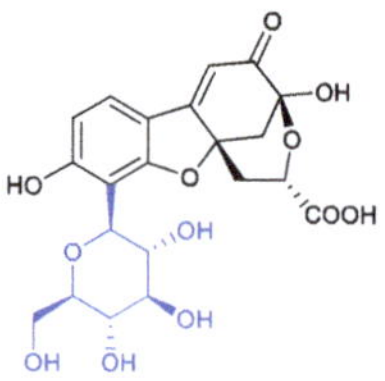

170

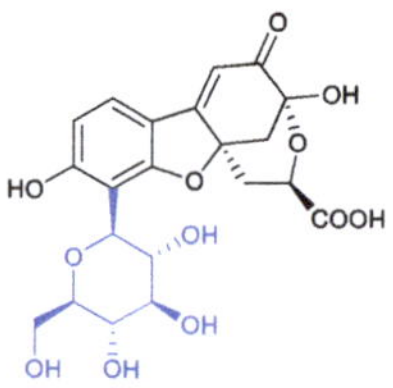

171

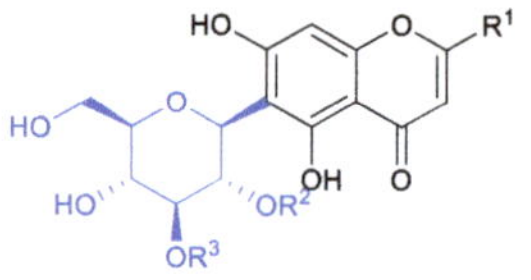

172, kunzeachromone A, R¹ = ⁱPr, R² = R³ = galloyl; **173**, kunzeachromone C, R¹ = Me, R² = galloyl, R³ = H; **174**, kunzeachromone E, R¹ = Me, R² = R³ = galloyl

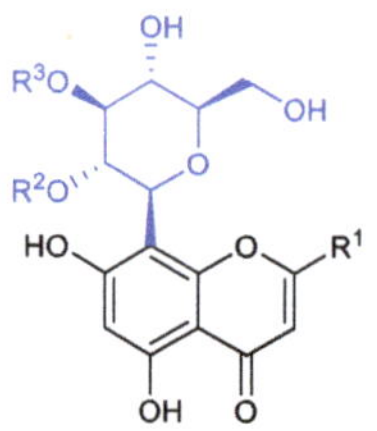

175, kunzeachromone B, R¹ = ⁱPr, R² = R³ = galloyl; **176**, kunzeachromone D, R¹ = Me, R² = galloyl, R³ = H; **177**, kunzeachromone F, R¹ = Me, R² = R³ = galloyl

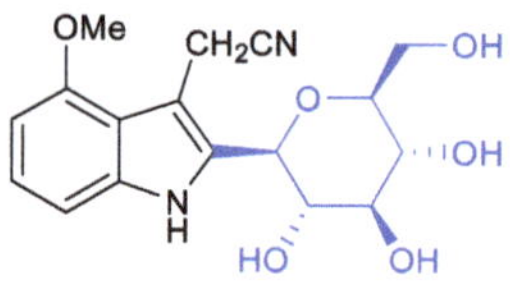

178

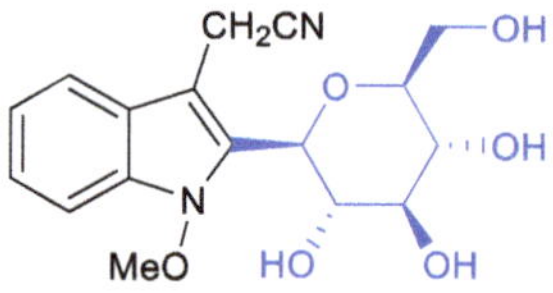

179

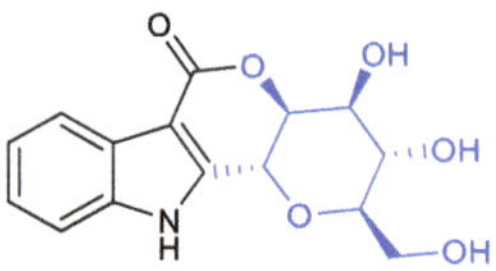

180, neopetrosin A

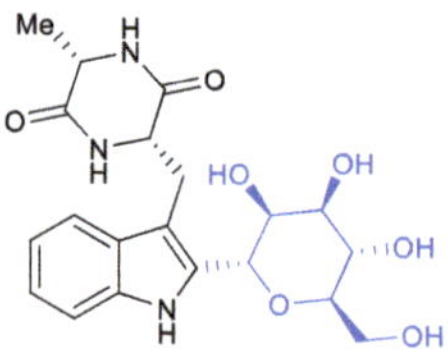

181, neopetrosin B

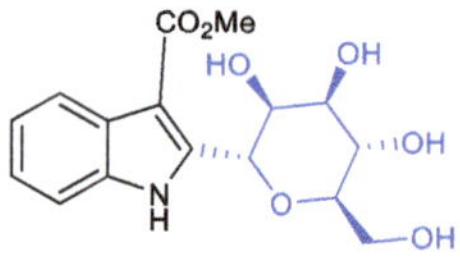

182, neopetrosin C

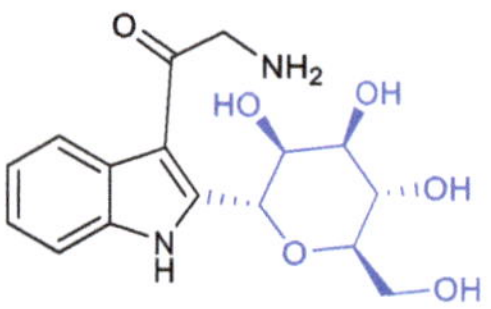

183, neopetrosin D

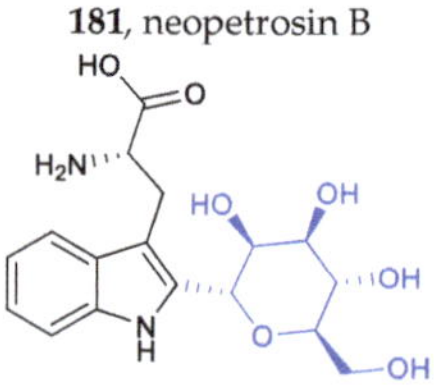

184, *C*-mannosyl tryptophan

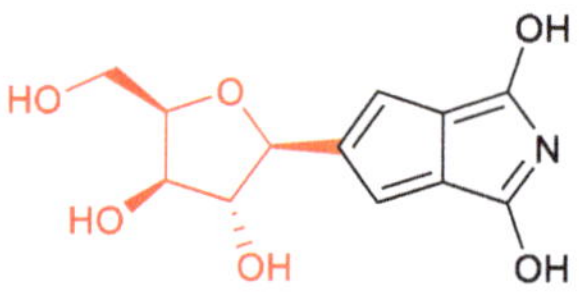

185

Figure 6. *Cont.*

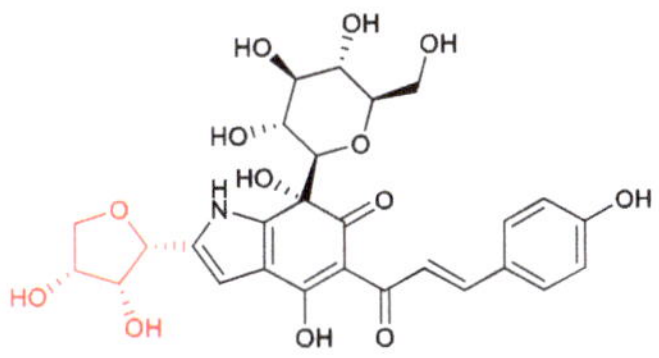

186, isocartormin

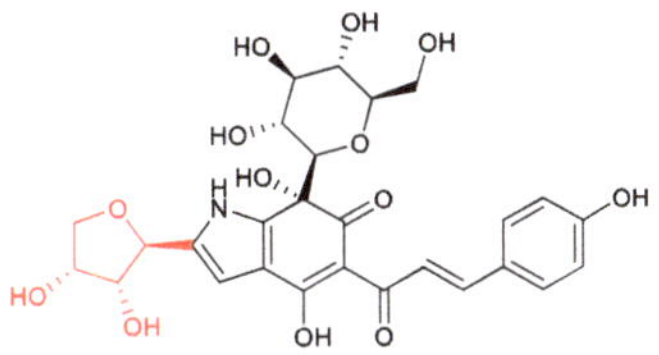

187, cartormin

Figure 6. Chemical structures (**170–187**) of heteroaryl C-glycosides.

2.6. Other Aryl-C-Glycosides

Other aryl-C-glycosides include aryl-C-glycosides of amino sugars and other rare sugars [30–36]. Compounds **190–197** (4-aminosugar) and compound **198–199** (3-aminosugar), which are 2-deoxylaminoglycoside antibiotics, resemble the well-known pluramycins. Among them, monacyclione G (**190**) possesses a unique scaffold featuring a xanthone core linked to the aminodeoxysugar ossamine, and monacycliones H–J (**194–195**, **192**) are rare examples of natural angucyclines with an S-methyl group (Figure 7).

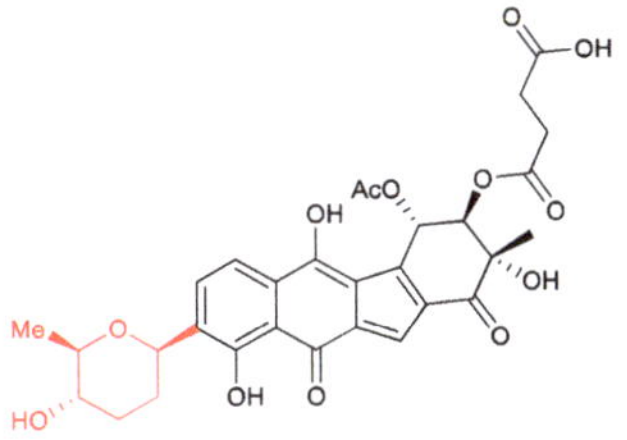

188, konamycin A

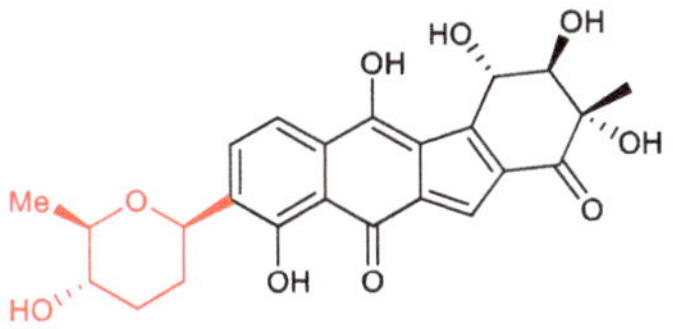

189, konamycin B

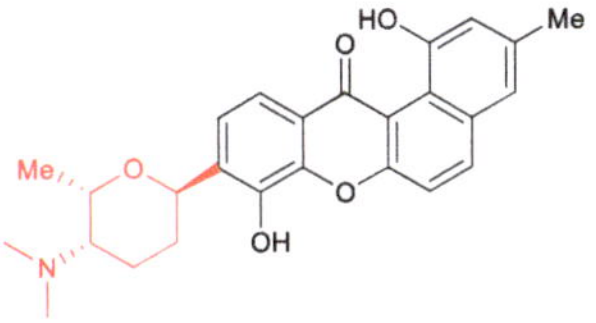

190, monacyclione G

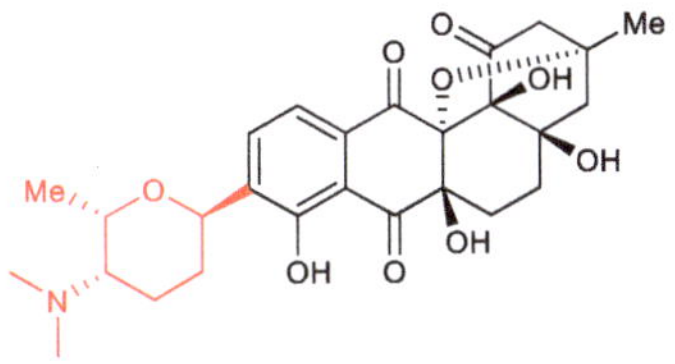

191, monacyclione K

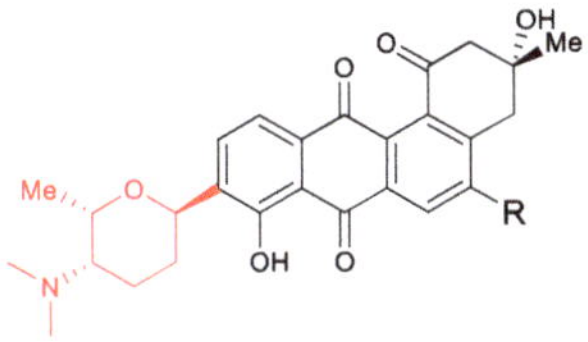

192, monacyclinone J, R = SMe
193, frigocyclinone, R = H

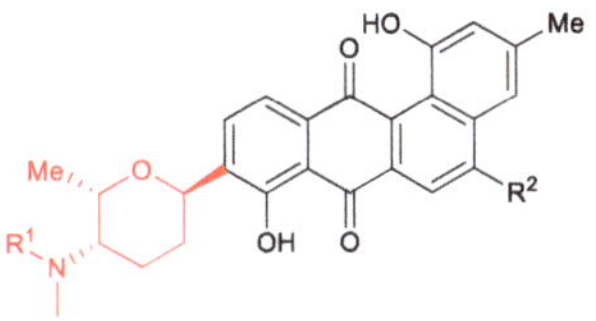

194, monacyclinone H, R¹ = Me, R² = SMe;
195, monacyclinone I, R¹ = H, R² = SMe

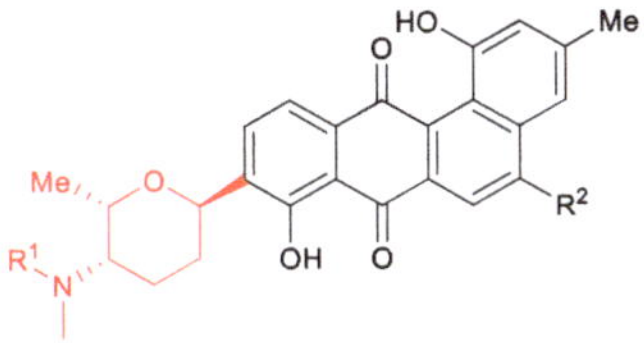

196, monacyclione A, R¹ = Me, R² = H; **197**, monacyclione B, R¹ = Me, R² = OH

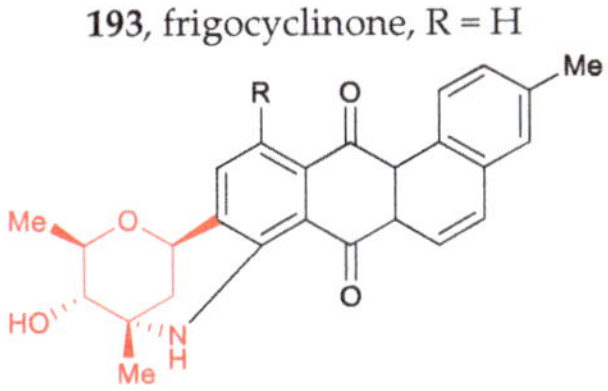

198, marmycins A, R = H
199, marmycins B, R = Cl

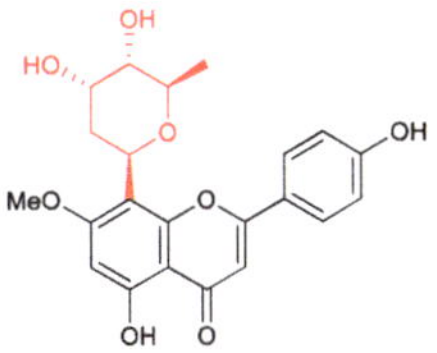

200, aciculatin

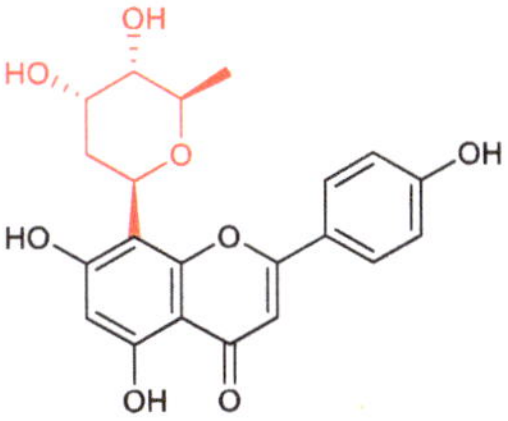

201

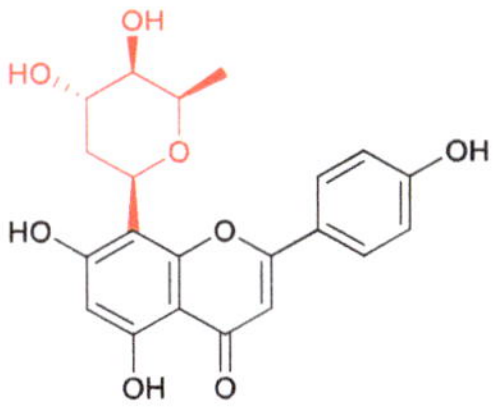

202

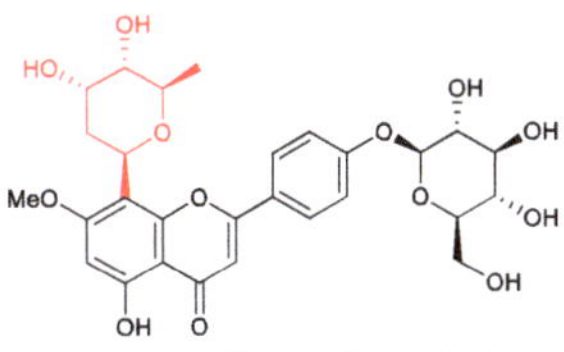

203, 4′-*O*-glucosylaciculatin

Figure 7. *Cont.*

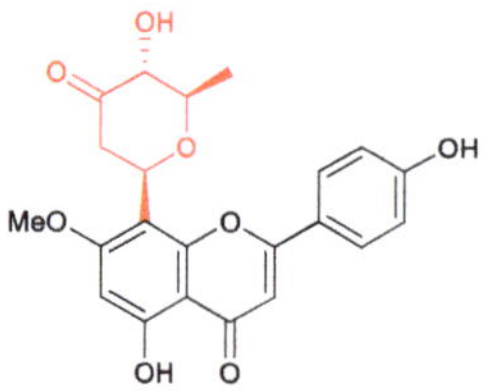

204, aciculatinone

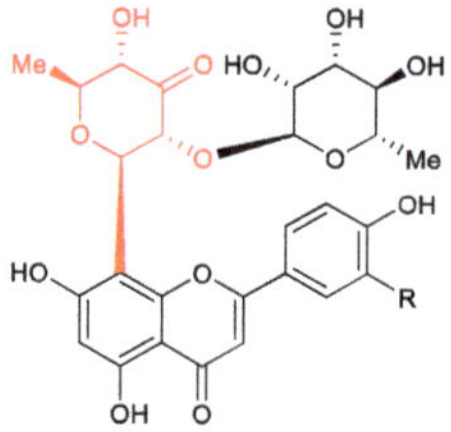

205, R = H
206, isocassiaoccidentalin B, R = OH

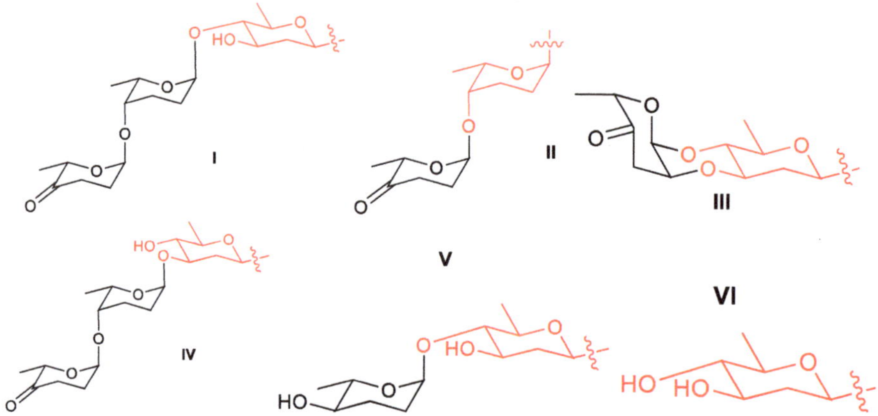

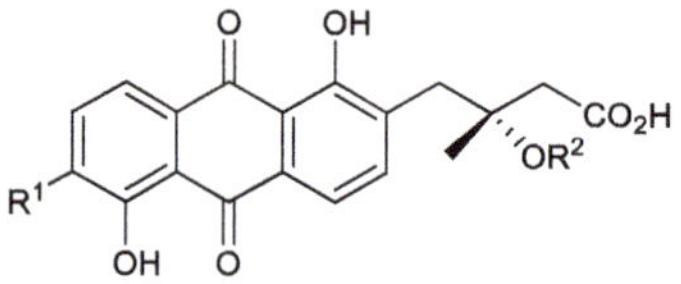

207, grincamycin B, R¹ = I, R² = II
208, grincamycin C, R¹ = I, R² = H
209, grincamycin D, R¹ = III, R² = II

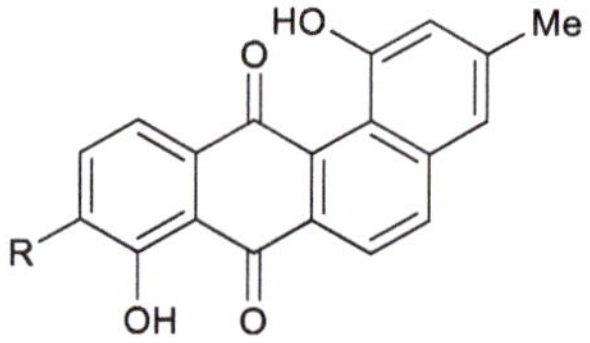

210, marangucycline A, R = V
211, marangucycline B, R = III
212, dehydroxyaquayamycin, R = VI

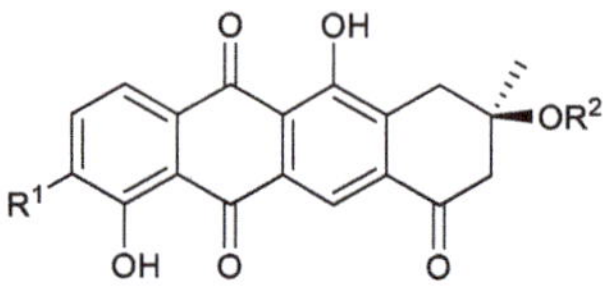

213, grincamycin E, R¹ = I, R² = II

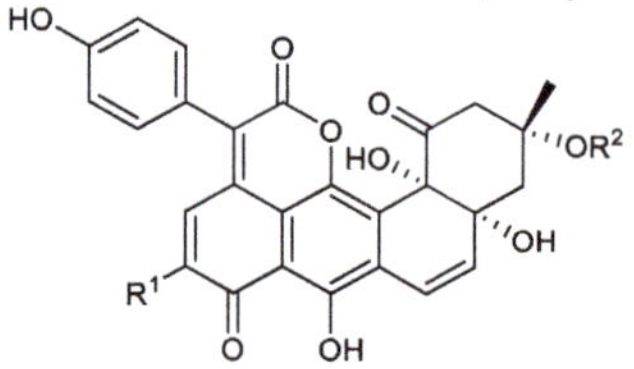

214, grincamycin F, R¹ = IV, R² = II

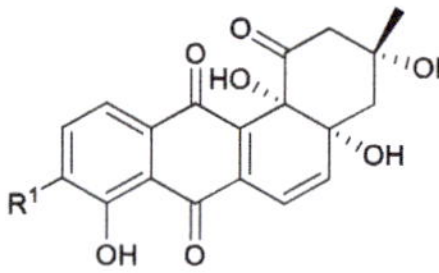

215, grincamycin, R¹ = I, R² = II

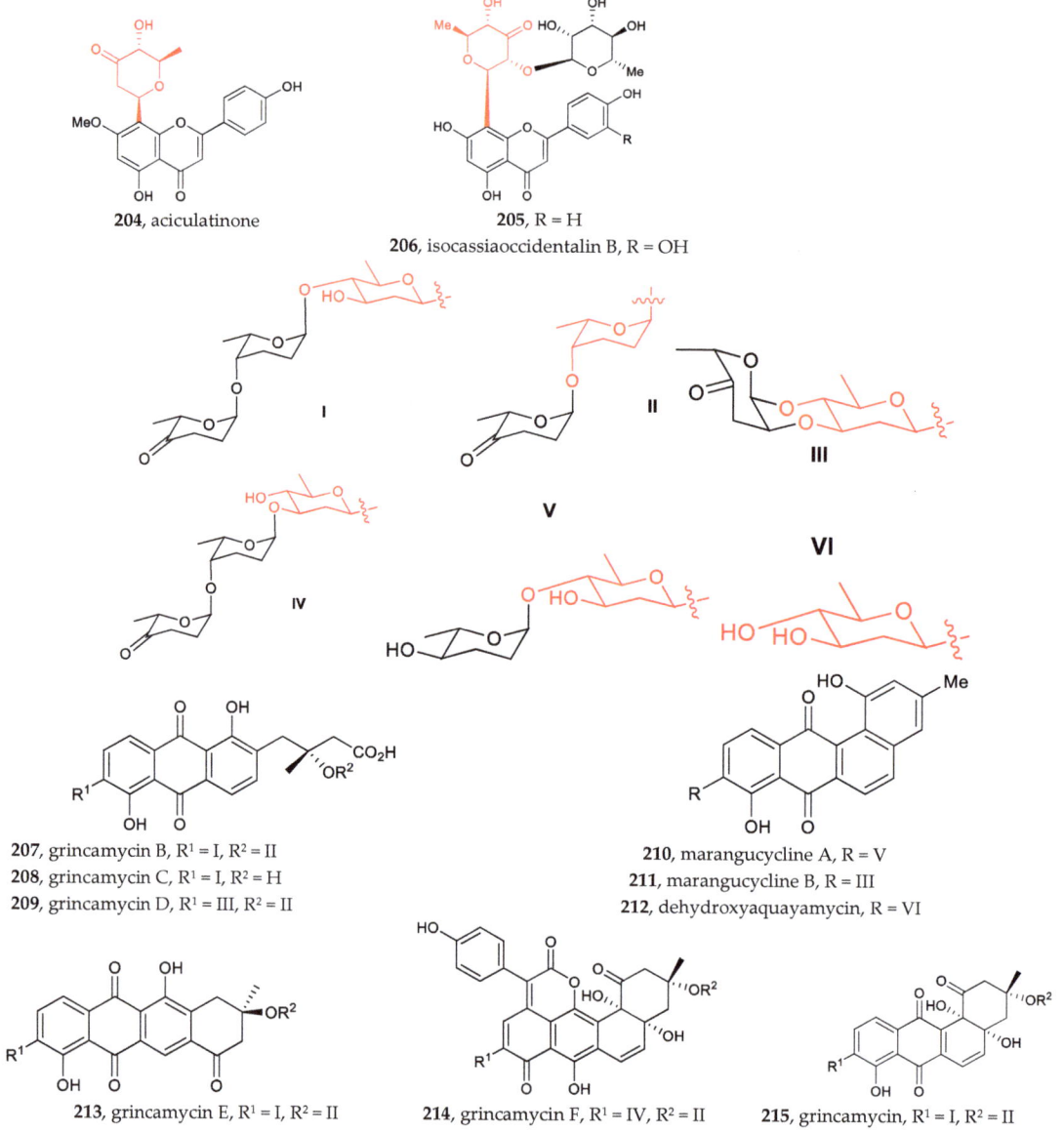

Figure 7. Chemical structures (**188–215**) of aryl *C*-glycosides.

3. Pharmacological Activity

Numerous researchers have investigated the pharmacological activities of various aryl-*C*-glycosides. Table 1 summarizes the pharmacological features of recently discovered aryl-*C*-glycoside molecules. They include, but are not limited to, anticancer, anti-inflammatory, antioxidant, antiviral activities, glycation inhibitory activity, other pharmacological activities such as neuroprotective effects, hepatoprotective activity, and antipyretic activity. These pharmacological activities are summarized below.

3.1. Anticancer Activity

The indole C-glucopyranoside **178** exhibited significant cytotoxic activity against human myeloid leukemia cells HL-60 and human liver cancer cells HepG2, with IC50 of 1.3 ± 0.1 and 2.1 ± 0.3 μM, respectively. The indole C-glucopyranoside **179** showed potential cytotoxic activity against HL-60 and human myeloid leukemia Mata cells, with IC_{50} of 5.1 ± 0.4 and 12.1 ± 0.8 μM, respectively [85]. Monacycliones I (**195**) and J (**192**), isolated from the marine-derived *Streptomyces* sp. HDN15129, showed cytotoxic activity against multiple human cancer cell lines, with IC_{50} values ranging from 3.5 to 10 μM [31]. Marmycin A (**198**), isolated from the culture broth of a marine sediment-derived actinomycete related to the genus *Streptomyces*, displayed significant cytotoxicity against several cancer cell lines, some at nanomolar concentrations, while marmycin B (**199**) was less potent. For marmycin A (**198**), tumor cell cytotoxicity appeared to coincide with a modest induction of apoptosis and the arrest in the G1 phase of the cell cycle [32]. Li et al. reported that compounds **34–36**, isolated from the small flowering aquatic plant *Lemna japonica*, exhibited weak cytotoxicity against HepG-2, SW-620, and A-549 cell lines, with IC_{50} values between 42.5 and 19.2 μg/mL [11]. Isoorientin, recently isolated from leaf and root methanolic extracts of *Petrorhagia Velutina*, a Mediterranean herbaceous plant, significantly reduced the proliferation of HepG2 cells, as determined by the complete conversion of a tetrazolium probe into formazan after 48 h of exposure [67]. Aciculatin **200**, **201**, **202**, **204**, isolated from an ethanolic extract of *Chrysopogon aciculatis*, showed differential potency on different cancer cell lines. Noticeably, aciculatin and **201** indicated specificity of cytotoxicity in MCF-7 and CEM cell lines [33]. Grincamycins B-E (**207–209**, **213**), isolated from *Streptomyces lusitanus* SCSIO LR32, exhibited in vitro cytotoxicity against the human cancer cell lines HepG2, SW-1990, HeLa, NCI-H460, and MCF-7 and the mouse melanoma cell line B16, with IC50 values ranging from 1.1 to 31 μM [35]. Marangucycline B (**211**), isolated from the deep-sea-derived *Streptomyces* sp. SCSIO 11594, displayed in vitro cytotoxicity against four cancer cell lines, i.e., A594, CNE2, HepG2, and MCF-7, superior to that obtained with cisplatin, used as a positive control. Notably, marangucycline B bearing a keto-sugar displayed significant cytotoxicity against various cancer cell lines, with IC_{50} values ranging from 0.24 to 0.56 μM. An IC_{50} value of 3.67 μM was found when using the non-cancerous hepatic cell line HL7702, demonstrating the cancer cell selectivity of marangucycline B [36].

3.2. Anti-Inflammatory Activity

Compound **68**, isolated from the rhizomes of *Cyperus rotundus*, showed moderate inhibitory activity against MRB, with an IC_{50} value of approximately 56.03 μM [15]. Nervilifordin J (**94**), isolated from a 60% EtOH extract of the aerial parts of *Nervilia fordii*, showed interesting inhibitory effects on nitric oxide production in lipopolysaccharide-activated RAW264.7 macrophages, with EC_{50} values of 14.80 μM [17]. Vicenin-2 was isolated and identified from an ethanol extract of the aerial parts of *Urtica circularis*. This crude extract was found to possess significant anti-inflammatory activity in a carrageenan-induced rat hind paw edema model (41.5% inhibition at a dose of 300 mg/kg). In cultured murine macrophages, this compound modified LPS-induced total nitrite and TNF-α production, in addition to promoting the LPS-induced translocation of nuclear factor NF-κB [87].

3.3. Antioxidant Activity

Shamimoside (**135**), isolated from a methanolic extract of the leaves of Bombax ceiba, showed antioxidant potential (IC_{50} = 150 μg/mL) [22]. Nelumboside B (**40**) exhibited strong scavenging activity (SC_{50} = 14.12 μM, ABTS assay), compared with the positive control L-ascorbic acid (SC_{50} = 26.15 μM, ABTS assay) [12]. Isocassiaoccidentalin B (**206**), isolated from whole *Cassia nomame* (SIEBER) HONDA plants, showed significant free-radical scavenging activity [34]. Compound **42**, isolated from *Gentiana piasezkii*, showed significant free-radical scavenging activity ($IC_{20} = 5.20 \pm 0.10$ μM) in the DPPH assay [13]. Compound **113**, isolated from the methanolic extracts of *Dtps*. Tinny Ribbon × *Dtps*. Plum Rose (*Phalaenopsis hybrids*), exhibited moderate α, α-diphenyl-β-picrylhydrazyl free-radical scavenging activity, with half-maximal inhibitory concentration (IC_{50}) values of 27.3 μM, compared to the reference compound vitamin E (IC_{50} 12.5 μM) [71].

3.4. Antiviral Activity

Speciflavoside A (**32**), isolated from a 70% methanolic extract of *Lilium speciosum* var. *gloriosoides* Baker, showed potent antiviral activity against RSV, with an IC_{50} value of 2.9 μg/mL, comparable to that of ribavirin, an approved drug for the treatment of RSV infections in humans [10].

3.5. Glycation Inhibitory Activity

In 2003, Okuyama T. et al. reported that chrysoeriol 6-C-β-fucopyranoside **109**, isolated from the style of *Zea mays* L. showed an inhibitory effect on glycation, with a percent inhibition value greater than that of aminoguanidine, a known glycation inhibitor [19]. Compounds **130–131**, which contains the rare sugar boivinose, exhibited a glycation inhibitory activity similar to that of aminoguanidine [20].

3.6. Other Pharmacological Effects

Other pharmacological effects include neuroprotective effects, hepatoprotective activity, HIV inhibitory activity, antipyretic activity, transcriptional inhibitory activity of RXRα, cytotoxicity activity, and other activities.

Glomexanthones A–C (**131**, **129**, **132**), isolated from an ethanol extract of *Polygala glomerata*, showed moderate neuroprotective effects on L-glutamic acid-induced cellular damage in human neuroblastoma SK-N-SH cells [21]. Compound **136**, isolated from the entire plant of *Swertia punicea*, exhibited potent neuroprotective activity against H_2O_2-induced PC12 cell damage [23]. Neopetrosins A, B, D (**180–181**, **183**) were isolated from the marine sponge *Neopetrosia chaliniformis* collected off Xisha Island in the South China Sea. They exhibited in vivo hepatoprotective activity in a zebrafish model at a concentration of 20 μM [29]. Ardimerin digallate (**169**) was isolated from the whole plant of *Ardisia japonica* and was shown to inhibit HIV-1 and HIV-2 RNase H in vitro, with IC_{50} values of 1.5 and 1.1 μM, respectively [27]. The compound 3,5-di-C-β-D-glucopyranosyl phloroacetophenone (**155**), isolated from the edible leaves of *Melicope pteleifolia*, was found to be responsible for the antipyretic activity of *M. pteleifolia* based on in vivo experiments [26]. Calophymembranside C (**141**), isolated from the stems of *Calophyllum membranaceum*, showed transcriptional inhibitory activity towards RXRα, with 50% inhibitory concentration (IC_{50}) values of 29.95 ± 1.08 [24]. Arenicolin A (**152**), isolated from *Penicillium arenicola*, exhibited cytotoxicity toward mammalian cell lines, including colorectal carcinoma (HCT-116), neuroblastoma (IMR-32), and ductal carcinoma (BT-474) cells, with IC_{50} values of 7.3, 6.0, and 9.7 μM, respectively [73]. Apigenosylide B (**16**), isolated from an EtOH extract of the leaves of *Machilus japonica* var. *kusanoi*, possesses moderate inhibitory activity against α-glucosidase [9]. Compounds **45–46** inhibited complement activation in the classic pathway in vitro, with IC_{50} values ranging from 0.88 to 4.02 mM. This may suggest the application of the herb for the treatment of acute respiratory distress syndrome, etc. [14]. Diandraflavone (**81**), isolated from *Drymaria diandra*, showed significantly selective inhibition on superoxide anion generation from human neutrophils stimulated by fMLP/CB, with an IC_{50} value of 10.0 μg/mL [16].

4. Synthesis

In nature, most C-glycosides are derived from plants. Aryl C-glycosides are biosynthetically prepared by the catalysis of C-glycosyltransferases (CGTs). However, while a large family of O-glycosyltransferases (OGTs) is known, a limited number of CGTs have been discovered in plants [88]. Many detailed reviews on the chemical synthesis of aryl-C-glycoside have been published [8,89–93]. Herein, a brief historical review of the total synthesis of natural aryl-C-glycosides is presented. From a retrosynthetic viewpoint, the strategy of synthesis of aryl-C-glycoside usually includes two protocols, as described below.

4.1. C-Glycosylation of Arenes and De Novo Construction of the Aromatic Moiety

This protocol exploits the installation of a sugar moiety on a full or partial aromatic structure. For a partial aryl-C-glycoside structure, it uses a simple C-glycoside as the starting material, which is functionalized, providing a complex aryl unit. The classical electrophilic aromatic substitution approach was often applied. The sugar portion includes glycosyl trichloroacetimidate (Scheme 1) [94], glycosyl acetate (Scheme 2) [95], glycosyl fluoride (Scheme 3) [96], glycosyl lactone (Scheme 4) [97], glycosyl thioglycoside (Scheme 5) [98], and other sugars. This strategy was successfully applied to the total synthesis of pluramycins [99], aciculatin [98], vineomycin B_2 and its methyl ester [100–102], angucycline C5 glycosides [103], paecilomycin B [104,105], aspalathin [106], nothofagin [106], chrysomycin A [107], vicenin-2 [108], 3,3′-Di-O-methyl Ardimerin [109], deacetylravidomycin M [110], aquayamycin [111], a precursor of kendomycin [112], and anthraquinone-based aryl-C-glycosides [113].

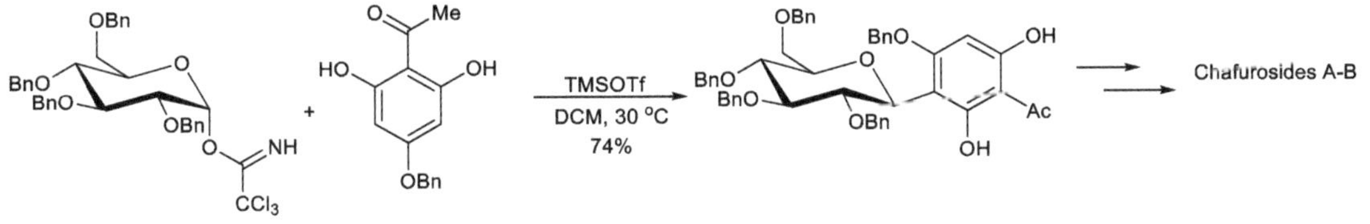

Scheme 1. Concise synthesis of chafurosides A and B.

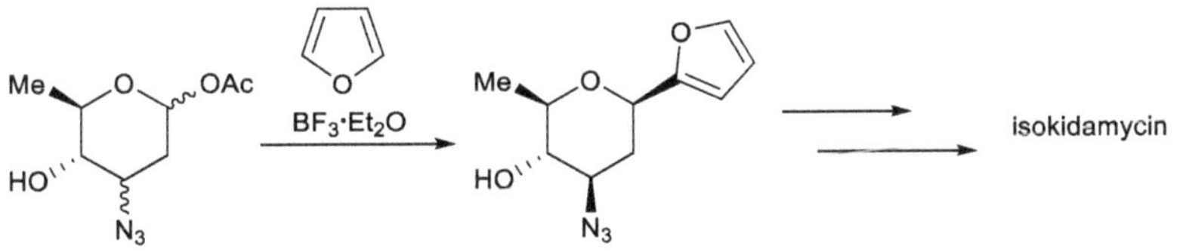

Scheme 2. Total synthesis of isokidamycin.

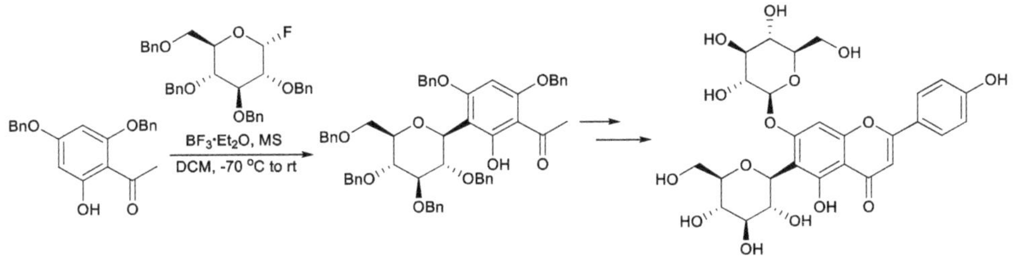

Scheme 3. Total synthesis of saponarin.

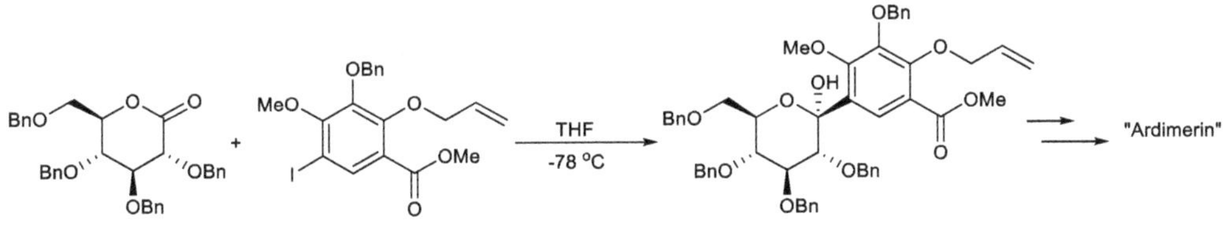

Scheme 4. Total synthesis of the proposed structure of ardimerin.

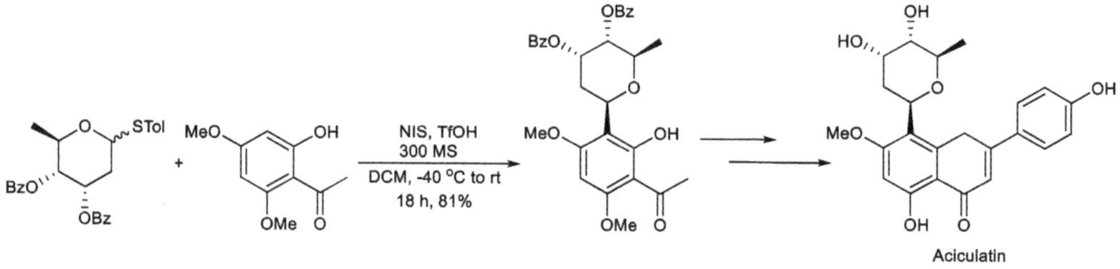

Scheme 5. Synthesis of aciculatin.

Toshiyuki Kan et al. reported the regioselective synthesis of chafurosides A (**107**) and B (**108**) using a novel protecting-group strategy. The construction of the dihydrofuran ring was achieved via an intramolecular Mitsunobu reaction. The key step in the C-glycosylation is the O→C rearrangement of the phenolic glycoside formed by TMSOTf-catalyzed glycosylation (Scheme 1) [94].

Martin et al. reported the total synthesis of isokidamycin, which features the use of a silicon tether as a disposable regiocontrol element in an intramolecular Diels–Alder reaction between a substituted naphthyne and a glycosyl furan and a subsequent *O*→*C*-glycoside rearrangement (Scheme 2) [95].

Sato et al. reported the total synthesis of saponarin. Saponarin was efficiently synthesized via 11 steps from 2, 4-*O*-dibenzylphloroacetophenone, with an overall yield of 37%. The key step also features a glycosylation involving per-*O*-benzylglucosyl α-fluoride and 2, 4-*O*-dibenzylphloroacetophenone, applying the *O*→*C* glycoside rearrangement method (Scheme 3) [96].

Suzuki K. et al. reported the total synthesis of the proposed structure of ardimerin, whose key step includes the β-selective formation of the crucial *C*-glycoside linkage by the reaction between aryl iodide, through a halogen–metal exchange reaction, and lactone (Scheme 4) [97].

Lee et al. reported the total synthesis of aciculatin. The key step is the glycosylation of the digitoxosyl thioglycoside with an electron-rich phenol activated by NIS/TfOH, which afforded the β-D-digitoxopyranoside (Scheme 5) [98].

Besides the nucleophilic attack reaction of the aryl portions to provide aryl-C-glycoside molecules, the transition metal-catalyzed cross-coupling reactions to form the C_{sp3}-C_{sp2} bond of aryl-C-glycosides are becoming more and more powerful. The Negeshi [114], Kumada [115], Stille [116–118], Heck [119,120], Sonogashira [121], Hiyama [122], and radical [123–125] cross-coupling reactions with different metals such as Pd [126,127], Ni [128,129], Fe [124,125], Co [115], Ir [130] as catalysts, as well as C-H activation [131,132] were successfully applied to the construction of aryl-C-glycoside scaffolds.

4.2. *De Novo Construction of the Sugar Moiety*

This approach uses the reactions of assembled aryl units based on methods for the construction of a sugar, including, but not limited to (a) the hetero-Diels–Alder reaction, (b) the 1,3-dipolar cycloaddition of nitrile oxides, (c) the ring-closing olefin metathesis [8], the de novo asymmetric approach [133–135], and the ring opening-ring closure strategy [136]. It was successfully applied to the synthesis of *O*-spiro-*C*-aryl glycosides [137] and 2-deoxy-β-*C*-aryl glycosides [138].

Hauser et al. reported that de novo synthesis of *C*-aryl glycosides based on cycloaddition of an aryl nitrile oxide with 4-pentyn-2-ol, which was straightforwardly converted to the pyranone through sequential hydrogenolysis of the N-O bond of the isoxazole followed by acid-catalyzed intramolecular cyclization. (Scheme 6) [139].

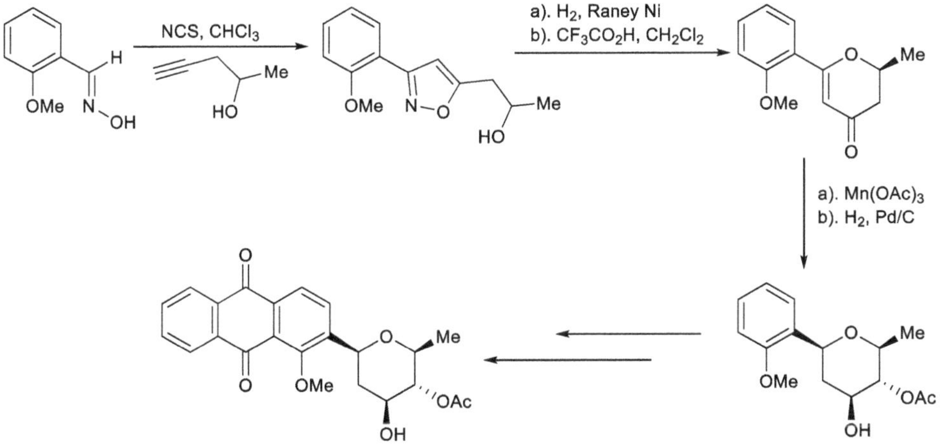

Scheme 6. Total synthesis of naturally occurring *C*-aryl glycosides.

Johann Mulzer et al. reported the development of a convergent and concise route to an advanced precursor of kendomycin by applying an S_N1 ring cyclization as a key step. The sugar moiety formed by the acid-catalyzed intramolecular etherification reaction of aryl-substituted 1, 3, 5-triol (Scheme 7) [140].

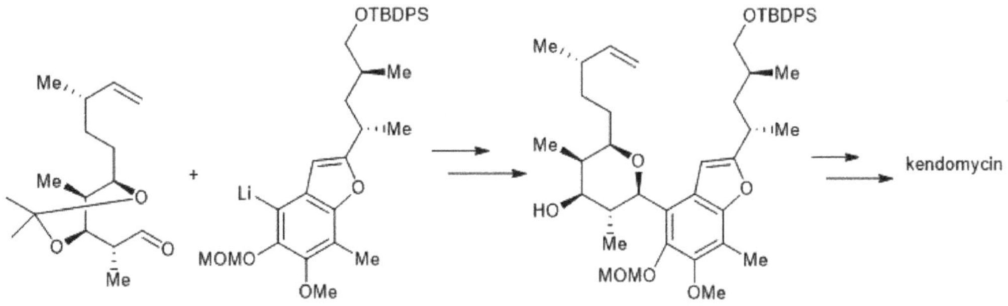

Scheme 7. Concise synthesis of kendomycin.

5. Conclusions and Perspectives

In this review, the recently discovered aryl-C-glycoside structures were listed and their biological activities as well as their synthetical approaches were summarized. The diverse structures of natural aryl-C-glycosides and their multiple pharmacological effects suggest significant medicinal applications. This review presents a summary of studies published from 2002 to date on this promising compounds. The core structure of aryl-C-glycosides, which is glycosylated and/or acylated, exhibits a great number of molecular entities and possesses various pharmaceutical effects. It is expected that more and more aryl-C-glycoside scaffolds including natural and synthetical molecules will be disclosed, and their medicinal application will come true in the near future to benefit human health.

Funding: This research was funded by National Natural Science Foundation of China, grant number [21762004].

Institutional Review Board Statement: Not applicable.

Informed Consent Statement: Not applicable.

Acknowledgments: The research was financially supported by the National Natural Science Foundation of China (21762004).

Conflicts of Interest: The authors declare no conflict of interest.

References

1. Levy, D.E.; Tang, C. *The Chemistry of C-Glycosides*; Pergamon: Tarrytown, NY, USA, 1995.
2. Pałasz, A.; Cież, D.; Trzewik, B.; Miszczak, K.; Tynor, G.; Bazan, B. In the Search of Glycoside-Based Molecules as Antidiabetic Agents. *Top. Curr. Chem.* **2019**, *37*, 19. [CrossRef] [PubMed]
3. Bokor, É.; Kun, S.; Goyard, D.; Tóth, M.; Praly, J.-P.; Vidal, S.; Somsák, L. C-Glycopyranosyl Arenes and Hetarenes: Synthetic Methods and Bioactivity Focused on Antidiabetic Potential. *Chem. Rev.* **2017**, *117*, 1687–1764. [CrossRef] [PubMed]
4. Sezaki, M.; Kondo, S.; Maeda, K.; Umezawa, H.; Ohno, M. The structure of aquayamycin. *Tetrahedron* **1970**, *26*, 5171–5190. [CrossRef]
5. Bililign, T.; Griffith, B.R.; Thorson, J.S. Structure, activity, synthesis and biosynthesis of aryl-C-glycosides. *Nat. Prod. Rep.* **2005**, *22*, 742–760. [CrossRef]
6. Kharel, M.K.; Pahari, P.; Shepherd, M.D.; Tibrewal, N.; Nybo, S.E.; Shaaban, K.A.; Rohr, J. Angucyclines: Biosynthesis, mode-of-action, new natural products, and synthesis. *Nat. Prod. Rep.* **2012**, *29*, 264–325. [CrossRef]
7. Bajracharya, G.B. Diversity, pharmacology and synthesis of bergenin and its derivatives: Potential materials for therapeutic usages. *Fitoterapia* **2015**, *101*, 133–152. [CrossRef] [PubMed]
8. Kitamura, K.; Ando, Y.; Matsumoto, T.; Suzuki, K. Total Synthesis of Aryl C-Glycoside Natural Products: Strategies and Tactics. *Chem. Rev.* **2018**, *118*, 1495–1598. [CrossRef]
9. Lee, S.-S.; Lin, Y.-S.; Chen, C.-K. Three Adducts of Butenolide and Apigenin Glycoside from the Leaves of *Machilus japonica*. *J. Nat. Prod.* **2009**, *72*, 1249–1252. [CrossRef]
10. Chen, W.; Zhang, H.; Wang, J.; Hu, X. Flavonoid Glycosides from the Bulbs of *Lilium speciosum* var. gloriosoides and their Potential Antiviral Activity Against RSV. *Chem. Nat. Compd.* **2019**, *55*, 461–464.
11. Bai, H.-H.; Wang, N.-N.; Mi, J.; Yang, T.; Fang, D.-M.; Wu, L.-W.; Zhao, H.; Li, G.-Y. Hydroxycinnamoylmalated flavone C-glycosides from *Lemna japonica*. *Fitoterapia* **2018**, *124*, 211–216. [CrossRef]
12. Jiang, X.-L.; Wang, L.; Wang, E.-J.; Zhang, G.-L.; Chen, B.; Wang, M.-K.; Li, F. Flavonoid glycosides and alkaloids from the embryos of *Nelumbo nucifera* seeds and their antioxidant activity. *Fitoterapia* **2018**, *125*, 184–190. [CrossRef] [PubMed]
13. Wu, Q.-X.; Li, Y.; Shi, Y.-P. Antioxidant phenolic glucosides from *Gentiana piasezkii*. *J. Asian Nat. Prod. Res.* **2006**, *8*, 391–396. [CrossRef] [PubMed]
14. Liu, J.-Y.; Li, S.-Y.; Feng, J.-Y.; Sun, Y.; Cai, J.-N.; Sun, X.-F.; Yang, S.-L. Flavone C-glycosides from the flowers of *Trollius chinensis* and their anti-complementary activity. *J. Asian Nat. Prod. Res.* **2013**, *15*, 325–331. [CrossRef] [PubMed]
15. Zhou, Z.; Zhang, T.; Xiao, H.; Zhou, X.; Wu, H. A New C-Glycosylflavone from the Rhizomes of *Cyperus rotundus*. *Chem. Nat. Compd.* **2015**, *51*, 640–642. [CrossRef]
16. Hsieh, P.-W.; Chang, F.-R.; Lee, K.-H.; Hwang, T.-L.; Chang, S.-M.; Wu, Y.-C. A New Anti-HIV Alkaloid, Drymaritin, and a New C-Glycoside Flavonoid, Diandraflavone, from Drymaria diandra. *J. Nat. Prod.* **2004**, *67*, 1175–1177. [CrossRef]
17. Qiu, L.; Jiao, Y.; Xie, J.-Z.; Huang, G.-K.; Qiu, S.-L.; Miao, J.-H.; Yao, X.-S. Five new flavonoid glycosides from *Nervilia fordii*. *J. Asian Nat. Prod. Res.* **2013**, *15*, 589–599. [CrossRef]
18. Iwao, Y.; Ishida, H.; Kimura, S.I.; Wakimoto, T.; Kondo, H.; Itai, S.; Noguchic, S. Crystal Structures of Flavone C-Glycosides from Oolong Tea Leaves: Chafuroside A Dihydrate and Chafuroside B Monohydrate. *Chem. Pharm. Bull.* **2019**, *67*, 935–939. [CrossRef]
19. Suzuki, R.; Okada, Y.; Okuyama, T. A New Flavone C-Glycoside from the Style of *Zea mays* L. with Glycation Inhibitory Activity. *Chem. Pharm. Bull.* **2003**, *51*, 1186–1188. [CrossRef]
20. Suzuki, R.; Okada, Y.; Okuyama, T. Two Flavone C-Glycosides from the Style of *Zea mays* with Glycation Inhibitory Activity. *J. Nat. Prod.* **2003**, *66*, 564–565. [CrossRef]
21. Li, C.-J.; Yang, J.-Z.; Yu, S.-S.; Zhao, C.-Y.; Peng, Y.; Wang, X.-L.; Zhang, D.-M. Glomexanthones A-C, three xanthonolignoid C-glycosides from *Polygala glomerata* Lour. *Fitoterapia* **2014**, *93*, 175–181.
22. Faizi, S.; Zikr-ur-Rehman, S.; Naz, A.; Versiani, M.A.; Dar, A.; Naqvi, S. Bioassay-guided studies on *Bombax ceiba* leaf extract: Isolation of shamimoside, a new antioxidant xanthone C-glucoside. *Chem. Nat. Compd.* **2012**, *48*, 774–779. [CrossRef]
23. Du, X.-G.; Wang, W.; Zhang, S.-P.; Pu, X.-P.; Zhang, Q.-Y.; Ye, M.; Zhao, Y.-Y.; Wang, B.-R.; Khan, I.A.; Guo, D.-A. Neuroprotective Xanthone Glycosides from *Swertia punicea*. *J. Nat. Prod.* **2010**, *73*, 1422–1426. [CrossRef] [PubMed]
24. Ming, M.; Zhang, X.; Chen, H.-F.; Zhu, L.-J.; Zeng, D.-Q.; Yang, J.; Wu, G.-X.; Wu, Y.-Z.; Yao, X.-S. RXR alpha transcriptional inhibitors from the stems of *Calophyllum membranaceum*. *Fitoterapia* **2016**, *108*, 66–72. [CrossRef] [PubMed]
25. Perlatti, B.; Lan, N.; Earp, C.E.; AghaAmiri, S.; Vargas, S.H.; Azhdarinia, A.; Bills, G.F.; Gloer, J.B. Arenicolins: C-Glycosylated Depsides from *Penicillium arenicola*. *J. Nat. Prod.* **2020**, *83*, 668–674. [CrossRef]
26. Lee, B.-W.; Park, J.-G.; Ha, T.K.Q.; Pham, H.T.T.; An, J.-P.; Noh, J.-R.; Lee, C.-H.; Oh, W.-K. Constituents of the Edible Leaves of *Melicope pteleifolia* with Potential Analgesic Activity. *J. Nat. Prod.* **2019**, *82*, 2201–2210. [CrossRef]
27. Dat, N.T.; Bae, K.; Wamiru, A.; McMahon, J.B.; Grice, S.F.J.L.; Bona, M.; Beutler, J.A.; Kim, Y.H. A Dimeric Lactone from *Ardisia japonica* with Inhibitory Activity for HIV-1 and HIV-2 Ribonuclease H. *J. Nat. Prod.* **2007**, *70*, 839–841. [CrossRef]

28. Ito, H.; Kasajima, N.; Tokuda, H.; Nishino, H.; Yoshida, T. Dimeric Flavonol Glycoside and Galloylated C-Glucosylchromones from Kunzea ambigua. *J. Nat. Prod.* **2004**, *67*, 411–415. [CrossRef]
29. Zhang, D.; Li, Y.; Li, X.; Han, X.; Wang, Z.; Zhang, W.; Dou, B.; Lu, Z.; Li, P.; Li, G. Neopetrosins A–D and Haliclorensin D, Indole-C-Mannopyranosides and a Diamine Alkaloid Isolated from the South China Sea Marine Sponge Neopetrosia chaliniformis. *J. Nat. Prod.* **2022**, *85*, 1626–1633. [CrossRef]
30. Harunari, E.; Imada, C.; Igarashi, Y. Konamycins A and B and Rubromycins CA1 and CA2, Aromatic Polyketides from the Tunicate-Derived *Streptomyces hyaluromycini* MB-PO13^{T}. *J. Nat. Prod.* **2019**, *82*, 1609–1615. [CrossRef]
31. Chang, Y.; Xing, L.; Sun, C.; Liang, S.; Liu, T.; Zhang, X.; Zhu, T.; Pfeifer, B.A.; Che, Q.; Zhang, G.; et al. Monacycliones G–K and ent-Gephyromycin A, Angucycline Derivatives from the Marine-Derived *Streptomyces* sp. HDN15129. *J. Nat. Prod.* **2020**, *83*, 2749–2755. [CrossRef]
32. Martin, G.D.A.; Tan, L.T.; Jensen, P.R.; Dimayuga, R.E.; Fairchild, C.R.; Raventos-Suarez, C.; Fenical, W. Marmycins A and B, Cytotoxic Pentacyclic C-Glycosides from a Marine Sediment-Derived Actinomycete Related to the Genus *Streptomyces*. *J. Nat. Prod.* **2007**, *70*, 1406–1409. [CrossRef] [PubMed]
33. Shen, C.-C.; Cheng, J.-J.; Lay, H.-L.; Wu, S.-Y.; Ni, C.-L.; Teng, C.-M.; Chen, C.-C. Cytotoxic Apigenin Derivatives from *Chrysopogon aciculatis*. *J. Nat. Prod.* **2012**, *75*, 198–201. [CrossRef] [PubMed]
34. Syed, A.S.; Akram, M.; Bae, O.-N.; Kim, C.Y. Isocassiaoccidentalin B, A New C-Glycosyl Flavone Containing a 3-Keto Sugar, and Other Constituents from *Cassia nomame*. *Helv. Chim. Acta* **2016**, *99*, 691–695. [CrossRef]
35. Huang, H.; Yang, T.; Ren, X.; Liu, J.; Song, Y.; Sun, A.; Ma, J.; Wang, B.; Zhang, Y.; Huang, C.; et al. Cytotoxic Angucycline Class Glycosides from the Deep-Sea Actinomycete *Streptomyces lusitanus* SCSIO LR32. *J. Nat. Prod.* **2012**, *75*, 202–208. [CrossRef] [PubMed]
36. Song, Y.; Liu, G.; Li, J.; Huang, H.; Zhang, X.; Zhang, H.; Ju, J. Cytotoxic and Antibacterial Angucycline- and Prodigiosin-Analogues from the Deep-Sea Derived *Streptomyces* sp. SCSIO 11594. *Mar. Drugs* **2015**, *13*, 1304–1316. [CrossRef] [PubMed]
37. McNally, D.J.; Wurms, K.V.; Labbé, C.; Quideau, S.; Bélanger, R.R. Complex C-Glycosyl Flavonoid Phytoalexins from *Cucumis sativus*. *J. Nat. Prod.* **2003**, *66*, 1280–1283. [CrossRef]
38. Ancheeva, E.; Daletos, G.; Muharini, R.; Lin, W.H.; Teslov, L.; Proksch, P. Flavonoids from *Stellaria nemorum* and *Stellaria holostea*. *Nat. Prod. Commun.* **2015**, *10*, 437–440. [CrossRef]
39. Zhang, Y.-Q.; Luo, J.-G.; Han, C.; Xu, J.-F.; Kong, L.-Y. Bioassay-guided preparative separation of angiotensin-converting enzyme inhibitory C-flavone glycosides from Desmodium styracifolium by recycling complexation high-speed counter-current chromatography. *J. Pharmaceut. Biomed.* **2015**, *102*, 276–281. [CrossRef]
40. Elbandy, M.; Miyamoto, T.; Lacaille-Dubois, M.-A. Sulfated Lupane Triterpene Derivatives and a Flavone C-Glycoside from *Gypsophila repens*. *Chem. Pharm. Bull.* **2007**, *55*, 808–811. [CrossRef]
41. Olennikov, D.N.; Chirikova, N.K. New C, O-Glycosylflavones from *Melandrium divaricatum*. *Chem. Nat. Compd.* **2019**, *55*, 1032–1038. [CrossRef]
42. Olennikov, D.N.; Kashchenko, N.I. New C, O-Glycosylflavones from the Genus *Silene*. *Chem. Nat. Compd.* **2020**, *56*, 1026–1034. [CrossRef]
43. Obmann, A.; Werner, I.; Presser, A.; Zehl, M.; Swoboda, Z.; Purevsuren, S.; Narantuya, S.; Kletter, C.; Glasl, S. Flavonoid C- and O-glycosides from the Mongolian medicinal plant Dianthus versicolor Fisch. *Carbohyd. Res.* **2011**, *346*, 1868–1875. [CrossRef] [PubMed]
44. Chen, J.-M.; Wei, L.-B.; Lu, C.-L.; Zhou, G.-X. A flavonoid 8-C-glycoside and a triterpenoid cinnamate from *Nervilia fordii*. *J. Asian Nat. Prod. Res.* **2013**, *15*, 1088–1093. [CrossRef]
45. Devkota, H.P.; Fukusako, K.; Ishiguro, K.; Yahara, S. Flavone C-Glycosides from *Lychnis senno* and their Antioxidative Activity. *Nat. Prod. Commun.* **2013**, *8*, 1413–1414. [CrossRef]
46. Xu, K.-J.; Xu, X.-M.; Deng, W.-L.; Zhang, L.; Wang, M.-K.; Ding, L.-S. Three new flavone C-glycosides from the aerial parts of *Paraquilegia microphylla*. *J. Asian Nat. Prod. Res.* **2011**, *13*, 409–416. [CrossRef] [PubMed]
47. Alqahtani, J.; Formisano, C.; Chianese, G.; Luciano, P.; Stornaiuolo, M.; Perveen, S.; Taglialatela-Scafati, O. Glycosylated Phenols and an Unprecedented Diacid from the Saudi Plant *Cissus rotundifolia*. *J. Nat. Prod.* **2020**, *83*, 3298–3304. [CrossRef] [PubMed]
48. Li, M.; Wang, Y.; Tsoi, B.; Jin, X.-J.; He, R.-R.; Yao, X.-J.; Dai, Y.; Kurihara, H.; Yao, X.-S. Indoleacetic acid derivatives from the seeds of *Ziziphus jujuba* var. *spinosa*. *Fitoterapia* **2014**, *99*, 48–55. [CrossRef] [PubMed]
49. Olennikov, D.N.; Chirikova, N.K. New Compounds from Siberian *Gentiana* Species. II. Xanthone and C, O-Glycosylflavone. *Chem. Nat. Compd.* **2021**, *57*, 681–684. [CrossRef]
50. Xie, Y.-Y.; Xu, Z.-L.; Wang, H.; Kano, Y.; Yuan, D. A novel spinosin derivative from Semen Ziziphi Spinosae. *J. Asian Nat. Prod. Res.* **2011**, *13*, 1151–1157. [CrossRef]
51. Song, Z.; Hashi, Y.; Sun, H.; Liang, Y.; Lan, Y.; Wang, H.; Chen, S. Simultaneous determination of 19 flavonoids in commercial trollflowers by using high-performance liquid chromatography and classification of samples by hierarchical clustering analysis. *Fitoterapia* **2013**, *91*, 272–279. [CrossRef]
52. Wu, L.-Z.; Wu, H.-F.; Xu, X.-D.; Yang, J.-S. Two New Flavone C-Glycosides from *Trollius ledebourii*. *Chem. Pharm. Bull.* **2011**, *59*, 1393–1395. [CrossRef] [PubMed]
53. Zou, J.-H.; Yang, J.-S.; Dong, Y.-S.; Zhou, L.; Lin, G. Flavone C-glycosides from flowers of *Trollius ledebouri*. *Phytochemistry* **2005**, *66*, 1121–1125. [CrossRef] [PubMed]

54. Song, Z.; Wang, H.; Ren, B.; Zhang, B.; Hashi, Y.; Chen, S. On-line study of flavonoids of Trollius chinensis Bunge binding to DNA with ethidium bromide using a novel combination of chromatographic, mass spectrometric and fluorescence techniques. *J. Chromatogr. A* **2013**, *1282*, 102–112. [CrossRef] [PubMed]
55. Kanchanapoom, T. Aromatic diglycosides from *Cladogynos orientalis*. *Phytochemistry* **2007**, *68*, 692–696. [CrossRef] [PubMed]
56. Le Moullec, A.; Juvik, O.J.; Fossen, T. First identification of natural products from the African medicinal plant *Zamioculcas zamiifolia*—A drought resistant survivor through millions of years. *Fitoterapia* **2015**, *106*, 280–285. [CrossRef]
57. Tang, L.; Xu, X.-M.; Rinderspacher, K.A.; Cai, C.-Q.; Ma, Y.; Long, C.-L.; Feng, J.-C. Two new compounds from *Comastoma pedunlulatum*. *J. Asian Nat. Prod. Res.* **2011**, *13*, 895–900. [CrossRef]
58. Ebrahimi, S.N.; Gafner, F.; Dell'Acqua, G.; Schweikert, K.; Hamburger, M. Flavone 8-C-Glycosides from *Haberlea rhodopensis* Friv. (Gesneriaceae). *Helv. Chim. Acta* **2011**, *94*, 38–45. [CrossRef]
59. Li, Z.-L.; Li, D.-Y.; Hua, H.-M.; Chen, X.-H.; Kim, C.-S. Three new acylated flavone C-glycosides from the flowers of *Trollius chinensis*. *J. Asian Nat. Prod. Res.* **2009**, *11*, 426–432. [CrossRef]
60. Zou, J.-H.; Yang, J.; Zhou, L. Acylated Flavone C-Glycosides from *Trollius ledebouri*. *J. Nat. Prod.* **2004**, *67*, 664–667. [CrossRef]
61. Dong, F.-Y.; Guan, L.-N.; Zhang, Y.-H.; Cui, Z.-H.; Wang, L.; Wang, W. Acylated flavone C-glycosides from *Hemistepta lyrate*. *J. Asian Nat. Prod. Res.* **2010**, *12*, 776–780. [CrossRef]
62. Chen, Y.; Yan, X.; Lu, F.; Jiang, X.; Friesen, J.B.; Pauli, G.F.; Chen, S.-N.; Li, D.-P. Preparation of flavone di-C-glycoside isomers from Jian-Gu injection (*Premna fulva* Craib.) using recycling counter-current chromatography. *J. Chromatogr. A* **2019**, *1599*, 180–186. [CrossRef] [PubMed]
63. Dong, Q.; Huang, Y.; Qiao, S.-Y. Studies on chemical constituents from Stellaria media I. *China J. Chin. Mater. Med.* **2007**, *32*, 1048–1051.
64. Feng, X.; Jiang, D.; Shan, Y.; Dai, T.; Dong, Y.; Cao, W. New flavonoid-C-Glycosides from *Triticum aestivum*. *Chem. Nat. Compd.* **2008**, *44*, 171–173. [CrossRef]
65. Pan, Y.-X.; Zhou, C.-X.; Zhang, S.-L.; Zheng, X.-X.; Zhao, Y. Constituents from *Ranunculus sieboldii* Miq. *J. Chin. Pharm. Sci.* **2004**, *13*, 92–96.
66. Zheleva-Dimitrova, D.; Nedialkov, P.; Giresser, U. A Validated HPLC Method for Simultaneous Determination of Caffeoyl Phenylethanoid Glucosides and Flavone 8-C-glycosides in *Haberlea rhodopensis*. *Nat. Prod. Commun.* **2016**, *11*, 791–792. [CrossRef]
67. Pacifico, S.; Scognamiglio, M.; D'Abrosca, B.; Piccolella, S.; Tsafantakis, N.; Gallicchio, M.; Ricci, A.; Fiorentino, A. Spectroscopic Characterization and Antiproliferative Activity on HepG2 Human Hepatoblastoma Cells of Flavonoid C-Glycosides from *Petrorhagia velutina*. *J. Nat. Prod.* **2010**, *73*, 1973–1978. [CrossRef]
68. Zheng, J.-X.; Zheng, Y.; Dai, Y.; Wang, N.-L.; Fang, Y.-X.; Du, Z.-Y.; Zhao, S.-Q.; Zhang, K.; Wu, L.-Y.; Fan, M. Flavone Di-C-Glycosides from *Selaginella uncinate* and Their Antioxidative Activities. *Chem. Nat. Compd.* **2016**, *52*, 306–308. [CrossRef]
69. Xie, C.; Veitch, N.C.; Houghton, P.J.; Simmonds, M.S.J. Flavone C-Glycosides from *Viola yedoensis* MAKINO. *Chem. Pharm. Bull.* **2003**, *51*, 1204–1207. [CrossRef]
70. Anh, N.H.; Yen, D.T.H.; Cuong, N.T.; Tai, B.H.; Yen, P.H.; Chinh, P.T.; Cuong, P.V.; Nam, N.H.; Kiem, P.V.; Cho, S.-H.; et al. Three new chromanes and one new flavone C-glycoside from *Mallotus apelta*. *J. Asian Nat. Prod. Res.* **2022**, *24*, 1–9. [CrossRef]
71. Lam, S.-H.; Hung, H.-Y.; Yang, M.-L.; Chen, H.-H.; Kuo, P.-C.; Wu, T.-S. Chemical Constituents from *Phalaenopsis* Hybrids and Their Bioactivities. *Nat. Prod. Commun.* **2019**, *14*. [CrossRef]
72. Hu, H.-B.; Zhu, J.-H. Flavonoid Constituents from the Roots of *Acanthopanax brachypus*. *Chem. Pharm. Bull.* **2011**, *59*, 135–139. [CrossRef] [PubMed]
73. Li, G.-H.; Zhang, Q.-W.; Wang, L.; Zhang, X.-Q.; Ye, W.-C.; Wang, Y.-T. New Isoflavone C-Glycosides from *Pueraria lobata*. *Helv. Chim. Acta* **2011**, *94*, 423–428. [CrossRef]
74. Ukida, K.; Doi, T.; Sugimoto, S.; Matsunami, K.; Otsuka, H.; Takeda, Y. Schoepfiajasmins A–H: C-Glycosyl Dihydrochalcones, Dihydrochalcone Glycoside, C-Glucosyl Flavanones, Flavanone Glycoside and Flavone Glycoside from the Branches of *Schoepfia jasminodora*. *Chem. Pharm. Bull.* **2013**, *61*, 1136–1142. [CrossRef] [PubMed]
75. Zheng, X.-K.; Cao, Y.-G.; Ke, Y.-Y.; Zhang, Y.-L.; Li, F.; Gong, J.-H.; Zhao, X.; Kuang, H.-X.; Feng, W.-S. Phenolic constituents from the root bark of *Morus alba* L. and their cardioprotective activity in vitro. *Phytochemistry* **2017**, *135*, 128–134. [CrossRef]
76. Ateba, S.B.; Njamen, D.; Gatterer, C.; Scherzer, T.; Zehl, M.; Kählig, H.; Krenn, L. Rare phenolic structures found in the aerial parts of *Eriosema laurentii* De Wild. *Phytochemistry* **2016**, *128*, 5–11. [CrossRef]
77. Shimokawa, Y.; Akao, Y.; Hirasawa, Y.; Awang, K.; Hadi, A.H.A.; Sato, S.; Aoyama, C.; Takeo, J.; Shiro, M.; Morita, H. Gneyulins A and B, Stilbene Trimers, and Noidesols A and B, Dihydroflavonol-C-Glucosides, from the Bark of *Gnetum gnemonoides*. *J. Nat. Prod.* **2010**, *73*, 763–767. [CrossRef]
78. Tsujimoto, T.; Nishihara, M.; Osumi, Y.; Hakamatsuka, T.; Goda, Y.; Uchiyama, N.; Ozekia, Y. Structural Analysis of Polygalaxanthones, C-Glucosyl Xanthones of *Polygala tenuifolia* Roots. *Chem. Pharm. Bull.* **2019**, *67*, 1242–1247. [CrossRef]
79. Abdel-Mageed, W.M.; Bayoumi, S.A.H.; Chen, C.; Vavricka, C.J.; Li, L.; Malik, A.; Dai, H.; Song, F.; Wang, L.; Zhang, J.; et al. Benzophenone C-glucosides and gallotannins from mango tree stem bark with broad-spectrum anti-viral activity. *Bioorg. Med. Chem.* **2014**, *22*, 2236–2243. [CrossRef]
80. Achari, B.; Dutta, P.K.; Roy, S.K.; Chakraborty, P.; Sengupta, J.; Bandyopadhyay, D.; Maity, J.K.; Khan, I.A.; Ding, Y.; Ferreira, D. Fluorescent Pigment and Phenol Glucosides from the Heartwood of *Pterocarpus marsupium*. *J. Nat. Prod.* **2012**, *75*, 655–660. [CrossRef]

81. Zou, J.; Jin, D.; Chen, W.; Wang, J.; Liu, Q.; Zhu, X.; Zhao, W. Selective Cyclooxygenase-2 Inhibitors from *Calophyllum membranaceum*. *J. Nat. Prod.* **2005**, *68*, 1514–1518. [CrossRef]
82. Zhuravleva, O.I.; Afiyatullov, S.S.; Denisenko, V.A.; Ermakova, S.P.; Slinkina, N.N.; Dmitrenok, P.S.; Kim, N.Y. Secondary metabolites from a marine-derived fungus *Aspergillus carneus* Blochwitz. *Phytochemistry* **2012**, *80*, 123–131. [CrossRef] [PubMed]
83. Zhu, L.-J.; Yi, S.; Li, X.; Chen, H.-F.; Ming, M.; Zhang, X.; Yao, X.-S. *C*-glycosides from the stems of *Calophyllum membranaceum*. *J. Asian Nat. Prod. Res.* **2018**, *20*, 49–54. [CrossRef] [PubMed]
84. Bringmann, G.; Lang, G.; Steffens, S.; Günther, E.; Schaumann, K. Evariquinone, isoemericellin, and stromemycin from a sponge derived strain of the fungus *Emericella variecolor*. *Phytochemistry* **2003**, *63*, 437–443. [CrossRef]
85. Wu, Y.; Zhang, Z.-X.; Hu, H.; Li, D.; Qiu, G.; Hu, X.; He, X. Novel indole *C*-glycosides from *Isatis indigotica* and their potential cytotoxic activity. *Fitoterapia* **2011**, *82*, 288–292. [CrossRef] [PubMed]
86. Li, F.; He, Z.; Ye, Y. Isocartormin, a novel quinochalcone *C*-glycoside from Carthamus tinctorius. *Acta Pharm. Sin. B* **2017**, *7*, 527–531. [CrossRef]
87. Marrassini, C.; Davicino, R.; Acevedo, C.; Anesini, C.; Gorzalczany, S.; Ferraro, G. Vicenin-2, a Potential Anti-inflammatory Constituent of Urtica circularis. *J. Nat. Prod.* **2011**, *74*, 1503–1507. [CrossRef] [PubMed]
88. Zhang, M.; Li, F.-D.; Li, K.; Wang, Z.-L.; Wang, Y.-X.; He, J.-B.; Su, H.-F.; Zhang, Z.-Y.; Chi, C.-B.; Shi, X.-M.; et al. Functional Characterization and Structural Basis of an Efficient Di-C-glycosyltransferase from *Glycyrrhiza glabra*. *J. Am. Chem. Soc.* **2020**, *142*, 3506–3512. [CrossRef]
89. Bennett, C.S.; Galan, M.C. Methods for 2-Deoxyglycoside Synthesis. *Chem. Rev.* **2018**, *118*, 7931–7985. [CrossRef]
90. Yang, Y.; Yu, B. Recent Advances in the Chemical Synthesis of *C*-Glycosides. *Chem. Rev.* **2017**, *117*, 12281–12356. [CrossRef]
91. Liao, H.; Ma, J.; Yao, H.; Liu, X.-W. Recent progress of *C*-glycosylation methods in the total synthesis of natural products and pharmaceuticals. *Org. Biomol. Chem.* **2018**, *16*, 1791–1806. [CrossRef]
92. Lee, D.Y.W.; He, M. Recent Advances in Aryl *C*-Glycoside Synthesis. *Curr. Top. Med. Chem.* **2005**, *5*, 1333–1350. [CrossRef] [PubMed]
93. Suzuki, K. Lessons from Total Synthesis of Hybrid Natural Products. *Chem. Rec.* **2010**, *10*, 291–307. [CrossRef] [PubMed]
94. Furuta, T.; Nakayama, M.; Suzuki, H.; Tajimi, H.; Inai, M.; Nukaya, H.; Wakimoto, T.; Kan, T. Concise Synthesis of Chafurosides A and B. *Org. Lett.* **2009**, *11*, 2233–2236. [CrossRef] [PubMed]
95. O'Keefe, B.M.; Mans, D.M.; Kaelin, D.E.; Martin, S.F. Total Synthesis of Isokidamycin. *J. Am. Chem. Soc.* **2010**, *132*, 15528–15530. [CrossRef] [PubMed]
96. Misawa, K.; Takahashi, Y.; Sato, S. First Synthesis of Saponarin, 6-*C*- and 7-*O*-Di-β-d-glucosylapigenin. *Chem. Pharm. Bull.* **2013**, *61*, 776–780. [CrossRef] [PubMed]
97. Nakayama, R.; Tanzer, E.-M.; Kusumi, T.; Ohmori, K.; Suzuki, K. Total Synthesis of the Proposed Structure of Ardimerin, and Proposal for its Structural Revision. *Helv. Chim. Acta* **2016**, *99*, 944–960. [CrossRef]
98. Kitamura, K.; Ando, Y.; Matsumoto, T.; Suzuki, K. Synthesis of the Pluramycins 1: Two Designed Anthrones as Enabling Platforms for Flexible Bis-*C*-Glycosylation. *Angew. Chem. Int. Ed.* **2014**, *53*, 1258–1261. [CrossRef]
99. Yao, C.-H.; Tsai, C.-H.; Lee, J.-C. Total Synthesis of the Naturally Occurring Glycosylflavone Aciculatin. *J. Nat. Prod.* **2016**, *79*, 1719–1723. [CrossRef]
100. Kusumi, S.; Tomono, S.; Okuzawa, S.; Kaneko, E.; Ueda, T.; Sasaki, K.; Takahashi, D.; Toshima, K. Total Synthesis of Vineomycin B_2. *J. Am. Chem. Soc.* **2013**, *135*, 15909–15912. [CrossRef]
101. Chen, Q.; Zhong, Y.; O'Doherty, G.A. Convergent de novo synthesis of vineomycinone B2 methyl ester. *Chem. Commun.* **2013**, *49*, 6806–6808. [CrossRef]
102. Chen, C.-L.; Sparks, S.M.; Martin, S.F. *C*-Aryl Glycosides via Tandem Intramolecular Benzyne–Furan Cycloadditions. Total Synthesis of Vineomycinone B2 Methyl Ester. *J. Am. Chem. Soc.* **2006**, *128*, 13696–13697. [CrossRef] [PubMed]
103. Mitra, P.; Behera, B.; Maiti, T.K.; Mal, D. Angucycline C5 Glycosides: Regio- and Stereocontrolled Synthesis and Cytotoxicity. *J. Org. Chem.* **2013**, *78*, 9748–9757. [CrossRef]
104. Ohba, K.; Nakata, M. Total Synthesis of Paecilomycin B. *Org. Lett.* **2015**, *17*, 2890–2893. [CrossRef] [PubMed]
105. Ohba, K.; Nataka, M. Convergent Total Synthesis of Paecilomycin B and 6′-epi-Paecilomycin B by a Barbier-Type Reaction Using 2,4,6-Triisopropylphenyllithium. *J. Org. Chem.* **2018**, *83*, 7019–7032. [CrossRef] [PubMed]
106. Yepremyan, A.; Salehani, B.; Minehan, T.G. Concise Total Syntheses of Aspalathin and Nothofagin. *Org. Lett.* **2010**, *12*, 1580–1583. [CrossRef] [PubMed]
107. Wu, F.; Zhang, J.; Song, F.; Wang, S.; Guo, H.; Wei, Q.; Dai, H.; Chen, X.; Xia, X.; Liu, X.; et al. Chrysomycin A Derivatives for the Treatment of Multi-Drug-Resistant Tuberculosis. *ACS Cent. Sci.* **2020**, *6*, 928–938. [CrossRef]
108. Ho, T.C.; Kamimura, H.; Ohmori, K.; Suzuki, K. Total Synthesis of (+)-Vicenin-2. *Org. Lett.* **2016**, *18*, 4488–4490. [CrossRef]
109. Mavlan, M.; Ng, K.; Panesar, H.; Yepremyan, A.; Minehan, T.G. Synthesis of 3,3′-Di-O-methyl Ardimerin and Exploration of Its DNA Binding Properties. *Org. Lett.* **2014**, *16*, 2212–2215. [CrossRef]
110. Ben, A.; Hsu, D.-S.; Matsumoto, T.; Suzuki, K. Total synthesis and structure revision of deacetylravidomycin M. *Tetrahedron* **2011**, *67*, 6460–6468. [CrossRef]
111. Acharya, P.P.; Khatri, H.R.; Janda, S.; Zhu, J. Synthesis and antitumor activities of aquayamycin and analogues of derhodinosylurdamycin A. *Org. Biomol. Chem.* **2019**, *17*, 2691–2704. [CrossRef]

112. Yuan, Y.; Men, H.; Lee, C. Total Synthesis of Kendomycin: A Macro-C-Glycosidation Approach. *J. Am. Chem. Soc.* **2004**, *126*, 14720–14721. [CrossRef] [PubMed]
113. Anand, N.; Upadhyaya, K.; Ajay, A.; Mahar, R.; Shukla, S.K.; Kumar, B.; Tripathi, R.P. A Strategy for the Synthesis of Anthraquinone-Based Aryl-C-glycosides. *J. Org. Chem.* **2013**, *78*, 4685–4696. [CrossRef]
114. Gong, H.; Gagné, M.R. Diastereoselective Ni-Catalyzed Negishi Cross-Coupling Approach to Saturated, Fully Oxygenated C-Alkyl and C-Aryl Glycosides. *J. Am. Chem. Soc.* **2008**, *130*, 12177–12183. [CrossRef]
115. Nicolas, L.; Angibaud, P.; Stansfield, I.; Bonnet, P.; Meerpoel, L.; Reymond, S.; Cossy, J. Diastereoselective Metal-Catalyzed Synthesis of C-Aryl and C-Vinyl Glycosides. *Angew. Chem. Int. Ed.* **2012**, *51*, 11101–11104. [CrossRef] [PubMed]
116. Zhu, F.; Rourke, M.J.; Yang, T.; Rodriguez, J.; Walczak, M.A. Highly Stereospecific Cross-Coupling Reactions of Anomeric Stannanes for the Synthesis of C-Aryl Glycosides. *J. Am. Chem. Soc.* **2016**, *138*, 12049–12052. [CrossRef] [PubMed]
117. Zhu, F.; Rodriguez, J.; Yang, T.; Kevlishvili, I.; Miller, E.; Yi, D.; O'Neill, S.; Rourke, M.J.; Liu, P.; Walczak, M.A. Glycosyl Cross-Coupling of Anomeric Nucleophiles: Scope, Mechanism, and Applications in the Synthesis of Aryl C-Glycosides. *J. Am. Chem. Soc.* **2017**, *139*, 17908–17922. [CrossRef] [PubMed]
118. Yi, D.; Zhu, F.; Walczak, M.A. Glycosyl Cross-Coupling with Diaryliodonium Salts: Access to Aryl C-Glycosides of Biomedical Relevance. *Org. Lett.* **2018**, *20*, 1936–1940. [CrossRef]
119. Xiong, D.-C.; Zhang, L.-H.; Ye, X.-S. Oxidant-Controlled Heck-Type C-Glycosylation of Glycals with Arylboronic Acids: Stereoselective Synthesis of Aryl 2-Deoxy-C-glycosides. *Org. Lett.* **2009**, *11*, 1709–1712. [CrossRef]
120. Li, H.-H.; Ye, X.-S. Regio- and stereo-selective synthesis of aryl 2-deoxy-C-glycopyranosides by palladium-catalyzed Heck coupling reactions of glycals and aryl iodides. *Org. Biomol. Chem.* **2009**, *7*, 3855–3861. [CrossRef]
121. Yepremyan, A.; Minehan, T.G. Total synthesis of indole-3-acetonitrile-4-methoxy-2-C-β-D-glucopyranoside. Proposal for structural revision of the natural product. *Org. Biomol. Chem.* **2012**, *10*, 5194–5196. [CrossRef]
122. Denmark, S.E.; Regens, C.S.; Kobayashi, T. Total Synthesis of Papulacandin D. *J. Am. Chem. Soc.* **2007**, *129*, 2774–2776. [CrossRef] [PubMed]
123. Wei, Y.; Ben-zvi, B.; Diao, T. Diastereoselective Synthesis of Aryl C-Glycosides from Glycosyl Esters via C-O Bond Homolysis. *Angew. Chem. Int. Ed.* **2021**, *60*, 9433–9438. [CrossRef]
124. Wang, Q.; Sun, Q.; Jiang, Y.; Zhang, H.; Yu, L.; Tian, C.; Chen, G.; Koh, M.J. Iron-catalysed reductive cross-coupling of glycosyl radicals for the stereoselective synthesis of C-glycosides. *Nat. Synth.* **2022**, *1*, 235–244. [CrossRef]
125. Adak, L.; Kawamura, S.; Toma, G.; Takenaka, T.; Isozaki, K.; Takaya, H.; Orita, A.; Li, H.C.; Shing, T.K.M.; Nakamura, M. Synthesis of Aryl C-Glycosides via Iron-Catalyzed Cross Coupling of Halosugars: Stereoselective Anomeric Arylation of Glycosyl Radicals. *J. Am. Chem. Soc.* **2017**, *139*, 10693–10701. [CrossRef] [PubMed]
126. Tang, S.; Zheng, Q.; Xiong, D.-C.; Jiang, S.; Li, Q.; Ye, X.-S. Stereocontrolled Synthesis of 2-Deoxy-C-glycopyranosyl Arenes Using Glycals and Aromatic Amines. *Org. Lett.* **2018**, *20*, 3079–3082. [CrossRef] [PubMed]
127. Singh, A.K.; Kanaujiya, V.K.; Tiwari, V.; Sabiah, S.; Kandasamy, J. Development of Routes for the Stereoselective Preparation of β-Aryl-C-glycosides via C-1 Aryl Enones. *Org. Lett.* **2020**, *22*, 7650–7655. [CrossRef]
128. Liu, J.; Gong, H. Stereoselective Preparation of α-C-Vinyl/Aryl Glycosides via Nickel-Catalyzed Reductive Coupling of Glycosyl Halides with Vinyl and Aryl Halides. *Org. Lett.* **2018**, *20*, 7991–7995. [CrossRef]
129. Mou, Z.-D.; Wang, J.-X.; Zhang, X.; Niu, D. Stereoselective Preparation of C-Aryl Glycosides via Visible-Light-Induced Nickel-Catalyzed Reductive Cross-Coupling of Glycosyl Chlorides and Aryl Bromides. *Adv. Synth. Catal.* **2021**, *363*, 3025–3029. [CrossRef]
130. Yu, C.; Liu, Y.; Xie, X.; Hu, S.; Zhang, S.; Zeng, M.; Zhang, D.; Wang, J.; Liu, H. Ir(I)-Catalyzed C-H Glycosylation for Synthesis of 2-Indolyl-C-Deoxyglycosides. *Adv. Synth. Catal.* **2021**, *363*, 4926–4931. [CrossRef]
131. Ghouilem, J.; Tran, C.; Grimblat, N.; Retailleau, P.; Alami, M.; Gandon, V.; Messaoudi, S. Diastereoselective Pd-Catalyzed Anomeric C(sp^3) −H Activation: Synthesis of α-(Hetero)aryl C-Glycosides. *ACS Catal.* **2021**, *11*, 1818–1826. [CrossRef]
132. Liu, M.; Niu, Y.; Wu, Y.-F.; Ye, X.-S. Ligand-Controlled Monoselective C-Aryl Glycoside Synthesis via Palladium-Catalyzed C–H Functionalization of N-Quinolyl Benzamides with 1-Iodoglycals. *Org. Lett.* **2016**, *18*, 1836–1839. [CrossRef] [PubMed]
133. Ahmed, M.M.; O'Doherty, G.A. De novo synthesis of a *galacto*-papulacandin moiety via an iterative dihydroxylation strategy. *Tetrahedron Lett.* **2005**, *46*, 4151–4155. [CrossRef]
134. Balachari, D.; O'Doherty, G.A. Enantioselective Synthesis of the Papulacandin Ring System: Conversion of the Mannose Diastereoisomer into a Glucose Stereoisomer. *Org. Lett.* **2000**, *2*, 4033–4036. [CrossRef] [PubMed]
135. Balachari, D.; O'Doherty, G.A. Sharpless Asymmetric Dihydroxylation of 5-Aryl-2-vinylfurans: Application to the Synthesis of the Spiroketal Moiety of Papulacandin D. *Org. Lett.* **2000**, *2*, 863–866. [CrossRef]
136. Liu, C.-F.; Xiong, D.-C.; Ye, X.-S. "Ring Opening–Ring Closure" Strategy for the Synthesis of Aryl-C-glycosides. *J. Org. Chem.* **2014**, *79*, 4676–4686. [CrossRef]
137. Mainkar, P.S.; Johny, K.; Rao, T.P.; Chandrasekhar, S. Synthesis of O-Spiro-C-Aryl Glycosides Using Organocatalysis. *J. Org. Chem.* **2012**, *77*, 2519–2525. [CrossRef]
138. Moral, J.A.; Moon, S.-J.; Rodriguez-Torres, S.; Minehan, T.G. A Sequential Indium-Mediated Aldehyde Allylation/Palladium-Catalyzed Cross-Coupling Reaction in the Synthesis of 2-Deoxy-β-C-Aryl Glycosides. *Org. Lett.* **2009**, *11*, 3734–3737. [CrossRef]

139. Hauser, F.M.; Hu, X. A New Route to *C*-Aryl Glycosides. *Org. Lett.* **2002**, *4*, 977–978. [CrossRef]
140. Pichlmair, S.; Marques, M.M.B.; Green, M.P.; Martin, H.J.; Mulzer, J. A Novel Approach toward the Synthesis of Kendomycin: Selective Synthesis of a *C*-Aryl Glycoside as a Single Atropisomer. *Org. Lett.* **2003**, *5*, 4657–4659. [CrossRef]

molecules

MDPI

Communication

Acid Catalyzed Stereocontrolled Ferrier-Type Glycosylation Assisted by Perfluorinated Solvent

Zhiqiang Lu [1,2,*,†], Yanzhi Li [1,2,†], Shaohua Xiang [3,*], Mengke Zuo [1], Yangxing Sun [2], Xingxing Jiang [1,2], Rongkai Jiao [1], Yinghong Wang [1] and Yuqin Fu [1,*]

1 College of Chemistry and Chemical Engineering and Henan Key Laboratory of Function-Oriented Porous Materials, Luoyang Normal University, Luoyang 471934, China
2 Hubei Key Laboratory of Natural Products Research and Development, Key Laboratory of Functional Yeast (China National Light Industry), College of Biological and Pharmaceutical Sciences, China Three Gorges University, Yichang 443002, China
3 Academy for Advanced Interdisciplinary Studies, Southern University of Science and Technology, Shenzhen 518055, China
* Correspondence: zqlu2000@lynu.edu.cn (Z.L.); xiangsh@sustech.edu.cn (S.X.); lyfyq@lynu.edu.cn (Y.F.)
† These authors contributed equally to this work.

Abstract: Described herein is the first application of perfluorinated solvent in the stereoselective formation of *O*-/*S*-glycosidic linkages that occurs via a Ferrier rearrangement of acetylated glycals. In this system, the weak interactions between perfluoro-*n*-hexane and substrates could augment the reactivity and stereocontrol. The initiation of transformation requires only an extremely low loading of resin-H^+ and the mild conditions enable the accommodation of a broad spectrum of glycal donors and acceptors. The 'green' feature of this chemistry is demonstrated by low toxicity and easy recovery of the medium, as well as operational simplicity in product isolation.

Keywords: glycosylation; Ferrier rearrangement; perfluorinated solvent; high stereoselectivity; reusability

Citation: Lu, Z.; Li, Y.; Xiang, S.; Zuo, M.; Sun, Y.; Jiang, X.; Jiao, R.; Wang, Y.; Fu, Y. Acid Catalyzed Stereocontrolled Ferrier-Type Glycosylation Assisted by Perfluorinated Solvent. *Molecules* **2022**, *27*, 7234. https://doi.org/10.3390/molecules27217234

Academic Editors: Jian Yin, Jing Zeng, De-Cai Xiong and Philippe Compain

Received: 24 September 2022
Accepted: 18 October 2022
Published: 25 October 2022

1. Introduction

Facile and stereoselective construction of glycosidic linkages has always been one of the major focal points in the carbohydrate research community. Among these, the 2,3-unsaturated *O*-glycosides have attracted great attention because of their wide occurrence in bioactive molecules (Figure 1a) [1,2] and the potential for rapid functionalization [3,4]. Over the past several decades, various efficient methods have been established for forging such core scaffolds with a Ferrier rearrangement [5,6] that employs readily accessible glycals, and *O*-nucleophiles emerging as the most robust strategy [7–10]. Owing to the mild conditions and short reaction times, Lewis acids are the catalyst class of choice to promote this type of transformation [7–10], while Brønsted acids [7–10] and transition metal catalysts were also found to be effective [7–12]. Alternatively, a Ferrier-type *O*-glycosylation could be mediated by single-electron transfer reagents [13,14] via a radical pathway. These developments notwithstanding, a predominant α-selectivity in the formation of *O*-glycosidic linkages is normally dictated by multiple factors, including the conformation of glycal, anomeric effect, as well as the solvent effect in most cases [10,15,16] (Figure 1b).

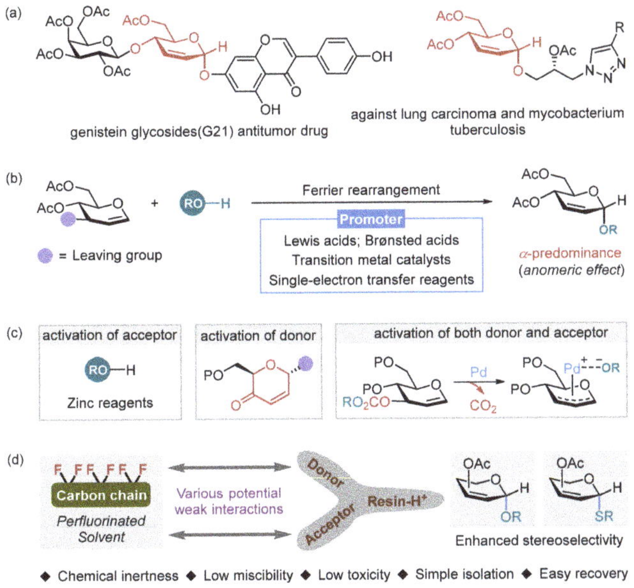

Figure 1. Motivation and reaction design for acid-catalyzed stereocontrolled Ferrier-type glycosylation assisted by perfluorinated solvent. (**a**) Representative bioactive molecules with 2,3-unsaturated *O*-glycoside scaffold; (**b**) Conventional approaches to access 2,3-unsaturated *O*-glycoside scaffold; (**c**) Strategies to activate donor or/and acceptor for Pd-catalyzed *O*-glycosylation; (**d**) This work: acid-catalyzed stereocontrolled *O*-glycosylation assisted by perfluorinated solvent.

In this context, palladium-catalyzed *O*-glycosylation with glycals as donors offers complementary and more programmable access by which excellent stereocontrol could be governed through the rational selection of the leaving group [17,18], ligand [19], or palladium source [20]. In this paradigm, tactics such as the addition of zinc reagent to render a softer acceptor [17,19], modification of glycal to activate the donor [21,22], or application of decarboxylative pathway to formally activate both reactants [23,24] are invoked to improve the performance of these reactions (Figure 1c). A review of these systems suggested that by incorporating a catalyst that could bring the donor and acceptor together through noncovalent interactions, the reaction might be catalytically mediated via a stereoselective manifold. Inspired by the recent advance in stereoselective *O*-glycosylation by means of bifunctional H-bond catalysis with *O*-acceptor [25], we envisioned devising a novel catalytic system to mimic this activation mode with other less-explored weak interactions.

Perfluorinated hydrocarbons displaying low chemical activity, low toxicity and low miscibility with common organic solvents have been recognized as a class of useful reaction mediums in various research fields [26,27], particularly in molecular-oxygen-involved aerobic oxidation reactions [28–34]. Wide application potential is also found in biphase catalysis by virtue of their unique physical properties [35,36]. Moreover, fluorous solvents could engage in diverse weak interactions such as π–π_F, C–F···H hydrogen bond, C–F···C=O, and anion-π_F, which play essential roles in the promotion of chemical transformations by enhancing reactivity and stereoselectivity as well as the design of functional materials [26,27,37–41]. In carbohydrate chemistry, it has been found that introducing a perfluorinated solvent could improve the reaction outcome [42–45]. These findings led us to postulate that the weak interactions stemming from perfluorinated solvent could be leveraged to improve the acid-catalyzed Ferrier-type glycosylation reaction (Figure 1d). On account of the weak acidic condition compared to traditional acid-catalyzed Ferrier rearrangement, the translation of this design into an effective process would further enable stereocontrol and broadens the substrate generality. Notably, the use of perfluorinated solvent could additionally ease the isolation of the glycoside products and the recovery of the reaction medium and assistor.

2. Results and Discussion

2.1. Optimization of Reaction Conditions

Based on these design criteria, the study on this stereocontrolled glycosylation commenced by employing tri-*O*-acetylated glucal **1a** as the donor, while ethanol **2a** serves as both the acceptor and solvent (Table 1). TFE (trifluoroethanol) was first attempted as the additive, which might promote glycosylation through acidic proton or/and other noncovalent weak interactions with **2a** **[46]**. Encouragingly, the *O*-glycosidic product **3a** was provided in 45% yield after 6 h at 100 °C (entry 1). The use of PFD (1H,1H,2H,2H-perfluoro-1-decanol) with a longer perfluorinated alkyl chain improved the yield to 55%, indicating the dominant role of the fluorine effect (entry 2). This speculation was further corroborated by the enhanced chemical yield when PFH (perfluoro-*n*-hexane) without an acidic proton was used as the catalyst (entry 3). Nonetheless, a significant decrease in conversion was observed when PFTEA (perfluoro-triethylamine) [47] was utilized, implying that the basic environment could retard the progress of this transformation (entry 4). It should be noted that high α-selectivity was detected for the generated *O*-glycosidic product for all these reactions (α:β > 20:1). Unsurprisingly, less than 10% yield and poor stereoselectivity (α:β = 5:1) was obtained in the absence of additive (entry 5). These results illustrated the positive effect of weak interactions on both efficiency and stereocontrol. As more complex glycosyl acceptors may not be accessed as easily and well-suited for use in solvent quantities, the reaction using stoichiometric glycosyl acceptors was evaluated in PFH due to the environmental friendliness and recyclability. However, under this set of conditions, only a trace amount of **3a** was detected (entry 6). Exogenous proton was introduced, and notably, 0.6 wt% of H^+ type sulfonic resin (resin-H^+) was sufficient to deliver a quantitative amount of glycosylated α-**3a** (entry 7). Meanwhile, when CH_2Cl_2 was used as the solvent, low yield (16%) and poor stereoselectivity (α:β =1.5:1) were delivered (entry 8). Similarly, the stereoselectivity was decreased (α:β = 7:1) when PFH was substituted by ethanol (entry 9), and no **3a** was obtained with less amount of PFH (10%) and *n*-hexane as a solvent, further affirming our hypothesis (entry 10). Other solvents were also screened, but no satisfactory results could be observed (entries 11–13). Lowering the temperature to 80 °C led to appreciable erosion of chemical yield (entry 14), whereas a prolonged reaction time of 14 h led again to a good yield (entry 15). A trace amount of **3a** was detected when the temperature was further decreased to 60 °C (entry 16). The absolute configuration of **3a** was determined by X-ray crystallographic analysis.

Table 1. Optimization of the reaction conditions.

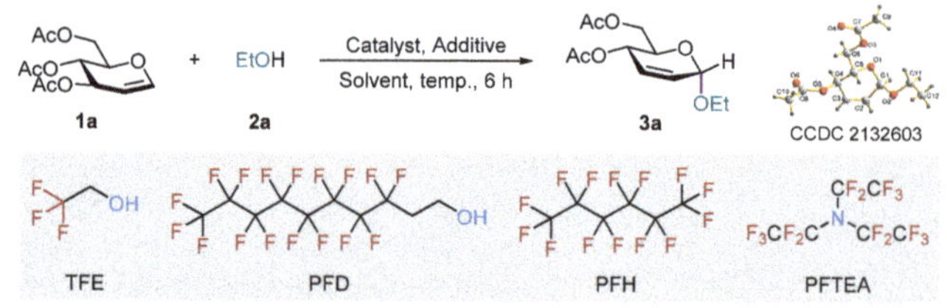

Entry [a]	Additive	Catalyst [b]	Solvent	Temp. (°C)	Yield (%) [c]	Stereoselectivity (α:β)
1	TFE	-	Ethanol	100	45	>20:1
2	PFD	-	Ethanol	100	55	>20:1
3	PFH	-	Ethanol	100	60	>20:1

Table 1. *Cont.*

Entry [a]	Additive	Catalyst [b]	Solvent	Temp. (°C)	Yield (%) [c]	Stereoselectivity (α:β)
4	PFTEA	-	Ethanol	100	15	>20:1
5	-	-	Ethanol	100	<10	5:1
6	-	-	PFH	100	trace	-
7	-	resin-H$^+$	PFH	100	96	>20:1
8	-	resin-H$^+$	CH_2Cl_2	100	16	1.5:1
9	-	resin-H$^+$	Ethanol	100	85	7:1
10	PFH	resin-H$^+$	Hexane	100	-	-
11	PFH	resin-H$^+$	Toluene	100	trace	-
12	PFH	resin-H$^+$	DCE	100	trace	-
13	PFH	resin-H$^+$	DMF	100	-	-
14	-	resin-H$^+$	PFH	80	55	>20:1
15 [d]	-	resin-H$^+$	PFH	80	95	>20:1
16	-	resin-H$^+$	PFH	60	trace	-

[a] Unless otherwise specified, all reactions were performed with **1a** (0.138 mmol, 1 equiv), **2a** (1.2 equiv), additive (10 mol%), catalyst (0.2 mg, 0.6 wt%) for 6 h under N_2 in 0.5 mL solvent. [b] Resin-H$^+$: sulfonic polystyrene type resin. [c] Isolated yields. [d] 14 h. DCE: dichloride ethane, DMF: *N, N*-dimethylformamide.

2.2. Substrate Scope

With the optimized conditions in hand, the substrate generality with respect to glycosyl acceptors was evaluated using glucal **1a** as the standard donor. As depicted in Scheme 1a, various types of glycosyl acceptors, including alkyl, allyl, benzyl, and propargyl alcohols, could give the desired glycosidic products in excellent yield with high stereocontrol at the anomeric center (**3b**-**3o**, α:β > 20:1). It is noteworthy that sterically hindered (**3f** and **3j**) and structurally rigid (**3o**) alcohols that are unreactive reactants for conventional Ferrier rearrangement approaches could convert efficiently to respective *O*-glycosylation products. Subsequently, phenols with different substituents and substitution patterns were examined, and the glycosidic **3p**-**3ac** was synthesized smoothly (Scheme 1b). Compared to aliphatic alcohol acceptors, the yields and stereoselectivities deteriorated in most cases, probably due to the strong background reaction catalyzed by an acidic hydroxyl group of phenols. Apart from *O*-nucleophiles, *S*-nucleophiles were also applicable for this reaction (Scheme 1c). Although all the tested substrates reacted well with **1a** to give compounds **4a**-**4e** in good yields, the stereochemical outcome varied greatly. For instance, a 1:1 α:β mixture was detected for **4a** (from *n*-butylthiol) while **4b** (from *t*-butylthiol) was generated with an α:β ratio > 20:1. Likewise, thiophenol with electron-withdrawing group delivered *S*-glycosidic **4c** in poor stereocontrol while **4d** with an electron-donating group on thiophenol was obtained with α:β ratio of 10:1. When 2-methylbenzenethiol was utilized, the desired glycosylation product **4e** was formed in 75% yield with 6:1 α:β selectivity. Additionally, C-3 substitution products **4c′** and **4e′** were isolated alongside 6% and 8% yields, respectively. The absolute configurations of **3aa**, **3ab**, **4e**, and **4e′** were determined by X-ray crystallographic analysis, and those of other products in this scheme were assigned by analogy. Water also functioned well as an acceptor in the developed reaction, giving α-**5** an 87% yield.

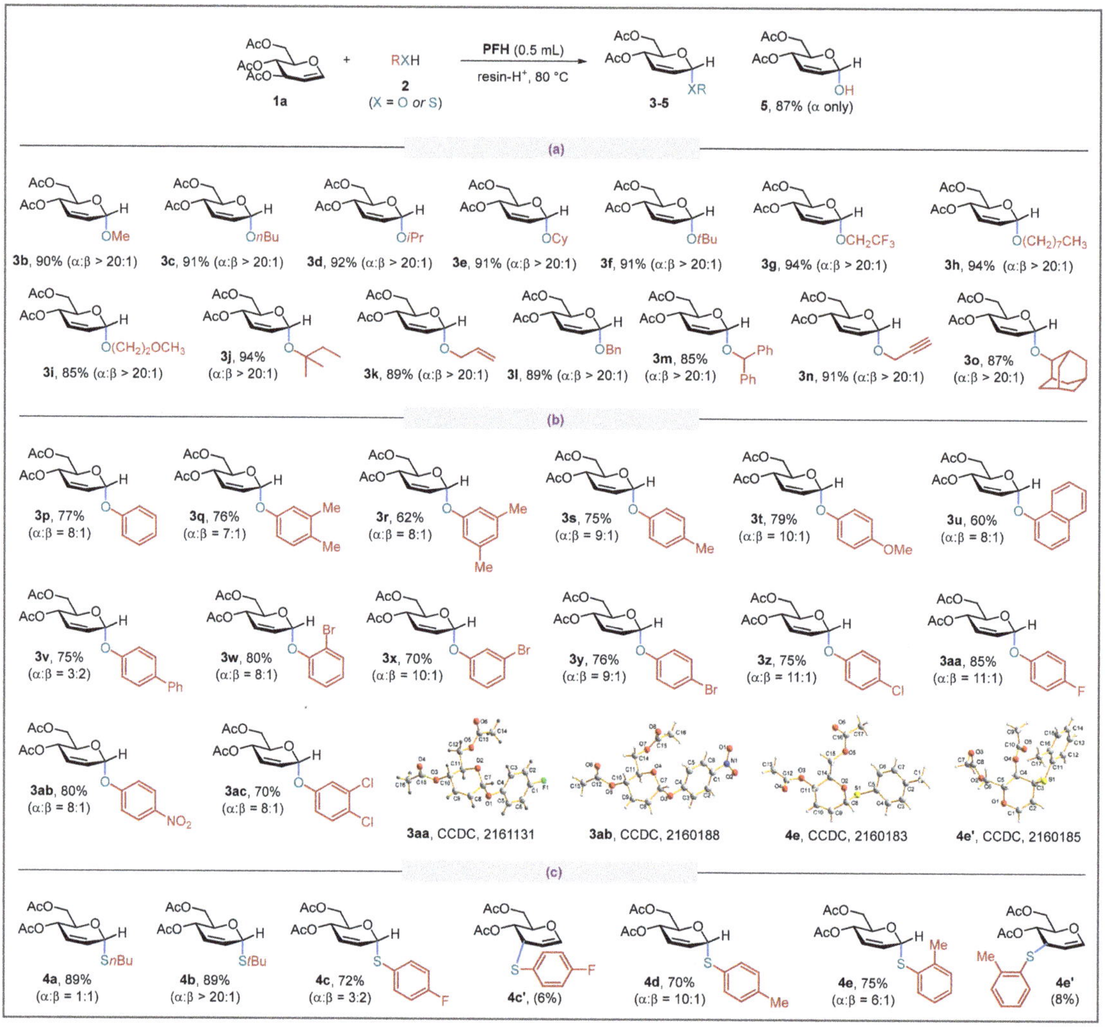

Scheme 1. Substrate generality with respect to glycosyl acceptors. (**a**) Substrate scope with respect to alcohols; (**b**) Substrate scope with respect to phenols; (**c**) Substrate scope of *S*-glycosyl acceptors.

Subsequently, the generality of this glycosylation method was studied with other types of glycal donors (Scheme 2). Firstly, D-galactal **1b**, C-4 epimer of **1a** was employed, and the results were summarized in Scheme 2a. A series of alcohols were examined, and these reactions invariably gave only **6a-6e** in excellent yields and α:β > 20:1. Phenols, thiols, and thiophenols were also applicable to afford **6f-6i** in good yields and stereoselectivities. As a C-3 epimer of **1a**, the combination of D-allal **1c** with selected glycosyl acceptors forged the corresponding products in more than 80% yield (**3a**, **3aa**, **4b**, and **4c**). Interestingly, remarkable α-selectivities were detected for all of these reactions, same with the case for glucal **1a** (Scheme 2b). L-Rhamnal **1d** was also verified to be a competent donor for this transformation, and **7a-7d** was established with excellent outcomes (Scheme 2c). However, when the pentose substrates were employed in this procedure, such as *D*-xylal **1e** or D-arabinal **1f** (a pair of C-3 epimers) as glycosyl donors, poor α:β ratios were observed for these reactions (Scheme S1, **8a-8d**), indicating the direct significance of C-5 substitution in stereoinduction.

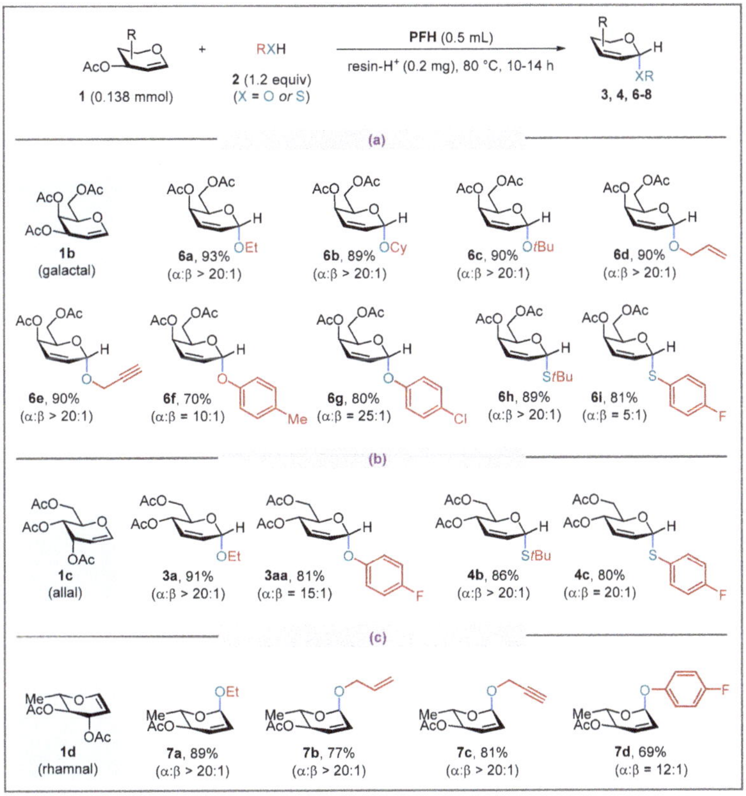

Scheme 2. Substrate generality with respect to glycosyl donors. (**a**) Substrate scope with respect ot galactal; (**b**) Substrate scope with respect to allal; (**c**) Substrate scope with respect to rhamnal.

To demonstrate the practicality of the developed glycosylation strategy, the reactions of **1a** with an array of functional molecules as acceptors were investigated (Scheme 3a). First, glycosylated product **9a** with a long alkyl chain was prepared in 90% chemical yield with α:β > 20:1, indicating the potential utility in lipidosome assembly. A fluorous tag containing long-chain linear perfluorocarbon was well tolerated to afford **9b** with the same level of outcome. Glycosylation with sugar alcohol delivered disaccharide **9c** in 80% yield with α:β selectivity of 12:1. When phenol derived from tetraphenylethylene with aggregation-induced emission attribute was reacted, **9d** could be generated in moderate yield with α:β = 9:1. Furthermore, the reaction operated smoothly on bioactive diosgenin to generate the C-O bond formation product **9e** with perfect stereochemical control.

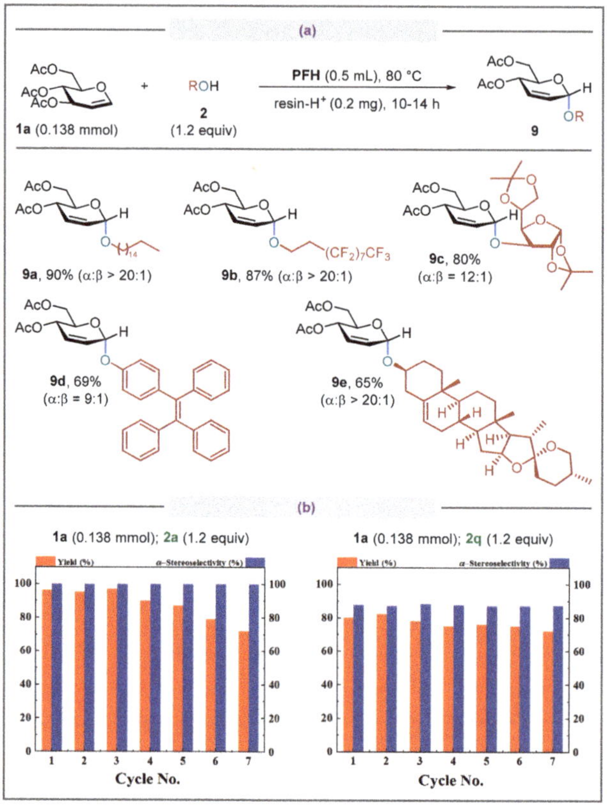

Scheme 3. Reactions with representative functional molecules as glycosyl acceptors and recycling experiments. (**a**) Functional molecules as acceptors; (**b**) Recycling experiment investigation.

A gram-scale reaction between **1a** and **2a** was also implemented under the standard conditions, in which the synthetic efficiency and stereocontrol observed for the small-scale reaction were perfectly preserved (Scheme S2). Additionally, given the ease of isolation and good recyclability of organofluorine solvent, the recycling experiments were conducted to reinforce the utility of this strategy. After the completion of each reaction, the target product was easily isolated by phase separation, and the recovered reaction system (bottom phase) was reused successively. As summarized in Scheme 3b, when ethanol **2a** was used to react with donor **1a**, the stereoselectivity (α:β > 20:1) was perfectly preserved, and the chemical yield was maintained at a good level (>70%) even after a repetition of this procedure for seven times. Similar results were obtained by using 3,4-dimethylphenol **2q** as a glycosyl acceptor for the recycling experiment.

3. Materials and Methods

The detailed procedure of the synthesis and characterization of the products are given in Supplementary Materials.

4. Conclusions

In conclusion, an acid-catalyzed stereoselective Ferrier-type glycosylation assisted by perfluorinated solvent has been established. A wide range of glycal donors and glycosyl acceptors are well accommodated to provide structurally diverse *O*- and *S*-glycosylated linkages products in good efficiency for most cases. The utilization of perfluoro-*n*-hexane as the solvent improves the reaction conditions, increases the yield, and enhances the stereocontrol at the anomeric center. Notably, the turnover of this procedure is achieved with a minimal amount of resin-H^+. Aside from experimental ease in isolating products, the use of low toxic and recyclable perfluorinated solvent highlights the environmental friendliness of the developed method.

Supplementary Materials: The following are available online at https://www.mdpi.com/article/10.3390/molecules27217234/s1, CCDC 2132603, 2160183, 2160185, 2160188, and 2161131 contain the supplementary crystallographic data for this paper. These data can be obtained free of charge via www.ccdc.cam.ac.uk/data_request/cif (accessed on 22 March 2022), by emailing da-ta_request@ccdc.cam.ac.uk, or by contacting The Cambridge Crystallographic Data Centre, 12 Union Road, Cambridge CB2 1EZ, UK; fax: +44 1223 336033, synthesis and characterization of all compounds described in this paper. References [48–68] are cited in the supplementary materials.

Author Contributions: Conceptualization, Z.L. and S.X.; methodology, Z.L.; software, Y.L.; validation, Y.L. and M.Z.; formal analysis, Y.L., M.Z., Y.S., X.J., R.J. and Y.W.; investigation, Y.L., M.Z., Y.S., X.J., R.J. and Y.W.; resources, Y.L. and Y.S.; data curation, Y.L.; writing—original draft preparation, Z.L. and S.X.; writing—review and editing, Z.L., S.X. and Y.F.; visualization, Y.L.; supervision, Y.F.; project administration, Z.L.; funding acquisition, Z.L. All authors have read and agreed to the published version of the manuscript.

Funding: We thank the financial support from Shenzhen Special Funds (JCYJ20180305123508258), the National Natural Science Foundation of China (U1304206, 21801112), the Natural Science Foundation of Henan Province (212300410374), the Science and Technology Project of Henan Province (212102210549), and the Key Scientific Research Project of Higher Education of Henan Province (18A150012, 19A150003, 19A150004, 13A150799).

Institutional Review Board Statement: Not applicable.

Informed Consent Statement: Not applicable.

Data Availability Statement: The data presented in this study are available in the Supplementary Materials.

Conflicts of Interest: The authors declare no conflict of interest.

Sample Availability: Samples of the compounds are not available from the authors.

References

1. Polkowski, K.; Popiołkiewicz, J.; Krzeczyński, P.; Ramza, J.; Pucko, W.; Zegrocka-Stendel, O.; Boryski, J.; Skierski, J.S.; Ma-zurek, A.P.; Grynkiewicz, G. Cytostatic and Cytotoxic Activity of Synthetic Genistein Glycosides Against Human Cancer Cell Lines. *Cancer Lett.* **2004**, *203*, 59–69. [CrossRef] [PubMed]
2. Popiołkiewicz, J.; Polkowski, K.; Skierski, J.S.; Mazurek, A.P. In Vitro Toxicity Evaluation in the Development of New Anticancer Drugs–Genistein Glycosides. *Cancer Lett.* **2005**, *229*, 67–75. [CrossRef] [PubMed]
3. Babu, R.S.; Zhou, M.; O'Doherty, G.A. De Novo Synthesis of Oligosaccharides Using a Palladium-Catalyzed Glycosylation Reaction. *J. Am. Chem. Soc.* **2004**, *126*, 3428–3429. [CrossRef] [PubMed]
4. Ghosh, A.K.; Veitschegger, A.M.; Nie, S.; Relitti, N.; MacRae, A.J.; Jurica, M.S. Enantioselective Synthesis of Thailanstatin A Methyl Ester and Evaluation of *in Vitro* Splicing Inhibition. *J. Org. Chem.* **2018**, *83*, 5187–5198. [CrossRef] [PubMed]
5. Ferrier, R.J. Unsaturated Carbohydrates. Part II. Three Reactions Leading to Unsaturated Glycopyranosides. *J. Chem. Soc.* **1964**, 5443–5449. [CrossRef]
6. Ferrier, R.J.; Prasad, N. Unsaturated Carbohydrates. Part XI. Isomerisation and Dimerisation of Tri-*O*-acetyl-D-glucal. *J. Chem. Soc. C* **1969**, 581–586. [CrossRef]
7. Gómez, A.M.; Lobo, F.; Uriel, C.; López, J.C. Recent Developments in the Ferrier Rearrangement. *Eur. J. Org. Chem.* **2013**, *32*, 7221–7262. [CrossRef]
8. Gómez, A.M.; Lobo, F.; Miranda, S.; López, J.C. A Survey of Recent Synthetic Applications of 2,3-Dideoxy-Hex-2-enopyranosides. *Molecules* **2015**, *20*, 8357–8394. [CrossRef]
9. Jiang, N.; Wu, Z.; Dong, Y.; Xu, X.; Liu, X.; Zhang, J. Progress in the Synthesis of 2,3-Unsaturated Glycosides. *Curr. Org. Chem.* **2020**, *24*, 184–199. [CrossRef]
10. Bennett, C.S.; Galan, M.C. Methods for 2-Deoxyglycoside Synthesis. *Chem. Rev.* **2018**, *118*, 7931–7985. [CrossRef]
11. McKay, M.J.; Nguyen, H.M. Recent Advances in Transition Metal-Catalyzed Glycosylation. *ACS Catal.* **2012**, *2*, 1563–1595. [CrossRef] [PubMed]

12. Bauer, E.B. Transition Metal Catalyzed Glycosylation Reactions—An Overview. *Org. Biomol. Chem.* **2020**, *18*, 9160–9180. [CrossRef] [PubMed]
13. Gómez, A.M.; Valverde, S.; Fraser-Reid, B. A Route to Unsaturated Spiroketals from Phenylthio Hex-2-enopyranosides via Sequential alkylation, Allylic Rearrangement and Intramolecular Glycosidation. *J. Chem. Soc. Chem. Commun.* **1991**, 1207–1208. [CrossRef]
14. Rafiee, E.; Tangestaninejad, S.; Habibi, M.H.; Mirkhani, V. A Mild, Efficient and α-Selective Glycosidation by Using Potassium Dodecatungstocobaltate Trihydrate as Catalyst. *Bioorg. Med. Chem. Lett.* **2004**, *14*, 3611–3614. [CrossRef]
15. Leng, W.-L.; Yao, H.; He, J.-X.; Liu, X.-W. Venturing beyond Donor-Controlled Glycosylation: New Perspectives toward Anomeric Selectivity. *Acc. Chem. Res.* **2018**, *51*, 628–639. [CrossRef]
16. Alabugin, I.V.; Kuhn, L.; Medvedev, M.G.; Krivoshchapov, N.V.; Vil', V.A.; Yaremenko, I.A.; Mehaffy, P.; Yarie, M.; Terent'ev, A.O.; Zolfigol, M.A. Stereoelectronic Power of Oxygen in Control of Chemical Reactivity: The Anomeric Effect is Not Alone. *Chem. Soc. Rev.* **2021**, *50*, 10253–10345. [CrossRef]
17. Schuff, B.P.; Mercer, G.J.; Nguyen, H.M. Palladium-Catalyzed Stereoselective Formation of α-*O*-Glycosides. *Org. Lett.* **2007**, *9*, 3173–3176. [CrossRef] [PubMed]
18. Xiang, S.-H.; Hoang, K.L.; He, M.J.; Tan, Y.-J.; Liu, X.-W. Reversing the Stereoselectivity of a Palladium-Catalyzed *O*-Glycosylation through an Inner-sphere or Outer-sphere Pathway. *Angew. Chem. Int. Ed.* **2015**, *54*, 604–607.
19. Kim, H.; Men, H.; Lee, C. Stereoselective Palladium-Catalyzed *O*-Glycosylation Using Glycals. *J. Am. Chem. Soc.* **2004**, *126*, 1336–1337. [CrossRef] [PubMed]
20. Yao, H.; Zhang, S.; Leng, W.-L.; Leow, M.-L.; Xiang, S.; He, J.; Liao, H.; Hoang, K.L.M.; Liu, X.-W. Catalyst-Controlled Stereoselective *O*-Glycosylation: Pd(0) vs Pd(II). *ACS Catal.* **2017**, *7*, 5456–5460. [CrossRef]
21. Comely, A.C.; Eelkema, R.; Minnaard, A.J.; Feringa, B.L. De Novo Asymmetric Bio- and Chemocatalytic Synthesis of Saccharides–Stereoselective Formal *O*-Glycoside Bond Formation Using Palladium Catalysis. *J. Am. Chem. Soc.* **2003**, *125*, 8714–8715. [CrossRef]
22. Babu, R.S.; O'Doherty, G.A. A Palladium-Catalyzed Glycosylation Reaction: The de Novo Synthesis of Natural and Unnatural Glycosides. *J. Am. Chem. Soc.* **2003**, *125*, 12406–12407. [CrossRef] [PubMed]
23. Xiang, S.; Lu, Z.; He, J.; Hoang, K.L.M.; Zeng, J.; Liu, X.-W. β-Type Glycosidic Bond Formation by Palladium-Catalyzed Decarboxylative Allylation. *Chem. Eur. J.* **2013**, *19*, 14047–14051. [CrossRef]
24. Wang, Q.; Lai, M.; Luo, H.; Ren, K.; Wang, J.; Huang, N.; Deng, Z.; Zou, K.; Yao, H. Stereoselective *O*-Glycosylation of Glycals with Arylboronic Acids Using Air as the Oxygen Source. *Org. Lett.* **2022**, *24*, 1587–1592. [CrossRef]
25. Loh, C.C.J. Exploiting Non-Covalent Interactions in Selective Carbohydrate Synthesis. *Nat. Rev. Chem.* **2021**, *5*, 792–815. [CrossRef]
26. Betzemeier, B.; Knochel, P. Modern Solvents in Organic Synthesis: Perfluorinated Solvents–A Novel Reaction Medium in Organic Chemistry. *Top. Curr. Chem.* **1999**, *206*, 60–78.
27. Berger, R.; Resnati, G.; Metrangolo, P.; Weberd, E.; Hulliger, J. Organic Fluorine Compounds: A Great Opportunity for Enhanced Materials Properties. *Chem. Soc. Rev.* **2011**, *40*, 3496–3508. [CrossRef] [PubMed]
28. Klement, I.; Knochel, P. Selective Oxidation of Zinc Organometallics to Hydroperoxides Using Oxygen in Perfluorohexanes. *Synlett* **1995**, *27*, 1113–1114. [CrossRef]
29. Brown, H.C.; Negishi, E. Bisborolane. Highly Elusive Bisboracyclane. *J. Am. Chem. Soc.* **1971**, *93*, 6682–6683. [CrossRef]
30. Brown, H.C.; Midland, M.M.; Kabalka, G.W. Stoichiometrically Controlled Reaction of Organoboranes with Oxygen Under very Mild Conditions to Achieve Essentially Quantitative Conversion into Alcohols. *J. Am. Chem. Soc.* **1971**, *93*, 1024–1025. [CrossRef]
31. Barton, D.H.R.; Jang, D.O.; Jaszberenyi, J.C. An Improved Radical Chain Procedure for the Deoxygenation of Secondary and Primary Alcohols using Diphenylsilane as Hydrogen Atom Donor and Triethylborane-Air as Initiator. *Tetrahedron Lett.* **1990**, *31*, 4681–4684. [CrossRef]
32. Klement, I.; Lütjens, H.; Knochel, P. Transition Metal Catalyzed Oxidations in Perfluorinated Solvents. *Angew. Chem. Int. Ed.* **1997**, *36*, 1454–1456. [CrossRef]
33. Tada, N.; Cui, L.; Ishigami, T.; Ban, K.; Miura, T.; Uno, B.; Itoh, A. Facile Aerobic Photooxidative Oxylactonization of Oxocarboxylic Acids in Fluorous Solvents. *Green Chem.* **2012**, *14*, 3007–3009. [CrossRef]
34. Karimi, M.; Sadeghi, S.; Mohebali, H.; Azarkhosh, Z.; Safarifard, V.; Mahjoub, A.; Heydari, A. Fluorinated Solvent-Assisted Photocatalytic Aerobic Oxidative Amidation of Alcohols *via* Visible-Light-Mediated HKUST-1/Cs-POMoW Catalysis. *New J. Chem.* **2021**, *45*, 14024–14035. [CrossRef]
35. Horváth, I.T.; Rábi, J. Facile Catalyst Separation Without Water: Fluorous Biphases Hydroformylation of Olefins. *Science* **1994**, *266*, 72–75. [CrossRef]
36. Maayan, G.; Fish, R.H.; Neumann, R. Polyfluorinated Quaternary Ammonium Salts of Polyoxometalate Anions: Fluorous Biphasic Oxidation Catalysis with and without Fluorous Solvents. *Org. Lett.* **2003**, *5*, 3547–3550. [CrossRef]
37. O'Hagan, D. Understanding Organofluorine Chemistry. An Introduction to the C–F Bond. *Chem. Soc. Rev.* **2008**, *37*, 308–319. [CrossRef]
38. Champagne, P.A.; Desroches, J.; Paquin, J.-F. Organic Fluorine as a Hydrogen-Bond Acceptor: Recent Examples and Applications. *Synthesis* **2015**, *47*, 306–322.
39. Yu, J.-S.; Liu, Y.-L.; Tang, J.; Wang, X.; Zhou, J. Highly Efficient "On Water" Catalyst-Free Nucleophilic Addition Reactions Using Difluoroenoxysilanes: Dramatic Fluorine Effects. *Angew. Chem. Int. Ed.* **2014**, *53*, 9512–9516. [CrossRef]

40. Cao, Z.; Wang, W.; Liao, K.; Wang, X.; Zhou, J.; Ma, J. Catalytic Enantioselective Synthesis of Cyclopropanes Featuring Vicinal All-Carbon Quaternary Stereocenters with a CH_2F Group; Study of Influence of C-F···H-N Interactions on Reactivity. *Org. Chem. Front.* **2018**, *5*, 2960–2968. [CrossRef]
41. Myers, K.E.; Kumar, K. Fluorophobic Acceleration of Diels-Alder Reactions. *J. Am. Chem. Soc.* **2000**, *122*, 12025–12026. [CrossRef]
42. Piscelli, B.A.; Sanders, W.; Yu, C.; Maharik, N.A.; Lebl, T.; Cormanich, R.A.; O'Hagan, D. Fluorine-Induced Pseudo-Anomeric Effects in Methoxycyclohexanes through Electrostatic 1,3-Diaxial Interactions. *Chem. Eur. J.* **2020**, *26*, 11989–11994. [CrossRef] [PubMed]
43. Misbahi, K.; Lardic, M.; Ferrières, V.; Noiret, N.; Kerbal, A.; Plusquellec, D. Unexpected fluorous solvent effect on oxidation of 1-thioglycosides. *Tetrahedron Asymmetry* **2001**, *12*, 2389–2393. [CrossRef]
44. Oikawa, M.; Tanak, T.; Fukud, N.; Kusumoto, S. One-Pot Preparation and Activation of Glycosyl Trichloroacetimidates: Operationally Simple Glycosylation Induced by Combined Use of Solid-Supported, Reactivity-Opposing Reagents. *Tetrahedron Lett.* **2004**, *45*, 4039–4042. [CrossRef]
45. Farrán, A.; Cai, C.; Sandoval, M.; Xu, Y.; Liu, J.; Hernáiz, M.J.; Linhardt, R.J. Green Solvents in Carbohydrate Chemistry: From Raw Materials to Fine Chemicals. *Chem. Rev.* **2015**, *115*, 6811–6853. [CrossRef]
46. Di Salvo, A.; David, M.; Crousse, B.; Bonnet-Delpon, D. Self-Promoted Nucleophilic Addition of Hexafluoro-2-propanol to Vinyl Ethers. *Adv. Synth. Catal.* **2006**, *348*, 118–124. [CrossRef]
47. Nakano, H.; Kitazume, T. Organic Reactions without an Organic Medium–Utilization of Perfluorotriethylamine as a Reaction Medium. *Green Chem.* **1999**, *1*, 21–22. [CrossRef]
48. Gorityala, B.K.; Lorpitthaya, R.; Bai, Y.; Liu, X.-W. ZnCl2/Alumina Impregnation Catalyzed Ferrier Rearrangement: An Expedient Synthesis of Pseudoglycosides. *Tetrahedron* **2009**, *65*, 29–30. [CrossRef]
49. Zhou, J.; Chen, H.; Shan, J.; Li, J.; Yang, G.; Chen, X.; Xin, K.; Zhang, J.; Tang, J. $FeCl_3{\cdot}6H_2O$/C: An Efficient and Recyclable Catalyst for the Synthesis of 2,3-Unsaturated *O*- and S-Glycosides. *J. Carbohydr. Chem.* **2014**, *33*, 313–325. [CrossRef]
50. Santra, A.; Guchhait, G.; Misra, A. Nitrosyl Tetrafluoroborate Catalyzed Preparation of 2,3-Unsaturated Glycosides and 2-Deoxyglycosides of Hindered Alcohols, Thiols, and Sulfonamides. *Synlett* **2013**, *24*, 581–586. [CrossRef]
51. Ruan, Z.; Dabideen, D.; Blumenstein, M.; Mootoo, D.R.A. Modular Synthesis of the Bis-Tetrahydrofuran Core of Rolliniastatin from Pyranoside Precursors. *Tetrahedron* **2000**, *56*, 9203–9211. [CrossRef]
52. Yadav, J.S.; Reddy, B.V.S.; Pandey, S.K. Ceric (IV) Ammonium Nitrate-Catalyzed Glycosidation of Glycals: A Facile Synthesis of 2,3-Unsaturated Glycosides. *New J. Chem.* **2001**, *25*, 538–540. [CrossRef]
53. Srinivas, B.; Reddy, T.R.; Radha Krishna, P.; Kashyap, S. Copper (II) Triflate as a Mild and Efficient Catalyst for Ferrier Glycosylation: Synthesis of 2,3-Unsaturated O-Glycosides. *Synlett* **2014**, *25*, 1325–1330. [CrossRef]
54. Saeeng, R.; Siripru, O.; Sirion, U. IBr-Catalyzed O-Glycosylation of D-Glucals: Facile Synthesis of 2,3-Unsaturated-O-Glycosides. *Heterocycles* **2015**, *91*, 849–861.
55. Bound, D.J.; Bettadaiah, B.K.; Srinivas, P. ZnBr2-Catalyzed and Microwave-Assisted Synthesis of 2,3-Unsaturated Glucosides of Hindered Phenols and Alcohols. *Synth. Commun.* **2014**, *44*, 2565–2576. [CrossRef]
56. Frappa, I.; Sinou, D. An Easy and Efficient Preparation of Aryl α-O-Δ2−Glycosides. *Synth. Commun.* **1995**, *25*, 2941–2951. [CrossRef]
57. Sun, G.; Qiu, S.; Ding, Z.; Chen, H.; Zhou, J.; Wang, Z.; Zhang, J. Magnetic Core-Shell Fe3O4@C-SO3H as an Efficient and Renewable 'Green Catalyst' for the Synthesis of O-2,3-Unsaturated Glycopyranosides. *Synlett* **2017**, *28*, 347–352.
58. Babu, B.S.; Balasubramanian, K.K. Indium Trichloride Catalyzed Glycosidation. An Expeditious Synthesis of 2,3-Unsaturated Glycopyranosides. *Tetrahedron Lett.* **2000**, *41*, 1271–1274. [CrossRef]
59. Yadav, J.S.; Reddy, B.V.S.; Murthy, C.V.S.R.; Kumar, G.M. Scandium Triflate Catalyzed Ferrier Rearrangement: An Efficient Synthesis of 2,3-Unsaturated Glycopyranosides. *Synlett* **2000**, *S11210*, 1450–1451. [CrossRef]
60. Stevanović, D.; Pejović, A.; Damljanović, I.; Vukićević, R.D.; Vukić ević, M.; Bogdanovic, G.A. Anodic Generation of a Zirconium Catalyst for Ferrier Rearrangement and Hetero Michael Addition. *Tetrahedron Lett.* **2012**, *53*, 6257–6260. [CrossRef]
61. Stevanović, D.; Pejović, A.; Damljanović, I.; Minić, A.; Bogdanović, G.A.; Vukićević, M.; Radulović, N.S.; Vukićević, R.D. Ferrier Rearrangement Promoted by an Electrochemically Generated Zirconium Catalyst. *Carbohydr. Res.* **2015**, *407*, 111–121. [CrossRef]
62. Bhagavathy, S.; Ajay, K.B.; Kalpattu, K.B. Microwave-induced, Montmorillonite K10-Catalyzed Ferrier Rearrangement of Tri-O-Acetyl-D-Galactal: Mild, Eco-Friendly, Rapid Glycosidation with Allylic Rearrangement. *Tetrahedron Lett.* **2002**, *43*, 6795–6798.
63. Grynkiewicz, G.; Priebe, W.; Zamojski, A. Synthesis of Alkyl 4, 6-di-O-Acetryl-2,3-Dideoxy-a-D-Three-hex-2-Enopyranosides from 3,4,6-tri-O-Acetyl-1,5-Anhydro-2-Droxy D-lyxo-hex-1-enitol(3,4,6-tri-O-acetyl-e-galactal). *Carbohydr. Res.* **1979**, *68*, 33–41. [CrossRef]
64. Chen, P.; Zhang, D.D. Sm(OTf)3 as a Highly Effificient Catalyst for the Synthesis of 2,3-Unsaturated O- and S-Pyranosides from Glycals and the Temperature-Dependent Formation of 4-O-Acetyl-6-Deoxy-2,3-Unsaturated S-Pyranosides and 4-O-Acetyl-6-Deoxy-3-Alkylthio Glycals. *Tetrahedron* **2014**, *70*, 8505–8510. [CrossRef]
65. Tatina, M.B.; Mengxin, X.; Peilin, R.; Judeh, Z.M.A. Robust Perfluorophenylboronic Acid-Catalyzed Stereoselective Synthesis of 2,3-Unsaturated O-, C-, N- and S-linked Glycosides. *Beilstein J. Org. Chem.* **2019**, *15*, 1275–1280. [CrossRef] [PubMed]
66. Kim, B.H.; Jacobs, P.B.; Elliott, R.L.; Curran, D.P. A Gentral- Synthetic Approach to S113 Optically Active Iridoid Aglpoaes. The Total Synthesis of Beta-Ethyl Descarbomethoxyverbenalol, Ethyl Catalpot, and (-)-Specionin. *Tetrahedron* **2014**, *44*, 3079–3092.

67. Bartlett, M.J.; Peter, T.; Northcote, P.T.; Lein, M.; Harvey, J.E. ^{13}C NMR Analysis of 3,6-Dihydro-2*H*-pyrans: Assignment of Remote Stereochemistry Using Axial Shielding Effects. *J. Org. Chem.* **2014**, *79*, 5521–5532. [CrossRef]
68. Khan, A.T.; Sidick Basha, R.S.; Lal, M. Bromodimethylsulfonium Bromide (BDMS) Catalyzed Synthesis of 2,3-Unsaturated-*O*-Glycosides via Ferrier Rearrangement. *Arkivoc* **2012**, *2013*, 201–212. [CrossRef]

molecules

MDPI

Review

Not Just Anticoagulation—New and Old Applications of Heparin

Lixuan Zang [1,2], Haomiao Zhu [1,3], Kun Wang [1], Yonghui Liu [4,*], Fan Yu [1,*] and Wei Zhao [1]

1 State Key Laboratory of Medicinal Chemical Biology, College of Life Sciences, College of Pharmacy, Nankai University, 38 Tongyan Road, Jinnan District, Tianjin 300350, China
2 National Glycoengineering Research Center and Shandong Key Laboratory of Carbohydrate Chemistry and Glycobiology, Shandong University, Qingdao 266237, China
3 Department of Pharmacy, Qilu Hospital, Shandong University, 107 Cultural West Road, Jinan 250012, China
4 School of Chemistry, Tiangong University, Tianjin 300387, China
* Correspondence: yhliu@mail.nankai.edu.cn (Y.L.); fanyu@nankai.edu.cn (F.Y.)

Abstract: In recent decades, heparin, as the most important anticoagulant drug, has been widely used in clinical settings to prevent and treat thrombosis in a variety of diseases. However, with in-depth research, the therapeutic potential of heparin is being explored beyond anticoagulation. To date, heparin and its derivatives have been tested in the protection against and repair of inflammatory, antitumor, and cardiovascular diseases. It has also been explored as an antiangiogenic, preventive, and antiviral agent for atherosclerosis. This review focused on the new and old applications of heparin and discussed the potential mechanisms explaining the biological diversity of heparin.

Keywords: heparin; applications of heparin; challenges of heparin therapy

Citation: Zang, L.; Zhu, H.; Wang, K.; Liu, Y.; Yu, F.; Zhao, W. Not Just Anticoagulation—New and Old Applications of Heparin. *Molecules* **2022**, *27*, 6968. https://doi.org/10.3390/molecules27206968

Academic Editors: Jian Yin, Jing Zeng and De-Cai Xiong

Received: 25 August 2022
Accepted: 12 October 2022
Published: 17 October 2022

1. Introduction

A sulfated glycosaminoglycan, heparin was named after its initial isolation from liver tissue a century ago [1,2]. It exists in the lung, vascular wall, intestinal mucosa, and so on. Due to its unique pentasaccharide sequence, heparin exerts anticoagulant activity after binding with antithrombin, inhibits the activation of factors Xa and IIa in the coagulation cascade, and finally exerts anticoagulant activity. Heparin is a natural anticoagulant in animals and is the most widely used anticoagulant treatment [3,4]. Heparin is usually used to prevent or treat thrombosis-related diseases, such as embolic diseases and myocardial infarction, or for blood anticoagulation during surgeries such as cardiovascular surgery, cardiac catheterization, cardiopulmonary bypass, and hemodialysis. In recent years, with the development of synthetic low-molecular-weight heparin (LMWH) and heparin derivatives, these compounds have been proven through various activity studies to have many pharmacological effects, such as anti-inflammatory, antiangiogenic, antitumor, and antimetastatic effects, in addition to anticoagulant effects [5–7]. At the same time, as a natural water-soluble polysaccharide, heparin has good biocompatibility, so there have also been studies on combining heparin with nanomaterials to give it wider medical application value.

In particular, the infection caused by severe acute respiratory syndrome coronavirus 2 (SARS-CoV-2, COVID-19) spread rapidly all over the world in 2019. Patients with severe COVID-19 may eventually develop systemic thrombovasculitis, leading to severe organ dysfunction [8]. Heparin can reduce the systemic symptoms of patients, which indicates that heparin may have an undeveloped effect in the treatment of COVID-19 [9,10]. Therefore, in this review, we briefly summarized the new and old applications and related mechanisms of heparin and its derivatives (Figure 1).

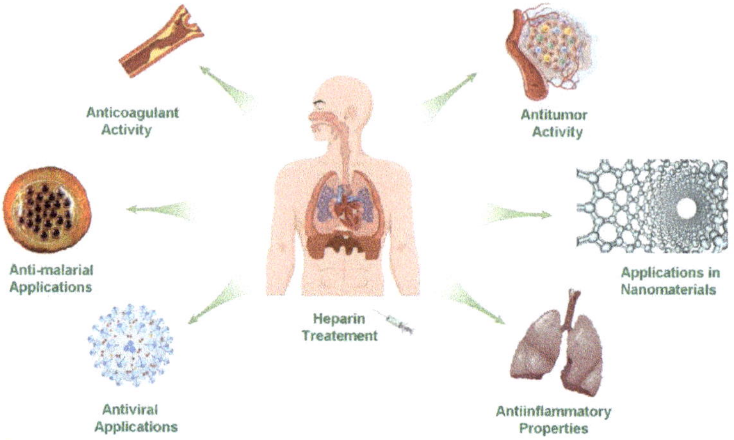

Figure 1. The applications of heparin.

2. The Structure of Heparin

Heparin belongs to the glycosaminoglycans (GAGs), which are composed of a linear set of negatively charged polysaccharides. According to differences in the polysaccharide unit, GAGs include heparin sulfate, chondroitin sulfate, hyaluronic acid, dermal sulfate, keratin sulfate, etc. There are three common and commercially available anticoagulant heparins: unfractionated heparin (UFH), low-molecular-weight heparin (LMWH), and ultralow-molecular-weight heparin (ULMWH). UFH is a natural heparin extracted from animal tissues, with a molecular weight of 3000 to 30,000 Da [7,11,12]. The heparin commonly used in treatment is mainly composed of the trisulfated disaccharide L-iduronic acid-2-sulfate and D-glucosamine-N,6-disulfate. Moreover, it is worth noting that these regular sequences are interrupted by undersulfated (occasionally persulfated) sequences containing D-glucosamine acid and N-acetylated D-glucosamine.

Various enzymes are involved in the biosynthetic pathway of heparin, such as synthetase, epimerase, etc. Moreover, heparin is further modified in vivo by N-deacetylation/N-sulfation of the glucosamine units, C-5 epimerization of the glucuronic acid, and O-sulfation at different sites of the chain [13,14]. These modifications lead to the heterogenization of heparin structure, and realize the regulation of multiple cellular mechanisms in organisms.

About 70% of the heparin polymer is highly sulfated and consists of repeated trisulfated disaccharide units [15]. The remainder is low-sulfated heparin, in which some sites are desulfated, such as 6-O-desulfated glucosamine, N-acetylglucosamine instead of N-sulfated glucosamine, and glucuronic acid instead of iduronic acid. Different extraction sources produce different types of heparin, for example, in the heparin from bovine lung, the content of N-acetyl glucosamine is lower and sulfation is higher [16]. However, there also exist conserved sequence forms among different type of heparins. Around one third of UFH has a specific five-sugar sequence of which the central glycogen is 3-O-sulfated glucosamine. This sequence realizes anticoagulant activity by binding antithrombin (AT) [17,18].

As part of the rapid development of research related to heparin, especially the study of the molecular mechanism of anticoagulation, the LMWHs, which are produced by incomplete depolymerization of UFH by chemical or enzymatic methods, have become a new research field [19]. LMWH is more uniform, with a chain between 2000 and 8000 Da (average 4500 Da) [20]. In addition, the pharmacokinetics and anticoagulant effects of LMWHs are different, as the depolymerization method is different. With this characteristic, the administration of LMWHs could be personalized according to individual patients in the prevention and treatment of venous thrombosis and pulmonary embolism, making LMWHs the first choice for many indications [21–23]. Moreover, LMWHs not only play an important role in thrombus treatment, but can also be used in a variety of other indications, such as maintaining vascular patency during hemodialysis and arterial bypass grafting [24], and the prevention of acute bronchial asthma contractions [25].

3. Anticoagulant Activity

Coagulation is an extremely complex process in organisms. When a series of coagulation factors are continuously activated in a specific order and finally form insoluble polymers, a coagulation reaction occurs. As one of the most widely used anticoagulants for the prevention and treatment of thromboembolic diseases, the anticoagulant effect of heparin is mainly mediated by antithrombin III (AT-III) [26]. Heparin combines with AT-III lysine residues to form a reversible complex, which changes the configuration of AT-III, fully exposes the active site of arginine, and quickly combines with the serine active centers of factor IIa (thrombin) and IXa, Xa, Xia, and XIIa to accelerate the inactivation of coagulation factors, effectively preventing the formation of blood clots and playing an anticoagulant role. When inactivating IXa/IIa, heparin must combine with AT-III and coagulation factor to form a ternary complex, while when inactivating Xa, it only needs to combine with AT-III [27–29]. Once the heparin–AT-III coagulation factor complex is formed, heparin can be dissociated from the complex and reused (Figure 2). To form the heparin–AT-III thrombin ternary complex, the chain length of the heparin molecule needs at least 18 monosaccharide units [30].

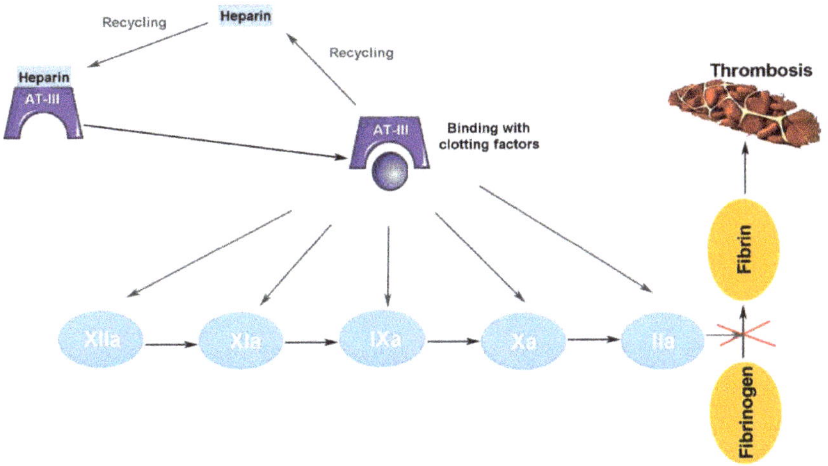

Figure 2. The interactions of heparin with the anticoagulant systems.

Heparin plays an important role in the treatment and prevention of venous thrombosis (VTE). When it occurs as hypercoagulation in pregnancy, VTE can lead to higher maternal morbidity. Heparin is safer than other anticoagulants when used during pregnancy [31–35]. Due to their animal origin and biosynthesis, GAGs have highly variable chain and sulfation patterns, which ultimately prevent the perfect purification of UFH. This became fatal during the "heparin contamination crisis" of 2007 and 2008 [36]. LMWH was developed in the 1980s and is produced by chemical or enzymatic degradation of heparin. The carbohydrate chain contains an average of 15 monosaccharide units and has a molecular weight of 3–8 kDa. The purpose of splitting heparin into LMWH is to reduce the length of the glycosaminoglycan chain to make these preparations easier to absorb, especially when delivered through subcutaneous injection. LMWH exhibits better subcutaneous bioavailability and a longer half-life (3–6 h) due to its low affinity for plasma proteins, endothelial cells, and blood cells; it can even be used once or twice daily without laboratory monitoring. Since the 1990s, LMWH has been recommended for the prevention and treatment of thromboembolic events because of its association with fewer adverse events than UFH [37,38]. At present, the common commercial ULMWH is fondaparin, which is a chemically synthesized pentosan methyl derivative (active fragment of UFC and LMWH) with a molecular weight of 1728 Da [39,40]. Due to its production characteristics, the product of fondaparin has stable properties, uniform structure, and easy derivation and modification of functional groups. It is the focus of heparin drug research at present (Figure 3).

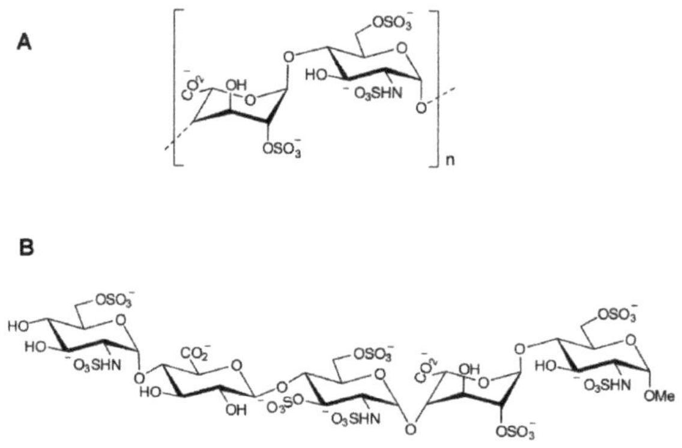

Figure 3. (**A**) The repeated sulfate disaccharides structure of heparin. (**B**) The pentasaccharide responsible for the anticoagulant activity of fondaparin.

4. Antitumor Activity

As early as the 20th century, there were reports on the use of heparin in antitumor research. First, because the blood of patients with advanced cancer is usually in a hypercoagulable state, it is necessary to use heparin for anticoagulant treatment while fighting cancer. Several clinical reports have observed that the use of heparin or heparin derivatives in the treatment of cancer-related thromboembolic diseases seems to prolong the survival time of cancer patients, which has aroused interest in the antimetastatic properties of heparin [41,42]. This characteristic seems to be related to a variety of potential anticoagulant and nonanticoagulant mechanisms.

Cancer metastasis is closely related to angiogenesis and cell adhesion. Some authors believe that LMWH and recombinant tissue factor pathway inhibitors (TFPIs) prevent angiogenesis induced by various angiogenesis factors (such as VEGF). These findings suggest that the interaction between LMWH and TFPI plays a key role in the regulation of angiogenesis. Meanwhile, Mousa and Norrby et al. also proved the potential of LMWH for angiogenesis of new tumors using a chorioallantoic membrane assay (CAM) angiogenesis model [43,44]. The regulation of the interactions between chemokines and their receptors constitutes another effect of heparin on cancer metastasis. Among these, the interaction between chemokine CXCL12 and its receptor CXCR4 is the basis of the metastasis of breast cancer. LMWH inhibited the interaction between CXCL12 and CXCR4, thereby reducing the metastatic spread of breast cancer cells in mice [45]. In animal cancer models, UFH also reduced tumor cell adhesion and LMWH reduced metastatic burden and primary tumor growth, but LMWH did not increase overall survival in patients with solid tumors [46,47]. Galectin 3 is a beta-galactoside-binding protein that is commonly overexpressed in most types of cancer. It has also recently been shown that it can be inhibited by heparin, thereby inhibiting tumor cell metastasis. Although the mechanism of this inhibition of metastasis is still unclear, the linkage of heparin has a significant inhibitory effect on galectin-3-mediated cancer cell metastasis [48].

In addition, heparin can also help to inhibit the proliferation of several cell types. It mainly exerts this antiproliferative effect by modifying the protein kinase C dependent signal transduction pathway and inhibiting some protooncogenes such as c-myc and c-fos. Heparin has been shown to inhibit the phosphorylation (i.e., activation) of MAPK as part of the protein kinase C signaling cascade [49]. Heparin can also affect invasive cancer metastasis by inhibiting the activity of heparanase, which affects the integrity of the extracellular matrix (ECM) and the basement membrane of the vascular wall [50,51]. At present, clinical results show that the inhibition by heparin and LMWH of tumor tissue-related clots is conducive to the better efficacy of radiotherapy and chemotherapy drugs [52–54]. The current clinical guidelines state that LMWH is the first choice for antithrombotic therapy in cancer patients.

5. Anti-Inflammatory Properties

Inflammation is a series of defensive responses to harmful stimuli, often involving the local vascular system and immune system. Since endogenous heparin is only stored in mast cell granules, it is not surprising that drug-grade heparin has immunomodulatory properties. Filkin et al. first studied the protective effect of heparin on lipopolysaccharide (LPS)-induced shock in 1968 and effectively reduced the mortality of LPS model mice [55]. However, apart from the high-affinity binding of antithrombin to heparin through a unique pentasaccharide sequence [56], there is no conclusive evidence describing other examples of "specific binding". Therefore, the mechanism of the anti-inflammatory effect of heparin is complex and not fully understood. It is known that heparin exerts anti-inflammatory effects through a variety of mechanisms [57]. Heparin can not only inhibit the specific functions of neutrophils [58,59] and reduce the migration of eosinophils and vascular permeability, thereby alleviating inflammatory reactions, but also interact with cytokines such as NF-κB, NF-α, IL-6, IL-8, and IL-1β in the vascular endothelium [60,61], preventing activation of the innate immune system. In addition, heparin can inhibit the proliferation of vascular smooth muscle cells [57]. The anti-inflammatory effect of heparin is regulated by many factors, such as source, length, and structure, which cause changes in the anti-inflammatory effect [11].

At present, some studies suggest that heparin can combine with cytokines, chemokines, and acute-phase proteins, including IL-8, platelet growth factor 4 (PGF4), matrix-derived factor 1a, neutrophil elastase, and P- and L-selectin, to exert anti-inflammatory effects [31,62,63].

Several recent studies have shown that heparin and its derivatives have good anti-inflammatory effects in asthma, chronic obstructive pulmonary disease (COPD), acute lung injury, and sepsis [64]. In particular, the good therapeutic effect of nonanticoagulant heparin on sepsis is very exciting. Sepsis is a life-threatening organ dysfunction caused by an imbalance in the body's response to infection. It has become one of the most important causes of death in patients with clinically critical illnesses, with the "three high characteristics" of high prevalence, high mortality, and high treatment costs. Histone is one of the regulatory mediators in sepsis. As a negatively charged, highly sulfated polysaccharide structure, heparin can hinder the interaction between positively charged histones and platelets and may represent a potential solution for sepsis through regulation of the inflammatory response [65]. In addition, through the glycocalyx, heparin can also participate in the cross-endothelial channels of leukocytes and the endothelial and cross-endothelial effects of inflammatory cytokines [66], inhibit the inflammatory response, repair damaged endothelial cells, and rebuild vascular barrier function. It is well known that the surface of endothelial cells is covered by a macromolecular reticular structure called the glycocalyx, which is a glycoprotein with syndecan-1 as the core [67]. In recent years, some studies have shown that UFH, as a HS analogue, can participate in the mobilization of glycocalyx core protein syndecan-1 to reconstruct the glycocalyx left on the cell surface and protect it from shedding, so as to achieve the integrity of the cell surface and ensure a good vascular barrier [68,69].

Hala et al. studied the preparation and characterization of a polyelectrolyte multilayer (PEM) coating which combined the anti-inflammatory activity of heparin as a polyanion and the potential release of naproxen. PEM containing heparin was shown to reduce cell adhesion and IL-β as a substitute for polyanions to form multilayers [70].

6. Antiviral Application

Research on heparin as an antiviral treatment has surged again following the recent outbreak of COVID-19. Amongst the long history of heparin, there were early studies on the inhibition of herpes simplex virus by heparin in vitro as early as the mid-20th century [71]. Heparin inhibits herpes simplex virus as a natural inhibitor with a unique sulfonation mode [72,73]. Subsequently, there were studies on the inhibitory potential of heparin on a variety of RNA and DNA viruses, such as HIV, in vitro [74–76].

Heparan sulfate (HS) on the cell surface can be used as a coreceptor by viruses, helping many viruses attach to or enter target cells. Modhiran et al. showed that a heparan sulfate analogue (PG545) can be used as an inhibitor of virus–cell adhesion, which can effectively prevent the transmission of dengue virus in clinical settings [77,78]. Similarly, by affecting cell adhesion, HS promotes the adhesion and entry of rabies virus (RABV) into target cells [79]. It should be noted that heparin is not necessary for cell adhesion of Zika virus, but it participates in virus replication and induces apoptosis of infected cells [80].

In recent years, due to the frequent outbreaks of severe acute respiratory syndrome (SARS), studies on the effects of heparin and its analogues as antiviral treatments have been ongoing. It was recently shown in an in vitro study that SARS-CoV-2 first attaches to heparan sulfate proteoglycans before interacting with ACE2 [81]. Similarly, the study of Clausen et al. [82] also showed that the spike protein of SARS-CoV-2 can bind to the cell surface through both HS and ACE2 protein receptors. All of the abovementioned results indicate that HS is required for the attachment of SARS-CoV-2 to the cell surface. Therefore, it is now generally believed that in severe acute respiratory syndrome (SARS)-coronavirus (COV)-2 infection, heparin can be used as a bait receptor to bind the SARS-CoV-2 spike protein, inhibiting the binding of the virus to HS and reducing the infectivity of the virus [83,84]. In addition, in patients with severe COVID-19 infection, platelet activation, the increase in blood viscosity caused by high fibrinogen levels, and the increase in heparanase expression [85] all further increase the risk of vascular disease [86]. Although previous anticoagulant treatments have failed in critical illnesses, convincing observations of coagulation dysfunction and high venous thromboembolism rates in COVID-19 increase the possibility that heparin may benefit the prognosis of patients. Moreover, heparin has beneficial effects on inflammation, which is associated with COVID-19. Heparin binds and regulates the activity of many inflammatory proteins, including IL-8, platelet growth factor 4, neutrophils, elastase control, CD11b/CD18, etc. [85]. In a randomized clinical trial by Alex et al., therapeutic-dose LMWH reduced major thromboembolism and death compared with institutional standard heparin thromboprophylaxis among inpatients with COVID-19 with very elevated D-dimer levels [87].

According to the study of Hidesaku et al., COVID-19 shows some characteristics similar to disseminated intravascular coagulation (DIC). Nafamostat mesylate (NM) is a drug used for treating DIC, and is also expected to be a drug used for treating COVID-19. However, the anticoagulant effect of NM is weak. Heparin can make up for this shortcoming and can be used in combination to treat COVID-19. Therefore, heparin as a combined drug in COVID-19 may be a future research direction [88].

7. Application of Heparin in Malaria

Malaria is caused by infected with Plasmodium, which is an infectious disease. Malaria can be spread by mosquito bites and then propagate in the liver of patients, leading to infection, causing the destruction of red blood cells, and then continuing to reproduce and be destroyed.

At present, with global warming, globalization, and the living environment continuing to deteriorate in some areas due to war, the incidence rate of malaria is continuing to increase. Although it has attracted attention globally, because the existing first-line drugs cannot effectively treat malaria and there has been no breakthrough in vaccine research, there is an urgent need for new treatment schemes and new mechanisms to improve treatment efficiency.

GAGs can bind to Plasmodium parasite red blood cells (PBRC), thus blocking the binding of PRBC to various host-cell surface receptors [89–92]. Negatively charged polysaccharides, such as heparin and chondroitin sulfate, have the ability to bind to PBRC, thus achieving antimalaria effects. Xuerong Dong et al. constructed an intraerythrocytic parasite-targeted nanostructured lipid carrier (NLC) that was developed for potentiation of artemether (ARM) by combination with PPIX and iron-loaded transferrin (holo-Tf). ARM and PPIX were co-loaded into NLCs with high entrapment efficiency. A targeting ligand heparin (HP) was then electrostatically adsorbed onto the periphery of the NLCs, followed by conjugation with holo-Tf to obtain the final formulation, Tf-HP-NLC/ARM/PPIX. This delivery system showed increased inhibitory activity against Plasmodium falciparum in culture [93].

However, heparin has anticoagulant and bleeding properties, so it will increase the risk of infection [94–97], which is a disadvantage in the treatment of malaria. Heparin has been shown to bind merozoites inside late-stage pRBCs. This finding can be used as a new direction in future research into antimalarial drugs.

8. Application of Heparin in Nanomaterials

Heparin nanocomposites first appeared in 2008. The authors of that first study synthesized hollow capsules based on an iron heparin complex and multilayer films, which significantly increased anticoagulation time. The capsules can be used as injectable anticoagulant carriers for parenteral injection to treat iron deficiency [98]. Since then, heparin nanomaterial composites have been continuously explored and innovated in the field of medical applications.

Nanomaterials and heparin are connected by covalent bonds or electrostatic interaction. Heparin, as a natural water-soluble polysaccharide, has good biocompatibility and a variety of biological functions. When combined with nanomaterials, it shows increased stability and water solubility, and can improve the targeting of molecules. Studies have shown that heparin nanoparticles have strong growth-factor-loading capacity and are cytokines that maintain activity for a long time [99–101]. Heparin nanoparticles have strong potential value in the field of drug delivery. Heparin nanomaterials also show significant value in the treatment of other diseases. Qi Tan et al. constructed a heparin chitosan nanoparticle-immobilized scaffold, which showed high-efficiency vascular endothelial growth factor (VEGF) localization and release ability in vitro and significantly increased fibroblast infiltration and extracellular matrix generation, and accelerated angiogenesis in the subcutaneous implantation model of mice [102]. Yasutaka et al. constructed LMWH protamine nanoparticles (lmwh-h/P NPs) as a carrier for heparin-binding growth factor. Fibroblast growth factor (FGF-2) combined with nanoparticles significantly extended the biological half-life of FGF-2 [103]. Jeong et al. constructed a PEG-LHT7/TRAIL/Protamine nanocomplex within a PEG-LMWH-taurocholate conjugate (LHT7). The selective cytotoxicity of tumor necrosis factor-related apoptosis-inducing ligand (TRAIL) to cancer cells, but not to normal cells, makes it an attractive candidate for cancer therapeutics. However, the disadvantages of TRAIL, such as physicochemical instability and short half-life, limit its further clinical applications. This heparin nanoparticle improved the short life of this drug and not only had anti-angiogenesis effect in the treatment of tumors, but also uniformly induced tumor cell apoptosis [104].

In conclusion, heparin nanocomposites, with their superior pharmacokinetic properties and biological functions, have prompted scientists to explore new drug delivery systems for improved therapeutic effects. Combining the advantages of the two can also overcome the inherent limitations of the structures of heparin and nanomaterials.

9. Challenges of Heparin Therapy

The effects of heparin are not exclusively positive. Some adverse reactions reported during heparin treatment are closely related to the biological activity of heparin itself. Firstly, among its own characteristics, heparin has a short half-life which requires frequent administration in clinical applications, leading to poor compliance of patients. At present, attempts are being made to extend the half-life using slow-release preparations [105].

Heparin is the most effective anticoagulant in clinical use, but it also carries a certain risk of bleeding [106,107]. Especially in elderly individuals or those with renal insufficiency, this needs special attention. Bleeding-related complications from heparin can range in severity from mild symptoms such as injection site hematomas to potentially fatal events such as intracranial hemorrhage. Some researchers have also observed in clinical settings that hematoma at the puncture site can occur when heparin is used for epidural anesthesia or spinal cord puncture; this symptom is indeed extremely dangerous because it carries the risk of causing paralysis. In addition, with increasing chain length, heparin has enhanced binding ability with positively charged molecules such as platelet 4 (PF4), which can form new antigen complexes to destroy receptors on platelets and endothelial cells, leading to thrombocytopenia (HIT). HIT can also cause skin damage or even necrosis at the heparin injection site. Moreover, HIT is a very noteworthy problem because of the ease of relapse [108]; immune memory is cleared several months after the onset of the disease. After the reintroduction of heparin, the risk of HIT recurrence is the same as that of patients not affected by heparin. In addition, some studies have shown that heparin can bind to bone protein, affect osteoblast bone synthesis, increase osteoblast absorption, and finally reduce bone mass, even leading to osteoporosis [109]. This is also one of the common side effects of long-term use of heparin and LMWH [110,111]. Although some studies have found that short-term use of LWMH (3–6 months) does not affect bone mineral density, this should be noted in pregnant women, the elderly, and children as the effect is long-lasting and irreversible [112].

Increases in eosinophils, hyperkalemia, and other side effects are not common and can recover with the cessation of heparin treatment. Some patients with special diseases, such as patients with chronic renal failure, may have adverse reactions such as calcium deposition at the injection site of heparin due to abnormalities in calcium and phosphorus in the body.

10. Conclusions

With the discovery of new indications and new possibilities, the role of heparin continues to develop. Although heparin is well known by the public as a mainstream anticoagulant drug, with the deepening of research, heparin and heparin derivatives are showing an edge in various fields. At present, LMWH and UFH are commonly used, but heparin has a complex structure and can be modified, for example, by sulfation and acetylation. At present, the specific efficacy of heparins with different molecular weights has not been studied thoroughly. With the maturing of enzymatic synthesis of sugar chains, high-purity oligoheparin may have broad application prospects. In addition, many proteins can bind to heparin and heparin can produce various pharmacological effects, such as anti-inflammatory, antiviral, antiangiogenic, antitumor, and antimetastatic effects, by interacting with a variety of proteases, protease inhibitors, and chemokines. In this review, we comprehensively discussed the old and new applications of heparin and explained its therapeutic mechanism to a certain extent. Due to the coexistence of clinical benefits and adverse reactions, efforts should be made to optimize the medication regimen or modify the structure–activity relationship to improve the therapeutic effect and reduce the possibility of adverse reactions. At the same time, more clinical studies are necessary in order to provide effective, accurate, and reliable evidence. In short, heparin, as a classic drug that has been used for a century, has gone beyond its traditional role of anticoagulation, and achieved new development potential under the optimization of current research and development.

Author Contributions: Conceptualization, L.Z., H.Z., Y.L., F.Y. and W.Z.; writing—original draft preparation, L.Z., H.Z. and K.W.; writing—review and editing, L.Z., H.Z., Y.L., F.Y. and W.Z.; supervision, Y.L., F.Y. and W.Z.; project administration, L.Z., H.Z. and K.W.; funding acquisition, W.Z. All authors have read and agreed to the published version of the manuscript.

Funding: This research was funded by the National Natural Science Foundation of China (grant number 22077068) and the Fundamental Research Funds for the Central Universities.

Institutional Review Board Statement: Not applicable.

Informed Consent Statement: Not applicable.

Data Availability Statement: Not applicable.

Conflicts of Interest: The authors declare that they have no known competing financial interests or personal relationships that could have appeared to influence the work reported in this paper.

References

1. Zhang, C.; Yang, B.-C.; Liu, W.-T.; Li, Z.-Y.; Song, Y.-J.; Zhang, T.-C.; Luo, X.-G. Structure-based engineering of heparinase I with improved specific activity for degrading heparin. *BMC Biotechnol.* **2019**, *19*, 59. [CrossRef] [PubMed]
2. Malavaki, C.J.; Theocharis, A.D.; Lamari, F.N.; Kanakis, I.; Tsegenidis, T.; Tzanakakis, G.N.; Karamanos, N.K. Heparan sulfate: Biological significance, tools for biochemical analysis and structural characterization. *Biomed. Chromatogr.* **2010**, *25*, 11–20. [CrossRef] [PubMed]
3. Oduah, E.I.; Linhardt, R.J.; Sharfstein, S.T. Heparin: Past, Present, and Future. *Pharmaceuticals* **2016**, *9*, 38. [CrossRef] [PubMed]
4. Hemker, H.C. A century of heparin: Past, present and future. *J. Thromb. Haemost.* **2016**, *14*, 2329–2338. [CrossRef] [PubMed]
5. Lazrak, H.H.; René, E.; Elftouh, N.; Leblanc, M.; Lafrance, J.-P. Safety of low-molecular-weight heparin compared to unfractionated heparin in hemodialysis: A systematic review and meta-analysis. *BMC Nephrol.* **2017**, *18*, 187. [CrossRef]
6. Biran, R.; Pond, D. Heparin coatings for improving blood compatibility of medical devices. *Adv. Drug Deliv. Rev.* **2017**, *112*, 12–23. [CrossRef]
7. Lima, M.; Rudd, T.; Yates, E. New Applications of Heparin and Other Glycosaminoglycans. *Molecules* **2017**, *22*, 749. [CrossRef]
8. Mazilu, L.; Katsiki, N.; Nikolouzakis, T.K.; Aslanidis, M.I.; Lazopoulos, G.; Kouretas, D.; Tsatsakis, A.; Suceveanu, A.-I.; Stoian, A.-P.; Parepa, I.-R.; et al. Thrombosis and Haemostasis challenges in COVID-19—Therapeutic perspectives of heparin and tissue-type plasminogen activator and potential toxicological reactions-a mini review. *Food Chem. Toxicol.* **2021**, *148*, 111974. [CrossRef]
9. Magnani, H.N. Rationale for the Role of Heparin and Related GAG Antithrombotics in COVID-19 Infection. *Clin. Appl. Thromb.* **2021**, *27*, 1076029620977702. [CrossRef]
10. Tandon, R.; Sharp, J.S.; Zhang, F.; Pomin, V.H.; Ashpole, N.M.; Mitra, D.; McCandless, M.G.; Jin, W.; Liu, H.; Sharma, P.; et al. Effective Inhibition of SARS-CoV-2 Entry by Heparin and Enoxaparin Derivatives. *J. Virol.* **2021**, *95*, e01987-20. [CrossRef]
11. Liu, H.; Zhang, Z.; Linhardt, R.J. Lessons learned from the contamination of heparin. *Nat. Prod. Rep.* **2009**, *26*, 313–321. [CrossRef] [PubMed]
12. Fang, G.; Tang, B. Advanced delivery strategies facilitating oral absorption of heparins. *Asian J. Pharm. Sci.* **2020**, *15*, 449–460. [CrossRef] [PubMed]
13. Rabenstein, D.L. Heparin and heparan sulfate: Structure and function. *Nat. Prod. Rep.* **2002**, *19*, 312–331. [CrossRef]
14. Mizumoto, S.; Kitagawa, H.; Sugahara, K. *Biosynthesis of Heparin and Heparan Sulfate*; Elsevier: Amsterdam, The Netherlands, 2005; pp. 203–243.
15. Casu, B.; Naggi, A.; Torri, G. Re-visiting the structure of heparin. *Carbohydr. Res.* **2015**, *403*, 60–68. [CrossRef]
16. Naggi, A.; Gardini, C.; Pedrinola, G.; Mauri, L.; Urso, E.; Alekseeva, A.; Casu, B.; Cassinelli, G.; Guerrini, M.; Iacomini, M.; et al. Structural peculiarity and antithrombin binding region profile of mucosal bovine and porcine heparins. *J. Pharm. Biomed. Anal.* **2016**, *118*, 52–63. [CrossRef] [PubMed]
17. Petitou, M.; van Boeckel, C.A. A synthetic antithrombin III binding pentasaccharide is now a drug! What comes next? *Angew. Chem. Int. Ed. Engl.* **2004**, *43*, 3118–3133. [CrossRef]
18. Huntington, J.A. Heparin activation of serpins. In *Chemistry and Biology of Heparin and Heparan Sulfate*; Garg, H.G., Linhardt, R.J., Hales, C.A., Eds.; Elsevier: Amsterdam, The Netherlands, 2005; pp. 367–398.
19. Wang, P.; Chi, L.; Zhang, Z.; Zhao, H.; Zhang, F.; Linhardt, R.J. Heparin: An old drug for new clinical applications. *Carbohydr. Polym.* **2022**, *295*, 119818. [CrossRef]
20. Ibrahim, S.S.; Osman, R.; Awad, G.A.S.; Mortada, N.D.; Geneidy, A.-S. Low molecular weight heparins for current and future uses: Approaches for micro- and nano-particulate delivery. *Drug Deliv.* **2016**, *23*, 2661–2667. [CrossRef]
21. Lima, M.A.; de Farias, E.H.; Rudd, T.; Ebner, L.F.; Gesteira, T.F.; Mendes, A.; Bouças, R.I.; Martins, J.R.M.; Hoppensteadt, D.; Fareed, J.; et al. Low molecular weight heparins: Structural differentiation by spectroscopic and multivariate approaches. *Carbohydr. Polym.* **2011**, *85*, 903–909. [CrossRef]
22. Bisio, A.; Vecchietti, D.; Citterio, L.; Guerrini, M.; Raman, R.; Bertini, S.; Eisele, G.; Naggi, A.; Sasisekharan, R.; Torri, G. Structural features of low-molecular-weight heparins affecting their affinity to antithrombin. *Thromb. Haemost.* **2009**, *102*, 865–873. [CrossRef]
23. Mulloy, B.; Hogwood, J.; Gray, E.; Lever, R.; Page, C.P. Pharmacology of Heparin and Related Drugs. *Pharmacol. Rev.* **2015**, *68*, 76–141. [CrossRef] [PubMed]
24. Patel, R.P.; Narkowicz, C.; Jacobson, G.A. Investigation of the Effect of Heating on the Chemistry and Antifactor Xa Activity of Enoxaparin. *J. Pharm. Sci.* **2009**, *98*, 1700–1711. [CrossRef] [PubMed]

25. Campo, C.; Molinari, J.F.; Ungo, J.; Ahmed, T. Molecular-weight-dependent effects of nonanticoagulant heparins on allergic airway responses. *J. Appl. Physiol.* **1999**, *86*, 549–557. [CrossRef]
26. Kopterides, P. What Is the Appropriate Anticoagulation Therapy in Patients with a History of Heparin-Induced Thrombocytopenia? *Anesth. Analg.* **2005**, *101*, 1885. [CrossRef]
27. Li, W.; Johnson, D.J.D.; Esmon, C.T.; A Huntington, J. Structure of the antithrombin–thrombin–heparin ternary complex reveals the antithrombotic mechanism of heparin. *Nat. Struct. Mol. Biol.* **2004**, *11*, 857–862. [CrossRef]
28. Wagenvoord, R.; Al Dieri, R.; van Dedem, G.; Béguin, S.; Hemker, H.C. Linear diffusion of thrombin and factor Xa along the heparin molecule explains the effects of extended heparin chain lengths. *Thromb. Res.* **2008**, *122*, 237–245. [CrossRef]
29. Al Dieri, R.; Wagenvoord, R.; Van Dedem, G.W.K.; Beguin, S.; Hemker, C. The inhibition of blood coagulation by heparins of different molecular weight is caused by a common functional motif-the C-domain. *J. Thromb. Haemost.* **2003**, *1*, 907–914. [CrossRef]
30. Crush, J.; Seah, M.; Chou, D.; Rawal, J.; Hull, P.; Carrothers, A. Sequential low molecular weight heparin and rivaroxaban for venous thromboprophylaxis in pelvic and acetabular trauma. *Arch. Orthop. Trauma. Surg.* **2022**, *142*, 3271–3277. [CrossRef] [PubMed]
31. Hao, C.; Xu, H.; Yu, L.; Zhang, L. Heparin: An essential drug for modern medicine. *Prog. Mol. Biol. Transl. Sci.* **2019**, *163*, 1–19. [CrossRef]
32. Kher, A.; Bauersachs, R.; Nielsen, J.D. The management of thrombosis in pregnancy: Role of low-molecular-weight heparin. *Thromb. Haemost.* **2007**, *97*, 505–513. [CrossRef]
33. Lussana, F.; Coppens, M.; Cattaneo, M.; Middeldorp, S. Pregnancy-related venous thromboembolism: Risk and the effect of thromboprophylaxis. *Thromb. Res.* **2012**, *129*, 673–680. [CrossRef] [PubMed]
34. Guimicheva, B.; Czuprynska, J.; Arya, R. The prevention of pregnancy-related venous thromboembolism. *Br. J. Haematol.* **2015**, *168*, 163–174. [CrossRef] [PubMed]
35. Chen, D. Heparin beyond anti-coagulation. *Curr. Res. Transl. Med.* **2021**, *69*, 103300. [CrossRef]
36. Szajek, A.Y.; Chess, E.; Johansen, K.; Gratzl, G.; Gray, E.; Keire, D.; Linhardt, R.J.; Liu, J.; Morris, T.; Mulloy, B.; et al. The US regulatory and pharmacopeia response to the global heparin contamination crisis. *Nat. Biotechnol.* **2016**, *34*, 625–630. [CrossRef] [PubMed]
37. Weitz, J.I. Low-molecular-weight heparins. *N. Engl. J. Med.* **1997**, *337*, 688–698. [CrossRef]
38. Hirsh, J.; Warkentin, T.E.; Raschke, R.; Granger, C.; Ohman, E.M.; Dalen, J.E. Heparin and low-molecular-weight heparin: Mechanisms of action, pharmacokinetics, dosing considerations, monitoring, efficacy, and safety. *Chest* **1998**, *114*, 489S–510S. [CrossRef]
39. Baytas, S.N.; Linhardt, R.J. Advances in the preparation and synthesis of heparin and related products. *Drug Discov. Today* **2020**, *25*, 2095–2109. [CrossRef]
40. Zhang, T.; Liu, X.; Li, H.; Wang, Z.; Chi, L.; Li, J.-P.; Tan, T. Characterization of epimerization and composition of heparin and dalteparin using a UHPLC-ESI-MS/MS method. *Carbohydr. Polym.* **2019**, *203*, 87–94. [CrossRef]
41. Sanford, D.; Naidu, A.; Alizadeh, N.; Lazo-Langner, A. The effect of low molecular weight heparin on survival in cancer patients: An updated systematic review and meta-analysis of randomized trials. *J. Thromb. Haemost.* **2014**, *12*, 1076–1085. [CrossRef]
42. Kuderer, N.M.; Khorana, A.A.; Lyman, G.H.; Francis, C.W. A meta-analysis and systematic review of the efficacy and safety of anticoagulants as cancer treatment: Impact on survival and bleeding complications. *Cancer Am. Cancer Soc.* **2007**, *110*, 1149–1161. [CrossRef]
43. Mousa, S.; Mohamed, S. Anti-angiogenic mechanisms and efficacy of the low molecular weight heparin, tinzaparin: Anti-cancer efficacy. *Oncol. Rep.* **2004**, *12*, 683–688. [CrossRef] [PubMed]
44. Norrby, K. Low-molecular-weight heparins and angiogenesis. *APMIS* **2006**, *114*, 79–102. [CrossRef] [PubMed]
45. Harvey, J.R.; Mellor, P.; Eldaly, H.; Lennard, T.W.; Kirby, J.A.; Ali, S. Inhibition of CXCR4-mediated breast cancer metastasis: A potential role for heparinoids? *Clin. Cancer Res.* **2007**, *13*, 1562–1570. [CrossRef] [PubMed]
46. Montroy, J.; Lalu, M.M.; Auer, R.C.; Grigor, E.; Mazzarello, S.; Carrier, M.; Kimmelman, J.; Fergusson, D.A. The Efficacy and Safety of Low Molecular Weight Heparin Administration to Improve Survival of Cancer Patients: A Systematic Review and Meta-Analysis. *Thromb. Haemost.* **2020**, *120*, 832–846. [CrossRef] [PubMed]
47. Ripsman, D.; Fergusson, D.A.; Montroy, J.; Auer, R.C.; Huang, J.W.; Dobriyal, A.; Wesch, N.; Carrier, M.; Lalu, M.M. A systematic review on the efficacy and safety of low molecular weight heparin as an anticancer therapeutic in preclinical animal models. *Thromb. Res.* **2020**, *195*, 103–113. [CrossRef]
48. Sindrewicz, P.; Yates, E.A.; Turnbull, J.E.; Lian, L.-Y.; Yu, L.-G. Interaction with the heparin-derived binding inhibitors destabilizes galectin-3 protein structure. *Biochem. Biophys. Res. Commun.* **2020**, *523*, 336–341. [CrossRef]
49. Atallah, J.; Khachfe, H.H.; Berro, J.; Assi, H.I. The use of heparin and heparin-like molecules in cancer treatment: A review. *Cancer Treat. Res. Commun.* **2020**, *24*, 100192. [CrossRef]
50. Pfankuchen, D.B.; Stölting, D.P.; Schlesinger, M.; Royer, H.-D.; Bendas, G. Low molecular weight heparin tinzaparin antagonizes cisplatin resistance of ovarian cancer cells. *Biochem. Pharmacol.* **2015**, *97*, 147–157. [CrossRef]
51. Shute, J.K.; Puxeddu, E.; Calzetta, L. Therapeutic use of heparin and derivatives beyond anticoagulation in patients with bronchial asthma or COPD. *Curr. Opin. Pharmacol.* **2018**, *40*, 39–45. [CrossRef]

52. Smorenburg, S.M.; Van Noorden, C.J. The complex effects of heparins on cancer progression and metastasis in experimental studies. *Pharmacol. Rev.* **2001**, *53*, 93–105.
53. Phillips, P.G.; Yalcin, M.; Cui, H.; Abdel-Nabi, H.; Sajjad, M.; Bernacki, R.; Veith, J.; A Mousa, S. Increased tumor uptake of chemotherapeutics and improved chemoresponse by novel non-anticoagulant low molecular weight heparin. *Anticancer Res.* **2011**, *31*, 411–419. [PubMed]
54. Pan, Y.; Li, X.; Duan, J.; Yuan, L.; Fan, S.; Fan, J.; Xiaokaiti, Y.; Yang, H.; Wang, Y.; Li, X. Enoxaparin sensitizes human non-small-cell lung carcinomas to gefitinib by inhibiting DOCK1 expression, vimentin phosphorylation, and Akt activation. *Mol. Pharmacol.* **2015**, *87*, 378–390. [CrossRef] [PubMed]
55. Filkins, J.; Di Luzio, N. Heparin protection in endotoxin shock. *Am. J. Physiol.* **1968**, *214*, 1074–1077. [CrossRef] [PubMed]
56. Lindahl, U.; Thunberg, L.; Bäckström, G.; Riesenfeld, J.; Nordling, K.; Björk, I. Extension and structural variability of the antithrombin-binding sequence in heparin. *J. Biol. Chem.* **1984**, *259*, 12368–12376. [CrossRef]
57. Poterucha, T.J.; Libby, P.; Goldhaber, S.Z. More than an anticoagulant: Do heparins have direct anti-inflammatory effects? *Thromb. Haemost.* **2017**, *117*, 437–444. [CrossRef]
58. Wakefield, T.W.; Greenfield, L.J.; Rolfe, M.W.; DeLucia, A., 3rd; Strieter, R.M.; Abrams, G.D.; Kunkel, S.L.; Esmon, C.T.; Wrobleski, S.K.; Kadell, A.M.; et al. Inflammatory and procoagulant mediator interactions in an experimental baboon model of venous thrombosis. *Thromb. Haemost.* **1993**, *69*, 164–172. [CrossRef]
59. Tichelaar, Y.I.G.V.; Kluin-Nelemans, J.C.; Meijer, K. Infections and inflammatory diseases as risk factors for venous thrombosis. A systematic review. *Thromb. Haemost.* **2012**, *107*, 827–837. [CrossRef]
60. Etulain, J.; Martinod, K.; Wong, S.L.; Cifuni, S.M.; Schattner, M.; Wagner, D.D. P-selectin promotes neutrophil extracellular trap formation in mice. *Blood* **2015**, *126*, 242–246. [CrossRef]
61. Demers, M.; Wagner, D.D. NETosis: A New Factor in Tumor Progression and Cancer-Associated Thrombosis. *Semin. Thromb. Hemost.* **2014**, *40*, 277–283. [CrossRef]
62. Young, E. The anti-inflammatory effects of heparin and related compounds. *Thromb. Res.* **2008**, *122*, 743–752. [CrossRef]
63. Li, X.; Li, L.; Shi, Y.; Yu, S.; Ma, X. Different signaling pathways involved in the anti-inflammatory effects of unfractionated heparin on lipopolysaccharide-stimulated human endothelial cells. *J. Inflamm.* **2020**, *17*, 1–9. [CrossRef] [PubMed]
64. Abd-Elaty, N.M.; Elprince, M.; El-Salam, M.A. Efficacy of inhaled heparin is effective in the treatment of acute exacerbation of asthma. *World Allergy Organ. J.* **2007**, *62*, S42–S43. [CrossRef]
65. Alhamdi, Y.; Abrams, S.T.; Lane, S.; Wang, G.; Toh, C.-H. Histone-Associated Thrombocytopenia in Patients Who Are Critically Ill. *JAMA* **2016**, *315*, 817–819. [CrossRef]
66. Rao, N.V.; Argyle, B.; Xu, X.; Reynolds, P.R.; Walenga, J.M.; Prechel, M.; Prestwich, G.D.; MacArthur, R.B.; Walters, B.B.; Hoidal, J.R.; et al. Low anticoagulant heparin targets multiple sites of inflammation, suppresses heparin-induced thrombocytopenia, and inhibits interaction of RAGE with its ligands. *Am. J. Physiol. Cell Physiol.* **2010**, *299*, C97–C110. [CrossRef]
67. Henrich, M.; Gruss, M.; Weigand, M.A. Sepsis-Induced Degradation of Endothelial Glycocalix. *Sci. World J.* **2010**, *10*, 917–923. [CrossRef] [PubMed]
68. Nelson, A.; Berkestedt, I.; Schmidtchen, A.; Ljunggren, L.; Bodelsson, M. Increased levels of glycosaminoglycans during septic shock: Relation to mortality and the antibacterial actions of plasma. *Shock* **2008**, *30*, 623–627. [CrossRef] [PubMed]
69. Yini, S.; Heng, Z.; Xin, A.; Xiaochun, M. Effect of unfractionated heparin on endothelial glycocalyx in a septic shock model. *Acta Anaesthesiol. Scand.* **2015**, *59*, 160–169. [CrossRef]
70. Al-Khoury, H.; Espinosa-Cano, E.; Aguilar, M.R.; Román, J.S.; Syrowatka, F.; Schmidt, G.; Groth, T. Anti-inflammatory Surface Coatings Based on Polyelectrolyte Multilayers of Heparin and Polycationic Nanoparticles of Naproxen-Bearing Polymeric Drugs. *Biomacromolecules* **2019**, *20*, 4015–4025. [CrossRef]
71. Nahmias, A.J.; Kibrick, S. INHIBITORY EFFECT OF HEPARIN ON HERPES SIMPLEX VIRUS. *J. Bacteriol.* **1964**, *87*, 1060–1066. [CrossRef]
72. Copeland, R.; Balasubramaniam, A.; Tiwari, V.; Zhang, F.; Bridges, A.; Linhardt, R.J.; Shukla, D.; Liu, J. Using a 3-*O*-Sulfated Heparin Octasaccharide To Inhibit the Entry of Herpes Simplex Virus Type 1. *Biochemistry* **2008**, *47*, 5774–5783. [CrossRef]
73. Ramos-Kuri, M.; Barron Romero, B.L.; Aguilar-Setien, A. Inhibition of three alphaherpesviruses (herpes simplex 1 and 2 and pseudorabies virus) by heparin, heparan and other sulfated polyelectrolytes. *Arch. Med. Res.* **1996**, *27*, 43–48.
74. Rider, C. The potential for heparin and its derivatives in the therapy and prevention of HIV-1 infection. *Glycoconj. J.* **1997**, *14*, 639–642. [CrossRef] [PubMed]
75. Shukla, D.; Liu, J.; Blaiklock, P.; Shworak, N.W.; Bai, X.; Esko, J.D.; Cohen, G.H.; Eisenberg, R.J.; Rosenberg, R.D.; Spear, P.G. A Novel Role for 3-O-Sulfated Heparan Sulfate in Herpes Simplex Virus 1 Entry. *Cell* **1999**, *99*, 13–22. [CrossRef]
76. Liu, J.; Shriver, Z.; Pope, R.M.; Thorp, S.C.; Duncan, M.B.; Copeland, R.J.; Raska, C.S.; Yoshida, K.; Eisenberg, R.J.; Cohen, G.; et al. Characterization of a Heparan Sulfate Octasaccharide That Binds to Herpes Simplex Virus Type 1 Glycoprotein D. *J. Biol. Chem.* **2002**, *277*, 33456–33467. [CrossRef] [PubMed]
77. Buijsers, B.; Yanginlar, C.; de Nooijer, A.; Grondman, I.; Maciej-Hulme, M.L.; Jonkman, I.; Janssen, N.A.F.; Rother, N.; de Graaf, M.; Pickkers, P.; et al. Increased Plasma Heparanase Activity in COVID-19 Patients. *Front. Immunol.* **2020**, *11*, 575047. [CrossRef] [PubMed]
78. Modhiran, N.; Gandhi, N.; Wimmer, N.; Cheung, S.; Stacey, K.; Young, P.R.; Ferro, V.; Watterson, D. Dual targeting of dengue virus virions and NS1 protein with the heparan sulfate mimic PG545. *Antivir. Res.* **2019**, *168*, 121–127. [CrossRef] [PubMed]

79. Sasaki, M.; Anindita, P.D.; Ito, N.; Sugiyama, M.; Carr, M.; Fukuhara, H.; Ose, T.; Maenaka, K.; Takada, A.; Hall, W.W.; et al. The Role of Heparan Sulfate Proteoglycans as an Attachment Factor for Rabies Virus Entry and Infection. *J. Infect. Dis.* **2018**, *217*, 1740–1749. [CrossRef] [PubMed]
80. Gao, H.; Lin, Y.; He, J.; Zhou, S.; Liang, M.; Huang, C.; Li, X.; Liu, C.; Zhang, P. Role of heparan sulfate in the Zika virus entry, replication, and cell death. *Virology* **2019**, *529*, 91–100. [CrossRef]
81. Bermejo-Jambrina, M.; Eder, J.; Kaptein, T.M.; Helgers, L.C.; Geijtenbeek, T. Infection and transmission of SARS-CoV-2 depend on heparan sulfate proteoglycans. *EMBO J.* **2021**, *40*, e106765. [CrossRef]
82. Clausen, T.M.; Sandoval, D.R.; Spliid, C.B.; Pihl, J.; Perrett, H.R.; Painter, C.D.; Narayanan, A.; Majowicz, S.A.; Kwong, E.M.; McVicar, R.N.; et al. SARS-CoV-2 Infection Depends on Cellular Heparan Sulfate and ACE2. *Cell* **2020**, *183*, 1043–1057.e15. [CrossRef]
83. Hendricks, G.L.; Velazquez, L.; Pham, S.; Qaisar, N.; Delaney, J.C.; Viswanathan, K.; Albers, L.; Comolli, J.C.; Shriver, Z.; Knipe, D.M.; et al. Heparin octasaccharide decoy liposomes inhibit replication of multiple viruses. *Antivir. Res.* **2015**, *116*, 34–44. [CrossRef] [PubMed]
84. Hippensteel, J.A.; LaRiviere, W.B.; Colbert, J.F.; Langouët-Astrié, C.J.; Schmidt, E.P. Heparin as a therapy for COVID-19: Current evidence and future possibilities. *Am. J. Physiol. Lung Cell Mol. Physiol.* **2020**, *319*, L211–L217. [CrossRef] [PubMed]
85. Qiu, M.; Huang, S.; Luo, C.; Wu, Z.; Liang, B.; Huang, H.; Ci, Z.; Zhang, D.; Han, L.; Lin, J. Pharmacological and clinical application of heparin progress: An essential drug for modern medicine. *Biomed. Pharmacother.* **2021**, *139*, 111561. [CrossRef] [PubMed]
86. Maier, C.L.; Truong, A.D.; Auld, S.C.; Polly, D.M.; Tanksley, C.L.; Duncan, A. COVID-19-associated hyperviscosity: A link between inflammation and thrombophilia? *Lancet* **2020**, *395*, 1758–1759. [CrossRef]
87. Spyropoulos, A.C.; Goldin, M.; Giannis, D.; Diab, W.; Wang, J.; Khanijo, S.; Mignatti, A.; Gianos, E.; Cohen, M.; Sharifova, G.; et al. Efficacy and Safety of Therapeutic-Dose Heparin vs Standard Prophylactic or Intermediate-Dose Heparins for Thromboprophylaxis in High-risk Hospitalized Patients With COVID-19: The HEP-COVID Randomized Clinical Trial. *JAMA Intern. Med.* **2021**, *181*, 1612–1620. [CrossRef]
88. Asakura, H.; Ogawa, H. Potential of heparin and nafamostat combination therapy for COVID-19. *J. Thromb. Haemost.* **2020**, *18*, 1521–1522. [CrossRef]
89. Andrews, K.T.; Klatt, N.; Adams, Y.; Mischnick, P.; Schwartz-Albiez, R. Inhibition of Chondroitin-4-Sulfate-Specific Adhesion of *Plasmodium falciparum* -Infected Erythrocytes by Sulfated Polysaccharides. *Infect. Immun.* **2005**, *73*, 4288–4294. [CrossRef]
90. Clark, D.L.; Su, S.; A Davidson, E. Saccharide anions as inhibitors of the malaria parasite. *Glycoconj. J.* **1997**, *14*, 473–479. [CrossRef]
91. Xiao, L.; Yang, C.; Patterson, P.S.; Udhayakumar, V.; A Lal, A. Sulfated polyanions inhibit invasion of erythrocytes by plasmodial merozoites and cytoadherence of endothelial cells to parasitized erythrocytes. *Infect. Immun.* **1996**, *64*, 1373–1378. [CrossRef]
92. Adams, Y.; Freeman, C.; Schwartz-Albiez, R.; Ferro, V.; Parish, C.R.; Andrews, K.T. Inhibition of *Plasmodium falciparum* Growth In Vitro and Adhesion to Chondroitin-4-Sulfate by the Heparan Sulfate Mimetic PI-88 and Other Sulfated Oligosaccharides. *Antimicrob. Agents Chemother.* **2006**, *50*, 2850–2852. [CrossRef]
93. Dong, X.; Zhang, X.; Wang, M.; Gu, L.; Li, J.; Gong, M. Heparin-decorated nanostructured lipid carriers of artemether-protoporphyrin IX-transferrin combination for therapy of malaria. *Int. J. Pharm.* **2021**, *605*, 120813. [CrossRef] [PubMed]
94. Smitskamp, H.; Wolthuis, F.H. New concepts in treatment of malignant tertian malaria with cerebral involvement. *Br. Med. J.* **1971**, *1*, 714–716. [CrossRef] [PubMed]
95. Jaroonvesama, N. INTRAVASCULAR COAGULATION IN FALCIPARUM MALARIA. *Lancet* **1972**, *1*, 221–223. [CrossRef]
96. Munir, M.; Tjandra, H.; Rampengan, T.H.; Mustadjab, I.; Wulur, F.H. Heparin in the treatment of cerebral malaria. *Paediatr. Indones.* **1980**, *20*, 47–50. [PubMed]
97. Rampengan, T.H. Cerebral malaria in children. Comparative study between heparin, dexamethasone and placebo. *Paediatr. Indones.* **1991**, *31*, 59–66. [PubMed]
98. Yu, L.; Gao, Y.; Yue, X.; Liu, S.; Dai, Z. Novel Hollow Microcapsules Based on Iron−Heparin Complex Multilayers. *Langmuir* **2008**, *24*, 13723–13729. [CrossRef] [PubMed]
99. Costalat, M.; Alcouffe, P.; David, L.; Delair, T. Controlling the complexation of polysaccharides into multi-functional colloidal assemblies for nanomedicine. *J. Colloid Interface Sci.* **2014**, *430*, 147–156. [CrossRef]
100. Xiong, G.M.; Yap, Y.Z.; Choong, C. Single-step synthesis of heparin-doped polypyrrole nanoparticles for delivery of angiogenic factor. *Nanomedicine* **2016**, *11*, 749–765. [CrossRef]
101. La, W.-G.; Yang, H.S. Heparin-Conjugated Poly(Lactic-Co-Glycolic Acid) Nanospheres Enhance Large-Wound Healing by Delivering Growth Factors in Platelet-Rich Plasma. *Artif. Organs* **2015**, *39*, 388–394. [CrossRef]
102. Tan, Q.; Tang, H.; Hu, J.; Hu, Y.; Zhou, X.; Tao, R.; Wu, Z. Controlled release of chitosan/heparin nanoparticle-delivered VEGF enhances regeneration of decellularized tissue-engineered scaffolds. *Int. J. Nanomed.* **2011**, *6*, 929–942. [CrossRef]
103. Mori, Y.; Nakamura, S.; Kishimoto, S.; Kawakami, M.; Suzuki, S.; Matsui, T.; Ishihara, M. Preparation and characterization of low-molecular-weight heparin/protamine nanoparticles (LMW-H/P NPs) as FGF-2 carrier. *Int. J. Nanomed.* **2010**, *5*, 147–155. [CrossRef] [PubMed]
104. Choi, J.U.; Kim, J.-Y.; Chung, S.W.; Lee, N.K.; Park, J.; Kweon, S.; Cho, Y.S.; Kim, H.R.; Lim, S.M.; Park, J.W.; et al. Dual mechanistic TRAIL nanocarrier based on PEGylated heparin taurocholate and protamine which exerts both pro-apoptotic and anti-angiogenic effects. *J. Control. Release* **2021**, *336*, 181–191. [CrossRef]

105. Yang, X.; Wang, Q.; Zhang, A.; Shao, X.; Liu, T.; Tang, B.; Fang, G. Strategies for sustained release of heparin: A review. *Carbohydr. Polym.* **2022**, *294*, 119793. [CrossRef]
106. Cossette, B.; Pelletier, M.; Carrier, N.; Turgeon, M.; LeClair, C.; Charron, P.; Echenberg, D.; Fayad, T.; Farand, P. Evaluation of Bleeding Risk in Patients Exposed to Therapeutic Unfractionated or Low-Molecular Weight Heparin: A Cohort Study in the Context of a Quality Improvement Initiative. *Ann. Pharmacother.* **2010**, *44*, 994–1002. [CrossRef] [PubMed]
107. Nieuwenhuis, H.K.; Albada, J.; Banga, J.D.; Sixma, J.J. Identification of risk factors for bleeding during treatment of acute venous thromboembolism with heparin or low molecular weight heparin. *Blood* **1991**, *78*, 2337–2343. [CrossRef] [PubMed]
108. Wu, W.; Wang, M.; Zhou, W.; Wang, Y. Heparin-induced thrombocytopenia with hematoma necrosis and persistent high fever after gastric cancer surgery: A case report. *Asian J. Surg.* **2020**, *43*, 387–388. [CrossRef]
109. Garcia, D.A.; Baglin, T.P.; Weitz, J.I.; Samama, M.M. Parenteral Anticoagulants: Antithrombotic Therapy and Prevention of Thrombosis, 9th ed: American College of Chest Physicians Evidence-Based Clinical Practice Guidelines. *Chest* **2012**, *141* (Suppl. 2), e24S–e43S. [CrossRef]
110. Pettilä, V.; Leinonen, P.; Markkola, A.; Hiilesmaa, V.; Kaaja, R. Postpartum Bone Mineral Density in Women Treated for Thromboprophylaxis with Unfractionated Heparin or LMW Heparin. *Thromb. Haemost.* **2002**, *87*, 182–186. [CrossRef]
111. Rajgopal, R.; Bear, M.; Butcher, M.K.; Shaughnessy, S.G. The effects of heparin and low molecular weight heparins on bone. *Thromb. Res.* **2008**, *122*, 293–298. [CrossRef]
112. Gajic-Veljanoski, O.; Phua, C.W.; Shah, P.S.; Cheung, A.M. Effects of Long-Term Low-Molecular-Weight Heparin on Fractures and Bone Density in Non-Pregnant Adults: A Systematic Review With Meta-Analysis. *J. Gen. Intern. Med.* **2016**, *31*, 947–957. [CrossRef]

molecules

MDPI

Article

Detailed Structural Analysis of the Immunoregulatory Polysaccharides from the Mycobacterium Bovis BCG

Lan Luo [1,†], Xuemei Song [1,2,†], Xiao Chang [1,†], Sheng Huang [3,4], Yunxi Zhou [3], Shengmei Yang [3], Yan Zhu [3], Lanyan Zhang [3], Yongsheng Wu [3,4], Junyin Zhang [1], Zhipeng Zhou [1] and Mingyi Wu [1,2,*]

1 State Key Laboratory of Phytochemistry and Plant Resources in West China, Kunming Institute of Botany, Chinese Academy of Sciences, Kunming 650201, China
2 University of Chinese Academy of Sciences, Beijing 100049, China
3 Hunan Jiuzhitang Siqi Biological Pharmaceutical Co., Ltd., Changsha 410329, China
4 Jiuzhitang Co., Ltd., Changsha 410205, China
* Correspondence: wumingyi@mail.kib.ac.cn
† These authors contributed equally to this work.

Abstract: Bacillus Calmette-Guérin polysaccharide and nucleic acid (BCG-PSN), extracted from *Mycobacterium bovis*, is an immunoregulatory medicine commonly used in clinic. However, the structural characteristics and potential pharmacological efficacy of the polysaccharides from BCG-PSN remain unclear. Herein, two polysaccharides (BCG-1 and BCG-2) were purified and their structures were characterized. Monosaccharide composition analysis combined with methylation analysis and NMR data indicated that BCG-1 and BCG-2 were an α-D-(1→4)-mannan with (1→2)-linked branches, and an α-D-(1→4)-glucan with (1→6)-linked branches, respectively. Herein, the mannan from BCG-PSN was first reported. Bioactivity assays showed that BCG-1 and BCG-2 dose-dependently and potently increased the production of inflammatory mediators (NO, TNF-α, IL-6, IL-1β, and IL-10), as well as their mRNA expressions in RAW264.7 cells; both have similar or stronger effects compared with BCG-PSN injection. These data suggest that BCG-1 and BCG-2 are very likely the active ingredients of BCG-PSN.

Keywords: active ingredients; polysaccharide purification; glucan; mannan; immunomodulator

Citation: Luo, L.; Song, X.; Chang, X.; Huang, S.; Zhou, Y.; Yang, S.; Zhu, Y.; Zhang, L.; Wu, Y.; Zhang, J.; et al. Detailed Structural Analysis of the Immunoregulatory Polysaccharides from the Mycobacterium Bovis BCG. *Molecules* **2022**, *27*, 5691. https://doi.org/10.3390/molecules27175691

Academic Editors: Jian Yin, Jing Zeng and De-Cai Xiong

Received: 6 August 2022
Accepted: 31 August 2022
Published: 3 September 2022

1. Introduction

Bacillus Calmette-Guérin polysaccharide and nucleic acid (BCG-PSN) from *Mycobacterium bovis* is an immunoregulatory medicine [1]. As a strong immunomodulator, BCG-PSN can orchestrate the immune system to resist virus infections and to ameliorate autoimmune diseases [2]. BCG-PSN injection is approved by State Food and Drug Administration (China) for the prevention and treatment of chronic bronchitis, asthma and cold, and commonly used in clinic. Additionally, BCG-PSN is also effective in treating recurrent oral aphthosis [3], idiopathic urticaria [4], melanoma [5], and lichen planus [6], and it may be an effective immunoadjuvant in some immunotherapy strategies [7].

Plenty of research is conducted to explore the underlying mechanism of action of BCG-PSN. BCG-PSN injection can elicit the Th1-mediated immune response by facilitating T cell differentiation to the Th1 subset in vivo, as well as activate toll-like receptor (TLR) signaling pathway, and induce the secretion of Th1-type cytokines, such as IL-2 and IFN-γ [1]. In patients with chronic urticaria, BCG-PSN can promote IL-2 production by Th1 and inhibit IL-10 production by Th2; it can also decrease the release of β-hexosaminidase and regulate IgE-mediated mast cell activation [4,8]. In pulmonary tuberculosis patients, BCG-PSN can significantly enhance the phagocytic capacity of alveolar macrophage [9]. Macrophages serve as the indispensable first line of defense in the innate immune system and participate in the adaptive immune system; upon activation, they can phagocytize intracellular pathogens or promote the immune response by generating nitric oxide (NO) and inflammatory cytokines such as TNF-α, IFN-γ, IL-10, and IL-6 [10].

BCG-PSN is mainly composed of two components (polysaccharides and nucleic acids), in which the high-content and bioactive polysaccharides are macromolecules with long chains and complex chemical structures [2]. Thus, the exact structure and pharmacological characteristics of polysaccharides in BCG-PSN remain to be clarified. To date, six companies have registered for producing BCG injection under the quality criterion of Chinese Pharmacopeia. However, quality control of the injection has been challenging during preparation.

Herein, we purified the polysaccharides from BCG-PSN, characterized their chemical structures by analyzing their monosaccharide composition, methylated derivatives, and NMR data, and studied their immunomodulatory activity by testing their effects on cytokine production and mRNA expression in RAW264.7 cells. These data further contribute to the understanding of the chemical structure and bioactivity of BCG-PSN polysaccharide components, including a mannan that has not been reported, which may provide reference data for the industrial quality control and the pharmacological mechanism of BCG-PSN injection.

2. Materials and Methods

2.1. Materials

The BCG-PSN powder and BCG-PSN injection were provided by Hunan Jiuzhitang Siqi Biological Pharmaceutical Co., Ltd., China. Sephadex G-100 and Sephadex G-75 were purchased from GE Healthcare. Monosaccharides D-(+)mannose (Man), D-(-)ribose (Rib), and D-(-)arabinose (Ara) were from Alfa Aeasr (AR, Haverhill, MA, USA). D-(+)glucose (Glc), D-(+)galactose (Gal), and lipopolysaccharide (LPS) were from Sigma-Aldrich (MO, USA). Trifluoroacetic acid (TFA), 1-phenyl-3-methyl-5-pyrazolone (PMP), and α-amylase (A109181) were from Aladdin (Shanghai, China). Dulbecco's Modified Eagle's Medium (DMEM) culture medium was from Shanghai BasalMedia (Shanghai, China). Fetal bovine serum (FBS), streptomycin, and penicillin were from Biological Industries (Kibbutz Beit-Haemek, Israel). Cell counting kit-8 (CCK-8) was from Biosharp (Hefei, China). NO detecting kit was from Beyotime (Shanghai, China). Mouse IL-10, IL-6, and IL-1β enzyme-linked immunosorbent assay (ELISA) kits were from Multisciences (Hangzhou, China). Mouse TNF-α ELISA kit was from R&D (Minneapolis, MN, USA). E.Z.N.A. Total RNA Kit II was from Omega Bio-tek (Norcross, GA, USA). 5 × All-In-One MasterMix (with AccuRT Genomic DNA Removal Kit) and Blastaq^{TM} 2 × qPCR MasterMix were from Applied Biological Materials (Richmond, VAN, Canada). Other reagents were all commercial and of analytical grade.

2.2. Isolation and Purification of Polysaccharides

BCG-PSN powder (200 mg) was dissolved in distilled water; after centrifugation, the supernatant was subjected to a DEAE-52 cellulose column (3.0 cm × 13.0 cm), followed by gradient elution with sodium chloride solution (0, 0.5, 3.9 M) at a flow rate of 2 mL/min, and the eluate (10 mL/tude) was collected in sequence [11]. Then, total polysaccharides without nucleic acid were eluted by distilled water and named BCG. Alternatively, polysaccharides were also extracted from the BCG strains, which were suspended in PBS (1.0 g/mL) supplemented with Tyloxapol (0.1%). The capsular polysaccharides were extracted by vigorous shaking and centrifugation (3500× *g* 15 min), and the intracellular polysaccharides by ultrasonic cell disruption and centrifugation (3500× *g* 10 min). The supernatants were collected and filtered (0.2 μm) to remove the remaining bacterial cells. Then the filtrate was concentrated and precipitated by adding ethanol to 80% (*v*/*v*), and centrifugated (3500× *g* 15 min) to obtain the crude polysaccharides (BCG), which were dissolved in deionized water, dialyzed, and lyophilized to get the white powder.

The crude polysaccharides were further separated by Sephadex G-75 column (2 cm × 120 cm) or Sephadex G-100 column (2 cm × 120 cm), and eluted with 0.1 M NaCl (contained 0.02% NaN_3) at a flow rate of 0.5 mL/min. The collected fractions were detected by the phenol-sulfuric acid method [12]. Based on the retention time in HPGPC, three polysaccharides (BCG-1/18.756 min, BCG-2/16.6032 min, and BCG-3/17.5032 min) were obtained, concentrated, and lyophilized. The retention time and NMR spectroscopy of the polysaccharides from BCG-PSN powder were consistent with that of polysaccharides from BCG strains. Our preliminary results showed that BCG-3 has a complicated structure. Its structural analysis requires plenty of work, thus, it will be reported alone, and the following study focuses on BCG-1 and BCG-2.

To test the presence and composition of α-glucan, BCG-PSN powder (2 g in 100 mL distilled water) was incubated with 0.8 mL α-amylase on a rotary shaker (200 rpm, pH 6.5, 70 °C) to degrade the α-glucan.

2.3. Homogeneity and Molecular Weight Determination

The molecular weights of BCG-1 and BCG-2 were determined by high-performance gel permeation chromatography (HPGPC) using an Agilent technologies 1260 series (Agilent, Santa Clara, CA, USA), which was equipped with RID, DAD detectors, and a Shodex OH-pak SB-804 HQ column (8 mm × 300 mm), using the eluent of 0.1 M NaCl solution (0.5 mL/min, 35 °C). The average molecular weight of BCG-1 and BCG-2 was calculated according to the standard curve of dextrans (2700, 5250, 9750, 13,050, 36,800, 64,650, 135,350 Da), using PL Cirrus GPC/SEC Software (Agilent, CA, USA), and the homogeneity was evaluated based on the HPGPC profile. Chromatographic conditions and procedures were performed according to the previous method [13].

Permethylated derivatization is a common method for in-depth analysis of glycans as it provides more mass spectra information in glycosidic linkages [14,15]. The conversion of glycans to hydrophobic derivatives such as permethylation enhances their signal strengths [14]. Thus, the polysaccharides BCG-1 and BCG-2 were permethylated using the KOH/dimethyl sulfoxide/methyl iodide for MALDI-TOF MS profiling [16–18].

The M_n and M_w were calculated from the equations [19]:

$$M_n = \left(\sum m_i N_i\right) / \left(\sum N_i\right) \tag{1}$$

$$M_w = \left(\sum m_i^2 N_i\right) / \left(\sum m_i N_i\right) \tag{2}$$

m_i and N_i represent mass and intensity of the ion i, taking into account the proportionality among intensity of the ith peak, N_i and the number of chains with mass m_i.

2.4. Qualitative Analysis of Monosaccharide Composition

Monosaccharide composition of the polysaccharide was analyzed by reverse-phase HPLC after PMP derivatization [2,18,20,21]. To hydrolyze polysaccharides, 300 μL polysaccharide solution (BCG-1 or BCG-2, 1 mg/mL in deionized water) was mixed with 300 μL TFA (4 M) at 110 °C for 4 h in the sealed COD tube. Then, the reaction solution was evaporated to dryness at 70 °C to remove the residual TFA. Then, 100 μL of the hydrolyzed sample solution was incubated with 100 μL sodium hydroxide (0.6 M) and 200 μL PMP (0.5 M in methanol) at 70 °C for 60 min. After reaction, the solution pH was adjusted to 7.0 by adding HCI (0.3 M). Then, the solution was added and mixed with 2 mL of chloroform, and the chloroform layer was discarded; this step was repeated for at least six times and the top aqueous layer was collected for HPLC analysis. The standard monosaccharides (Man, Rib, Ara, Glc, and Gal) were processed by the same procedures.

The analysis of the PMP-labelled monosaccharides was carried out using an Agilent technologies 1260 series (Agilent, CA, USA) which was equipped with DAD detectors and a ZORBAX SB C18 column (4.6 mm × 150 mm, 5 μm). Mobile phase A and B (v/v, 83:17) were ammonium acetate (0.1 M, pH 5.5) and acetonitrile, respectively, at a flow rate of 1 mL/min, and UV absorbance of the effluent was monitored at 250 nm.

2.5. Methylation and GC-MS Analysis

Glycosidic linkage of the polysaccharides was analyzed after methylation according to the literature with minor modification [22,23]. The polysaccharide BCG-1 or BCG-2 was incubated with dimethyl sulfoxide, NaOH, and methyl iodide under the protection of nitrogen and ultrasonic conditions for 1 h. The reaction solution was added with ultrapure water to decompose methyl iodide. Then, a methylated sample was extracted with trichloromethane and evaporated to dryness. Complete hydrolysis of the methylated polysaccharides was performed by heating at 120 °C with TFA (2 M) for 4 h. The methylated sample was reduced and acetylated for GC-MS analysis. The GC-MS analysis was performed on a HP6890GC/5973 MS system (Agilent, CA, USA) equipped with an ion trap MS detector and a DB-5MS quartz capillary column (30 m × 0.25 mm, 0.25 μm). The temperature program was set as follows: the initial temperature was 80 °C, increased to 250 °C at 5 °C/min, holding for 5 min; injection temperature was 270 °C; and the ion source of the mass spectrometer was set at 230 °C. The injection volume was 1 μL and injector split ratio was 10:1.

2.6. UV, IR, and NMR Analysis

The UV-vis absorption spectra of the polysaccharides were recorded using a UV-2600 spectrophotometer (Shimadzu, Kyoto, Japan) in the wavelength range of 190–800 nm. The Fourier transform infrared spectroscopy (FT-IR) spectra of the polysaccharides (KBr pellets) were recorded by a Tensor-27 (Bruker, Karlsruhe, Germany) in 400–4000 cm^{-1} at room temperature.

The polysaccharide sample was dissolved in D_2O for NMR analysis, which was performed in a Bruker Avance spectrometer of 600 or 800 MHz (Bruker, Karlsruhe, Germany), equipped with a $^{13}C/^{1}H$ dual probe in FT mode [22]. $^{1}H/^{1}H$ correlated spectroscopy (COSY), total correlation spectroscopy (TOCSY), rotating frame overhauser effect spectroscopy (ROESY), heteronuclear single-quantum correlation–total correlation spectroscopy (HSQC-TOCSY), and heteronuclear multiple bond coherence (HMBC) spectra were recorded using state-time proportion phase incrementation for quadrature detection in the indirect dimension. TMSP-2,2,3,3-D4 (D, 98%) was used as internal and external reference for the resonance measurements of BCG-1 and BCG-2, respectively.

2.7. In Vitro Immunomodulatory Activity Assay

2.7.1. Cell Culture

The RAW 264.7 cell line from National Collection of Authenticated Cell Cultures was cultured in DMEM medium supplemented with 10% FBS (*v*/*v*), penicillin (100 U/mL), and streptomycin (100 μg/mL), under humidified conditions with 5% CO_2 at 37 °C.

2.7.2. Cell Viability Assays

Cell viability was evaluated by CCK8 assay using RAW 264.7 cells. The cells in logarithmic growth phase were adjusted to a concentration of 7.5×10^4 cells/mL. Cell suspension was added to a 96-well plate and cultured at 37°C for 24 h. The cells were subsequently treated with LPS (100 ng/mL), BCG-PSN, BCG-1 or BCG-2 at different concentrations for another 24 h. The serum-free culture media was used as a control. Then, 10 μL of CCK8 solution was pipetted into each of the wells for the next four hours. Then the absorbance at 450 nm was determined by a Flexstation3 microplate reader (Molecular Device). The cell viability was calculated as following:

Cell viability (%) = $A_1/A_0 \times 100\%$, where A_1 was the absorbance of treated group and the A_0 was the absorbance of control group.

2.7.3. NO and Cytokine Production

The RAW264.7 cells (4×10^5 cells/mL) in logarithmic growth phase were seeded into 24-well plates (500 μL/well) and incubated for 24 h. The cells were subsequently treated with LPS (100 ng/mL), BCG-PSN injection (dilute 4 times with serum-free DMEM medium, total polysaccharides was at 87.5 μg/mL), BCG-1 or BCG-2 (1, 10, 50, and 100 μg/mL), or control (serum-free DMEM medium) for another 24 h, then the culture supernatant of cells was collected for subsequent analysis. The macrophage NO content was determined based on the Griess method [24], and the levels of TNF-α, IL-6, IL-1β, and IL-10 were detected by ELISA kits according to the manufacturer's protocols.

2.7.4. RT-qPCR Analysis

The RAW 264.7 cells with different treatments were collected and washed three times with cold PBS for RNA extraction. The total RNA was isolated using Total RNA kit II and reversed to cDNA using a 5 $\times$ All-In-One MasterMix kit (with AccuRT Genomic DNA Removal). Then, amplification of the cDNA was carried out in a total volume of 20 μL containing 10 μL Blastaq^{TM} 2 $\times$ qPCR MasterMix, 0.4 μL specific primer (10 μM), 2 μL cDNA, and 7.2 μL nuclease-free H_2O. PCR was performed in multiple cycles using a QuantStudio 7 Flex (Thermo Scientific, Waltham, MA, USA) with the following program: denaturation at 95 °C for 3 min, annealing for 15 s, and elongation at 60 °C for 1 min. The nucleotide sequences of primers were shown in (Table S1, in the supplementary materials). β-actin gene was used as the internal reference. The expression levels of mRNA were calculated by $2^{-\Delta\Delta Ct}$ method.

2.8. Statistical Analysis

Statistical analyses were performed using GraphPad Prism 9.0 software. All the data were presented as mean $\pm$ SD. Normal distribution was determined by Shapiro-Wilk tests. Data were analyzed using one-way analysis of variance (ANOVA) followed by Dunnett's test, p values less than 0.05 were considered statistically significant (* $p < 0.05$, ** $p < 0.01$, or *** $p < 0.001$).

3. Results

3.1. Isolation and Purification

In HPGPC profiles of both the BCG-PSN and total polysaccharide (BCG), a wide peak can be observed (Figure 1A and Figure S1, in the supplementary materials), suggesting that they contain distinct polysaccharides with different molecular weights. Three pure polysaccharides (BCG-1, BCG-2, and BCG-3) were further isolated from total polysaccharide BCG, their HPGPC profiles showed a single peak and symmetrical peak (Figure 1A), indicating that they are homogeneous polysaccharides. Furthermore, all the polysaccharide fractions (BCG, BCG-1, BCG-2, and BCG-3) had no absorption at 280 or 260 nm in UV spectra, indicating the absence of nucleic acids (Figure 1B). Additionally, in the RID profile of BCG-PSN, the peak in about 16 min disappeared after hydrolysis by α-amylase (Figure S1, in the supplementary materials); this indicates that BCG-PSN is rich in α-glucan as in previous reports [2,11]. The following study focused on polysaccharides BCG-1 and BCG-2 because our preliminary results, as well as the literature [25,26], showed that BCG-3 has a complicated structure (its structural analysis requires plenty of work and will be reported alone in future).

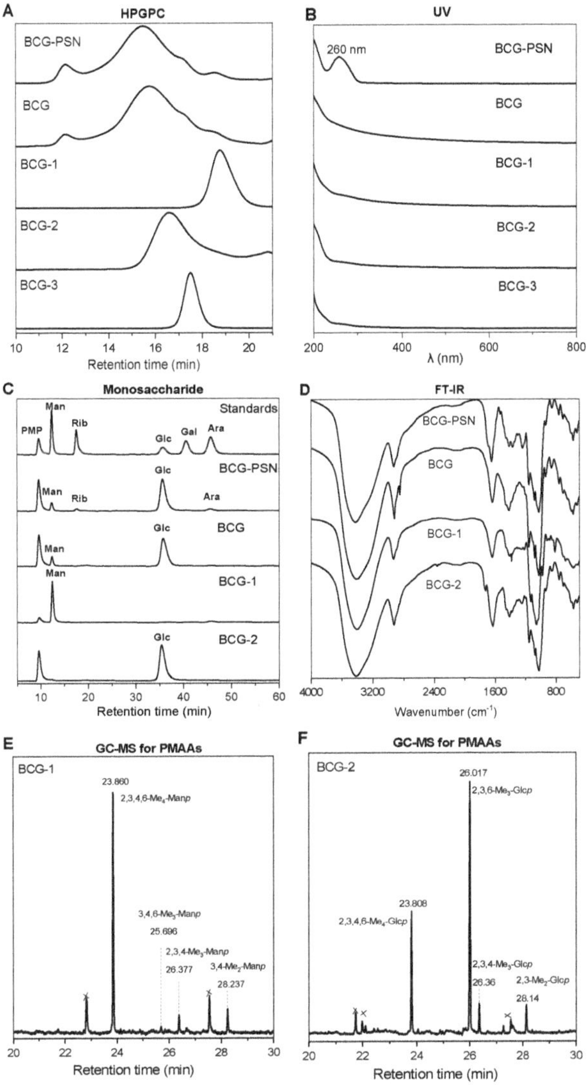

Figure 1. Purification and structural characterization of polysaccharides from BCG-PSN. (**A**) HPGPC profiles of BCG-PSN, BCG, BCG-1, BCG-2, and BCG-3; (**B**) UV absorption spectra of BCG-PSN, BCG, BCG-1, BCG-2, and BCG-3; (**C**) HPLC-DAD profiles of PMP derivatives of standard monosaccharides and the monosaccharides from polysaccharides; (**D**) FT-IR spectra of BCG-PSN, BCG, BCG-1, and BCG-2; (**E**) The TIC profile of PMAAs from the polysaccharide BCG-1; (**F**) The TIC profile of PMAAs from the polysaccharide BCG-2; noncarbohydrate signals are marked with ×.

3.2. Molecular Weight and Monosaccharide Composition Analysis

Based on the calibration curve of standard dextrans (LogM = 11.24 − 0.501 x + 0.006049 x^2 − 4.485e − 005x^3), the weight-average molecular weights (Mw) of BCG-1 and BCG-2 were determined to be 4600 Da and 23,588 Da, respectively (Table S2, in the supplementary materials). Their molecular weight distributions (Mw/Mn) were 1.14 and 1.49, respectively, indicating that both polysaccharides were homogeneous.

The molecular weights of these polysaccharides were also confirmed by MALDI-TOF-MS spectrometry analysis [18,19]. In the MALDI-TOF-MS profiles of permethylated BCG-1, a mass gap between two neighboring peaks was 204.1 Da (Figure S2, in the supplementary materials). The permethylated derivative was ionized as $[M + Na]^+$ ions (its relative error between experiment and theory is lower than that of $[M + K]^+$ [14]), and the mass was calculated as (219.1232 (a terminal sugar) + 204.0998 (2,4,6-trimethyl glucosyl unit) $\times$ n + Na^+31.0184 (OMe at reducing end residue)) Da [27] (Table S3, in the supplementary materials). The M_n and M_w of BCG-1 permethylated derivative were calculated to be 4158 Da and 5210 Da, respectively, the *Mw was* slightly higher than that measured by HPGPC (Table S4, in the supplementary materials). It is reported that the *Mw* upper limit of MALDI-TOF-MS spectrometry by current technology was about 5000 Da [19]. Due to the larger molecular weight of BCG-2, its signals were inadequate [15], hence further work is required to achieve optimal conditions to analyze BCG-2 [18].

Monosaccharide compositions of the polysaccharides were analyzed by reverse-phase HPLC after PMP precolumn derivatization (Figure 1C). BCG-PSN contained five monosaccharides (mannose, ribose, glucose, arabinose, and traces of galactose), among them glucose was in the highest content. Combined with the result of α-amylase hydrolysis, this indicates that BCG-PSN contains a large portion of glucan (Figure S1, in the supplementary materials). The total polysaccharide BCG was composed of mannose, glucose, and a small amount of arabinose, whereas, the purified polysaccharides BCG-1 and BCG-2 consisted exclusively of mannose and glucose, respectively (Figure 1C).

3.3. *FT-IR Analysis*

The functional groups of polysaccharides were analyzed by FT-IR spectroscopy. In the spectrum of BCG-1 (Figure 1D), the intense bands in the 3405 cm^{-1} region were ascribed to the stretching vibration of O-H, and those at 2933 cm^{-1} were produced by C-H stretching vibration [28]. The signal at 1640 cm^{-1} might be caused by the vibration of crystallized water [22,29]. A strong band between 975 and 1130 cm^{-1} was assigned to the stretching vibration of pyranose ring. The sharp peak at 814 cm^{-1} was attributed to the characteristic absorption of α-glycosidic bond of mannan [30,31], which was consistent with the results of monosaccharide composition.

Likewise, the FT-IR profile of the polysaccharide BCG-2 was also identified (Figure 1D). The signals at 3416 cm^{-1} and 2928 cm^{-1} were due to the stretching vibration of O-H and C-H, respectively. The peaks at around 1017, 1059, and 1150 cm^{-1} indicated that the monosaccharide in BCG-2 belonged to pyranose ring. Signals at 930, 840, and 760 cm^{-1} were ascribed to the characteristic absorption of α-D-glycosidic bond of glucose.

3.4. *Methylation Analysis*

For linkage analysis, methylation was conducted, and the total ion chromatograms (TIC) of the partially methylated alditol acetates (PMAAs) from BCG-1 and BCG-2 were studied. GC-MS analysis revealed that both BCG-1 and BCG-2 derivatives contained four PMAAs (Figure 1E,F). The data on glycosidic linkage and molar ratio were summarized in Table 1. According to the ESI-MS information (Figure S3, in the supplementary materials) and the GC-EIMS databases of PMAAs (Complex Carbohydrate Research Center, University of Georgia), the peaks in TIC of BCG-1 were identified as 1,5-di-*O*-acetyl-2,3,4,6-tetra-*O*- methyl-D-mannitol (2,3,4,6-Me_4-Man), 1,2,5-tri-*O*-acetyl-3,4,6-tri-O-methyl-D-mannitol (3,4,6-Me_3-Man), 1, 5,6-tri-*O*-acetyl-2,3,4-tri-*O*-methyl-D-mannitol (2,3,4-Me_3-Man), and 1,2,5,6-tri-*O*-acetyl-3,4-di-*O*-methyl-D-mannitol (3,4-Me_2-Man), with the molar ratio of 35.86:1.00:3.10:1.52 based on the peak areas (Table 1). The results indicated the presence of terminal 2-, 6-, 2,6-linked Man*p* pyranoside residues.

Table 1. GC-MS of alditol acetate derivatives of the BCG-1 and BCG-2 methylated products.

PMAA Derivatives [a]	Type of Linkage	Relative Retention Time [b]	Molar Ratio	Mass Fragments (m/z)
		BCG-1		
2,3,4,6-Me_4- Man*p*	t-Man*p*	1.00	35.86	71, 87, 101, 129, 145, 162, 205
3,4,6-Me_3-Man*p*	2- Man*p*	1.08	1.00	87, 129, 161, 189
2,3,4-Me_3-Man*p*	6- Man*p*	1.11	3.10	71, 87, 101, 118, 129, 161, 189
3,4-Me_2- Man*p*	2,6- Man*p*	1.18	3.52	87, 99, 129, 189
		BCG-2		
2,3,4,6-Me_4-Glc*p*	t-Glc*p*	1.00	5.16	87, 102, 118, 129, 145, 162, 205
2,3,6- Me_3- Glc*p*	4- Glc*p*	1.09	10.38	87, 100, 118, 129, 233
2,3,4-Me_3- Glc*p*	6- Glc*p*	1.17	1.13	87, 100, 118, 129, 161, 189
2,3-Me_2- Glc*p*	4,6- Glc*p*	1.18	1.00	100, 117, 129, 261

[a] 2,3,4,6-Me_4-Man = 1,5-di-*O*-acetyl-2,3,4,6-tetra-*O*-methyl-mannose, etc. [b] Relative retention times of the corresponding alditol acetate derivatives compared with 1,5-di-*O*-acetyl-2,3,4,6-tetra-*O*-methyl-D-mannitol (for BCG-1) and 1,5-di-*O*-acetyl-2,3,4,6-tetra-*O*-methyl-D-glucitol (for BCG-2).

Similarly, the PMAAs of BCG-2 were identified as 1,5-di-*O*-acetyl-2,3,4,6-tetra-*O*-methyl-D-glucitol (2,3,4,6-Me_4-Glc), 1,4,5-tri-*O*-acetyl-2,3,6-tri-*O*-methyl-D-glucitol (2,3,6-Me_3-Glc), 1,5,6-tri-*O*-acetyl- 2,3,4-tri-*O*-methyl-D-glucitol (2,3,4-Me_3-Glc), and 1,4,5,6-tri-*O*-acetyl-2,3-di-*O*-methyl-D-glucitol (2,3-Me_2-Glc), with the molar ratio of 5.16:10.38:1.13:1.00 (Figure 1F, Table 1, and Figure S4, in the supplementary materials). These results indicated the presence of terminal 4-,6-, 4,6-linked Glc*p* residues.

3.5. NMR Analysis

The structural features of BCG-1 and BCG-2 were further elucidated by ^{1}H and ^{13}C NMR analysis. Their chemical shifts were assigned according to 1D (^{1}H, ^{13}C) and 2D COSY, ROESY, HSQC-TOCSY, and HMBC data (Figures 2 and 3 and Table 2).

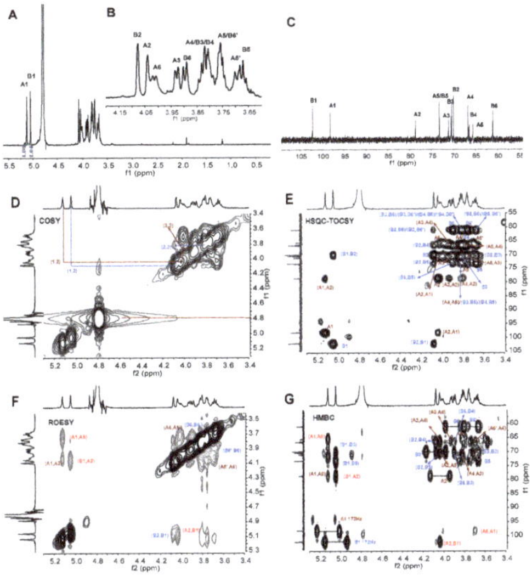

Figure 2. ^{1}H (**A**,**B**), ^{13}C (**C**), COSY (**D**), HSQC-TOCSY (**E**), ROESY (**F**), HMBC (**G**) NMR spectra of BCG-1. Chemical shifts are relative to internal trimethylsilylpropionic acid sodium at 0 ppm. The letters A (maroon) and B (blue) represented 2,6-*O*-α-Man*p* (Residue A) and t-α-Man*p* (Residue B), respectively.

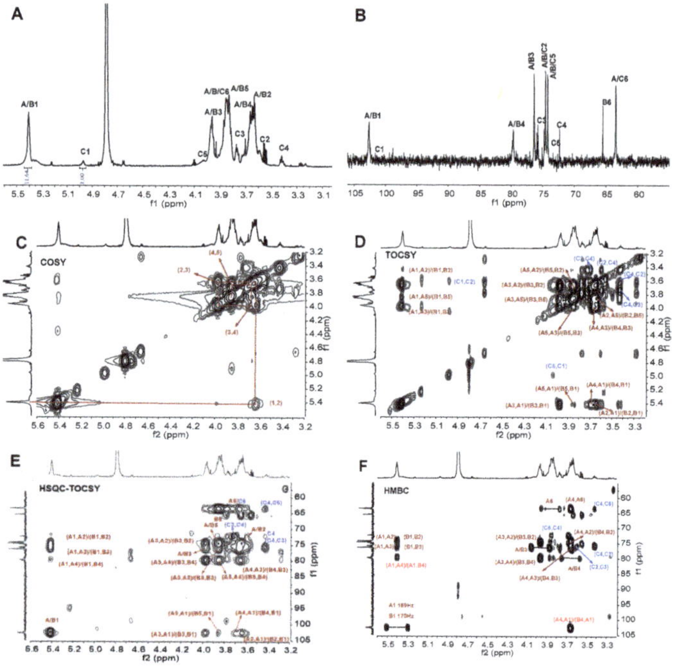

Figure 3. ^{1}H (**A**), ^{13}C (**B**), COSY (**C**), TOCSY (**D**), HSQC-TOCSY (**E**), HMBC (**F**) NMR spectra of BCG-2. Chemical shifts are relative to external trimethylsilylpropionic acid sodium at 0 ppm. The letters A/B (maroon) represented →4)-α-Glc→1 (residue A)/→4,6)-α-Glc(1→ (residue B). The letters C (blue) represented t-α-Glc (1→ (residue C).

Table 2. ^{1}H and ^{13}C NMR chemical shifts of the BCG-1 and BCG-2 in D_2O.

Sugar Residues		**Chemical Shifts (ppm)**						
		1	**2**	**3**	**4**	**5**	**6a**	**6b**
				BCG-1				
2,6-*O*-α-Man*p* (A)	H	5.12	4.04	3.94	3.82	3.77	4.01	3.68
	C	98.2	78.6	70.5	66.4	73.1	65.6	-
t-α-Man (B)	H	5.04	4.08	3.82	3.82	3.68	3.90	3.76
	C	102.2	70.0	70.6	66.4	72.8	61.0	-
				BCG-2				
→4)-α-Glc→1 (A)	H	5.41	3.63	3.96	3.66	3.86	3.88	3.76
	C	102.6	74.5	76.3	79.8	74.1	63.4	-
→4,6)-α-Glc(1→ (B)	H	5.39	3.63	3.96	3.66	3.83	3.86	3.85
	C	102.6	74.5	76.3	79.8	74.1	65.5	-
t-α-Glc(1→ (C)	H	4.96	3.61	3.71	3.41	4.04	3.85	
	C	101.6	74.5	75.7	72.3	73.4	63.4	

In the ^{1}H NMR spectrum of BCG-1, an anomeric zone (δ_H 4.9–5.5 ppm) exhibited two signals at 5.12 and 5.04 ppm (Figure 2A,B), which were unambiguously assigned as the anomeric proton resonance of residue A and B, respectively. The molar proportion of A and B residues was about 1:1.04 according to the signal integration. The ^{13}C NMR spectrum also displayed two signals in the anomeric region at 98.2 ppm and 102.2 ppm (Figure 2C), which were correlated with anomeric protons [31]. We thus deduced that BCG-1 may contain two mannosyl units, 2,6-*O*-α- Man*p* (Residue A), and t-α-D-Man*p* (Residue B). The detailed proton signals from H-2 to H-6 of the two residues were assigned using the ^{1}H-^{1}H COSY (Figure 2D) and HSQC-TOCSY (Figure 2E) data. The chemical shifts from δ_H

3.3 ppm to δ_H 4.0 ppm, showing overlapping peaks, were assigned to the protons of C-2 to C-6 in glycosidic rings [32,33]. For residue A, the typical signals of C-2 at ~78.6 ppm were shifted ~6 ppm to low-field of $\delta_{C\text{-}2}$ for α-mannose, and the H-2 signals at ~4.04 ppm were shifted ~0.2 ppm to $\delta_{H\text{-}2}$ for α-mannose, indicating that residue A was *O*-substituted at 2 position. In addition, the C-6 signals at ~65.6 ppm for residue A were shifted ~4.6 ppm to low-field of $\delta_{C\text{-}6}$, supporting the high substitution at C-6 of residue A. Combining the results of methylation analysis (Table 1) and NMR data, it suggested that BCG-1 has (1→2) and (1→6)-linkages. The linkages of the glycosyl residues were further confirmed by 2D ROESY (Figure 2F) and HMBC analysis (Figure 2G). The H-1 of residue A had a strong inter-residue ROE connected to H-6 of adjacent residue in addition to intra-residue ROE, indicating that residue A was linked to C-6 of adjacent residue. Similarly, the H-1 of residue B had a strong inter-residue ROE with H-2 of residue A, indicating that residue B was linked to C-2 of residue A. In addition, the HMBC spectrum (Figure 2G) also showed the sequence-defining C-1 of residue A and H-6 of adjacent residue. Accordingly, BCG-1 should have the linear backbone of →6-A-1→6-A-1→ and the side chain of B-1→2-A. To further confirm the configurations at the glycosidic linkages, the direct coupling constants ($^1J_{C\text{–}H}$) of C-1 of each saccharide from the HMBC spectrum were analyzed (Figure 2G). The large values of ~170 Hz for these mannose residues indicated that the protons at C-1 were equatorial [34], therefore the configurations of C-1 were determined to be α-D-mannose. This was consistent with the strongly specific rotation (+84°) and proton shifts of anomeric signals of mannan residue (Table 2). Taken together, the structure of BCG-1 was proposed in Figure 4A.

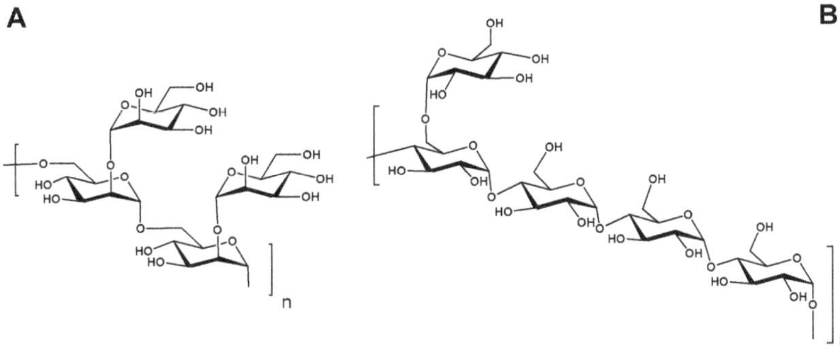

Figure 4. Proposed structures of BCG-1 (**A**) and BCG-2 (**B**) purified from BCG-PSN.

For BCG-2, its chemical shifts in 1D and 2D NMR spectra were also assigned (Table 2 and Figure 3). In the ^{1}H NMR spectrum (Figure 3A), the chemical shifts at 5.41 ppm, 5.39 ppm, and 4.96 ppm were anomeric proton signals of Residues A, B, and C, respectively [35], which corresponded to two low-field signals at 102.6 and 101.6 ppm in its ^{13}C spectrum (Figure 3B). The molar ratio of A and B to C was about 4:1, according to the integral area. The signals at δ_H 3.4–4.1 ppm were assigned to protons in other locations of saccharide rings [11,22,35,36]. Based on COSY and TOCSY spectra (Figure 3C,D), the chemical shifts of H2~H6 were in 3.63, 3.96, 3.66, 3.86, 3.88, and 3.76 ppm, respectively. The carbon signals were assigned based on the assignment of the protons in HSQC spectrum. The HSQC-TOCSY and HMBC spectra clearly showed the related signals between C-H on the saccharide ring (Figure 3E,F). In the ^{13}C NMR spectrum (Figure 3B), the downfield signals at 102.6 ppm and 101.6 ppm were attributed to anomeric carbons of A/B and C, respectively. The typical C-4 signals at ~79.8 ppm for A and B residues were shifted ~8 ppm to low-field of $\delta_{C\text{-}4}$, demonstrating that A and B residues were substituted at 4-position [22,35,37]. The signals of *O*-substituted C-6 and unsubstituted C-6 were at 65.4 ppm and 63.4 ppm, respectively [22]. Furthermore, the methylation analysis (Table 1) and HMBC spectrum (Figure 2G) indicated that BCG-2 has the (1→4) linked glucan main chain and (1→6)-linked branches (Figure 4B).

The branching point presented on the average of every tetrasaccharide fragment, indicating a high branching degree of the α-glucan. The structure proposed for BCG-2 was similar to that of the glucans from *M. bovis* [2,11] and *M. tuberculosis* [38], while BCG-2 exhibited different saccharide sequences from these reported glucans.

3.6. Effects of BCG-1 and BCG-2 on the NO Stimulation

To study the immunomodulatory activity of polysaccharides BCG-1 and BCG-2, their effects on NO production and inducible nitric oxide synthase (iNOS) expression were tested using RAW264.7 cells. NO is known as a signaling molecule released by activated macrophages, which is critical to defense against microbe invasion and tumor cells [39]. Additionally, iNOS is a crucial enzyme responsible for NO generation [40]. After stimulation with LPS or BCG-PSN injection, RAW264.7 cells significantly increased NO secretion, compared with control (Figure 5A). Both BCG-1 and BCG-2 dose-dependently increased NO production by macrophages, at 10 μg/mL or above their effects were significant, and at 50 μg/mL or above stronger than BCG-PSN. Consistently, BCG-1 and BCG-2 at 1~100 μg/mL also potently enhanced the mRNA expression of iNOS (Figure 5B). These results suggested that BCG-1 and BCG-2 can enhance NO production from macrophages.

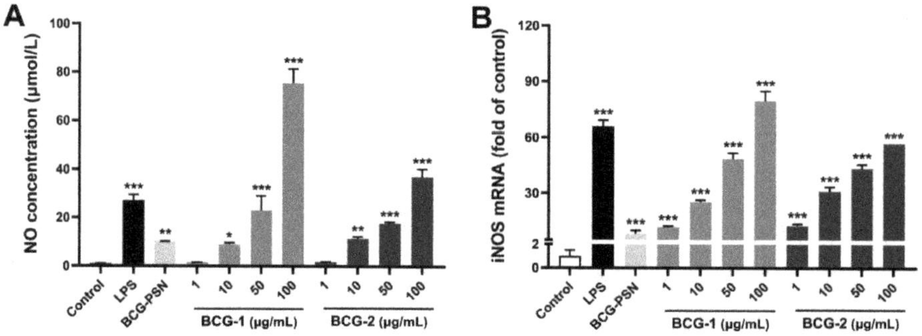

Figure 5. Effects of LPS, BCG-PSN, BCG-1, and BCG-2 on the production of NO (**A**) and mRNA expressions of *iNOS* (**B**) in RAW264.7 cells. LPS was at 100 ng/mL, BCG-PSN was at 87.5 μg/mL. * $p < 0.05$, ** $p < 0.01$, and *** $p < 0.001$ vs. control, one-way ANOVA, Dunnett's multiple comparisons test.

3.7. Effects of BCG-1 and BCG-2 on the Production of Cytokines

Activated macrophages can release inflammation cytokines, such as TNF-α, IL-6, IL-1β, and IL-10 [41]. In turn, the levels of these cytokines can be used to evaluate the activation degree of macrophage. At the tested concentration, BCG-1, BCG-2, and BCG-PSN injection could potently promote the release of TNF-α from RAW264.7 cells (Figure 6A), BCG-1 and BCG-2 at 1 μg/mL increased TNF-α to more than four folds of the control, comparable to the effect of LPS (100 ng/mL), and the production of TNF-α seemed to plateau without further increase with higher doses. BCG-1, BCG-2, and BCG-PSN injection also significantly increased the production of IL-6 to more than 1000 folds of the control, with the activity comparable to LPS (100 ng/mL) (Figure 6B). Although the activity of BCG-1 and BCG-2 in promoting IL-1β secretion was much weaker (exhibiting only significant effects at 10 μg/mL or above), it is also comparable to the potency of LPS (Figure 6C). Our results showed that BCG-1 and BCG-2 can also stimulate the production of IL-10 in a dose-dependent manner (Figure 6D). These results are consistent with previous reports of the immunomodulatory effects of polysaccharides [42–44]. TNF-α, IL-6, IL-1β, and IL-10 are essential immunomodulatory cytokines and play critical roles in innate and adaptive immune responses [45,46]. The results suggested that BCG-1 and BCG-2 exhibit potent immunomodulatory effects, comparable to or stronger than BCG-PSN, indicating that they may be the major active ingredients of BCG-PSN injection.

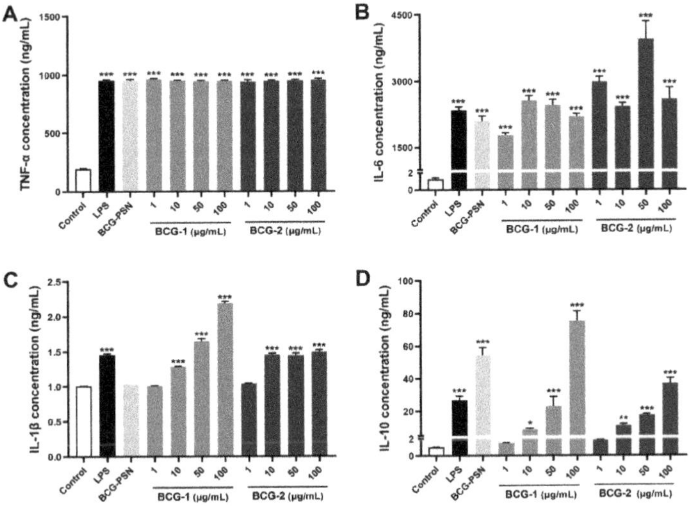

Figure 6. Effects of LPS, BCG-PSN, BCG-1, and BCG-2 on cytokines TNF-α (**A**), IL-6 (**B**), IL-1β (**C**), IL-10 (**D**) produced by the RAW 264.7 cells. LPS was at 100 ng/mL, BCG-PSN was at 87.5 μg/mL. * $p < 0.05$, ** $p < 0.01$, and *** $p < 0.001$ vs. control, one-way ANOVA, Dunnett's multiple comparisons test.

3.8. Effects of BCG-1 and BCG-2 on the Expression of Inflammation Genes

Since cytokine productions are related to their gene expressions, we next examined the effects of BCG-1 and BCG-2 on their relevant mRNA expression using RT-qPCR tests (Figure 7). Obviously, untreated RAW264.7 cells expressed low mRNA levels of all inflammatory cytokines (TNF-α, IL-6, IL-1β, and IL-10). After stimulation with BCG-PSN injection, the mRNA expression of cytokines TNF-α, IL-1β, and IL10 significantly increased except for IL-6. BCG-1 and BCG-2 at 1 μg/mL or above markedly up-regulated the expression levels of TNF-α and IL-1β, while it required 10 μg/mL or above to significantly increase IL-6 and IL-10. Taken together, it suggested that BCG-1 and BCG-2 increased the production of inflammation cytokines from RAW264.7 cells, at least partly by enhancing their expression at the transcriptional level.

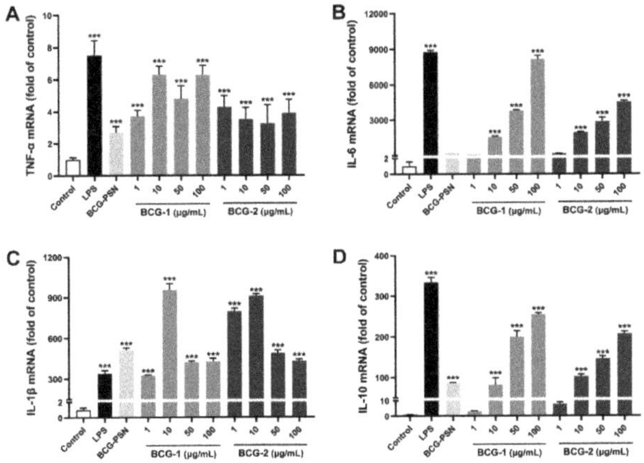

Figure 7. Effects of LPS, BCG-PSN, BCG-1, and BCG-2 on the mRNA expressions of TNF-α (**A**), IL-6 (**B**), IL-1β (**C**), IL-10 (**D**) in the RAW 264.7 cells. LPS was at 100 ng/mL, BCG-PSN was at 87.5 μg/mL. *** $p < 0.001$ vs. control, one-way ANOVA, Dunnett's multiple comparisons test.

In summary, the purified polysaccharides BCG-1 and BCG-2 showed obvious immunomodulatory effects by promoting the production of NO, TNF-α, IL-6, IL-1β, and IL-10 from macrophages, similar to BCG-PSN injection.

4. Discussion

Clinical trials have demonstrated that BCG-PSN is effective in the treatment of asthma, atopic dermatitis, chronic urticaria, oral, and cutaneous lichen planus [1,47]. The polysaccharide components are believed to be the active ingredients of BCG-PSN injection. Our present study indicated that BCG-PSN was composed of three polysaccharides, mostly α-D-glucan, a small amount of α-D-mannan, and arabinomannan. The chemical structures of two purified polysaccharides, BCG-1 and BCG-2, from BCG-PSN was characterized. The polysaccharide BCG-1 was an α-D-(1→4) mannan with (1→2)-linked branches, and BCG-2 was a glucan with α-D-(1→4)-linked backbone and (1→6)-linked branches.

In the past decades, lipomannan (LM), arabinomannan (AM), and mannosylated lipoarabinomannan (ManLAM) from *M. bovis* BCG were reported [25,26], while the structure–function relationship of these polysaccharides from BCG-PSN injection was less studied. A polysaccharide BDP and a water-soluble glucan (BCG-PASW) from BCG-PSN were reported [2,11], but their biological activity remains unclear. The present study has extended our knowledge in the structural diversity and immunostimulatory capability of polysaccharides from *Mycobacterium* species. Moreover, our results have important implications both for the understanding of active ingredients and quality control of BCG-PSN.

Macrophage activation is primarily regulated by recognizing pathogens through pattern recognition receptors such as Toll-like receptors (TLRs) [39,48]. TLRs could induce MyD88- and TRIF- dependent pathways, resulting in the activation of transcription factors NF-κB (nuclear factor kappa-B) and IRF3 (interferon regulatory factor 3), and the up-regulation of co-stimulatory markers and proinflammatory gene expression such as iNOS, TNF-α, IL-6, IL-1, and IL-8 [48,49]. Furthermore, TLR-MyD88 pathway is involved in dendritic cell maturation, thereby bridging the innate and adaptive immune response [50]. It is reported that BCG-PSN can enhance adaptive immunity via activating TLR signaling pathways and inducing the secretion of proinflammatory cytokines [1]. In our study, macrophages stimulated by BCG-1, BCG-2, or BCG-PSN injection increased the gene expression of proinflammatory cytokines TNF-α, IL-6, IL-1β, as well as inflammatory mediator iNOS (Figure 7). Consequently, these polysaccharides all substantially increased proinflammatory cytokines (NO, TNF-α, IL-6, and IL-1β) produced by macrophages (Figure 6). Therefore, the active ingredients in BCG-PSN, such as BCG-1 and BCG-2, may strengthen the immune responses, owing to activating the TLR pathway and inducing the production of inflammatory cytokines.

Additionally, BCG-PSN can induce macrophages to kill bladder cancer cells, possibly by expression of and synergy with Th1-stimulating cytokines [51,52]. IL-10 is a Th2 cytokine with anti-inflammatory and immunosuppressive properties; it may prevent the secretion of proinflammatory cytokines (TNF-α and IL-1β) and NO by macrophages [53,54]. Some studies have suggested that blockade of IL-10 may improve the therapeutic efficacy of the BCG vaccine in the treatment of bladder cancer patients [55,56]. Our in vitro data showed that although BCG-1 and BCG-2 increased IL-10 production, their activities were weaker than the positive control LPS. It is suggested that the polysaccharide chondroitin sulfate may possess both pro-inflammatory and anti-inflammatory effects [44]. However, the exact in vivo effect of these polysaccharides requires further study.

Taken together, BCG-1 and BCG-2 are most likely the functional components of BCG-PSN. However, given the difference of isolation, purification, and structural characterization, there may be other active ingredients that have not been purified and identified in BCG-PSN. For instance, in our study, we obtained another polysaccharide (arabinomannan, BCG-3), but its complex structure remains to be clarified in our future work. Since the bioactivity of polysaccharides depends on their structural features, different polysaccharide components or proportions may impact the immunomodulatory activity of BCG-PNS. Thus, it is essential to elucidate the structure–activity relationship of polysaccharides in BCG-PSN.

5. Conclusions

In conclusion, two polysaccharides, BCG-1 and BCG-2, were obtained from BCG-PSN. Chemical characteristic analysis revealed that BCG-1 was an α-D-(1→4) mannan with (1→2)-linked branches, and BCG-2 was an α-D-(1→4) glucan with (1→6)-linked branches. Both polysaccharides showed potent immunomodulatory effects, similar to BGC-PSN injection, significantly inducing the production of NO, TNF-α, IL-6, IL-1β, and IL-10, and their mRNA expression by RAW 264.7 cells. These results suggested that BCG-1 and BCG-2 were most likely the functional ingredients of BCG-PSN injection.

Supplementary Materials: The following supporting information can be downloaded at: https://www.mdpi.com/article/10.3390/molecules27175691/s1 Figure S1: HPGPC profiles of BCG-PSN before (A) and after (B) α-amylase hydrolysis, Figure S2: MALDI-TOF MS spectrum of BCG-1, Figure S3: MS fragments of PMAAs of BCG-1 and its deduced residues, Figure S4: MS fragments of PMAAs of BCG-2 and its deduced residues, Figure S5: Effects of BCG-PSN, BCG-1 and BCG-2 on the cell viability, Table S1: Primer sequences for RT-qPCR analysis, Table S2: Molecular weights of two polysaccharides, Table S3: MALDI-TOF MS data of BCG-1, Table S4: Weight average (Mw), number average (Mn) and polydispersity index (PI) of BCG-1 determined by HPGPC and MALDI-TOF MS.

Author Contributions: Methodology, Investigation, Writing-Original draft: L.L., X.S. and X.C.; Investigation, Supervision: S.H.; Conceptualization, Supervision, Funding acquisition: Y.Z. (Yunxi Zhou); Investigation: S.Y., Y.Z. (Yan Zhu) and L.Z.; Conceptualization, Supervision: Y.W.; Investigation, Writing-Reviewing and Editing: J.Z.; Investigation: Z.Z.; Conceptualization, Supervision, Writing-Reviewing and Editing, Funding acquisition, Data curation: M.W. All authors have read and agreed to the published version of the manuscript.

Funding: This work was funded by the Special Funding for the Construction of Innovative Provinces in Hunan Province (2020SK2043), Youth Innovation Promotion Association of Chinese Academy of Sciences (Y2021104), CAS "Light of West China" Program (E1264211W1), the Ten-thousand Talents Program of Yunnan Province (YNWR-QNBJ-2018-271), and Yunnan Fundamental Research Projects (grant NO. 202201AS070073 and 202101AT070152).

Institutional Review Board Statement: Not applicable.

Informed Consent Statement: Not applicable.

Data Availability Statement: Not applicable.

Conflicts of Interest: The authors state no conflict of interest.

Abbreviations

BCG-PSN: Bacillus Calmette-Guérin polysaccharide and nucleic acid; COSY, correlated spectroscopy; DMEM, dulbecco's modified eagle's medium; ESI-MS, electrospray ionization mass spectrometry; FTIR, fourier-transform infrared spectroscopy; GC-MS, gas chromatograph-mass spectrometer; HMBC, heteronuclear multiple bond coherence; HPGPC, high-performance gel permeation chromatography; HPLC, high performance liquid chromatography; HSQC–TOCSY, heteronuclear single-quantum correlation–total correlation spectroscopy; IL, interleukin; IR, infrared; iNOS, Inducible nitric oxide synthase; LPS, lipopolysaccharide; MALDI-TOF MS, matrix-assisted laser desorption/ionization time of flight mass spectrometry; Mn, number-average molecular weight; Mw, weight-average molecular weight; MyD88, myeloid differentiation factor 88; NMR, nuclear magnetic resonance; PMAAs, partially methylated alditol acetates; PMP, 1-phenyl-3-methyl-5-pyrazolone; qPCR, quantitative PCR; ROESY, rotating frame overhauser effect spectroscopy; RT-PCR, real-time polymerase chain reaction; TFA, trifluoroacetic acid; TIC, total ion chromatograms; TLR, toll-like receptor; TNF-α, tumor necrosis factor-α; TOCSY, total correlation spectroscopy.

References

1. Sun, J.; Hou, J.; Li, D.; Liu, Y.; Hu, N.; Hao, Y.; Fu, J.; Hu, Y.; Shao, Y. Enhancement of HIV-1 DNA vaccine immunogenicity by BCG-PSN, a novel adjuvant. *Vaccine* **2013**, *31*, 472–479. [CrossRef]
2. Liu, W.; Wang, H.; Yu, J.; Liu, Y.; Lu, W.; Chai, Y.; Liu, C.; Pan, C.; Yao, W.; Gao, X. Structure, chain conformation, and immunomodulatory activity of the polysaccharide purified from Bacillus Calmette Guerin formulation. *Carbohydr. Polym.* **2016**, *150*, 149–158.
3. Sharquie, K.E.; Hayani, R.K. BCG as a new therapeutic and prophylactic agent in patients with severe oral aphthosis. *Clin. Exp. Rheumatol.* **2005**, *23*, 914.
4. Li, N.; Cao, N.; Niu, Y.-D.; Bai, X.-H.; Lu, J.; Sun, Y.; Yu, M.; Sun, L.-X.; Duan, X.-S. Effects of the polysaccharide nucleic acid fraction of bacillus Calmette-Guérin on the production of interleukin-2 and interleukin-10 in the peripheral blood lymphocytes of patients with chronic idiopathic urticaria. *Biomed. Rep.* **2013**, *1*, 713–718. [PubMed]
5. Sloot, S.; Rashid, O.M.; Sarnaik, A.A.; Zager, J.S. Developments in intralesional therapy for metastatic melanoma. *Cancer Control* **2016**, *23*, 12–20. [CrossRef] [PubMed]
6. Gharib, K.; Kandil, A.; Marie, A.; Mounir, H. Efficacy of intralesional Bacillus Calmette-Guérin polysaccharide nucleic acid and vitamin D injections in the treatment of *Lichen Planus*. *J. Clin. Aesthet. Dermatol.* **2021**, *14*, 40–45.
7. Zhang, F.; Shi, L.; Liu, P.; Zhang, L.; Wu, Q.; Wang, B.; Zhang, G.; Wang, P.; Zhou, F.; Chen, W.R.; et al. A novel cosmetic and clinically practicable laser immunotherapy for facial verruca plana: Intense pulsed light combined with BCG-PSN. *Photodiagnosis Photodyn. Ther.* **2018**, *22*, 86–90. [CrossRef]
8. Yan, S.; Liu, R.; Mao, M.; Liu, Z.; Zhang, W.; Zhang, Y.; Li, J.; Peng, C.; Chen, X. Therapeutic effect of Bacillus Calmette—Guérin polysaccharide nucleic acid on mast cell at the transcriptional level. *PeerJ* **2019**, *7*, e7404. [CrossRef] [PubMed]
9. Li, M.C.; Zhang, H.S.; Tang, W.X.; Deng, F. Effects of BCG-PSN on pulmonary alveolar macrophage function in pulmonary tuberculosis patients. *Acta Acad. Med. Mil. Tertiae* **2004**, *26*, 1762–1764.
10. Beutler, B. Innate immunity: An overview. *Mol. Immunol.* **2004**, *40*, 845–859. [CrossRef]
11. Xu, X.; Gu, Z.X.; Liu, S.; Gao, N.; He, X.Z.; Xin, X. Purification and characterization of a glucan from Bacillus Calmette Guérin and the antitumor activity of its sulfated derivative. *Carbohydr. Polym.* **2015**, *128*, 138–146. [CrossRef] [PubMed]
12. Dubois, M.; Gilles, K.A.; Hamilton, J.K.; Rebers, P.A.; Smith, F. Colorimetric method for determination of sugars and related substances. *Anal. Chem.* **1956**, *28*, 350–356. [CrossRef]
13. Wu, Y.; Zhou, Z.; Luo, L.; Tao, M.; Chang, X.; Yang, L.; Huang, X.; Hu, L.; Wu, M. A non-anticoagulant heparin-like snail glycosaminoglycan promotes healing of diabetic wound. *Carbohydr. Polym.* **2020**, *247*, 116682. [CrossRef]
14. Zaia, J. Mass spectrometry of oligosaccharides. *Mass Spectrom. Rev.* **2004**, *23*, 161–227. [CrossRef] [PubMed]
15. Hung, W.-T.; Wang, S.-H.; Chen, Y.-T.; Yu, H.-M.; Chen, C.-H.; Yang, W.-B. MALDI-TOF MS analysis of native and permethylated or benzimidazole-derivatized polysaccharides. *Molecules* **2012**, *17*, 4950–4961. [CrossRef]
16. Ciucanu, I.; Kerek, F. A simple and rapid method for the permethylation of carbohydrates. *Carbohydr. Res.* **1984**, *131*, 209–217. [CrossRef]
17. Zhang, J.; Khoo, K.-H.; Wu, S.-W.; Chatterjee, D. Characterization of a distinct arabinofuranosyltransferase in Mycobacterium smegmatis. *J. Am. Chem. Soc.* **2007**, *129*, 9650–9662. [CrossRef] [PubMed]
18. Zhang, J.; Chen, H.; Luo, L.; Zhou, Z.; Wang, Y.; Gao, T.; Yang, L.; Peng, T.; Wu, M. Structures of fructan and galactan from *Polygonatum cyrtonema* and their utilization by probiotic bacteria. *Carbohydr. Polym.* **2021**, *267*, 118219. [CrossRef]
19. López-García, M.; García, M.S.D.; Vilariño, J.M.L.; Rodríguez, M.V.G. MALDI-TOF to compare polysaccharide profiles from commercial health supplements of different mushroom species. *Food Chem.* **2016**, *199*, 597–604. [CrossRef]
20. Fu, D.T.; ONeill, R.A. Monosaccharide composition analysis of oligosaccharides and glycoproteins by high-performance liquid chromatography. *Anal. Biochem.* **1995**, *227*, 377–384. [CrossRef]
21. Honda, S.; Akao, E.; Suzuki, S.; Okuda, M.; Kakehi, K.; Nakamura, J. High-performance liquid chromatography of reducing carbohydrates as strongly ultraviolet-absorbing and electrochemically sensitive 1-phenyl-3-methyl-5-pyrazolone derivatives. *Anal. Biochem.* **1989**, *180*, 351–357. [CrossRef]
22. Liu, J.; Shang, F.; Yang, Z.; Wu, M.; Zhao, J. Structural analysis of a homogeneous polysaccharide from *Achatina fulica*. *Int. J. Biol. Macromol.* **2017**, *98*, 786–792. [CrossRef] [PubMed]
23. Sims, I.M.; Carnachan, S.M.; Bell, T.J.; Hinkley, S.F.R. Methylation analysis of polysaccharides: Technical advice. *Carbohydr. Polym.* **2018**, *188*, 1–7. [CrossRef] [PubMed]
24. Griess, P. bemerkungen zu der abhandlung der hh. weselsky und benediktueber einige azoverbindungen. *Ber. Dtsch. Chem. Ges.* **1879**, *12*, 426–428. [CrossRef]
25. Angala, S.K.; Palčeková, Z.; Belardininelli, J.M.; Jackson, M. Covalent modifivations of polysaccharides in mycobacteria. *Nat. Chem. Biol.* **2018**, *14*, 193–198. [CrossRef]
26. Venisse, A.; Rivière, M.; Vercauteren, J.; Puzo, G. Structural analysis of the mannan region of lipoarabinomannan from *Mycobacterium bovis* BCG: Heterogeneity in phosphorylation state. *J. Biol. Chem.* **1995**, *270*, 15012–15021. [CrossRef]
27. Hung, W.-T.; Wang, S.-H.; Chen, C.-H.; Yang, W.-B. Structure determination of beta-glucans from *Ganoderma lucidum* with matrix-assisted laser desorption/ionization (MALDI) mass spectrometry. *Molecules* **2008**, *13*, 1538–1550. [CrossRef]
28. Luo, L.; Wu, M.; Xu, L.; Lian, W.; Xiang, J.; Lu, F.; Gao, N.; Xiao, C.; Wang, S.; Zhao, J. Comparison of physicochemical characteristics and anticoagulant activities of polysaccharides from three sea cucumbers. *Mar. Drugs* **2013**, *11*, 399–417. [CrossRef]

29. Lijour, Y.; Gentric, E.; Deslandes, E.; Guezennec, J. Estimation of the sulfate content of hydrothermal vent bacterial polysaccharides by fourier-transform infrared-spectroscopy. *Anal. Biochem.* **1994**, *220*, 244–248. [CrossRef]
30. Kath, F.; Kulicke, W.-M. Polymer analytical characterization of glucan and mannan from yeast *Saccharomyces cerevisiae*. *Macromol. Mater. Eng.* **1999**, *268*, 69–80.
31. Liu, Y.; Huang, G.; Lv, M. Extraction, characterization and antioxidant activities of mannan from yeast cell wall. *Int. J. Biol. Macromol.* **2018**, *118*, 952–956. [CrossRef] [PubMed]
32. Gilleron, M.; Nigou, J.; Cahuzac, B.; Puzo, G. Structural study of the lipomannans from *Mycobacterium bovis* BCG: Characterisation of multiacylated forms of the phosphatidyl-myo-inositol anchor. *J. Mol. Biol.* **1999**, *285*, 2147–2160. [CrossRef]
33. Kosaka, A.; Aida, M.; Katsumoto, Y. Reconsidering the activation entropy for anomerization of glucose and mannose in water studied by NMR spectroscopy. *J. Mol. Struct.* **2015**, *1093*, 195–200. [CrossRef]
34. Bubb, W.A. NMR spectroscopy in the study of carbohydrates: Characterizing the structural complexity. *Concepts Magn. Reson. Part A* **2003**, *19A*, 1–19. [CrossRef]
35. Dinadayala, P.; Lemassu, A.; Granovski, P.; Cérantola, S.; Winter, N.; Daffé, M. Revisiting the structure of the anti-neoplastic glucans of *Mycobacterium bovis* Bacille Calmette-Guérin: Structural analysis of the extracellular and boiling water extract-derived glucans of the vaccine substrains. *J. Biol. Chem.* **2004**, *279*, 12369–12378. [CrossRef]
36. Jarlier, V.; Nikaido, H. Mycobacterial cell wall: Structure and role in natural resistance to antibiotics. *FEMS Microbiol. Lett.* **1994**, *123*, 11–18. [CrossRef]
37. Lemassu, A.; Daffé, M. Structural features of the exocellular polysaccharides of *Mycobacterium tuberculosis*. *Biochem. J.* **1994**, *297*, 351–357. [CrossRef]
38. Ortalo-Magne, A.; Dupont, M.-A.; Lemassu, A.; Andersen, A.B.; Gounon, P.; Daffe, M. Molecular composition of the outermost capsular material of the tubercle bacillus. *Microbiology* **1995**, *141*, 1609–1620. [CrossRef]
39. Bogdan, C. Nitric oxide synthase in innate and adaptive immunity: An update. *Trends Immunol.* **2015**, *36*, 161–178. [CrossRef]
40. Aktan, F. iNOS-mediated nitric oxide production and its regulation. *Life Sci.* **2004**, *75*, 639–653. [CrossRef]
41. Peasura, N.; Laohakunjit, N.; Kerdchoechuen, O.; Vongsawasdi, P.; Chao, L.K. Assessment of biochemical and immunomodulatory activity of sulphated polysaccharides from *Ulva intestinalis*. *Int. J. Biol. Macromol.* **2016**, *91*, 269–277. [CrossRef] [PubMed]
42. Fang, Q.; Wang, J.-F.; Zha, X.-Q.; Cui, S.-H.; Cao, L.; Luo, J.-P. Immunomodulatory activity on macrophage of a purified polysaccharide extracted from *Laminaria japonica*. *Carbohydr. Polym.* **2015**, *134*, 66–73. [CrossRef] [PubMed]
43. Sun, H.; Zhang, J.; Chen, F.; Chen, X.; Zhou, Z.; Wang, H. Activation of RAW264.7 macrophages by the polysaccharide from the roots of *Actinidia eriantha* and its molecular mechanisms. *Carbohydr. Polym.* **2015**, *121*, 388–402. [CrossRef] [PubMed]
44. Wu, F.; Zhou, C.; Zhou, D.; Ou, S.; Liu, Z.; Huang, H. Immune-enhancing activities of chondroitin sulfate in murine macrophage RAW 264.7 cells. *Carbohydr. Polym.* **2018**, *198*, 611–619. [CrossRef]
45. Moore, K.W.; de Waal Malefyt, R.; Coffman, R.L.; O'Garra, A. Interleukin-10 and the interleukin-10 receptor. *Annu. Rev. Immunol.* **2001**, *19*, 683–765. [CrossRef]
46. Commins, S.P.; Borish, L.; Steinke, J.W. Immunologic messenger molecules: Cytokines, interferons, and chemokines. *J. Allergy Clin. Immunol.* **2010**, *125*, S53–S72. [CrossRef]
47. Wang, L.-H.; Ye, Y.; Zhang, Y.-Q.; Xiao, T. Curative effect of BCG-polysaccharide nuceic acid on atopic dermatitis in mice. *Asian Pac. J. Trop. Med.* **2014**, *7*, 913–917. [CrossRef]
48. Akira, S.; Uematsu, S.; Takeuchi, O. Pathogen recognition and innate immunity. *Cell* **2006**, *124*, 783–801. [CrossRef]
49. Medzhitov, R. Toll-like receptors and innate immunity. *Nat. Rev. Immunol.* **2001**, *1*, 135–145. [CrossRef]
50. Iwasaki, A.; Medzhitov, R. Toll-like receptor control of the adaptive immune responses. *Nat. Immunol.* **2004**, *5*, 987–995. [CrossRef]
51. Luo, Y.; Yamada, H.; Evanoff, D.P.; Chen, X. Role of Th1-stimulating cytokines in bacillus Calmette—Guérin (BCG)-induced macrophage cytotoxicity against mouse bladder cancer MBT-2 cells. *Clin. Exp. Immunol.* **2006**, *146*, 181–188. [CrossRef] [PubMed]
52. Yamada, H.; Kuroda, E.; Matsumoto, S.; Matsumoto, T.; Yamada, T.; Yamashita, U. Prostaglandin E2 down-regulates viable Bacille Calmette-Guérin-induced macrophage cytotoxicity against murine bladder cancer cell MBT-2 in vitro. *Clin. Exp. Immunol.* **2002**, *128*, 52–58. [CrossRef] [PubMed]
53. Cenci, E.; Romani, L.; Mencacci, A.; Spaccapelo, R.; Schiaffella, E.; Puccetti, P.; Bistoni, F. Interleukin-4 and interleukin-10 inhibit nitric oxide-dependent macrophage killing of *Candida albicans*. *Eur. J. Immunol.* **1993**, *23*, 1034–1038. [CrossRef]
54. Fiorentino, D.F.; Zlotnik, A.; Mosmann, T.R.; Howard, M.; O'Garra, A. IL-10 inhibits cytokine production by activated macrophages. *J. Immunol.* **1991**, *147*, 3815–3822. [PubMed]
55. Luo, Y.; Yamada, H.; Chen, X.; Ryan, A.A.; Evanoff, D.P.; Triccas, J.A.; O'Donnell, M.A. Recombinant *Mycobacterium bovis* bacillus Calmette—Guérin (BCG) expressing mouse *IL-18* augments Th1 immunity and macrophage cytotoxicity. *Clin. Exp. Immunol.* **2004**, *137*, 24–34. [CrossRef] [PubMed]
56. Luo, Y.; Han, R.; Evanoff, D.P.; Chen, X. Interleukin-10 inhibits *Mycobacterium bovis* bacillus Calmette—Guérin (BCG)-induced macrophage cytotoxicity against bladder cancer cells. *Clin. Exp. Immunol.* **2010**, *160*, 359–368. [CrossRef]

molecules

MDPI

Article

Hyaluronic Acid Oligosaccharide Derivatives Alleviate Lipopolysaccharide-Induced Inflammation in ATDC5 Cells by Multiple Mechanisms

Hesuyuan Huang [1,2,†], Xuyang Ding [3,†], Dan Xing [1,2], Jianjing Lin [1,2], Zhongtang Li [3,*] and Jianhao Lin [1,2,*]

1 Arthritis Clinic & Research Center, Peking University People's Hospital, Peking University, Beijing 100044, China
2 Arthritis Institute, Peking University, Beijing 100044, China
3 State Key Laboratory of Natural and Biomimetic Drugs, School of Pharmaceutical Sciences, Peking University, Beijing 100191, China
* Correspondence: lizhongtang@bjmu.edu.cn (Z.L.); linjianhao@pkuph.edu.cn (J.L.)
† These authors contributed equally to this work.

Abstract: High molecular weight hyaluronic acids (HMW-HAs) have been used for the palliative treatment of osteoarthritis (OA) for decades, but the pharmacological activity of HA fragments has not been fully explored due to the limited availability of structurally defined HA fragments. In this study, we synthesized a series glycosides of oligosaccharides of HA (o-HAs), hereinafter collectively referred to as o-HA derivatives. Their effects on OA progression were examined in a chondrocyte inflammatory model established by the lipopolysaccharide (LPS)-challenged ATDC5 cells. Cell Counting Kit-8 (CCK-8) assays and reverse transcription-quantitative polymerase chain reaction (RT-qPCR) showed that o-HA derivatives ($\leq$100 μg/mL) exhibited no cytotoxicity and pro-inflammatory effects. We found that the o-HA and o-HA derivatives alleviated LPS-induced inflammation, apoptosis, autophagy and proliferation-inhibition of ATDC5 cells, similar to the activities of HMW-HAs. Moreover, Western blot analysis showed that different HA derivatives selectively reversed the effects of LPS on the expression of extracellular matrix (ECM)-related proteins (MMP13, COL2A1 and Aggrecan) in ATDC5 cells. Our study suggested that o-HA derivatives may alleviate LPS-induced chondrocyte injury by reducing the inflammatory response, maintaining cell proliferation, inhibiting apoptosis and autophagy, and decreasing ECM degradation, supporting a potential oligosaccharides-mediated therapy for OA.

Keywords: osteoarthritis; hyaluronic acid; hyaluronic acid oligosaccharide derivatives; ATDC5; lipopolysaccharide

check for updates

Citation: Huang, H.; Ding, X.; Xing, D.; Lin, J.; Li, Z.; Lin, J. Hyaluronic Acid Oligosaccharide Derivatives Alleviate Lipopolysaccharide-Induced Inflammation in ATDC5 Cells by Multiple Mechanisms. *Molecules* **2022**, 27, 5619. https://doi.org/10.3390/molecules27175619

Academic Editors: Jian Yin, William Blalock, Jing Zeng and De-Cai Xiong

Received: 23 May 2022
Accepted: 17 August 2022
Published: 31 August 2022

1. Introduction

OA is a chronic disorder characterized by articular cartilage degradation, periarticular bone sclerosis, cysts, and inflammation. It is a major cause of inactivity, pain, and disability in middle-aged and elderly populations globally [1–3]. The disease affects the whole articular organ and seriously increases the public health burden, and has no effective treatment except for terminal patients to undergo joint replacement surgery [4–6]. OA also can affect multiple tissues surrounding the joints including articular cartilage, and chondrocytes are the major cellular component in the cartilage tissues [7,8]. The shift of chondrocyte phenotypes has been well documented to be involved in the onset and development of OA [8,9].

Lipopolysaccharide (LPS), a crucial metabolite and biomarker of gut microbiota, has been identified to cause the inflammation of OA [10,11]. ATDC5, a mouse teratocarcinoma-derived chondrogenic cell line, is regarded as an excellent in vitro cell model to explore the molecular mechanisms underlying chondrocyte biology, chondrogenesis, and skeletal development because it can recapitulate the main aspects of chondrocyte differentiation, cartilage extracellular matrix (ECM)-processing machinery, and synthesis of cartilage matrix [12–14]. LPS-treated ATDC5 cells have been used as an in vitro model to investigate the molecular basis of OA-related cartilage damage [15,16].

HA is a long-chain polysaccharide found in connective tissue, skin, eyes, cartilage, bone and synovial fluid. It consists of repeated disaccharide units [17,18]. In the joint, HA mainly concentrates on the surface of articular cartilage and synovial superficial layer, forming a semi-permeable barrier cartilage between synovial fluid and the synovial membrane. HA depletion is one of the pathologic factors of OA [19,20]. Moreover, the decrease of HA concentration and molecular weight (MW) in the joints is related to OA onset and progression [21,22]. Certain exogenous HA-based formulations have been demonstrated to be effective in the amelioration of OA-related symptoms (e.g., pain) by several mechanisms including the increase of HA content, inhibition of pro-inflammatory responses, and restoration of viscoelasticity of intra-articular fluid [22–24]. The molecular weight of HA ranges from 10^3 to 10^7 Da, and the complete hyaluronic acid chain can contain up to 25,000 disaccharide repeat units. Based on molecular weight, HA can be divided into several groups: HA oligosaccharide (o-HA), low molecular weight HA (LMW-HA), medium molecular weight HA (MMW-HA), high molecular weight HA (HMW-HA) and very high molecular weight hyaluronic acid (vHMW-HA) [25].

The size of HA decreases under pathological conditions, which increases the mechanical load of cartilage and induces cartilage damage [26,27]. In previous studies, HA was found to reduce the rate of chondrocyte apoptosis in OA patients [28]. In addition, it may promote not only chondrocyte proliferation and metabolism, but also promote extracellular matrix formation [19,20,29]. However, the underlying mechanisms are not well understood. Similarly, the effect of changes in the molecular mass of HA in the extracellular environment on the inflammatory processes is not fully understood [30]. Previous studies suggest that HMW-HA preparations are generally considered to have more positive pharmacological effects and stronger effects than LMW-HA or MMW-HA [27,31]. HMW-HA is considered as a physiological protective agent of cells and can act as a scaffold to assemble a proteoglycan matrix. It binds and aggregates its cell-surface receptors [32–34]. HA fragments (MW ranging from LMW-HA to o-HA) produced by hyaluronidases are formed in the inflammatory microenvironment and have been reported to induce defensive or pro-inflammatory responses in many cell types [35–37]. Endogenous HA oligosaccharides are recognized as the inflammatory response inducer by interacting with CD44 and toll-like receptor 4 (TLR-4) and 2 (TLR-2) in different types of cells [38–40]. However, several studies published in recent years have reported that HA fragments (including HA oligosaccharides) may fail to induce proinflammatory factors or other signaling effects in some cell types, especially in the area of OA [30,41,42].

The effects of o-HAs on the inflammation of articular chondrocytes are still unclear. Despite advances in the chemical synthesis of o-HAs [43–45], obtaining structurally defined derivatives is difficult. We developed a semi-synthesis route starting from acidolysis and enzymolysis of HAs to obtain HA oligosaccharides, which were further modified at the terminal position to to obtain methylation and azidation oligosaccharides. And their anti-inflammatory effects were examined in ATDC5 cells. Our results show that o-HA derivatives alleviated LPS-induced inflammation by multiple potential mechanisms.

2. Results

2.1. Study Design

Figure 1 briefly introduces the synthetic routes and molecular biological function detection of six glycosides of HA oligosaccharides

2.2. Synthesis of o-HA Derivatives

2.2.1. Synthesis of o-HA_2OMe (**6**) and o-HA_2N_3 (**7**)

The HA disaccharides modified by β-methyl and β-azide were synthesized (Figure 2) using HA polysaccharide (MW. ~500 kDa) as the starting material. Disaccharide **1** was obtained by acid hydrolysis [46]. Acetylation in pyridine (Pyr)/acetic anhydride (Ac_2O) resulted in medium to low yield due to the presence of a large amount of inorganic salts. A more efficient procedure for acetylation of **1** using Ac_2O/dichloromethane (DCM) as the solvent and dimethyl aminopyridine (DMAP) as the catalyst was used to obtain **2** in 31% yield with α/β = 2.5/1 [47]. The oxazoline intermediate **3** was generated from **2** in 82% yield [48], and the disaccharides **4** and **5** were obtained through a copper(II)-mediated ring-opening reaction from the intermediate **3** in 85% and 83% yield [49], respectively. The disaccharides **6** and **7** were provided by deprotection of the disaccharides **4** and **5** through a standard procedure in 86% and 85% yield, respectively [50] (Figure 2).

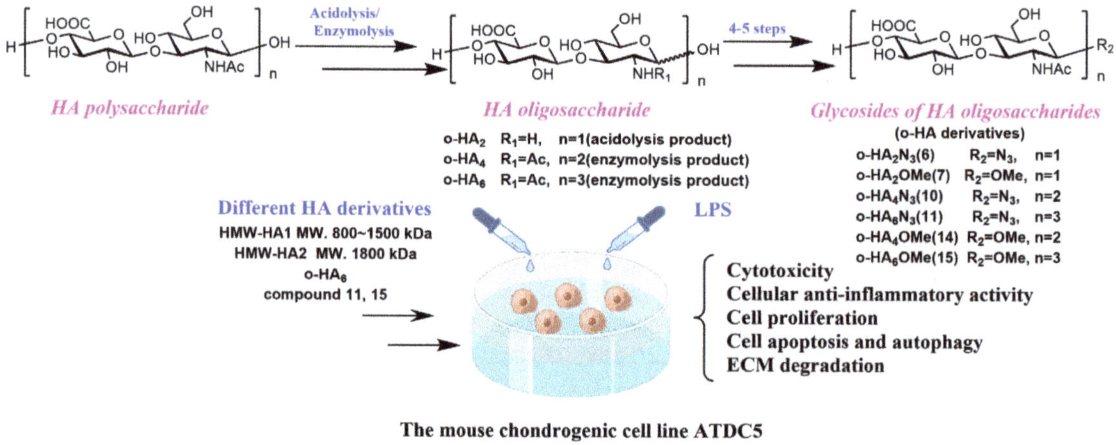

Figure 1. Synthesis of glycosides of HA oligosaccharides and evaluation of their biological activity in ATDC5 cells.

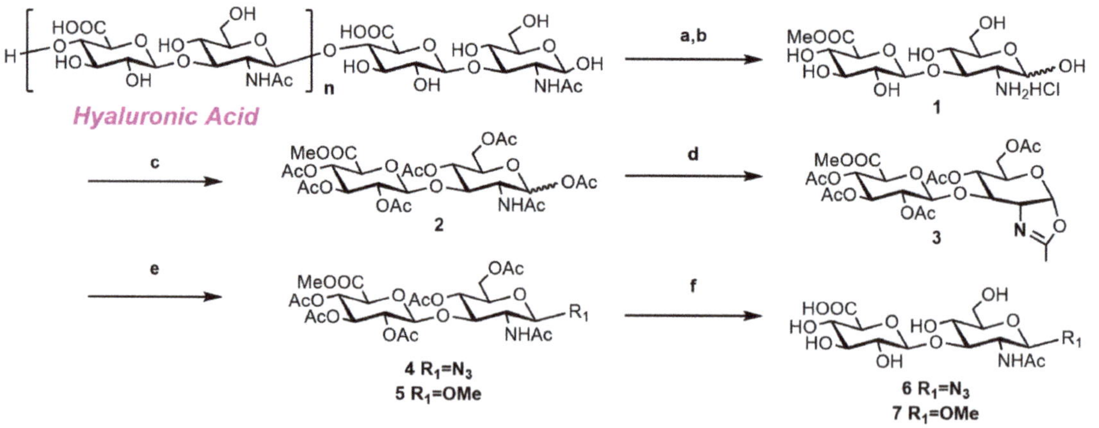

Figure 2. Synthesis of o-HA_2OMe (**6**) and o-HA_2N_3 (**7**). Reagents and conditions: (**a**) 0.5 M HCl, 80 °C, 2 d; (**b**) 0.06 M HCl, MeOH, 4 °C, 4 d; (**c**) Ac_2O, Et_3N, DMAP, DCM, 2 h, α/β = 2.5/1, 31% for 3 steps; (**d**) TMSOTf, DCM, 4 °C to room temperature (rt), 82%; (**e**) $CuCl_2$, MeOH/$TMSN_3$, $CHCl_3$, reflux, 82% for **4**, 85% for **5**; (**f**) LiOH/H_2O_2, THF/H_2O, −5 °C to rt, then 4 M NaOH, MeOH, 0 °C to rt, 85% for **6**, 86% for **7**.

2.2.2. Synthesis of o-HA$_4$N$_3$ (**11**) and o-HA$_6$N$_3$ (**12**)

As shown in Figure 2, HA polysaccharide (MW. ~500 kDa) was digested with bovine testis (or testicular) hyaluronidase (BTH) (200 mg for 10 g HA polysaccharide) in NaCl/NaOAc buffer (pH = 5.0) for 7 days to generate HA tetrasaccharide and hexasaccharide mixture [51]. The mixture was converted into β-azide derivatives through a (2-chloro-N,N′-1, 3-dimethylimidazolium chloride)-mediated nucleophilic reaction with NaN$_3$ in water without protecting groups [52]. Through a continuous esterification method by sequentially methylating and acetylating the degradation mixture, the tetrasaccharide **8** and hexasaccharide **9** were isolated via chromatography in a three-step overall yield of 42% and 34%, respectively. The tetrasaccharide **10** and hexasaccharide **11** were provided by the deprotection of tetrasaccharide **8** and hexasaccharide **9** through a standard procedure in 85% and 82% yield, respectively (Figure 3).

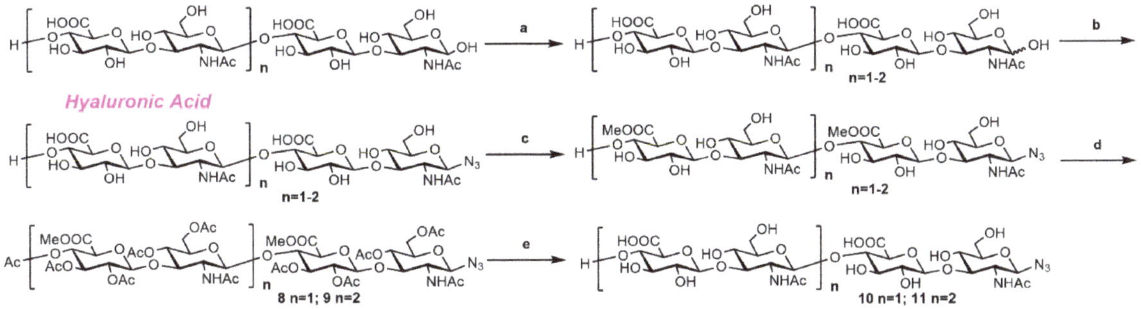

Figure 3. The synthesis of o-HA$_4$N$_3$ (**11**) and o-HA$_6$N$_3$ (**12**). Reagents and conditions: (**a**) 2.5% bovine testis hyaluronidase, NaOAc/NaCl buffer, pH 5.0, 37 °C, 7 days; (**b**) DMC, N-methylmorpholine, NaN$_3$, H$_2$O, 0 °C to rt; (**c**) 0.06 M HCl, MeOH, 4 °C, 4 d; (**d**) Ac$_2$O, Et$_3$N, DMAP, DCM, 2 h, 42% for **8**, 34% for **9**; (**e**) LiOH/H$_2$O$_2$, THF/H$_2$O, −5 °C to rt, then 4 M NaOH, MeOH, 0 °C to rt, 85% for **10**, 86% for **11**.

2.2.3. Synthesis of o-HA$_4$OMe (**14**) and o-HA$_6$OMe (**15**)

As shown in Figure 4, the HA tetrasaccharide and hexasaccharide mixture digested by BTH was dissolved in a high concentration of HCl/MeOH (0.5 M) solution to convert into β-methyl derivatives, following a standard procedure of acetylation. The β-methyl derivatives **12** and **13** were separated in a three-step overall yield of 21% and 19%, respectively (Figure 4). The tetrasaccharide **14** and hexasaccharide **15** were provided by the deprotection of tetrasaccharide **12** and hexasaccharide **13** in 85% and 82% yield, respectively.

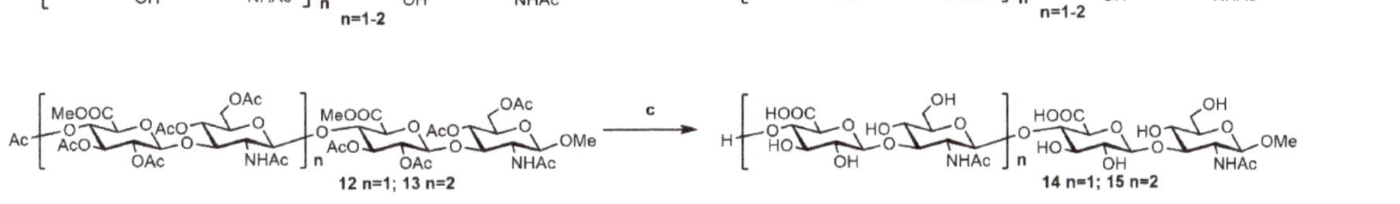

Figure 4. Synthesis of o-HA$_4$OMe (**14**) and o-HA$_6$OMe (**15**). Reagents and conditions: (**a**) 0.5 M HCl, MeOH, 0 °C to rt, overnight; (**b**) Ac$_2$O, Et$_3$N, DMAP, DCM, 2 h, 21% for **12**, 19% for **13**. (c) LiOH/H$_2$O$_2$, THF/H$_2$O, −5 °C to rt, then 4 M NaOH, MeOH, 0 °C to rt, 85% for **14**, 82% for **15**.

2.3. Evaluation of the Cytotoxicity and Pro-Inflammatory Activity of o-HA Derivatives in ATDC5 Cells

Before evaluating anti-inflammatory activity, we tested the cytotoxicity of the synthetic o-HA derivatives in ATDC5 cells using the CCK-8 assay. The cell inhibitions of the six o-

HA derivatives at different doses (0, 1, 10, 100 or 1000 μg/mL) were examined after 24 h. o-HA_2N_3 and o-HA_4N_3 showed little effect on cell viability (Figure 5A). o-HA_6N_3 and o-HA_6OMe demonstrated good safety at concentrations of less than 1000 μg/mL. The other two derivatives o-HA_2OMe and o-HA_6OMe showed a dose-dependent inhibition of ATDC5 cell viability. This showed that the β-azide derivatives exhibited better safety compared to β-methyl derivatives and the safe concentrations between 0–100 μg/mL of the β-azide derivatives and o-HA_6OMe were used in the following experiments.

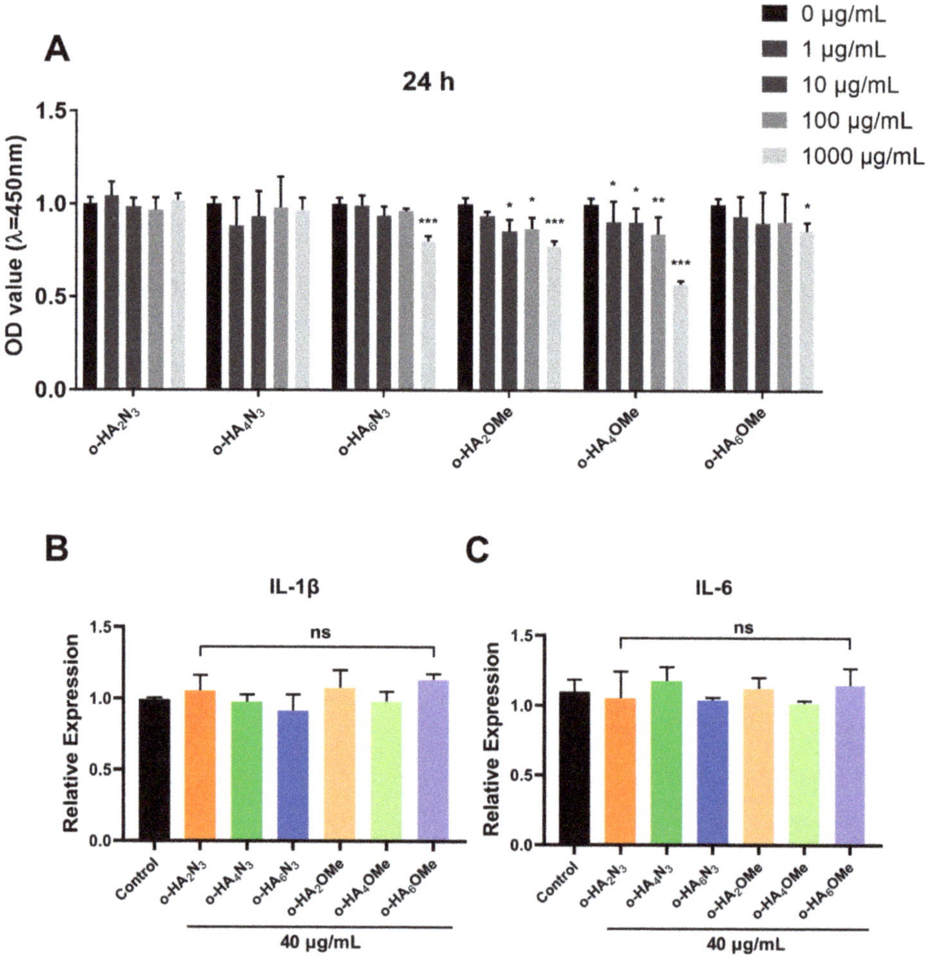

Figure 5. Evaluation of the cytotoxicity and pro-inflammation of o-HA derivatives in ATDC5 cells. (**A**) ATDC5 cells were seeded into 96-well plates. After 12 h of incubation, cells were treated with different concentrations (0, 1, 10, 100, 1000 μg/mL) of o-HA derivatives for 24 h. Next, cell viability was examined by the CCK-8 assay. (**B**,**C**) RT-qPCR was used to confirm the expression levels of pro-inflammatory cytokines (IL-1β, IL-6) after co-culture with ATDC5 cell of o-HA derivatives (40 μg/mL) for 24h. * $p < 0.05$, ** $p < 0.01$, *** $p < 0.001$ vs. each 0 μg/mL group; ns = not statistically significant vs Control group.

In the past study [36], HA oligosaccharides (40 μg/mL) were reported to induce inflammatory response of mouse chondrocytes. We tested the related activity on ATDC5 cells after treatment with the synthetic o-HA derivatives at 40 μg/mL. After treatment with o-HA derivatives for 24 h, RT-qPCR for IL-1β (Figure 5B) and IL-6 (Figure 5C) was performed to confirm the inflammatory effect. Results showed that all six derivatives had no inflammation causing effect on the mRNA levels of these two inflammatory cytokines at 40 μg/mL. Moreover, to verify whether the o-HA derivatives exhibit inflammatory activity, we added the experiment related to THP-1 cells and obtained similar results (Figure S2).

2.4. Evaluation of the Protective Effect of Different HA Derivatives on LPS-Induced Inflammatory Injury of ATDC5 Cells

To investigate the anti-inflammatory potential of the synthetic o-HA derivatives, we tested their effects on the expression of pro-inflammatory cytokines IL-1β and IL-6 in LPS challenged ATDC5 cells. As shown in Figure 6A,D, compared with the control group, the mRNA levels of IL-1β and IL-6 were significantly elevated by LPS. All the synthetic derivatives groups showed decreased mRNA levels of both pro-inflammatory markers (IL-1β and IL-6) compared with the LPS group. The results also suggested that the disaccharides and hexasaccharides exhibited higher potency against LPS-induced inflammation. Considering the lower cytotoxicity of hexasaccharide derivatives, o-HA_6N_3 and o-HA_6OMe were selected for further investigation. RT-qPCR was used to investigate the anti-inflammatory effects of o-HA hexasaccharides, HMW-HAs (HMW-HA1 with MW. 800~1500 kDa and HMW-HA2 with MW. 1800 kDa), and o-HA_6 which were obtained from the enzymolysis of HA. We found that the mRNA levels of two HMW-HA groups were significantly decreased compared with the control group, and o-HA_6 could also reduce the levels of IL-1β and IL-6 mRNA (Figure 6B,E). Compared with o-HA_6, the other two synthetic o-HA_6 derivatives showed higher anti-inflammatory activity at a similar level of HMW-HA.

To further verify the effect on inflammatory response, enzyme-linked immunosorbent assay (ELISA) was used to determine the release of pro-inflammatory cytokines (IL-1β and IL-6) (Figure 6C,F) after treatment with different HA derivatives (HMW-HA1/2, o-HA_6, o-HA_6N_3 and o-HA_6OMe). It was found that the release of inflammatory cytokines decreased significantly after 24 h treatment with different HA derivatives. For the inhibition of IL-1β, all derivatives showed high potency and had no significant difference, except that the difference between HWM-HA2 and o-HA_6N_3 was significant ($p < 0.05$) (Figure 6C). For the inhibition of IL-6, o-HA_6 and o-HA_6 derivatives showed superior activity to HMW-HA2 with statistically significant difference ($p < 0.05$, $p < 0.01$), but showed no difference among the three o-HA_6 derivatives (Figure 6F). The anti-inflammatory activity of different HA derivatives was also confirmed by Western blot analysis (Figure S3). We also examined the anti-inflammatory activity of different HA derivatives in THP-1 cells by RT-qPCR, and the result confirmed the anti-inflammatory activities of different HA derivatives (Figure S4). However, a different activity pattern among the tested group was observed compared with that in ATDC5 cells, which may be related to the different cell line and experimental condition. Further experimental study is needed.

2.5. The Proliferation Effect of Different HA Derivatives on LPS Challenged ATDC5 Cells

To confirm the proliferation effects of different HA derivatives on LPS-challenged ATDC5 cells, flow cytometry was used to evaluate the 5-Ethynyl-2′-deoxyuridine (EdU)-positive cell rate. As shown in Figure 7A,B, the percentage EdU-positive cells decreased significantly after LPS was added. In contrast, treatments with different HA derivatives (HMW-HA1, HMW-HA2, o-HA_6, o-HA_6N_3 and o-HA_6OMe) for 24 h significantly increased the percentage of EdU-positive ATDC5 cells, indicating that various HA derivatives can restore cell proliferation ability (Figure 7C–G). However, according to the statistical results of Figure 7H, there was no significant difference in the alleviating effects of different HA derivatives on LPS-induced inflammatory injury.

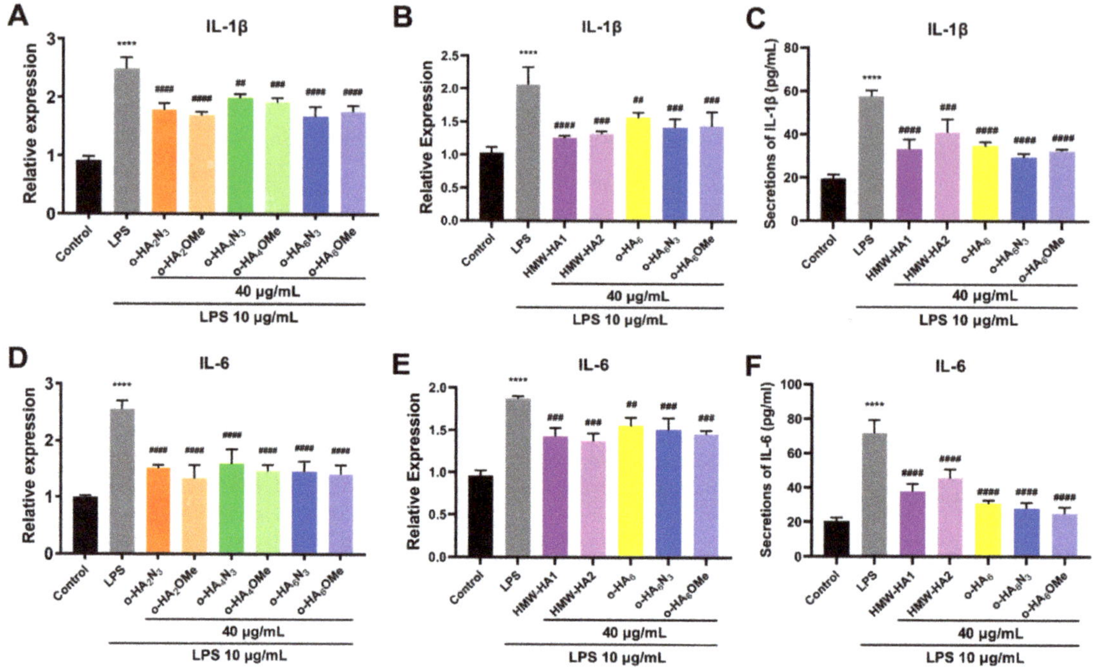

Figure 6. Effects of different HA derivatives on LPS-induced inflammatory responses in ATDC5 cells. ATDC5 cells in the LPS group were stimulated with LPS (10 μg/mL) for 24 h and then cultured in a medium containing 10% FBS for another 24 h. For different HA derivatives groups, ATDC5 cells were treated with LPS (10 μg/mL) for 24 h followed by treatment with corresponding samples (40 μg/mL) for an additional 24 h, respectively. (**A–C**) IL-1β mRNA expression levels and cytokine level were measured by RT-qPCR and ELISA. (**D–F**) IL-6 mRNA expression levels and cytokine level were measured by RT-qPCR and ELISA. **** $p < 0.0001$ vs. Control group; $^{\#}$ $p < 0.05$, $^{\#\#}$ $p < 0.01$, $^{\#\#\#}$ $p < 0.001$, $^{\#\#\#\#}$ $p < 0.0001$ vs. LPS group.

2.6. Anti-Apoptotic Effect of Different HA Derivatives on LPS Challenged ATDC5 Cells

To study the effects of different HA derivatives on apoptosis, the TUNEL (TdT mediated dUTP Nick End Labeling) assay was used to analyze apoptosis of ATDC5 cells after LPS treatment. TUNEL assay showed that LPS-induced apoptosis of ATDC5 cells with a significantly increased apoptotic cell ratio. All HA derivatives (HMW-HA1, HMW-HA2, o-HA$_6$, o-HA$_6$N$_3$ and o-HA$_6$OMe) significantly inhibited LPS-induced apoptosis, among which HMW-HA1 showed the most potent effect. o-HA$_6$N$_3$ and o-HA$_6$OMe showed moderate alleviating effect on apoptosis, and the difference between o-HA$_6$N$_3$ and o-HA$_6$OMe was statistically significant ($p < 0.05$) (Figure 8A,B). The Bcl-2 gene is known as an apoptosis-suppressing gene. Bcl-2 can homodimerize and form heterodimers with Bax, in which Bcl-2 acts as the apoptosis inhibitor and Bax acts as the apoptosis promoter. Therefore, we also analyzed the protein expression of Bcl-2 and Bax by Western blot assay. The result showed that the expression of Bcl-2 was selectively increased and that of Bax was decreased, indicating increased cell resistance to apoptosis due to the protective effect of different HA derivatives. Compared with the control group, Bcl-2 expression was significantly decreased and Bax expression was significantly increased in the LPS group. After treatment with different HA derivatives in the LPS group, the expression of Bax decreased significantly. However, o-HA or o-HA derivatives did not significantly alter the expression of Bcl-2 in the LPS group, but HMW-HA treatment restored the expression of Bcl-2 (Figure 8C).

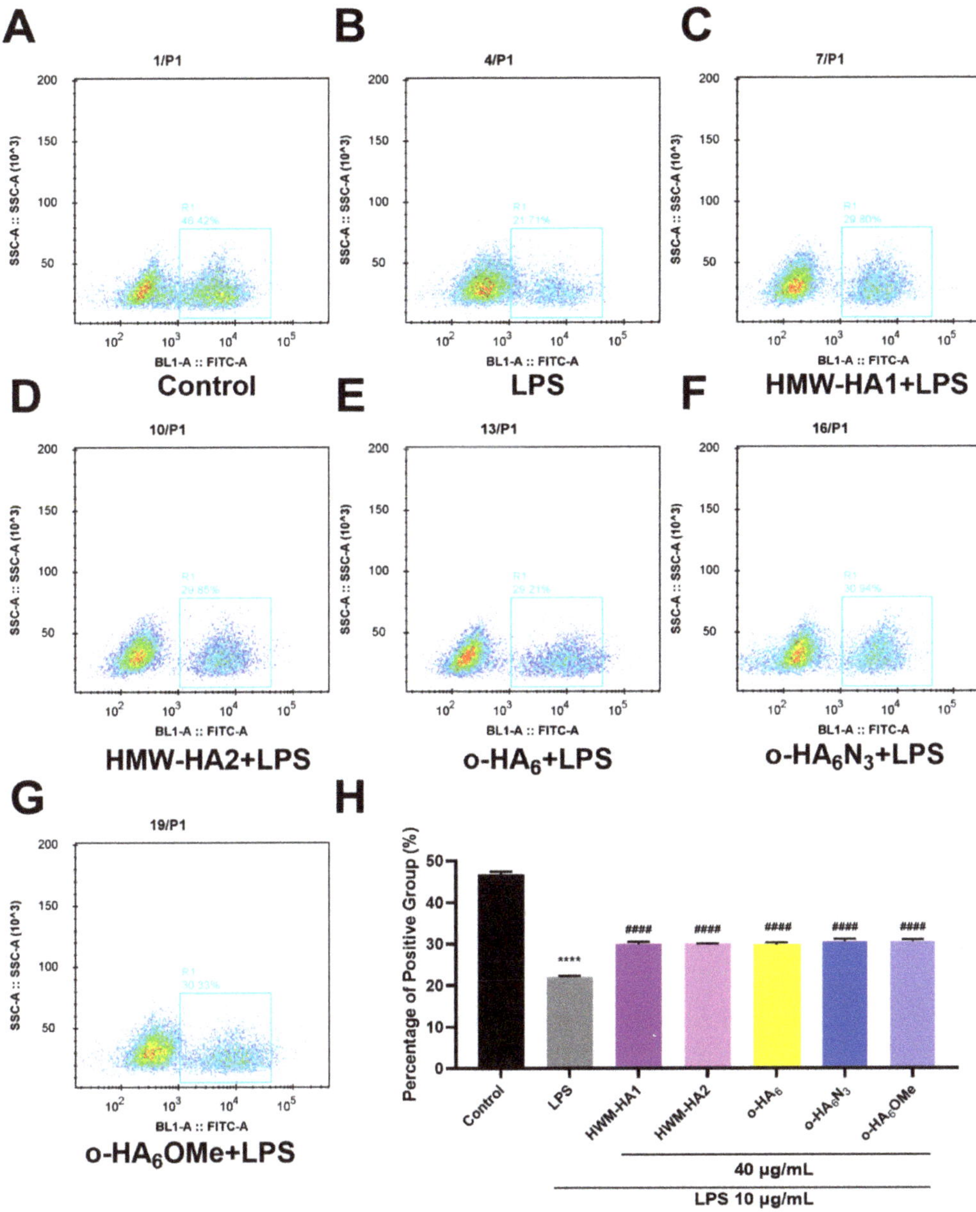

Figure 7. Effects of different HA derivatives on cell proliferation after inflammatory injury. (**A–H**) ATDC5 cells in the LPS group were stimulated with LPS (10 μg/mL) for 24 h and then cultured in the medium containing 10% FBS for another 24 h. For different HA derivatives groups, ATDC5 cells were treated with LPS (10 μg/mL) for 24 h and different HA derivatives (40 μg/mL) for an additional 24 h, respectively. (**A–G**) The positive cell ratio was measured by EdU flow cytometry assay. (**H**) Histogram quantification was performed for the triplicate results. **** $p < 0.0001$ vs. Control group; #### $p < 0.0001$ vs. LPS group.

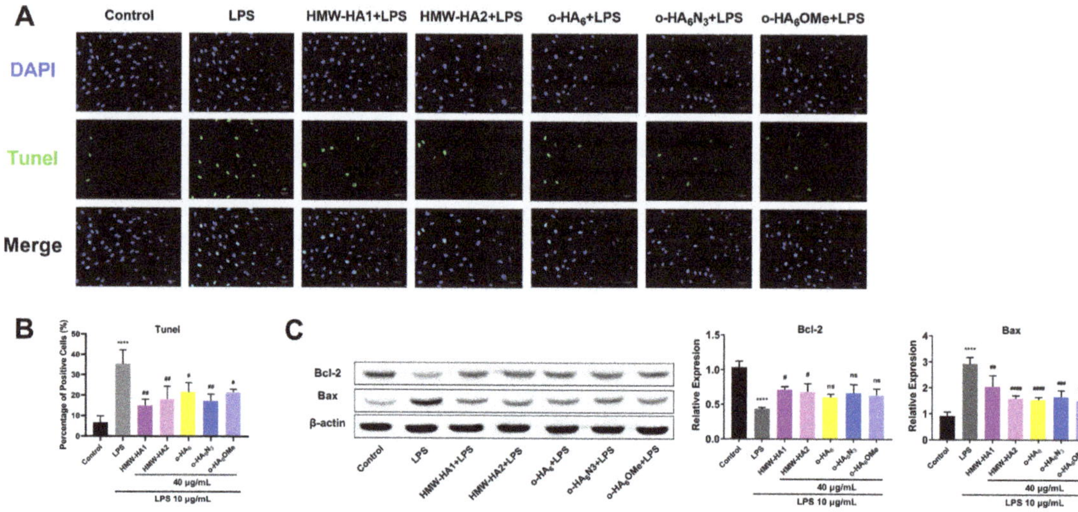

Figure 8. Anti-apoptotic effects of different HA derivatives on ATDC5 inflammatory injury. ATDC5 cells in the LPS group were stimulated with LPS (10 μg/mL) for 24 h and then cultured in a medium containing 10% FBS for another 24 h. For different HA derivatives groups, ATDC5 cells were treated with LPS (10 μg/mL) for 24 h and different HA derivatives (40 μg/mL) for an additional 24 h, respectively. (**A**) Cell apoptotic pattern was examined by TUNEL assay. (**B**) Positive cell counts were plotted for each treatment group for three independent replicates. (**C**) Western blot assay was used to detect the protein expression levels of Bcl-2 and Bax after treatment with different HA derivatives. **** $p < 0.0001$ vs. Control group; # $p < 0.05$, ## $p < 0.01$, ### $p < 0.001$, #### $p < 0.0001$, ns = not statistically significant vs. LPS group.

2.7. Anti-Autophagy Effect of Different HA Derivatives on LPS Challenged ATDC5 Cells

The most commonly used indicator for autophagy detection is the microtubule-associated protein light chain 3 (LC3) [53]. LC3 conversion (LC3-I to LC3-II) can be detected by immunofluorescence analysis because the level of LC3-II correlates with the number of autophagosomes. Our results showed that compared with control group LPS promoted the autophagy process of ATDC5 cells (Figure 9A,B). Most of the autophagy structures appeared at the outside of the cell membrane after LPS injury, showing the highest fluorescence intensity. After treatment with the HA derivatives, the aggregation of autophagy around the cell membrane was alleviated significantly with decreased fluorescence intensity. Different HA derivatives showed prominent anti-autophagy effects revealed by the decreased immunofluorescence intensity, but the differences among them were not statistically significant. The endogenous LC3 protein exhibits two bands expressed from the same mRNA, LC-I with MW of 18 kDa and LC-II with MW of 16 kDa. The ratio of LC3-II/LC3-I is closely correlated with the number of autophagosomes, serving as a reliable indicator of autophagosome formation [54]. Therebefore, we used Western blot assay to measure the LC3-II/LC3-I ratio in the study. The ratio of LC3-II/LC3-I increased significantly after LPS treatment (Figure 9C). In comparison, after treatment with 40 μg/mL of different HA derivatives for 24 h, the ratio of LC3-II/LC3-I decreased significantly. The results of o-HA and o-HA derivatives suggest that they may be more beneficial in inhibiting autophagy in LPS-stimulated ATDC5 cells. In addition, the autophagy-inhibiting effect of HA derivatives was investigated in the presence of an autophagy inhibitor (Figure S5). The results showed that in the absence of the autophagy inhibitor, o-HA derivative (o-HA_6OMe) and HMW-HA (HMW-HA1) had similar decreasing effects on the LC3-II/LC3-I ratio increase caused by LPS ($p < 0.0001$). However, after the addition of the inhibitor, the decreasing effect on the ratio was weakened. HMW-HA1 showed a decreasing effect that was not statistically significant, and o-HA_6OMe showed a statistically significant ($p < 0.05$) decreasing effect.

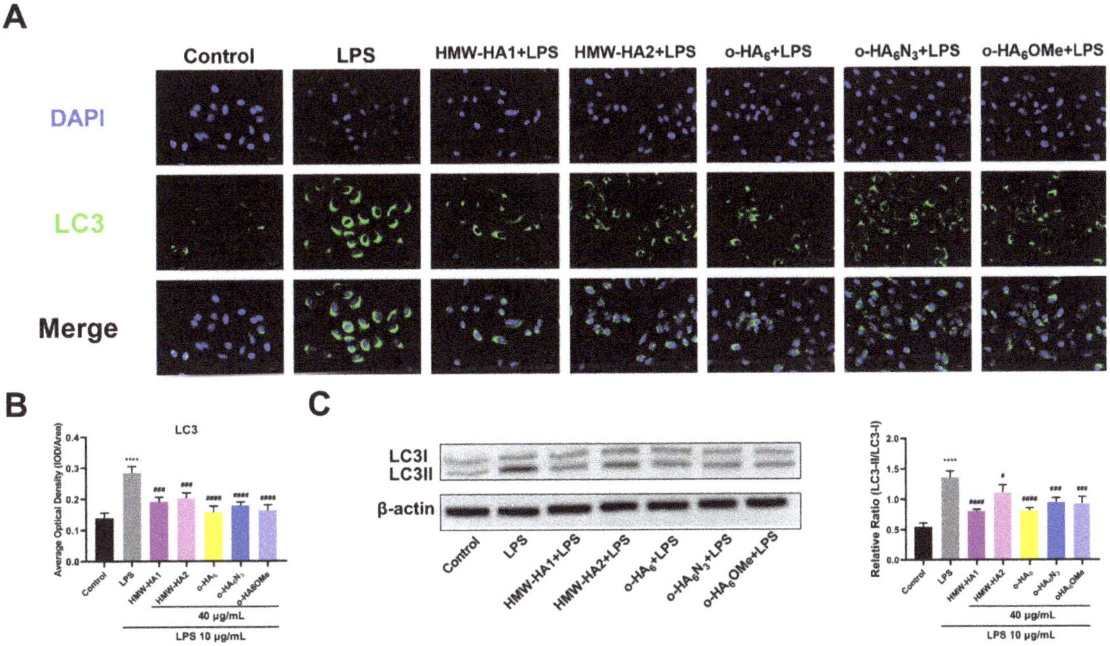

Figure 9. Anti-autophagy effect of different HA derivatives on ATDC5 cells after inflammatory injury. ATDC5 cells in the LPS group were stimulated with LPS (10 μg/mL) for 24 h and then cultured in the medium containing 10% FBS for another 24 h. For different HA derivatives groups, ATDC5 cells were treated with LPS (10 μg/mL) for 24 h and different HA derivatives (40 μg/mL) for an additional 24 h, respectively. (**A**) Cell autophagy pattern was assessed by LC3 immunofluorescence analysis. (**B**) Relative fluorescence intensity was plotted for each treatment group for three independent replicates. (**C**) LC3 protein expression level and LC3-II/LC3-I ratio were detected by Western Blot assay. **** $p < 0.0001$ vs. Control group; # $p < 0.05$, ### $p < 0.001$, #### $p < 0.0001$ vs. LPS group.

2.8. Alleviation of the ECM Degradation by Different HA Derivatives

ECM degradation is an important pathogenic mechanism of OA, inducing a degenerative cycle of inflammation and degradation of cartilage by proteinases such as matrix metalloproteinases (MMPs). MMP13 plays a key role in cartilage degradation in OA, destroying not just type II collagen, but proteoglycan, type IV and type IX collagen, as well as osteonectin and basement membrane [55]. Proteoglycans (such as Aggrecan), glycoproteins (such as COMP), polysaccharides and fibrins (such as collagen type II alpha 1 (COL2A1) and elastin) synthesized by chondrocytes are the main components in ECM, and they are closely related to the progress of OA [56–58]. To evaluate the effects of different HA derivatives on ECM degradation we measured the levels of MMP13, COL2A1 and Aggrecan by Western blot to determine the protein indexes of cell matrix and collagen. Results of Western blot assay showed that stimulated by LPS (Figure 10A), MMP13 was significantly increased, and the different HA derivatives reduced the increased level; especially, o-HA_6OMe showed high potency (Figure 10B). COL2A1 and Aggrecan of ATDC5 cells decreased significantly under LPS stimulation. Moreover, the treatment of different HA derivatives (HMW-HA1, HMW-HA2, o-HA_6 and o-HA_6N_3, except o-HA_6OMe group) reversed the decreasing trend of COL2A1 at different degrees. However, for Aggrecan, almost all the different HA derivatives did not show protective effects except for o-HA_6N_3, which showed a slightly significant protective effect (Figure 10C,D).

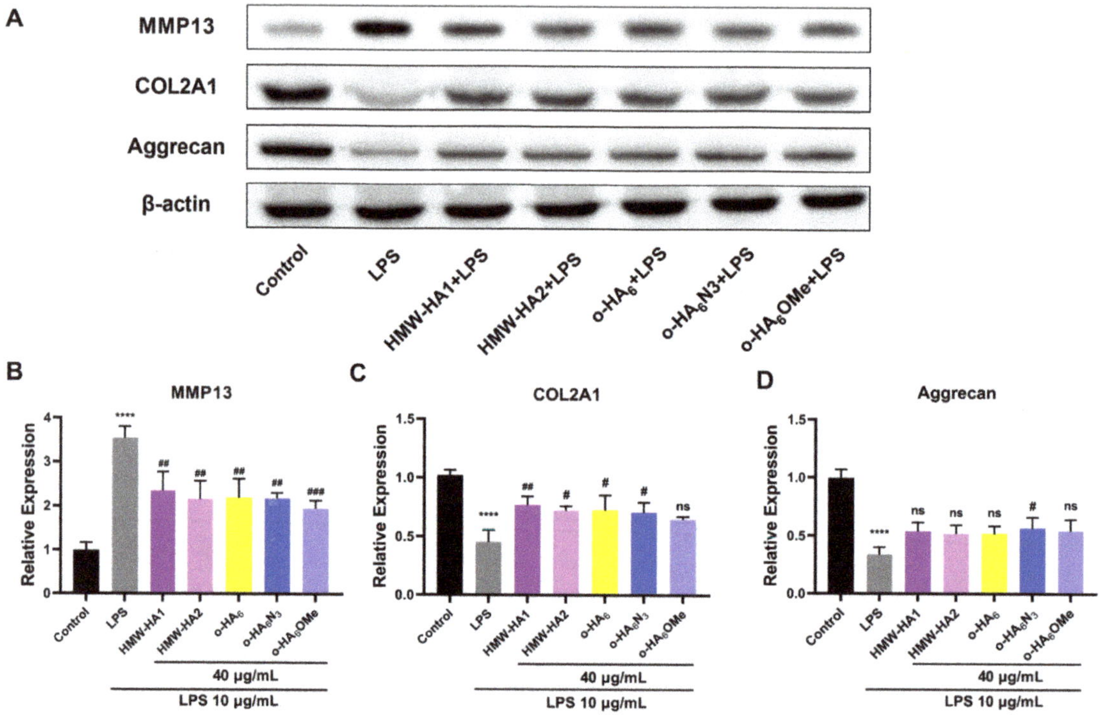

Figure 10. Effects of different HA derivatives on the degradation of ECM caused by ATDC5 inflammatory injury. ATDC5 cells in the LPS group were stimulated with LPS (10 μg/mL) for 24 h and then cultured in the medium containing 10% FBS for another 24 h. For different HA derivatives groups, ATDC5 cells were treated with LPS (10 μg/mL) for 24 h and different HA derivatives (40 μg/mL) for an additional 24 h, respectively. (**A–D**) The protein expression levels of ECM degradation related indicators (MMP13, Aggrecan and COL2A1) were detected by Western blot assay. **** $p < 0.0001$ vs. Control group; # $p < 0.05$, ## $p < 0.01$, ### $p < 0.001$, ns = not statistically significant vs. LPS group.

3. Discussion

OA is a chronic degenerative disease of the joints that leads to pain and loss of mobility [59]. Intra-articular injection of high molecular weight HA to replace synovial fluid that has lost its viscoelastic properties is widely used in the treatment for OA [60]. HA plays a dual role in the process of inflammation and damage of cartilage, as a pro-inflammatory molecule or an anti-inflammatory molecule [31] It basically depends on the molecular weight of HA. In addition, a study of rheumatoid arthritis have shown that HA modulates inflammation based on its molecular weight, and HMW-HA was reported to inhibit the production of pro-inflammatory mediators and down-regulate NF-kB by binding to ICAM-1 [61]. It is well known that the immune system is essential in protecting the body from environmental threats and pathogens [62]. However, excessive activation of the immune system can lead to a variety of chronic and acute inflammation and cause a variety of diseases. As members of the pattern recognition receptor (PRRs) family, 13 Toll-like receptors (TLRs) have been identified in laboratory mice and 10 functional TLRs identified in humans. These recognize various PAMPs (pathogen-associated molecular patterns), including lipopolysaccharides (LPS) [63]. LPS-induced cartilage inflammation model is often used to mimic OA in humans, which has several advantages including convenient operation and inflammatory effects consistent with the OA disease. Multiple studies have shown that LPS exposure can trigger marked up-regulation in the expression levels of pro-inflammatory cytokines (e.g., IL-1β, IL-6), pro-apoptotic protein Bax, and a noticeable reduction in the expression of anti-apoptotic protein Bcl-2 in ATDC5 cells [64]. However, although HMW-HAs had been demonstrated to be effective in reducing the inflammation induced by LPS, the effects of HA fragments such as o-HAs or their derivatives on LPS-induced cartilage inflammation are still not clear.

Great advances have been made in the chemical synthesis of HA oligosaccharides in recent years, but the extremely long routes, low yields, small scale and high cost limit the biological activity evaluations of HA oligosaccharides [43–45]. In addition, hyaluronic acid can be degraded into oligosaccharides by acids and enzymes, but the main limitation of inhomogeneity of degradation products remains [46,65,66]. In our study, HA oligosaccharides were prepared in only four to five steps after acidolysis and enzymolysis of the HA polysaccharide. In addition to obtaining structurally defined oligosaccharides, HA oligosaccharides were modified by β-azide and β-methyl at the terminal position. The effects of these o-HA compounds on OA development were investigated in the LPS-induced ATDC5 cell model. Our results showed that the six o-HA derivatives had similar anti-inflammatory activity, but some differences remained. The following general conclusions were obtained: (1) hyaluronic acid oligosaccharides derivatives exhibited anti-inflammatory activity in a manner as effective as HMW-HAs in an LPS-challenged inflammatory model; (2) the cytotoxicity of methylation derivates seemed to be higher than that of azide derivatives; (3) the modifications at the terminal group might benefit the anti-inflammatory activity.

We found there were significant differences in cell viability at 24 h between the cells treated with o-HAs modified with β-methyl group or with β-azide group. Methylated disaccharides and tetrasaccharides showed a certain cytotoxicity, while the cytotoxicity of methylated hexasaccharides and azide-modified hexasaccharides remained low. It is generally believed that the longer the sugar chain of HA, the more stable the compound and the less toxic it is to the cells. To be consistent, we selected two hexasaccharides with terminal modifications for activity comparison with HMW-HA and unmodified o-HA_6.

Although different cell sources and models were used, our results are consistent with a previous report that various HA fragments with low endotoxin content (including o-HAs) cannot induce or enhance proinflammatory cytokine release from chondrocytes [42]. Another study reported that exogenous HA fragments up to about 40 kDa in molecular weight do not show pro-inflammatory activity in human articular chondrocytes [30]. Contrary to most studies in this field in recent years, this suggests that the accumulation of HA-fragment in inflammatory tissues may be the result of inflammation rather than a driver of it. This conclusion is challenged by concerns of the purity and endotoxin content of earlier samples. Our study has now confirmed that the HA oligosaccharide derivatives exhibited anti-inflammatory activity comparable to HMW-HA. A recent study showed that the enzymatically obtained HA disaccharide, ΔHA2, inhibited LPS-induced inflammation in different cell lines (including THP-1 cells). In addition, hyaluronic acid oligosaccharides (2mer–8mer) prepared by acid hydrolysis can reduce LPS-induced inflammatory damage to varying degrees in RAW 264.7 cells [67]. Our experiment verified the activity of different HA derivatives against LPS-induced inflammatory injury in THP-1 cells. However, evidence for the therapeutic application of hyaluronic acid oligosaccharides and their derivatives in the treatment of osteoarthritis is still lacking. Admittedly, our study of the ATDC5 cell line is only the first step in advancing the application of o-HAs and their derivatives in OA.

Chondrocyte apoptosis and autophagy have been implicated in the pathogenesis of OA [68,69]. Cell apoptosis is mediated by multiple pathways, including caspase family members, of which anti-apoptotic protein Bcl-2 and pro-apoptotic protein Bax regulate the release of apoptotic activators such as cytochrome C by controlling mitochondrial membrane permeability. The Bax dimer opens channels in the membrane and increases permeability. Bcl-2 and Bax form a heteromer and reduce permeability. Therefore, the increased expression of Bcl-2 and the decreased expression of Bax indicate that the cells are more resistant to apoptosis [68]. Our study showed that LPS facilitated cell apoptosis, while these effects were inhibited by different HA derivatives. The expression of Bax decreased significantly after treatment with different HA derivatives in the LPS group. However, HMW-HAs restored the protein expression level of Bcl-2. And o-HA_6 or o-HA_6 derivatives did not change the expression. This result may be related to the different molecular mechanisms underlying the actions of oligosaccharides and HWM-HA. The TUNEL assay indicated HMW-HA had the most significant anti-apoptotic effect compared with o-HA_6 derivatives, followed by o-HA_6N_3 and o-HA_6OMe. The effect of o-HA_6N_3 was higher than that of o-HA_6OMe.

During autophagy, the target structures are sequestered by phagophores, matured into autophagosomes, and finally delivered to lysosomes for degradation. Autophagy is involved in the pathophysiology of many diseases, and the study of its regulation helps understand the outcome of many diseases. Lysosomal inhibitors that primarily block lysosomal degradation, such as Bafilomycin A1 (Baf-A1), protease inhibitors, and chloroquine (CQ) [70], are used to block the final step of the autophagy process. We found that the anti-autophagy effects of HA derivatives were weakened in the presence of the autophagy inhibitor. After administration of lysosomal inhibitor, o-HA_6OMe showed stronger inhibition. Compared with HMW-HA1, the number of autophagosomes was significantly reduced. (Figure S5). Inflammation can induce the remodeling of ECM of cartilage, which is closely associated with OA development [71]. ECM, a complex molecular network surrounding the cells, plays vital roles in the maintenance and regulation of cell phenotypes, cell/tissue functions, and tissue homeostasis [72,73]. Type II Collagen (COL2) is secreted by chondrocytes in the form of soluble collagen. Its supramolecular structure is generally composed of three identical alpha chains (triple helices), which is held together by interchain or intra-chain hydrogen bonds [74]. Aggrecans are the most abundant proteoglycans in cartilage, which is mainly composed of anionic sulfated glycoaminoglycan (GAG), chondroitin sulfate (CS) and keratin sulfate (KS) covalently bound with the core protein backbone. Aggrecans only exist as aggregates within the ECM. For example, the components mentioned above are non-covalently linked to another long GAG chain, hyaluronic acid, to form larger protein aggregates via "link proteins" [75]. Our results of Western blot assay suggested that the effect of HA oligosaccharide derivatives on ECM supramolecular structure in OA may depend on the down-regulation of MMP, a degradation factor, and restoration of the three-dimensional structure of ECM (up-regulation of COL2A1). The abnormal remodeling of ECM is related to the pathogenesis of multiple diseases [72,76]. Moreover, the destruction and disorganization of cartilage ECM is a common feature of OA [77]. Chondrocytes play vital roles in the production, maintenance, and remodeling of cartilage ECM by inflammatory mediators, growth factors, and enzymes [78]. In OA, anomalous cartilage ECM signals can be transmitted to chondrocytes and lead to the abnormality in the phenotypes and behaviors of chondrocytes, which further triggers ECM remodeling and cartilage disturbance, and facilitates OA progression [71,77]. Moreover, previous studies have shown that HA can exert its biological and pharmacological activities by regulating the chemical and physical properties of ECM [79,80]. Given the impact of the o-HA derivates on inflammation, we further demonstrated that these o-HA derivates could exert their protective functions by altering ECM composition and structure. Consistent with a previous report [64], our study also demonstrated that the expression levels of aggrecan and COL2A1 were notably reduced, and ECM degradation-related protein MMP13 was markedly increased, in ATDC5 cells following LPS stimulation. In conclusion, the different HA derivatives selectively alleviated LPS-induced ECM destruction in ATDC5 cells by abating the expression of ECM-related proteins.

The o-HA derivatives with β-azide or β-methyl modification were synthesized with a shortened procedure; these compounds can be easily produced at a large scale. These o-HA derivatives inhibited LPS-induced pro-inflammatory responses and cell apoptosis and autophagy in ATDC5 cells. They also restored the inhibition of cell proliferation caused by LPS inflammatory injury. Moreover, these o-HA derivatives partially reversed the effects of LPS on the expression of ECM proteins (i.e., COL2A1, aggrecan) and ECM degradation-related proteins (i.e., MMP13) in ATDC5 cells. These data suggest that o-HA derivatives could alleviate LPS-induced chondrocyte injury by reducing inflammation and altering ECM components, suggesting the potential therapeutic value of these HA oligosaccharides in OA management.

4. Materials and Methods

4.1. Material and Reagents

LPS (Escherichia coli 055: B5) was obtained from Sigma-Aldrich (St. Louis, MO, USA). HMW-HA (0.8–1.5 $\times$ 10^6 Da, 1.8 $\times$ 10^6 Da) was obtain from Meilunbio (Dalian, China). CCK-8 and One Step TUNEL Apoptosis Assay Kit were purchased from Beyotime (Shanghai, China). DAPI solution and FITC conjugated Goat Anti-Rabbit IgG (H + L) was obtained from Servicebio (Wuhan, China). All primary antibodies used in this study were from Proteintech Group (Chicago, IL, USA) except the anti-β-actin (Medical Discovery Leader (MDL) Biotechnology, Beijing, China), anti-IL-1β (Abcam, Cambridge, UK) and anti-COL2A1 (Santa Cruz Biotechnology, Dallas, TX, USA). Bafilomycin A1 (Baf-A1) was purchased from Selleck (Houston, TX, USA). The reagents used for synthesis of o-HA derivatives were of analytical grade.

4.2. Chemical Synthesis

The detailed synthesis procedures of HA oligosaccharides are shown in Figures 2–4 and Materials and Methods S1. The ^{1}H NMR, ^{13}C NMR, HRMS and HPLC spectra of the compounds are presented in Figure S1.

4.3. Cell Culture and Treatment

The mouse chondrogenic cell line ATDC5 obtained from FuHeng Cell Center (Shanghai, China). ATDC5 cell line was cultured in DMEM medium (Hyclone, Thermo Scientific, Waltham, MA, USA) supplemented with 10% fetal bovine serum (Gibco, Thermo Scientific) and penicillin/streptomycin solution (Solarbio Science & Technology, Beijing, China) at 37 °C in the humidified atmosphere of 5% CO_2 in air. For ATDC5 inflammation damage, the medium was replaced with medium containing 10% FBS and LPS Supplement (final concentration 10 μg/mL) for inflammatory injury of chondrocytes.

Human monocyte cell line THP-1 cells were obtained from Procell Life Science & Technology (Wuhan, China). RPMI-1640 was a suitable medium for THP-1 cells and was obtained from Gibco. THP-1 cells in FBS concentration and CO_2 incubator conditions were the same as ATDC5 cells. PMA (Phorbol 12-myristate 13-acetate, 25 ng/mL, Meilunbio, China) induced THP-1 monocytes to differentiate into macrophages. After 48 h of induction, THP-1 cells were incubated in RPMI-1640 containing LPS (final concentration 200 ng/mL) or LPS plus different HA derivates (40 μg/mL) for 24 h.

4.4. CCK-8 Assay

The CCK-8 assay was performed using the CCK-8 kit (Beyotime) to measure the cytotoxicity of six o-HA derivatives in ATDC5 cells. Briefly, cells were seeded into 96-well plates and then treated with different doses of HA oligosaccharides. At 24 or 72 h after treatment, 10 μL of CCK-8 solution was added to each well. After 1 h of incubation, the absorbance was measured at 450 nm. Three replicates were performed for each group.

4.5. RT-qPCR Assay

RNA was extracted from ATDC5 and THP-1 cells using the Trizol reagent (Thermo Scientific) following the protocols of the manufacturer. RNA was reversely transcribed into cDNA using the SuperScript III First-Strand Synthesis SuperMix for RT-qPCR (Thermo Scientific). Real-time quantitative PCR reactions were performed on the Applied Biosystems StepOne Real-time PCR system (Thermo Scientific) using the SYBR Select Master Mix

(Thermo Scientific) and corresponding quantitative PCR primers under the reaction conditions: 95 °C for 5 min and 40 cycles of 95 °C for 10 s, 58 °C for 20 s, and 72 °C for 20 s. The quantitative PCR primers for ATDC5 were as follows: 5′-CTCCTGAGCGCAAGTACTCT-3′ (forward) and 5′-TACTCCTGCTTGCTGATCCAC-3′ (reverse) for β-actin; 5′-GTGAAATGC CACCTTTTGACA-3′ (forward) and 5′-GATTTGAAGCTGGATGCTCT-3′ (reverse) for IL-1β; 5′-CTTCCATCCAGTTGCCTT-3′ (forward) and 5′-CTGTGAAGTCTCCTCTCCG-3′ (reverse) for IL-6. The quantitative PCR primers for THP-1 were as follows: 5′-TCCTCCTGAGC GCAAGTACTCC-3′ (forward) and 5′-CATACTCCTGCTTGCTGATCCAC-3′ (reverse) for β-actin; 5′-CTCTCTCCTTTCAGGGCCAA-3′ (forward) and 5′-GCGGTTGCTCATCAGAATGT-3′ (reverse) for IL-1β; 5′-ACTCACCTCTTCAGAACGAATTG-3′ (forward) and 5′-CCATCTT TGGAAGGTTCAGGTTG-3′ (reverse) for IL-6.

4.6. ELISA

After treating the cells according to the experimental requirements, cell culture supernatants were collected, and the expression of IL-1β and IL-6 was detected by ELISA using Mouse IL-1β (BOSTER, Wuhan, China) and Mouse IL-6 ELISA kits (BOSTER, Wuhan, China) according to the manufacturer's instructions.

4.7. Western Blot Assay

Protein was extracted from ATDC5 cells using the Protein Extraction Solution (MDL) supplemented with protease inhibitor (MDL). A BCA protein analysis kit (MDL) was used to quantify protein concentration. Protein was separated by 10% sodium dodecyl sulfate polyacrylamide gel electrophoresis (SDS-PAGE) and transferred to polyvinylidene fluoride membranes (0.22 μm; Millipore, Bedford, USA). After blocking non-specific interactions, the membranes were incubated overnight at 4 °C with primary antibody against β-actin (MDL), Aggrecan, Bax, Bcl-2, LC3 (Proteintech), and COL2A1(Santa Cruz). Next, the membranes were incubated for 1 h with corresponding HRP*Polyclonal Goat Anti-Rabbit IgG (H + L) (MDL) conjugated with horseradish peroxidase at room temperature. Finally, the proteins were stained and captured using Pierce ECL Western Blotting Substrate (Thermo Scientific) on the Bio-Rad ChemiDoc MP imaging system (Bio-Rad Laboratories, Hercules, CA, USA).

4.8. TUNEL Assay

Cells were seeded into 24-well plates containing the chamber slides and cultured overnight. Next, cells were treated with LPS (10 μg/mL) for 24 h alone or in combination with different HA derivatives (40 μg/mL) for an additional 24 h. The cell apoptotic pattern was detected using the One Step TUNEL Apoptosis Assay Kit (Beyotime) according to the protocols of the manufacturer. Briefly, the cells were fixed with 4% paraformaldehyde in phosphate buffered saline for 30 min, rinsed with PBS, and permeabilized with 0.3% TritonX-100 for 5 min on ice followed by TUNEL for 1 h at 37 °C in the dark. After rinsing f3 times with a PBS solution containing 0.1% Triton X-100 and 5 mg/mL BSA, cells were stained with DAPI solution (Servicebio) for 5 min at room temperature. The cells with green fluorescence were detected as apoptotic cells. After washing with PBS again for three times, cells were observed under a fluorescence microscope (Leica, Wetzlar, Germany). TUNEL-positive apoptotic cells were labeled with FITC (green), and cell nuclei were stained with DAPI (blue).

4.9. Immunofluorescence (IF) Assay

Cells were seeded into 24-well plates containing the chamber slides and cultured overnight. Cells were treated with 10 μg/mL of LPS, followed by 40 μg/mL of different HA derivatives for 24 h. The cells growing on the chamber slides (cell-climbing films) were fixed with 4% paraformaldehyde and blocked with goat serum (Beyotime) for 1 h at 37 °C. Next, cells were incubated overnight at 4 °C with anti-LC3 primary antibody (Proteintech). Subsequently, cells were incubated with fluorescence-labeled goat anti-rabbit secondary antibody (MDL) for 1 h in the dark at 37 °C. Then, cells were stained with DAPI solution

(Servicebio) for 10 min in the dark. Finally, cells were imaged under the Leica DM3000 fluorescent microscope.

4.10. EdU Flow Cytometry Assay

Cell proliferative activity was examined using the Cell-Light EdU Apollo567 In Vitro Flow Cytometry Kit (Ribo Biotechnology, Guangzhou, China) according to the instructions of the manufacturer. Briefly, cells were treated with 10 μg/mL of LPS, then 40 μg/mL of different HA derivatives for 24 h. Next, 50 μM EdU solution was added to cells. After 2 h of incubation, cells were fixed with 4% paraformaldehyde for 20 min and treated with 2 mg/mL glycine for 5 min. The cells were then incubated with 0.5% TritonX-100 solution for 10 min and stained with Apollo staining solution for an additional 10 min in the dark at room temperature. After rinsing for 3 times with PBS, the cells were suspended again and analyzed by flow cytometry.

4.11. Statistical Analysis

Data were analyzed using the GraphPad Prism software Version 8.3.0 (San Diego, CA, USA). Results are shown as mean ± standard deviation. The difference among groups was analyzed using one-way or two-way ANOVA and the Turkey test. A statistically significant difference was defined at $p < 0.05$.

Supplementary Materials: The following supporting information can be downloaded at: https://www.mdpi.com/article/10.3390/molecules27175619/s1, Supplementary Materials: Materials and Methods S1 The detailed synthesis procedures of glycosides of HA oligosaccharides and its derivatives. Figure S1 ^{1}H NMR, ^{13}C NMR HRMS and HPLC spectra of the compounds **2–15**. Table S1 HPLC of compounds **6**, **7**, **10**, **11**, **14** and **15**. Figure S2 o-HA derivatives had no pro-inflammatory effect on THP-1 cells. Figure S3 Different HA derivatives down-regulated IL-1β and IL-6 protein levels in LPS-induced inflammatory injury in ATDC5 cells. Figure S4 Different HA derivatives alleviated the inflammatory injury of THP-1 cells induced by LPS. Figure S5 The ratio of LC3-II/LC3-I was determined by Western blot analysis after using autophagy inhibitor.

Author Contributions: Conceptualization, Z.L. and J.L. (Jianhao Lin); data curation, Z.L. and J.L. (Jianhao Lin); formal analysis, H.H. and X.D.; funding acquisition, Z.L. and J.L. (Jianhao Lin); investigation, H.H. and X.D.; methodology, H.H., X.D., D.X. and Z.L.; project administration, J.L. (Jianhao Lin); resources, J.L. (Jianhao Lin); software, H.H. and X.D.; supervision, D.X. and Z.L.; validation, H.H., X.D., J.L. (Jianjing Lin) and Z.L.; visualization, H.H. and X.D.; writing—original draft, H.H. and X.D.; writing—review & editing, H.H., X.D., D.X., J.L. (Jianjing Lin) and Z.L. All authors have read and agreed to the published version of the manuscript.

Funding: This work was supported by grants from the National Natural Science Foundation of China (NSFC) (81973606, 82151233) and National Key Research and Development Program of China (2020YFC2004904).

Institutional Review Board Statement: Not applicable.

Informed Consent Statement: Not applicable.

Data Availability Statement: The data presented in this study are available on request from the corresponding author. The data are not publicly available due to data protection protocol among researchers.

Conflicts of Interest: The authors declare no conflict of interest.

References

1. Hunter, D.J.; Bierma-Zeinstra, S. Osteoarthritis. *Lancet* **2019**, *393*, 1745–1759. [CrossRef]
2. Sofat, N. Analysing the role of endogenous matrix molecules in the development of osteoarthritis. *Int. J. Exp. Pathol.* **2009**, *90*, 463–479. [CrossRef]
3. Primorac, D.; Molnar, V. Knee Osteoarthritis: A Review of Pathogenesis and State-Of-The-Art Non-Operative Therapeutic Considerations. *Genes* **2020**, *11*, 854. [CrossRef]
4. From the Centers for Disease Control and Prevention. Prevalence and impact of arthritis among women—United States, 1989–1991. *JAMA* **1995**, *273*, 1820. [CrossRef]

5. Hunter, D.J.; Schofield, D.; Callander, E. The individual and socioeconomic impact of osteoarthritis. *Nat. Rev. Rheumatol.* **2014**, *10*, 437–441. [CrossRef]
6. Global, regional, and national incidence, prevalence, and years lived with disability for 354 diseases and injuries for 195 countries and territories, 1990–2017: A systematic analysis for the Global Burden of Disease Study 2017. *Lancet* **2018**, *392*, 1789–1858. [CrossRef]
7. Varela-Eirin, M.; Loureiro, J.; Fonseca, E.; Corrochano, S.; Caeiro, J.R.; Collado, M.; Mayan, M.D. Cartilage regeneration and ageing: Targeting cellular plasticity in osteoarthritis. *Ageing Res. Rev.* **2018**, *42*, 56–71. [CrossRef]
8. Charlier, E.; Deroyer, C.; Ciregia, F.; Malaise, O.; Neuville, S.; Plener, Z.; Malaise, M.; de Seny, D. Chondrocyte dedifferentiation and osteoarthritis (OA). *Biochem. Pharmacol.* **2019**, *165*, 49–65. [CrossRef]
9. Zheng, L.; Zhang, Z.; Sheng, P.; Mobasheri, A. The role of metabolism in chondrocyte dysfunction and the progression of osteoarthritis. *Ageing Res. Rev.* **2021**, *66*, 101249. [CrossRef]
10. Huang, Z.Y.; Stabler, T.; Pei, F.X.; Kraus, V.B. Both systemic and local lipopolysaccharide (LPS) burden are associated with knee OA severity and inflammation. *Osteoarthr. Cartil.* **2016**, *24*, 1769–1775. [CrossRef]
11. Binvignat, M.; Sokol, H.; Mariotti-Ferrandiz, E.; Berenbaum, F.; Sellam, J. Osteoarthritis and gut microbiome. *Jt. Bone Spine* **2021**, *88*, 105203. [CrossRef] [PubMed]
12. Yao, Y.; Wang, Y. ATDC5: An excellent in vitro model cell line for skeletal development. *J. Cell. Biochem.* **2013**, *114*, 1223–1229. [CrossRef]
13. Wilhelm, D.; Kempf, H.; Bianchi, A.; Vincourt, J.B. ATDC5 cells as a model of cartilage extracellular matrix neosynthesis, maturation and assembly. *J. Proteom.* **2020**, *219*, 103718. [CrossRef] [PubMed]
14. Santoro, A.; Conde, J.; Scotece, M.; Abella, V.; López, V.; Pino, J.; Gómez, R.; Gómez-Reino, J.J.; Gualillo, O. Choosing the right chondrocyte cell line: Focus on nitric oxide. *J. Orthop. Res.* **2015**, *33*, 1784–1788. [CrossRef]
15. Zhang, Q.; Bai, X.; Wang, R.; Zhao, H.; Wang, L.; Liu, J.; Li, M.; Chen, Z.; Wang, Z.; Li, L.; et al. 4-octyl Itaconate inhibits lipopolysaccharide (LPS)-induced osteoarthritis via activating Nrf2 signalling pathway. *J. Cell Mol. Med.* **2022**, *26*, 1515–1529. [CrossRef]
16. Conde, J.; Gomez, R.; Bianco, G.; Scotece, M.; Lear, P.; Dieguez, C.; Gomez-Reino, J.; Lago, F.; Gualillo, O. Expanding the adipokine network in cartilage: Identification and regulation of novel factors in human and murine chondrocytes. *Ann. Rheum. Dis.* **2011**, *70*, 551–559. [CrossRef]
17. Weindl, G.; Schaller, M.; Schäfer-Korting, M.; Korting, H.C. Hyaluronic acid in the treatment and prevention of skin diseases: Molecular biological, pharmaceutical and clinical aspects. *Ski. Pharmacol. Physiol.* **2004**, *17*, 207–213. [CrossRef]
18. Rohrich, R.J.; Ghavami, A.; Crosby, M.A. The role of hyaluronic acid fillers (Restylane) in facial cosmetic surgery: Review and technical considerations. *Plast Reconstr. Surg.* **2007**, *120*, 41s–54s. [CrossRef]
19. Neustadt, D.H. Intra-articular injections for osteoarthritis of the knee. *Cleve Clin. J. Med.* **2006**, *73*, 897–911.
20. Am McGrath, A.F.M. A Comparison of Intra-Articular Hyaluronic Acid Competitors in the Treatment of Mild to Moderate Knee Osteoarthritis. *J. Arthritis* **2013**, *2*, 1–5. [CrossRef]
21. Belcher, C.; Yaqub, R.; Fawthrop, F.; Bayliss, M.; Doherty, M. Synovial fluid chondroitin and keratan sulphate epitopes, glycosaminoglycans, and hyaluronan in arthritic and normal knees. *Ann. Rheum. Dis.* **1997**, *56*, 299–307. [CrossRef] [PubMed]
22. Moreland, L.W. Intra-articular hyaluronan (hyaluronic acid) and hylans for the treatment of osteoarthritis: Mechanisms of action. *Arthritis. Res. Ther.* **2003**, *5*, 54–67. [CrossRef] [PubMed]
23. Cooper, C.; Rannou, F.; Richette, P.; Bruyère, O.; Al-Daghri, N.; Altman, R.D.; Brandi, M.L.; Collaud Basset, S.; Herrero-Beaumont, G.; Migliore, A.; et al. Use of Intraarticular Hyaluronic Acid in the Management of Knee Osteoarthritis in Clinical Practice. *Arthritis Care Res.* **2017**, *69*, 1287–1296. [CrossRef]
24. Bowman, S.; Awad, M.E.; Hamrick, M.W.; Hunter, M.; Fulzele, S. Recent advances in hyaluronic acid based therapy for osteoarthritis. *Clin. Transl. Med.* **2018**, *7*, 6. [CrossRef]
25. Tavianatou, A.G.; Caon, I.; Franchi, M.; Piperigkou, Z.; Galesso, D.; Karamanos, N.K. Hyaluronan: Molecular size-dependent signaling and biological functions in inflammation and cancer. *Febs J.* **2019**, *286*, 2883–2908. [CrossRef]
26. Aggarwal, A.; Sempowski, I.P. Hyaluronic acid injections for knee osteoarthritis. Systematic review of the literature. *Can. Fam. Physician* **2004**, *50*, 249–256.
27. Ghosh, P.; Guidolin, D. Potential mechanism of action of intra-articular hyaluronan therapy in osteoarthritis: Are the effects molecular weight dependent? *Semin Arthritis Rheum* **2002**, *32*, 10–37. [CrossRef]
28. Barreto, R.B.; Sadigursky, D.; de Rezende, M.U.; Hernandez, A.J. Effect of hyaluronic acid on chondrocyte apoptosis. *Acta Ortop Bras* **2015**, *23*, 90–93. [CrossRef]
29. Patti, A.M.; Gabriele, A.; Vulcano, A.; Ramieri, M.T.; Della Rocca, C. Effect of hyaluronic acid on human chondrocyte cell lines from articular cartilage. *Tissue Cell* **2001**, *33*, 294–300. [CrossRef]
30. Cowman, M.K.; Shortt, C.; Arora, S.; Fu, Y.; Villavieja, J.; Rathore, J.; Huang, X.; Rakshit, T.; Jung, G.I.; Kirsch, T. Role of Hyaluronan in Inflammatory Effects on Human Articular Chondrocytes. *Inflammation* **2019**, *42*, 1808–1820. [CrossRef]
31. Gupta, R.C.; Lall, R.; Srivastava, A.; Sinha, A. Hyaluronic Acid: Molecular Mechanisms and Therapeutic Trajectory. *Front Vet Sci* **2019**, *6*, 192. [CrossRef] [PubMed]
32. Evanko, S.P.; Tammi, M.I.; Tammi, R.H.; Wight, T.N. Hyaluronan-dependent pericellular matrix. *Adv. Drug Deliv. Rev.* **2007**, *59*, 1351–1365. [CrossRef] [PubMed]

33. Yang, C.; Cao, M.; Liu, H.; He, Y.; Xu, J.; Du, Y.; Liu, Y.; Wang, W.; Cui, L.; Hu, J.; et al. The high and low molecular weight forms of hyaluronan have distinct effects on CD44 clustering. *J. Biol. Chem.* **2012**, *287*, 43094–43107. [CrossRef] [PubMed]
34. Knudson, W.; Ishizuka, S.; Terabe, K.; Askew, E.B.; Knudson, C.B. The pericellular hyaluronan of articular chondrocytes. *Matrix Biol.* **2019**, *78–79*, 32–46. [CrossRef] [PubMed]
35. Campo, G.M.; Avenoso, A.; D'Ascola, A.; Scuruchi, M.; Prestipino, V.; Calatroni, A.; Campo, S. 6-Mer hyaluronan oligosaccharides increase IL-18 and IL-33 production in mouse synovial fibroblasts subjected to collagen-induced arthritis. *Innate Immun* **2012**, *18*, 675–684. [CrossRef]
36. Campo, G.M.; Avenoso, A.; D'Ascola, A.; Prestipino, V.; Scuruchi, M.; Nastasi, G.; Calatroni, A.; Campo, S. Hyaluronan differently modulates TLR-4 and the inflammatory response in mouse chondrocytes. *Biofactors* **2012**, *38*, 69–76. [CrossRef]
37. Taylor, K.R.; Yamasaki, K.; Radek, K.A.; Nardo, A.D.; Goodarzi, H.; Golenbock, D.; Beutler, B.; Gallo, R.L. Recognition of hyaluronan released in sterile injury involves a unique receptor complex dependent on Toll-like receptor 4, CD44, and MD-2. *J. Biol. Chem.* **2007**, *282*, 18265–18275. [CrossRef]
38. Campo, G.M.; Avenoso, A.; D'Ascola, A.; Scuruchi, M.; Calatroni, A.; Campo, S. Beta-arrestin-2 negatively modulates inflammation response in mouse chondrocytes induced by 4-mer hyaluronan oligosaccharide. *Mol. Cell Biochem.* **2015**, *399*, 201–208. [CrossRef]
39. D'Ascola, A.; Scuruchi, M.; Ruggeri, R.M.; Avenoso, A.; Mandraffino, G.; Vicchio, T.M.; Campo, S.; Campo, G.M. Hyaluronan oligosaccharides modulate inflammatory response, NIS and thyreoglobulin expression in human thyrocytes. *Arch. Biochem. Biophys.* **2020**, *694*, 108598. [CrossRef]
40. Scuruchi, M.; D'Ascola, A.; Avenoso, A.; Campana, S.; Abusamra, Y.A.; Spina, E.; Calatroni, A.; Campo, G.M.; Campo, S. 6-Mer Hyaluronan Oligosaccharides Modulate Neuroinflammation and α-Synuclein Expression in Neuron-Like SH-SY5Y Cells. *J. Cell Biochem.* **2016**, *117*, 2835–2843. [CrossRef]
41. Dong, Y.; Arif, A.; Olsson, M.; Cali, V.; Hardman, B.; Dosanjh, M.; Lauer, M.; Midura, R.J.; Hascall, V.C.; Brown, K.L.; et al. Endotoxin free hyaluronan and hyaluronan fragments do not stimulate TNF-α, interleukin-12 or upregulate co-stimulatory molecules in dendritic cells or macrophages. *Sci. Rep.* **2016**, *6*, 36928. [CrossRef] [PubMed]
42. Olsson, M.; Bremer, L.; Aulin, C.; Harris, H.E. Fragmented hyaluronan has no alarmin function assessed in arthritis synovial fibroblast and chondrocyte cultures. *Innate Immun.* **2018**, *24*, 131–141. [CrossRef]
43. Lu, X.; Kamat, M.N.; Huang, L.; Huang, X. Chemical synthesis of a hyaluronic acid decasaccharide. *J. Org. Chem.* **2009**, *74*, 7608–7617. [CrossRef] [PubMed]
44. Blatter, G.; Jacquinet, J.-C. The use of 2-deoxy-2-trichloroacetamido-d-glucopyranose derivatives in syntheses of hyaluronic acid-related tetra-, hexa-, and octa-saccharides having a methyl β-d-glucopyranosiduronic acid at the reducing end. *Carbohydr. Res.* **1996**, *288*, 109–125. [CrossRef]
45. Furukawa, T.; Hinou, H.; Shimawaki, K.; Nishimura, S.-I. A potential glucuronate glycosyl donor with 2-O-acyl-6,3-lactone structure: Efficient synthesis of glycosaminoglycan disaccharides. *Tetrahedron Lett.* **2011**, *52*, 5567–5570. [CrossRef]
46. Tokita, Y.; Okamoto, A. Hydrolytic degradation of hyaluronic acid. *Polym. Degrad. Stab.* **1995**, *48*, 269–273. [CrossRef]
47. Song, Z.; Meng, L.; Xiao, Y.; Zhao, X.; Fang, J.; Zeng, J.; Wan, Q. Calcium hypophosphite mediated deiodination in water: Mechanistic insights and applications in large scale syntheses of d-quinovose and d-rhamnose. *Green Chem.* **2019**, *21*, 1122–1127. [CrossRef]
48. Kobayashi, S.; Fujikawa, S.-i.; Ohmae, M. Enzymatic Synthesis of Chondroitin and Its Derivatives Catalyzed by Hyaluronidase. *J. Am. Chem. Soc.* **2003**, *125*, 14357–14369. [CrossRef]
49. Wipf, P.; Eyer, B.R.; Yamaguchi, Y.; Zhang, F.; Neal, M.D.; Sodhi, C.P.; Good, M.; Branca, M.; Prindle, T.; Lu, P.; et al. Synthesis of anti-inflammatory α-and β-linked acetamidopyranosides as inhibitors of toll-like receptor 4 (TLR4). *Tetrahedron Lett.* **2015**, *56*, 3097–3100. [CrossRef]
50. Zhang, X.; Liu, H.; Yao, W.; Meng, X.; Li, Z. Semisynthesis of Chondroitin Sulfate Oligosaccharides Based on the Enzymatic Degradation of Chondroitin. *J. Org. Chem.* **2019**, *84*, 7418–7425. [CrossRef]
51. Sha, M.; Yao, W.; Zhang, X.; Li, Z. Synthesis of structure-defined branched hyaluronan tetrasaccharide glycoclusters. *Tetrahedron Lett.* **2017**, *58*, 2910–2914. [CrossRef]
52. Tanaka, T.; Nagai, H.; Noguchi, M.; Kobayashi, A.; Shoda, S.-I. One-step conversion of unprotected sugars to β-glycosyl azides using 2-chloroimidazolinium salt in aqueous solution. *Chem. Commun.* **2009**, 3378–3379. [CrossRef] [PubMed]
53. Tanida, I.; Ueno, T.; Kominami, E. LC3 and Autophagy. *Methods Mol. Biol.* **2008**, *445*, 77–88. [CrossRef] [PubMed]
54. Kabeya, Y.; Mizushima, N.; Ueno, T.; Yamamoto, A.; Kirisako, T.; Noda, T.; Kominami, E.; Ohsumi, Y.; Yoshimori, T. LC3, a mammalian homologue of yeast Apg8p, is localized in autophagosome membranes after processing. *Embo J.* **2000**, *19*, 5720–5728. [CrossRef] [PubMed]
55. Huang, H.; Xing, D.; Zhang, Q.; Li, H.; Lin, J.; He, Z.; Lin, J. LncRNAs as a new regulator of chronic musculoskeletal disorder. *Cell Prolif* **2021**, *54*, e13113. [CrossRef]
56. Carballo, C.B.; Nakagawa, Y.; Sekiya, I.; Rodeo, S.A. Basic Science of Articular Cartilage. *Clin. Sports Med.* **2017**, *36*, 413–425. [CrossRef]
57. Gao, Y.; Liu, S.; Huang, J.; Guo, W.; Chen, J.; Zhang, L.; Zhao, B.; Peng, J.; Wang, A.; Wang, Y.; et al. The ECM-cell interaction of cartilage extracellular matrix on chondrocytes. *Biomed. Res. Int.* **2014**, *2014*, 648459. [CrossRef]

58. Shahid, M.; Manchi, G.; Slunsky, P.; Naseer, O.; Fatima, A.; Leo, B.; Raila, J. A systemic review of existing serological possibilities to diagnose canine osteoarthritis with a particular focus on extracellular matrix proteoglycans and protein. *Pol. J. Vet. Sci.* **2017**, *20*, 189–201. [CrossRef]
59. Sun, A.R.; Friis, T.; Sekar, S.; Crawford, R.; Xiao, Y.; Prasadam, I. Is Synovial Macrophage Activation the Inflammatory Link Between Obesity and Osteoarthritis? *Curr. Rheumatol. Rep.* **2016**, *18*, 57. [CrossRef]
60. Bauer, C.; Niculescu-Morzsa, E.; Jeyakumar, V.; Kern, D.; Späth, S.S.; Nehrer, S. Chondroprotective effect of high-molecular-weight hyaluronic acid on osteoarthritic chondrocytes in a co-cultivation inflammation model with M1 macrophages. *J. Inflamm.* **2016**, *13*, 31. [CrossRef]
61. Hiramitsu, T.; Yasuda, T.; Ito, H.; Shimizu, M.; Julovi, S.M.; Kakinuma, T.; Akiyoshi, M.; Yoshida, M.; Nakamura, T. Intercellular adhesion molecule-1 mediates the inhibitory effects of hyaluronan on interleukin-1beta-induced matrix metalloproteinase production in rheumatoid synovial fibroblasts via down-regulation of NF-kappaB and p38. *Rheumatol* **2006**, *45*, 824–832. [CrossRef] [PubMed]
62. Wang, X.; Majumdar, T.; Kessler, P.; Ozhegov, E.; Zhang, Y.; Chattopadhyay, S.; Barik, S.; Sen, G.C. STING Requires the Adaptor TRIF to Trigger Innate Immune Responses to Microbial Infection. *Cell Host Microbe* **2016**, *20*, 329–341. [CrossRef] [PubMed]
63. Vijay, K. Toll-like receptors in immunity and inflammatory diseases: Past, present, and future. *Int. Immunopharmacol.* **2018**, *59*, 391–412. [CrossRef] [PubMed]
64. Zhang, H.; Lu, Y.; Wu, B.; Xia, F. Semaphorin 3A mitigates lipopolysaccharide-induced chondrocyte inflammation, apoptosis and extracellular matrix degradation by binding to Neuropilin-1. *Bioengineered* **2021**, *12*, 9641–9654. [CrossRef] [PubMed]
65. Kakizaki, I.; Ibori, N.; Kojima, K.; Yamaguchi, M.; Endo, M. Mechanism for the hydrolysis of hyaluronan oligosaccharides by bovine testicular hyaluronidase. *FEBS J.* **2010**, *277*, 1776–1786. [CrossRef]
66. Kobayashi, S.; Morii, H.; Itoh, R.; Kimura, S.; Ohmae, M. Enzymatic Polymerization to Artificial Hyaluronan: A Novel Method to Synthesize a Glycosaminoglycan Using a Transition State Analogue Monomer. *J. Am. Chem. Soc.* **2001**, *123*, 11825–11826. [CrossRef]
67. Han, W.; Lv, Y.; Sun, Y.; Wang, Y.; Zhao, Z.; Shi, C.; Chen, X.; Wang, L.; Zhang, M.; Wei, B.; et al. The anti-inflammatory activity of specific-sized hyaluronic acid oligosaccharides. *Carbohydr Polym* **2022**, *276*, 118699. [CrossRef]
68. Hwang, H.S.; Kim, H.A. Chondrocyte Apoptosis in the Pathogenesis of Osteoarthritis. *Int. J. Mol. Sci.* **2015**, *16*, 26035–26054. [CrossRef]
69. Duan, R.; Xie, H.; Liu, Z.Z. The Role of Autophagy in Osteoarthritis. *Front Cell Dev. Biol.* **2020**, *8*, 608388. [CrossRef]
70. Mauthe, M.; Orhon, I.; Rocchi, C.; Zhou, X.; Luhr, M.; Hijlkema, K.J.; Coppes, R.P.; Engedal, N.; Mari, M.; Reggiori, F. Chloroquine inhibits autophagic flux by decreasing autophagosome-lysosome fusion. *Autophagy* **2018**, *14*, 1435–1455. [CrossRef]
71. Maldonado, M.; Nam, J. The role of changes in extracellular matrix of cartilage in the presence of inflammation on the pathology of osteoarthritis. *Biomed. Res. Int.* **2013**, *2013*, 284873. [CrossRef] [PubMed]
72. Theocharis, A.D.; Manou, D.; Karamanos, N.K. The extracellular matrix as a multitasking player in disease. *FEBS J.* **2019**, *286*, 2830–2869. [CrossRef] [PubMed]
73. Walker, C.; Mojares, E.; Del Río Hernández, A. Role of Extracellular Matrix in Development and Cancer Progression. *Int. J. Mol. Sci.* **2018**, *19*, 3028. [CrossRef]
74. Shoulders, M.D.; Raines, R.T. Collagen structure and stability. *Annu. Rev. Biochem.* **2009**, *78*, 929–958. [CrossRef] [PubMed]
75. Roughley, P.J.; Mort, J.S. The role of aggrecan in normal and osteoarthritic cartilage. *J. Exp. Orthop.* **2014**, *1*, 8. [CrossRef] [PubMed]
76. Bonnans, C.; Chou, J.; Werb, Z. Remodelling the extracellular matrix in development and disease. *Nat. Reviews. Mol. Cell Biol.* **2014**, *15*, 786–801. [CrossRef]
77. Peng, Z.; Sun, H.; Bunpetch, V.; Koh, Y.; Wen, Y.; Wu, D.; Ouyang, H. The regulation of cartilage extracellular matrix homeostasis in joint cartilage degeneration and regeneration. *Biomaterials* **2021**, *268*, 120555. [CrossRef]
78. Sanchez, C.; Bay-Jensen, A.C.; Pap, T.; Dvir-Ginzberg, M.; Quasnichka, H.; Barrett-Jolley, R.; Mobasheri, A.; Henrotin, Y. Chondrocyte secretome: A source of novel insights and exploratory biomarkers of osteoarthritis. *Osteoarthr. Cartil.* **2017**, *25*, 1199–1209. [CrossRef]
79. Marinho, A.; Nunes, C.; Reis, S. Hyaluronic Acid: A Key Ingredient in the Therapy of Inflammation. *Biomolecules* **2021**, *11*, 1518. [CrossRef]
80. Amorim, S.; Reis, C.A.; Reis, R.L.; Pires, R.A. Extracellular Matrix Mimics Using Hyaluronan-Based Biomaterials. *Trends Biotechnol* **2021**, *39*, 90–104. [CrossRef]

molecules

MDPI

Review

Multivalent Pyrrolidine Iminosugars: Synthesis and Biological Relevance

Yali Wang [1], Jian Xiao [1], Aiguo Meng [2] and Chunyan Liu [1,*]

[1] College of Pharmacy, North China University of Science and Technology, Tangshan 063000, China
[2] Affiliated Hospital, North China University of Science and Technology, Tangshan 063000, China
[*] Correspondence: chunyanliu@ncst.edu.cn

Abstract: Recently, the strategy of multivalency has been widely employed to design glycosidase inhibitors, as glycomimetic clusters often induce marked enzyme inhibition relative to monovalent analogs. Polyhydroxylated pyrrolidines, one of the most studied classes of iminosugars, are an attractive moiety due to their potent and specific inhibition of glycosidases and glycosyltransferases, which are associated with many crucial biological processes. The development of multivalent pyrrolidine derivatives as glycosidase inhibitors has resulted in several promising compounds that stand out. Herein, we comprehensively summarized the different synthetic approaches to the preparation of multivalent pyrrolidine clusters, from total synthesis of divalent iminosugars to complex architectures bearing twelve pyrrolidine motifs. Enzyme inhibitory properties and multivalent effects of these synthesized iminosugars were further discussed, especially for some less studied therapeutically relevant enzymes. We envision that this comprehensive review will help extend the applications of multivalent pyrrolidine iminosugars in future studies.

Keywords: iminosugar; pyrrolidine; multivalent effect; glucosidase inhibitors

Citation: Wang, Y.; Xiao, J.; Meng, A.; Liu, C. Multivalent Pyrrolidine Iminosugars: Synthesis and Biological Relevance. *Molecules* **2022**, 27, 5420. https://doi.org/10.3390/molecules27175420

Academic Editors: Jian Yin, Jing Zeng and De-Cai Xiong

Received: 22 July 2022
Accepted: 21 August 2022
Published: 24 August 2022

1. Introduction

Iminosugars, containing an endocycling nitrogen atom that effectively mimics carbohydrates by facilitating the reversible and competitive inhibition of their processing enzymes, have generated much attention in recent years as targets in the treatment of a wide range of illnesses (e.g., diabetes, cancer, tuberculosis, and lysosomal storage disorders, etc.) [1–6]. Given the enormous range of biochemical events in which carbohydrate processing enzymes are implicated, iminosugars have enormous potential to be developed as inhibitors of glycosidases (glycoside hydrolases), glycosyltransferases (glycoside synthases), metalloproteinases, and nucleoside-processing enzymes [7–11]. It is worth noting that structural modifications to find potent inhibitors of the above enzymes among two of the most studied classes of iminosugars, polyhydroxylated pyrrolidines and piperidines, arouse great interest because of the wide range of their biological properties, such as glycosidase inhibition, shown over the past five decades [12–16].

The most famous representative iminosugars belonging to piperidine are derivatives of 1-deoxynojirimycin (DNJ, Figure 1). Since DNJ's isolation from white mulberry root bark in 1976, hundreds of artificial iminosugars based on DNJ have been synthesized and their bioactivities evaluated [17–21]. Two *N*-alkylated DNJ derivatives are approved drugs, *N*-hydroxyethyl-1-deoxynojirimycin (Miglitol, Figure 1) to treat type II diabetes, and *N*-butyl-1-deoxynojirimycin (Miglustat, Figure 1) to treat lysosomal storage disorders (e.g., Gaucher disease). Similarly, five-membered iminocyclitols, also known as pyrrolidine iminosugars, exhibited excellent inhibition toward glycosidases. For example, 2,5-dihydroxymethyl-3,4-dihydroxypyrrolidine (DMDP, Figure 1), the first pyrrolidine iminosugar extracted from the leaves of *Derris elliptica* in 1976 [22], proved to be a potent glycosidase inhibitor; subsequently, its analogs were also found to have significant effects on glycosidases [23–25]. 1,4-Dideoxy-1,4-imino-D-arabinitol (DAB, Figure 1), isolated from the fruit of *Angylocalyx boutiqueanus*, exhibited strong inhibition of glycogen phosphorylase, and is currently being explored for the treatment of type II diabetes [26,27]. 1,4-Dideoxy-1,4-imino-L-arabinitol (LAB, Figure 1), the enantiomer of DAB, displayed more potent specific glycosidase inhibition. A new α-glucosidase inhibitor based on LAB was reported by Kato et al. in 2012, which showed huge potential in reducing elevated plasma glucose after food intake when tested in vivo with a carbohydrate load at doses approximately ten times lower than the required dose of miglitol [27–30]. In addition, Radicamines A and B have attracted extensive interest because of their potent inhibition of α-glucosidases and potential pharmaceutical applications [31–33]. Overall, ample evidence has established that iminosugars have anti-diabetic effects [34–38]. Both classes of iminosugar derivatives would be promising drug candidates, and therefore the development of synthetic strategies and the evaluation of bioactivities are of decisive importance.

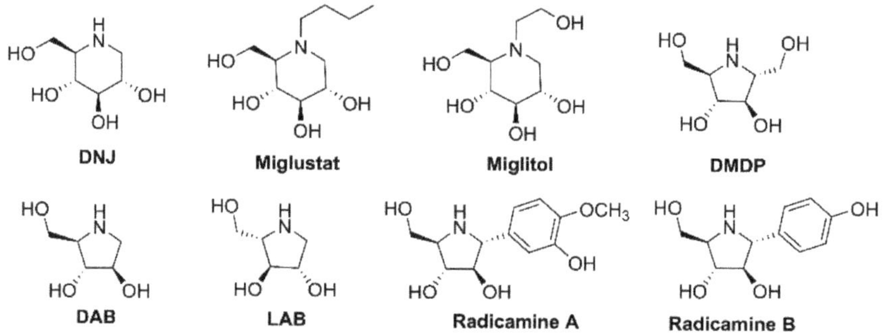

Figure 1. Examples of natural and synthetic iminosugars.

However, as previously introduced, only a few drugs are on the market. New strategies for developing iminosugar-based glycosidase inhibitors to understand vital biological processes or as clinical candidates are therefore major challenges in both academia and the pharmaceutical industry [39–41]. In the last decade, a multivalent glycosidase inhibition effect, which has been extensively used in developing lectin inhibitors to seek new therapeutic opportunities for carbohydrate-related diseases, was found and rapidly developed [42–49]. Johns and Johnson first reported the synthesis of divalent iminosugars and explored the contribution of the multivalent strategy to biological activity in 1998 [50]. Probably due to the reported multivalent compounds, which could not exhibit the expected inhibition ability to glycosidase enzymes, the design of glycosidase inhibitors has not been able to attract the interest of researchers [51,52]. The factors that hampered the application of multivalency to iminosugars may be as follows: Firstly, there is an intrinsic structural difference between glycosidases and lectins. The surface of lectins shows multiple carbohydrate-binding sites, while glycosides and other carbohydrate-processing enzymes are usually monomeric and therefore bind relatively weakly to multivalent substrates [53–55]. Secondly, the synthesis of multivalent iminosugars is challenging. Since the click reaction was immature before 2001, it was particularly hard to graft several monomers to a skeleton simultaneously [56,57]. In addition, the experimental results obtained were not encouraging [50–52]. It was not a rapidly emerging area with exciting potential until the discovery of a small but quantitative multivalent effect in α-mannosidase inhibition [58]. Based on the extensive literature involving lectins and glycoclusters, several potential interactions have been proposed to explain the multivalent effect. The "bind-and-recapture" process is a classical mode due to the increased concentration of active molecules concentration in proximity to the binding site (Figure 2a). The chelate effect can occur when the enzyme presents more than one active site (Figure 2b). In addition, stronger interactions will occur when some non-catalytic subsites interact with glycoclusters (Figure 2c). Moreover, cross-linking and aggregation processes may prevail with glycoclusters if the enzyme possesses a multimetric nature (Figure 2d) [59,60].

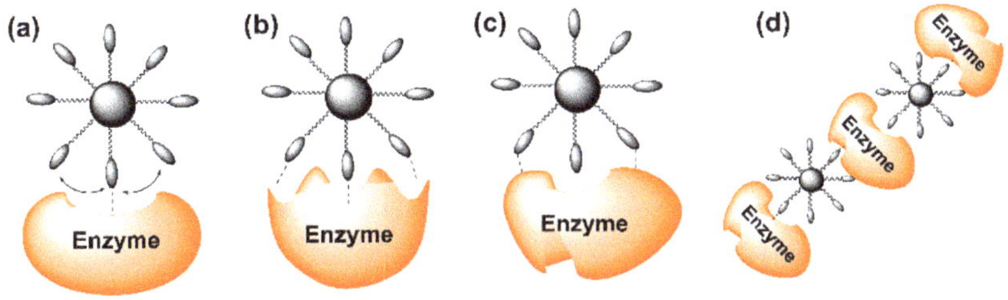

Figure 2. Proposed binding models accounting for the multivalent effect: (**a**) Bind and recapture; (**b**) Chelate process; (**c**) Subsite binding; (**d**) Cross-linking and aggregation.

The construction of multivalent iminosugars follows conventional strategies, including modifications and protecting group chemistry of the iminosugars [50,61,62], coupling reactions using click chemistry [45,48,63], and recently developed supramolecular self-assembly based on π–π stacking or hydrophobic interactions [64–66]. Multimerization of the pyrrolidines using these strategies resulted in some interesting results. For example, multimeric pyrrolidine iminosugars were reported to be the first example of multivalent enhancers of human α-galactosidase A (α-Gal A), an enzyme involved in Fabry disease [49]. Multivalent dendrimers decorated with the DAB exhibited a relevant multivalent effect toward the lysosomal enzyme *N*-acetylgalactosamine-6-sulfatase (GALNS), which is involved in a rare metabolic disorder [67]. Other multivalent effects of multivalent pyrrolidine iminosugars were obtained with carbohydrate-active enzymes such as Golgi α-mannosidase II [68], β-*N*-acetylglucosaminidase [69], and α-L-fucosidase [70]. This unique class of compounds may provide new pharmaceutical opportunities to treat diseases involving carbohydrate-processing enzymes.

Several reviews on the topic of multivalent iminosugars have been published; however, large parts deal with the synthesis and biological properties of multivalent piperidine iminosugars rather than pyrrolidine [12,14,71,72]. Moreover, the latter research field was scarcely reviewed in the decade of its rapid development (2012–2022) [73]. The present review illustrates the detailed synthesis and multivalent effects of all multivalent pyrrolidine iminosugars that have been assayed against various glycosidases and provide an overview of the main achievements made to date.

2. Syntheses of Multivalent Pyrrolidine Iminosugars

2.1. Synthesis of Di- and Trivalent Iminosugars

The first synthesis of tethered di- and trivalent pyrrolidine iminosugars to interfere with carbohydrate processing enzymes was reported by Robina and co-workers in 2013 [70]. At that time, the advantages of the multivalent effect for glycosidase inhibition over the corresponding monomer were gradually realized (e.g., for α- and β-glucosidases [74], for β-galactosidases [75]). Taking advantage of their experience in designing glycosidase inhibitors [76,77], the authors investigated the multivalent approach by comparing the α-L-fucosidase inhibitory activities of multi- and mono-pyrrolidine iminosugars. For this purpose, four di- and trivalent pyrrolidine derivatives (**1**–**4**) were synthesized based on *fuco*-configured 1,4-imino-cyclitols **5** and **6**, which displayed good inhibitory activity towards α-L-fucosidase (Figure 3). The benzylamino pyrrolidine **5**, designed as a monovalent reference for dimer **1** and trimers **2** and **3**, was synthesized for the first time, while the furyl-substituted pyrrolidine **6**, previously reported by the same research group [78], was selected for comparison with trimer **4**.

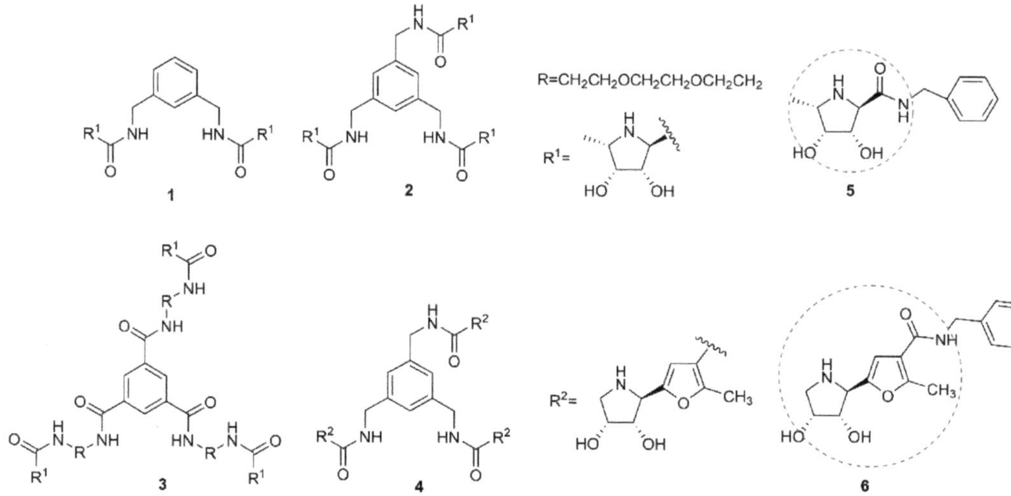

Figure 3. Structures of multivalent pyrrolidine iminosugars **1–4** and monovalent α-L-fucosidase inhibitors **5** and **6**.

The authors employed a classical amide coupling reaction to synthesize the desired iminosugars, starting from commercially available benzylamine and the *O*- and *N*-protected carboxylic acid **7** [70], using PyBOP as the coupling agent and DIPEA as base gave protected intermediates. Then, the excess benzylamine was easily separated by column chromatography. A similar method was used to remove excess amines for the preparations of **1–4**. Finally, isopropylidene deprotection with HCl and subsequent catalytic hydrogenation with H_2/Pd/C gave the corresponding target product **5** in 84% yield. Di- and trivalent iminosugars were obtained with commercially available *m*-xylylenediamine **9** and triamine **10** as the scaffolds for bi- and trivalent glycomimetics, respectively. Moreover, the long-tethered triamine **11** reported by the same research group [79] was chosen as C-3 symmetric template to yield the long spacer trivalent iminosugar **3**. Amide coupling reactions between scaffolds and compound **7** under the same condition above gave the corresponding target products **1–3** in moderate-to-good yields. Similarly, trimer **4** was obtained by the coupling reaction between a previously synthesized pyrrolidine-furan carboxylic acid **12** **[78]** and template **10**, followed by standard reductive hydrogenation in 46% yield (Scheme 1).

To further understand the complicated multivalent effect on α-fucosidase inhibition, Behr, Robina, and co-workers reported a library of divalent pyrrolidine iminosugars **13–17** using polyamine and triazole benzene as spacers to evaluate the contributions of the length and rigidity of the bridge, the number of nitrogen atoms present, and the moieties close to the pyrrolidine to the biological activity of divalent inhibitors [80]. Since there is no report of the monovalent references **18–20**, their synthesis routes were also introduced. The inhibitory effect of chemically diverse spacers in dimers on α-fucosidase was systematically investigated, and a potent and specific α-fucosidase inhibitor (compound **17**, Ki = 3.7 nM) was thus discovered (Figure 4).

The target monovalent inhibitors **18** and **19** were synthesized from the known allyl-pyrrolidines **21** and **24** **[81]**. Starting from the (2*R*)-configured **21**, after dihydroxylation, oxidative cleavage with $NaIO_4$ gave a stable intermediate **22**. Pyrrolidinyl ethanol **18a** was obtained in 37% yield by reducing **22** with sodium borohydride, followed by deprotection with hydrogen and acidification in three steps. The congener **18b** was obtained by reacting **23** with benzylamine followed by deprotection with H_2/Pd–C (10%), in 43% yield. However, the stereoisomers **19a** and **19b** of compounds **18a** and **18b** could not be obtained by the same synthetic route through (2*S*)-configured **24**, mainly due to the key intermediate

25 after reaction with **24**. The authors reported that the aldehyde **25** would go through an epimerization process, which spontaneously opened the pyrrolidine ring to form the conjugated aldehyde **22**, impeding the synthetic purpose [81,82]. Alternatively, protection of the amino group by switching from Bn to Boc solved this problem and gave the clean and stable (2*S*)-configured **26**. Target iminosugars **19a** and **19b** were afforded by reduction and reductive amination under the same conditions as introduced above (Scheme 2).

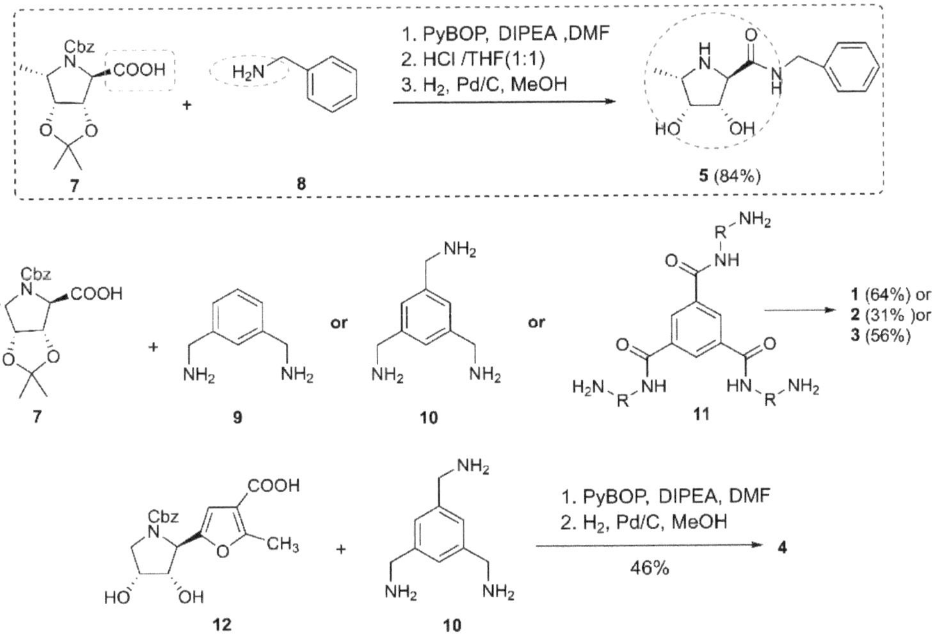

Scheme 1. Synthesis of di- and trivalent pyrrolidine iminosugars **1–4**.

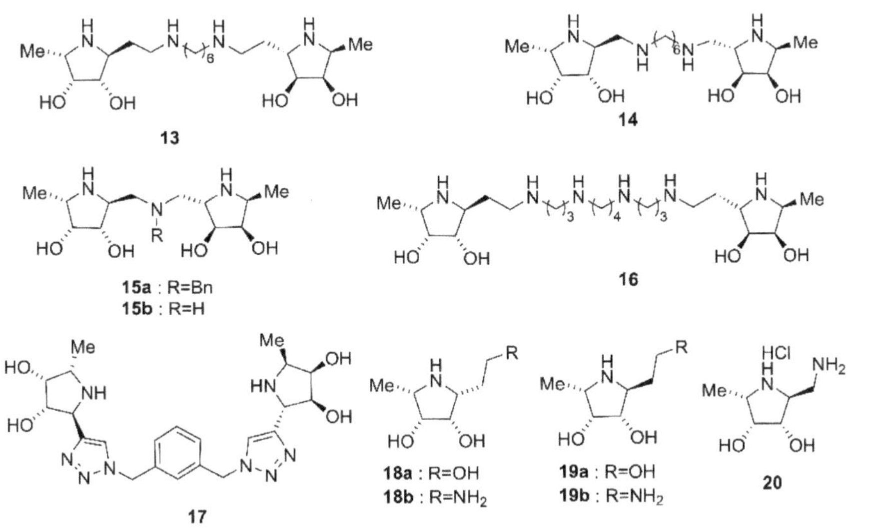

Figure 4. Structures of divalent iminosugars **13–17** and monovalent references **18–20**.

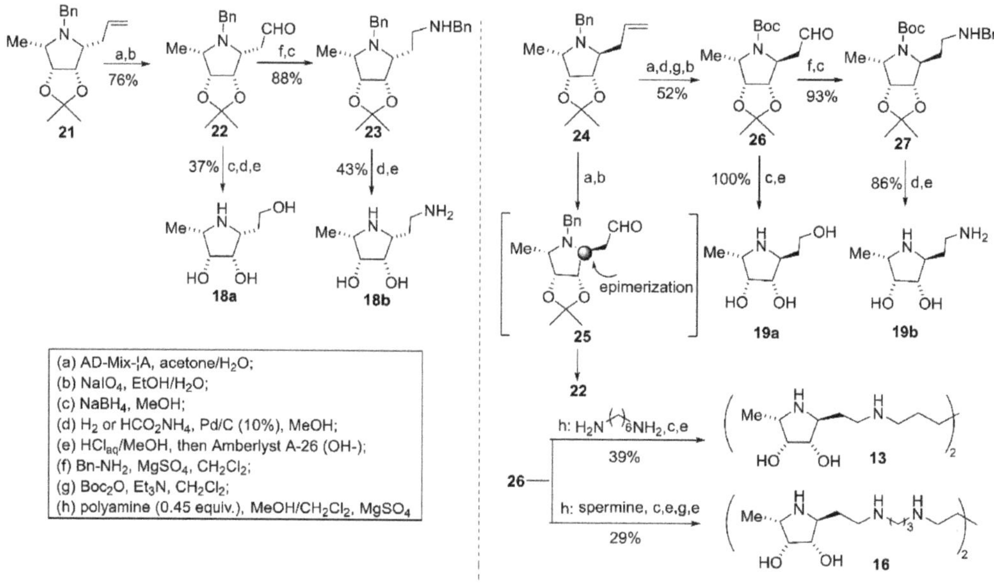

Scheme 2. Synthesis of divalent iminosugars **13**, **16**, and the corresponding monomers.

The synthetic routes for dimers **13** and **16** were similar to those of their corresponding monomers. An excess of compound **26** (2.2-fold) reacted with hexamethylene-diamine or spermine to produce the corresponding di-imines, which were then reduced via sodium borohydride. Silica gel chromatography was employed to separate the excess **26** and yield the corresponding protected dimers **13** and **16**, which were further deprotected to give the target products. It is worth noting that under this method, homologue **16** with a spermine bridge contained some impurities. Hence, a further sequence of Boc protection/purification/deprotection (MeOH: HClaq) was required to obtain pure **16** in 29% yield (Scheme 2).

The synthesis of the divalent iminosugars **14** and **15** was started from the known compound **28** [83], which was reacted with benzylamine and sodium triacetoxyborohydride, followed by acidification and deprotection to afford compound **15a** in 26% yield. Dimer **15b** was obtained by debenzylation of **15a** under the H_2/Pd/C system in 65% yield. Reacting **28** (2-fold excess) with hexamethylenediamine by similar methods ($MgSO_4$ then $NaBH_4$ or amine then $NaBH(OAc)_3$) both gave dimer **29** in a low yield (25%) with an undesired trivalent product **31** (27%). Gratifyingly, the yield of **29** could be increased to 49% by reacting **28** with ethylenediamine in the presence of sodium borohydride and 2,2,2-trifluoroethanol. For exploration, the amino-protected dimer **29** and trimer **30** were both deprotected under hydrochloric acid to generate the target products **14** and **31**, which were likewise tested towards α-fucosidase. Dimer **17** was synthesized due to the good inhibitory activities of (pyrrolidin-2-yl)triazoles shown by the researchers previously [83]. Thus, after the reduction of **28** to **32**, target dimer **17** was generated through a two-step reaction by treating excess **32** with 1,3-bis(azidomethyl)benzene under the catalysts of CuI and DIPEA, followed by acidification. Column chromatography was carried out to remove unreacted **32** and yield **17** in 44% yield (Scheme 3). Increasing results began to highlight the advantages of multivalent effects. However, some contradictory experimental results were still reported. Elucidating the specific binding mechanisms of multi-ligands with enzymes is urgent and challenging. Behr and co-workers reported three stereoisomeric pyrrolidine dimers in 2016 to explore the divalent effect on fucosidase inhibition [84]. The divalent iminosugars (**33**, ***ent*-33**, and ***meso*-33**) were constructed based on a known fucosidase inhibitor **34** reported by Steensma [85]. The monovalent iminosugars **35**, ***ent*-35**, and **36** were also synthesized as referenced (Figure 5).

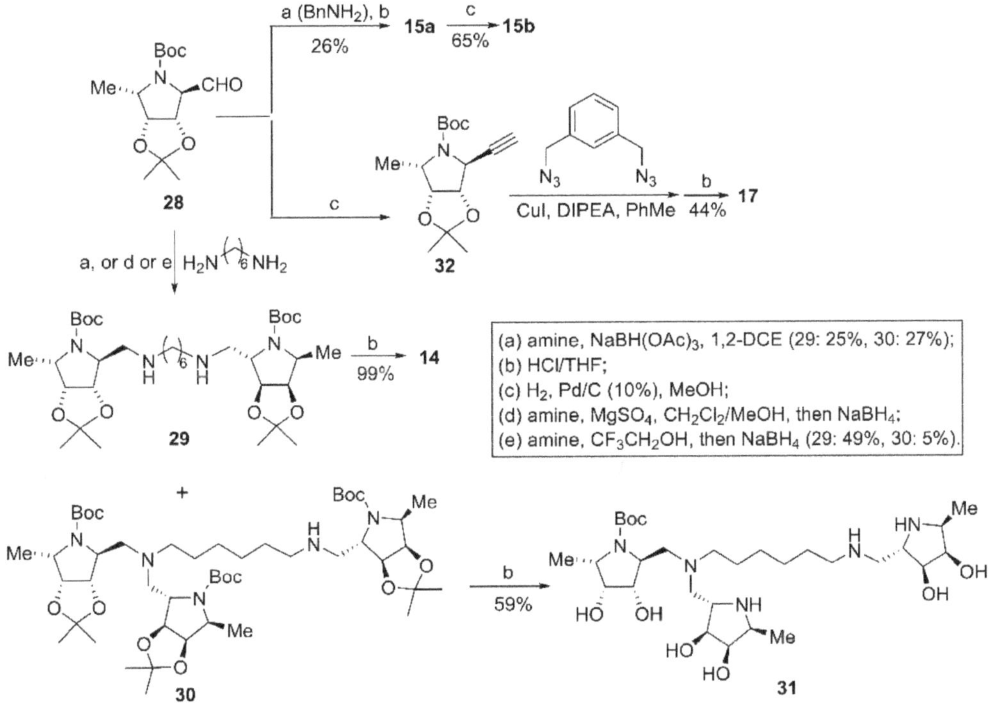

Scheme 3. Synthesis of divalent iminosugars **14**, **15**, and **17**.

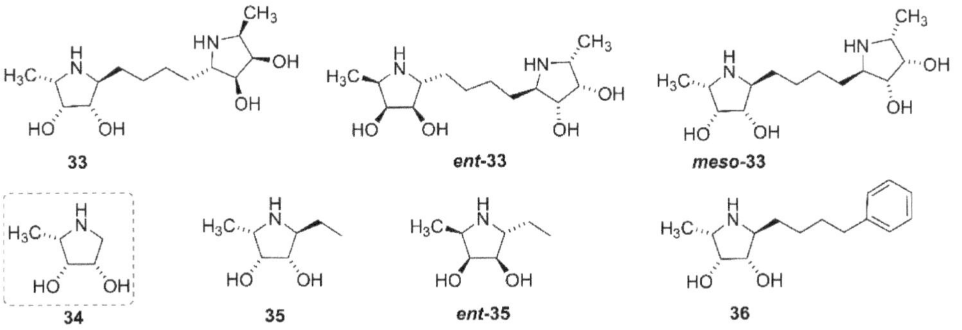

Figure 5. Structures of divalent inhibitors and the corresponding monomers by Behr.

Known compound **37** and its enantiomer ***ent*-37** were used as starting materials for synthesizing homodimer **33** and its enantiomer ***ent*-33**, respectively [86]. Hemiacetal ***ent*-37** was converted to **38** by amination with benzylamine. The intermediate **38** was then subjected to a highly stereoselective ethynylmagnesium bromide-mediated nucleophilic addition to aminoalcohol ***ent*-39** in 70% yield for two steps [87]. Then, azide ***ent*-40** was obtained in 68% yield from ***ent*-39** upon intramolecular nucleophilic reaction in the presence of MsCl, which was employed to activate the secondary hydroxyl to invert the configuration at C(OH). Homodimerization was carried out simply via the oxidation coupling of pyrrolidine ***ent*-40** using $Pd(PPh_3)_2Cl_2$ and CuI as catalysts in the presence of i-$PrNH_2$ to generate the diyne ***ent*-41** in 84% yield. Finally, the target homodimer ***ent*-33** was prepared in 38% yield by alkyne reduction, hydrogenolysis of the benzyl groups, and acidolysis of ***ent*-41**. Monomer ***ent*-35** was readily prepared by the same reduction/acidolysis sequence from ***ent*-40**. The same synthetic route was applied to the known **37** to prepare compounds **33** and **35** (Scheme 4).

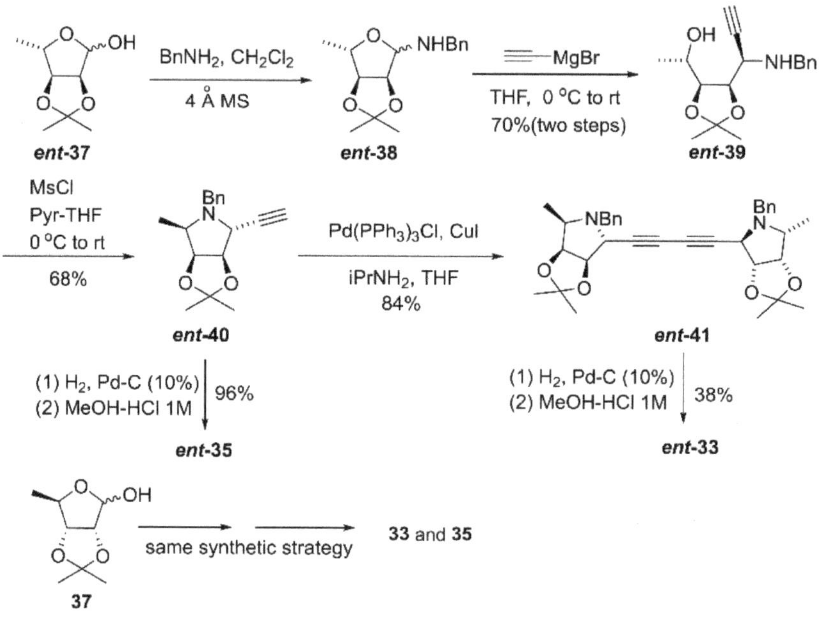

Scheme 4. Synthesis of divalent iminosugars **33**, ***ent*-33** and monovalent references **35**, ***ent*-35**.

The *meso* analog ***meso*-33** cannot be obtained by coupling ***ent*-40** with its enantiomer directly, due to the formation of hard-to-remove mixture ***ent*-41/41**. In order to avoid reaction monitoring and isolation problems, ***ent*-40** and known *N*-allyl protected enantiomer **42** [88] were employed to obtain ***meso*-33** through the same way used to generate **41** described above. An excess of enantiomer **42** was necessary to decrease the production of compounds ***ent*-41** and **43**. As expected, the target hetero-diyne **44** was obtained and isolated in high yield (61%). Cleavage of the *N*-allyl group from **44** in the presence of NDMBA and $Pd(PPh_3)_4$, followed by hydrogenolysis of the benzyl groups and final acidolysis, afforded the target dimer ***meso*-33**. Monomer **36**, an analog of **33** whose second pyrrolidine moiety was replaced by a phenyl group, was prepared from the known diyne **46** [53] using classic hydrogenation (H_2, Pd/C, MeOH) in 93% yield (Scheme 5).

Two years later, Moreno Vargas and co-workers pioneered a valuable methodology for rapid, efficient screening of the divalent inhibitors to α-fucosidases and β-galactosidase, as well as studying the multivalent approach in the inhibition of glycosidases [89]. The Cu(I)-catalyzed alkyne-azide cycloaddition (CuAAC) reaction, a fantastic chemical reaction based on Huisgen 1,3-dipolar cycloaddition chemistry [90,91] and then developed by Meldal [56] and Sharpless [57], was employed to generate three libraries of divalent iminosugars (**47a–l, 48a–l, and 49a–l**) between alkynyl pyrrolidines **47–49** and the set of diazides **a–i**. Due to the high efficiency of the CuAAC reaction, the obtained crude products could be directly screened for enzyme inhibitors without purification. It is worth noting that the discovery of the CuAAC reaction extensively promoted the development of the multivalent approach (Figure 6).

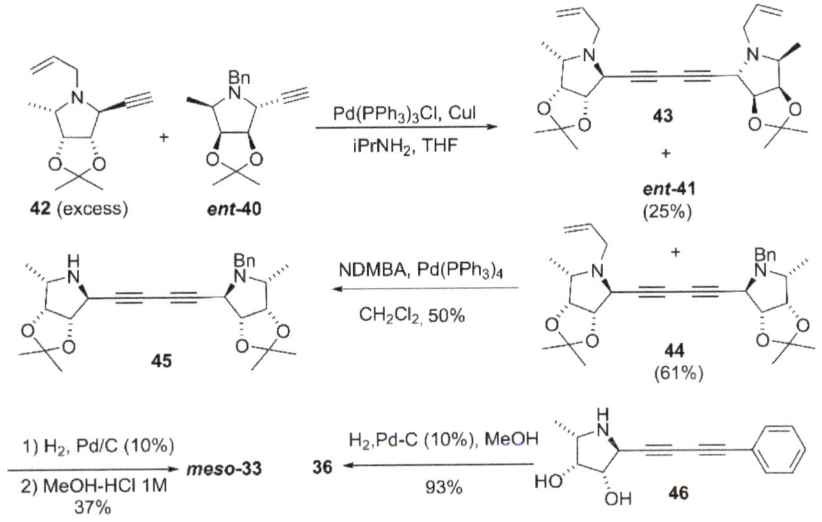

Scheme 5. Synthesis of divalent iminosugar ***meso*-33** and monovalent reference **36**.

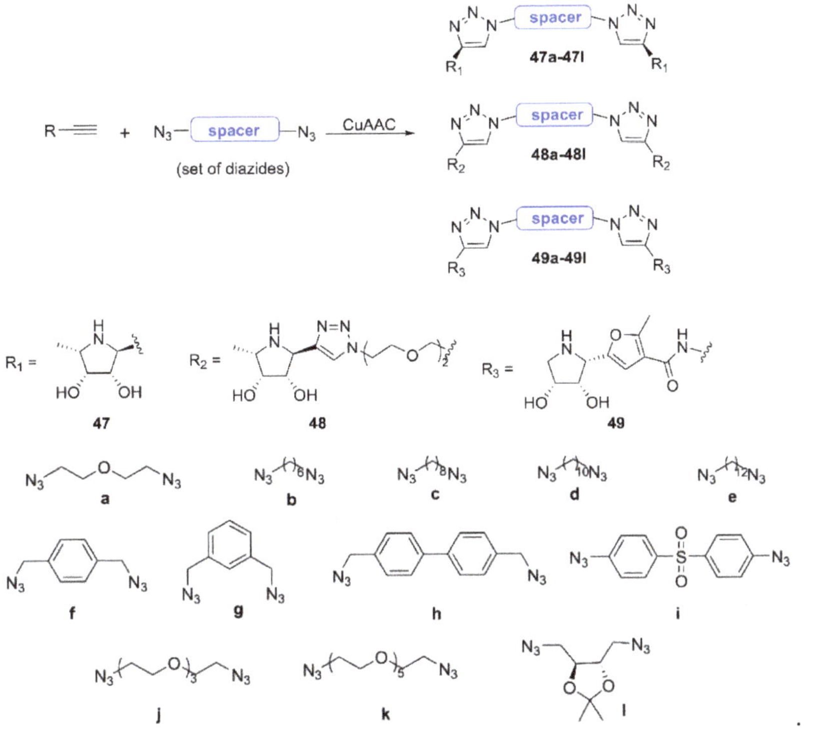

Figure 6. Generation of libraries of dimeric iminosugars (**47a–l**, **48a–l**, and **49a–l**) for in situ screening.

Alkynyl pyrrolidines **47–49** were selected as the skeletons because their analogs (pyrrolidin-2-yl)triazole and (pyrrolidin-2-yl)furans were previously shown to exhibit significant glycosidase inhibition to α-fucosidases and β-galactosidases by the same group [83,92]. The initial step was to prepare the different tethered alkynyl pyrrolidine derivatives **47–49**.

Known compounds **32** [**84**] and **53** [93], previously prepared by the same group from D-lyxose and D-glucose, were employed as starting materials. Alkynyl pyrrolidine **47** was readily prepared from **32** via Boc-deprotection in TFA in 87% yield. As expected, the CuAAC coupling reaction was carried out between alkynyl pyrrolidine **32** and azide **50** [94] to yield quantitative triazole **51**. Then, propargylation of triazole **51** with NaH followed by acidic deprotection of derivative **52** quantitatively provided the desired alkynyl derivative **48**. Classic amide coupling conditions (PyBOP/propargylamine) were employed to form the epimers **54a** and **54b** from compound **53**, further separated by chromatography. Based on the previous report, *cis*-configured epimer **54b** was chosen for deprotection to yield the target pyrrolidine-furan hybrid **49**, since *trans*-configured epimer **54a** exhibited weak α-fucosidase inhibition. Finally, the diazides **a–l**, another part for CuAAC coupling, were prepared according to the design of the spacer. (Pyrrolidin-2-yl)triazole libraries were generated by parallel CuAAC couplings between alkynyl functionalized pyrrolidines **47–49** and diazides **a–l** under the catalysts of CuI or $CuSO_4$. Due to the high efficiency of CuAAC coupling, granting almost quantitative yields with no side reactions, the desired products were all processed and directly screened for enzyme inhibition testing (Scheme 6).

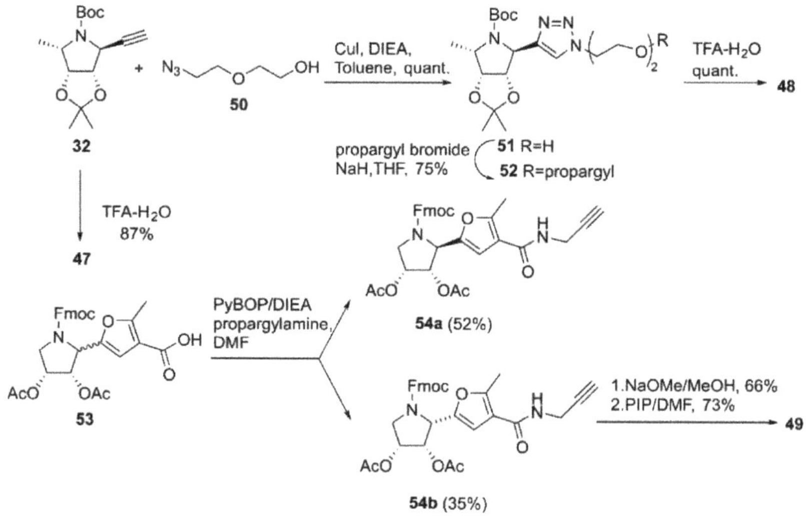

Scheme 6. Synthesis of alkynyl pyrrolidines **47–49**.

Crude screening indicated that dimer **47i** was the best inhibitor of α-fucosidases from the bovine kidney (k_i = 0.15 nM), and that dimer **49e** was the best inhibitor of β-galactosidase from the bovine liver (k_i = 5.8 μM). Hence, compounds **47i** and **49e** were scaled up for detailed and complete analysis. To evaluate the multivalent effect on enzyme inhibition, monovalent references **56–59** were prepared. The synthesis routes adopted the same reaction conditions with compound **51** and CuAAC cycloaddition, and therefore will not be repeated here (Scheme 7).

The generation and in situ bio-screening of compound libraries mediated by efficient chemical reactions such as click reactions has proven to be an economical, rapid and efficient screening method for enzyme inhibitors, mainly in the context that the spatial structure of most enzymes is still unknown. More recently, Moreno and co-workers continued their work by screening a library of divalent pyrrolidine iminosugars to find inhibitors of human hexosaminidase [69]. A nanomolar and remarkably selective inhibitor of human nucleocytoplasmic β-*N*-acetylglucosaminidase was thus discovered.

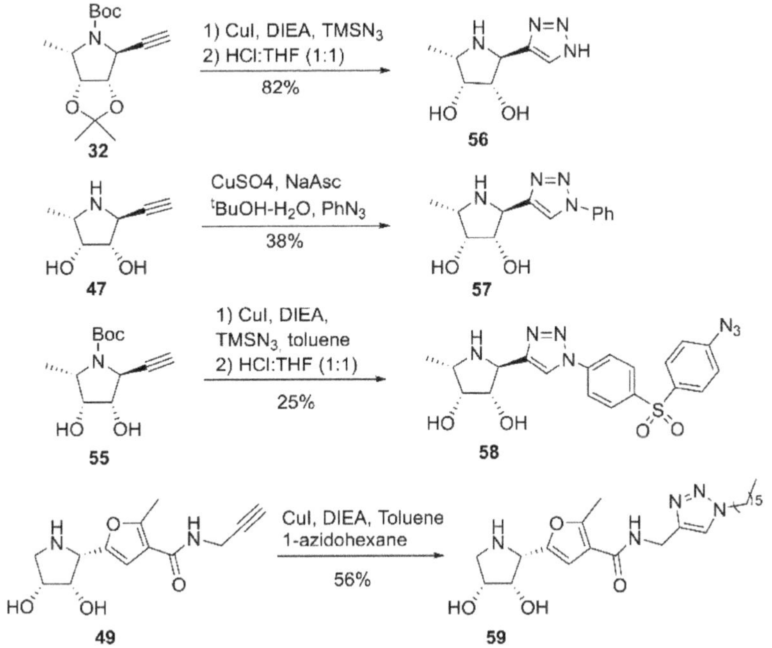

Scheme 7. Synthesis of monovalent references **56–59**.

The authors selected pyrrolidine derivative **60** as the skeleton of the divalent iminosugar libraries. Compound **60** was proved to be an outstanding inhibitor of β-*N*-acetylhexosaminidase in 2001 by Wong's group [95], which is consistent with the purpose of this research. Amino and azido functional groups were introduced based on **60** through molecular modification to carry out the subsequent click reaction. Azide **63** was obtained via reduction of cyanide in the known compound **61** **[96]**, followed by acylation and reductive amination with 6-azidohexanal using $NaBH_3CN$ as catalyst in 49% yield (Scheme 8) [97]. Catalytic hydrogenation of **63** by $H_2/Pd/C$ gave the amine **64** in 76% yield (Scheme 8).

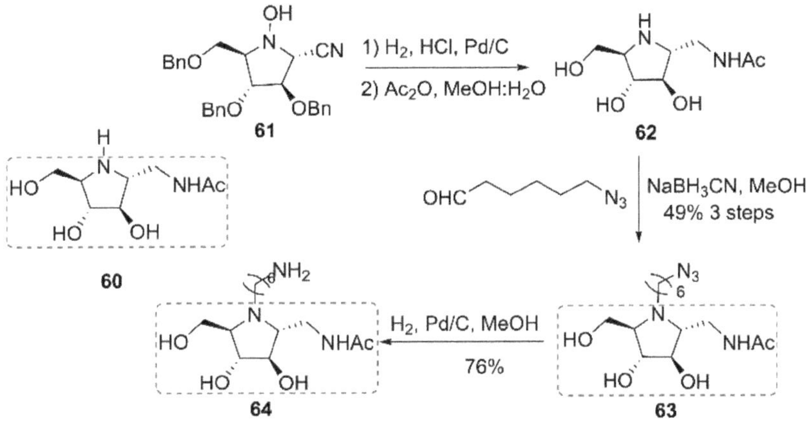

Scheme 8. Synthesis of azide **63** and amine **64**.

A sub-library **I** was generated via CuAAC reaction between **63** (2.4–2.5 equiv.) and dialkynes **a–e** (1.0 equiv.) in the presence of $CuSO_4{\cdot}5H_2O$ (0.14 equiv.), sodium ascorbate (0.44 equiv.), and *t*-BuOH/H_2O with high yield (Scheme 9). In parallel, (thio)urea-bond forming reactions between compound **64** (2.4–2.5 equiv.) and diisothiocyanates **A–E** (1.0 equiv.) were carried out in solvent DMSO to give a sub-library **II** (Scheme 10). Finally, the crude divalent iminosugars **63a–e** and **64A–E** were assayed as β-*N*-acetylglucosaminidase inhibitors, and thus compounds **63A** and **64D**, with the highest inhibitory potency, were screened and studied in detail. Similar to the previous protocol [89], the inhibition potency of divalent iminosugars was compared with corresponding monomers to evaluate the multivalent effect. Compound **65**, the reference of **63A**, was synthesized by CuAAC cycloaddition between **63** and methyl propargyl ether in 61% yield. Similarly, compound **66**, as control of **64D**, was generated through (thio)urea-bond forming reactions between **64** and phenyl isothiocyanate in 62% yield (Scheme 11).

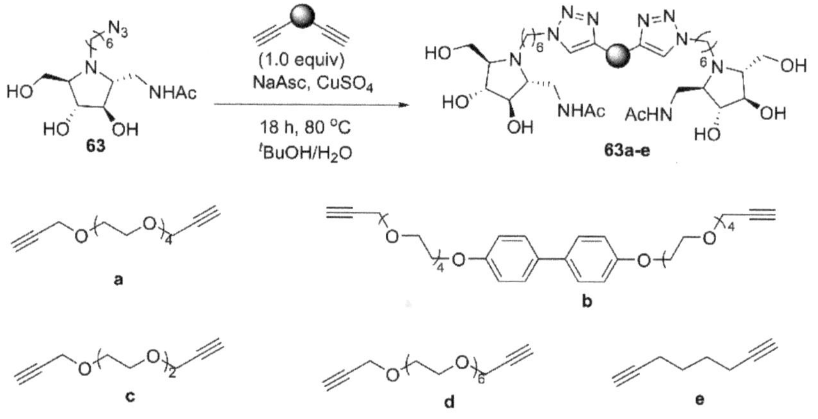

Scheme 9. Synthesis of divalent iminosugars **63a–e**.

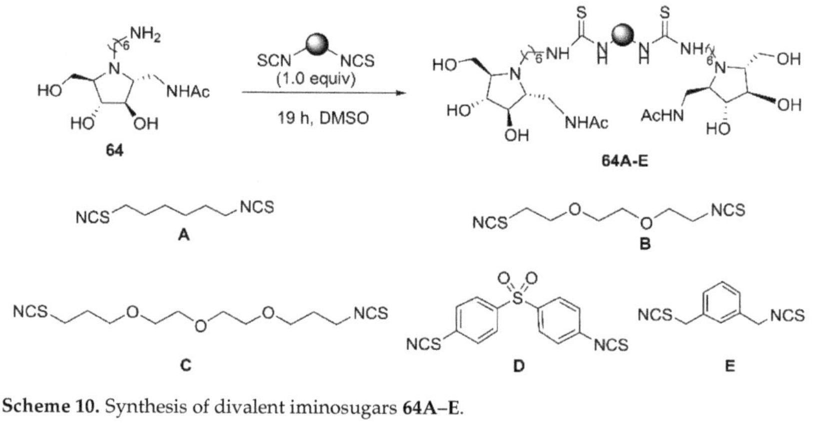

Scheme 10. Synthesis of divalent iminosugars **64A–E**.

As described above, several examples of multivalent pyrrolidine iminosugars were successfully prepared and used for biological activity exploration through the efficient CuAAC reaction. However, this reaction also brings some problems. For example, the catalyst copper ion required has a high chance of complexing with multiple nitrogen atoms in the triazole produced by the reaction, increasing the risk of metal ion contamination [98]. Moreover, the CuAAC reaction is usually carried out in the last step between the monovalent skeleton and scaffold, which limits the choice of monomer part of the final iminosugars. Therefore, developing new strategies without metal catalysts for the preparation of bio-related iminosugars is highly desirable [61].

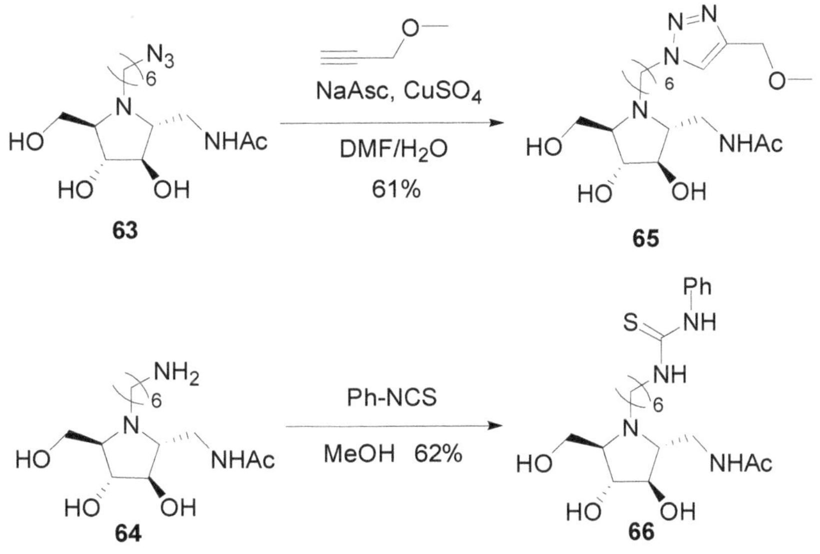

Scheme 11. Synthesis of monovalent references **65–66**.

Cardona and co-workers reported an alternative way to synthesize multivalent pyrrolidine iminosugars without metals in 2019 [62]. The synthesis relies on iminosugar pyrrolidine DAB and three selective and high-yielding steps (1,3-dipole cycloaddition with nitrone **67**, *N–O* bond cleavage of the adduct, and selective *N*- and/or *O*-allylation), and allows the preparation of different topologies in the DAB clusters. Nitrone **67**, obtained from the commercially available tribenzylated D-arabinose [99], and allyl benzyl ether **68 [100]** were employed to prepare *exo-anti* isoxazolidine **69** in 85% yield through 1,3-dipole cycloaddition (1,3 DC). The 1,3 DC process is a crucial step since a high degree of stereoselectivity in the reaction must be guaranteed to reduce isomer formation. Previous research showed that high *exo-anti* selectivity is ascribed to the *tans-trans* configurated nitrone **67**, whose C-3 and C-5 substituents on the same face are opposite to C-4. Thus, the *exo* mode was preferred to avoid repulsive steric interactions with a substituent at C-4 [99,101,102]. Then, the cleavage of the *N–O* bond of **69** in the presence of 10 equiv. of Zn afforded **70** quantitatively. Note that compound **70** is a key intermediate since the selective *N*-and/or *O*-allylation would generate new dipolarophiles **71–73**, which would introduce a second or third DAB moiety by 1,3 DC with nitrone **67**. MW-assisted selective *N*-allylation of **70** was carried out using electrophile allyl bromide to afford intermediate **71** in 86% yield. Protecting the amine of **70** using benzyl bromide, followed by selective *O*-allylation in the presence of allyl bromide and K_2CO_3, gave intermediate **72** in 71% yield. The *N,O*-bis allylated pyrrolidine **73** was obtained by treating **70** with a high excess of allyl bromide (6 equiv.) and NaH (8 equiv.) in 81% yield (Scheme 12).

The synthesis of multivalent DAB iminosugars was carried out as initially designed. 1,3 DC reaction between **71** and nitrone **67**, followed by catalytic hydrogenation with Pd/C in MeOH/HCl, gave the bis-pyrrolidinium hydrochloride **75** in quantitative yield, which was submitted to the ion exchange resin Dowex 50WX8–200, followed by treatment with the strongly basic Ambersep 900-OH resin to afford the divalent DAB-based iminosugar **76** in 40% yield (Scheme 13). A similar approach was applied to the preparation of dimer **79** and trimer **82** from corresponding intermediates *O*-allylated **72** and *N,O*-allylated **73**

(Scheme 14). After three steps—1,3 DC, catalytic hydrogenation, and ion exchange resin—DAB-based iminosugars **79** and **82** were generated in 95% and 60% yield.

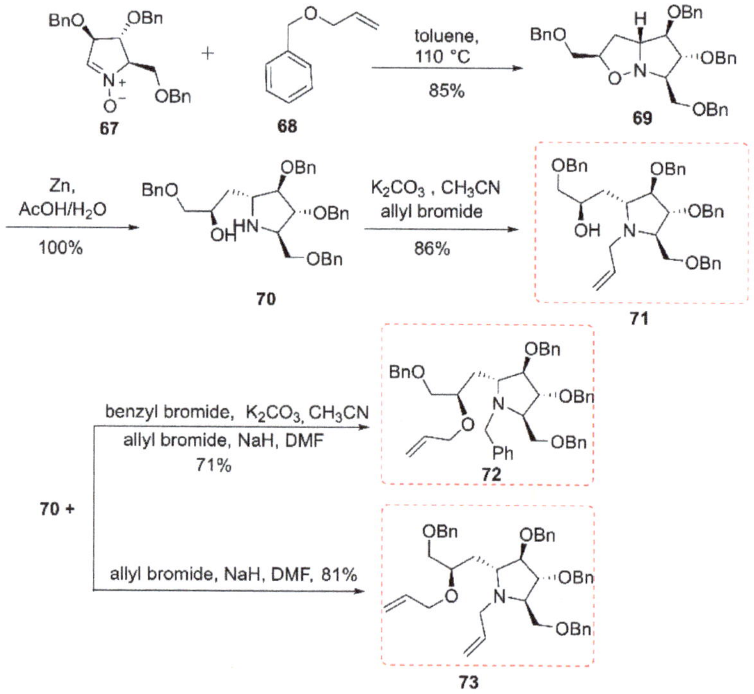

Scheme 12. Synthesis of crucial intermediates **71–73**.

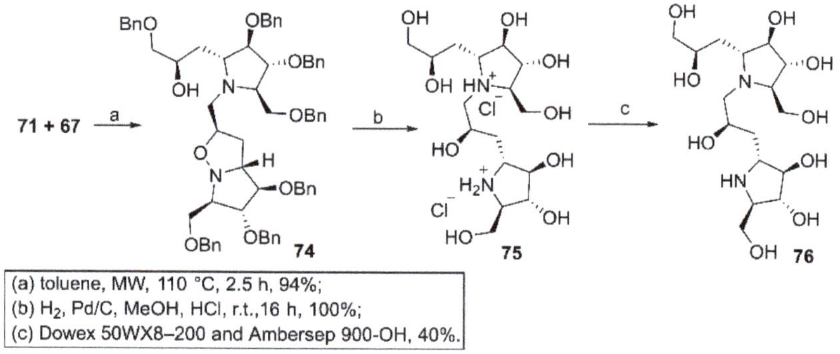

Scheme 13. Synthesis of DAB-based iminosugar **76**.

2.2. *Synthesis of Multivalent Iminosugars*

In 2016, Cardona and co-workers explored the inhibition of sulfatases using the first two examples of pyrrolidine clusters [67]. Nonavalent pyrrolidine iminosugars **83** and **84** were obtained from DAB and 1,4-dideoxy-1,4-imino-D-ribitol (**86**), prepared by the deprotection of the starting nitrones **67 [103]** and **85 [104]**. Upon selective *N*-alkylation with 1-azido-6-bromohexane [105] in the presence of K_2CO_3 under MW irradiation, DAB gave the deprotected azide **87** in 92% yield. Reacting **87** with the nonadiyne scaffold **89** [106] using the standard CuAAC cycloaddition condition ($CuSO_4$ (30 mol%), sodium ascorbate (60 mol%), THF/H_2O) afforded a mixture of unreacted **87** and desired product **83**, which was further purified by column chromatography and size exclusion chromatography Sephadex LH-20, obtaining pure **83** in 81% yield. Similar synthetic routes were applied for the synthesis of target iminosugar **84**. Starting from a different configuration of bioactive 1,4-dideoxy-1,4-imino-D-ribitol (**86**), selective *N*-alkylation afforded **88**, and CuAAC coupling gave ribose configured **84** in 71% yield (Scheme 15).

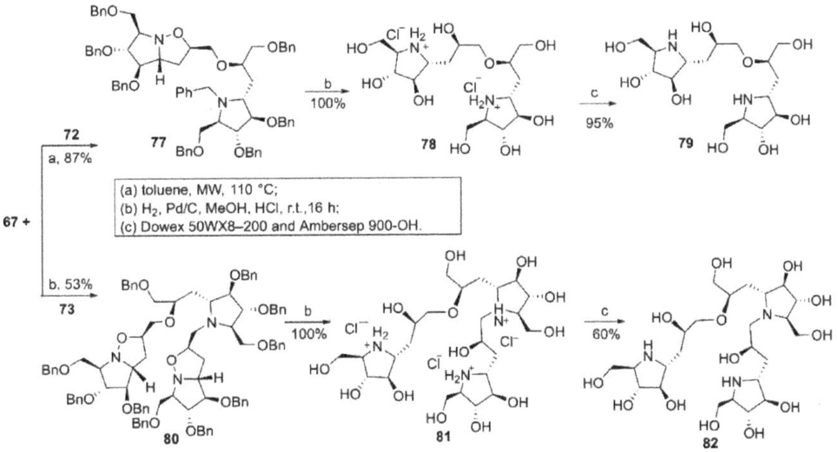

Scheme 14. Synthesis of DAB-based iminosugars **79** and **82**.

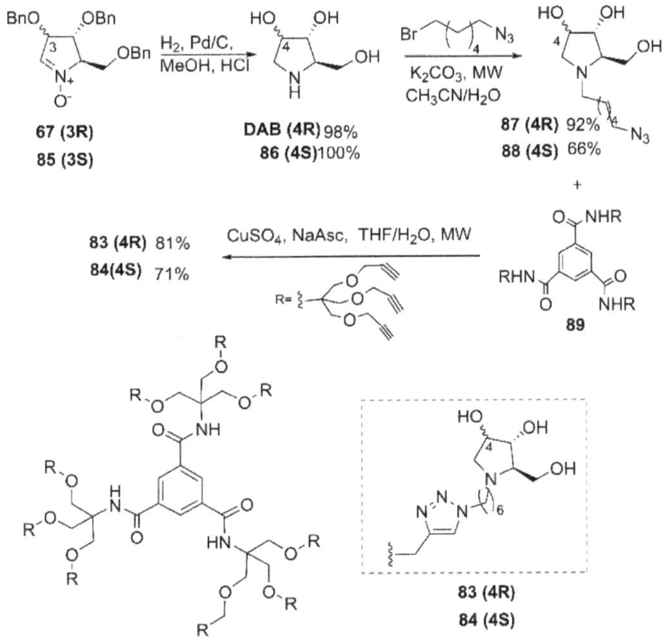

Scheme 15. Synthesis of nonavalent iminosugars **83** and **84**.

The monovalent references **90** and **91** corresponding to compounds **83** and **84** were synthesized to explore the multivalent effect further (Scheme 16). The reference **90** was synthesized from known D-arabinose derived nitrone **67**. Two-step reduction by $NaBH_4$ followed by Zn in AcOH gave amine **92** in 98% yield, which was treated with 1-azido-6-

bromohexane in basic conditions under microwave irradiation to afford azide **93** in 88% yield. The CuAAC coupling between azide **93** and 3-butyn-1-ol was carried out under MW irradiation to give **94** (93%), which was further deprotected using H_2 and Pd/C in acidic MeOH to give the final product **90** in 74% yield. Compound **91**, generated through one-step synthesis, was obtained by cycloaddition of azide **88** to 3-butyn-1-ol in the presence of $CuSO_4$ and sodium ascorbate under MW irradiation in 89% yield (Scheme 16).

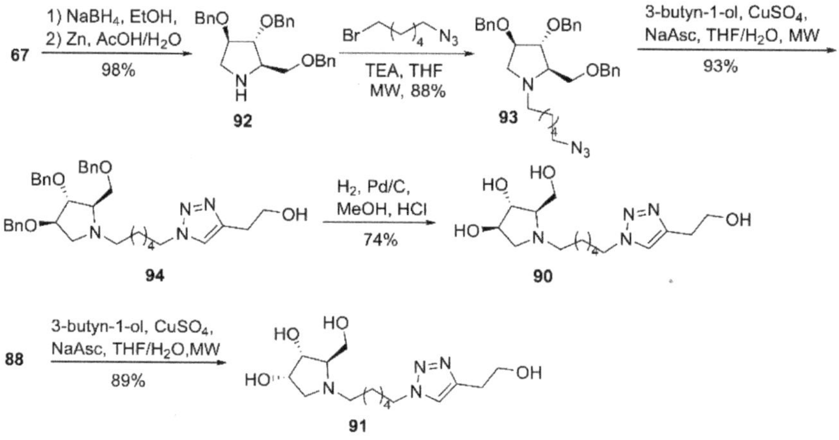

Scheme 16. Synthesis of monovalent references **90** and **91**.

Results showed that the nonavalent pyrrolidine iminosugar **83** exhibited impressive inhibition of *N*-acetylgalactosamine-6-sulfatase (GALNS). Considering that the DAB motif in **83** is a widely available glycosidase inhibitor, and the fact that GALNS and α-mannosidases both have dimer properties, the same research group continued to explore the interaction of α-mannosidases with different multivalent architectures based on iminosugar DAB in 2017 [107].

Similar to the method for synthesizing compound **83**, the CuAAC cycloaddition of intermediate azide **87** and multivalent alkyne scaffolds was exploited to generate new tetra- and trivalent pyrrolidine iminosugars **96** and **98** (Scheme 17). Microwave-assisted CuAAC of azide **87** (3.5 or 4.5 equiv.) with trivalent scaffold tris[(propargyloxy)methyl]aminomethane **97** and tetravalent scaffold **95** (1.0 equiv.) in the presence of $CuSO_4$ (0.3 equiv.) and sodium ascorbate (0.6 equiv.) in THF/H_2O a, followed by purification through flash column chromatography and size-exclusion chromatography Sephadex LH-20 (H_2O) to separate the excess azide **87**, gave the pyrrolidine iminosugar clusters **98** (48%) and **96** (76%) in good yields. In addition, inhibitory performance against a panel of glycosidases was evaluated among the nona-, tetra-, and trivalent iminosugars **83**, **96**, and **98**, as well as monovalent references DAB and its derivative **90**.

Two years later, Moreno Vargas and co-workers prepared four multivalent pyrrolidine iminosugars for GH1 β-glucosidases A and B (BglA and BglB) to continue their study on binding modes and key determinants responsible for the inhibitory effect displayed by pyrrolidine-based clusters [108]. Therefore, as before, the CuAAC click reaction between pyrrolidine-azide derivatives (**99** and **101**) and two different tri- or hexavalent alkynyl spacers (**97** and **102**) was exploited to give the target clusters (**103**–**106**). It was found that spacers containing aromatic moieties in multivalent inhibitors showed excellent inhibition against octameric BglA (μM range) compared to the similar monomeric BglB. Moreover, a modest multivalent effect was detected for the hexavalent inhibitor **106**.

Starting from the azidomethyl pyrrolidine **107** reported by the same group, the protected azide **99** was obtained in good yield by reacting with the aromatic alkyne **108** [109] through CuAAC and then treating with NaN_3 in 93% yield. Then, classic conditions

(HCl/THF) were used to get rid of the protecting group to give the unprotected derivative **100** quantitatively. The same reactions performed in **107** and alkyne **109** [95] via a sequence of CuAAC coupling, nucleophilic displacement with NaN_3, and acidic deprotection afforded **101** in high yield. The synthesis of azides **100** and **101** was not only to obtain the final multivalent clusters but also as monomers for bioactivity control. However, due to the poor solubility of azide **100**, protected azide **99** was selected in the subsequent synthesis and then deprotected again with hydrochloric acid (Scheme 18).

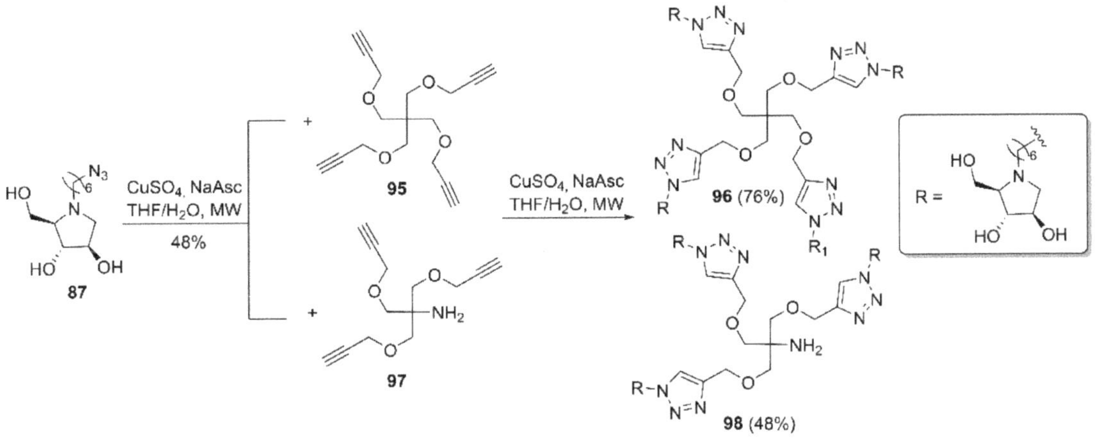

Scheme 17. Synthesis of DAB-based iminosugars **96** and **98**.

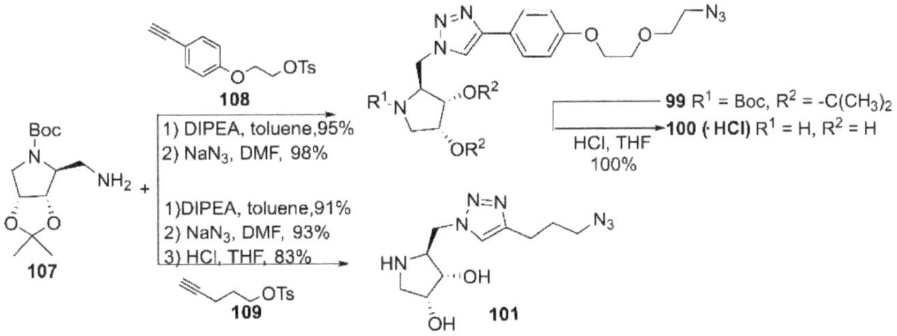

Scheme 18. Synthesis of monovalent references **99–101**.

Finally, microwave-assisted CuAAC cycloaddition of azide **99** or **101** with scaffolds **97** [110] and **102** [108] in the presence of $CuSO_4$, sodium ascorbate in THF/H_2O (2:1), followed by checking reactions through ^1H NMR spectra of the crude mixtures, gave the pyrrolidine-based iminosugar clusters **103–106** in good yields (57–93%, Scheme 19). It is worth noting that the resulting crudes, mixed with excess azide **99** or **101** and desired products, were purified by stirring with Quadrasil® MP followed by chromatography column (silica gel or Sephadex LH-20).

Moreno Vargas and co-workers have long worked on the design and synthesis of mono- and multivalent iminosugars to evaluate their inhibition activities toward various disease-related enzymes. Because two enzymes, β-glucocerebrosidase (GCase) and α-galactosidase (α-Gal A), are involved in Fabry and Gaucher diseases, respectively, and combined with the experimental results that pyrrolidine-3,4-diol skeleton-based iminosugars exhibit bioactivity to human lysosomal GCase reported by his group [111,112], exploring the multivalent effect on these two enzymes became their target. The author reported four sets of multivalent pyrrolidine iminosugars with different valency, configuration, and spacers to perform a systematic analysis of the inhibition of the lysosomal glycosidases in 2020 [49].

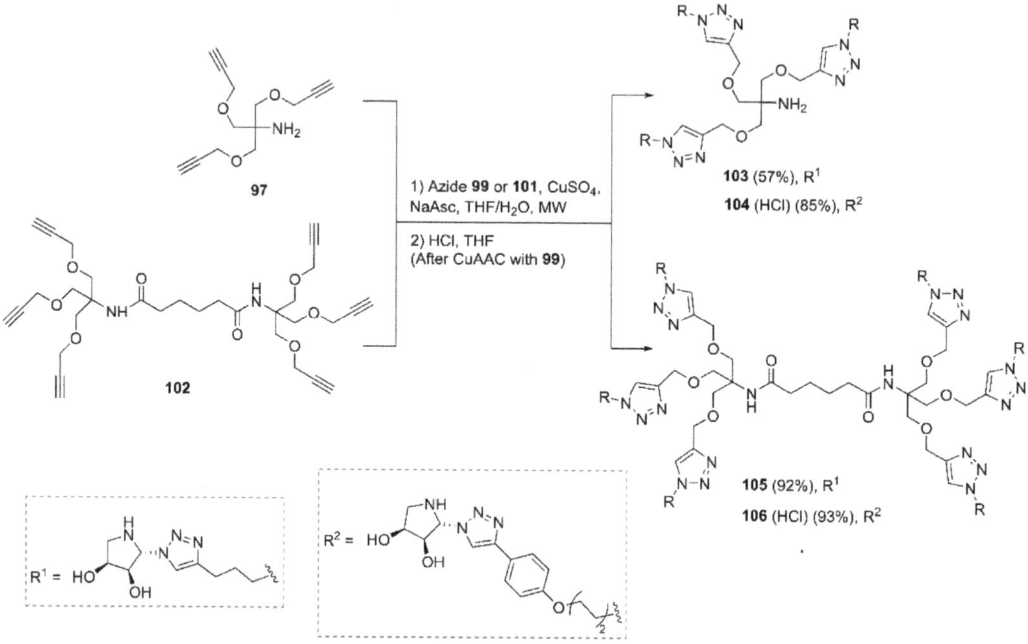

Scheme 19. Synthesis of multivalent iminosugars **103–106**.

The six azidoalkyl pyrrolidines shown in Figure 7 were selected as monovalent references as well as anchoring moieties for CuAAC coupling reactions. Compounds **100, 101, 110,** and **111** were all known compounds reported by the same group [108,111], while **112** and **113** were newly synthesized. The synthetic routes of **112** and **113** were the same as the preparation for their epimers **101** and **100** described in Scheme 18 [108]. The known scaffolds shown in Figure 8 were selected to synthesize the tri-, tetra-, hexa- and nonavalent iminosugars via CuAAC coupling with azido derivatives.

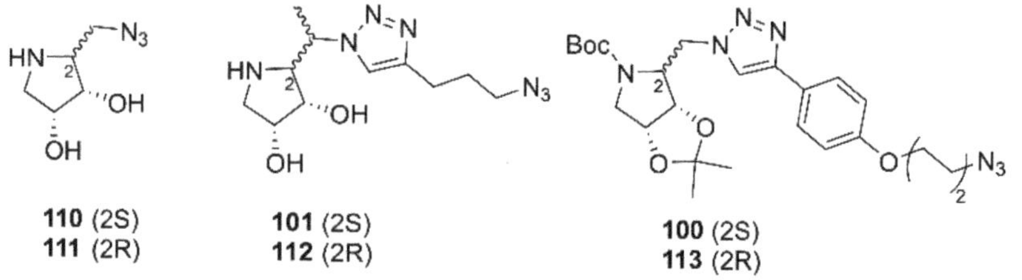

Figure 7. Azidoalkyl pyrrolidines used as anchoring moieties.

General reaction conditions of the click reaction involved using $CuSO_4$ and sodium ascorbate as catalysts and in THF–H_2O under microwave irradiation at 80 °C. Similar to the methods for compounds **103–106**, chromatography column (silica gel or Sephadex LH-20) was employed to separate the unreacted azidoalkyl pyrrolidines. Multimeric derivatives (**114–133**) were generated in high yields (55–99%). The monovalent references (**134–139**) were prepared via CuAAC reaction between monomers (**110** and **111**) and the corresponding alkynes in high yields (79–91%), which were submitted to evaluate inhibition against human GCase and α-Gal A along with the corresponding multivalent iminosugars **114–133** (Figure 9, Scheme 20).

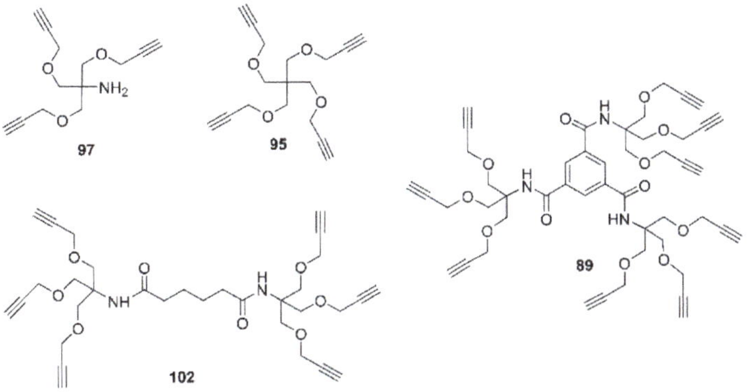

Figure 8. Alkynyl scaffolds for CuAAC reaction.

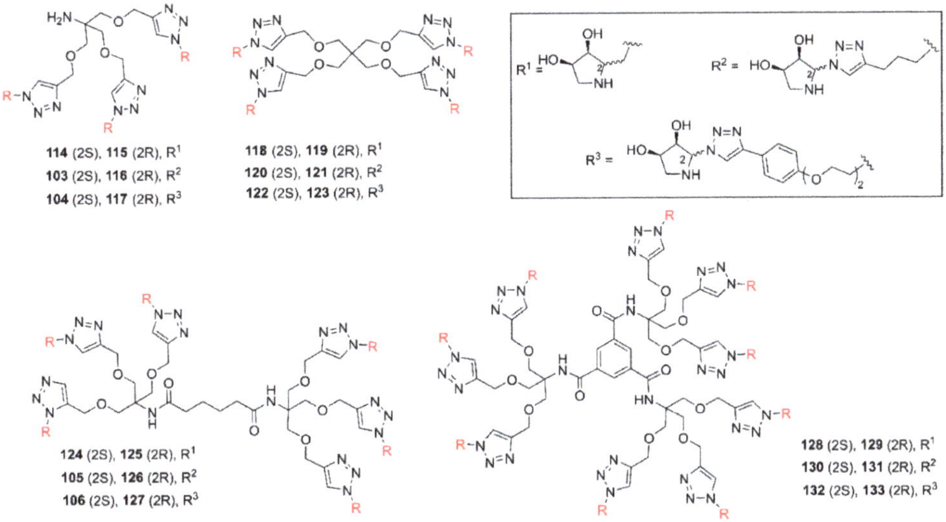

Figure 9. Tri-, tetra-, hexa- and nonavalent iminosugars **114–133**.

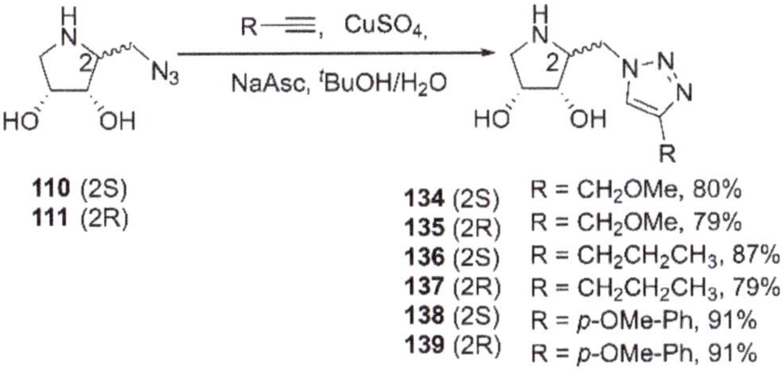

Scheme 20. Synthesis of monovalent references **134–139**.

More recently, Gaeta, Cardona, and co-workers constructed pyrrolidine-based multivalent clusters, employing the less researched scaffold resorcinarene [113], which exhibited conformationally mobile ability [114–116], to explore the role of both conformability and valency in the inhibition of therapeutically relevant enzyme Golgi α-mannosidase IIb (GMIIb) [68].

Resorcinarene **140–142** [113,116] are macrocycles consisting of four to six rings of resorcinol obtained by resorcinol/aldehyde acid-catalyzed condensation reaction (Scheme 21). Interestingly, each aromatic moiety of the macrocycle contains two hydroxyl functional groups, making it suitable for constructing multivalent iminosugars. Azides **87** and **143**, belonging to the DAB-derived pyrrolidine family, were selected as skeletons (Scheme 21). Azide **87 [67]** was reported by the same group previously, and new azido-ending ligand **143**, which possessed a more hydrophilic linker, was newly synthesized to explore the role of the nature of linkers in bioactivity [68].

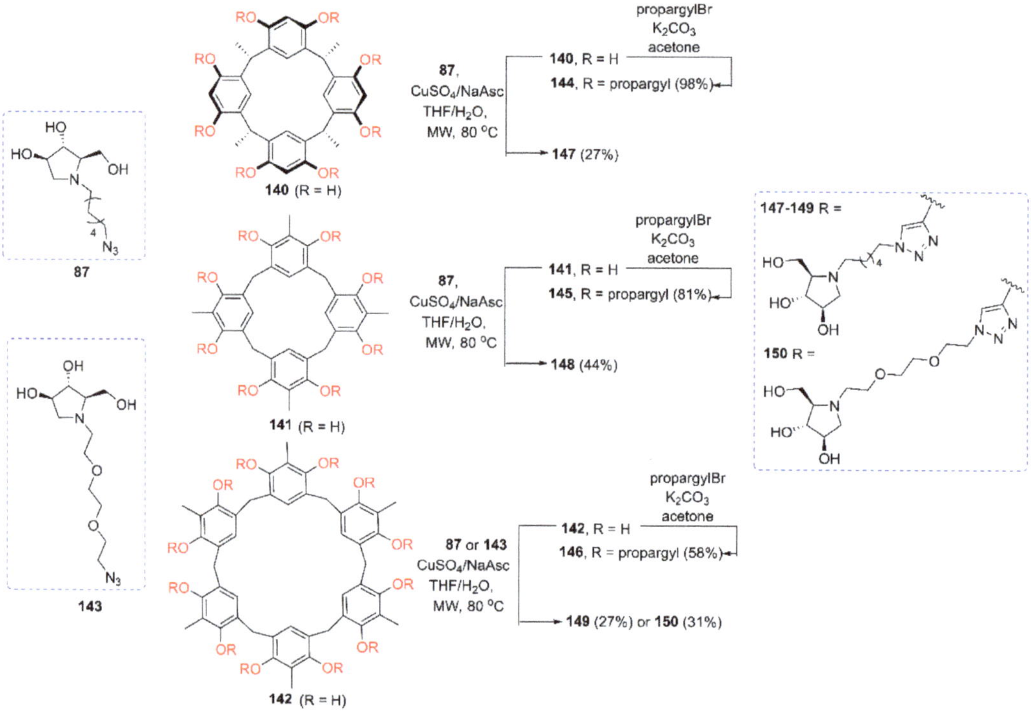

Scheme 21. Synthesis of multivalent iminosugars **147–150**.

Different scaffolds C-methyl-resorcin [4] arene **140**, resorcin [4] arenes **141**, and resorcin [6] arenes **142** were allowed to react with propargyl bromide 2 equivalents per hydroxyl in acetone by treatment with excess K_2CO_3 to give alkyne-ending scaffolds **144–146** in high yields (58–98%). Then, target resorcinarene-based iminosugars **147–149** were obtained by reacting azide **87** with scaffolds **144–146** through the CuAAC click reaction in moderate yields (27–44%). The unsubstituted azides **87** and **143** were purified through chromatographic column (silica gel, gradient: from MeOH to ammonia solution 4 M in MeOH). Finally, the compound **150**, featuring a more hydrophilic linker, was obtained by CuAAC reaction between azide **143** and scaffold **146** (Scheme 21). It is worth pointing out that, although the valency of **147** and **148** was the same, the scaffold resorcin [4] arene (**140**) was conformationally blocked in a cone conformation, thanks to the presence of CH_3CH bridges between aromatic rings [117]. As a result, iminosugar **147** was more flexible than **148**.

3. Biological Activity of Multivalent Pyrrolidine Iminosugars

3.1. Inhibition of α-Fucosidases

α-Fucosidases (AFU) are lysosomal acid hydrolase enzymes that catalyze the hydrolysis of α-fucose units located on the cell surface oligosaccharides and participate in various biological processes, including immune response, signal transduction, and antigenic determination [118–120]. Changes in the activity of AFU in serum or tissue significantly correlate with the occurrence of tumors, such as hepatocellular carcinoma [121], colon adenocarcinoma [122,123], and gastric cancer [124]. Since the discovery of pyrrolidine 1,4-iminocyclitols as potent inhibitors of AFU [77,78], Robina and co-workers first explored the AFU-inhibitory activity of di- (**1**) and trivalent pyrrolidine iminosugars **2–4** in 2013 (Figure 3) [70]. Results showed that all the newly synthesized compounds displayed high AFU inhibition (IC_{50}: 1.6–17 μM) and excellent selectivity. However, compared with the monomer references **5** and **6** (Figure 3), the effect of multivalency was not convincing, except for the trivalent iminosugar **2** (K_i = 0.3 μM, Table 1), which showed seven-fold more potent inhibition activity than monovalent reference **5** (K_i = 2.1 μM). Compounds **2** and **3** (K_i = 0.4 μM, Table 1) displayed almost equivalent activities, indicating that the increase in the length of multivalent iminosugars was not clearly linked to the inhibitory properties of the enzyme. Divalent iminosugars with more diverse spacers were also reported subsequently [80]. Polyamino and triazole-benzyl bridged iminosugars (**13–17**, Figure 4) were constructed to develop potential inhibitors for AFU. Dimers **13**, **14**, and **16** showed stronger inhibition than their corresponding monomers, while compounds **14**, **15**, and **31** yielded the opposite results. Triazole-benzyl bridged iminosugar **17** showed excellent enzyme inhibition to AFU (IC_{50} = 74 nM, K_i = 3.7 nM, Table 1) while dimer **13** (IC_{50} = 1.2 μM, Table 1) indicated the existence of multivalency compared with its control **19b** (IC_{50} = 13 μM, Figure 4), a 10.8-fold potency enhancement. The result that compound **17** exhibited excellent inhibition toward AFU was consistent with the fact that the presence of an additional aromatic or heteroaromatic binding component close to the five membered iminocyclitols notably increases their inhibitory activity to AFU, which was shown by Robina [83], Behr [125], and Wong [126]. To further explore the ligand-enzyme binding modes, stereoisomeric pyrrolidine dimers (**33**, ***ent*-33**, and ***meso*-33**, Figure 5) with short and flexible space were synthesized [84]. Dimer **33** showed potential inhibition of AFU (IC_{50} = 0.108 μM, K_i = 23 nM, Table 1) when compared to its monovalent reference **35** (IC_{50} = 2.0 μM, K_i = 0.18 μM, Figure 5), which to some extent confirmed the existence of the multivalent effect. The divalent ***meso*-33** also showed potential inhibition of AFU (IC_{50} = 0.365 μM, K_i = 0.051 μM), while compound ***ent*-33** was significantly less potent (IC_{50} = 84 μM, K_i = 12 μM). Through detailed controlled trials and structural analysis, the authors suggested that the inhibition enhancement obtained with divalent compounds could be explained by additional interactions of the hydrophobic moiety with a lipophilic binding pocket other than the active site. This hypothesis was confirmed by the 3-D structure of the bacterial fucosidase *Bt*Fuc2970 complexed with the best divalent inhibitor **33**. However, other mechanisms such as rebinding could not be completely ruled out. In 2018, Moreno Vargas and co-workers successfully screened a batch of AFU inhibitors through the CuAAC click reaction followed by in situ biological screening and identified one of the most effective enzyme inhibitors, **47i** (IC_{50} = 48 nM, K_i = 15 nM, Figure 6, Table 1) [89]. The higher inhibition shown by dimer **47i** compared to its analogue **48i** could be argued to be due to non-specific interactions of the diphenylsulfone spacer in the loop regions near the GH29 family's enzymatic active site. Due to controversy over the reference selection, a valid multivalent effect could not be given, but the discovery of compound **47i** proved the rapidity and efficiency of the methodology, which should be highlighted in the screening of enzyme inhibitors. To some extent, these results indicated that the multivalent effect of iminosugars on α-fucosidases is probably due to the additional unspecific interactions with a noncatalytic subsite, which would be beneficial to medicinal chemists in the rational design of α-fucosidase inhibitors.

Table 1. Inhibition activities (K_i or IC_{50} [μM]) and relative inhibition potencies (rp and rp/n) of some selected pyrrolidine iminosugars described in the previous section of the present review.

Enzyme	Compound	Valency (*n*)	K_i [a]	IC_{50} [b]	Rp [c]	Rp/n [d]	Ref.
α-fucosidase [e]	**2**	3	0.3	1.6	7.0	2.3	[70]
"	**3**	3	0.4	3.8	5.3	1.8	[70]
"	**13**	2	-	1.2	10.8	5.4	[80]
"	**17**	2	0.0037	0.074	4.1	2.1	[80]
"	**33**	2	0.023	0.108	7.8	3.9	[84]
"	**47i**	2	0.48×10^{-3}	0.15×10^{-3}	-	-	[89]
α-mannosidase [f]	**83**	9	-	0.095	13,684	1520	[107]
"	**147**	8	-	5.3	245	31	[68]
"	**148**	8	-	14.8	88	11	[68]
"	**149**	12	-	1.2	1083	90	[68]
"	**150**	12	-	10.5	124	10	[68]
Golgi α-mannosidase IIb [g]	**147**	8	–	3.7	47	6	[68]
"	**148**	8		5.3	33	4.1	[68]
"	**149**	12	-	0.7	250	21	[68]
"	**150**	12	-	28.5	6	0.5	[68]
N-acetylgalactosamine-6-	**83**	9	-	47	83	9.2	[67]
sulfatase [h]	**84**	9	-	85	59	6.5	[67]
iduronate-2-sulfatase [i]	**83**	9	-	140	23	2.5	[67]
"	**84**	9	-	31	177	19.7	[67]
β-*N*-acetylhexosaminidase [j]	**64D**	2	168	-	1.9	0.96	[69]
β-*N*-acetylglucosaminidase [k]	**64D**	2	0.0061	-	7.8	3.9	[69]
α-galactosidase A [l]	**133**	9	0.2	1.2	378	42	[49]

[a] Glycosidase inhibition constant (K_i, μM). The K_i on glycosidases was calculated from the measured IC_{50} value using the Cheng–Prusoff equation. [b] Half maximal inhibitory concentration (IC_{50}, μM). [c] Relative inhibitory potency: K_i(monovalent reference)/K_i(glycocluster) or IC_{50}(monovalent reference)/IC_{50} (glycocluster). [d] Inhibitory potency per iminosugar unit. [e] Bovine kidney. [f] Jack bean. [g] *Drosophila melanogaster*. [h] Human leukocytes. [i] Human leukocytes. [j] Human recombinant enzyme (*pichia pastoris*). [k] Human recombinant enzyme (*escherichia coli*). [l] Human lysosome.

3.2. Inhibition of α-Mannosidases

α-Mannosidases are mainly involved in the biosynthesis and catabolism of *N*-glycans in cells. Such processes are, for instance, involved in the treatment of cancers and lysosomal diseases [127–129]. The first evidence of the multivalent effect on iminosugars was gained through the interaction between Jack bean α-mannosidase (JBMan) and a trivalent DNJ conjugate [58]. Due to the successful analysis of its crystal structure and the ease of purchase, JBMan has become the most investigated enzyme for multimeric inhibition studies [48]. Novel tri-, tetra-, and nonavalent pyrrolidine iminosugars (**98**, **96**, and **83**, Schemes 15 and 17) were constructed by Cardona and co-workers to investigate the binding modes to α-mannosidases [107]. A large multivalent effect was observed from the three iminosugars (rp/n >> 1). The DAB-based nonavalent iminosugar **83** (Scheme 15) was the best inhibitor of JBMan (IC_{50} = 95 nM, Table 1), with a 13,684-fold (rp/n = 1520) stronger inhibitory potency than the corresponding the monovalent reference **90** (IC_{50} = 1300 μM, Scheme 16). The trivalent compound **98** and the tetravalent **96** also showed good multivalent effects towards JBMan, with rp/n values of 46 and 10, respectively. Transmission electron microscope (TEM) analysis, nuclear magnetic resonance (NMR), and molecular dynamic studies were carried out to elucidate the binding mode of the multivalent iminosugars and α-mannosidases. NMR studies showed the existence of specific interactions of the multivalent ligands with JBMan, which presumably take place within the enzyme active site. TEM studies indicated that the binding mode would probably be intermolecular cross-linking, due to the formation of ligand–JBMan aggregates. It is worth noting that a remarkable selectivity of iminosugars (**83**, **96**, and **98**) for Golgi α-mannosidase IIb (GMIIb) over lysosomal α-mannosidase II (LManII), two biologically relevant enzymes (GMIIb: tumor growth and cell metastasis; LManII: disorder mannosidosis), was observed. The interesting selectivity appeared particularly relevant for selective application of multivalent compounds in anticancer therapy without the undesirable side effect of mannosidosis syndrome. Subsequently, scaffold resorcinarene was employed to explore the role of both the conformability and the valency of multivalent iminosugars to therapeutically relevant target GMIIb [68]. Similarly, both the 8-valent (**147**, **148**, Scheme 21) and 12-valent iminosugars (**149**, **150**, Scheme 21) exhibited greater selectivity to JBMan and GMIIb over LManII. Biological assay indicated that 12-valent **149** had stronger inhibition, for example, towards GMIIb (IC_{50} = 0.7 μM, Table 1) than 8-valent **147** (IC_{50} = 3.7 μM, Table 1) and **148** (IC_{50} = 5.3 μM, Table 1), which further showed that the inhibitory activity of resorcinarene-based conjugates was related to their valency. The 12-valent iminosugar **150**, possessing a more hydrophilic group, showed weaker inhibition (GMIIb, IC_{50} = 28.5 μM, Table 1) than the same valent **149** (GMIIb, IC_{50} = 0.7 μM, Table 1). This was ascribed to the unfavorable repulsions between oxygen atoms on the linker with electron-rich atoms of the amino acid residues of the GMIIb protein. In addition, the 12-valent **149** showed a remarkable multivalent effect towards JBMan (IC_{50} = 1.2 μM) compared to its monovalent reference **90** (Scheme 16, IC_{50} = 1300 μM, rp/n = 90). Computational studies suggest that the binding mode should be the rebinding process, since the resorcinarene ligands bind the dimer of the JBMan by coordination of one Zn ion at a time. From these results, we can know that the multivalent effect of pyrrolidine iminosugars on α-mannosidases has a great relationship with valency. Generally, higher valency iminosugars usually exhibit better inhibitory activities. The multivalent effect is also affected by the type of linker and the conformation of the scaffold. In addition, the proposed binding modes—cross-linking and aggregation, and bind and recapture (Figure 2)—are more likely involved in better responses of multivalent pyrrolidine iminosugars toward α-mannosidases. However, a binding mode that involves both the active site and non-catalytic subsites cannot be completely excluded.

3.3. *Inhibition of Other Disease-Related Glycosidases*

Besides the α-fucosidases and α-mannosidases introduced above, Cardona and coworkers explored the impact of multivalency on sulfatases involved in lysosomal storage disorders (LSD) for the first time [67]. A decrease in two lysosomal enzymes, *N*-acetylgalactosamine-6-sulfatase (GALNS) and iduronate-2-sulfatase (IDS) could cause diseases of mucopolysaccharidoses: Morquio A syndrome and Hunter disease, respectively [130–132]. Nonavalent DAB-based iminosugars **83** (GALNS: IC_{50} = 47 μM, IDS: IC_{50} = 140 μM, Scheme 15, Table 1) and **84** (GALNS: IC_{50} = 85 μM, IDS: IC_{50} = 31 μM, Scheme 15, Table 1) exhibited strong inhibition to both enzymes compared to the negligible monovalent references **90** (GALNS: IC_{50} = 3900 μM, IDS: IC_{50} = 3200 μM, Scheme 16) and **91** (GALNS: IC_{50} = 5000 μM, IDS: IC_{50} = 5500 μM, Scheme 16). The results demonstrated that a good multivalent effect was achieved with pyrrolidine-based clusters towards sulfatases. For example, **84** showed a remarkable multivalent effect toward IDS (rp/n = 19.7, Table 1). However, detailed kinetic studies and proposed binding modes were not given. On the basis of this result, DAB-based iminosugars **79** (Scheme 14) and **82** (Scheme 14) with different ligand topologies were synthesized for GALNS inhibition two years later by changing CuAAC coupling to a new strategy which avoided contamination with copper ions [62]. Dimer **79** and trimer **82** showed IC_{50} to GALNS in the low micromolar range (0.3 and 0.2 μM, respectively), confirming that multimerization of DAB epitopes generates potent GALNS inhibitors. Comparing the inhibitory activities of **79** and **82** with **83** and **84**, we can learn that the ligand topology strongly affected the affinity of the DAB-based multivalent iminosugars for GALNS. Divalent iminosugar **64D** (Scheme 10) was discovered to be a potent inhibitor of human hexosaminidases, the potential pharmacological targets for drug development, via the screening of two libraries of divalent pyrrolidine iminosugars [69]. The results showed that compound **64D** exhibited remarkable inhibition of human β-*N*-acetylglucosaminidase (hOGA) in the nanomolar range (K_i = 6.1 nM, Table 1) compared to the monovalent reference **66** (K_i = 47.6 nM, Scheme 11). No significant multivalent effect was observed in the inhibition of any of the hexosaminidases by dimers

63 and **64**. However, compound **64D** displayed excellent selectivity towards hOGA compared with human lysosomal β-*N*-acetylhexosaminidases (hHexB, K_i = 168 μM, Table 1), with an approximately 27500-fold enzyme affinity enhancement. It was observed very clearly from the result that multivalency could also be a promising tool to modulate the inhibition selectivity of multivalent iminosugars. Similarly, the (2*R*)-nonavalent iminosugar **133** (Figure 9) was screened by Moreno-Vargas and co-workers for human α-galactosidase A (α-Gal A), which is involved a common lysosomal storage disorder, Fabry disease [49]. Compound **133** displayed remarkably potent inhibition and multivalent effect (K_i = 0.2 μM, rp/n = 42, Table 1) towards α-Gal A, being a 375-fold more potent inhibitor than the monovalent reference **139** (K_i = 75 μM, Scheme 20). The author suggested that the multivalent effect was probably due to the involvement of interaction mechanisms such as statistical rebinding, additional binding with allosteric sites, and/or aggregative processes. More importantly, the activity enhancement effect of compound **133** towards α-Gal A in Fabry fibroblasts constitutes the first evidence of the potential of multivalent iminosugars to act as pharmacological chaperones in the treatment of this LSD.

4. Conclusions

The last decade has witnessed the rapid development of multivalent effects in glycosidase inhibition and drug discovery. In this review, we systematically summarized the process of fabricating multivalent iminosugars based on pyrrolidine in terms of design strategies, synthesis routes, and glycosidase inhibition investigations. Up to 12-valent pyrrolidine iminosugars were synthesized through classic click reactions, and thus several outstanding inhibitors were discovered. For example, nonavalent inhibitors based on DAB and one of its epimers demonstrated the existence of the multivalent effect in sulfatases for the first time [67]. Moreover, nonavalent iminosugars based on pyrrolidine-triazole moieties exhibited a remarkable multivalent effect on one important therapeutic enzyme, human α-galactosidase A, and constitute the first evidence of a multivalent enzyme activity enhancer for Fabry disease [49]. Despite advances in the design and investigation of multivalent iminosugars based on pyrrolidines, some problems and challenges remain.

The enzymes used for studying the multivalent approach are mostly limited to the more researched models, such as α-mannosidase and α-fucosidase, which means the importance of some therapeutically relevant glycosidases is overlooked. The complex and confusing enzyme–ligand binding mechanism is not a negligible issue when developing new relevant multivalent inhibitors. Elucidating the binding mode(s) would improve glycosidase inhibition efficiency and selectivity, two major problems currently existing. Despite the many challenges, we hope that, with the information presented in this review, researchers in this field will continue to explore multivalent effects based on pyrrolidine for developing new glycosidase inhibitors, as well as for candidates for advanced clinical trials or markets.

Author Contributions: Conceptualization, C.L. Original draft preparation, Y.W. and J.X. Review and editing, A.M. and C.L. All authors have read and agreed to the published version of the manuscript.

Funding: This research was funded by the National Science Foundation of Hebei province, China (no. H2020209288 and H2021209056).

Institutional Review Board Statement: Not applicable.

Conflicts of Interest: The authors declare no conflict of interest.

References

1. Zamoner, L.O.B.; Aragão-Leoneti, V.; Carvalho, I. Iminosugars: Effects of stereochemistry, ring size, and n-substituents on glucosidase activities. *Pharmaceuticals* **2019**, *12*, 108. [CrossRef] [PubMed]
2. Ferhati, X.; Matassini, C.; Fabbrini, M.G.; Goti, A.; Morrone, A.; Cardona, F.; Moreno-Vargas, A.J.; Paoli, P. Dual targeting of ptp1b and glucosidases with new bifunctional iminosugar inhibitors to address type 2 diabetes. *Bioorganic Chem.* **2019**, *87*, 534–549. [CrossRef] [PubMed]

3. Liu, X.; Li, F.; Su, L.; Wang, M.; Jia, T.; Xu, X.; Li, X.; Wei, C.; Luo, C.; Chen, S.; et al. Design and synthesis of novel benzimidazole-iminosugars linked a substituted phenyl group and their inhibitory activities against β-glucosidase. *Bioorganic Chem.* **2022**, *127*, 106016. [CrossRef] [PubMed]
4. Liu, C.; Hou, L.; Meng, A.; Han, G.; Zhang, W.; Jiang, S. Design, synthesis and bioactivity evaluation of galf mimics as antitubercular agents. *Carbohydr. Res.* **2016**, *429*, 135–142. [CrossRef]
5. Liu, C.; Kang, H.; Wightman, R.H.; Jiang, S. Stereoselective synthesis of a novel galf-disaccharide mimic: β-D-galactofuranosyl-(1-5)-β-D-galactofuranosyl motif of mycobacterial cell walls. *Tetrahedron Lett.* **2013**, *54*, 1192–1194. [CrossRef]
6. Evans DeWald, L.; Starr, C.; Butters, T.; Treston, A.; Warfield, K.L. Iminosugars: A host-targeted approach to combat flaviviridae infections. *Antivir. Res.* **2020**, *184*, 104881. [CrossRef]
7. Conforti, I.; Marra, A. Iminosugars as glycosyltransferase inhibitors. *Org. Biomol. Chem.* **2021**, *19*, 5439–5475. [CrossRef]
8. Huonnic, K.; Linclau, B. The synthesis and glycoside formation of polyfluorinated carbohydrates. *Chem. Rev.* **2022**. [CrossRef]
9. Uhrig, M.L.; Mora Flores, E.W.; Postigo, A. Approaches to the synthesis of perfluoroalkyl-modified carbohydrates and derivatives: Thiosugars, iminosugars, and tetrahydro(thio)pyrans. *Chem. Eur. J.* **2021**, *27*, 7813–7825. [CrossRef]
10. Compain, P.; Martin, O.R. *Iminosugars: From Synthesis to Therapeutic Applications*; John Wiley & Sons Ltd.: Chichester, UK, 2007; ISBN 978-0-470-03391-3.
11. Stutz, A.E. *Iminosugars as Glycosidase Inhibitors: Nojirimycin and Beyond*; Wiley-VCH: Chichester, Germany, 1999; ISBN 3-527-29544-5.
12. Wang, H.; Shen, Y.; Zhao, L.; Ye, Y. Deoxynojirimycin and its derivatives: A mini review of the literature. *Curr. Med. Chem.* **2021**, *28*, 628–643. [CrossRef]
13. Matassini, C.; Warren, J.; Wang, B.; Goti, A.; Cardona, F.; Morrone, A.; Bols, M. Imino- and azasugar protonation inside human acid β-glucosidase, the enzyme that is defective in gaucher disease. *Angew. Chem. Int. Ed.* **2020**, *59*, 10466–10469. [CrossRef] [PubMed]
14. Iftikhar, M.; Lu, Y.; Zhou, M. An overview of therapeutic potential of *N*-alkylated 1-deoxynojirimycin congeners. *Carbohydr. Res.* **2021**, *504*, 108317. [CrossRef] [PubMed]
15. Kar, S.; Sanderson, H.; Roy, K.; Benfenati, E.; Leszczynski, J. Green chemistry in the synthesis of pharmaceuticals. *Chem. Rev.* **2022**, *122*, 3637–3710. [CrossRef] [PubMed]
16. Simone, M.I.; Wood, A.; Campkin, D.; Kiefel, M.J.; Houston, T.A. Recent results from non-basic glycosidase inhibitors: How structural diversity can inform general strategies for improving inhibition potency. *Eur. J. Med. Chem.* **2022**, *235*, 114282. [CrossRef]
17. Ye, X.-S.; Sun, F.; Liu, M.; Li, Q.; Wang, Y.; Zhang, G.; Zhang, L.-H.; Zhang, X.-L. Synthetic iminosugar derivatives as new potential immunosuppressive agents. *J. Med. Chem.* **2005**, *48*, 3688–3691. [CrossRef]
18. Wang, G.-N.; Xiong, Y.-L.; Ye, J.; Zhang, L.-H.; Ye, X.-S. Synthetic *N*-alkylated iminosugars as new potential immunosuppressive agents. *ACS Med. Chem. Lett.* **2011**, *2*, 682–686. [CrossRef]
19. Yu, W.; Gill, T.; Wang, L.; Du, Y.; Ye, H.; Qu, X.; Guo, J.-T.; Cuconati, A.; Zhao, K.; Block, T.M.; et al. Design, synthesis, and biological evaluation of *N*-alkylated deoxynojirimycin (DNJ) derivatives for the treatment of dengue virus infection. *J. Med. Chem.* **2012**, *55*, 6061–6075. [CrossRef]
20. Du, Y.; Ye, H.; Gill, T.; Wang, L.; Guo, F.; Cuconati, A.; Guo, J.-T.; Block, T.M.; Chang, J.; Xu, X. *N*-alkyldeoxynojirimycin derivatives with novel terminal tertiary amide substitution for treatment of bovine viral diarrhea virus (BVDV), dengue, and tacaribe virus infections. *Bioorganic Med. Chem. Lett.* **2013**, *23*, 2172–2176. [CrossRef]
21. Cipolla, L.; Sgambato, A.; Forcella, M.; Fusi, P.; Parenti, P.; Cardona, F.; Bini, D. *N*-bridged 1-deoxynojirimycin dimers as selective insect trehalase inhibitors. *Carbohydr. Res.* **2014**, *389*, 46–49. [CrossRef]
22. Welter, A.; Jadot, J.; Dardenne, G.; Marlier, M.; Casimir, J. 2,5-dihydroxymethyl 3,4-dihydroxypyrrolidine dans les feuilles de derris elliptica. *Phytochemistry* **1976**, *15*, 747–749. [CrossRef]
23. Pinto, A.J.; Stewart, D.; van Rooijen, N.; Morahan, P.S. Selective depletion of liver and splenic macrophages using liposomes encapsulating the drug dichloromethylene diphosphonate: Effects on antimicrobial resistance. *J. Leukoc. Biol.* **1991**, *49*, 579–586. [CrossRef] [PubMed]
24. Yan, X.; Shimadate, Y.; Kato, A.; Li, Y.-X.; Jia, Y.-M.; Fleet, G.W.J.; Yu, C.-Y. Synthesis of pyrrolidine monocyclic analogues of pochonicine and its stereoisomers: Pursuit of simplified structures and potent β-*N*-acetylhexosaminidase inhibition. *Molecules* **2020**, *25*, 1498. [CrossRef] [PubMed]
25. Martínez, R.F.; Jenkinson, S.F.; Nakagawa, S.; Kato, A.; Wormald, M.R.; Fleet, G.W.J.; Hollinshead, J.; Nash, R.J. Isolation from stevia rebaudiana of DMDP acetic acid, a novel iminosugar amino acid: Synthesis and glycosidase inhibition profile of glycine and β-alanine pyrrolidine amino acids. *Amino Acids* **2019**, *51*, 991–998. [CrossRef]
26. Nash, R.J.; Arthur Bell, E.; Michael Williams, J. 2-hydroxymethyl-3,4-dihydroxypyrrolidine in fruits of angylocalyx boutiqueanus. *Phytochemistry* **1985**, *24*, 1620–1622. [CrossRef]
27. Wang, J.-Z.; Cheng, B.; Kato, A.; Kise, M.; Shimadate, Y.; Jia, Y.-M.; Li, Y.-X.; Fleet, G.W.J.; Yu, C.-Y. Design, synthesis and glycosidase inhibition of *C*-4 branched LAB and DAB derivatives. *Eur. J. Med. Chem.* **2022**, *233*, 114230. [CrossRef]
28. Kato, A.; Hayashi, E.; Miyauchi, S.; Adachi, I.; Imahori, T.; Natori, Y.; Yoshimura, Y.; Nash, R.J.; Shimaoka, H.; Nakagome, I.; et al. A-L-C-butyl-1,4-dideoxy-1,4-imino-L-arabinitol as a second-generation iminosugar-based oral α-glucosidase inhibitor for improving postprandial hyperglycemia. *J. Med. Chem.* **2012**, *55*, 10347–10362. [CrossRef]

29. Trapero, A.; Llebaria, A. A prospect for pyrrolidine iminosugars as antidiabetic α-glucosidase inhibitors. *J. Med. Chem.* **2012**, *55*, 10345–10346. [CrossRef]
30. Mena-Barragán, T.; García-Moreno, M.I.; Nanba, E.; Higaki, K.; Concia, A.L.; Clapés, P.; García Fernández, J.M.; Ortiz Mellet, C. Inhibitor versus chaperone behaviour of D-fagomine, DAB and LAB sp^2-iminosugar conjugates against glycosidases: A structure–activity relationship study in Gaucher fibroblasts. *Eur. J. Med. Chem.* **2016**, *121*, 880–891. [CrossRef]
31. Tsou, E.-L.; Chen, S.-Y.; Yang, M.-H.; Wang, S.-C.; Cheng, T.-R.R.; Cheng, W.-C. Synthesis and biological evaluation of a 2-aryl polyhydroxylated pyrrolidine alkaloid-based library. *Biorg. Med. Chem.* **2008**, *16*, 10198–10204. [CrossRef]
32. Li, Y.-X.; Iwaki, R.; Kato, A.; Jia, Y.-M.; Fleet, G.W.J.; Zhao, X.; Xiao, M.; Yu, C.-Y. Fluorinated radicamine A and B: Synthesis and glycosidase inhibition. *Eur. J. Org. Chem.* **2016**, *2016*, 1429–1438. [CrossRef]
33. Liu, C.; Gao, J.; Yang, G.; Wightman, H.R.; Jiang, S. Enantiospecific synthesis of (-)-radicamine B. *Lett. Org. Chem.* **2007**, *4*, 556–558. [CrossRef]
34. Nash, R.J.; Kato, A.; Yu, C.-Y.; Fleet, G.W.J. Iminosugars as therapeutic agents: Recent advances and promising trends. *Future Med. Chem.* **2011**, *3*, 1513–1521. [CrossRef] [PubMed]
35. Chennaiah, A.; Dahiya, A.; Dubbu, S.; Vankar, Y.D. A stereoselective synthesis of an imino glycal: Application in the synthesis of (−)-1-epi-adenophorine and a homoimindosugar. *Eur. J. Org. Chem.* **2018**, *2018*, 6574–6581. [CrossRef]
36. Chennaiah, A.; Bhowmick, S.; Vankar, Y.D. Conversion of glycals into vicinal-1,2-diazides and 1,2-(or 2,1)-azidoacetates using hypervalent iodine reagents. Application in the synthesis of *N*-glycopeptides, pseudo-trisaccharides and an iminosugar. *RSC Adv.* **2017**, *7*, 41755–41762. [CrossRef]
37. Verma, A.K.; Dubbu, S.; Chennaiah, A.; Vankar, Y.D. Synthesis of di- and trihydroxy proline derivatives from D-glycals: Application in the synthesis of polysubstituted pyrrolizidines and bioactive *C*-aryl/alkyl pyrrolidines. *Carbohydr. Res.* **2019**, *475*, 48–55. [CrossRef]
38. Horne, G.; Wilson, F.X.; Tinsley, J.; Williams, D.H.; Storer, R. Iminosugars past, present and future: Medicines for tomorrow. *Drug Discovery Today* **2011**, *16*, 107–118. [CrossRef]
39. Ghani, U. Re-exploring promising α-glucosidase inhibitors for potential development into oral anti-diabetic drugs: Finding needle in the haystack. *Eur. J. Med. Chem.* **2015**, *103*, 133–162. [CrossRef]
40. Singh, A.; Mhlongo, N.; Soliman, E.S.M. Anti-cancer glycosidase inhibitors from natural products: A computational and molecular modelling perspective. *Anticancer Agents Med. Chem.* **2015**, *15*, 933–946. [CrossRef]
41. Vanni, C.; Bodlenner, A.; Marradi, M.; Schneider, J.P.; Ramirez, M.d.l.A.; Moya, S.; Goti, A.; Cardona, F.; Compain, P.; Matassini, C. Hybrid multivalent jack bean α-mannosidase inhibitors: The first example of gold nanoparticles decorated with deoxynojirimycin inhitopes. *Molecules* **2021**, *26*, 5864. [CrossRef]
42. Brissonnet, Y.; Ladevèze, S.; Tezé, D.; Fabre, E.; Deniaud, D.; Daligault, F.; Tellier, C.; Šesták, S.; Remaud-Simeon, M.; Potocki-Veronese, G.; et al. Polymeric iminosugars improve the activity of carbohydrate-processing enzymes. *Bioconjugate Chem.* **2015**, *26*, 766–772. [CrossRef]
43. González-Cuesta, M.; Ortiz Mellet, C.; García Fernández, J.M. Carbohydrate supramolecular chemistry: Beyond the multivalent effect. *Chem. Commun.* **2020**, *56*, 5207–5222. [CrossRef] [PubMed]
44. Lepage, M.L.; Schneider, J.P.; Bodlenner, A.; Meli, A.; De Riccardis, F.; Schmitt, M.; Tarnus, C.; Nguyen-Huynh, N.-T.; Francois, Y.-N.; Leize-Wagner, E.; et al. Iminosugar-cyclopeptoid conjugates raise multivalent effect in glycosidase inhibition at unprecedented high levels. *Chem. Eur. J.* **2016**, *22*, 5151–5155. [CrossRef] [PubMed]
45. Brissonnet, Y.; Ortiz Mellet, C.; Morandat, S.; Garcia Moreno, M.I.; Deniaud, D.; Matthews, S.E.; Vidal, S.; Sestak, S.; El Kirat, K.; Gouin, S.G. Topological effects and binding modes operating with multivalent iminosugar-based glycoclusters and mannosidases. *J. Am. Chem. Soc.* **2013**, *135*, 18427–18435. [CrossRef] [PubMed]
46. Li, R.-F.; Yang, J.-X.; Liu, J.; Ai, G.-M.; Zhang, H.-Y.; Xu, L.-Y.; Chen, S.-B.; Zhang, H.-X.; Li, X.-L.; Cao, Z.-R.; et al. Positional isomeric effects on the optical properties, multivalent glycosidase inhibition effect, and hypoglycemic effect of perylene bisimide-deoxynojirimycin conjugates. *J. Med. Chem.* **2021**, *64*, 5863–5873. [CrossRef] [PubMed]
47. Compain, P. Multivalent effect in glycosidase inhibition: The end of the beginning. *Chem. Rec.* **2020**, *20*, 10–22. [CrossRef]
48. Howard, E.; Cousido-Siah, A.; Lepage, M.L.; Schneider, J.P.; Bodlenner, A.; Mitschler, A.; Meli, A.; Izzo, I.; Alvarez, H.A.; Podjarny, A.; et al. Structural basis of outstanding multivalent effects in jack bean α-mannosidase inhibition. *Angew. Chem. Int. Ed.* **2018**, *57*, 8002–8006. [CrossRef]
49. Martínez-Bailén, M.; Carmona, A.T.; Cardona, F.; Matassini, C.; Goti, A.; Kubo, M.; Kato, A.; Robina, I.; Moreno-Vargas, A.J. Synthesis of multimeric pyrrolidine iminosugar inhibitors of human β-glucocerebrosidase and α-galactosidase a: First example of a multivalent enzyme activity enhancer for Fabry disease. *Eur. J. Med. Chem.* **2020**, *192*, 112173. [CrossRef]
50. Johns, B.A.; Johnson, C.R. Scaffolded bis-azasugars: A dual warhead approach to glycosidase inhibition. *Tetrahedron Lett.* **1998**, *39*, 749–752. [CrossRef]
51. Wennekes, T.; van den Berg, R.J.B.H.N.; Bonger, K.M.; Donker-Koopman, W.E.; Ghisaidoobe, A.; van der Marel, G.A.; Strijland, A.; Aerts, J.M.F.G.; Overkleeft, H.S. Synthesis and evaluation of dimeric lipophilic iminosugars as inhibitors of glucosylceramide metabolism. *Tetrahedron Asymmetry* **2009**, *20*, 836–846. [CrossRef]
52. Lohse, A.; Jensen, K.B.; Lundgren, K.; Bols, M. Synthesis and deconvolution of the first combinatorial library of glycosidase inhibitors. *Bioorganic Med. Chem.* **1999**, *7*, 1965–1971. [CrossRef]

53. Gestwicki, J.E.; Cairo, C.W.; Strong, L.E.; Oetjen, K.A.; Kiessling, L.L. Influencing receptor−ligand binding mechanisms with multivalent ligand architecture. *J. Am. Chem. Soc.* **2002**, *124*, 14922–14933. [CrossRef] [PubMed]
54. Wittmann, V.; Pieters, R.J. Bridging lectin binding sites by multivalent carbohydrates. *Chem. Soc. Rev.* **2013**, *42*, 4492–4503. [CrossRef] [PubMed]
55. Kiessling, L.L.; Gestwicki, J.E.; Strong, L.E. Synthetic multivalent ligands in the exploration of cell-surface interactions. *Curr. Opin. Chem. Biol.* **2000**, *4*, 696–703. [CrossRef]
56. Tornøe, C.W.; Christensen, C.; Meldal, M. Peptidotriazoles on solid phase: [1,2,3]-triazoles by regiospecific copper(I)-catalyzed 1,3-dipolar cycloadditions of terminal alkynes to azides. *J. Org. Chem.* **2002**, *67*, 3057–3064. [CrossRef]
57. Rostovtsev, V.V.; Green, L.G.; Fokin, V.V.; Sharpless, K.B. A stepwise huisgen cycloaddition process: Copper(I)-catalyzed regioselective "ligation" of azides and terminal alkynes. *Angew. Chem. Int. Ed.* **2002**, *41*, 2596–2599. [CrossRef]
58. Diot, J.; García-Moreno, M.I.; Gouin, S.G.; Ortiz Mellet, C.; Haupt, K.; Kovensky, J. Multivalent iminosugars to modulate affinity and selectivity for glycosidases. *Org. Biomol. Chem.* **2009**, *7*, 357–363. [CrossRef]
59. Gouin, S.G. Multivalent inhibitors for carbohydrate-processing enzymes: Beyond the "lock-and-key" concept. *Chem. Eur. J.* **2014**, *20*, 11616–11628. [CrossRef]
60. Compain, P.; Bodlenner, A. The multivalent effect in glycosidase inhibition: A new, rapidly emerging topic in glycoscience. *Chembiochem* **2014**, *15*, 1239–1251. [CrossRef]
61. Zelli, R.; Bartolami, E.; Longevial, J.-F.; Bessin, Y.; Dumy, P.; Marra, A.; Ulrich, S. A metal-free synthetic approach to peptide-based iminosugar clusters as novel multivalent glycosidase inhibitors. *RSC. Adv.* **2016**, *6*, 2210–2216. [CrossRef]
62. Matassini, C.; D'Adamio, G.; Vanni, C.; Goti, A.; Cardona, F. Studies for the multimerization of DAB-L-based iminosugars through iteration of the nitrone cycloaddition/ring-opening/allylation sequence. *Eur. J. Org. Chem.* **2019**, *2019*, 4897–4905. [CrossRef]
63. Zamoner, L.O.B.; Aragao-Leoneti, V.; Mantoani, S.P.; Rugen, M.D.; Nepogodiev, S.A.; Field, R.A.; Carvalho, I. Cuaac click chemistry with *N*-propargyl 1,5-dideoxy-1,5-imino-D-gulitol and n-propargyl 1,6-dideoxy-1,6-imino-D-mannitol provides access to triazole-linked piperidine and azepane pseudo-disaccharide iminosugars displaying glycosidase inhibitory properties. *Carbohydr. Res.* **2016**, *429*, 29–37. [CrossRef] [PubMed]
64. Bonduelle, C.; Huang, J.; Mena-Barragán, T.; Ortiz Mellet, C.; Decroocq, C.; Etamé, E.; Heise, A.; Compain, P.; Lecommandoux, S. Iminosugar-based glycopolypeptides: Glycosidase inhibition with bioinspired glycoprotein analogue micellar self-assemblies. *Chem. Commun.* **2014**, *50*, 3350–3352. [CrossRef] [PubMed]
65. Li, J.-J.; Wang, K.-R.; Li, R.-F.; Yang, J.-X.; Li, M.; Zhang, H.-X.; Cao, Z.-R.; Li, X.-L. Synthesis, self-assembly behaviours and multivalent glycosidase inhibition effects of a deoxynojirimycin modified perylene bisimide derivative. *J. Mater. Chem. B* **2019**, *7*, 1270–1275. [CrossRef] [PubMed]
66. Yang, J.-X.; Li, J.-J.; Yin, F.-Q.; Wang, G.-Y.; Wei, W.-T.; Li, X.-L.; Wang, K.-R. Multivalent glucosidase inhibitors based on perylene bisimide and iminosugar conjugates. *Eur. J. Med. Chem.* **2022**, *241*, 114621. [CrossRef] [PubMed]
67. D'Adamio, G.; Matassini, C.; Parmeggiani, C.; Catarzi, S.; Morrone, A.; Goti, A.; Paoli, P.; Cardona, F. Evidence for a multivalent effect in inhibition of sulfatases involved in lysosomal storage disorders. *RSC. Adv.* **2016**, *6*, 64847–64851. [CrossRef]
68. Della Sala, P.; Vanni, C.; Talotta, C.; Di Marino, L.; Matassini, C.; Goti, A.; Neri, P.; Šesták, S.; Cardona, F.; Gaeta, C. Multivalent resorcinarene clusters decorated with DAB-1 inhitopes: Targeting golgi α-mannosidase from drosophila melanogaster. *Org. Chem. Front.* **2021**, *8*, 6648–6656. [CrossRef]
69. Pingitore, V.; Martinez-Bailen, M.; Carmona, A.T.; Meszaros, Z.; Kulik, N.; Slamova, K.; Kren, V.; Bojarova, P.; Robina, I.; Moreno-Vargas, A.J. Discovery of human hexosaminidase inhibitors by in situ screening of a library of mono- and divalent pyrrolidine iminosugars. *Bioorganic Chem.* **2022**, *120*, 105650. [CrossRef]
70. Moreno-Clavijo, E.; Carmona, A.T.; Moreno-Vargas, A.J.; Molina, L.; Wright, D.W.; Davies, G.J.; Robina, I. Exploring a multivalent approach to α-L-fucosidase inhibition. *Eur. J. Org. Chem.* **2013**, *2013*, 7328–7336. [CrossRef]
71. Prichard, K.; Campkin, D.; O'Brien, N.; Kato, A.; Fleet, G.W.J.; Simone, M.I. Biological activities of 3,4,5-trihydroxypiperidines and their *N*- and *O*-derivatives. *Chem. Biol. Drug Des.* **2018**, *92*, 1171–1197. [CrossRef]
72. Dhara, D.; Dhara, A.; Bennett, J.; Murphy, P.V. Cyclisations and strategies for stereoselective synthesis of piperidine iminosugars. *Chem. Rec.* **2021**, *21*, 2958–2979. [CrossRef]
73. Stocker, B.L.; Dangerfield, E.M.; Win-Mason, A.L.; Haslett, G.W.; Timmer, M.S.M. Recent developments in the synthesis of pyrrolidine-containing iminosugars. *Eur. J. Org. Chem.* **2010**, *2010*, 1615–1637. [CrossRef]
74. Decroocq, C.; Rodríguez-Lucena, D.; Russo, V.; Mena Barragán, T.; Ortiz Mellet, C.; Compain, P. The multivalent effect in glycosidase inhibition: Probing the influence of architectural parameters with cyclodextrin-based iminosugar click clusters. *Chem. Eur. J.* **2011**, *17*, 13825–13831. [CrossRef]
75. Cagnoni, A.J.; Varela, O.; Gouin, S.G.; Kovensky, J.; Uhrig, M.L. Synthesis of multivalent glycoclusters from L-thio-β-D-galactose and their inhibitory activity against the β-galactosidase from *E. coli*. *J. Org. Chem.* **2011**, *76*, 3064–3077. [CrossRef]
76. Moreno-Clavijo, E.; Carmona, A.T.; Vera-Ayoso, Y.; Moreno-Vargas, A.J.; Bello, C.; Vogel, P.; Robina, I. Synthesis of novel pyrrolidine 3,4-diol derivatives as inhibitors of α-L-fucosidases. *Org. Biomol. Chem.* **2009**, *7*, 1192–1202. [CrossRef] [PubMed]
77. Moreno-Clavijo, E.; Carmona, A.T.; Moreno-Vargas, A.J.; Rodríguez-Carvajal, M.A.; Robina, I. Synthesis and inhibitory activities of novel *C*-3 substituted azafagomines: A new type of selective inhibitors of α-L-fucosidases. *Bioorganic Med. Chem.* **2010**, *18*, 4648–4660. [CrossRef] [PubMed]

78. Robina, I.; Moreno-Vargas, A.J.; Fernández-Bolaños, J.G.; Fuentes, J.; Demange, R.; Vogel, P. New leads for selective inhibitors of α-L-fucosidases. Synthesis and glycosidase inhibitory activities of [(2*R*,3*R*,4*R*)-3,4-dihydroxypyrrolidin-2-yl]furan derivatives. *Bioorganic Med. Chem. Lett.* **2001**, *11*, 2555–2559. [CrossRef]
79. Szczepanska, A.; Espartero, J.L.; Moreno-Vargas, A.J.; Carmona, A.T.; Robina, I.; Remmert, S.; Parish, C. Synthesis and conformational analysis of novel trimeric maleimide cross-linking reagents. *J. Org. Chem.* **2007**, *72*, 6776–6785. [CrossRef]
80. Hottin, A.; Carrión-Jiménez, S.; Moreno-Clavijo, E.; Moreno-Vargas, A.J.; Carmona, A.T.; Robina, I.; Behr, J.-B. Expanding the library of divalent fucosidase inhibitors with polyamino and triazole-benzyl bridged bispyrrolidines. *Org. Biomol. Chem.* **2016**, *14*, 3212–3220. [CrossRef]
81. Hottin, A.; Dubar, F.; Steenackers, A.; Delannoy, P.; Biot, C.; Behr, J.-B. Iminosugar–ferrocene conjugates as potential anticancer agents. *Org. Biomol. Chem.* **2012**, *10*, 5592–5597. [CrossRef]
82. Bergeron-Brlek, M.; Goodwin-Tindall, J.; Cekic, N.; Roth, C.; Zandberg, W.F.; Shan, X.; Varghese, V.; Chan, S.; Davies, G.J.; Vocadlo, D.J.; et al. A convenient approach to stereoisomeric iminocyclitols: Generation of potent brain-permeable OGA inhibitors. *Angew. Chem. Int. Ed.* **2015**, *54*, 15429–15433. [CrossRef]
83. Elías-Rodríguez, P.; Moreno-Clavijo, E.; Carmona, A.T.; Moreno-Vargas, A.J.; Robina, I. Rapid discovery of potent α-fucosidase inhibitors by in situ screening of a library of (pyrrolidin-2-yl)triazoles. *Org. Biomol. Chem.* **2014**, *12*, 5898–5904. [CrossRef] [PubMed]
84. Hottin, A.; Wright, D.W.; Moreno-Clavijo, E.; Moreno-Vargas, A.J.; Davies, G.J.; Behr, J.-B. Exploring the divalent effect in fucosidase inhibition with stereoisomeric pyrrolidine dimers. *Org. Biomol. Chem.* **2016**, *14*, 4718–4727. [CrossRef]
85. Wong, C.-H.; Provencher, L.; Porco, J.A.; Jung, S.-H.; Wang, Y.-F.; Chen, L.; Wang, R.; Steensma, D.H. Synthesis and evaluation of homoaza sugars as glycosidase inhibitors. *J. Org. Chem.* **1995**, *60*, 1492–1501. [CrossRef]
86. Srihari, P.; Prem Kumar, B.; Subbarayudu, K.; Yadav, J.S. A convergent approach for the total synthesis of (−)-synrotolide diacetate. *Tetrahedron Lett.* **2007**, *48*, 6977–6981. [CrossRef]
87. Rao, M.V.; Rao, B.V.; Ramesh, B. Total synthesis of AI-77-B. *Tetrahedron Lett.* **2014**, *55*, 5921–5924. [CrossRef]
88. Hottin, A.; Wright, D.W.; Davies, G.J.; Behr, J.-B. Exploiting the hydrophobic terrain in fucosidases with aryl-substituted pyrrolidine iminosugars. *Chembiochem* **2015**, *16*, 277–283. [CrossRef] [PubMed]
89. Carmona, A.T.; Carrion-Jimenez, S.; Pingitore, V.; Moreno-Clavijo, E.; Robina, I.; Moreno-Vargas, A.J. Harnessing pyrrolidine iminosugars into dimeric structures for the rapid discovery of divalent glycosidase inhibitors. *Eur. J. Med. Chem.* **2018**, *151*, 765–776. [CrossRef] [PubMed]
90. Huisgen, R.; Szeimies, G.; Möbius, L. 1.3-dipolare cycloadditionen, XXXII. Kinetik der additionen organischer azide an cc-mehrfachbindungen. *Chem. Ber.* **1967**, *100*, 2494–2507. [CrossRef]
91. Huisgen, R. 1,3-dipolar cycloadditions. Past and future. *Angew. Chem. Int. Ed.* **1963**, *2*, 565–598. [CrossRef]
92. Moreno-Vargas, A.J.; Demange, R.; Fuentes, J.; Robina, I.; Vogel, P. Synthesis of [(2*S*,3*S*,4*R*)-3,4-dihydroxypyrrolidin-2-yl]-5-methylfuran-4-carboxylic acid derivatives: New leads as selective β-galactosidase inhibitors. *Bioorganic Med. Chem. Lett.* **2002**, *12*, 2335–2339. [CrossRef]
93. Moreno-Vargas, A.J.; Robina, I.; Demange, R.; Vogel, P. Synthesis and glycosidase inhibitory activities of 5-(1′,4′-dideoxy-1′,4′-imino-D-erythrosyl)-2-methyl-3-furoic acid (=5-[(3*S*,4*R*)-3,4-dihydroxypyrrolidin-2-yl]-2-methylfuran-3-carboxylic acid) derivatives: New leads as selective α-L-fucosidase and β-galactosidase inhibitors. *Helv. Chim. Acta* **2003**, *86*, 1894–1913.
94. Aucagne, V.; Valverde, I.E.; Marceau, P.; Galibert, M.; Dendane, N.; Delmas, A.F. Towards the simplification of protein synthesis: Iterative solid-supported ligations with concomitant purifications. *Angew. Chem. Int. Ed.* **2012**, *51*, 11320–11324. [CrossRef]
95. Liu, J.; Shikhman, A.R.; Lotz, M.K.; Wong, C.-H. Hexosaminidase inhibitors as new drug candidates for the therapy of osteoarthritis. *Chem. Biol.* **2001**, *8*, 701–711. [CrossRef]
96. Tsou, E.-L.; Yeh, Y.-T.; Liang, P.-H.; Cheng, W.-C. A convenient approach toward the synthesis of enantiopure isomers of DMDP and ADMDP. *Tetrahedron* **2009**, *65*, 93–100. [CrossRef]
97. Shaikh, T.M.; Sudalai, A. Enantioselective synthesis of (+)-α-conhydrine and (−)-sedamine by L-proline-catalysed α-aminooxylation. *Eur. J. Org. Chem.* **2010**, *2010*, 3437–3444. [CrossRef]
98. Ornelas, C.; Broichhagen, J.; Weck, M. Strain-promoted alkyne azide cycloaddition for the functionalization of poly(amide)-based dendrons and dendrimers. *J. Am. Chem. Soc.* **2010**, *132*, 3923–3931. [CrossRef] [PubMed]
99. Martella, D.; D'Adamio, G.; Parmeggiani, C.; Cardona, F.; Moreno-Clavijo, E.; Robina, I.; Goti, A. Cycloadditions of sugar-derived nitrones targeting polyhydroxylated indolizidines. *Eur. J. Org. Chem.* **2016**, *2016*, 1588–1598. [CrossRef]
100. Ilardi, E.A.; Stivala, C.E.; Zakarian, A. Hexafluoroisopropanol as a unique solvent for stereoselective iododesilylation of vinylsilanes. *Org. Lett.* **2008**, *10*, 1727–1730. [CrossRef]
101. Ishikawa, T.; Tajima, Y.; Fukui, M.; Saito, S. Synthesis and asymmetric [3 + 2] cycloaddition reactions of chiral cyclic nitrone: A novel system providing maximal facial bias for both nitrone and dipolarophile. *Angew. Chem. Int. Ed.* **1996**, *35*, 1863–1864. [CrossRef]
102. Tamayo, J.A.; Franco, F.; Lo Re, D.; Sánchez-Cantalejo, F. Synthesis of pentahydroxylated pyrrolizidines and indolizidines. *J. Org. Chem.* **2009**, *74*, 5679–5682. [CrossRef]
103. Merino, P.; Delso, I.; Tejero, T.; Cardona, F.; Marradi, M.; Faggi, E.; Parmeggiani, C.; Goti, A. Nucleophilic additions to cyclic nitrones en route to iminocyclitols—total syntheses of DMDP, 6-deoxy-DMDP, DAB-1, CYB-3, nectrisine, and radicamine B. *Eur. J. Org. Chem.* **2008**, *2008*, 2929–2947. [CrossRef]

104. Parmeggiani, C.; Catarzi, S.; Matassini, C.; D'Adamio, G.; Morrone, A.; Goti, A.; Paoli, P.; Cardona, F. Human acid β-glucosidase inhibition by carbohydrate derived iminosugars: Towards new pharmacological chaperones for gaucher disease. *Chembiochem* **2015**, *16*, 2054–2064. [CrossRef] [PubMed]
105. Coutrot, F.; Busseron, E. Controlling the chair conformation of a mannopyranose in a large-amplitude [2]rotaxane molecular machine. *Chem. Eur. J.* **2009**, *15*, 5186–5190. [CrossRef] [PubMed]
106. Chabre, Y.M.; Giguère, D.; Blanchard, B.; Rodrigue, J.; Rocheleau, S.; Neault, M.; Rauthu, S.; Papadopoulos, A.; Arnold, A.A.; Imberty, A.; et al. Combining glycomimetic and multivalent strategies toward designing potent bacterial lectin inhibitors. *Chem. Eur. J.* **2011**, *17*, 6545–6562. [CrossRef]
107. Mirabella, S.; D'Adamio, G.; Matassini, C.; Goti, A.; Delgado, S.; Gimeno, A.; Robina, I.; Moreno-Vargas, A.J.; Šesták, S.; Jiménez-Barbero, J.; et al. Mechanistic insight into the binding of multivalent pyrrolidines to α-mannosidases. *Chem. Eur. J.* **2017**, *23*, 14585–14596. [CrossRef]
108. Martínez-Bailén, M.; Jiménez-Ortega, E.; Carmona, A.T.; Robina, I.; Sanz-Aparicio, J.; Talens-Perales, D.; Polaina, J.; Matassini, C.; Cardona, F.; Moreno-Vargas, A.J. Structural basis of the inhibition of gh1 β-glucosidases by multivalent pyrrolidine iminosugars. *Bioorganic Chem.* **2019**, *89*, 103026. [CrossRef]
109. Veliks, J.; Seifert, H.M.; Frantz, D.K.; Klosterman, J.K.; Tseng, J.-C.; Linden, A.; Siegel, J.S. Towards the molecular borromean link with three unequal rings: Double-threaded ruthenium(II) ring-in-ring complexes. *Org. Chem. Front.* **2016**, *3*, 667–672. [CrossRef]
110. Chabre, Y.M.; Contino-Pépin, C.; Placide, V.; Shiao, T.C.; Roy, R. Expeditive synthesis of glycodendrimer scaffolds based on versatile tris and mannoside derivatives. *J. Org. Chem.* **2008**, *73*, 5602–5605. [CrossRef]
111. Martinez-Bailen, M.; Carmona, A.T.; Moreno-Clavijo, E.; Robina, I.; Ide, D.; Kato, A.; Moreno-Vargas, A.J. Tuning of β-glucosidase and α-galactosidase inhibition by generation and in situ screening of a library of pyrrolidine-triazole hybrid molecules. *Eur. J. Med. Chem.* **2017**, *138*, 532–542. [CrossRef]
112. Martinez-Bailen, M.; Carmona, A.T.; Patterson-Orazem, A.C.; Lieberman, R.L.; Ide, D.; Kubo, M.; Kato, A.; Robina, I.; Moreno-Vargas, A.J. Exploring substituent diversity on pyrrolidine-aryltriazole iminosugars: Structural basis of β-glucocerebrosidase inhibition. *Bioorganic Chem.* **2019**, *86*, 652–664. [CrossRef]
113. Tunstad, L.M.; Tucker, J.A.; Dalcanale, E.; Weiser, J.; Bryant, J.A.; Sherman, J.C.; Helgeson, R.C.; Knobler, C.B.; Cram, D.J. Host-guest complexation. 48. Octol building blocks for cavitands and carcerands. *J. Org. Chem.* **1989**, *54*, 1305–1312. [CrossRef]
114. Gaeta, C.; Della Sala, P.; Talotta, C.; De Rosa, M.; Soriente, A.; Brancatelli, G.; Geremia, S.; Neri, P. A tetrasulfate-resorcin [6]arene cavitand as the host for organic ammonium guests. *Org. Chem. Front.* **2016**, *3*, 1276–1280. [CrossRef]
115. Brancatelli, G.; Geremia, S.; Gaeta, C.; Della Sala, P.; Talotta, C.; De Rosa, M.; Neri, P. Solid-state assembly of a resorcin [6]arene in twin molecular capsules. *CrystEngComm* **2016**, *18*, 5045–5049. [CrossRef]
116. Della Sala, P.; Gaeta, C.; Navarra, W.; Talotta, C.; De Rosa, M.; Brancatelli, G.; Geremia, S.; Capitelli, F.; Neri, P. Improved synthesis of larger resorcinarenes. *J. Org. Chem.* **2016**, *81*, 5726–5731. [CrossRef] [PubMed]
117. Timmerman, P.; Verboom, W.; Reinhoudt, D.N. Resorcinarenes. *Tetrahedron* **1996**, *52*, 2663–2704. [CrossRef]
118. Moloney, D.J.; Shair, L.H.; Lu, F.M.; Xia, J.; Locke, R.; Matta, K.L.; Haltiwanger, R.S. Mammalian notch1 is modified with two unusual forms of *O*-linked glycosylation found on epidermal growth factor-like modules. *J. Biol. Chem.* **2000**, *275*, 9604–9611. [CrossRef] [PubMed]
119. Schneider, M.; Al-Shareffi, E.; Haltiwanger, R.S. Biological functions of fucose in mammals. *Glycobiology* **2017**, *27*, 601–618. [CrossRef]
120. Listinsky, J.J.; Siegal, G.P.; Listinsky, C.M. A-l-fucose: A potentially critical molecule in pathologic processes including neoplasia. *Am. J. Clin. Pathol.* **1998**, *110*, 425–440. [CrossRef]
121. Lopes, G.; Ho, C.K.; Liau, K.H.; Chung, A.; Cheow, P.; Chang, A.Y. Gefitinib in the adjuvant treatment of hepatocellular carcinoma: A pilot study by the singapore hepatocellular carcinoma consortium. *J. Clin. Oncol.* **2010**, *28*, 210. [CrossRef]
122. Ayude, D.; de la Cadena, M.P.; Cordero, O.J.; Nogueira, M.; Ayude, J.; Fernández-Briera, A.; Rodríguez-Berrocal, F.J. Clinical interest of the combined use of serum CD26 and alpha-L-fucosidase in the early diagnosis of colorectal cancer. *Dis. Markers* **2004**, *19*, 834309. [CrossRef]
123. Ayude, D.; Páez de la Cadena, M.; Martínez-Zorzano, V.S.; Fernández-Briera, A.; Rodríguez-Berrocal, F.J. Preoperative serum alpha-L-fucosidase activity as a prognostic marker in colorectal cancer. *Oncology* **2003**, *64*, 36–45. [CrossRef] [PubMed]
124. Liu, T.-W.; Ho, C.-W.; Huang, H.-H.; Chang, S.-M.; Popat Shide, D.; Wang, Y.-T.; Wu, M.-S.; Chen, Y.-J.; Lin, C.-H. Role for α-L-fucosidase in the control of helicobacter pylori-infected gastric cancer cells. *Proc. Natl. Acad. Sci. USA* **2009**, *106*, 14581–14586. [CrossRef] [PubMed]
125. Kotland, A.; Accadbled, F.; Robeyns, K.; Behr, J.-B. Synthesis and fucosidase inhibitory study of unnatural pyrrolidine alkaloid 4-epi-(+)-codonopsinine. *J. Org. Chem.* **2011**, *76*, 4094–4098. [CrossRef]
126. Wu, C.-Y.; Chang, C.-F.; Chen, J.S.-Y.; Wong, C.-H.; Lin, C.-H. Rapid diversity-oriented synthesis in microtiter plates for in situ screening: Discovery of potent and selective α-fucosidase inhibitors. *Angew. Chem. Int. Ed.* **2003**, *42*, 4661–4664. [CrossRef] [PubMed]
127. Gnanesh Kumar, B.S.; Pohlentz, G.; Schulte, M.; Mormann, M.; Siva Kumar, N. Jack bean α-mannosidase: Amino acid sequencing and *N*-glycosylation analysis of a valuable glycomics tool. *Glycobiology* **2014**, *24*, 252–261. [CrossRef] [PubMed]
128. Daniel, P.F.; Winchester, B.; Warren, C.D. Mammalian α-mannosidases—multiple forms but a common purpose? *Glycobiology* **1994**, *4*, 551–566. [CrossRef]

129. Berardi, A.C.; Manieri, P.; Ciraci, E.; Tribuzi, R.; Di Girolamo, I.; Cavalieri, C.; Isacchi, G.; Emiliani, C.; Bottazzo, G.; Orlacchio, A.; et al. Lysosomal glycohydrolase activities in dendritic cells: Is it a function of hematopoietic stem cells differentiation process? *Blood* **2004**, *104*, 4193. [CrossRef]
130. Solanki, G.A.; Martin, K.W.; Theroux, M.C.; Lampe, C.; White, K.K.; Shediac, R.; Lampe, C.G.; Beck, M.; Mackenzie, W.G.; Hendriksz, C.J.; et al. Spinal involvement in mucopolysaccharidosis IVA (morquio-brailsford or morquio A syndrome): Presentation, diagnosis and management. *J. Inherit. Metab. Dis.* **2013**, *36*, 339–355. [CrossRef]
131. Sanford, M.; Lo, J.H. Elosulfase alfa: First global approval. *Drugs* **2014**, *74*, 713–718. [CrossRef]
132. Platt, F.M. Sphingolipid lysosomal storage disorders. *Nature* **2014**, *510*, 68–75. [CrossRef]

molecules

MDPI

Review

Rhamnose-Containing Compounds: Biosynthesis and Applications

Siqiang Li [1,2], Fujia Chen [1,2], Yun Li [1,2], Lizhen Wang [3], Hongyan Li [1], Guofeng Gu [4,*] and Enzhong Li [1,2,*]

1 School of Biological and Food Processing Engineering, Huanghuai University, Zhumadian 463000, China
2 Institute of Agricultural Products Fermentation Engineering and Application, Huanghuai University, Zhumadian 463000, China
3 Biology Institute, Qilu University of Technology (Shandong Academy of Sciences), Jinan 250100, China
4 National Glycoengineering Research Center, Shandong Key Laboratory of Carbohydrate Chemistry and Glycobiology, Shandong University, 72 Binhai Road, Qingdao 266237, China
* Correspondence: guofenggu@sdu.edu.cn (G.G.); enzhongli@163.com (E.L.)

Abstract: Rhamnose-associated molecules are attracting attention because they are present in bacteria but not mammals, making them potentially useful as antibacterial agents. Additionally, they are also valuable for tumor immunotherapy. Thus, studies on the functions and biosynthetic pathways of rhamnose-containing compounds are in progress. In this paper, studies on the biosynthetic pathways of three rhamnose donors, i.e., deoxythymidinediphosphate-L-rhamnose (dTDP-Rha), uridine diphosphate-rhamnose (UDP-Rha), and guanosine diphosphate rhamnose (GDP-Rha), are firstly reviewed, together with the functions and crystal structures of those associated enzymes. Among them, dTDP-Rha is the most common rhamnose donor, and four enzymes, including glucose-1-phosphate thymidylyltransferase RmlA, dTDP-Glc-4,6-dehydratase RmlB, dTDP-4-keto-6-deoxy-Glc-3,5-epimerase RmlC, and dTDP-4-keto-Rha reductase RmlD, are involved in its biosynthesis. Secondly, several known rhamnosyltransferases from *Geobacillus stearothermophilus*, *Saccharopolyspora spinosa*, *Mycobacterium tuberculosis*, *Pseudomonas aeruginosa*, and *Streptococcus pneumoniae* are discussed. In these studies, however, the functions of rhamnosyltransferases were verified by employing gene knockout and radiolabeled substrates, which were almost impossible to obtain and characterize the products of enzymatic reactions. Finally, the application of rhamnose-containing compounds in disease treatments is briefly described.

Keywords: rhamnose; deoxythymidinediphosphate-L-rhamnose; guanosine diphosphate rhamnose; uridine diphosphate-rhamnose; rhamnosyltransferase; rhamnose biosynthesis

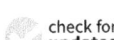

Citation: Li, S.; Chen, F.; Li, Y.; Wang, L.; Li, H.; Gu, G.; Li, E. Rhamnose-Containing Compounds: Biosynthesis and Applications. *Molecules* **2022**, *27*, 5315. https://doi.org/10.3390/molecules27165315

Academic Editors: Jian Yin, Jing Zeng and De-Cai Xiong

Received: 9 July 2022
Accepted: 15 August 2022
Published: 20 August 2022

1. Introduction

Glycans are important components of various glycoconjugates, such as glycoproteins, glycolipids, and proteoglycans, and play pivotal roles in many biological processes, including intracellular trafficking, cell adhesion and development, cancer progression, host–pathogen interactions, and immune responses [1]. For a detailed structure–activity relationship analysis of functional glycans, it is necessary to obtain molecules in structurally homogeneous forms, which is not easy to achieve via the isolation of natural products from biological sources. Therefore, the total synthesis of polysaccharides and their oligomeric analogs has become a hot research topic. Rhamnose (Rha)-containing compounds (RCCs) are especially interesting due to their potential applications, including antibacterial vaccines and killing tumors [2,3]. Additionally, Rha is a common component of various bacterial polysaccharides, such as lipopolysaccharides (LPSs) [4], extracellular polysaccharides (EPSs) [5], capsular polysaccharides (CPSs) [6], and cell wall polysaccharides [7]. In addition to bacteria, Rha is also found in viruses [8], fungi [9], plants [10], and lower animals [11]. Interestingly, Rha has not been found in humans or other mammals. In recent years, more evidence has emerged about its essential roles in many pathogenic bacteria, making it a potentially attractive therapeutic target. Furthermore, RCCs are also candidates for vaccines, antitumor drugs, and antibacterial drugs [3,12]. Thus, there is a keen desire to obtain and characterize RCCs. However, due to the complexity of the target molecules and the difficulty in constructing certain glycosidic linkages, such as β-linked Rha, via chemical glycosylation, enzymatic synthesis is particularly attractive [13]. For the enzymatic synthesis of RCCs or their conjugates, rhamnosyl donors are the key substrates [14,15], which are utilized by rhamnosyltransferases (Rha-Ts) and attached to sugar acceptors [16]. Therefore, the catalysis of Rha-Ts and the preparation of Rha donors and acceptors in vitro are hot topics [13,17,18]. In this paper, the biosynthetic pathways of Rha donors are reviewed, and the development of Rha-Ts and their medical perspectives are also explored. Such knowledge expands our understandings of the biosynthetic pathways of RCCs and could facilitate their enzymatic synthesis.

2. Biosynthetic Pathways of Donors of RCCs

Three sugar nucleotides, including deoxythymidinediphosphate-L-rhamnose (dTDP-Rha), guanosine diphosphate rhamnose (GDP-Rha), and uridine diphosphate-rhamnose (UDP-Rha), can serve as Rha donors in reactions catalyzed by Rha-Ts. dTDP-Rha and GDP-Rha are present in bacteria and fungi, whereas UDP-Rha is only found in plants. There are probably other Rha donors involving in Rha biosynthetic pathways in *Mycoplasma* [19]. The biosynthetic pathways of these three Rha donors and structural, mechanistic, and biochemical aspects of the key enzymes involved are reviewed below.

2.1. Biosynthetic Pathways of dTDP-Rha

The dTDP-Rha is one of the most important sugar precursors. Four enzymes, glucose-1-phosphate thymidylyltransferase (RmlA), dTDP-D-glucose 4,6-dehydratase (RmlB), dTDP-4-keto-6-deoxy-D-glucose3,5-epimerase (RmlC), and dTDP-4-keto-L-Rha reductase (RmlD), are responsible for the formation of dTDP-Rha (Figure 1) [20]. The *rmlA, rmlB, rmlC,* and *rmlD* genes are usually located in biosynthetic gene clusters of polysaccharides in conserved gene orders with few exceptions [20]. Below, we discuss what is known about the steps involved in the biosynthesis of dTDP-Rha, as well as the functions, physicochemical properties, and crystal structures of RmlA, RmlB, RmlC, and RmlD [21].

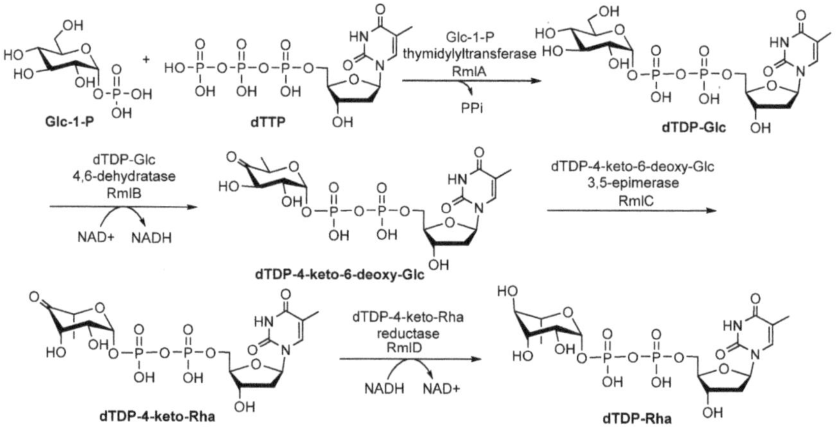

Figure 1. The biosynthetic pathway of dTDP-Rha from Glc-1-P in bacteria [22,23].

RmlA is a nucleotidyltransferase that catalyzes the first reaction to form dTDP-glucose (dTDP-Glc) by transferring a deoxythymidine triphosphate (dTTP) to glucose-1-phosphate (Glc-1-P) via a single sequential displacement mechanism [24]. Based on the reverse reaction, RmlA is also known to be a pyrophosphorylase [24]. RmlA has attracted more attention because it displays unusual promiscuity toward both sugar-1-phosphates and nucleotide triphosphate substrates, which could be harnessed in glycorandomization [25,26]. The inherent sugar-1-phosphate and/or nucleotide triphosphate (NTP) promiscuity of RmlA was further expanded by mutation studies. For example, L89T [27], E162D, Y177F [27], T201A, and W224H [28] mutants increased its sugar-1-phosphate tolerance and conversion [27], whereas Q24S [29] and Q83D/S [30] mutants altered the preference for the NTP of wildtype RmlA (also called the inherent NTP purine/pyrimidine bias). RmlA and its variants can utilize 57 sugar-1-phosphates ranging from all epimers [26], substituted compounds (amino [30], N-acetyl [31], methyl [32], azido [32], thiol [32], and alkyl [33]), deoxy sugars of D-glucose [32], two anomers of L-fucose [34], to pentofuranosyl-1-phosphate [34]. In addition to sugar-1-phosphate substrates, Moretti et al. reported that RmlA could recognize all eight natural NTPs as substrates despite its reduced activity toward purine NTPs [35]. Furthermore, Cps2L (a RmlA homolog) can act on deoxythymidine 5-tetraphosphate (p4dT) and Glc-1-P to form dTDP-Glc and triphosphate (PPPi) [32]. To date, 154 (d) nucleotide diphosphate (NDP)-sugars have been produced by RmlA and its variants (Figure 2 and Supplementary information Table S1) [36–42].

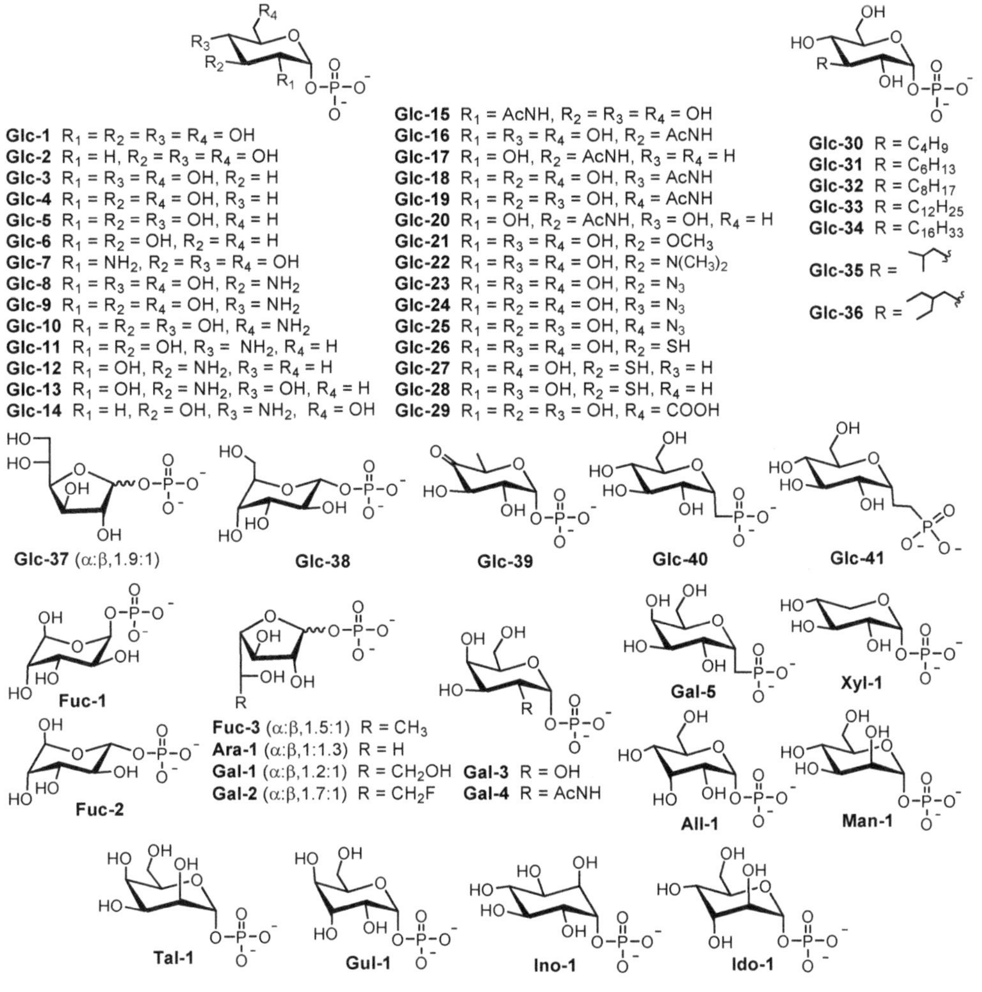

Figure 2. Sugar-1-phosphates recognized by RmlA [22].

The activity of RmlA is inhibited by dTDP-Glc, inorganic pyrophosphate (PPi) and dTDP-Rha [43], and dTDP-Rha is both a competitive and a noncompetitive inhibitor [43]. Although the mechanism of noncompetitive inhibition of RmlA by dTDP-Rha proposed by Mmot et al. remains unclear and needs to be further studied [43], the mechanism of competitive inhibition is well understood [44], which involved: (1) dTDP-Rha occupies the same site as dTDP-Glc; E161 of RmlA interacts with O2′ and O3′ of Rha through a bidentate hydrogen, similar to the dTDP-Glc complex (Figure 3D); (2) The two phosphates of dTDP-Rha move into the active site and form strong salt bridges with R194, which is absent in the dTDP-Glc complex; (3) A hydrogen bond between ribose O3 of dTDP-Rha and the side chain of D110 is likely replaced by the α-phosphate, resulting in the decomposition of dTDP-Rha. Thus, it was concluded that targeting these sites could provide a potential basis for inhibitor design. In addition, the R15 loop probably affects catalytic activity because it is different in the active site of the dTDP-Rha complex [44]. Crystal structures of RmlA from *Pseudomonas aeruginosa* [44], *Escherichia coli* [45], and *Salmonella typhimurium* [45] showed that RmlA is a homotetramer (Figure 3A).

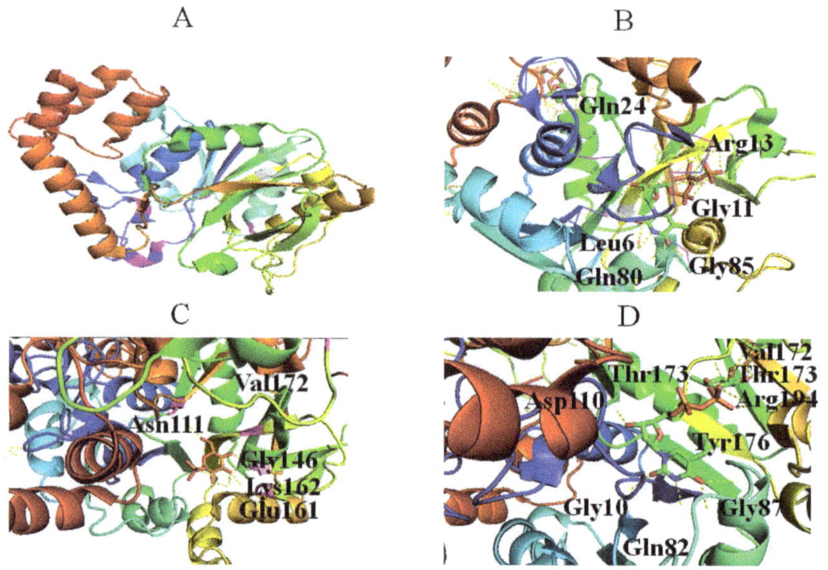

Figure 3. Stereo views of RmlA (Protein Data Bank (PDB) entry 4HO3) (**A**), dTTP (PDB entry 4HO3) (**B**), Glc-1-P (PDB entry 1G23) (**C**), and dTDP-Rha (PDB entry 1G3L) (**D**) bound to RmlA. Hydrogen bonds are shown as red lines. Helices, sheets, and loops of RmlA are colored blue, purple, and beige, respectively. C, N, O, and P elements of ligands are shown in green, blue, red, and brown, respectively. RmlA is shown in cartoon representation, and ligands are shown as sticks [22].

The active center of RmlA lies in a deep pocket formed by core and sugar-binding domains [45]. G11, Q80, and G85 form hydrogen bonds with the thymine: N3 and O4 of thymine engage in hydrogen bonds with Q80; O4 of thymine also forms hydrogen bonds with the N atom of G85; O2 of the thymine base engages in hydrogen bonds with G11 (Figure 3B) [45]. Neither methyl group of the pyrimidine ring nor the 2-OH of ribose interacts with RmlA, which explains why RmlA can accept UTP and dTTP as substrates.

The 3-hydroxyl group of ribose contacts Q24 (Figure 3B) [45], and glucose residue interacts with RmlA via hydrogen bonds. Specifically, O2 and O3 of glucose form hydrogen bonds with E161; O2, O3, and O4 of glucose form hydrogen bonds with G146 and L172; O6 of glucose forms hydrogen bonds with N111 (Figure 3C). In addition, Q26, G11, S13, and two water molecules bind magnesium. Therefore, the crystal structures of RmlA helped to reveal the reaction mechanism and provide a basis for active site engineering of RmlA [35].

The second step in the dTDP-Rha biosynthetic pathway is the dehydration of dTDP-Glc to form dTDP-4-keto-6-deoxy-D-glucose (dT4k6dG), which is catalyzed by RmlB. Four steps have been proposed during the reaction: (1) NAD^+ extracts a hydride from C4 of the glucose ring; (2) Glu135 removes a C5 proton; (3) elimination of a water molecule between C5 and C6 generates 4-keto-5,6-glucosene as an intermediate; and (4) a hydride is transferred from NADH to C6 of the glucose ring [46]. The substrate tolerance of RmlB is more limited compared with that of RmlA, probably because it catalyzes the committal step in the dTDP-Rha biosynthetic pathway [47]. The crystal structure of RmlB from *Salmonella enterica* serovar Typhimurium showed that it functions as a homodimer (Figure 4A). RmlB has two domains: a larger N-terminal domain consisting of seven β-strands and ten α-helices to bind the nucleotide cofactor NAD^+; and a smaller C-terminal domain composed of four β-strands and six α-helices to bind dTDP-Glc [46]. The two domains create a deep cavity in the enzyme to form the active site (Figure 4A) [46]. The key residues interacting with NAD^+ include (1) a hydrogen bond (Asp62) and a hydrophobic crevice consisting of Ile21, Ala57, Ile59, Val77, Ala81, and Leu107 binding to the adenine portion of NAD^+, and (2) Asp37, Tyr161, and Lys171 forming hydrogen bonds with the ribose sugar (Figure 4B) [46]. In addition, Thr133, Asp134, Glu135, Asn196, Arg231, and Asn266 make contacts with dTDP-Glc (Figure 4C) [46]. Specifically, Thr133, Glu135, and Asp134 bind to the 4, 6-hydroxyl groups of the glucose ring (Figure 4C), while Asn196 and Arg231 interact with the phosphoryl oxygen atom (Figure 4C), and Asn266 hydrogen binds to the 3-hydroxyl group of the ribose sugar (Figure 4D) [46]. Notably, Asn266 may also control the selectivity for the deoxy-nucleotide sugar substrate in the binding site [46].

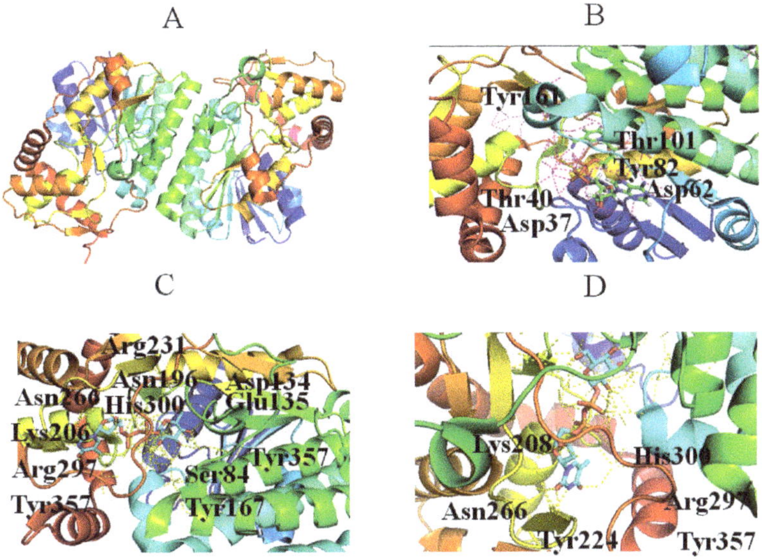

Figure 4. Stereo view of RmlB (PDB entry 1KEP) (**A**), NAD^+ (PDB entry 1KEP) (**B**), and dTDP-Glc (PDBentry 1KEU) (**C**) bound to RmlB, and the position of the Asn266 side chain as the sugar moves to stack against Tyr224 (**D**). Hydrogen bonds are shown as red lines. Helices, sheets, and loops of RmlB are colored blue, purple, and beige, respectively. C, N, O, and P elements of ligands are green, blue, red, and brown, respectively. RmlB is shown in cartoon representation, and ligands are shown as sticks [22].

RmlC catalyzes the third step in the dTDP-Rha biosynthetic pathway, in which the C3 and C5 positions of dT4k6dG are epimerized to generate dTDP-4-keto-Rha [48]. The catalytic mechanism of this catalytic reaction is proposed as follows: (1) a proton is abstracted from C5 of glucose of dT4k6dG accompanied by epimerization, then proton donation to C5, resulting in a mono-epimerized intermediate; (2) a proton from C3 of glucose is abstracted accompanied by epimerization, followed by proton donation to C3; (3) a ring flip occurs [49]. These reactions need strict stereo control and a cofactor is not required [50]. RmlC and/or RmlC co-complex structures have been obtained with dTDP-phenol, dTDP, dTDP-Glc and dTDP-D-xylose [49,51]. RmlC functions as a homodimer (Figure 5A). The monomer consists of 11 β-strands and seven α-helices that can be divided into three parts, including an N-terminal portion, a core active site, and a C-terminal portion. A His-Asp dyad (Figure 5B) in the active site is crucial in the RmlC catalytic mechanism because a conserved His65 residue from the His-Asp dyad extracts C5 and C3 protons (Figure 5B). Moreover, Tyr134 is essential for epimerization and for proton incorporation at C5. However, a water molecule may replace Tyr134 to facilitate C3 proton incorporation (Figure 5B) [49].

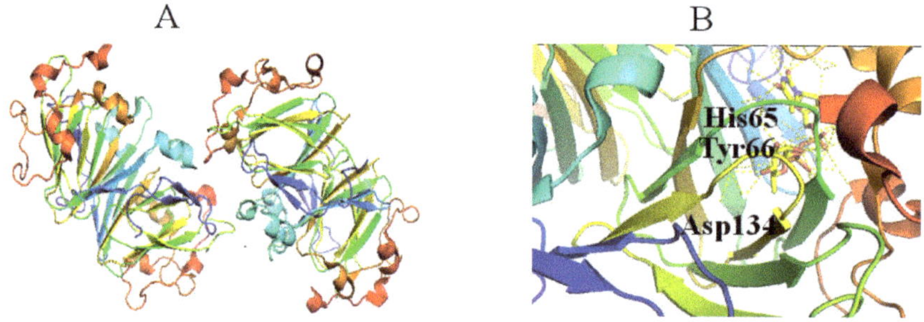

Figure 5. Stereo view of the RmlC (PDB entry 2IXL) (**A**) and His-Asp dyad of RmlC (PDB entry 2IXT) (**B**). Helix, sheet, and loop of RmlC are shown as blue, purple, and beige, respectively. C, N, O, and P elements of ligands are green, blue, red, and brown, respectively. RmlC is shown in cartoon representation, and ligands are shown as sticks [22].

RmlD catalyzes the last step in the dTDP-Rha biosynthetic pathway, in which the C4 keto group of dTDP-4-keto-Rha is reduced to a hydroxyl group to produce dTDP-Rha (Figure 1) [52,53]. During the reaction, proton transferred from the nicotinamide ring of the cofactor to the C4 keto group requires the assistance of Mg^{2+} [52]. RmlD is a homodimer, and the monomer consists of two domains: an N-terminal domain that binds NAD(H), and a C-terminal domain that binds substrate [52]. Various residues are involved in interactions with NAD(P)H, including (1) a ribose moiety located in the space formed by Ala62, Ala63, Gly7 and Gly10, in which the 2′- and 3′-hydroxyl groups of the ribose ring and Lys132 from the conserved YXXXK motif engage in two hydrogen bonds (Figure 6A); (2) the adenine ring of the cofactor located in a pocket formed by Val31, Asp39, Phe40, Ala62, Ala63, Leu80, and Phe40, in which Asp39 interacts with adenine via hydrogen bonds (Figure 6A); (3) Gln11 and Thr 65 interact with diphosphate (Figure 6A) [52]. Three glutamic acids (Glu15, Glu190, and Glu292) of two monomers bind to Mg^{2+} [52], and dTDP-Rha binds in a pocket of RmlD built from the hydrophobic parts of the side chains of Thr65, Tyr106, Tyr128, and Val67, together with the nicotinamide ring of the cofactor [52]. Additionally, Thr104, 105, Trp153, the carboxamide group of the cofactor, and a water molecule interact with L-Rha (Figure 6B) [52].

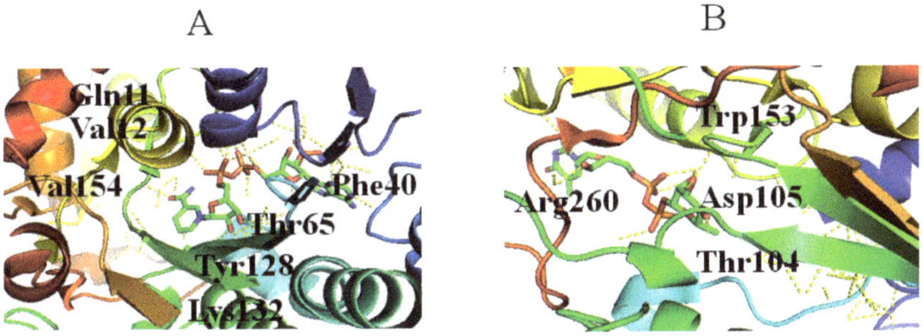

Figure 6. Stereo view of RmlD bound to NADH (PDB entry 1KC3) (**A**) and dTDP-Rha (PDB entry 1KC3) (**B**). Hydrogen bonds are shown as red lines. Helices, sheets, and loops of RmlD are colored blue, purple, and beige, respectively. C, N, O, and P elements of ligands are green, blue, red, and brown, respectively. RmlD is shown in cartoon representation, and ligands are shown as sticks [22].

2.2. Biosynthetic Pathway of GDP-Rha

D-Rha is a rare 6-deoxy monosaccharide found in the LPS of pathogenic bacteria [54]. GDP-Rha is the precursor for the biosynthesis of D-Rha-containing compounds, and it is synthesized in two steps: (1) GDP-mannose-4,6-dehydratase (GMD) catalyzes the conversion of GDP-D-mannose (GDP-Man) to GDP-4-keto-6-deoxy-D-Man; (2) GDP-6-deoxy-D-lyxo-hexos-4-ulose-4-reductase (RMD) catalyzes the production of GDP-Rha (Figure 7). Both GMD and RMD are members of the short-chain dehydrogenase/reductase (SDR) family. GMD is homologous to RmlB, while RMD is homologous to RmlD. However, GMD and RMD cannot catalyze the conversion of dT4k6dG to dTDP-Rha, indicating that enzymes involved in the GDP-D-Rha biosynthesis pathway possess strict substrate specificity. The functions of GMD and RMD from *Aneurinibacillus thermoaerophilus* strain L420-91 (T) [55] and *Pseudomonas aeruginosa* [56] have been confirmed in vitro.

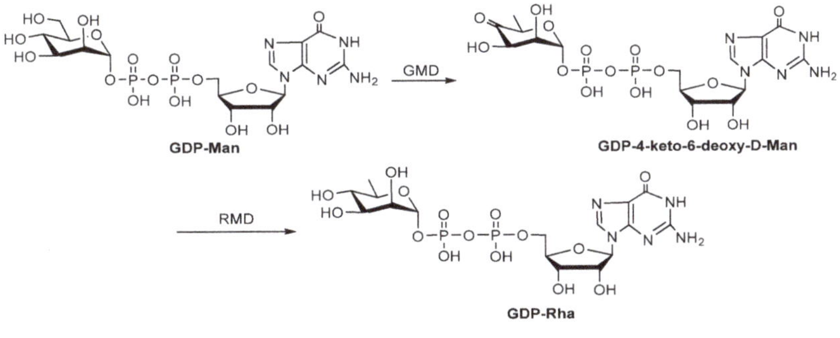

Figure 7. Biosynthesis pathway of GDP-Rha [22].

GMD is present in bacteria [57], plants [58], and animals [59], and its production serves as a branch point for several different deoxyhexoses, such as GDP-Rha, GDP-L-fucose, GDP-6-deoxy-D-talose, and the GDP-dideoxy amino sugars [56]. GMD functions as a homodimer [60] or a homotetramer [56] in cells. In particular, PBCV-1 GMD behaves as a bifunctional enzyme, displaying not only dehydratase activity but also a strong NAD(P)H-dependent reductase activity toward GDP-4-keto-6-deoxy-D-Man (the dehydration product), leading to the formation of GDP-Rha [61]. The crystal structures of GMD from *E. coli* [62], *Arabidopsis thaliana* [58], *P. aeruginosa* [56], and *Paramecium bursaria Chlorella virus 1* (PBCV-1) [63] have been reported. The GMD monomer folds into two domains: a N-terminal cofactor-binding domain and a C-terminal substrate-binding domain. Residues of GMDs that contact the GDP moiety are highly conserved, including Val190, Asn179,

Lys193, Arg218, Arg279, and Glu282. However, the hexose moiety has not been successfully crystallized. The crystal structure of RMD from *Aneurinibacillus thermoaerophilus* was reported in 2008, but the quality of the crystal structure was not good [64].

2.3. *Biosynthesis Pathway of UDP-Rha*

UDP-Rha is found in fungi and plants, and its biosynthesis pathway involves dehydration, epimerization, and reduction, similar to dTDP-Rha (Figure 8) [9]. Three isoenzymes (UDP-Rha synthases, RHMs) RHM1, RHM2/RHM4 and RHM3 convert UDP-Glc to UDP-Rha via UDP-4-keto-6-deoxy-glucose (U4k6dG) as an intermediate [65]. All the RHMs function as the activities of UDP-D-Glc 4,6-dehydratase, UDP-4-keto-6-deoxy-D-Glc 3,5-epimerase, and UDP-4-keto-L-Rha4-keto-reductase, respectively [66]. Other enzymes, such as a bifunctional enzyme named nucleotide-Rha synthase/epimerase-reductase (NRS/ER), can also act on the intermediate U4k6dG to form UDP-Rha [56,65]. Notably, it is not known what substrates are utilized by Rha-Ts as donors in plants because there are two rhamnose donors in plants (UDP-Rha and dTDP-Rha).

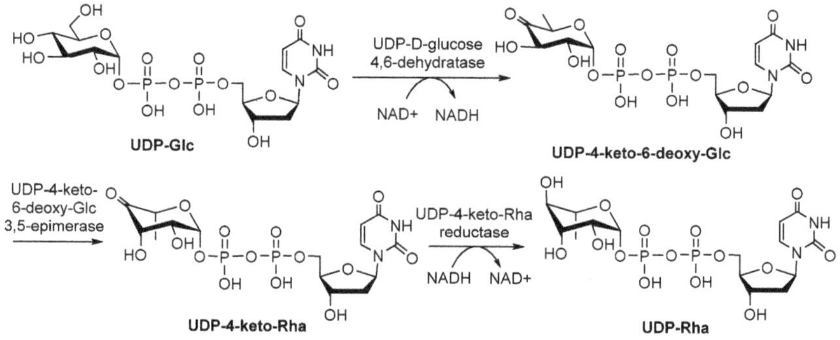

Figure 8. Biosynthesis pathway of UDP-Rha [22].

3. Rha-Ts Generating RCCs in Bacteria

Glycosyltransferases (GTs) are a large family of enzymes that catalyze the transfer of saccharide moieties from glycosyl donors to a broad range of acceptor substrates, including monosaccharides, oligosaccharides and polysaccharides, lipids, proteins, nucleic acids, and small organic molecules, to form complex carbohydrates and glycoconjugates that are essential to many fundamental biological processes [1]. There are three main methods for the classification of GTs: Firstly, based on the anomeric configuration of reactants and products, GTs are classified as inverting or retaining enzymes; Secondly, GT-A, GT-B, and GT-C topologies of GTs are divided based on Rossmann-fold domains and the locations of donors and acceptors; Thirdly, according to sequence similarity, GTs are divided into 114 different families, as listed in the carbohydrate-active enzymes (CAZy) database (http://www.cazy.org accessed on 14 March 2022). Rha-Ts are GTs that generate RCCs, which are universally present in bacteria [67,68]. However, biochemical knowledge on Rha-Ts is still limited.

3.1. *Rha-Ts from Geobacillus Stearothermophilus*

The S-layer protein of *Geobacillus stearothermophilus* NRS 2004/3a serves as a model for investigating O-glycosylation pathways in bacteria, and the glycans of this protein are 2-OMe-α-L-Rha-(1→3)-β-L-Rha-(1→2)-α-L-Rha-(1→[2)-α-L-Rha-(1→3)-β-L-Rha-(1→2)-α-L-Rha-(1→]$_{n=13-18}$[2)-α-L-Rha-(1→]$_{n=1-2}$3)-α-L-Rha-(1→3)-β-D-Gal-(1→Protein (Figure 9) [69,70].

The polycistronic S-layer glycosylation (*slg*) gene cluster encodes four GTs, of which three Rha-Ts (WsaC, WsaD and WsaF) catalyze the biosynthesis of the glycan [70]. The biosynthesis pathway for this glycan is initiated by the transfer of a galactose residue to a membrane-associated lipid carrier, followed by two steps catalyzed by α-1,3-Rha-Ts

(WsaC and WsaD) that add Rha to build up the [2)-α-L-Rha-(1→]$_{n=1-2}$3)-α-L-Rha-(1→3) linker, and two Rha-Ts (WsaE and WsaF) then extend the glycan chain by adding the repeating trisaccharide motif [→2)-α-L-Rha-(1→3)-β-L-Rha-(1→2)-α-L-Rha-(1→]$_{n=13-18}$ through addition of Rha. The complete glycan chain is thereafter transported across the membrane by an ATP-binding cassette transporter (ABC transporter) and transferred to the S-layer protein by oligosaccharyltransferase WsaB [70]. The functions of WsaC−WsaF were proved by using the chemically synthesized β-D-Gal-(1→O)-octyl as substrate. Among them, both WsaC and WsaD are transmembrane proteins, and the activity of WsaC requires membranes, while WsaD can only recognize the natural substrate; WsaE is a multifunctional enzyme, and the N-terminal domain of WsaE possesses methylase activity, whereas the central and C-terminal domains of WsaE possess Rha-Ts activity, generating α-1,2 and α-1,3 linkages; WsaF is a β-1,2-Rha-T enzyme (Figure 10) [70].

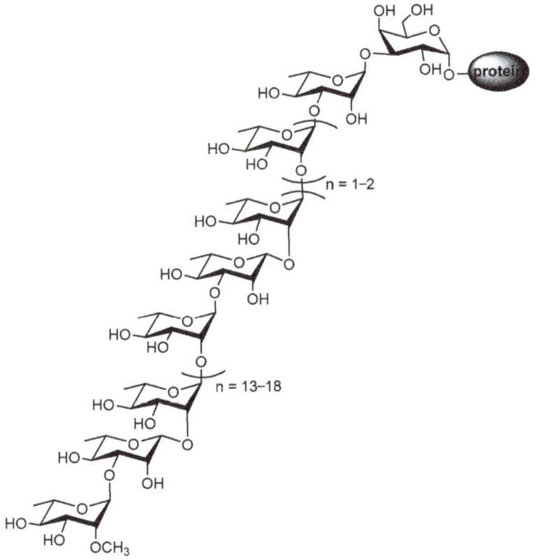

Figure 9. The glycan structure of the *Geobacillus stearothermophilus* S-layer glycoprotein.

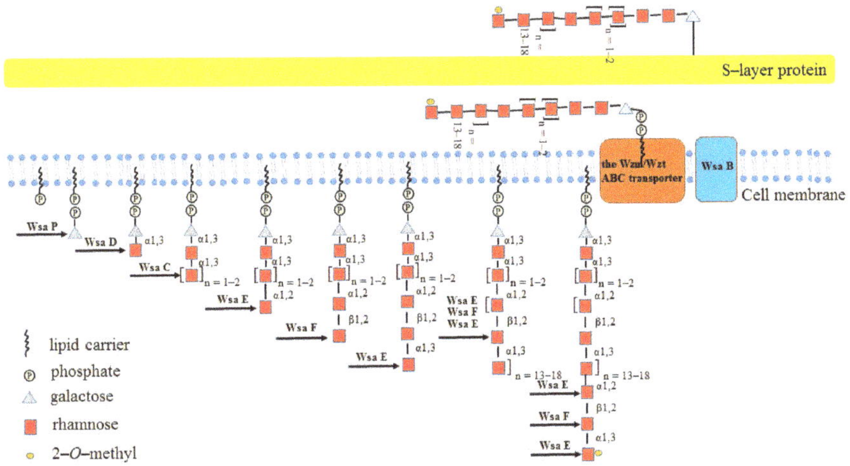

Figure 10. Diagram of the biosynthesis pathway of the S-layer glycoprotein glycan of *Geobacillus stearothermoph ilus* NRS 2004/3a.

WsaF is a dimer formed by two monomers that consist of two GT-B-fold domains and a cleft between the two domains [69]. dTDP-Rha interacts with WsaF, and the dTDP-WsaF and dTDP-Rha-WsaF complex structures revealed that thymidine contacts with K302 and L303, while V282 and G283 interact with thymidine via van der Waals forces, pyrophosphate binds to G63, R249 and K302 through hydrogen bonds, Rha contacts N227, K225 and E333 directly, and Y329 engages in a stacking interaction with the hydrophobic face of Rha [69]. The crystal structure of the WsaF–acceptor complex has not been reported. However, the acceptor fragments of both α-L-Rha-(1-2)-α-L-Rha-(1-3)-α-L-Rha and α-L-Rha-(1-2)-α-L-Rha-(1-3)-α-L-Rha-(1-3)-α-D-Gal were modeled manually in the tunnel using PyMOL, suggesting that G63, I65, P54, S55, A140, Q170, D171, E173 and F176 would form van der Waals interactions with the acceptor fragments. This was confirmed by mutant studies [69].

3.2. *Spinosyn SpnG Rha-T from Saccharopolyspora Spinosa*

Spinosyn from *S. spinosa* is a type of macrocyclic lactone that has been used as an agricultural antibiotic [71,72]. The entire spinosyn biosynthetic gene cluster contains 19 genes, including 5 large genes (*spnA, spnB, spnC, spnD,* and *spnE*) encoding a type I polyketide synthase, 4 genes (*spnF, spnJ, spnL,* and *spnM*) encoding proteins involved in intramolecular C-C bond formation, 4 genes (*spnG, spnI, spnK,* and *spnH*) encoding proteins involved in Rha attachment and methylation, and 6 genes (*spnP, spnO, spnN, spnQ, spnR,* and *spnS*) encoding proteins involved in forosamine biosynthesis [72]. SpnG is known to be capable of transferring Rha from dTDP-Rha donors to spinosyn aglycone (AGL) and to display relaxed substrate specificity [73].

SpnG forms a C2-symmetric homodimer, and each monomer contains two domains connected by a long loop (residues 183–209) [71]. The C-terminal domain binds the donor substrate (dTDP-Rha) and the N-terminal domain binds the acceptor substrate [71,73]. Specifically, interactions between dTDP-Rha and SpnG include thymine contacting L279, V277, P257, and L279; α-phosphate forming hydrogen bonds with G296, T297, and T297; β-phosphate making hydrogen bonds with M227 and V228; the 3-OH group of deoxyribose directly forming a hydrogen bond with N202; and Rha contacting with D316, Q317, Y314, and W142 [71]. The crystal structure of a SpnG–acceptor complex has not been reported.

3.3. *WbbL from Mycobacterium Tuberculosis*

The cell wall of *M. tuberculosis*, essential for cell proliferation and growth, is composed of peptidoglycan, arabinogalactan, and mycolic acids [7]. The galactan of arabinogalactan combines with peptidoglycan via a disaccharide linker, α-L-Rha-(1→3)-α-D-GlcNAc-(1→P), to form the integrated mycobacterial cell wall [74]. The Rha-T enzyme WbbL forms the disaccharide linker by transferring Rha from dTDP-Rha to decaprenyldiphosphoryl-α-D-N-acetyl glucosamine (GlcNAc-PP-DP) [75]. The *wbbL* gene is essential for mycobacterial viability and is found in the genomes of all mycobacteria [76]; hence, it is an attractive target for antituberculosis therapeutics. Activity analysis of WbbL was performed using endogenous GlcNAc-PP-DP as a substrate, and a microtiter plate method was established [74]. The bioinformatics analysis of WbbL showed that it belongs to the GT2 family with a fold characteristic of the GT-A superfamily [74], members of which can utilize dTDP-β-Rha as a substrate and produce an α-Rha product. In addition, this protein has a N-terminal GT domain, no signal peptide or transmembrane helices, and it is located outside the membrane.

3.4. *Rha-Ts from Pseudomonas Aeruginosa*

Pseudomonas aeruginosa is a pathogen of plants and animals, and an opportunistic human pathogen that causes serious nosocomial infections [77]. LPSs are major virulence factors composed of three distinct regions, i.e., lipid A, core oligosaccharide (OS), and O polysaccharide (O antigen), which contain diverse repeating saccharide units. In this section, we focus on Rha found in OS and O antigens.

OS is divided into two types: one is capped (linked to O polysaccharides) with O antigen through an α-1,3-linked L-Rha, while the other is uncapped (devoid of O polysaccharides) and contains an α-1,6-linked L-Rha [78]. Gene knockout analysis showed that *migA* and *wapR* genes encode α-1,3 Rha-Ts and α-1,6 Rha-Ts, respectively, which are responsible for the biosynthesis of the α-1,3-linked L-Rha and α-1,6-linked L-Rha [78]. O antigen can also be divided into two types: heteropolymeric O antigen (formerly called B band), containing mannosuronic acid derivatives with *N*-acetyl-D-fucosamine (D-FucNAc), and an alternative LPS containing the common polysaccharide antigen (CPA; formerly called A band). There are repeating units of O-polysaccharides of A band, namely (→3-α-D-Rha-(1→2)-α-D-Rha (1→3)-α-D-Rha α-1→), containing a D-Rha moiety [79]. D-Rha-Ts (including WbpY, WbpX and WbpZ) catalyze the transfer of D-Rha to an acceptor [80,81].

Rhamnolipids are detergents composed of α-D-(α-D-hydroxyalkanoyloxy) alkanoic acids (HAA) derivatized with one or two Rha sugars (monorhamnolipids and dirhamnolipids; Figure 11), which are secreted by *P. aeruginosa* [82]. Rha-Ts I [83] and Rha-Ts II [84] generate rhamnolipids, and their mechanism has been determined: (1) Rha-Ts I are encoded by *rhlA* and *rhlB* genes, and gene knockout analysis of these genes indicated that RhlA forms HAA, while RhlB is a Rha-T enzyme [85], and the heterologous expression of RhlA and RhlB was achieved [86]; (2) gene knock-in assay proved that RhlC encodes Rha-Ts II, which transfers the second Rha to dirhamnolipids [84].

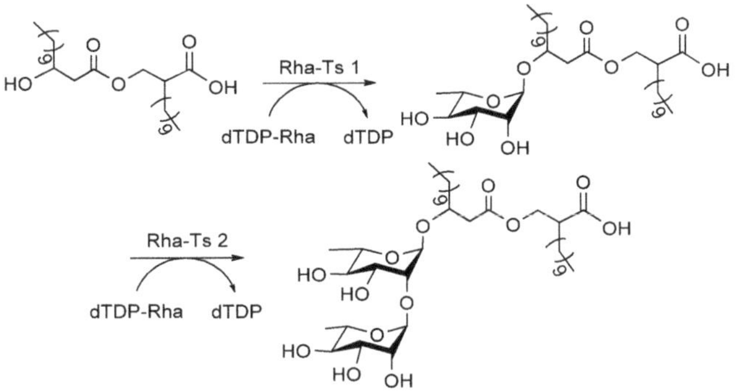

Figure 11. Rha-Ts involved in the biosynthesis of rhamnolipids [22].

Additionally, a Rha-T EarP derived from *P. aeruginosa* has been reported that transfers Rha from dTDP-Rha to Arg32 of the translation elongation factor P (EF-P) [87,88]. This rhamnosylation of Arg32 by EarP can activate the functions of EF-P, which is important in the process of protein translation in ribosome. EarP is also discovered in other clinically relevant bacteria [89,90], indicating that this type of post-translational modification strategy is crucial for protein translation and bacteria pathogenicity [87,88].

3.5. *Rha-Ts from Streptococcus Pneumoniae*

Capsular polysaccharides (CPSs) are produced by almost all isolates of *S. pneumoniae* recovered from cases of invasive disease, and they are major virulence factors and immunogens [91]. Rha-containing CPS has been identified in at least 27 serotypes. Rha-containing CPS of *S. pneumoniae* is particularly attractive: (1) L-Rha may increase the immunogenicity of CPS based on the immune analysis of 23F CPS, showing that α-(1→2)-linked L-Rha is a dominant antigen [92]; (2) modified L-Rha may increase the stability of CPS based on the analysis of a 19F CPS analog in which a residue of carba-L-Rha was inserted into the natural trisaccharide, and this increased the stability of CPS [93]; (3) Rha-Ts are the most prevalent GT genes in *S. pneumoniae* cps loci [94]. Therefore, studies on Rha of *S. pneumonia* CPS may provide a new strategy for developing novel drugs to treat anti-pneumococcal infections. However, new serotypes should be identified, and attempts to determine the structures of CPSs and Rha-Ts have been reported [95,96].

3.6. *Rha-Tss from Other Bacteria*

Although numerous Rha-Ts have been predicted, the in vitro biochemical knowledge of these enzymes is limited. Gene mutants have confirmed the functions of some Rha-Ts, including RgpF [97], WbgA [97], AceR [98], AntB [99], and GacB [100]. Heterologous expression has also been used to confirm the functions of Rha-Ts, as exemplified by HlpA/RtfA [101]. Rha-Ts from *Mycobacterium smegmatis* [93], *Streptococcus anginosus* [102], serotype VIII capsular polysaccharide (CPS) of Group B Streptococci (GBS) [103], and *Vibrio cholera* [104] have also been reported.

This review mainly overviews the research advance of the Rha-Ts derived from bacterial; however, there are also other enzymes involved in RCC biosynthesis that will not be described in detail here. For example, several Rha-Ts from plants have been reported [105–107]. Additionally, in recent years, α-L-rhamnosidase has been found to synthesize RCCs by a reverse hydrolyzing mechanism, which has attracted extensive attention [108,109].

4. Application of RCC in the Research and Development of New Drugs

4.1. *Rha Increases the Immunogenicity of Tumour-Associated Carbohydrate Antigen (TACA) Vaccines*

TACAs are carbohydrates expressed at high levels on the surface of tumor cells [110,111], and anti-TACA vaccines have been well developed [112]. However, the immunogenicity of TACAs is very low [113]. Saccharide conjugating to proteins can increase its immunogenicity, and this approach was then widely applied in conjugation vaccinations [114]. Although some glycoconjugate TACA cancer vaccines have shown promising therapeutic potential, no vaccine has yet achieved a satisfactory survival rate in clinical trials [115,116]. Guo group developed both positive and negative immunotherapies with unnatural TACAs for testing against cancers [117,118]. However, the quality control of reactions was difficult, and unexpected immune responses to proteins and linkages limited their application.

To solve these problems, two strategies have been developed: using a low-molecular-weight peptide (such as YAF) in place of proteins to increase immunogenicity of TACAs, and antigens targeting antigen-presenting cells (APCs) [117,118]. Additionally, saccharide binding to Rha can improve immunogenicity, as demonstrated by Oyelaran et al. who reported that human serum contains high levels of anti-Rha antibody [119]. Zhang et al. reported that L-Rha conjugated with truncated MAGE-A3 enhanced the immunogenicity of melanoma-associated antigen A3, thereby stimulating antitumor immune responses [120]. A study by Sarkar et al. showed that L-Rha binding to carbohydrate antigens enhanced antigenicity in mice [121]. In 2013, this team also successfully formulated a MUC1 VNTR TACA conjugate into a liposome-based anticancer vaccine, and the immunogenicity of the vaccine was further augmented by incorporating surface-displayed L-Rha epitopes onto liposomes [122]. Li et al. reported a strategy targeting tumor cells using ligand-incorporated Rha-functionalized liposomes [123]. Additionally, L-Rha epitopes can also enhance cellular immunogenicity. Partha et al. reported that the Rha-decorated liposomal Pam_3Cys-MUC_1-Tn vaccine showed higher cellular immunogenicity [2]. In addition, the immunogenicity of Rha-decorated liposomal Pam_3Cys-MUC_1-Tn was further augmented in mice when received human anti-Rha antibodies prior to its vaccination [124]. Additionally, Rha and sTn antigen, co-conjugated to bovine serum albumin (BSA), significantly enhanced antigen uptake through the involvement of Rha-specific antibodies [125]. Together, these studies showed that TACA vaccines containing Rha can increase immunogenicity. Compared with Galα1-3Galβ1-4GlcNAc-R (α-Gal epitope), Rha not only increases the immunogenicity of TACAs, but also can be used directly in wild-type mice [126].

In addition to TACA vaccines, the strategies of enhancing the monoclonal antibodies' (mAbs) efficacy were also developed by using high levels of anti-Rha antibody of the human serum [127,128]. MAbs are one of the most rapidly growing drug classes used for the clinical practice, such as cancer and infectious and autoimmune diseases. Complement-dependent cytotoxicity (CDC) and antibody-dependent cell-mediated cytotoxicity (ADCC) are effector functions for antibodies to deplete target cells [128]. Rituximab is one of the commercially available mAbs, which is site-specifically conjugated with the Rha hapten to generate rituximab–Rha conjugates, to recruit anti-Rha antibodies onto the cancer cell surface and further form an immune complex that leads to magnifying ADCC and CDC simultaneously [128]. Ou et al. reported an efficient chemoenzymatic synthesis of structurally well-defined conjugates of antibody–rhamnose clusters to recruit natural anti-rhamnose antibodies for the enhancement of the CDC effects [127]. In addition, Coen et al. reported on antibody-recruiting glycopolymers (ARGPs) that consist of polymeric copies of a rhamnose motif, which can bind anti-Rha antibody of the human serum, for the design of potent immunotherapeutics that mark target cells for destruction by the immune system through ADCC [129]. These studies developed general and cost-effective approaches to augment the mAb effector functions with the engagement of anti-Rha antibody of the human serum that may have broad applications.

4.2. Rha-Containing Tumor-Killing Agents

Many natural products are known to have human health benefits, such as saponins and tumor-killing agents. The relationships between biological activity and chemical structure of some tumor-killing agents indicate that Rha may play a crucial role in determining biological properties. For example, kaempferol-3-O-(3′′,4′′-di-O-acetyl-α-L-rhamnopyranoside; SL0101) from *Forsteronia refracta* can inhibit the activity of Ser/Thr protein kinases (RSKs) that are closely related to the proliferation and metastasis of many tumor cells [130]. During this process, acylation of the Rha moiety of SL0101 is required for high-affinity binding and selectivity [118]. In addition, the Rha moiety of solamargine and solasonine is a key factor in anticancer activity [131,132]. Lou group demonstrated why Rha plays an important role in the anticancer activity of solasodine-derived rhamnosides; they reported that Rha-binding lectins (RBLs) on the surface of tumor cells conjugated with Rha to mediate the transportation of rhamnosides [133].

Furthermore, due to specificity of the interactions between carbohydrates and cell receptors, a lectin-directed enzyme-activated prodrug therapy (LEAPT) strategy was developed [134]. Specifically, in the first phase of this strategy, a glycosylated enzyme is targeted to specific cell types or tissues; in the second phase, prodrugs capped with sugars are administered; the glycosylated enzyme is then able to activate the prodrugs at the site of interest by cleaving the prodrug linkage; the interaction of both prodrug and enzyme relies on their precise glycosylation, and Rha-doxorubicin and Rha-5-fluorouracil are effective examples [134]. Although the Rha of tumor-killing agents is a key factor in tumor killing, L-Rha cannot kill tumor cells directly because it does not affect energy metabolism [135].

4.3. Inhibitors of Rha Synthetases as Drug Targets

Many prevalent and opportunistic pathogens, including *M. tuberculosis*, *P. aeruginosa*, and *S. pneumoniae*, are particularly difficult to treat due to their intrinsic chemo-resistance and their ability to acquire further resistance mechanisms against antimicrobial agents. Rha biosynthesis pathways have been discovered in numerous bacteria and fungi, but they have not been discovered in humans, hence they might be potential therapeutic targets [136,137]. The first nanomolar inhibitors of RmlA from *P. aeruginosa* were thymine analogs, and some inhibitors also showed inhibitory activity against *M. tuberculosis* [138]. In addition, L-Rha-1-C-phosphonate is the best inhibitor of Cps2L, and a fluorine atom at C1 can increase inhibition by 25%, but two fluorine atoms at C1 had an adverse effect [139]. Furthermore, RmlC is the most promising therapeutic target because it possesses high substrate specificity and it does not require a cofactor [140].

5. Conclusions

RCCs are present in bacteria but not in humans and other mammals, making them valuable for tumor immunotherapy and treating antibacterial infections. To date, RCCs have been studied extensively, and produced a series of excellent results, i.e., the discovery of the biosynthetic pathways of three rhamnose donors, the discovery of Rha-Ts, and their application to the treatment of various diseases. In this review, the biosynthesis pathways and the properties of the related enzymes from three donor substrates, including dTDP-Rha, GDP-Rha, and UDP-Rha, were reviewed in detail, which is of great significance for the development of the strategies for the preparation of donor substrates of Rha-Ts in vitro. In addition, the functions and properties of Rha-Ts were also reviewed, which provides theoretical guidance for the development of Rha-Ts and the enzymatic synthesis of RCCs. It is important to note the complex and diverse structures of the receptor substrates of Rha-Ts, which need to be further studied. However, the research of the synthesis pathways of RCCs from different cells, the properties of related enzymes and their catalytic mechanisms is rather little; therefore, further studies on the biosynthesis and applications of RCCs are being carried out at present and subsequently via the latest biochemical technologies, such as molecular biology, structural biology, and computational biochemistry techniques.

Supplementary Materials: The following supporting information can be downloaded at: https://www.mdpi.com/article/10.3390/molecules27165315/s1, Table S1: Sugar-1-phosphate substrates of RmlA and their conversion into corresponding NDP-sugars.

Author Contributions: S.L. and L.W. made the bibliography, analyzed the literature, wrote the first drafts, and revised the final version of the manuscript. F.C., Y.L. and H.L. corrected the manuscript and wrote the final draft. G.G. and E.L. suggested the topic of the article, revised the manuscript, and adjusted it to be in a suitable form for publication. All authors have read and agreed to the published version of the manuscript.

Funding: This research was funded by the Natural Science Foundation of Henan Province (grant number 212300410203, 202300410279), the Key Science and Technology Innovation Demonstration Projects of Henan Province, (grant number 191110110600),the Scientific and Technological Research Project Foundation of Henan Provincial Scientific and Technological Department (grant number 202102310479, 222102110180), the Young Backbone Teachers Fundation of Huanghuai University, the Scientific Research Foundation for Advanced Talents of Huanghuai University, grant number 1000.12.01.1678.

Conflicts of Interest: The authors declare no conflict of interest.

References

1. McArthur, J.B.; Chen, X. Glycosyltransferase engineering for carbohydrate synthesis. *Biochem. Soc. Trans.* **2016**, *44*, 129–142. [CrossRef] [PubMed]
2. Karmakar, P.; Lee, K.; Sarkar, S.; Wall, K.A.; Sucheck, S.J. Synthesis of a Liposomal MUC1 Glycopeptide-Based Immunotherapeutic and Evaluation of the Effect of L-Rhamnose Targeting on Cellular Immune Responses. *Bioconjug. Chem.* **2016**, *27*, 110–120. [CrossRef] [PubMed]
3. Chen, W.X.; Cheng, L.; Pan, M.; Qian, Q.; Zhu, Y.L.; Xu, L.Y.; Ding, Q. D Rhamnose beta-Hederin against human breast cancer by reducing tumor-derived exosomes. *Oncol. Lett.* **2018**, *16*, 5172–5178. [PubMed]
4. Vanacore, A.; Vitiello, G.; Wanke, A.; Cavasso, D.; Clifton, L.A.; Mahdi, L.; Campanero-Rhodes, M.A.; Solis, D.; Wuhrer, M.; Nicolardi, S.; et al. Lipopolysaccharide O-antigen molecular and supramolecular modifications of plant root microbiota are pivotal for host recognition. *Carbohydr. Polym.* **2022**, *277*, 118839. [CrossRef] [PubMed]
5. Concordio-Reis, P.; Alves, V.D.; Moppert, X.; Guezennec, J.; Freitas, F.; Reis, M.A.M. Characterization and Biotechnological Potential of Extracellular Polysaccharides Synthesized by Alteromonas Strains Isolated from French Polynesia Marine Environments. *Mar. Drugs* **2021**, *19*, 522. [CrossRef]
6. Garcia-Vello, P.; Sharma, G.; Speciale, I.; Molinaro, A.; Im, S.H.; De Castro, C. Structural features and immunological perception of the cell surface glycans of *Lactobacillus plantarum*: A novel rhamnose-rich polysaccharide and teichoic acids. *Carbohydr. Polym.* **2020**, *233*, 115857. [CrossRef] [PubMed]
7. Wu, Q.; Zhou, P.; Qian, S.; Qin, X.; Fan, Z.; Fu, Q.; Zhan, Z.; Pei, H. Cloning, expression, identification and bioinformatics analysis of Rv3265c gene from Mycobacterium tuberculosis in *Escherichia coli*. *Asian Pac. J. Trop. Med.* **2011**, *4*, 266–270. [CrossRef]
8. De Castro, C.; Molinaro, A.; Piacente, F.; Gurnon, J.R.; Sturiale, L.; Palmigiano, A.; Lanzetta, R.; Parrilli, M.; Garozzo, D.; Tonetti, M.G.; et al. Structure of N-linked oligosaccharides attached to chlorovirus PBCV-1 major capsid protein reveals unusual class of complex N-glycans. *Proc. Natl. Acad. Sci. USA* **2013**, *110*, 13956–13960. [CrossRef]
9. Martinez, V.; Ingwers, M.; Smith, J.; Glushka, J.; Yang, T.; Bar-Peled, M. Biosynthesis of UDP-4-keto-6-deoxyglucose and UDP-rhamnose in pathogenic fungi *Magnaporthe grisea* and *Botryotinia fuckeliana*. *J. Biol. Chem.* **2012**, *287*, 879–892. [CrossRef]
10. Yu, Y.; Wen, Q.; Song, A.; Liu, Y.; Wang, F.; Jiang, B. Isolation and immune activity of a new acidic *Cordyceps militaris* exopolysaccharide. *Int. J. Biol. Macromol.* **2022**, *194*, 706–714. [CrossRef]

11. Allen, S.; Richardson, J.M.; Mehlert, A.; Ferguson, M.A. Structure of a complex phosphoglycan epitope from gp72 of *Trypanosoma cruzi*. *J. Biol. Chem.* **2013**, *288*, 11093–11105. [CrossRef] [PubMed]
12. Hossain, M.K.; Vartak, A.; Sucheck, S.J.; Wall, K.A. Synthesis and Immunological Evaluation of a Single Molecular Construct MUC1 Vaccine Containing L-Rhamnose Repeating Units. *Molecules* **2020**, *25*, 3137. [CrossRef] [PubMed]
13. Wagstaff, B.A.; Zorzoli, A.; Dorfmueller, H.C. NDP-rhamnose biosynthesis and rhamnosyltransferases: Building diverse glycoconjugates in nature. *Biochem. J.* **2021**, *478*, 685–701. [CrossRef]
14. Oh, J.; Lee, S.G.; Kim, B.G.; Sohng, J.K.; Liou, K.; Lee, H.C. One-pot enzymatic production of dTDP-4-keto-6-deoxy-D-glucose from dTMP and glucose-1-phosphate. *Biotechnol. Bioeng.* **2003**, *84*, 452–458. [CrossRef]
15. Glaser, L. The synthesis of thymidine-linked sugars. III. On the mechanism of thymidine diphosphate-L-rhamnose formation. *Biochim. Biophys. Acta* **1961**, *51*, 169–171. [CrossRef]
16. Thibodeaux, C.J.; Melancon, C.R.; Liu, H.W. Natural-product sugar biosynthesis and enzymatic glycodiversification. *Angew. Chem. Int. Ed. Engl.* **2008**, *47*, 9814–9859. [CrossRef]
17. Pal, D.; Mukhopadhyay, B. Chemical Synthesis of beta-L-Rhamnose Containing the Pentasaccharide Repeating Unit of the O-Specific Polysaccharide from a *Halophilic Bacterium Halomonas ventosae* RU5S2EL in the Form of Its 2-Aminoethyl Glycoside. *J. Org. Chem.* **2021**, *86*, 8683–8694. [CrossRef] [PubMed]
18. Cloutier, M.; Prevost, M.J.; Lavoie, S.; Feroldi, T.; Piochon, M.; Groleau, M.C.; Legault, J.; Villaume, S.; Crouzet, J.; Dorey, S.; et al. Total synthesis, isolation, surfactant properties, and biological evaluation of ananatosides and related macrodilactone-containing rhamnolipids. *Chem. Sci.* **2021**, *12*, 7533–7546. [CrossRef]
19. Jordan, D.S.; Daubenspeck, J.M.; Dybvig, K. Rhamnose biosynthesis in mycoplasmas requires precursor glycans larger than monosaccharide. *Mol. Microbiol.* **2013**, *89*, 918–928. [CrossRef]
20. Madduri, K.; Waldron, C.; Merlo, D.J. Rhamnose biosynthesis pathway supplies precursors for primary and secondary metabolism in *Saccharopolyspora spinosa*. *J. Bacteriol.* **2001**, *183*, 5632–5638. [CrossRef]
21. Yang, S.; An, X.; Gu, G.; Yan, Z.; Jiang, X.; Xu, L.; Xiao, M. Novel dTDP-L-Rhamnose Synthetic Enzymes (RmlABCD) From *Saccharothrix syringae* CGMCC 4.1716 for One-Pot Four-Enzyme Synthesis of dTDP-L-Rhamnose. *Front. Microbiol.* **2021**, *12*, 772839. [CrossRef] [PubMed]
22. Li, S. Studies and Application of Rhamnosyltransferase and Related Enzymes from *Streptococcus pneumonia* Serotype 23F. Ph.D. Thesis, Shandong University, Shandong, China, 2017.
23. Li, S.; Wang, H.; Ma, J.; Gu, G.; Chen, Z.; Guo, Z. One-pot four-enzyme synthesis of thymidinediphosphate-L-rhamnose. *Chem. Commun.* **2016**, *52*, 13995–13998. [CrossRef] [PubMed]
24. Pazur, J.H.; Anderson, J.S. Thymidine triphosphate: Alpha-D-galactose-L-phosphate thymj-L-phosphate thymidylyltransferase from *streptococcus faecalis* grown on d-galactose. *J. Biol. Chem.* **1963**, *238*, 3155–3160. [CrossRef]
25. Barton, W.A.; Lesniak, J.; Biggins, J.B.; Jeffrey, P.D.; Jiang, J.; Rajashankar, K.R.; Thorson, J.S.; Nikolov, D.B. Structure, mechanism and engineering of a nucleotidylyltransferase as a first step toward glycorandomization. *Nat. Struct. Biol.* **2001**, *8*, 545–551. [CrossRef]
26. Jiang, J.; Biggins, J.B.; Thorson, J.S. A General Enzymatic Method for the Synthesis of Natural and "Unnatural" UDP- and TDP-Nucleotide Sugars. *J. Am. Chem. Soc.* **2000**, *122*, 6803–6804. [CrossRef]
27. Barton, W.A.; Biggins, J.B.; Jiang, J.; Thorson, J.S.; Nikolov, D.B. Expanding pyrimidine diphosphosugar libraries via structure-based nucleotidylyltransferase engineering. *Proc. Natl. Acad. Sci. USA* **2002**, *99*, 13397–13402. [CrossRef]
28. Moretti, R.; Chang, A.; Peltier-Pain, P.; Bingman, C.A.; Phillips, G.J.; Thorson, J.S. Expanding the nucleotide and sugar 1-phosphate promiscuity of nucleotidyltransferase RmlA via directed evolution. *J. Biol. Chem.* **2011**, *286*, 13235–13243. [CrossRef]
29. Jakeman, D.L.; Young, J.L.; Huestis, M.P.; Peltier, P.; Daniellou, R.; Nugier-Chauvin, C.; Ferrieres, V. Engineering ribonucleoside triphosphate specificity in a thymidylyltransferase. *Biochemistry* **2008**, *47*, 8719–8725. [CrossRef]
30. Jiqing Jiang, J.B.B.A. Expanding the Pyrimidine Diphosphosugar Repertoire: The Chemoenzymatic Synthesis of Amino- and Acetamidoglucopyranosyl Derivatives. *Angew. Chem. Int. Ed. Engl.* **2001**, *8*, 1502–1505. [CrossRef]
31. Timmons, S.C.; Mosher, R.H.; Knowles, S.A.; Jakeman, D.L. Exploiting nucleotidylyltransferases to prepare sugar nucleotides. *Org. Lett.* **2007**, *9*, 857–860. [CrossRef]
32. Forget, S.M.; Smithen, D.A.; Jee, A.; Jakeman, D.L. Mechanistic evaluation of a nucleoside tetraphosphate with a thymidylyltransferase. *Biochemistry* **2015**, *54*, 1703–1707. [CrossRef] [PubMed]
33. Huestis, M.P.; Aish, G.A.; Hui, J.P.; Soo, E.C.; Jakeman, D.L. Lipophilic sugar nucleotide synthesis by structure-based design of nucleotidylyltransferase substrates. *Org. Biomol. Chem.* **2008**, *6*, 477–484. [CrossRef] [PubMed]
34. Timmons, S.C.; Hui, J.P.; Pearson, J.L.; Peltier, P.; Daniellou, R.; Nugier-Chauvin, C.; Soo, E.C.; Syvitski, R.T.; Ferrieres, V.; Jakeman, D.L. Enzyme-catalyzed synthesis of furanosyl nucleotides. *Org. Lett.* **2008**, *10*, 161–163. [CrossRef] [PubMed]
35. Moretti, R.; Thorson, J.S. Enhancing the latent nucleotide triphosphate flexibility of the glucose-1-phosphate thymidylyltransferase RmlA. *J. Biol. Chem.* **2007**, *282*, 16942–16947. [CrossRef]
36. Yang, J.; Hoffmeister, D.; Liu, L.; Fu, X.; Thorson, J.S. Natural product glycorandomization. *Bioorg. Med. Chem.* **2004**, *12*, 1577–1584. [CrossRef]
37. Zhang, C.; Moretti, R.; Jiang, J.; Thorson, J.S. The in vitro characterization of polyene glycosyltransferases AmphDI and NysDI. *ChemBioChem* **2008**, *9*, 2506–2514. [CrossRef]

38. Jiang, J.; Albermann, C.; Thorson, J.S. Application of the nucleotidylyltransferase Ep toward the chemoenzymatic synthesis of dTDP-desosamine analogues. *ChemBioChem* **2003**, *4*, 443–446. [CrossRef]
39. Albermann, C.; Soriano, A.; Jiang, J.; Vollmer, H.; Biggins, J.B.; Barton, W.A.; Lesniak, J.; Nikolov, D.B.; Thorson, J.S. Substrate specificity of NovM: Implications for novobiocin biosynthesis and glycorandomization. *Org. Lett.* **2003**, *5*, 933–936. [CrossRef]
40. Williams, G.J.; Gantt, R.W.; Thorson, J.S. The impact of enzyme engineering upon natural product glycodiversification. *Curr. Opin. Chem. Biol.* **2008**, *12*, 556–564. [CrossRef]
41. Thorson, J.S.; Barton, W.A.; Hoffmeister, D.; Albermann, C.; Nikolov, D.B. Structure-based enzyme engineering and its impact on in vitro glycorandomization. *ChemBioBhem* **2004**, *5*, 16–25. [CrossRef]
42. Beaton, S.A.; Huestis, M.P.; Sadeghi-Khomami, A.; Thomas, N.R.; Jakeman, D.L. Enzyme-catalyzed synthesis of isosteric phosphono-analogues of sugar nucleotides. *Chem. Commun.* **2009**, *8*, 238–240. [CrossRef] [PubMed]
43. Melo, A.; Glaser, L. The nucleotide specificity and feedback control of thymidine diphosphate D-glucose pyrophosphorylase. *J. Biol. Chem.* **1965**, *240*, 398–405. [CrossRef]
44. Blankenfeldt, W.; Giraud, M.F.; Leonard, G.; Rahim, R.; Creuzenet, C.; Lam, J.S.; Naismith, J.H. The purification, crystallization and preliminary structural characterization of glucose-1-phosphate thymidylyltransferase (RmlA), the first enzyme of the dTDP-L-rhamnose synthesis pathway from *Pseudomonas aeruginosa*. *Acta. Crystallogr. D. Biol. Crystallogr.* **2000**, *56 Pt 11*, 1501–1504. [CrossRef] [PubMed]
45. Sivaraman, J.; Sauve, V.; Matte, A.; Cygler, M. Crystal structure of Escherichia coli glucose-1-phosphate thymidylyltransferase (RffH) complexed with dTTP and Mg^{2+}. *J. Biol. Chem.* **2002**, *277*, 44214–44219. [CrossRef]
46. Allard, S.T.; Beis, K.; Giraud, M.F.; Hegeman, A.D.; Gross, J.W.; Wilmouth, R.C.; Whitfield, C.; Graninger, M.; Messner, P.; Allen, A.G.; et al. Toward a structural understanding of the dehydratase mechanism. *Structure* **2002**, *10*, 81–92. [CrossRef]
47. James, D.B.; Yother, J. Genetic and biochemical characterizations of enzymes involved in *Streptococcus pneumoniae* serotype 2 capsule synthesis demonstrate that Cps2T (WchF) catalyzes the committed step by addition of beta1-4 rhamnose, the second sugar residue in the repeat unit. *J. Bacteriol.* **2012**, *194*, 6479–6489. [CrossRef]
48. Dong, C.; Major, L.L.; Allen, A.; Blankenfeldt, W.; Maskell, D.; Naismith, J.H. High-resolution structures of RmlC from *Streptococcus suis* in complex with substrate analogs locate the active site of this class of enzyme. *Structure* **2003**, *11*, 715–723. [CrossRef]
49. Dong, C.; Major, L.L.; Srikannathasan, V.; Errey, J.C.; Giraud, M.F.; Lam, J.S.; Graninger, M.; Messner, P.; McNeil, M.R.; Field, R.A.; et al. RmlC, a C3′ and C5′ carbohydrate epimerase, appears to operate via an intermediate with an unusual twist boat conformation. *J. Mol. Biol.* **2007**, *365*, 146–159. [CrossRef]
50. Giraud, M.F.; Leonard, G.A.; Field, R.A.; Berlind, C.; Naismith, J.H. RmlC, the third enzyme of dTDP-L-rhamnose pathway, is a new class of epimerase. *Nat. Struct. Biol.* **2000**, *7*, 398–402.
51. Giraud, M.F.; Gordon, F.M.; Whitfield, C.; Messner, P.; McMahon, S.A.; Naismith, J.H. Purification, crystallization and preliminary structural studies of dTDP-6-deoxy-D-xylo-4-hexulose 3,5-epimerase (RmlC), the third enzyme of the dTDP-L-rhamnose synthesis pathway, from *Salmonella enterica* serovar typhimurium. *Acta Crystallogr. D Biol. Crystallogr.* **1999**, *55 Pt 3*, 706–708. [CrossRef]
52. Blankenfeldt, W.; Kerr, I.D.; Giraud, M.F.; McMiken, H.J.; Leonard, G.; Whitfield, C.; Messner, P.; Graninger, M.; Naismith, J.H. Variation on a theme of SDR. dTDP-6-deoxy-L- lyxo-4-hexulose reductase (RmlD) shows a new Mg^{2+}-dependent dimerization mode. *Structure* **2002**, *10*, 773–786. [CrossRef]
53. Graninger, M.; Nidetzky, B.; Heinrichs, D.E.; Whitfield, C.; Messner, P. Characterization of dTDP-4-dehydrorhamnose 3,5-epimerase and dTDP-4-dehydrorhamnose reductase, required for dTDP-L-rhamnose biosynthesis in *Salmonella enterica* serovar Typhimurium LT2. *J. Biol. Chem.* **1999**, *274*, 25069–25077. [CrossRef] [PubMed]
54. Ovod, V.; Rudolph, K.; Knirel, Y.; Krohn, K. Immunochemical characterization of O polysaccharides composing the alpha-D-rhamnose backbone of lipopolysaccharide of Pseudomonas syringae and classification of bacteria into serogroups O1 and O2 with monoclonal antibodies. *J. Bacteriol.* **1996**, *178*, 6459–6465. [CrossRef] [PubMed]
55. Kneidinger, B.; Graninger, M.; Adam, G.; Puchberger, M.; Kosma, P.; Zayni, S.; Messner, P. Identification of two GDP-6-deoxy-D-lyxo-4-hexulose reductases synthesizing GDP-D-rhamnose in *Aneurinibacillus thermoaerophilus* L420-91T. *J. Biol. Chem.* **2001**, *276*, 5577–5583. [CrossRef] [PubMed]
56. Webb, N.A.; Mulichak, A.M.; Lam, J.S.; Rocchetta, H.L.; Garavito, R.M. Crystal structure of a tetrameric GDP-D-mannose 4,6-dehydratase from a bacterial GDP-D-rhamnose biosynthetic pathway. *Protein Sci.* **2004**, *13*, 529–539. [CrossRef]
57. Ginsburg, V. Formation of guanosine diphosphate L-fucose from guanosine diphosphate D-mannose. *J. Biol. Chem.* **1960**, *235*, 2196–2201. [CrossRef]
58. Mulichak, A.M.; Bonin, C.P.; Reiter, W.D.; Garavito, R.M. Structure of the MUR1 GDP-mannose 4,6-dehydratase from Arabidopsis thaliana: Implications for ligand binding and specificity. *Biochemistry* **2002**, *41*, 15578–15589. [CrossRef]
59. Reitman, M.L.; Trowbridge, I.S.; Kornfeld, S. Mouse lymphoma cell lines resistant to pea lectin are defective in fucose metabolism. *J. Biol. Chem.* **1980**, *255*, 9900–9906. [CrossRef]
60. Bisso, A.; Sturla, L.; Zanardi, D.; De Flora, A.; Tonetti, M. Structural and enzymatic characterization of human recombinant GDP-D-mannose-4,6-dehydratase. *FEBS. Lett.* **1999**, *456*, 370–374. [CrossRef]
61. Tonetti, M.; Zanardi, D.; Gurnon, J.R.; Fruscione, F.; Armirotti, A.; Damonte, G.; Sturla, L.; De Flora, A.; Van Etten, J.L. Paramecium bursaria Chlorella virus 1 encodes two enzymes involved in the biosynthesis of GDP-L-fucose and GDP-D-rhamnose. *J. Biol. Chem.* **2003**, *278*, 21559–21565. [CrossRef]

62. Somoza, J.R.; Menon, S.; Schmidt, H.; Joseph-McCarthy, D.; Dessen, A.; Stahl, M.L.; Somers, W.S.; Sullivan, F.X. Structural and kinetic analysis of Escherichia coli GDP-mannose 4,6 dehydratase provides insights into the enzyme's catalytic mechanism and regulation by GDP-fucose. *Structure* **2000**, *8*, 123–135. [CrossRef]
63. Rosano, C.; Zuccotti, S.; Sturla, L.; Fruscione, F.; Tonetti, M.; Bolognesi, M. Quaternary assembly and crystal structure of GDP-D-mannose 4,6 dehydratase from *Paramecium bursaria Chlorella* virus. *Biochem. Biophys. Res. Commun.* **2006**, *339*, 191–195. [CrossRef] [PubMed]
64. King, J.D.; Poon, K.K.; Webb, N.A.; Anderson, E.M.; McNally, D.J.; Brisson, J.R.; Messner, P.; Garavito, R.M.; Lam, J.S. The structural basis for catalytic function of GMD and RMD, two closely related enzymes from the GDP-D-rhamnose biosynthesis pathway. *FEBS J.* **2009**, *276*, 2686–2700. [CrossRef]
65. Qian, L.; Han, X.; Zhang, L.; Liu, X. Structural and biochemical insights into nucleotide-rhamnose synthase/epimerase-reductase from *Arabidopsis thaliana*. *Biochim. Biophys. Acta* **2015**, *10 Pt A*, 1476–1486.
66. Oka, T.; Nemoto, T.; Jigami, Y. Functional analysis of Arabidopsis thaliana RHM2/MUM4, a multidomain protein involved in UDP-D-glucose to UDP-L-rhamnose conversion. *J. Biol. Chem.* **2007**, *282*, 5389–5403. [CrossRef] [PubMed]
67. Kenyon, J.J.; Kasimova, A.A.; Sviridova, A.N.; Shpirt, A.M.; Shneider, M.M.; Mikhaylova, Y.V.; Shelenkov, A.A.; Popova, A.V.; Perepelov, A.V.; Shashkov, A.S.; et al. Correlation of Acinetobacter baumannii K144 and K86 capsular polysaccharide structures with genes at the K locus reveals the involvement of a novel multifunctional rhamnosyltransferase for structural synthesis. *Int. J. Biol. Macromol.* **2021**, *193 Pt B*, 1294–1300. [CrossRef]
68. Kenyon, J.J.; Arbatsky, N.P.; Sweeney, E.L.; Zhang, Y.; Senchenkova, S.N.; Popova, A.V.; Shneider, M.M.; Shashkov, A.S.; Liu, B.; Hall, R.M.; et al. Involvement of a multifunctional rhamnosyltransferase in the synthesis of three related Acinetobacter baumannii capsular polysaccharides, K55, K74 and K85. *Int. J. Biol. Macromol.* **2021**, *166*, 1230–1237. [CrossRef]
69. Steiner, K.; Hagelueken, G.; Messner, P.; Schaffer, C.; Naismith, J.H. Structural basis of substrate binding in WsaF, a rhamnosyltransferase from *Geobacillus stearothermophilus*. *J. Mol. Biol.* **2010**, *397*, 436–447. [CrossRef]
70. Steiner, K.; Novotny, R.; Werz, D.B.; Zarschler, K.; Seeberger, P.H.; Hofinger, A.; Kosma, P.; Schaffer, C.; Messner, P. Molecular basis of S-layer glycoprotein glycan biosynthesis in *Geobacillus stearothermophilus*. *J. Biol. Chem.* **2008**, *283*, 21120–21133. [CrossRef]
71. Isiorho, E.A.; Liu, H.W.; Keatinge-Clay, A.T. Structural studies of the spinosyn rhamnosyltransferase, SpnG. *Biochemistry.* **2012**, *51*, 1213–1222. [CrossRef]
72. Huang, K.X.; Xia, L.; Zhang, Y.; Ding, X.; Zahn, J.A. Recent advances in the biochemistry of spinosyns. *Appl. Microbiol. Biotechnol.* **2009**, *82*, 13–23. [CrossRef] [PubMed]
73. Chen, Y.L.; Chen, Y.H.; Lin, Y.C.; Tsai, K.C.; Chiu, H.T. Functional characterization and substrate specificity of spinosyn rhamnosyltransferase by in vitro reconstitution of spinosyn biosynthetic enzymes. *J. Biol. Chem.* **2009**, *284*, 7352–7363. [CrossRef] [PubMed]
74. Grzegorzewicz, A.E.; Ma, Y.; Jones, V.; Crick, D.; Liav, A.; McNeil, M.R. Development of a microtitre plate-based assay for lipid-linked glycosyltransferase products using the mycobacterial cell wall rhamnosyltransferase WbbL. *Microbiology* **2008**, *154 Pt 12*, 3724–3730. [CrossRef] [PubMed]
75. Sivendran, S.; Jones, V.; Sun, D.; Wang, Y.; Grzegorzewicz, A.E.; Scherman, M.S.; Napper, A.D.; McCammon, J.A.; Lee, R.E.; Diamond, S.L.; et al. Identification of triazinoindol-benzimidazolones as nanomolar inhibitors of the Mycobacterium tuberculosis enzyme TDP-6-deoxy-d-xylo-4-hexopyranosid-4-ulose 3,5-epimerase (RmlC). *Bioorg. Med. Chem.* **2010**, *18*, 896–908. [CrossRef]
76. Mills, J.A.; Motichka, K.; Jucker, M.; Wu, H.P.; Uhlik, B.C.; Stern, R.J.; Scherman, M.S.; Vissa, V.D.; Pan, F.; Kundu, M.; et al. Inactivation of the mycobacterial rhamnosyltransferase, which is needed for the formation of the arabinogalactan-peptidoglycan linker, leads to irreversible loss of viability. *J. Biol. Chem.* **2004**, *279*, 43540–43546. [CrossRef] [PubMed]
77. Aguirre-Ramirez, M.; Medina, G.; Gonzalez-Valdez, A.; Grosso-Becerra, V.; Soberon-Chavez, G. The Pseudomonas aeruginosa rmlBDAC operon, encoding dTDP-L-rhamnose biosynthetic enzymes, is regulated by the quorum-sensing transcriptional regulator RhlR and the alternative sigma factor sigmaS. *Microbiology* **2012**, *158 Pt 4*, 908–916. [CrossRef]
78. Poon, K.K.; Westman, E.L.; Vinogradov, E.; Jin, S.; Lam, J.S. Functional characterization of MigA and WapR: Putative rhamnosyltransferases involved in outer core oligosaccharide biosynthesis of *Pseudomonas aeruginosa*. *J. Bacteriol.* **2008**, *190*, 1857–1865. [CrossRef]
79. Rocchetta, H.L.; Burrows, L.L.; Lam, J.S. Genetics of O-antigen biosynthesis in *Pseudomonas aeruginosa*. *Microbiol. Mol. Biol. Rev.* **1999**, *63*, 523–553. [CrossRef] [PubMed]
80. Rocchetta, H.L.; Burrows, L.L.; Pacan, J.C.; Lam, J.S. Three rhamnosyltransferases responsible for assembly of the A-band D-rhamnan polysaccharide in *Pseudomonas aeruginosa*: A fourth transferase, WbpL, is required for the initiation of both A-band and B-band lipopolysaccharide synthesis. *Mol. Microbiol.* **1998**, *28*, 1103–1119. [CrossRef]
81. Melamed, J.; Kocev, A.; Torgov, V.; Veselovsky, V.; Brockhausen, I. Biosynthesis of the *Pseudomonas aeruginosa* common polysaccharide antigen by D-Rhamnosyltransferases WbpX and WbpY. *Glycoconj. J.* **2022**, *39*, 393–411. [CrossRef]
82. Zhu, K.; Rock, C.O. RhlA converts beta-hydroxyacyl-acyl carrier protein intermediates in fatty acid synthesis to the beta-hydroxydecanoyl-beta-hydroxydecanoate component of rhamnolipids in *Pseudomonas aeruginosa*. *J. Bacteriol.* **2008**, *190*, 3147–3154. [CrossRef]
83. Ochsner, U.A.; Fiechter, A.; Reiser, J. Isolation, characterization, and expression in Escherichia coli of the *Pseudomonas aeruginosa* rhlAB genes encoding a rhamnosyltransferase involved in rhamnolipid biosurfactant synthesis. *J. Biol. Chem.* **1994**, *269*, 19787–19795. [CrossRef]

84. Rahim, R.; Ochsner, U.A.; Olvera, C.; Graninger, M.; Messner, P.; Lam, J.S.; Soberon-Chavez, G. Cloning and functional characterization of the Pseudomonas aeruginosa rhlC gene that encodes rhamnosyltransferase 2, an enzyme responsible for di-rhamnolipid biosynthesis. *Mol. Microbiol.* **2001**, *40*, 708–718. [CrossRef] [PubMed]
85. Deziel, E.; Lepine, F.; Milot, S.; Villemur, R. rhlA is required for the production of a novel biosurfactant promoting swarming motility in *Pseudomonas aeruginosa*: 3-(3-hydroxyalkanoyloxy)alkanoic acids (HAAs), the precursors of rhamnolipids. *Microbiology* **2003**, *149 Pt 8*, 2005–2013. [CrossRef] [PubMed]
86. Tavares, L.F.; Silva, P.M.; Junqueira, M.; Mariano, D.C.; Nogueira, F.C.; Domont, G.B.; Freire, D.M.; Neves, B.C. Characterization of rhamnolipids produced by wild-type and engineered *Burkholderia kururiensis*. *Appl. Microbiol. Biotechnol.* **2013**, *97*, 1909–1921. [CrossRef] [PubMed]
87. Pan, X.; Luo, J.; Li, S. Bacteria-Catalyzed Arginine Glycosylation in Pathogens and Host. *Front. Cell. Infect. Microbiol.* **2020**, *10*, 185. [CrossRef]
88. He, C.; Liu, N.; Li, F.; Jia, X.; Peng, H.; Liu, Y.; Xiao, Y. Complex Structure of Pseudomonas aeruginosa Arginine Rhamnosyltransferase EarP with Its Acceptor Elongation Factor P. *J. Bacteriol.* **2019**, *201*, e00128-19. [CrossRef]
89. Lassak, J.; Keilhauer, E.C.; Fürst, M.; Wuichet, K.; Gödeke, J.; Starosta, A.L.; Chen, J.M.; Søgaard-Andersen, L.; Rohr, J.; Wilson, D.N.; et al. Arginine-rhamnosylation as new strategy to activate translation elongation factor P. *Nat. Chem. Biol.* **2015**, *11*, 266–270. [CrossRef]
90. Yanagisawa, T.; Takahashi, H.; Suzuki, T.; Masuda, A.; Dohmae, N.; Yokoyama, S. Neisseria meningitidis Translation Elongation Factor P and Its Active-Site Arginine Residue Are Essential for Cell Viability. *PLoS ONE* **2016**, *11*, e0147907. [CrossRef]
91. Geno, K.A.; Gilbert, G.L.; Song, J.Y.; Skovsted, I.C.; Klugman, K.P.; Jones, C.; Konradsen, H.B.; Nahm, M.H. Pneumococcal Capsules and Their Types: Past, Present, and Future. *Clin. Microbiol. Rev.* **2015**, *28*, 871–899. [CrossRef]
92. Park, S.; Nahm, M.H. L-rhamnose is often an important part of immunodominant epitope for pneumococcal serotype 23F polysaccharide antibodies in human sera immunized with PPV23. *PLoS ONE* **2013**, *8*, e83810. [CrossRef] [PubMed]
93. Legnani, L.; Ronchi, S.; Fallarini, S.; Lombardi, G.; Campo, F.; Panza, L.; Lay, L.; Poletti, L.; Toma, L.; Ronchetti, F.; et al. Synthesis, molecular dynamics simulations, and biology of a carba-analogue of the trisaccharide repeating unit of *Streptococcus pneumoniae* 19F capsular polysaccharide. *Org. Biomol. Chem.* **2009**, *7*, 4428–4436. [CrossRef] [PubMed]
94. Aanensen, D.M.; Mavroidi, A.; Bentley, S.D.; Reeves, P.R.; Spratt, B.G. Predicted functions and linkage specificities of the products of the *Streptococcus pneumoniae* capsular biosynthetic loci. *J. Bacteriol.* **2007**, *189*, 7856–7876. [CrossRef]
95. James, D.B.; Gupta, K.; Hauser, J.R.; Yother, J. Biochemical activities of *Streptococcus pneumoniae* serotype 2 capsular glycosyltransferases and significance of suppressor mutations affecting the initiating glycosyltransferase Cps2E. *J. Bacteriol.* **2013**, *195*, 5469–5478. [CrossRef] [PubMed]
96. Wang, H.; Li, S.; Xiong, C.; Jin, G.; Chen, Z.; Gu, G.; Guo, Z. Biochemical studies of a beta-1,4-rhamnoslytransferase from *Streptococcus pneumonia* serotype 23F. *Org. Biomol. Chem.* **2019**, *17*, 1071–1075. [CrossRef] [PubMed]
97. Ardissone, S.; Noel, K.D.; Klement, M.; Broughton, W.J.; Deakin, W.J. Synthesis of the flavonoid-induced lipopolysaccharide of *Rhizobium Sp.* strain NGR234 requires rhamnosyl transferases encoded by genes rgpF and wbgA. *Mol. Plant-Microbe Interact.* **2011**, *24*, 1513–1521. [CrossRef] [PubMed]
98. Ishida, T.; Sugano, Y.; Shoda, M. Novel glycosyltransferase genes involved in the acetan biosynthesis of *Acetobacter xylinum*. *Biochem. Biophys. Res. Commun.* **2002**, *295*, 230–235. [CrossRef]
99. Dong, S.; McPherson, S.A.; Wang, Y.; Li, M.; Wang, P.; Turnbough, C.J.; Pritchard, D.G. Characterization of the enzymes encoded by the anthrose biosynthetic operon of *Bacillus anthracis*. *J. Bacteriol.* **2010**, *192*, 5053–5062. [CrossRef]
100. Zorzoli, A.; Meyer, B.H.; Adair, E.; Torgov, V.I.; Veselovsky, V.V.; Danilov, L.L.; Uhrin, D.; Dorfmueller, H.C. Group A, B, C, and G Streptococcus Lancefield antigen biosynthesis is initiated by a conserved α-d-GlcNAc-β-1,4-L-rhamnosyltransferase. *J. Biol. Chem.* **2019**, *294*, 15237–15256. [CrossRef]
101. Miyamoto, Y.; Mukai, T.; Naka, T.; Fujiwara, N.; Maeda, Y.; Kai, M.; Mizuno, S.; Yano, I.; Makino, M. Novel rhamnosyltransferase involved in biosynthesis of serovar 4-specific glycopeptidolipid from *Mycobacterium avium* complex. *J. Bacteriol.* **2010**, *192*, 5700–5708. [CrossRef]
102. Tsunashima, H.; Miyake, K.; Motono, M.; Iijima, S. Organization of the capsule biosynthesis gene locus of the oral *streptococcus Streptococcus anginosus*. *J. Biosci. Bioeng.* **2012**, *113*, 271–278. [CrossRef] [PubMed]
103. Liang, M.; Gong, W.; Sun, C.; Zhao, J.; Wang, H.; Chen, Z.; Xiao, M.; Gu, G. Sequential One-Pot Three-Enzyme Synthesis of the Tetrasaccharide Repeating Unit of Group B Streptococcus Serotype VIII Capsular Polysaccharide. *Chin. J. Chem.* **2022**, *40*, 1039–1044. [CrossRef]
104. Li, Q.; Hobbs, M.; Reeves, P.R. The variation of dTDP-L-rhamnose pathway genes in *Vibrio cholerae*. *Microbiology* **2003**, *149 Pt 9*, 2463–2474. [CrossRef] [PubMed]
105. Feng, K.; Chen, R.; Xie, K.; Chen, D.; Guo, B.; Liu, X.; Liu, J.; Zhang, M.; Dai, J. A regiospecific rhamnosyltransferase from Epimedium pseudowushanense catalyzes the 3-O-rhamnosylation of prenylflavonols. *Org. Biomol. Chem.* **2018**, *16*, 452–458. [CrossRef]
106. Zong, G.; Fei, S.; Liu, X.; Li, J.; Gao, Y.; Yang, X.; Wang, X.; Shen, Y. Crystal structures of rhamnosyltransferase UGT89C1 from Arabidopsis thaliana reveal the molecular basis of sugar donor specificity for UDP-β-L-rhamnose and rhamnosylation mechanism. *Plant J.* **2019**, *99*, 257–269.

107. Takenaka, Y.; Kato, K.; Ogawa-Ohnishi, M.; Tsuruhama, K.; Kajiura, H.; Yagyu, K.; Takeda, A.; Takeda, Y.; Kunieda, T.; Hara-Nishimura, I.; et al. Pectin RG-I rhamnosyltransferases represent a novel plant-specific glycosyltransferase family. *Nat. Plants* **2018**, *4*, 669–676. [CrossRef]
108. Lu, C.; Dong, Y.; Ke, K.; Zou, K.; Wang, Z.; Xiao, W.; Pei, J.; Zhao, L. Modification to increase the thermostability and catalytic efficiency of α-L-rhamnosidase from Bacteroides thetaiotaomicron and high-level expression. *Enzym. Microb. Technol.* **2022**, *158*, 110040. [CrossRef]
109. Xu, L.; Liu, X.; Li, Y.; Yin, Z.; Jin, L.; Lu, L.; Qu, J.; Xiao, M. Enzymatic rhamnosylation of anticancer drugs by an α-L-rhamnosidase from *Alternaria* sp. L1 for cancer-targeting and enzyme-activated prodrug therapy. *Appl. Microbiol. Biotechnol.* **2019**, *103*, 7997–8008. [CrossRef]
110. Hakomori, S. Tumor-associated carbohydrate antigens defining tumor malignancy: Basis for development of anti-cancer vaccines. *Adv. Exp. Med. Biol.* **2001**, *491*, 369–402.
111. Hakomori, S. Aberrant glycosylation in cancer cell membranes as focused on glycolipids: Overview and perspectives. *Cancer Res.* **1985**, *45*, 2405–2414.
112. Guo, Z.; Wang, Q. Recent development in carbohydrate-based cancer vaccines. *Curr. Opin. Chem. Biol.* **2009**, *13*, 608–617. [CrossRef] [PubMed]
113. Wilson, R.M.; Danishefsky, S.J. A Vision for Vaccines Built from Fully Synthetic Tumor-Associated Antigens: From the Laboratory to the Clinic. *J. Am. Chem. Soc.* **2013**, *135*, 14462–14472. [CrossRef]
114. Grabenstein, J.D.; Klugman, K.P. A century of pneumococcal vaccination research in humans. *Clin. Microbiol. Infect.* **2012**, *18* (Suppl. 5), 15–24. [CrossRef]
115. Holmberg, L.A.; Sandmaier, B.M. Vaccination with Theratope (STn-KLH) as treatment for breast cancer. *Expert. Rev. Vaccines* **2004**, *3*, 655–663. [CrossRef]
116. Franco, A. Glycoconjugates as vaccines for cancer immunotherapy: Clinical trials and future directions. *Anticancer Agents Med. Chem.* **2008**, *8*, 86–91. [CrossRef] [PubMed]
117. Wang, Q.; Zhou, Z.; Tang, S.; Guo, Z. Carbohydrate-monophosphoryl lipid a conjugates are fully synthetic self-adjuvanting cancer vaccines eliciting robust immune responses in the mouse. *ACS Chem. Biol.* **2012**, *7*, 235–240. [CrossRef] [PubMed]
118. Yu, S.; Wang, Q.; Zhang, J.; Wu, Q.; Guo, Z. Synthesis and Evaluation of Protein Conjugates of GM3 Derivatives Carrying Modified Sialic Acids as Highly Immunogenic Cancer Vaccine Candidates. *Medchemcomm* **2011**, *2*, 524–530. [CrossRef]
119. Oyelaran, O.; McShane, L.M.; Dodd, L.; Gildersleeve, J.C. Profiling human serum antibodies with a carbohydrate antigen microarray. *J. Proteome Res.* **2009**, *8*, 4301–4310. [CrossRef]
120. Zhang, H.; Wang, B.; Ma, Z.; Wei, M.; Liu, J.; Li, D.; Zhang, H.; Wang, P.G.; Chen, M. L-Rhamnose Enhances the Immunogenicity of Melanoma-Associated Antigen A3 for Stimulating Antitumor Immune Responses. *Bioconjug. Chem.* **2016**, *27*, 1112–1118. [CrossRef]
121. Sarkar, S.; Lombardo, S.A.; Herner, D.N.; Talan, R.S.; Wall, K.A.; Sucheck, S.J. Synthesis of a single-molecule L-rhamnose-containing three-component vaccine and evaluation of antigenicity in the presence of anti-L-rhamnose antibodies. *J. Am. Chem. Soc.* **2010**, *132*, 17236–17246. [CrossRef]
122. Sarkar, S.; Salyer, A.C.; Wall, K.A.; Sucheck, S.J. Synthesis and immunological evaluation of a MUC1 glycopeptide incorporated into L-rhamnose displaying liposomes. *Bioconjug. Chem.* **2013**, *24*, 363–375. [CrossRef] [PubMed]
123. Li, X.; Rao, X.; Cai, L.; Liu, X.; Wang, H.; Wu, W.; Zhu, C.; Chen, M.; Wang, P.G.; Yi, W. Targeting Tumor Cells by Natural Anti-Carbohydrate Antibodies Using Rhamnose-Functionalized Liposomes. *ACS Chem. Biol.* **2016**, *11*, 1205–1209. [CrossRef] [PubMed]
124. Hossain, M.K.; Vartak, A.; Karmakar, P.; Sucheck, S.J.; Wall, K.A. Augmenting Vaccine Immunogenicity through the Use of Natural Human Anti-rhamnose Antibodies. *ACS Chem. Biol.* **2018**, *13*, 2130–2142. [CrossRef] [PubMed]
125. Lin, H.; Hong, H.; Wang, J.; Li, C.; Zhou, Z.; Wu, Z. Rhamnose modified bovine serum albumin as a carrier protein promotes the immune response against sTn antigen. *Chem. Commun.* **2020**, *56*, 13959–13962. [CrossRef] [PubMed]
126. Chen, W.; Gu, L.; Zhang, W.; Motari, E.; Cai, L.; Styslinger, T.J.; Wang, P.G. L-rhamnose antigen: A promising alternative to alpha-gal for cancer immunotherapies. *ACS Chem. Biol.* **2011**, *6*, 185–191. [CrossRef]
127. Ou, C.; Prabhu, S.K.; Zhang, X.; Zong, G.; Yang, Q.; Wang, L.X. Synthetic Antibody-Rhamnose Cluster Conjugates Show Potent Complement-Dependent Cell Killing by Recruiting Natural Antibodies. *Chemistry* **2022**, *28*, e202200146. [CrossRef]
128. Zhou, K.; Hong, H.; Lin, H.; Gong, L.; Li, D.; Shi, J.; Zhou, Z.; Xu, F.; Wu, Z. Chemical Synthesis of Antibody-Hapten Conjugates Capable of Recruiting the Endogenous Antibody to Magnify the Fc Effector Immunity of Antibody for Cancer Immunotherapy. *J. Med. Chem.* **2022**, *65*, 323–332. [CrossRef]
129. De Coen, R.; Nuhn, L.; Perera, C.; Arista-Romero, M.; Risseeuw, M.D.P.; Freyn, A.; Nachbagauer, R.; Albertazzi, L.; Van Calenbergh, S.; Spiegel, D.A.; et al. Synthetic Rhamnose Glycopolymer Cell-Surface Receptor for Endogenous Antibody Recruitment. *Biomacromolecules* **2020**, *21*, 793–802. [CrossRef]
130. Hilinski, M.K.; Mrozowski, R.M.; Clark, D.E.; Lannigan, D.A. Analogs of the RSK inhibitor SL0101: Optimization of in vitro biological stability. *Bioorg. Med. Chem. Lett.* **2012**, *22*, 3244–3247. [CrossRef]
131. Chun, J.; Ha, I.J.; Kim, Y.S. Antiproliferative and apoptotic activities of triterpenoid saponins from the roots of *Platycodon grandiflorum* and their structure-activity relationships. *Planta Med.* **2013**, *79*, 639–645. [CrossRef]

132. Ha, I.J.; Kang, M.; Na, Y.C.; Park, Y.; Kim, Y.S. Preparative separation of minor saponins from *Platycodi Radix* by high-speed counter-current chromatography. *J. Sep. Sci.* **2011**, *34*, 2559–2565. [CrossRef] [PubMed]
133. Wang, Y.; Gao, J.; Gu, G.; Li, G.; Cui, C.; Sun, B.; Lou, H. In situ RBL receptor visualization and its mediated anticancer activity for solasodine rhamnosides. *Chembiochem* **2011**, *12*, 2418–2420. [CrossRef] [PubMed]
134. Garnier, P.; Wang, X.T.; Robinson, M.A.; van Kasteren, S.; Perkins, A.C.; Frier, M.; Fairbanks, A.J.; Davis, B.G. Lectin-directed enzyme activated prodrug therapy (LEAPT): Synthesis and evaluation of rhamnose-capped prodrugs. *J. Drug Target.* **2010**, *18*, 794–802. [CrossRef] [PubMed]
135. Tomsik, P.; Stoklasova, A.; Micuda, S.; Niang, M.; Suba, P.; Knizek, J.; Rezacova, M. Evaluation of the antineoplastic activity of L-rhamnose in vitro. A comparison with 2-deoxyglucose. *Acta Med.* **2008**, *51*, 113–119.
136. Ma, Y.; Pan, F.; McNeil, M. Formation of dTDP-rhamnose is essential for growth of mycobacteria. *J. Bacteriol.* **2002**, *184*, 3392–3395. [CrossRef]
137. Xiao, G.; Alphey, M.S.; Tran, F.; Pirrie, L.; Milbeo, P.; Zhou, Y.; Bickel, J.K.; Kempf, O.; Kempf, K.; Naismith, J.H.; et al. Next generation Glucose-1-phosphate thymidylyltransferase (RmlA) inhibitors: An extended SAR study to direct future design. *Bioorg. Med. Chem.* **2021**, *50*, 116477. [CrossRef]
138. Alphey, M.S.; Pirrie, L.; Torrie, L.S.; Boulkeroua, W.A.; Gardiner, M.; Sarkar, A.; Maringer, M.; Oehlmann, W.; Brenk, R.; Scherman, M.S.; et al. Allosteric competitive inhibitors of the glucose-1-phosphate thymidylyltransferase (RmlA) from *Pseudomonas aeruginosa*. *ACS Chem. Biol.* **2013**, *8*, 387–396. [CrossRef] [PubMed]
139. Loranger, M.W.; Forget, S.M.; McCormick, N.E.; Syvitski, R.T.; Jakeman, D.L. Synthesis and evaluation of L-rhamnose 1C-phosphonates as nucleotidylyltransferase inhibitors. *J. Org. Chem.* **2013**, *78*, 9822–9833. [CrossRef]
140. Kantardjieff, K.A.; Kim, C.Y.; Naranjo, C.; Waldo, G.S.; Lekin, T.; Segelke, B.W.; Zemla, A.; Park, M.S.; Terwilliger, T.C.; Rupp, B. Mycobacterium tuberculosis RmlC epimerase (Rv3465): A promising drug-target structure in the rhamnose pathway. *Acta Crystallogr. D Biol. Crystallogr.* **2004**, *60 Pt 5*, 895–902. [CrossRef]

molecules

MDPI

Article

Critical Quality Control Methods for a Novel Anticoagulant Candidate LFG-Na by HPSEC-MALLS-RID and Bioactivity Assays

Shunliang Zheng [1,2], Yi Wang [2], Jiashuo Wu [1], Siyao Wang [2], Huaifu Wei [2], Yongchun Zhang [2], Jianbo Zhou [2,*] and Yue Shi [1,*]

1 Institute of Medicinal Plant Development, Chinese Academy of Medical Sciences and Peking Union Medical College, Beijing 100193, China; zhengshunliang@163.com (S.Z.); wjs_implad@163.com (J.W.)
2 Mudanjiang Youbo Pharmceutical Co., Ltd., Mudanjiang 157013, China; realwangyi1221@163.com (Y.W.); wsiy11@126.com (S.W.); whf8369@163.com (H.W.); ybyyzyc@163.com (Y.Z.)
* Correspondence: zjbjianbo@163.com (J.Z.); yshi@implad.ac.cn (Y.S.); Tel.: +86-10-57833270 (Y.S.)

Abstract: A low molecular weight fucosylated glycosaminoglycan sodium (LFG-Na) is a novel anticoagulant candidate from the sea cucumber *Holothuria fuscopunctata* that selectively inhibits intrinsic tenase (iXase). The molecular weight, molecular weight distribution and bioactivities are the critical quality attributes of LFG-Na. The determination of these quality attributes of such an oligosaccharides mixture drug is challenging but critical for the quality control process to ensure its safety and efficacy in clinical use. Herein, the molecular weight and molecular weight distribution of LFG-Na were successfully determined using high performance size exclusion chromatography coupled with multi angle laser light scattering and refractive index detector (HPSEC-MALLS-RID). Comparing to the conventional method, HPSEC-MALLS-RID based on the refractive index increment (*dn/dc*) did not require the reference substances to establish the calibration curve. The acceptance criteria of LFG-Na were established, the weight-average molecular weight (M_w) should be 4000 to 6000 Da, the polydispersity (M_w/M_n) < 1.40, and the fraction with molecular weights of 1500 to 8000 Da should be no less than 80% of the total. HPSEC-MALLS-RID was also utilized for the determination of the starting material native fucosylated glycosaminoglycan (NFG) to choose a better manufacturing process. Furthermore, APTT assay was selected and the potency of anti-iXase, referring to the parallel line assay (PLA) method, was established to clarify the consistency of its biological activities. The results suggest that HPSEC-MALLS-RID and bioactivity assays are critical quality control methods for multi-component glycosaminoglycan LFG-Na. The methods also provide a feasible strategy to control the quality of other polysaccharide medicines.

Keywords: LFG-Na; sea cucumber; *Holothuria fuscopunctata*; molecular weight; HPSEC-MALLS-RID; bioactivity assay; quality control

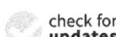

Citation: Zheng, S.; Wang, Y.; Wu, J.; Wang, S.; Wei, H.; Zhang, Y.; Zhou, J.; Shi, Y. Critical Quality Control Methods for a Novel Anticoagulant Candidate LFG-Na by HPSEC-MALLS-RID and Bioactivity Assays. *Molecules* **2022**, *27*, 4522. https://doi.org/10.3390/molecules27144522

Academic Editors: Jian Yin, Jing Zeng and De-Cai Xiong

Received: 3 July 2022
Accepted: 13 July 2022
Published: 15 July 2022

1. Introduction

Venous thromboembolism (VTE), such as deep vein thrombosis (DVT) and pulmonary embolism (PE), is a common underlying pathology of cardiovascular disease, which is a global health burden associated with high morbidity and mortality [1,2]. Anticoagulants, antiplatelets and thrombolytics are three types of antithrombotic drugs in great demand [3]. Among them, anticoagulants are effective in inhibiting the activity or synthesis of coagulation factors, which ultimately prevent or limit the formation of fibrin clots by breaking the coagulation cascade.

To date, numerous anticoagulant drugs have been developed. The classical management of VTE in adults consists of an initial treatment with adjusted-dose intravenous unfractionated heparin (UFH), body weight-adjusted subcutaneous low molecular weight heparin (LMWH), or body weight-adjusted subcutaneous fondaparinux followed by long-term treatment with a vitamin K antagonist (VKA) [4]. UFH and LMWH have been the clinical cornerstones of antithrombotic treatment and prophylaxis for over 80 years, and early in the 21st century, direct oral anticoagulants (DOACs) targeting thrombin (f.IIa) or factor Xa (f.Xa) were developed. However, the risk of haemorrhagic complications is still a major concern with their clinical application, and their therapeutic monitoring requirement is controversial [5–8]. Consequently, there is an unmet medical need in discovering safer anticoagulants for antithrombotic therapy.

The inhibitors of intrinsic coagulation pathway can inhibit pathological thrombosis without or slightly affecting hemostatic function and prevent thrombosis with negligible bleeding risks [9,10]. Intrinsic factor Xase complex (iXase) consisting of f.IXa-f.VIIIa is the last and rate-limiting enzyme of the intrinsic coagulation pathway [10–12], indicating that targeting iXase is a promising safer anticoagulant therapy with lower risk of bleeding.

Native fucosylated glycosaminoglycan (NFG) from sea cucumber is a unique glycosaminoglycan with fucose branches. The NFG extracted from the sea cucumber *Holothuria fuscopunctata* Jaeger mainly exhibits a chondroitin sulfate (CS) chain and 3,4-di-*O*-sulfated-fucose (Fuc_{3S4S}) branches [10,13]. Its β-eliminative depolymerized product, a low molecular weight fucosylated glycosaminoglycan sodium (LFG-Na), is a novel anticoagulant candidate to enter clinical trials permitted by the U.S. Food and Drug Administration. LFG-Na has clear chemical composition, selective anti-iXase activity, potent anticoagulation, antithrombosis with low bleeding tendency and predictable pharmacodynamic characteristics without the NFG's undesired effects of platelet aggregation and factor XII (f.XII) activation [12,14].

The biological and pharmacological effects of fucosylated glycosaminoglycans are closely related to their molecular weight, molecular weight distribution and sulfation patterns [15,16]. LFG-Na is composed of a series of oligosaccharides, and it is thus a complex multicomponent drug. Moreover, its pharmacological activities, such as anti-iXase, factor IXa-binding, anticoagulant and antithrombotic activities, result from its oligosaccharides in the terms of weighted average sum [12]. Therefore, the molecular weight, molecular weight distribution and bioactivities are the critical quality attributes of LFG-Na.

The current typical method for the determination of the molecular weight and molecular weight distribution of glycosaminoglycans is high performance gel permeation chromatography (HPGPC) [12,17]. The data are calculated using GPC software, so it is necessary to fit the calibration curve with a series of narrow standard reference standards with known molecular weight for calculation. Currently, it is temporarily impossible to obtain a series of narrow standard references that completely cover the maximum molecular weight range of glycosaminoglycans, including NFG and LFG-Na, due to the complicated and laborious procedures for their separation and purification [18]. The accuracy of HPGPC with standard curves is relatively poor. Additionally, it is time consuming, considering the injection of the standard references. Therefore, accurate and rapid determination of the molecular weight of LFG-Na and its starting material NFG is crucially important and urgently required to control quality of the new drug. Herein, high performance size exclusion chromatography coupled with multi angle laser light scattering and refractive index detector (HPSEC-MALLS-RID) was used to determine the molecular weight and molecular weight distribution of LFG-Na and compared to HPGPC. Furthermore, the chemical characteristics and bioactivities of multiple batches of LFG-Na were tested and compared to their consistency. Our results indicate that HPSEC-MALLS-RID and bioactivity assays are critical quality control methods for the multi-component anticoagulant candidate LFG-Na.

2. Results and Discussion

2.1. Determination of the Refractive Index Increment (*dn/dc*)

For the multi-angle light scattering coupled with size exclusion chromatography (SEC-MALLS), the *dn/dc* of the solution is required. The *dn/dc* of a solution is a constant that indicates the variation of the refractive index with the solute concentration. It is used in the multi-angle light scattering technique to determine the concentration and the weight-average molecular weight (M_w) of polymers [19]. Since the *dn/dc* appeared as an important parameter, the accurate value was therefore essential for the determination of the M_w of LFG-Na and its starting material NFG.

A representative chromatogram obtained from the determination of the *dn/dc* value of LFG-Na is shown in Figure 1, whereas the chromatograms of NFG samples are similar. The changes in concentration of glycosaminoglycan solutions were converted to changes in refractive index. The *dn/dc* values were obtained with ASTRA software (version 7.1.3, Wyatt Technology Co., Santa Barbara, CA, USA).

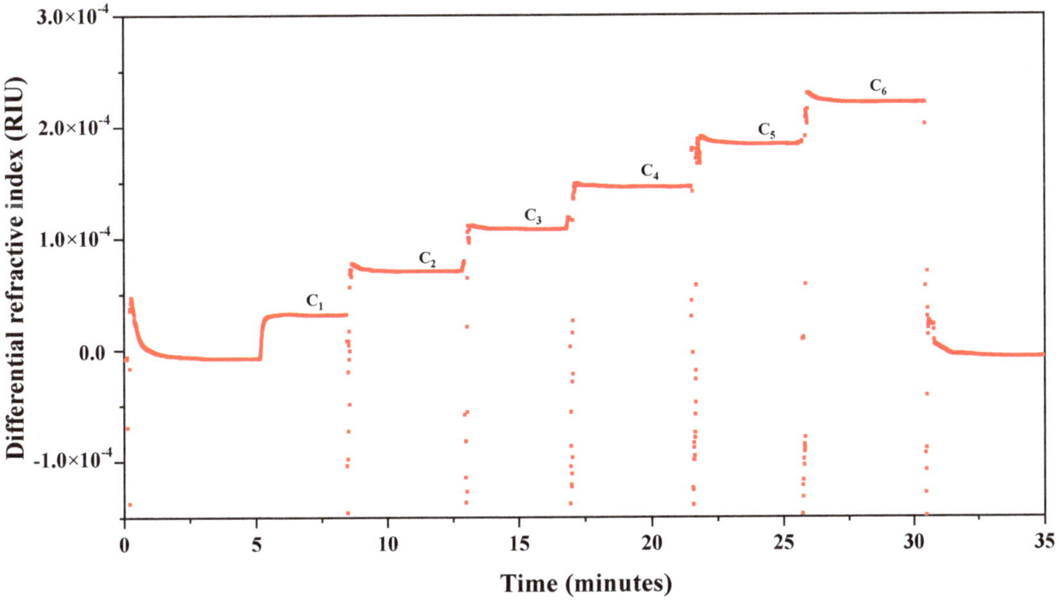

Figure 1. Chromatogram obtained from the determination of *dn/dc* value (LFG-Na, batch L3, concentrations of LFG-Na solutions-differential refractive index were fitted, the Fit R^2 = 1.0000 from ASTRA software (version 7.1.3, Wyatt Technology Co., Santa Barbara, CA, USA)).

Table 1 shows *dn/dc* values of LFG-Na and NFG in the solution of 0.1 mol/L sodium nitrate (containing 0.02% sodium azide). Five batches of LFG-Na samples had *dn/dc* values varying from 0.1173 to 0.1283 mL/g, with a mean value of 0.1248 mL/g. Six batches of NFG samples had *dn/dc* values varying from 0.1113 to 0.1192 mL/g, with a mean value of 0.1163 mL/g; while four batches of NFG-2.0M (obtained before the NFG manufacturing process optimization) samples had a mean *dn/dc* value of 0.1166 mL/g. The mean *dn/dc* value of LFG-Na was slightly larger than that of NFG. These mean *dn/dc* values of LFG-Na and NFG were used for the calculation of their M_w by the method of HPSEC-MALLS-RID.

2.2. The Starting Material NFG, NFG-2.0M and Their Molecular Weights

Six batches of NFG were extracted and purified from dried body wall of the sea cucumber *H. fuscopunctata* Jaeger. The yields of NFGs were about 0.85% by dry weight and the yields of another four batches of NFG-2.0M were about 1.20% by dry weight.

After being subjected to a Shodex Ohpak SB-804 HQ column, the HPSEC-MALLS-RID of NFG showed only one peak, while NFG-2.0M showed two peaks (Figure 2). As shown in Figure 2B, NFG-2.0M appeared to contain NFG (peak 2) and a small amount of sulfated fucan SF-II (peak 1) according to the studies reported previously [20]. The M_w and the polydispersity (M_w/M_n) of NFG and NFG-2.0M by HPSEC-MALLS-RID (only the NFG peak was involved in the calculation) are presented in Table 2. The M_w of NFG from different batches were determined to range from 58,270 to 74,280 Da, while that of NFG-2.0M were larger (from 77,680 to 86,950 Da). The M_w/M_n of NFG from different batches were determined to range from 1.105 to 1.241, while that of NFG-2.0M were higher (from 1.297 to 1.407). Furthermore, the labeled value (true value) of M_w of BSA was 66,430 Da, and the mean value of M_w from the determinations was 64,100 (the RSD was 0.67%, n = 5). The standard deviation is 3.51% of the true value of M_w, which indicates that the HPSEC-MALLS-RID has good accuracy and system suitability. The RSD of repeatability for NFG is 1.6% (n = 6), which suggests that the method for NFG has good repeatability. Meanwhile, the RSD of stability (0 h, 4 h, 8 h, 12 h, 24 h at room temperature) for NFG solutions was 1.27% (n = 5), which indicates that NFG was stable in 0.1 mol/L sodium nitrate (containing 0.02% sodium azide) at room temperature during the tested period. It was found that the M_w and the M_w/M_n of the NFG peak (peak 2) of the NFG-2.0M sample could be not precisely determined due to the relatively poor resolution of the column and the co-elution of the sulfated fucan SF-II (peak 1) [20].

Table 1. Measurement results of *dn/dc* values of LFG-Na and NFG using Optilab rEX refractometer.

Sample	Batch	*dn/dc* (mL/g)	Mean Value (mL/g)
LFG-Na			
	L1	0.1277	
	L2	0.1269	
	L3	0.1237	0.1248 ± 0.0046
	L4	0.1283	
	L5	0.1173	
NFG			
	N1	0.1168	
	N2	0.1154	
	N3	0.1187	
	N4	0.1113	0.1163 ± 0.0028
	N5	0.1162	
	N6	0.1192	
NFG-2.0M			
	NM1	0.1162	
	NM2	0.1172	
	NM3	0.1157	0.1166 ± 0.0008
	NM4	0.1173	

These data indicate that the purity and homogeneity of NFG were better than NFG-2.0M after the manufacturing process optimization. That is to say, the FPA98 ion-exchange resin elution process was the crucial step for ensuring the purity and homogeneity of NFG. As the starting material, NFG was more conducive to the quality control of LFG-Na. Therefore, the manufacturing process after optimization of NFG was selected. Based on multiple batches of measurement data, the acceptance criteria of the molecular weight of NFG were established. Considering that NFG was just the starting material of LFG-Na, the acceptance criteria could be relatively broad. Therefore, the M_w of NFG should be 50,000 to 10,000 Da, and $M_w/M_n < 1.50$.

2.3. Determination of Molecular Weight and Molecular Weight Distribution of LFG-Na Using HPSEC-MALLS-RID

HPSEC-MALLS-RID, an absolute method, has been proven as the powerful and efficient technique for analysis of the molecular weight and molecular weight distribution of polysaccharides in dilute polymer solution without using a series of standards [21]. In some countries, molecular weight has been considered as one of the indicators to control the quality of some drugs in pharmacopoeias, such as LMWH [22]. Therefore, the established HPSEC-MALLS-RID method was applied for the determination of the molecular weight and molecular weight distribution of LFG-Na.

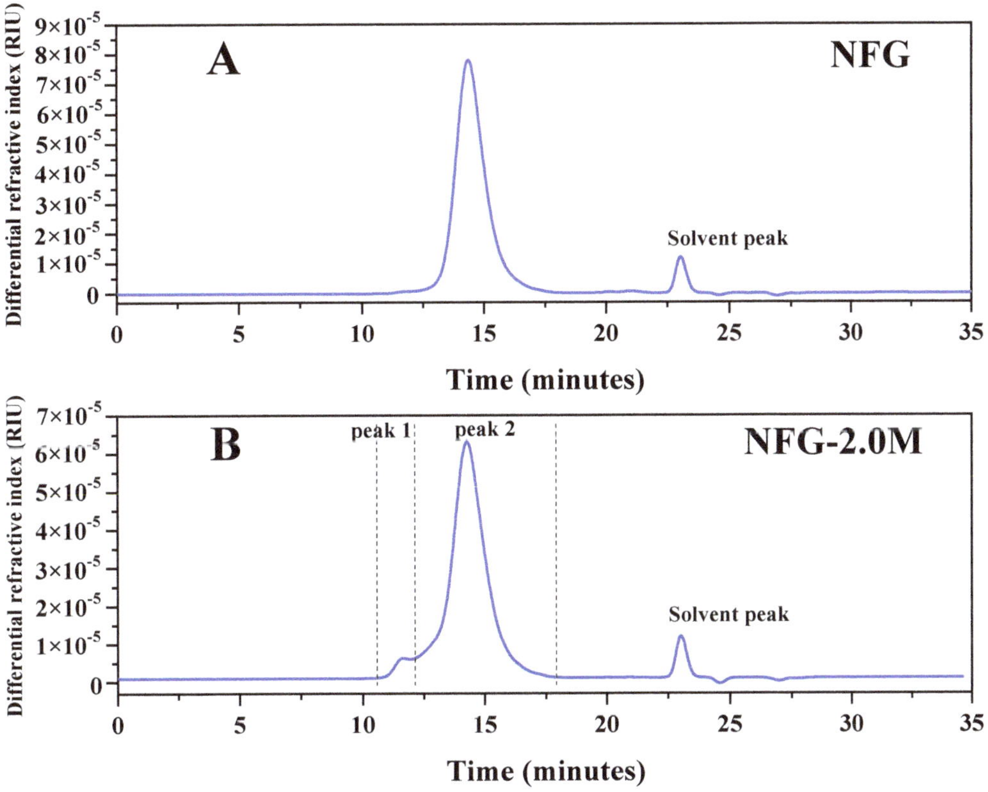

Figure 2. Representative HPSEC chromatogram of NFG (**A**) and NFG-2.0M (**B**) obtained before the NFG manufacturing process optimization from the determination of HPSEC-MALLS-RID.

Table 2. The M_w and the M_w/M_n of NFG and NFG-2.0M by HPSEC-MALLS-RID (only the NFG peak was involved in the calculation).

Sample	Batch	M_w(Da)	M_w/M_n
NFG			
	N1	60,000	1.184
	N2	70,940	1.239
	N3	61,350	1.180
	N4	58,270	1.175
	N5	74,280	1.241
	N6	64,260	1.105
NFG-2.0M			
	NM1	84,840	1.407
	NM2	86,950	1.360
	NM3	77,680	1.297
	NM4	84,320	1.360

Five batches of LFG-Na were prepared from NFG by β-eliminative depolymerization. The representative HPSEC-MALLS-RID chromatogram of LFG-Na is shown in Figure 3. Only one fraction of glycosaminoglycan was detected (dRI), whereas the peak of retention time from 22 to 24 min in the HPSEC-MALLS-RID chromatogram was the solvent peak. The result showed that LFG-Na was distributed from 15 to 19 min. Since there was no peak between the retention time of 10 to 15 min in the RI chromatogram (dRI), even if a peak was found in the MALLS chromatogram (LS), it could be considered that there was no substance from 10 to 15 min based on the different detection principles of MALLS and RI detectors.

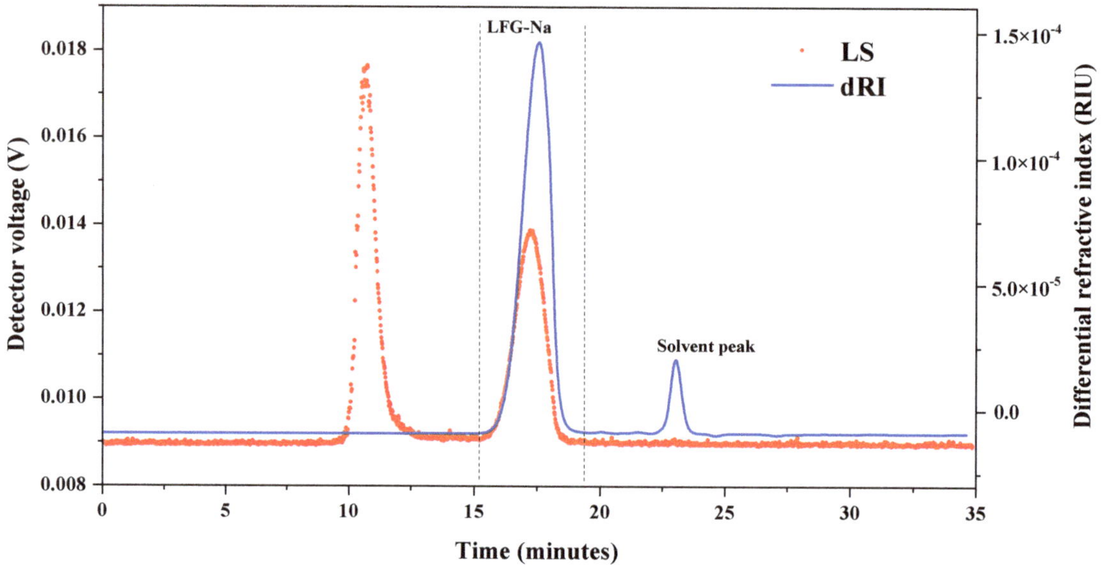

Figure 3. Representative HPSEC-MALLS-RID chromatogram of LFG-Na.

Using the Astra software (version 7.1.3, Wyatt Technology Co., Santa Barbara, CA, USA)), the results of molecular weight and molecular weight distribution for LFG-Na by HPSEC-MALLS-RID are summarized in Table 3. The M_w of LFG-Na from different batches were determined to range from 4596 to 5708 Da; it could be considered that the M_w of LFG-Na from different batches were similar. Sample L1 was observed as the lowest M_w among the test samples, while sample L5 was the highest one. The true value of M_w of BSA was 66,430 Da, and the mean value of M_w from the determinations was 64,100 (the RSD was 0.67%, n = 5). The standard deviation was 3.51% of the true value of M_w, which indicated that the HPSEC-MALLS-RID had good accuracy and system suitability. The RSDs of repeatability and intermediate precision for the M_w of LFG-Na were 2.13% (n = 6), and 2.87% (different days, n = 12), respectively. The data suggest that the HPSEC-MALLS-RID for LFG-Na has good accuracy, system suitability, repeatability and intermediate precision. The molecular weight distribution is used to measure the width of the molecular weight distribution, which represents the homogeneity and dispersibility of the polysaccharides. The M_w/M_n of LFG-Na from different batches were determined to range from 1.121 to 1.240, which suggest that the molecular weight distribution of each batch of LFG-Na was relatively narrow. Furthermore, the percentages of the fractions of LFG-Na with different molecular weights were obtained (Table 3). From the data of five batches of LFG-Na, the percentages of the fraction with molecular weights range from 1500 to 8000 Da (M1500~8000) were all more than 85%, and the percentages of the fraction with molecular weights greater than 8000 Da (M8000) were all less than 15%, while there was no fraction with molecular weights lower than 1500 Da (M1500).

Table 3. Test results of molecular weight and molecular weight distribution for LFG-Na by HPGPC and HPSEC-MALLS-RID.

Batch	HPGPC					HPSEC-MALLS-RID				
	Molecular Parameters		Molecular Weight Distribution			Molecular Parameters		Molecular Weight Distribution		
	M_w (Da)	M_w/M_n	M1500	M1500~8000	M8000	M_w (Da, ±Error)	M_w/M_n (±Error)	M1500	M1500~8000	M8000
L1	5657	1.438	1.25%	85.34%	13.41%	4596 (±0.546%)	1.141 (±0.897%)	0	94.192%	5.808%
L2	5241	1.281	0.57%	86.80%	12.63%	4629 (±0.519%)	1.121 (±0.851%)	0	94.539%	5.461%
L3	5227	1.280	0.53%	86.60%	12.86%	4870 (±0.583%)	1.128 (±0.921%)	0	93.024%	6.976%
L4	5082	1.387	2.52%	82.91%	14.57%	5361 (±0.479%)	1.240 (±0.699%)	0	87.329%	12.671%
L5	5531	1.420	1.91%	81.48%	16.61%	5708 (±0.541%)	1.204 (±0.854%)	0	85.055%	14.945%

M1500: the percentage of the fraction of LFG-Na with molecular weights lower than 1500 Da; M1500~8000: the percentage of the fraction of LFG-Na with molecular weights range from 1500 to 8000 Da; M8000: the percentage of the fraction of LFG-Na with molecular weights greater than 8000 Da.

Based on multiple batches of measurement data, the acceptance criteria of the molecular weight and molecular weight distribution of LFG-Na were established. The M_w of LFG-Na should be 4000 to 6000 Da, $M_w/M_n < 1.40$, the fraction with molecular weights of 1500 to 8000 Da should be no less than 80% of the total, and the fraction with molecular weights greater than 8000 should be no more than 20% of the total.

2.4. Determination of Molecular Weight and Molecular Weight Distribution of LFG-Na Using HPGPC

Size exclusion chromatography has been widely employed for separation of polysaccharides and the molecular weight and molecular weight distribution could be determined by HPGPC with suitable standards. Five batches of LFG-Na were also determined by HPGPC with five oligosaccharide standards HS5, HS8, HS11, HS14 and HS17. The HPGPC profiles of the representative batch of LFG-Na and oligosaccharide standards (Figure 4A) show that LFG-Na is the mixture of oligosaccharides with different degrees of polymerization.

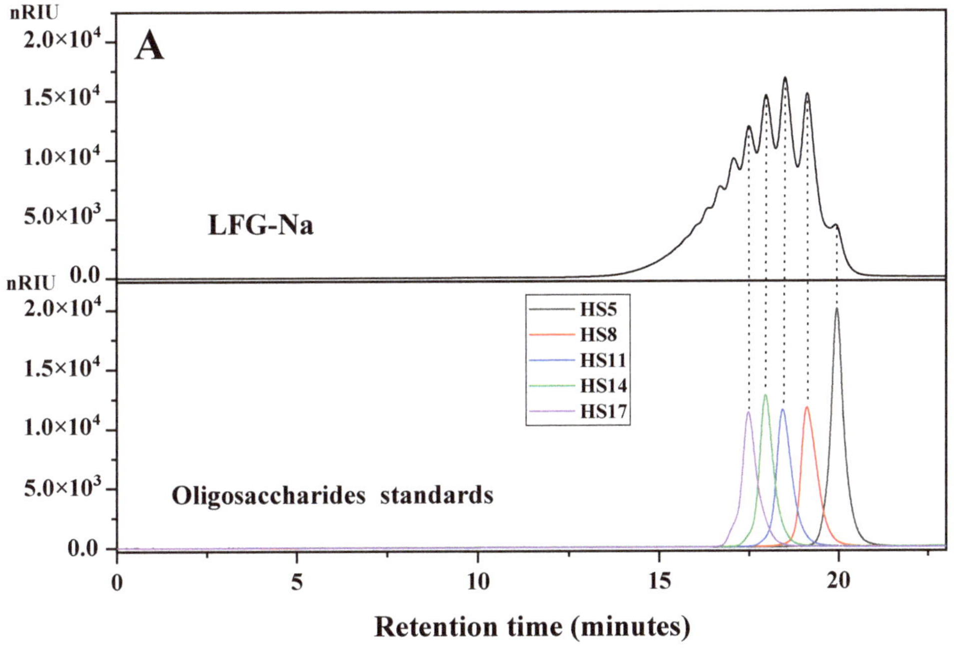

Figure 4. *Cont.*

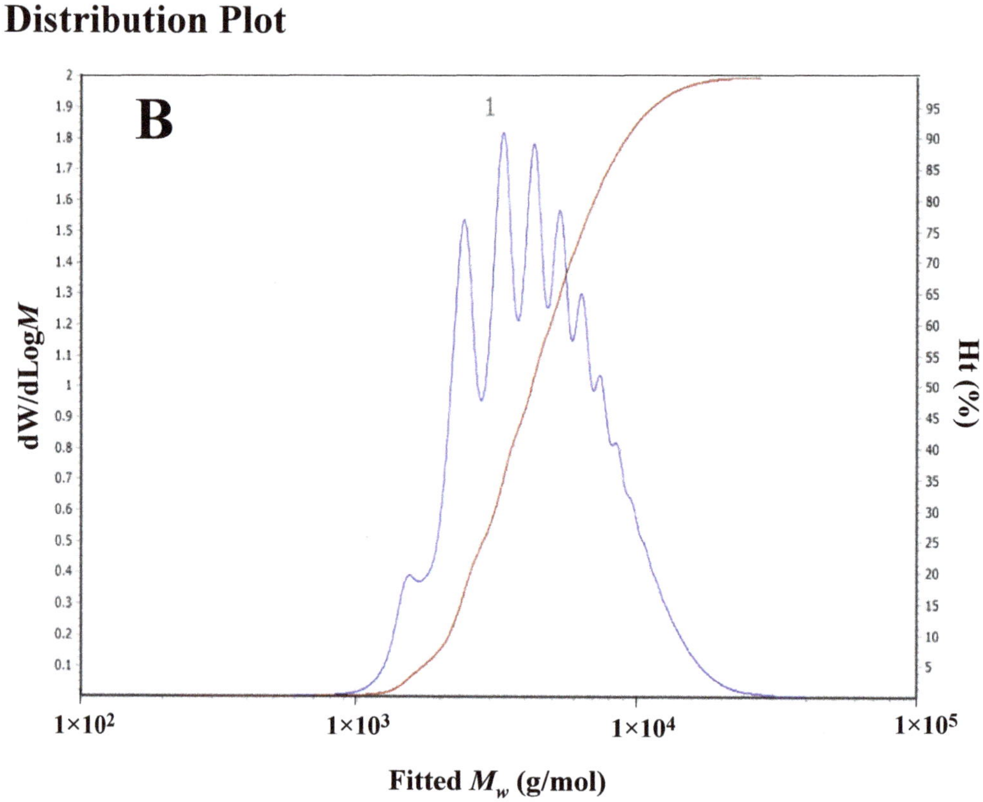

Figure 4. HPGPC profiles of LFG-Na and oligosaccharide standards HS5, HS8, HS11, HS14, HS17, which were determined by the differential refractive index detector (**A**). The distribution plot of LFG-Na was analyzed by using Agilent GPC/SEC software (version A.02.01, Agilent Technologies, Palo Alto, CA, USA) (**B**).

The M_w of LFG-Na was calculated according to the calibration curve from the data of oligosaccharide standards and the distribution plot of LFG-Na (Figure 4B) was analyzed using Agilent GPC/SEC software (version A.02.01, Agilent Technologies, Palo Alto, CA, USA). The results of molecular weight and molecular weight distribution for LFG-Na by HPGPC were also summarized in Table 3. The M_w of LFG-Na from different batches were determined to range from 5082 to 5657 Da. Sample L1 was observed as having the highest M_w among the test samples, while sample L4 had the lowest one. The M_w/M_n from different batches were determined to range from 1.280 to 1.438. From the data of five batches of LFG-Na by HPGPC, the percentages of the fraction of M1500, M1500~8000, and M8000 were <3%, >80% and <17%, respectively. The above results suggest that the molecular weight and molecular weight distribution of LFG-Na from different batches were similar.

2.5. Comparison of the Determination Using HPGPC and HPSEC-MALLS-RID

The analysis results of molecular weight and molecular weight distribution for LFG-Na using the developed HPSEC-MALLS-RID method were also compared with those of HPGPC analysis. As shown in Table 3 and Figure 5, the M_w of LFG-Na from different batches using the HPSEC-MALLS-RID method were relatively close to those of HPGPC analysis (Figure 5A), though the measured value of Sample L1 had a largest absolute error of 1061 Da as compared with that of the HPGPC method (4596 Da vs. 5657 Da), while the measured value of Sample L5 had a smaller absolute error of 177 Da (5708 Da vs. 5531 Da). However, Figure 5B shows that the M_w/M_n of LFG-Na from different batches using HPSEC-MALLS-RID were all obviously smaller than those obtained by HPGPC (average of 1.17 vs. 1.36). There were slight differences in the percentages of the fraction of M1500, M1500~8000 and M8000 between the two methods. The percentages of the fraction of M1500~8000 were higher by the HPSEC-MALLS-RID method.

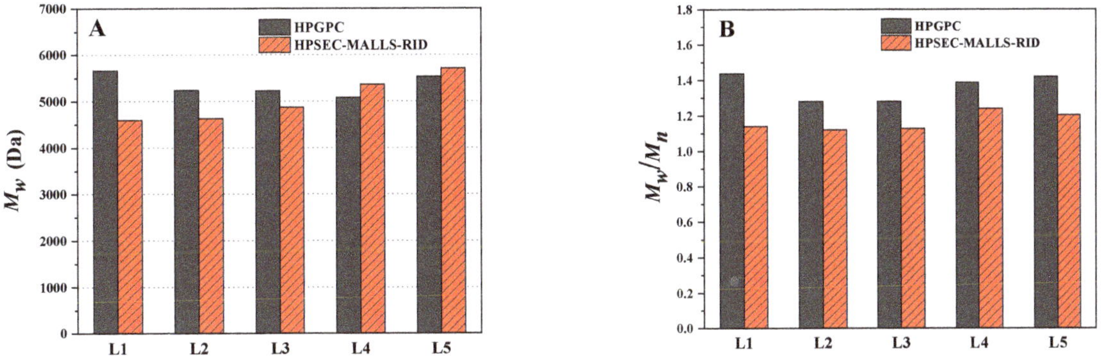

Figure 5. The results of molecular weight (**A**) and molecular weight distribution (**B**) for LFG-Na by HPGPC and HPSEC-MALLS-RID.

In conclusion, both HPSEC-MALLS-RID and HPGPC methods for the determination of LFG-Na included size exclusion chromatography, but were calculated in different ways. HPGPC was widely used and convenient, but the choice of different standards apparently affected the molecular weight result and the standards' structures should be similar to the samples for obtaining accurate and reliable results. In this study, because the highest molecular weight of oligosaccharide standards was 5328 Da (HS17), which did not cover the maximum molecular weight range of LFG-Na, the accuracy of HPGPC was poor. In addition, in view of the current technical means, it was temporarily impossible to obtain a series of narrow oligosaccharide standards that completely covers the maximum molecular weight range of LFG-Na due to the complicated and laborious procedures for their separation and purification. Compared to HPGPC, the method of HPSEC-MALLS-RID based on the *dn/dc* did not require the reference substances to establish the calibration curve, and the molecular weight and molecular weight distribution of NFG and its depolymerized fraction LFG-Na could be determined directly, efficiently and accurately. The limitation of the HPSEC-MALLS-RID method is that the *dn/dc* value of the sample should be determined or obtained before determining the molecular weight and molecular weight distribution. In addition, the accuracy and system suitability of the method should be assessed by standards, such as BSA or Dextran, which were used at the beginning of each run sequence.

2.6. Comparison of the Chemical Characteristics of LFG-Na

The monosaccharide composition analysis with the method of PMP (1-Phenyl-3-methyl-5-pyrazolone) precolumn derivatization-HPLC showed that LFG-Na was composed of glucuronic acid, N-acetyl galactosamine, and fucose (Figure 6). Five batches of LFG-Na (L1-L5) were also analyzed by the Similarity Evaluation System for Chromatographic Fingerprint of TCM software (Version 2012, Chinese Pharmacopoeia Commission, Beijing, China). The HPLC fingerprint similarities were more than 0.99, which showed the perfect correlation and similarity among them.

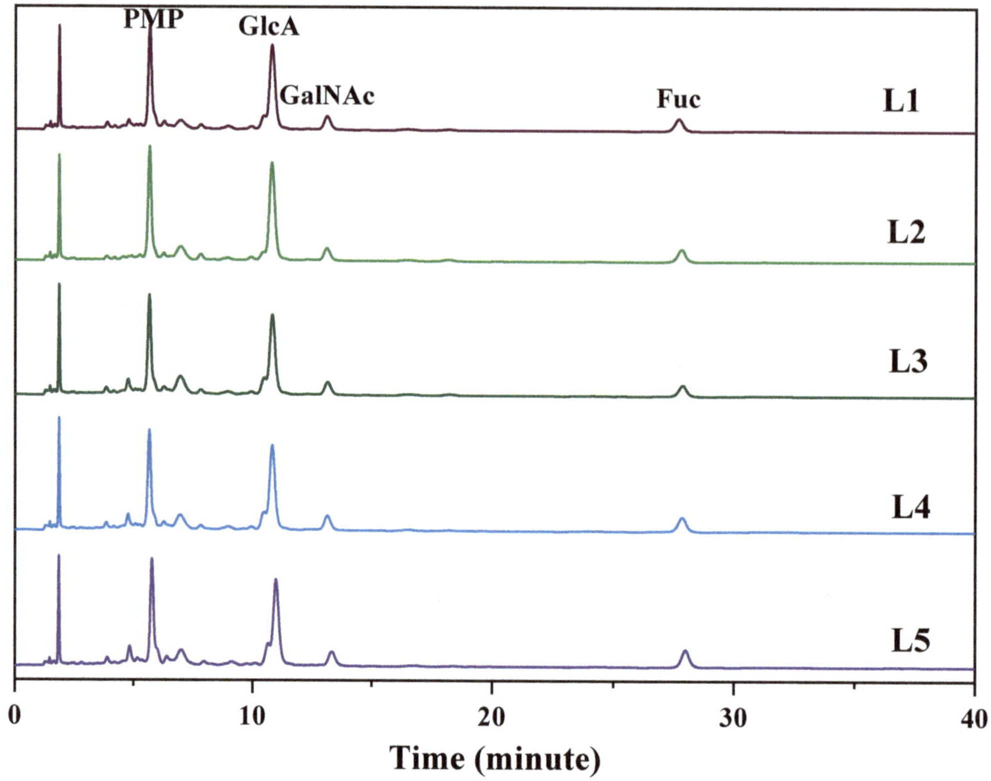

Figure 6. HPLC profiles of PMP derivatives from LFG-Na (L1–L5).

NMR spectroscopy was performed using a 600 MHz spectrometer, and the $^{1}H/^{13}C$ NMR chemical shifts and NMR spectra of LFG-Na were analyzed. The complete assignment of its signal peaks is given in Table 4. In the ^{1}H NMR spectrum, a strong signal at 5.760 ppm could be assigned to the proton H-4 of Δ4,5-unsaturated glucuronic acid (ΔU). There were three relatively strong signal peaks (at 5.356, 5.282, 5.116 ppm) from 5.0 to 5.6 ppm, which were the α-fucose terminal proton signals at different sites with the sulfation type of Fuc_{3S4S}. It could be judged in combination with the other spectra that the signals at 5.356, 5.282, and 5.116 ppm are connected to the glucuronic acid, the unsaturated glucuronic acid at the non-reducing end, and Fuc_{3S4S} terminal hydrogen of the GlcA-ol at the reducing end, respectively. The β-terminal proton signals of the main chain appeared at about 4.3 to 4.6 ppm and GlcA-ol was located at the reducing end with the proton H-1 (CH_2) at 3.800 and 3.761 ppm. Fuc methyl proton signals showed different chemical shifts due to their different locations. Specifically, the Fuc methyl protons at the reducing end (rF) and non-reducing end (dF) occurred at 1.30 to 1.35 ppm, Fuc in the sugar chain, connecting with glucuronic acid (F), occurred at about 1.4 ppm; and GalNAc acetyl protons occurred at about 1.9 to 2.1 ppm. In the ^{13}C NMR spectrum, the C-1 (dU and dA) peaks of GlcA and GalNAc at the non-reducing end of the main chain appeared at about 105.91 and 102.51 ppm, the C-1 (U and A) peaks of GlcA and GalNAc in the middle of the main chain appeared at about 106.71, and 102.34 ppm, and the C-1 (rU and rA) peaks of glucuronic acid alcohol and GalNAc at the reducing end appeared at 65.40 and 104.33 ppm, indicating the β-configuration of the bridgehead hydrogen was consistent with the GlcA-ol at the reducing end. C-1 of Fuc connected to different sites also showed different chemical shifts (dF:101.15 ppm, F:102.06 ppm and rF:104.33 ppm). C-3, C-4, and C-5 of ΔU at the non-reducing end appeared at 79.90, 109.54, and 149.79 ppm, respectively. With C-6 (rU) of GlcA-ol at reducing end and C-6 (ΔU) at non-reducing end ΔU appearing at 180.22 and 171.76 ppm as the exception, the carbonyl carbon C-6 (U6) in the GlcA carboxyl and the carbonyl carbon C-7 (rA/A/dA) in the GalNAc acetamido appeared at almost the same site (~177 ppm); Fuc methyl carbon signals appeared at about 18 to 19 ppm; the methyl carbon signal in GalNAc acetamido appeared at about 25 ppm. The C-2 signal of GalNAc appeared at about 54 ppm. The structure of LFG-Na is confirmed as shown in Figure 7A, which is consistent with our previous reports [10,12].

Table 4. NMR data and assignments of LFG-Na (Batch L1).

Sugar Ring	H	δ [a]	^{1}H-^{1}H Couplings [b]	Correlated Signals			C	δ	Correlated Signals	
				COSY	TOCSY	ROESY			HSQC	HMBC
rU	H-1	3.800	$J_{(1,1')}$ = 11.33	H1′, H2	H1′, H2, H3/4	H2, H3	C-1	65.40	H-1	H-1
	H-1′	3.761	$J_{(1,2)}$ = 6.12	H1, H2	H1, H2, H3/4	H3			H-1′	H-1′
	H-2	4.146	$J_{(1',2)}$ = 3.96	H1/1′, H3	H1/1′, H3/4	H1/1′, H3	C-2	72.67	H-2	–
	H-3	4.072	–	H2	H1/1′, H2	H2; rF1	C-3	82.54	H-3	–
	H-4	4.052	–	H5	H1/1′, H2	–	C-4	82.54	H-4	–
	H-5	4.329	–	H4	H4	–	C-5	74.78	H-5	–
	–	–	–	–	–	–	C-6	180.22	–	H5
rF	H-1	5.116	$J_{(1,2)}$ = 4.02	H2	H2, H3, H4	H2; rU3	C-1	104.33	H-1	H1, H5; rU3
	H-2	3.885	–	H1, H3	H1, H3, H4	H1, H3, H4	C-2	69.36	H-2	H2, H3, H4
	H-3	4.610	$J_{(3,4)}$ =2.94	H2, H4	H1, H2, H4	H2, H3, H4, H5	C-3	78.10	H-3	H1, H2, H4
	H-4	4.909	–	H3	H1, H2, H3	H3, H5, H6	C-4	81.83	H-4	H3, H5
	H-5	4.428	–	H6	H6	H3, H4, H6	C-5	69.71	H-5	H1, H4
	H-6	1.359	–	H5	H5	H4, H5	C-6	18.94	H-6	H4, H5, H6
rA	H-1	4.695	$J_{(1,2)}$ = 7.04	H2	H2, H3	H2, H3, H5; rU4	C-1	104.33	H-1	rU4
	H-2	3.974	–	H1, H3	H1, H3, H4	H3, H1	C-2	54.22	H-2	H3, H4
	H-3	4.173	–	H2, H4	H1, H2, H4	H1, H2, H4; U1	C-3	78.74	H-3	H1, H4; U1
	H-4	4.829	–	H3	H2, H3, H5	H3, H5, H6	C-4	79.25	H-4	H3, H5, H6′
	H-5	4.092	–	H6, H6′	H4, H6/6′	H1, H4, H6/6′	C-5	75.06	H-5	H6/6′
	H-6	4.276	–	H5, H6′	H5, H6′	H4, H5, H6′	C-6	70.52	H-6/6′	H5
	H-6′	4.199	–	H5, H6	H5, H6	H4, H5, H6	C-7	177.97	–	H2, H8
	H-8	2.090	–	–	–	–	C-8	25.26	H-8	H8
U	H-1	4.491	$J_{(1,2)}$ = 8.88	H2	H2, H3, H4, H5	H2, H3, H5; rA/A3	C-1	106.71	H-1	H2, H3; A3
	H-2	3.616	$J_{(2,3)}$ = 7.74	H1, H3	H1, H3, H4, H5	H4, H1	C-2	76.45	H-2	H2, H3
	H-3	3.690	–	H2, H4	H1, H2, H4, H5	H1, H2, H4; F1	C-3	82.00	H-3	H2, H4, H5; F1
	H-4	4.021	–	H3, H5	H1, H2, H3, H5	H2, H3/5; A1	C-4	78.07	H-4	H3, H5
	H-5	3.706	–	H4	H1, H2, H3, H4	H1, H3	C-5	79.84	H-5	H4, H5
	–	–	–	–	–	–	C-6	177.87	–	H4, H5
A	H-1	4.485	$J_{(1,2)}$ = 8.52	H2	H2, H3	H2, H3, H5; U4	C-1	102.34	H-1	H2; U4
	H-2	4.021	–	H1, H3	H1, H3, H4	H3, H1	C-2	54.24	H-2	H3, H4
	H-3	3.895	–	H2, H4	H1, H2, H4	H1, H2, H4; U1	C-3	79.11	H-3	H1, H4; U1
	H-4	4.751	–	H3	H2, H3, H5	H3, H5, H6	C-4	78.90	H-4	H3, H5, H6′

Table 4. *Cont.*

Sugar Ring	H	δ [a]	^{1}H-^{1}H Couplings [b]	Correlated Signals			C	δ	Correlated Signals	
				COSY	TOCSY	ROESY			HSQC	HMBC
	H-5	3.932	–	H6, H6′	H4, H6/6′	H1, H4, H6/6′	C-5	74.62	H-5	H6/6′
	H-6	4.287	–	H5, H6′	H5, H6′	H4, H5, H6′	C-6	69.88	H-6/6′	H5
	H-6′	4.183	–	H5, H6	H5, H6	H4, H5, H6	C-7	177.84	–	H2, H8
	H-8	1.998	–	–	–	–	C-8	25.38	H-8	H8
F	H-1	5.356	$J_{(1,2)}$ = 3.72	H2	H2, H3, H4	H2, H3, H4, H5; U3	C-1	102.06	H-1	H3; U3
	H-2	3.950	$J_{(2,3)}$ = 10.08	H1, H3	H1, H3, H4	H1, H3, H4	C-2	69.24	H-2	H2, H3, H4
	H-3	4.521	–	H2, H4	H1, H2, H4	H2, H3, H4, H5	C-3	78.28	H-3	H1, H2, H4
	H-4	5.040	–	H3	H1, H2, H3	H3, H5, H6	C-4	82.20	H-4	H3, H5
	H-5	4.849	–	H6	H6	H3, H4, H6	C-5	69.28	H-5	H1, H4
	H-6	1.404	–	H5	H5	H4, H5	C-6	18.89	H-6	H4, H5, H6
dA	H-1	4.571	$J_{(1,2)}$ = 8.52	H2	H2, H3	H2, H3, H5, U4	C-1	102.51	H-1	H2; U4
	H-2	4.163	–	H1, H3	H1, H3, H4	H3, H1	C-2	54.37	H-2	H3, H4
	H-3	4.166	–	H2, H4	H1, H2, H4	H1, H2, H4, dA1	C-3	78.69	H-3	H1, H4; dU1
	H-4	4.988	–	H3	H2, H3, H5	H3, H5, H6	C-4	78.91	H-4	H3, H5, H6′
	H-5	4.088	–	H6, H6′	H4, H6/6′	H1, H4, H6/6′	C-5	74.99	H-5	H6/6′
	H-6	4.371	–	H5, H6′	H5, H6′	H4, H5, H6′	C-6	70.16	H-6/6′	H5
	H-6′	4.263	–	H5, H6	H5, H6	H4, H5, H6	C-7	178.04	–	H2, H8
	H-8	2.065	–	–	–	–	C-8	25.38	H-8	H8
ΔU	H-1	4.922	$J_{(1,2)}$ = 7.20	H2	H2, H3, H4	H1, H3, H4, dA3	C-1	105.91	H-1	H1, H2; dA3
	H-2	3.911	$J_{(2,3)}$ = 8.22	H1, H3	H1, H3, H4	H3	C-2	73.10	H-2	H2, H3, H4
	H-3	4.491	–	H2, H4	H1, H2, H4	H1, H2, H4, dF1	C-3	79.90	H-3	H2; dF1
	H-4	5.760	–	H3	H1, H2, H3	H1, H3	C-4	109.54	H-4	H3, H4
	–	–	–	–	–	–	C-5	149.79	–	H4, H3
	–	–	–	–	–	–	C-6	171.76	–	H4
dF	H-1	5.282	$J_{(1,2)}$ = 3.96	H2	H2, H3, H4	H2, dU3	C-1	101.15	H-1	H5; dU3
	H-2	3.951	$J_{(2,3)}$ = 10.88	H1, H3	H1, H3, H4	H1, H3, H4	C-2	69.24	H-2	H2, H3, H4
	H-3	4.610	$J_{(3,4)}$ = 3.06	H2, H4	H1, H2, H4	H2, H3, H4, H5	C-3	78.10	H-3	H1, H2, H4
	H-4	4.909	–	H3	H1, H2, H3	H3, H5, H6	C-4	81.83	H-4	H3, H5
	H-5	4.356	–	H6	H6	H3, H4, H6	C-5	70.03	H-5	H1, H4
	H-6	1.301	–	H5	H5	H4, H5	C-6	18.71	H-6	H4, H5, H6

a δ: Chemical shift, ppm, internal standard: deuteration TSP. b J: Coupling constant, Hz.

A

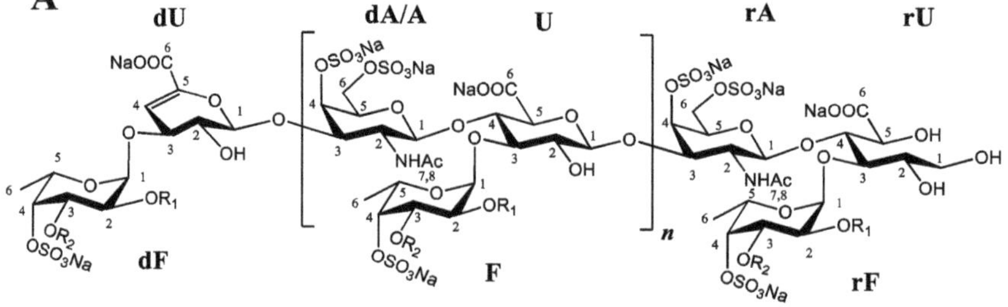

The mean value of ***n*** was about 4.

R_1: -H (~95%) or -SO_3Na (~5%); R_2: -SO_3Na (~85%) or -H (~15%); R_1 and R_2 should not be -SO_3Na simultaneously.

Figure 7. *Cont.*

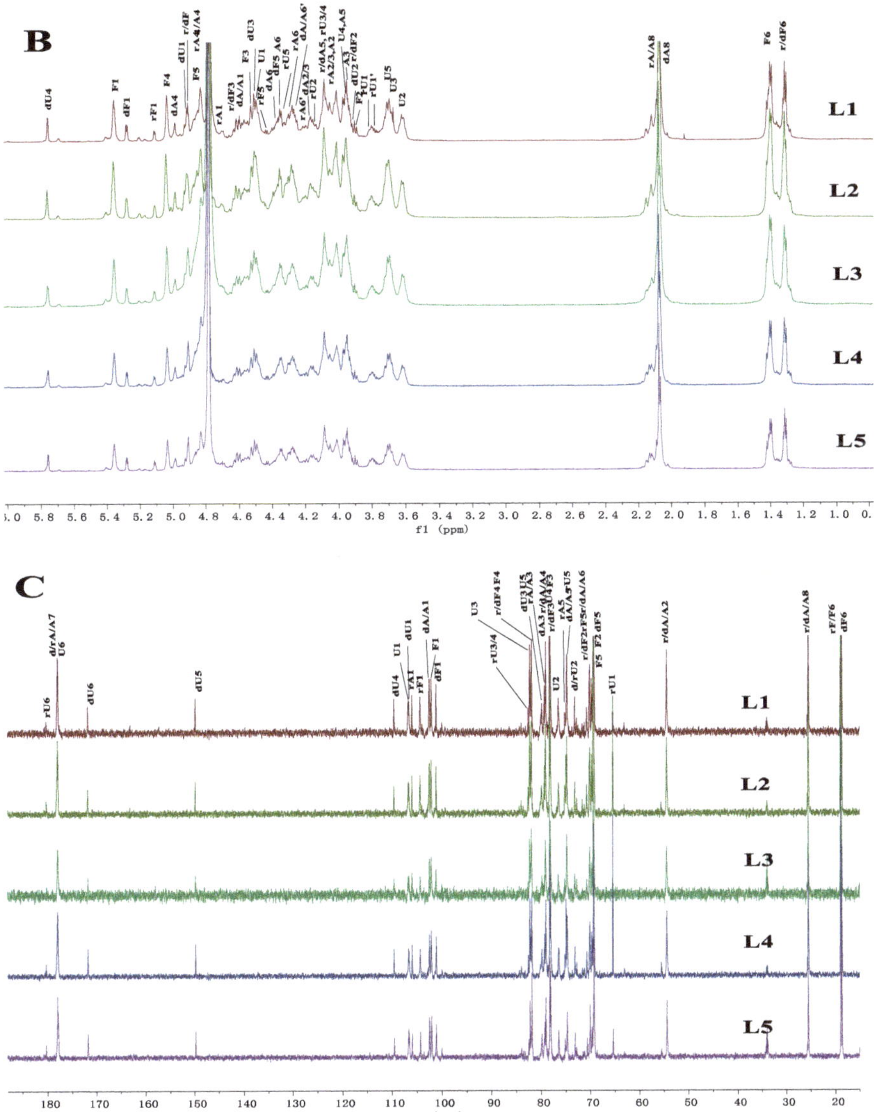

Figure 7. The structure of LFG-Na (**A**), 1D ^{1}H NMR spectra (**B**) and ^{13}C NMR spectra (**C**) from five batches of LFG-Na (L1–L5). The letters of (**A**) between parentheses are used as labels for assignments in the ^{1}H-NMR (**B**) and ^{13}C-NMR spectrum (**C**). They are the following: dU for the nonreducing terminal unsaturated uronic acid residue; U for internal glucuronic acid; rU for reducing terminal alcohol; dA for GalNAc$_{4S6S}$ linked to dU; A for internal GalNAc$_{4S6S}$; rA for GalNAc$_{4S6S}$ linked to rU; dF for Fuc$_{3S4S}$ residue linked to dU; F for Fuc$_{3S4S}$ residue linked to U; rF for Fuc$_{3S4S}$ residue linked to rU. 1D ^{1}H (**B**) and ^{13}C (**C**) NMR spectra are from five batches of LFG-Na (L1–L5).

As shown in Figures 6 and 7B,C, HPLC profiles of PMP derivatives and 1D (^{1}H/^{13}C) NMR spectra with the typical signal assignments from five batches of LFG-Na (L1-L5) were obtained. After comparing and evaluating carefully, the chemical characteristics of these five batches of LFG-Na were highly similar to each other.

2.7. Comparison of Biological Potency

The activity of FG derived from various sea cucumber species has been reported [10,23]. Herein, the APTT (activated partial thromboplastin time) assay was selected and the potency for anti-iXase, referring to the PLA (parallel line assay) method, was established. The APTT-prolonging activity of LFG-Na in human plasma indicated the potent inhibitory activity on the intrinsic coagulation pathway. The inhibition of APTT by LFG-Na was increased depending on the increasing concentration of LFG-Na in a certain range, and the bioactivity could be measured by measuring the concentrations of LFG-Na required to double APTT. The PLA method for estimating the potency for anti-iXase was established and validated with respect to specificity, linearity and range, repeatability, intermediate precision, accuracy and durability in our laboratory.

The activities of multiple batches of samples of LFG-Na are summarized in Table 5. Concentrations required to double the APTT of five batches samples were from 7.36 to 8.89 μg/mL. Sample L4 and L5 with higher M_w showed slightly stronger anticoagulant activity than L1-L3. In order to compare the potency for anti-iXase among different batches of LFG-Na, L5 with the highest M_w was selected as the reference substance, and the potency for anti-iXase was set as 100 U/mg. The results of potency for anti-iXase also indicated that L4 and L5 had slightly stronger activities of anti-iXase, whereas potency for anti-iXase of L1-L3 were 88.7, 85.9 and 89.8 U/mg. Basically, different batches of LFG-Na with the M_w ranging from 4000 to 6000 Da had similar activities, and the results of bioactivity assays in vitro indicated the stability and consistency of the preparation process of LFG-Na. Overall, the bioactivity assays also could be listed as the quality control methods of multi-component drugs to ensure the consistency of their efficacy in clinical use.

Table 5. Bioactivities and molecular weights of LFG-Na.

Batch	M_w [a] (Da)	APTT [b] (μg/mL)	Anti-iXase [c] (U/mg)
L1	4596	8.89 ± 0.08	88.7
L2	4629	8.84 ± 0.03	85.9
L3	4870	8.64 ± 0.06	89.8
L4	5361	7.92 ± 0.14	94.1
L5	5708	7.36 ± 0.11	100

[a] M_w data are from the result of HPSEC-MALLS-RID. [b] Concentrations required to double the clotting time ($n = 2$). [c] Sample L5 of LFG-Na was selected as the reference substance, and the potency for anti-iXase was set as 100 U/mg.

3. Materials and Methods

3.1. Materials and Chemicals

Dried sea cucumbers *H. fuscopunctata* Jaeger were purchased from local markets in Guangdong Province and Hainan Province, China. Five oligosaccharide standards from LFG-Na (HS5, pentasaccharide, M_w = 1506 Da; HS8, octasaccharide, M_w = 2462 Da; HS11, hendecasaccharide, M_w = 3417 Da; HS14, tetradecasaccharide, M_w = 4373 Da; HS17, heptadecasaccharide, M_w = 5328 Da) were obtained from Kunming Institute of Botany, Chinese Academy of Sciences. BSA standard was purchased from Sigma (St. Louis, MO, USA). The monosaccharides including L-fucose and *N*-acetyl-D-galactosamine were purchased from Sigma-Aldrich, and D-glucuronic acid was from J&K Scientific Ltd. (Beijing, China). 1-Phenyl-3-methyl-5-pyrazolone (PMP) was purchased from Xiya Reagnent Co. (Linyi, China). Amberlite FPA98Cl ion-exchange resin was purchased from Rohm and Haas Company (Philadelphia, PA, USA). Deuterium oxide (D_2O, 99.9% Atom D) was obtained from Sigma-Aldrich. The activated partial thromboplastin time (APTT) kits, $CaCl_2$ and standard human plasma were purchased from MDC Hemostasis (Neufahrn, Germany), BIOPHEN FVIII: C kit was from Hyphen Biomed (Neuville sur Oise, France). Human factor VIII was from Bayer HealthCare LLC. Deionized water was prepared by the Millipore Milli Q-Plus system (Millipore, Billerica, MA, USA). All other reagents were of analytical grade and obtained commercially.

3.2. Preparation of the Starting Material NFG

The NFG was extracted and purified from the sea cucumber *H. fuscopunctata* using a previously described procedure with minor modifications [12,24]. Briefly, the tissue of the dried body wall of the sea cucumber (30 kg) was digested by 1% papain (6000 U/mg) for 6 h and the polysaccharide components were released by 0.25 mol/L sodium hydroxide for 2 h. The crude polysaccharide was obtained by repeated salting-out with KOAc and precipitation by ethanol, which was further purified by strong anion-exchange chromatography using FPA98 ion-exchange resin (30 cm × 160 cm, 100 L resin), and the sample was eluted sequentially with distilled water, 0.5, 1.0 and 1.3 mol/L NaCl solutions. The fraction eluted by 1.3 mol/L NaCl solutions was collected, desalted by a Pellicon device with a 0.5 m^2 PES membrane cassette with a molecular weight cutoff (MWCO) of 10 kDa (Millipore) and lyophilized to obtain white powder. The NFG was the starting material, which was further used for chemical cleavage to prepare its low molecular weight fragments named 'LFG-Na'. Before expanding to pilot scale, six batches of NFG were obtained from the dried body wall of the sea cucumber (30 kg) according to the above procedures, based on which the FPA98 ion-exchange resin elution process was optimized.

Before the NFG manufacturing process optimization, we obtained another four batches of NFG (named 'NFG-2.0M'), which were from the fraction eluted by 2.0 mol/L NaCl solutions during the FPA98 ion-exchange resin elution process. In other words, after going through the processes of enzymolysis, alkaline hydrolysis, salting-out and alcohol precipitation, the crude polysaccharide was eluted sequentially with distilled water, 0.5, 1.0 and 2.0 mol/L NaCl solutions; then the fraction eluted by 2.0 mol/L NaCl solutions was collected during the FPA98 ion-exchange resin elution process.

3.3. Preparation of LFG

LFG-Na was subjected to glycosidic bond-selective β-eliminative cleavage of the starting material NFG through its activated benzyl ester derivative according to previous research with minor modifications [12,25]. The process of LFG-Na mainly included the following steps: quaternization, carboxyl esterification, elimination and depolymerization, saponification, terminal group reduction, ultrafiltration and freeze-drying. Briefly describing the preparation process of one batch of LFG-Na (batch L3), approximately 200 g of NFG was dissolved in 2.96 L of distilled water and transalification was completed with benzethonium salts. NFG benzethonium salts were obtained by precipitation and centrifugation and dried under vacuum conditions. The NFG benzethonium salts were dissolved in 2970 mL dimethyl formamide and esterified by 63.6 mL benzyl chloride under continual stirring at 35 °C for 24 h. Then the solution was cooled to 25 °C and depolymerized by adding 1010 mL freshly prepared 0.08 mol/L EtONa in ethanol and incubating for 0.5 h. The transalification of benzethonium salt to sodium salts was completed by adding about 4.1 L saturated sodium chloride solution and 41.2 L of ethanol successively to the reaction solution. The saponification procedure in alkaline solution was necessary to hydrolyze the residual benzyl esters, and the reducing ends were reduced to its alcoholic hydroxyl by $NaBH_4$ (31.9 g). After adjusting pH and precipitation by ethanol, the crude products of low molecular weight were obtained. A tangential flow ultrafiltration method with the MWCO of 3 or 10 kDa (Millipore) was selected for further purification. The fractions with M_w higher than 3 kDa and lower than 10 kDa were collected and freeze-dried; finally, about 59.0 g LFG-Na was obtained.

3.4. HPSEC-MALLS-RID Analysis

3.4.1. *dn/dc* Measurement of NFG and LFG-Na

The *dn/dc* value of NFG and LFG-Na were measured by using a refractive index detector (RID, Optilab rEX refractometer, DAWN EOS, Wyatt Technology Co., Santa Barbara, CA, USA). NFG or LFG-Na samples were dissolved in 0.1 mol/L sodium nitrate (containing 0.02% sodium azide), which the concentrations were about 1.8 mg/mL. From the solution, a series of solutions with six different concentrations (about 0.3, 0.6, 0.9, 1.2, 1.5, and 1.8 mg/mL) were prepared by successive dilutions. For the *dn/dc* measurements, the mobile phase was the same solvent used for the preparation of the solutions. The series of different concentrations were injected for each sample solution using a manual injector. The RID signals were analyzed and the *dn/dc* values were obtained with ASTRA software (version 7.1.3, Wyatt Technology Co., Santa Barbara, CA, USA).

3.4.1.1. The Method of HPSEC-MALLS-RID

The M_w and M_w/M_n of LFG-Na and the starting material NFG were measured using HPSEC-MALLS-RID. In brief, HPSEC-MALLS-RID measurements were carried out on a multi angle laser light scattering detector (MALLS, DAWN HELEOS-II, Wyatt Technology Co., Santa Barbara, CA, USA) with an Agilent 1200 series LC/DAD system (Agilent Technologies, Palo Alto, CA, USA) equipped with a Shodex Ohpak SB-804 HQ column (300 mm × 8.0 mm) at 35 °C. The MALLS instrument was equipped with a He-Ne laser (wavelength λ = 661.6 nm). A refractive index detector (RID, Optilab rEX refractometer, DAWN EOS, Wyatt Technology Co., Santa Barbara, CA, USA) was simultaneously connected. The MALLS was calibrated according to the manufacturer's recommended procedures by using HPLC grade toluene. BSA standard (M_w of 66430 Da, *dn/dc* value of 0.185 mL/g) was employed for normalization of the 18-angle light scattering detectors relative to the right angle detector. The signals of light scattering from MALLS, refractive index from RID and UV absorbance from DAD detector were also aligned by the BSA standard. The mobile phase was 0.1 mol/L sodium nitrate (containing 0.02% sodium azide) at a flow rate of 0.5 mL/min. Each LFG-Na sample was dissolved in the same solution as the mobile phase at a final concentration of about 10 mg/mL, while NFG was about 5 mg/mL. All standard solutions and sample solutions were filtered through a 0.22 μm membrane before analysis. The Light Scattering Model was Zimm. An injection volume of 100 μL was used. The accuracy and system suitability of the method were determined via repeating the analysis of BSA for five replicates and the mean value of M_w was compared to its labeled value (true value). The repeatability was confirmed with preparation and analysis of six parallel solutions of NFG (5 mg/mL) and LFG-Na (10 mg/mL), respectively. The intermediate precision was measured by the RSD based on a total of 12 determinations of LFG-Na on different days. The Astra software (version 7.1.3, Wyatt Technology Co., Santa Barbara, CA, USA) was utilized for data acquisition and analysis.

3.5. Determination of Molecular Weight and Homogeneity of LFG-Na by HPGPC

The molecular weight, including M_w, number-average molecular weight (M_n) and molecular weight distribution of LFG-Na were also determined by HPGPC using an Agilent 1200 series (Agilent Technologies, Palo Alto, CA, USA) apparatus with differential refraction detector equipped with a TSK-Gel G2000SWxl column (300 mm × 7.8 mm). According to the size-exclusion chromatography method, the chromatographic conditions and procedures are presented as follows. The measurements were carried out at 35 °C and the mobile phase was 0.1 mol/L NaCl aqueous solution at a flow rate of 0.5 mL/min. An amount of oligosaccharide standards from LFG-Na (HS5, HS8, HS11, HS14 and HS17) with mobile phase was dissolved to make the concentration 2 mg/mL, while the concentration of the test sample solution of LFG-Na was 10 mg/mL. All standard solutions and test sample solutions were filtered through a 0.22 μm membrane before analysis. An injection volume of 20 μL was used. Oligosaccharide standard retention time—the peak molecular weights curve was fitted by a third-order polynomial using Agilent GPC/SEC software, and the molecular weight and molecular weight distribution of LFG-Na were calculated using the same GPC/SEC software (version A.02.01, Agilent Technologies, Palo Alto, CA, USA).

3.6. Chemical Characteristics and NMR Analysis of LFG-Na

Monosaccharide composition of LFG-Na was analyzed by reverse-phase HPLC according to PMP derivatization procedures [26]. The LFG-Na sample (20 mg) was dissolved in 4 mL of 2 mol/L trifluoroacetic acid (TFA), then sealed and incubated at 110 °C for 4 h in a heating block. The reaction mixture was evaporated to dryness. The residue was then dissolved in 5 mL methanol and again evaporated to remove residual TFA for five times. The five batches of LFG-Na and the standard samples were prepared as described above. The above samples were reconstituted in 2 mL water. Then 50 μL of the sample solution, 50 μL of 0.6 mol/L sodium hydroxide and 100 μL of 0.5 mol/L PMP in methanol were mixed and incubated at 70 °C for 30 min. After adjusting the pH to 7, water was added to make the volume of solution to 1.0 mL, and then 5.0 mL of chloroform was added to extract PMP three times. The aqueous layer was collected for HPLC analysis. Analysis of the PMP-labeled polysaccharides was carried out using an Agilent technologies 1100 series LC/DAD system (Agilent Technologies, USA) and an Agilent Eclipse XDB C18 (150 mm × 4.6 mm). The flow rate was 1 mL/min, and UV absorbance of the effluent was monitored at 250 nm. Mobile phases A and B (*v/v*, 80:20) consisted of 0.1 mol/L ammonium acetate (pH 5.5) and acetonitrile, respectively.

NMR spectroscopy was performed at 298 K with a BRUKER-AVANCEIII-HD 600 MHz spectrometer equipped with a $^{13}C/^{1}H$ dual probe in FT mode according to a previously described method [27]. All samples of LFG-Na were dissolved in D_2O and lyophilized three times to replace exchangeable protons. The lyophilized samples were then dissolved in D_2O at a concentration of 20–30 g/L. The 1D ($^{1}H/^{13}C$) and 2D (^{1}H-^{1}H COSY, TOCSY, NOESY, ^{1}H-^{13}C HSQC, and HMBC) NMR spectra were recorded with HOD suppression by presaturation. All chemical shifts were relative to internal 3-(trimethylsilyl)-propionic-2,2,3,3-d4 acid sodium salt (TSP, δ_H and δ_C = 0.00). The NMR spectra were processed using a trial MestReNova software (v9.0.1-13254, MESTRELAB RESEARCH, S.L, Santiago de Compostela, Spain).

3.7. *Activity Assays of LFG-Na In Vitro*

3.7.1. APTT Assay

The APTT of LFG-Na was determined with a coagulometer (PRECIL C2000-4, China) using APTT kits and standard human plasma as previously described with minor modifications [23]. The LFG-Na samples were dissolved in 20 mmol/L Tris-HCl (pH 7.4) at various concentrations. LFG-Na solutions of 5 μL each were mixed with 45 μL of normal human plasma and incubated for 2 min at 37 °C. Then 50 μL of APTT reagent was added to the mixture, which was incubated for another 3 min at 37 °C. $CaCl_2$ (50 μL) was then added, and the clotting time was recorded. Linear regression was performed with LFG-Na concentration as abscissa and the APTT value of LFG-Na solution at each concentration as ordinate. The concentration of LFG-Na required to double APTT was calculated according to the linear regression equation.

3.7.2. Potency for Anti-iXase

The potency for anti-iXase, referred to as the PLA method, was established based on quantitative responses in the Chinese Pharmacopoeia 2020 Edition Volume IV General Principle 1431 and the methods using the reagents in the BIOPHEN FVIII:C Kit as previously described [23,27,28]. In order to compare the potency for anti-iXase among different batches of LFG-Na, sample L5 of LFG-Na was selected as the reference substance (LFG-Na RS), and the potency for its anti-iXase was set as 100 U/mg. Another four batches of LFG-Na were tested as follows. LFG-Na RS was dissolved in water and the solutions diluted 10 times with buffer solution (R4) in factor VIII kit to obtain the reference solutions with four concentrations of 0.01–0.1 U/mL (S1–S4). The test sample (T1–T4) was taken to prepare test solution in the same way. An amount of 30 μL of solutions prepared as above was added into the 96-well plate with a sequence of B1 (blank control 1), S1, S2, S3, S4, T1, T2, T3, T4, T1, T2, T3, T4, S1, S2, S3, S4, and B2. The solutions were incubated with 1 IU/mL factor VIII (50 μL), and activation reagent (R2, 50 μL) (containing human thrombin, calcium, and synthetic phospholipids) at 37 °C. The reaction was initiated by the addition of factor X (R1, 50 μL). An amount of 50 μL of factor Xa chromogenic substrate SXa-11 (R3) was added and incubated. Before detecting the absorbance (A) at 405 nm (recorded at 37 °C using a

Microplate Reader (Multiskan GO-1510, Thermo Fisher Scientific, Vantaa, Finland)), 50 μL of 30% acetic acid solution was added. A was converted according to the exponential equation $A' = e^A$, linear regression was performed with A′ as the ordinate and the concentration of reference solution (or test solution) as the abscissa, respectively, and the results were input into the statistical program of Biological Assay in Chinese Pharmacopoeia BS2000 for calculating the potency and average confidence limit (FL%) of the quantitative reaction by the PLA method (4 × 4). The FL% should not be greater than 20%.

4. Conclusions

LFG-Na, a low molecular weight fucosylated glycosaminoglycan sodium, is a novel anticoagulant candidate that selectively inhibits iXase. In this study, the HPSEC-MALLS-RID method was successfully developed to determine the molecular weight and molecular weight distribution of LFG-Na and compared to the conventional method of HPGPC. HPSEC-MALLS-RID can also be used to determine the molecular weight of the starting material NFG. Based on multiple batches of measurement data, the acceptance criteria of the molecular weight and molecular weight distribution of NFG and LFG-Na were established. Furthermore, APTT assay was selected and the potency for anti-iXase, referring to the parallel line assay (PLA) method, was established to clarify biological potency of LFG-Na. This work illustrated that HPSEC-MALLS-RID and bioactivity assays were critical quality control methods for multi-component glycosaminoglycan LFG-Na. The methods also provide a feasible strategy to control the quality of other polysaccharide medicines.

Author Contributions: Y.S. and J.Z. provided the way of thinking. Y.S., J.Z. and S.Z. designed the framework. S.Z. carried out the experiments, performed the data analyses, wrote and edited the manuscript. Y.W., J.W., S.W. and S.Z. conducted the preparation of samples, the determination of molecular weight and bioactivity assays. The study of monosaccharide composition was performed by H.W., Y.Z. and S.Z., Y.S. and J.Z. revised the paper. All authors have read and agreed to the published version of the manuscript.

Funding: This work was financially supported by the National Major Scientific and Technological Special Project for Significant New Drugs Development (2018ZX09711-001), the Transformation Project of Major Scientific and Technological Achievements of Heilongjiang Province (ZC2018SH0004), Special Funding for the Construction of Innovative Provinces in Hunan Province (2020SK2043), and the Industry Incubation and Cluster Innovation Project of Marine Economic Innovation and Development Demonstration City of Haikou (HHCF20180101).

Institutional Review Board Statement: Not applicable.

Data Availability Statement: Not applicable.

Acknowledgments: We would like to thank Jinhua Zhao of South-Central Minzu University (Wuhan, China) for his help in NMR analysis.

Conflicts of Interest: The authors declare no conflict of interest.

Sample Availability: Samples of LFG-Na and NFG are available from the authors.

Abbreviations

LFG-Na, low molecular weight fucosylated glycosaminoglycan sodium; NFG, native fucosylated glycosaminoglycan; f.IIa, activated factor II; f.Xa, activated factor X; iXase, intrinsic factor Xase; f.IXa, activated factor IX; f.VIIIa, activated factor VIII; f.XII, factor XII; Fuc_{3S4S}, 3,4-di-*O*-sulfated-fucose; Fuc, fucose; GlcA, glucuronic acid; GalNAc, N-acetyl galactosamine; HPGPC, high performance gel permeation chromatography; HPSEC-MALLS-RID, high performance size exclusion chromatography coupled with multi angle laser light scattering and refractive index detector; M_w, weight-average molecular weight; M_n, number-average molecular weight; LMWH, low molecular weight heparin; RSD, relative standard deviation; BSA, bovine serum albumin; KOAc, potassium acetate; DMF, dimethyl formamide; EtONa, sodium ethoxide; $NaBH_4$, sodium borohydride; DAD, diode array detector; MWCO, molecular weight cutoff; APTT, activated partial thromboplastin time; PLA, parallel line assay.

References

1. Mackman, N. Triggers, targets and treatments for thrombosis. *Nature* **2008**, *451*, 914–918. [CrossRef] [PubMed]
2. Timmis, A.; Townsend, N.; Gale, C.; Grobbee, R.; Maniadakis, N.; Flather, M.; Wilkins, E.; Wright, L.; Vos, R.; Bax, J.; et al. European Society of Cardiology: Cardiovascular Disease Statistics 2017. *Eur. Heart J.* **2018**, *39*, 508–579. [CrossRef] [PubMed]
3. Fan, P.; Gao, Y.; Zheng, M.; Xu, T.; Schoenhagen, P.; Jin, Z. Recent progress and market analysis of anticoagulant drugs. *J. Thorac. Dis.* **2018**, *10*, 2011–2025. [CrossRef] [PubMed]
4. Kearon, C.; Akl, E.A.; Comerota, A.J.; Prandoni, P.; Bounameaux, H.; Goldhaber, S.Z.; Nelson, M.E.; Wells, P.S.; Gould, M.K.; Dentali, F.; et al. Antithrombotic therapy for VTE disease: Antithrombotic Therapy and Prevention of Thrombosis, 9th ed: American College of Chest Physicians Evidence-Based Clinical Practice Guidelines. *Chest* **2012**, *141* (Suppl. S2), e419S–e496S. [CrossRef]
5. Eriksson, B.I.; Quinlan, D.J.; Eikelboom, J.W. Novel oral factor Xa and thrombin inhibitors in the management of thromboembolism. *Annu. Rev. Med.* **2011**, *62*, 41–57. [CrossRef]
6. Fareed, J.; Thethi, I.; Hoppensteadt, D. Old versus new oral anticoagulants: Focus on pharmacology. *Annu. Rev. Pharmacol. Toxicol.* **2012**, *52*, 79–99. [CrossRef]
7. Lesko, L.J. Anticoagulants: What is new and what is the standard? *Clin. Pharmacol. Ther.* **2016**, *100*, 126–128. [CrossRef]
8. Shoeb, M.; Fang, M.C. Assessing bleeding risk in patients taking anticoagulants. *J. Thromb. Thrombolysis* **2013**, *35*, 312–319. [CrossRef]
9. Lin, L.; Zhao, L.; Gao, N.; Yin, R.; Li, S.; Sun, H.; Zhou, L.; Zhao, G.; Purcell, S.W.; Zhao, J. From multi-target anticoagulants to DOACs, and intrinsic coagulation factor inhibitors. *Blood Rev.* **2020**, *39*, 100615. [CrossRef]
10. Zhou, L.; Gao, N.; Sun, H.; Xiao, C.; Yang, L.; Lin, L.; Yin, R.; Li, Z.; Zhang, H.; Ji, X.; et al. Effects of Native Fucosylated Glycosaminoglycan, Its Depolymerized Derivatives on Intrinsic Factor Xase, Coagulation, Thrombosis, and Hemorrhagic Risk. *Thromb. Haemost.* **2020**, *120*, 607–619. [CrossRef]
11. Ahmad, S.S.; Rawala-Sheikh, R.; Walsh, P.N. Components and assembly of the factor X activating complex. *Semin. Thromb. Hemost.* **1992**, *18*, 311–323. [CrossRef] [PubMed]
12. Sun, H.; Gao, N.; Ren, L.; Liu, S.; Lin, L.; Zheng, W.; Zhou, L.; Yin, R.; Zhao, J. The components and activities analysis of a novel anticoagulant candidate dHG-5. *Eur. J. Med. Chem.* **2020**, *207*, 112796. [CrossRef] [PubMed]
13. Pomin, V.H. Holothurian fucosylated chondroitin sulfate. *Mar. Drugs* **2014**, *12*, 232–254. [CrossRef] [PubMed]
14. Fonseca, R.J.; Santos, G.R.; Mourao, P.A. Effects of polysaccharides enriched in 2,4-disulfated fucose units on coagulation, thrombosis and bleeding. Practical and conceptual implications. *Thromb. Haemost.* **2009**, *102*, 829–836. [CrossRef]
15. Li, H.; Yuan, Q.; Lv, K.; Ma, H.; Gao, C.; Liu, Y.; Zhang, S.; Zhao, L. Low-molecular-weight fucosylated glycosaminoglycan and its oligosaccharides from sea cucumber as novel anticoagulants: A review. *Carbohydr. Polym.* **2021**, *251*, 117034. [CrossRef]
16. Yin, R.; Zhou, L.; Gao, N.; Li, Z.; Zhao, L.; Shang, F.; Wu, M.; Zhao, J. Oligosaccharides from depolymerized fucosylated glycosaminoglycan: Structures and minimum size for intrinsic factor Xase complex inhibition. *J. Biol. Chem.* **2018**, *293*, 14089–14099. [CrossRef]
17. Luo, L.; Wu, M.; Xu, L.; Lian, W.; Xiang, J.; Lu, F.; Gao, N.; Xiao, C.; Wang, S.; Zhao, J. Comparison of physicochemical characteristics and anticoagulant activities of polysaccharides from three sea cucumbers. *Mar. Drugs* **2013**, *11*, 399–417. [CrossRef]
18. Li, S.-P.; Wu, D.-T.; Lv, G.-P.; Zhao, J. Carbohydrates analysis in herbal glycomics. *TrAC Trends Anal. Chem.* **2013**, *52*, 155–169. [CrossRef]
19. Kim, C.; Deratani, A.; Bonfils, F. Determination of the Refractive Index Increment of Natural and Synthetic Poly(Cis-1,4-Isoprene) Solutions and Its Effect on Structural Parameters. *J. Liq. Chromatogr. Relat. Technol.* **2009**, *33*, 37–45. [CrossRef]
20. Gao, N.; Chen, R.; Mou, R.; Xiang, J.; Zhou, K.; Li, Z.; Zhao, J. Purification, structural characterization and anticoagulant activities of four sulfated polysaccharides from sea cucumber *Holothuria fuscopunctata*. *Int. J. Biol. Macromol.* **2020**, *164*, 3421–3428. [CrossRef]
21. Cheong, K.L.; Wu, D.T.; Zhao, J.; Li, S.P. A rapid and accurate method for the quantitative estimation of natural polysaccharides and their fractions using high performance size exclusion chromatography coupled with multi-angle laser light scattering and refractive index detector. *J. Chromatogr. A* **2015**, *1400*, 98–106. [CrossRef] [PubMed]
22. Li, H.; Gong, X.; Wang, Z.; Pan, C.; Zhao, Y.; Gao, X.; Liu, W. Multiple fingerprint profiles and chemometrics analysis of polysaccharides from Sarcandra glabra. *Int. J. Biol. Macromol.* **2019**, *123*, 957–967. [CrossRef] [PubMed]
23. Wu, M.; Wen, D.; Gao, N.; Xiao, C.; Yang, L.; Xu, L.; Lian, W.; Peng, W.; Jiang, J.; Zhao, J. Anticoagulant and antithrombotic evaluation of native fucosylated chondroitin sulfates and their derivatives as selective inhibitors of intrinsic factor Xase. *Eur. J. Med. Chem.* **2015**, *92*, 257–269. [CrossRef] [PubMed]
24. Li, X.; Luo, L.; Cai, Y.; Yang, W.; Lin, L.; Li, Z.; Gao, N.; Purcell, S.W.; Wu, M.; Zhao, J. Structural Elucidation and Biological Activity of a Highly Regular Fucosylated Glycosaminoglycan from the Edible Sea Cucumber Stichopus herrmanni. *J. Agric. Food. Chem.* **2017**, *65*, 9315–9323. [CrossRef]
25. Gao, N.; Lu, F.; Xiao, C.; Yang, L.; Chen, J.; Zhou, K.; Wen, D.; Li, Z.; Wu, M.; Jiang, J.; et al. beta-Eliminative depolymerization of the fucosylated chondroitin sulfate and anticoagulant activities of resulting fragments. *Carbohydr. Polym.* **2015**, *127*, 427–437. [CrossRef]
26. Liu, J.; Shang, F.; Yang, Z.; Wu, M.; Zhao, J. Structural analysis of a homogeneous polysaccharide from Achatina fulica. *Int. J. Biol. Macromol.* **2017**, *98*, 786–792. [CrossRef]

27. Shang, F.; Gao, N.; Yin, R.; Lin, L.; Xiao, C.; Zhou, L.; Li, Z.; Purcell, S.W.; Wu, M.; Zhao, J. Precise structures of fucosylated glycosaminoglycan and its oligosaccharides as novel intrinsic factor Xase inhibitors. *Eur. J. Med. Chem.* **2018**, *148*, 423–435. [CrossRef]
28. Liu, S.; Zhang, T.; Sun, H.; Lin, L.; Gao, N.; Wang, W.; Li, S.; Zhao, J. Pharmacokinetics and Pharmacodynamics of a Depolymerized Glycosaminoglycan from *Holothuria fuscopunctata*, a Novel Anticoagulant Candidate, in Rats by Bioanalytical Methods. *Mar. Drugs* **2021**, *19*, 212. [CrossRef]

molecules

MDPI

Article

Synthesis, Biological Evaluation and Docking Studies of Ring-Opened Analogues of Ipomoeassin F

Sarah O'Keefe [1,*,†,‡], Pratiti Bhadra [2,†], Kwabena B. Duah [3,†,§], Guanghui Zong [4,†], Levise Tenay [2,5], Lauren Andrews [3], Hayden Schneider [3], Ashley Anderson [2,5], Zhijian Hu [6], Hazim S. Aljewari [7], Belinda S. Hall [8], Rachel E. Simmonds [8], Volkhard Helms [2,*], Stephen High [1,*] and Wei Q. Shi [3,*]

1 School of Biological Sciences, Faculty of Biology, Medicine and Health, University of Manchester, Manchester M13 9PT, UK
2 Center for Bioinformatics, Saarland University, 66123 Saarbrucken, Germany; pratiti.bhadra@bioinformatik.uni-saarland.de (P.B.); levise.tenay@gmail.com (L.T.); aanderson8@bsu.edu (A.A.)
3 Department of Chemistry, Ball State University, Muncie, IN 47306, USA; kbduah@iu.edu (K.B.D.); landrews2@bsu.edu (L.A.); hoschneider@bsu.edu (H.S.)
4 Department of Chemistry and Biochemistry, University of Maryland, College Park, MD 20742, USA; gzong@umd.edu
5 Department of Biomedical Engineering, Biruni University, 34010 Istanbul, Turkey
6 Feinstein Institute for Medical Research, Northwell Health, 350 Community Dr., Manhasset, NY 11030, USA; zhu1@northwell.edu
7 Ralph E. Martin Department of Chemical Engineering, University of Arkansas, Fayetteville, AR 72701, USA; hsaljewa@uark.edu
8 Department of Microbial Sciences, School of Biosciences and Medicine, University of Surrey, Guildford GU2 7XH, UK; b.s.hall@surrey.ac.uk (B.S.H.); rachel.simmonds@surrey.ac.uk (R.E.S.)
* Correspondence: sarah.okeefe@helsinki.fi (S.O.); volkhard.helms@bioinformatik.uni-saarland.de (V.H.); stephen.high@manchester.ac.uk (S.H.); wqshi@bsu.edu (W.Q.S.)
† These authors contributed equally to this work.
‡ Current Address: Institute of Biotechnology, HiLIFE, University of Helsinki, 00014 Helsinki, Finland.
§ Current Address: Department of Chemistry, Indiana University, 800 E Kirkwood Ave, Bloomington, IN 47405, USA.

Citation: O'Keefe, S.; Bhadra, P.; Duah, K.B.; Zong, G.; Tenay, L.; Andrews, L.; Schneider, H.; Anderson, A.; Hu, Z.; Aljewari, H.S.; et al. Synthesis, Biological Evaluation and Docking Studies of Ring-Opened Analogues of Ipomoeassin F. *Molecules* **2022**, *27*, 4419. https://doi.org/10.3390/molecules27144419

Academic Editors: De-Cai Xiong, Jian Yin and Jing Zeng

Received: 13 June 2022
Accepted: 7 July 2022
Published: 10 July 2022

Abstract: The plant-derived macrocyclic resin glycoside ipomoeassin F (Ipom-F) binds to Sec61α and significantly disrupts multiple aspects of Sec61-mediated protein biogenesis at the endoplasmic reticulum, ultimately leading to cell death. However, extensive assessment of Ipom-F as a molecular tool and a therapeutic lead is hampered by its limited production scale, largely caused by intramolecular assembly of the macrocyclic ring. Here, using in vitro and/or in cellula biological assays to explore the first series of ring-opened analogues for the ipomoeassins, and indeed all resin glycosides, we provide clear evidence that macrocyclic integrity is not required for the cytotoxic inhibition of Sec61-dependent protein translocation by Ipom-F. Furthermore, our modeling suggests that open-chain analogues of Ipom-F can interact with multiple sites on the Sec61α subunit, most likely located at a previously identified binding site for mycolactone and/or the so-called lateral gate. Subsequent in silico-aided design led to the discovery of the stereochemically simplified analogue **3** as a potent, alternative lead compound that could be synthesized much more efficiently than Ipom-F and will accelerate future ipomoeassin research in chemical biology and drug discovery. Our work may also inspire further exploration of ring-opened analogues of other resin glycosides.

Keywords: resin glycosides; macrocyclic natural glycolipids; ring-opened analogues; cytotoxicity; protein translocation; Sec61 translocon; molecular docking

1. Introduction

Resin glycosides are a large family of plant-derived natural products unique to the morning glory family, Convolvulaceae [1,2]. The vast majority of them contain a macrocyclic ester ring with embedded carbohydrates and are considered active ingredients for many herbal medicines. Despite their distinctive structures and medicinal benefits, the pharmacological properties of most resin glycosides are underexplored. In 2005 and 2007, the ipomoeassin family of resin glycosides was isolated from the leaves of *Ipomoea squamosa* in the Suriname rainforest and exhibited potent cytotoxicity [3,4]. Following a great amount of effort on total synthesis [5–8] and medicinal chemistry [9–12], chemical proteomics studies discovered strong inhibition of Sec61-mediated protein translocation as the primary molecular mechanism for ipomoeassin F (Ipom-F, Figure 1) [13], the most potent member of the family. To date, ring expansion [14] and modifications at the *para* position of the cinnamate benzene ring [13] have afforded several analogues with biological activities comparable to or even better than Ipom-F.

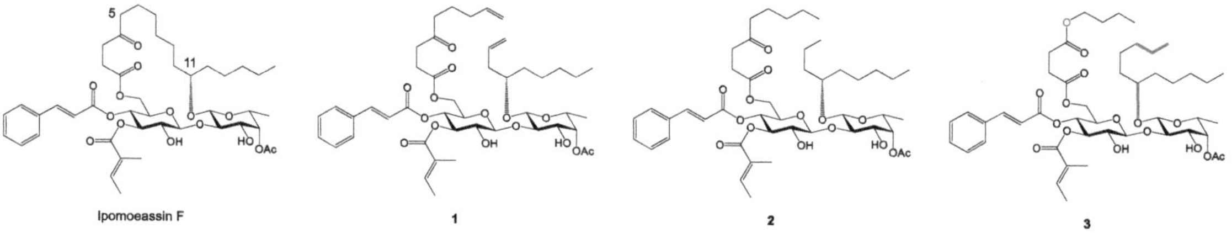

Figure 1. Structures of Ipom-F and its open-chain analogues (**1**–**3**).

The translocation of nascent polypeptides into and across the membrane of the endoplasmic reticulum (ER) is the first and decisive step during the biogenesis of many integral membrane and secretory proteins [15–17]. The Sec61 translocon [18,19] is the predominant protein conducting channel at the ER membrane, acting as a dynamic hub to coordinate the translocation of ~one third of the cellular proteome in eukaryotes [17]. While the fidelity of Sec61-mediated protein translocation is essential for proper cellular and organismal function [20,21], small molecule inhibitors that modulate this process have provided valuable insights into the mechanistic complexities of protein translocation at the ER [22,23] and also have potential therapeutic applications [24,25]. In the latter case, small molecule-mediated inhibition of Sec61-dependent protein production shows promise for the clinical treatment of solid tumors (ClinicalTrials.gov: NCT05047536) and presents an attractive, yet underexplored, strategy to mitigate the toxicity associated with the overexpression of *SEC* genes that have been linked to pathogenicity in kidney and liver diseases, diabetes and certain cancers [26].

Small molecule inhibitors typically bind to the central, Sec61α, subunit of the Sec61 translocon [19,22] and the subsequent blockade in Sec61-mediated protein translocation results in potent cytotoxicity that, ultimately, leads to cell death [13,22,27,28]. Besides Ipom-F, the current repertoire of small molecule Sec61 inhibitors includes several other structurally distinct classes of natural products: apratoxins [27], coibamide A [28], cotransins [29], decatransin [30], mycolactone [31,32] and derivatives thereof [24,25]. Although a limited number of synthetic Sec61 inhibitors, e.g., the eeyarestatins [33] and FMP-40139-3 [34], have been identified by library-based screening approaches, the natural products and their derivatives are substantially more potent.

Despite their structural diversity, a common feature shared amongst each of the natural product Sec61 inhibitors is a core, albeit differently sized, cyclic scaffold [24,25]. The integrity of most macrocyclic frameworks appears to confer an essential role for efficient Sec61 inhibition, particularly since two linear analogues of coibamide A showed a significant loss in activity when compared to the cyclic parent compound [35]. Hence, we were surprised when two ring-opened analogues (**1** and **2**, Figure 1) of Ipom-F were discovered to still be active in cytotoxicity assays [9]. In our initial characterization of Sec61α as the cellular target of Ipom-F [13], we additionally used a well-characterized cell-free assay to demonstrate that open-chain analogue **2** efficiently inhibits Sec61-mediated protein translocation in vitro [13]. These preliminary results raised an intriguing hypothesis that the macrocyclic ring may not be essential to the biological activities of Ipom-F, or perhaps resin glycosides in general.

In this report, we present unambiguous evidence using our established in vitro ER membrane translocation assay in combination with live-cell cytotoxicity screening to expand the current scope of ring-modified Ipom-F analogues [9,13,14] and prove that open-chain structures can act as authentic surrogates for Ipom-F. These conclusions are further supported by molecular docking of Ipom-F and analogues within the channel pore of Sec61α [22]. Our modeling also raises the intriguing possibility that several Ipom-F analogues can access multiple binding sites on the Sec61α subunit that likely reflect the stepwise interaction of signal peptides with the Sec61 translocon [36]. We further exploited our in silico modeling studies to design a new and stereochemically simpler open-chain Ipom-F analogue **3** (Figure 1) that we predicted to be an active Sec61 inhibitor. After its synthesis, **3** was experimentally validated to be almost as effective at inhibiting Sec61-mediated protein translocation as Ipom-F and the parent open-chain analogue **2**. Therefore, we present **3** as a new lead compound for the efficient generation of new small molecule inhibitors that can be used to explore the mechanisms of Sec61-mediated protein translocation and potential therapeutic applications.

2. Results and Discussion

2.1. *Synthesis and Cytotoxicity of Nine New Open Chain-Analogues*

To investigate the impact of ring-opening, we first submitted the open-chain analogue **1 [9]** to the National Cancer Institute (NCI) 60-cell line screening (Supplementary Material S1) and determined the cytotoxicity of nine newly synthesized open-chain analogues **4–12** (Figure 2, Supplementary Materials S2 and S3) against MDA-MB-231 cells (Table S1). When compared to Ipom-F, only a 4–5-fold (NCI 60-cell line: average GI_{50} ~135 nM versus ~30 nM) and 9-fold (MDA-MB-231 cells: IC_{50} ~60 nM versus ~7 nM) loss in potency was observed for the ring-opened analogue **1** and, while neither compound induced much cytotoxicity in most ovarian cell lines, both compounds showed a comparable cytotoxic potency towards the majority of breast and melanoma cell lines tested (Supporting information S1). In MDA-MB-231 cells, and when compared to **1** (IC_{50} ~59 nM), the methyl amide-modified analogue **5** (IC_{50} ~43 nM) showed slightly enhanced activity, analogues **4** (IC_{50} ~168 nM) and **6** (IC_{50} ~205 nM) showed only a marginal loss in activity (3–4 fold) and each of the other six open-chain analogues **7–12** showed substantially reduced cytotoxicity (μM range; >35-fold activity loss when compared to **1**). Furthermore, the majority of the nine new open-chain analogues were modestly less cytotoxic towards MDA-MB-231 cells when compared to their direct ring-closed counterparts (analogues **13–21** matching with **4–12**; Figures S1 and S2 and Table S1) [9,11–14]. Taken together, these data demonstrate that the macrocylic integrity of Ipom-F is not crucial for cytotoxic activity against the majority of cell lines tested and suggest that a similar mechanism likely underlies the cytotoxic activity of open- and closed-chain analogues.

2.2. *Open-Chain Analogues Selectively Inhibit Sec61-Mediated Protein Translocation In Vitro*

Having previously established that the principal molecular basis for the cytotoxicity of Ipom-F is its strong, yet reversible, binding to Sec61α and the resultant wide-ranging blockade of Sec61-mediated protein translocation at the ER [13], we sought evidence that this mechanism also underlies the cytotoxic effects of **1** and other open-chain analogues. Hence, based on our previous characterization of Ipom-F and the open-chain analogue **2** [13], we first used a well-established cell-free translation system supplemented with ER microsomes derived from canine pancreas [37] to study the effects of the open-chain analogues that were the most cytotoxic towards MDA-MB-231 cells (analogues **5** > **1** > **6**; Table S1) on Sec61-mediated protein translocation in vitro. For comparative purposes, we analyzed the effects of **2** [13] and two closed-chain compounds: Ipom-F and the diester-modified analogue **22** [11] in parallel (Figure 3A).

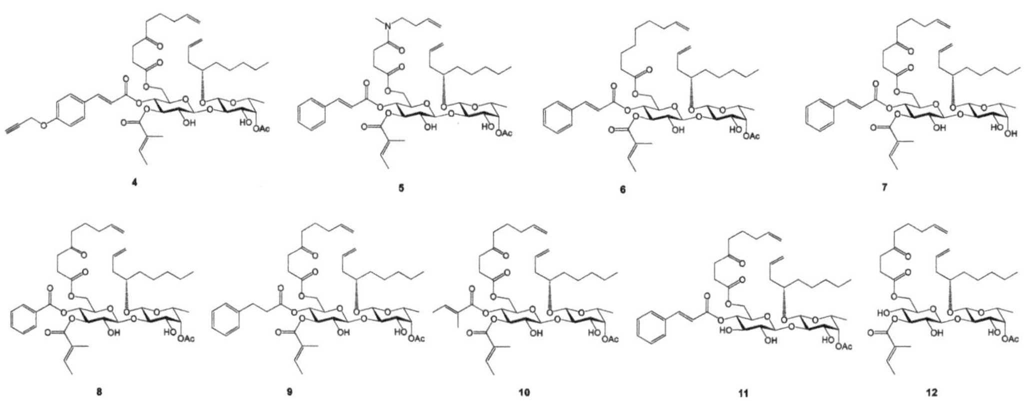

Figure 2. Structures of open-chain Ipom-F analogues (**4**–**12**).

Following the resolution of radiolabeled proteins by SDS-PAGE (Figure 3B), we used the efficiency with which the N-terminal domain of the type II integral membrane protein Ii (short form of HLA class II histocompatibility antigen gamma chain, isoform 1) was N-glycosylated inside the ER lumen as a robust reporter for the authentic membrane integration of this model Sec61-dependent protein client (Figure 3B, right upper panel, 0Gly versus 2 Gly forms) [13,14,33]. Based on the reduced levels of the N-glycosylated forms of Ii synthesized in the presence of 1 μM of each compound in comparison to the DMSO control (Figure 3B, right upper panel, lanes 3–8 versus lane 1), Ipom-F and analogues **1**, **2**, **5** and **22** efficiently, albeit variably, inhibited the in vitro membrane integration of Ii, while analogue **6** did not (Figure 3B,C).

To further analyze the apparent variations in the potency of ring-modified analogues (Figure 3B,C), we next synthesized Ii in the presence of decreasing (500 μM–5 nM) concentrations of each compound that efficiently inhibited the membrane integration of Ii at 1 μM (Figure 3D). Analyses of these in vitro titrations yielded estimated IC_{50} values in the mid-nanomolar range (Figure 1E), allowing us to rank order compound activity: Ipom-F (IC_{50}: 155 nM) > **2** (IC_{50}: 202 nM) > **5** (IC_{50}: 291 nM) > **22** (IC_{50}: 334 nM) > **1** (IC_{50}: 562 nM) > **6** (negligible inhibition at 1 μM). Given that the two closed-chain compounds, Ipom-F and **22** (Figure 3A), are, respectively, the most and second least potent active inhibitors of Sec61-mediated protein translocation in vitro, these data suggest that a combination of structural and chemical features, and not merely macrocyclic integrity, are important contributors to the potency of Sec61 inhibition.

2.3. Open-Chain Analogues Induce Cytotoxicity Via the Selective Inhibition of Sec61-Mediated Protein Translocation

To confirm that Sec61α is the primary target of open-chain analogues and that this interaction underlies their cytotoxic effects, we next used a resazurin-based cell viability assay (Figure 3F) to compare the effects of each compound on the growth of HCT-116 cells that were wild-type for Sec61α (HCT116 Sec61α-WT) or carrying a heterozygous point mutant in SEC61A1 (HCT116 Sec61α-G80W) that confers resistance to Ipom-F, and reduces binding of mycolactone to the Sec61 translocon by mechanism that involves an alteration in translocon dynamics [22].

Following 72 h treatment with 50 nM of each compound, Ipom-F and analogues **1**, **2**, **5** and **22** induced ~48–64% cell death in HCT116 Sec61α-WT cells, while, as observed in vitro, analogue **6** was the least potent compound tested (~20% cell death; Figure 3G, left). Strikingly, none of the compounds affected the viability of HCT116 Sec61α-G80W cells treated using the same concentration (Figure 3G, right), consistent with Sec61α being their primary molecular target.

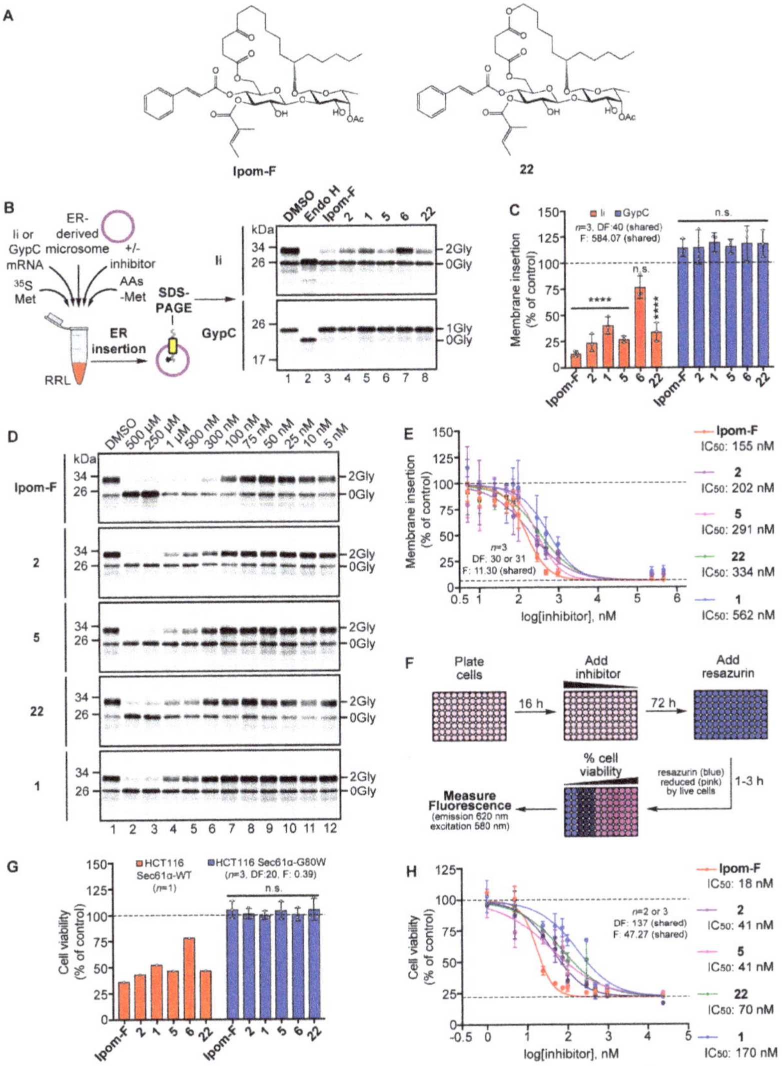

Figure 3. Open-chain Ipom-F analogues selectively inhibit Sec61-mediated protein translocation in vitro and in cellula. (**A**) Structures of ring-closed compounds; Ipom-F and its diester analogue **22**. (**B**) Phosphorimages of the membrane associated radiolabeled precursor proteins of a type II (Ii, top panel) and a type III integral membrane protein (GypC, bottom panel) synthesized in the presence or absence of 1 μM compound. (**C**) Quantification of the efficiency of membrane integration in B expressed relative to the control (set to 100%). (**D**) Phosphorimages of the membrane-associated products of Ii synthesized in the presence of 500 μM–5 nM concentrations of indicated compounds. (**E**) IC_{50} curves derived from the Ii insertion data in D. (**F**) Schematic of the resazurin-based cytotoxicity assay. (**G**) Cell viability of human HCT116 cells (Sec61α-WT; wild-type) and HCT116 cells heterozygous for a SEC61A1 missense mutation encoding G80W (Sec61α-G80W) [22] following 72 h treatment with 50 nM compound. (**H**) IC_{50} curves derived from the cell viability of HCT116 Sec61α-WT cells treated with 25 μM–1 nM concentrations of each compound. See methods for details of biological replicates and statistical analyses. Statistical significance is given as n.s., non-significant $p > 0.1$ and ****, $p < 0.0001$.

To further explore the observed variations in cytotoxicity, we performed cell viability assays using HCT116 Sec61α-WT cells and decreasing (25 μM–1 nM) concentrations of each compound that efficiently caused cell death at 50 nM. Analyses of these in cellula titrations yielded estimated IC_{50} values in the low-to-mid-nanomolar range (Figure 3H) allowing us to, once again, rank order compound activity. Although, as typical for Sec61 inhibitors, the IC_{50} values derived via cytotoxicity are consistently lower than those obtained by analyzing membrane insertion (see Figure S3) [13,14], the cytotoxic potency of each compound closely mirrored the rank order of activity that was observed in vitro. Taken together, these data strongly suggest that, similar to the closed-chain Ipom-F [13] and the diester analogue **22**, the interaction of active open-chain analogues (**1**, **2** and **5**) with Sec61α and the resultant inhibition of Sec61-mediated protein translocation at the ER underlies their cytotoxic effects.

2.4. Molecular Docking of Open-Chain Analogues within the Channel Pore of Sec61α Reveals Multiple Binding Sites

We have previously postulated that Ipom-F most likely occludes membrane access via the Sec61 lateral gate [23,38], as recently established for mycolactone [22]. Since the G80W mutation that is located in the transmembrane (TM) helix TM2 of Sec61α confers resistance to mycolactone [22], Ipom-F [22] and each of Ipom-F analogues tested here (Figure 3), while the Ipom-F-Sec61α binding site is yet to be elucidated, we used the cryogenic-electron microscopy (cryo-EM)-derived structure of the canine Sec61 channel bound to mycolactone as a model to explore the potential interaction sites and putative docking conformations of active closed-chain (Ipom-F and **22**) and open-chain analogues (**1**, **2** and **5**) in the previously defined inhibited state of the Sec61α channel pore [22].

These docking studies suggest that Ipom-F occupies the same groove between the TM helices TM2, TM7, TM8 and the cytosolic loop (CL) 4 of Sec61α (Figure 4B) that mycolactone was found to bind in [22], and that was approximately recovered in a previous docking analysis of mycolactone [39] using a similar docking protocol as used here. The predicted binding affinity of Ipom-F is -7.82 ± 0.2 kcal/mol and in this orientation Ipom-F may preferentially interact with the C-terminus of TM2 (Leu89-Ala97), the Gln170-Gly172 region of CL4 (loop between TM4-TM5) and the Trp379-Val382 region of CL8 (loop between TM8-TM9) (Figure S4A). The *D*-fucose region of Ipom-F most likely occupies the volume between the TM2 helix and CL4 (Figure 4B), since this region formed a hydrogen bond with side-chain or backbone atoms of Gln170 in CL4 in all of our independent final docking simulation results. Figure S5 illustrates the polar and non-polar contacts between the docked Ipom-F derivatives and Sec61α residues in their energetically most favorable binding poses.

Strikingly, and in contrast to Ipom-F, the closed-chain analogue **22** was predicted to bind in two different locations (Figure 4A): either in the same groove where Ipom-F and mycolactone bind or in the upper part of the lateral gate (Figure 4C), with predicted binding affinities of -8.81 ± 0.63 and -7.35 ± 0.23 kcal/mol, respectively. Similar to Ipom-F, analogue **22** interacts with the C-terminus of the TM2 helix, the CL4 region and CL8 (Figure S4A). However, when positioned in the upper part of the lateral gate, analogue **22** preferentially interacts with TM7 (Gln294-Val298) (Figure S4B).

Similar to the closed-chain analogue **22**, the open-chain analogue **2** was also predicted to bind in two different locations within the Sec61 translocon; namely, in the binding groove of mycolactone [22] and the middle of the lateral gate with very similar predicted binding affinities of -4.5 ± 0.75 and -4.4 ± 0.55 kcal/mol, respectively. As observed for Ipom-F and its closed-chain analogue **22**, the open-chain analogue **2** also strongly interacts with TM2, CL4 and CL8 when occupying the mycolactone binding site identified by cryo-EM (Figure S4A). Similar to Ipom-F, analogue **2** is also inclined to hydrogen bond with the CL4 region (Lys171), and this hydrogen bond was identified in four out of five independent docking simulations. However, when occupying the middle of the lateral gate, analogue **2** likely interacts with the plug region (Ile68), TM3 and TM7 (Figure S4B).

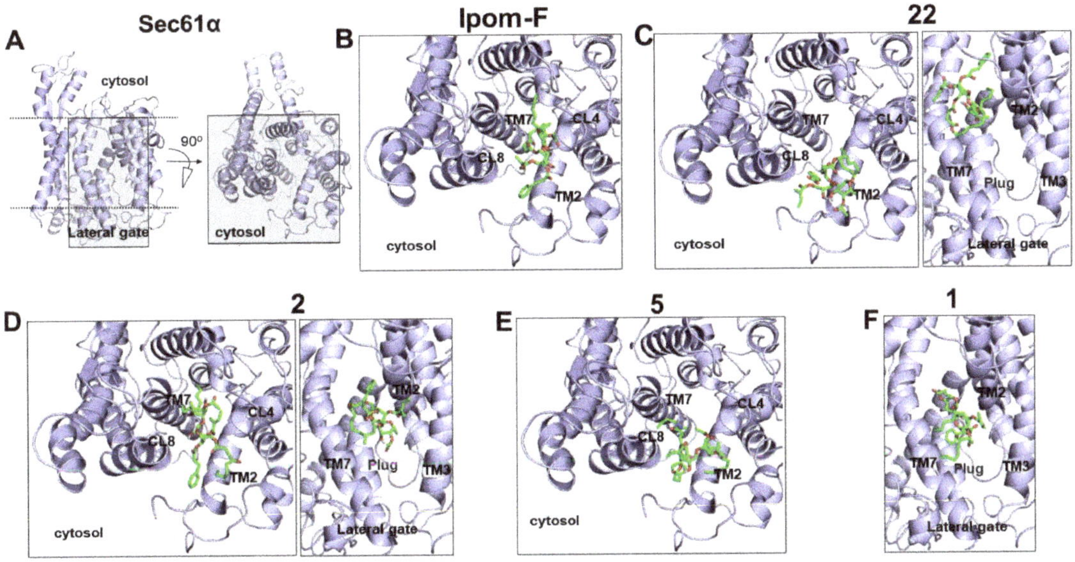

Figure 4. Homology model of human-Sec61α and molecular docking. (**A**) Homology model of human-Sec61α in the inhibited state. The two grey boxes indicate two different binding sites of analogues. The 'cytosol' view represents a top-view of Sec61α from the cytosolic side and focuses on the groove between TM2, TM7, CL8 and CL4 while the 'lateral gate' view represents a side-view from the lipid bilayer. The other panels show the most favorable docking conformations of closed-chain ((**B**) Ipom-F; (**C**) **22**) and open-chain ((**D**) **2**; (**E**) **5**; (**F**) **1**) analogues in the inhibited conformation of human-Sec61α.

In contrast, analogue **5** was predicted to bind only at the groove of the Ipom-F binding site with a predicted binding affinity of −3.28 ± 0.9 kcal/mol, where it interacts with the C-terminus of TM2, CL4 and CL8 (see Figure S4A). Likewise, the open-chain analogue **1** only binds in one position; at the lower part of the lateral gate (Figure 4F), with a predicted binding affinity of −4.25 ± 0.6 kcal/mol, where it strongly interacts with the plug region (Met65-Ile68), TM3 (Ile123-Gln127) and TM7 (Gln294-Val298) (see Figure S4B).

These docking studies offer several new insights into small molecule-mediated inhibition of Sec61. Firstly, they suggest that Ipom-F and the majority of the modeled Ipom-F analogues (**2**, **5** and **22**) likely bind within the same groove as mycolactone (between TM2, TM7, TM8 and CL4 of Sec61α) [22]. Secondly, since certain compounds also show favorable binding affinities at the middle (**2**) or upper part of the lateral gate (**22**) or an exclusive putative docking conformation at the lower part of the lateral gate (**1**), our studies reveal multiple putative binding sites for Ipom-F analogues. Since molecular docking of the (about 10–20 amino acids long) hydrophobic core-portions of signal peptides suggests that these regions prefer to bind in the vicinity of the lateral gate of Sec61α of the Sec61 complex [40], we speculate that Ipom-F analogues may perturb signal peptide binding at more than one site on the Sec61 translocon, and that the flexibility of open-chain analogues may even enhance their ability to perturb the interactions of signal peptides at multiple interaction sites within the Sec61α subunit, thereby influencing their potency and/or substrate selectivity. Finally, since the number of contact residues within the CL4 region reflected the rank order of compound activity observed in vitro and in cellula: Ipom-F = **2** ≥ **5** > **22** > **1** (CL4 contact residues: ~77%, 77%, 62%, 31%, 0% respectively; Figure S4C), we propose that the interaction of compounds with the CL4 region is important for the potency of Sec61 inhibition. We further propose that the potential of Ipom-F and **2** to form hydrogen bonds with residues Gln170 and Lys171 in the CL4 region (Figure S4C) may provide the molecular basis for their increased potency when compared to other analogues.

2.5. Biology-Directed and In Silico-Aided Design of Analogue 3

While the binding affinities of the open-chain Ipom-F analogues (**1**: -4.25 ± 0.6 kcal/mol; **2**: -4.45 ± 0.5 kcal/mol; **5**: -3.82 ± 0.6 kcal/mol) were predicted to be less favorable than those of the closed-chain Ipom-F (-7.82 ± 0.2 kcal/mol) and **22** (-8.09 ± 0.4 kcal/mol), it should be noted that it is difficult to compare compound docking affinities due to the different entropy changes between the open- and closed-chain analogues. Empirical docking scoring functions such as that used in Autodock 4.2 (Available online: https://autodocksuite.scripps.edu/autodock4/ (accessed on 2 June 2022)) typically approximate entropy penalties on binding using the number of rotatable bonds present in the ligand. Here, the difference in the docking score (~3 kcal/mol) is associated with the number of rotatable bonds present in the open-chain analogues **1**, **2** and **5** versus the closed-chain Ipom-F and **22** (respectively 31–32 rotatable bonds versus 15). In the present case, we, therefore, suggest that the docking score or binding affinity obtained from the docking software should not be considered as an accurate parameter to represent the experimental binding affinity. Adaptive biasing force (ABF)/metadynamics simulations may be helpful in future work to characterize the enthalpic and/or entropic contributions involved in the binding of closed-ring vs. open-ring Ipom-F derivatives.

Despite these caveats, we postulate that the disaccharide core is capable of controlling the overall conformation of ring-opened analogues that enables them to retain sufficient interaction with Sec61α. Therefore, we exploited our biology-directed and in silico-aided studies to design a new Ipom-F-derived lead compound that would be more synthetically accessible than compounds **1** and **2**, while also retaining a comparable level of biological activity. Since the alkene-reduced analogue **2** was the most potent open-chain analogue discovered to date, we first decided to remove both terminal double bonds from the new analogue. Second, and to avoid low-yielding Grignard reactions during the synthesis of the aglycones at the 6″-OH-Glu*p* and C-1′-Fuc*p* positions [8], we sought to replace 4-oxononanoic acid with mono-butyl succinate **28** (Scheme 1) in the synthetic route to the new analogue. Such a strategy, while synthetically attractive, additionally permits the well-tolerated bioisosteric replacement of the C-5 methylene with an oxygen atom (cf. **22**; Table 1) while retaining the C-4 carbonyl group, whose removal is detrimental to compound potency (cf. **6**; Table 1).

Table 1. Comparison of chemical properties, inhibition of in vitro protein translocation (IC_{50}, nM) and cytotoxicity in HCT116 Sec61α-WT cells (IC_{50}, nM) of Ipom-F and analogues.

Compound	Ring Integrity	Other Ring Modifications	C-11 Chirality	cLogP *	In Vitro Translocation Inhibition	Cytotoxicity HCT-116 Cells
Ipom-F	Closed	None	11*S*	5.97	155	18
2	Open	None	11*S*	8.72	202	41
5	Open	Dialkene, aza	11*S*	6.98	291	41
22	Closed	Diester	11*S*	4.94	334	70
1	Open	Dialkene	11*S*	7.75	562	170
6	Open	Dialkene, ketone removed	11*S*	9.39	Not determined	Not determined

* Calculated in ChemDraw.

Lastly, we decided to increase the lipophilicity of the new analogue for two reasons: (i) our earlier studies on how the ring size of closed-chain analogues affects compound potency revealed that ring expansion by two methylene units, and the concomitant increase in lipophilicity, is an advantageous feature [14]; and (ii) in this study, increased compound potency in vitro and in cellula for both the open-chain (**2** $\geq$ **5** > **1**) and closed-chain (Ipom-F > **22**) compounds appears to correlate with an increase in lipophilicity (based on cLogP calculations; Table 1).

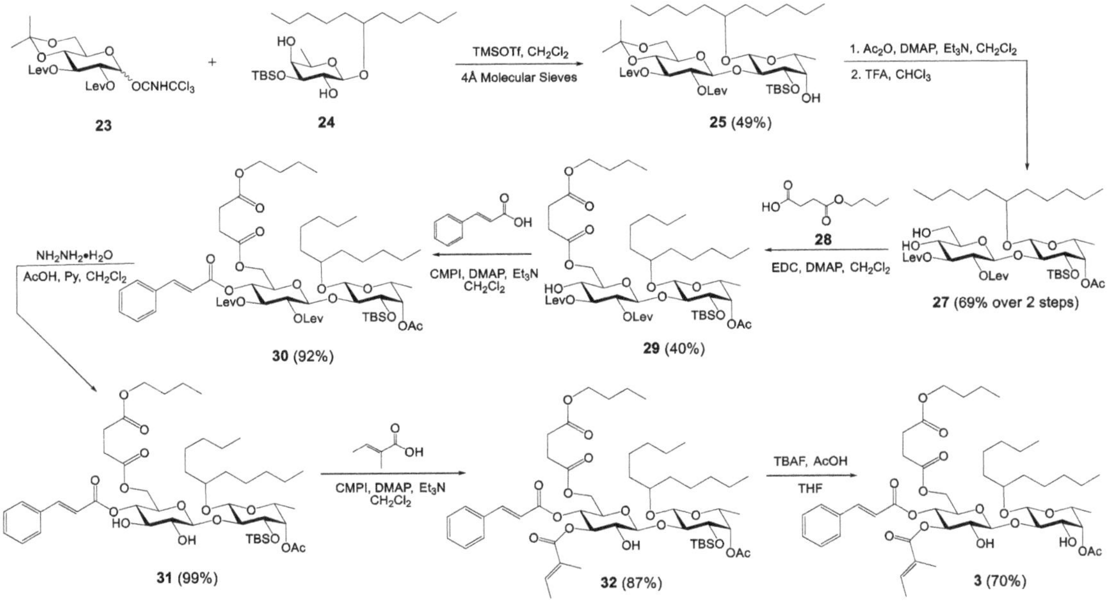

Scheme 1. Synthesis of the rationally designed, new open-chain analogue (**3**) of Ipom-F.

We, therefore, considered the possibility that the chiral starting material (*S*)-4-hydroxy-1-nonene (which requires a three-step synthesis from an expensive, chiral reagent) could be replaced with a simpler and cheaper alternative; the achiral, but two extra methylene unit-containing, 6-undecanol. While uncertain about how loss of the 11*S* chiral center may impact compound potency, we postulated that the increased lipophilicity from the extra two methylene units may compensate for the likely significant loss in activity following its removal [8].

To this end, and before embarking on its synthesis, we sought to use molecular docking to evaluate the potential interactions of the stereochemically simplified and synthetically more accessible open-chain analogue **3** in the inhibited state of the Sec61α channel pore (cf. Figures 3 and S4). Due to the limitation on the number of rotatable bonds that can be considered (maximum of 32 in AutoDock4.2), we docked the closely related compounds **3a** and **3b** instead (Figure 5A; respectively 32 and 31 rotatable bonds, with and without the 11*S* chiral center). Similar to the open-chain analogue 2, both analogues are predicted to bind both the mycolactone binding site and the middle of the lateral gate (Figure 5B). When positioned at the groove of mycolactone, **3a** and **3b** preferentially interact with the TM2, TM7, CL4 and CL8 regions of Sec61α (Figure S4A,C) with predicted binding affinities of -4.15 ± 0.6 kcal/mol and -4.78 ± 0.6 kcal/mol, respectively. Similar to Ipom-F and analogue **2**, **3b** preferentially forms hydrogen bonds with Gln170 and Lys171 in the CL4 region and a hydrogen bond was identified in four out of five independent docking simulations. In contrast to **3b**, hydrogen bonding between **3a** and the CL4 region (Gln170 and Lys171) was only observed in two out of five independent docking simulations. This suggests that **3b** interacts more strongly with CL4 than **3a**.

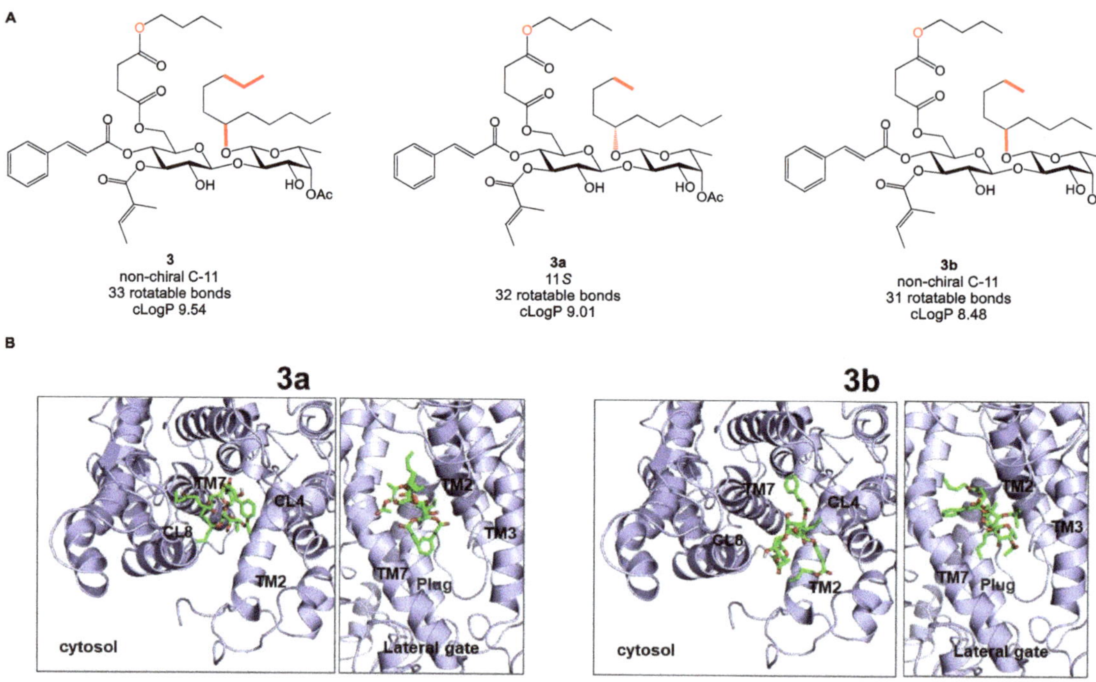

Figure 5. Molecular docking of **3a** and **3b**. (**A**) Structures of the closely related open-chain analogues **3**, **3a** and **3b**, which differ in C-11 chirality, number of rotatable bonds and lipophilicity. cLogP values were calculated in ChemDraw. (**B**) The most favorable docking conformations of **3a** and **3b** in the inhibited conformation of human-Sec61α. The 'cytosol' view represents a top-view of Sec61α from the cytosolic side and focuses on the groove between TM2, TM7, CL8 and CL4, while the 'lateral gate' view represents a side-view from the lipid bilayer. The homology model of human-Sec61α in the inhibited state used for docking is shown in Figure 4A.

When bound in the middle of the lateral gate, both analogues are predicted to be in contact with the plug region, TM2, TM3 and TM7 (Figure S4B,C) with binding affinities of -4.7 ± 0.2 kcal/mol and -5.26 ± 0.9 kcal/mol, respectively. Furthermore, both analogues form hydrogen bonds (observed in four out of five independent docking simulations) with Thr86 (TM2) and Gln127 (TM3) when bound at the lateral gate. In that position, the open-chain analogue **2** also likely forms a hydrogen bond with Thr86 (TM2), while this bond was not observed in any docking simulation for the closed-chain analogue **22**.

Taken together, our molecular docking studies of **3a** and **3b** suggested that removal of the 11*S* chiral center may result in a greater number of contact residues with the CL4 region (62% and 77% respectively) and an increased potential to hydrogen bond with residues Gln170 and Lys171 (Figure S4C). Thus, we elected to remove the 11*S* center, increase the lipophilicity of the fatty acid region and bioisosterically replace the C-5 methylene with an oxygen atom in our newly designed open-chain analogue **3**.

2.6. *Synthesis of Analogue* 3

Our synthesis of analogue **3** (Scheme 1) started with the known trichloroacetimidate glucosyl donor **23** [14] and the diol fucoside acceptor **24** (see Supporting information 2–3). Regioselective glycosylation on 2-OH-Fuc*p* afforded the monohydroxy intermediate **25** in moderate yield and, after acetylation of the free 4′-OH-Fuc*p* in **25** with acetic anhydride, the isopropylidene protecting group was removed by trifluoroacetic acid to give 4′′,6′′-diol **27**. EDC-mediated Steglich esterification of 6′′-OH-Glu*p* with mono-butyl succinate **28** (see Supplementary Materials S2) successfully produced a second monohydroxy intermediate **29**, despite somewhat poor regioselectivity. The cinnamate moiety was then introduced to 4-OH-Glu*p* by the Mukaiyama reagent, 2-chloro-1-methylpyridinium iodide (CMPI), to give the intermediate **30** in excellent yield. Subsequently, both levulinoyl (Lev) groups were removed by hydrazine under buffered conditions [14,41]. In the penultimate step, the tiglate moiety was selectively introduced to 3′′-OH-Glu*p* in **31** using CMPI to give the intermediate **32** in good yield. Finally, the TBS (*tert*-butyldimethylsilyl) group at 3′-OH-Fuc*p* was cleaved using TBAF and acetic acid in THF to give the target molecule **3**. In brief, **3** was synthesized in 7.5% yield (not optimized) over eight steps from two key monosaccharide building blocks **23** and **24**.

2.7. *Analogue* 3 *Inhibits Sec61-Mediated Protein Translocation with Potency and Selectivity Comparable to Ipom-F and* 2

Following the successful synthesis of analogue **3**, we first analyzed its effects on the in vitro membrane integration of our two model protein substrates, Ii and GypC (Figure 6A–C). In the first instance, 1 μM analogue **3** inhibited the membrane integration of Ii (Figure 6A, lane 4 versus lane 1) to a level comparable to that of Ipom-F, **2** and **5** (circa ~73–86% reduction in Ii membrane integration; Figure 3B,C). In the second instance, the same concentration of analogue **3** did not affect the membrane integration of GypC (Figure 6A, lanes 13–14 and Figure 6B), confirming the selective inhibition of Sec61-mediated protein translocation by analogue **3** in ER-derived microsomes. Thus, we proceeded to determine the estimated IC_{50} of analogue **3** on the in vitro membrane integration of Ii (Figure 6A, lanes 1–12, Figures 3D and S3), which allowed us to rank order analogue **3** as the third most potent compound of the seven tested in this study: Ipom-F (IC_{50}: 155 nM) > **2** (IC_{50}: 202 nM) > **3** (IC_{50}: 242 nM) > **5** (IC_{50}: 291 nM) > **22** (IC_{50}: 334 nM) > **1** (IC_{50}: 562 nM) > **6** (negligible inhibition at 1 μM).

When we analyzed analogue **3** in our resazurin-based cell viability assay (cf. Figure 3F) using HCT116 Sec61-WT cells (Figures 6D,E and S3), we found it to be the second most cytotoxic compound of the six that we performed IC_{50} analyses for: Ipom-F (IC_{50}: 18 nM) > **3** (IC_{50}: 40 nM) > **2** (IC_{50}: 41 nM) = **5** (IC_{50}: 41 nM) > **22** (IC_{50}: 70 nM) > **1** (IC_{50}: 170 nM). Furthermore, and as anticipated, cell death was not observed in resistance-conferring HCT116 Sec61-G80W mutant cells treated with the same concentration of analogue **3** that induced ~58% cell death in HCT116 Sec61-WT cells (Figure 6D). We, therefore, conclude that our chosen combination of advantageous chemical features permits the macrocyclic ring opening and removal of the 11*S* chiral center of Ipom-F without a significant loss in the potency or selectivity of the inhibition of Sec61-mediated protein translocation at the ER.

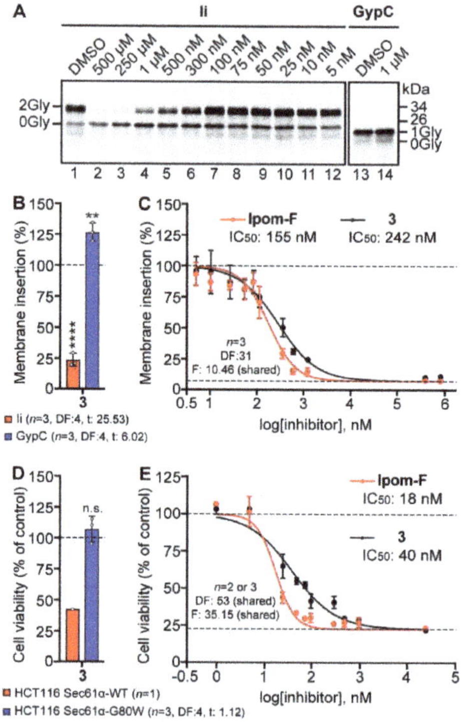

Figure 6. Analogue **3** inhibits Sec61-mediated protein translocation with potency and selectivity comparable to Ipom-F and **2**. (**A**) Phosphorimages of the membrane associated products of Ii (lanes 1–12) and GypC (lanes 13–14) synthesized in the presence and absence of 500 µM–5 nM concentrations of analogue **3**. (**B**) Quantification of the efficiency of membrane integration of Ii and GypC in the presence of 1 µM **3**, expressed relative to the control (100%). (**C**) IC_{50} curve derived from the Ii insertion data of **3** in A, compared to that of Ipom-F (also shown in Figure 3E). (**D**) Cell viability of HCT116 Sec61α-WT and HCT116 Sec61α-G80W cells treated with 50 nM **3**. (**E**) IC_{50} curve derived from the cell viability of HCT116 Sec61α-WT cells treated with 25 µM–1 nM concentrations of **3**, compared to that of Ipom-F (also shown in Figure 3F). See methods for details of biological replicates and statistical analyses. Statistical significance is given as n.s., non-significant n.s., non-significant $p > 0.1$; **, $p < 0.01$ and ****, $p < 0.0001$.

3. Materials and Methods

3.1. Chemical Synthesis General Methods

All reaction glassware was thoroughly washed and oven dried before any reactions were undertaken. Unless otherwise stated, all commercially obtained reagents were used without further purification and all reactions were conducted under argon atmosphere. Reaction progress was monitored by TLC using silica gel MF254 glass back plates with detection under UV lamp (254 nM) or charring with 5 % (v/v) H_2SO_4 (sulfuric acid) in EtOH (ethanol). Column chromatographic purifications were performed using silica gel (70–230 mesh) with a ratio that spanned from 100 to 50: 1 (w/w) between the silica gel and crude products. All ^{1}H nMR spectra were obtained in deuterated chloroform ($CDCl_3$), deuterated methanol (CD_3OD) or deuterated dimethyl sulfoxide (($CD_3)_2SO$ using chloroform ($CHCl_3$, δ = 7.27), methanol (CH_3OH, δ = 3.31) or dimethyl sulfoxide (($CH_3)_2SO$, δ = 2.50) as an internal reference for ^{1}H. All ^{13}C nMR spectra were proton decoupled and obtained in $CDCl_3$, CD_3OD or $(CD_3)_2SO$ with $CHCl_3$ (δ = 77.0), CH_3OH (δ = 49.9) or $CH_3)_2SO$ (δ = 40.4) as internal references for ^{13}C. nMR data are reported in the form:

chemical shifts (*δ*) in ppm, multiplicity, coupling constants (*J*) in Hz, and integrations. ^{1}H data are reported as though they were first order. An error less than 0.5 Hz is reported for coupling constants between two coupled protons. Other 1D and 2D nMR spectra, such as ^{135}DEPT, COSY, HMQC and HMBC, were collected in addition to ^{1}H and ^{13}C in the characterization of new compounds. Low-resolution mass spectra (LRMS) were obtained on a LTQ XL mass spectrometer equipped with an electrospray ion source using the positive ion mode and connected to a linear ion trap mass analyzer. Purity was analyzed using a Waters HPLC with a photodiode array (PDA) detector, a DIONEX Acclaim® 120 reverse phase column (C18, 5 μm, 120Å, 4.6 × 150 mm) and an isocratic mobile phase of 83% acetonitrile in water at a flow rate of 1.5 mL/min.

3.1.1. Synthesis of Compound **25**

The fucoside diol acceptor **24** (339.9 mg, 0.79 mmol), known glucoside trichloroacetimidate donor **23** [14] (478.1 mg, 0.85 mmol, 1.08 eq.) and crushed activated 4Å molecular sieves (1 g) were suspended in anhydrous CH_2Cl_2 (7 mL). The mixture was stirred under an argon atmosphere for ~30 min at room temperature and then cooled to −78 °C. TMSOTf (14.2 μL, 0.079 mmol, 0.1 eq.) was added dropwise via a syringe and the reaction mixture was gradually warmed to −20 °C over ~1 h. The reaction mixture was then quenched by the addition of Et_3N and filtered through a pad of celite and the filtrate concentrated. The resulting residue was purified by column chromatography (15:1→6:1 hexanes–EtOAc) to acquire pure **25** as a colorless oil (315.1 mg, 49%): R_f 0.56 (4:1 hexanes–EtOAc); ^{1}H nMR (400 MHz, $CDCl_3$, δ_H) 5.10 (d, *J* = 7.8 Hz, 1H), 5.03 (t, *J* = 9.4 Hz, 1H), 4.83 (dd, *J* = 10.4 Hz, *J* = 7.9 Hz, 1H), 4.21 (d, *J* = 7.5 Hz, 1H), 3.91 (dd, *J* = 10.9 Hz, *J* = 5.3 Hz, 1H), 3.65–3.79 (m, 3H), 3.62 (t, *J* = 9.6 Hz, 1H), 3.42–3.57 (m, 3H), 3.21 (qd, *J* = 9.7 Hz, *J* = 5.5 Hz, 1H), 2.45–2.81 (m, 9H), 2.13 (s, 6H), 1.10–1.57 (m, 25H), 0.80–0.95 (m, 15H), 0.12 (s, 3H), 0.07 (s, 3H); ^{13}C nMR (100 MHz, $CDCl_3$, δ_C) 206.5 (C=O), 206.4 (C=O), 172.1 (C=O), 171.3 (C=O), 101.1 (O_2CH), 99.8 (O_2C), 99.6 (O_2CH), 79.8 (OCH), 76.0 (OCH), 75.6 (OCH), 73.9 (OCH), 72.8 (OCH), 72.7 (OCH), 71.5 (OCH), 69.6 (OCH), 67.5 (OCH), 62.3 (OCH_2), 37.8 (CH_2), 37.7 (CH_2), 34.5 (CH_2), 33.7 (CH_2), 32.3 (CH_2), 32.1 (CH_2), 29.9(0) (CH_3), 29.8(9) (CH_3), 29.0 (CH_3), 28.0 (CH_2), 27.9 (CH_2), 26.1 ($C(\underline{C}H_3)_3$), 24.9 (CH_2), 24.6 (CH_2), 22.9 (CH_2), 22.7 (CH_2), 18.9 (CH_3), 18.1 ($Si\underline{C}(CH_3)_3$), 16.5 (CH_3), 14.3 (CH_3), 14.2 (CH_3), −4.1 ($SiCH_3$), −4.5 ($SiCH_3$).

3.1.2. Synthesis of Compound **26**

To an ice-cold solution of **25** (315.1 mg, 0.38 mmol), DMAP (4.6 mg, 0.1 eq.) and Et_3N (157.6 μL, 1.14 mmol, 3.0 eq.) in DCM (3 mL), Ac_2O was added (44.8 μL, 0.47 mmol, 1.25 eq.). The mixture was warmed to room temperature overnight and the reaction was then quenched by the addition of methanol. The resulting solution was extracted with DCM and the organic fractions washed with 1N HCl and saturated $NaHCO_3$. The collected organic layer was dried over Na_2SO_4, filtered and concentrated under vacuum to yield the crude product as a pale yellow syrup (288.3 mg), which was used directly for the next step without further purification.

3.1.3. Synthesis of Compound **27**

To a solution of **26** (288.3 mg, 0.33 mmol) in $CHCl_3$ (5 mL) trifluoroacetic acid (TFA) was added (126.9 μL, 1.65 mmol, 5.0 eq.). The mixture was stirred at room temperature for 4 h and then quenched by Et_3N. After evaporation, the residue was purified by column chromatography (8:1→2:1, hexanes–EtOAc) to afford **27** as a colorless syrup (217.6 mg, 69% over two steps): R_f 0.43 (2:1 hexanes–EtOAc); ^{1}H nMR (400 MHz, $CDCl_3$, δ_H) 4.92–5.02 (m, 3H), 4.83 (dd, *J* = 9.7 Hz, *J* = 7.8 Hz, 1H), 4.20 (d, *J* = 7.7 Hz, 1H), 3.79–3.93 (m, 2H), 3.62–3.79 (m, 3H), 3.43–3.61 (m, 3H), 3.32–3.40 (m, 1H), 2.68–2.80 (m, 3H), 2.41–2.68 (m, 6H), 2.13 (s, 3H), 2.12 (s, 3H), 2.09 (s, 3H), 1.13–1.61 (m, 16H), 1.07 (d, *J* = 6.4 Hz, 3H), 0.75–0.94 (m, 15H), 0.09 (s, 3H), 0.05 (s, 3H); ^{13}C nMR (100 MHz, $CDCl_3$, δ_C) 208.3 (C=O), 206.3 (C=O), 173.1 (C=O), 171.4 (C=O), 171.0 (C=O), 101.5 (O_2CH), 98.8 (O_2CH), 81.5 (OCH), 76.4 (OCH), 75.0 (OCH), 74.9 (OCH), 73.7 (OCH), 73.4 (OCH), 72.3 (OCH), 69.8 (OCH),

68.8 (OCH), 62.1 (OCH_2), 38.4 (CH_2), 37.8 (CH_2), 34.6 (CH_2), 34.0 (CH_2), 32.2 (CH_2), 32.0 (CH_2), 29.8(8) (CH_3), 29.8(5) (CH_3), 28.1 (CH_2), 28.0 (CH_2), 25.9 ($C(\underline{C}H_3)_3$), 24.8 ($2xCH_2$), 22.8 (CH_2), 22.7 (CH_2), 21.1 (CH_3), 17.8 ($Si\underline{C}(CH_3)_3$), 16.6 (CH_3), 14.3 (CH_3), 14.2 (CH_3), −4.2 ($SiCH_3$), −4.5 ($SiCH_3$).

3.1.4. Synthesis of Compound **29**

To an ice-cold solution of **27** (217.6 mg, 0.26 mmol), EDC (125.1 mg, 0.65 mmol, 2.5 eq.) and DMAP (8.0 mg, 0.065 mmol, 0.25 eq.) in DCM (7 mL), **28** was added (52.5 mg, 0.30 mmol, 1.16 eq.), and the reaction was warmed to room temperature overnight. The solution was then washed with 1N HCl and saturated $NaHCO_3$ and the collected organic fractions were dried over Na_2SO_4, filtered and concentrated under vacuum. The resulting residue was then purified by column chromatography (8:1→2:1, hexanes–EtOAc) to afford **29** as a colorless syrup (104.0 mg, 40%): R_f 0.52 (1:1 hexanes–EtOAc); ^{1}H nMR (400 MHz, $CDCl_3$, δ_H) 5.03 (d, J = 7.8 Hz, 1H), 4.91–5.01 (m, 2H), 4.83 (dd, J = 9.7 Hz, J = 7.8 Hz, 1H), 4.41 (dd, J = 8.1 Hz, J = 4.2 Hz, 1H), 4.29 (dd, J = 12.0 Hz, J = 2.3 Hz, 1H), 4.23 (d, J = 7.6 Hz, 1H), 4.04 (t, J = 6.7 Hz, 2H), 3.85 (dd, J = 9.1 Hz, J = 7.6 Hz, 1H), 3.76 (dd, J = 9.4 Hz, J = 3.5 Hz, 1H), 3.40–3.66 (m, 5H), 2.41–2.85 (m, 12H), 2.13 (s, 3H), 2.11 (s, 3H), 2.07 (s, 3H), 1.14–1.64 (m, 20H), 1.07 (d, J = 6.4 Hz, 3H), 0.77–0.96 (m, 18H), 0.09 (s, 3H), 0.05 (s, 3H); ^{13}C nMR (100 MHz, $CDCl_3$, δ_C) 207.9 (C=O), 206.3 (C=O), 173.0 (C=O), 172.6 (C=O), 172.5 (C=O), 171.3 (C=O), 170.9 (C=O), 101.4 (O_2CH), 98.9 (O_2CH), 81.0 (OCH), 76.0 (OCH), 75.1 (OCH), 73.7(5) (OCH), 73.7(0) (OCH), 73.6(6) (OCH), 72.5 (OCH), 69.1 (OCH), 68.8 (OCH), 64.8 (OCH_2), 63.3 (OCH_2), 38.3 (CH_2), 37.8 (CH_2), 34.5 (CH_2), 33.9 (CH_2), 32.2 (CH_2), 32.1 (CH_2), 30.6 (CH_2), 29.9 (CH_3), 29.8 (CH_3), 29.2 (CH_2), 29.0 (CH_2), 28.1 (CH_2), 28.0 (CH_2), 25.9 ($C(\underline{C}H_3)_3$), 24.8 (CH_2), 24.7 (CH_2), 22.9 (CH_2), 22.7 (CH_2), 21.0 (CH_3), 19.1 (CH_2), 17.8 ($Si\underline{C}(CH_3)_3$), 16.6 (CH_3), 14.3 (CH_3), 14.2 (CH_3), 13.8 (CH_3), −4.3 ($2xSiCH_3$).

3.1.5. Synthesis of Compound **30**

To an ice-cold solution of **29** (104.0 mg, 0.105 mmol), cinnamic acid (23.4 mg, 0.158 mmol, 1.5 eq.), CMPI (53.7 mg, 0.210 mmol, 2.0 eq.) and DMAP (12.8 mg, 0.105 mmol, 1.0 eq.) in DCM (4 mL), Et_3N was added (72.9 µL, 0.525 mmol, 5.0 eq.). The mixture was warmed to room temperature overnight. After evaporation, the resulting residue was then purified by column chromatography (10:1→4:1, hexanes–EtOAc) to afford **30** as a colorless syrup (108.1 mg, 92%): R_f 0.46 (2:1 hexanes–EtOAc); ^{1}H nMR (400 MHz, $CDCl_3$, δ_H) 7.62 (d, J = 16.0 Hz, 1H), 7.43–7.53 (m, 2H), 7.29–7.38 (m, 3H), 6.33 (d, J = 16.0 Hz, 1H), 5.06–5.25 (m, 3H), 4.91–5.00 (m, 2H), 4.20–4.28 (m, 2H), 4.16 (dd, J = 12.1 Hz, J = 2.9 Hz, 1H), 4.01 (t, J = 6.7 Hz, 2H), 3.89 (dd, J = 9.3 Hz, J = 7.6 Hz, 1H), 3.80 (dd, J = 9.4 Hz, J = 3.5 Hz, 1H), 3.45–3.70 (m, 3H), 2.35–2.80 (m, 12H), 2.12 (s, 3H), 2.08 (s, 3H), 2.03 (s, 3H), 1.15–1.61 (m, 20H), 1.08 (d, J = 6.4 Hz, 3H), 0.80–0.98 (m, 18H), 0.13 (s, 3H), 0.08 (s, 3H); ^{13}C nMR (100 MHz, $CDCl_3$, δ_C) 206.4 (C=O), 206.2 (C=O), 172.2 (C=O), 172.0 (2xC=O), 171.3 (C=O), 170.9 (C=O), 165.4 (C=O), 146.5 (=CH), 134.1 (=C), 130.7 (=CH), 129.0 (2x=CH), 128.4 (2x=CH), 116.7 (=CH), 101.4 (O_2CH), 98.9 (O_2CH), 81.3 (OCH), 75.1 (OCH), 73.8 (OCH), 73.6 (OCH), 73.0 (OCH), 72.5 (OCH), 71.8 (OCH), 68.9 (OCH), 68.8 (OCH), 64.6 (OCH_2), 62.8 (OCH_2), 37.8(3) (CH_2), 37.7(6) (CH_2), 34.5 (CH_2), 34.0 (CH_2), 32.3 (CH_2), 32.1 (CH_2), 30.6 (CH_2), 29.9 (CH_3), 29.6 (CH_3), 29.1 (CH_2), 29.0 (CH_2), 28.0 (CH_2), 27.9 (CH_2), 26.0 ($C(\underline{C}H_3)_3$), 25.0 (CH_2), 24.7 (CH_2), 23.0 (CH_2), 22.7 (CH_2), 21.0 (CH_3), 19.1 (CH_2), 17.8 ($Si\underline{C}(CH_3)_3$), 16.6 (CH_3), 14.3 (CH_3), 14.2 (CH_3), 13.8 (CH_3), −4.3 ($SiCH_3$), −4.4 ($SiCH_3$).

3.1.6. Synthesis of Compound **31**

To an ice-cold solution of **30** (108.1 mg, 0.0966 mmol) in DCM (3 mL), a buffer solution of hydrazine monohydrate was added (25.8 µL, 0.532 mmol, 5.5 eq.) in acetic acid (354.1 µL) and pyridine (531.1 µL). The mixture was warmed to room temperature and stirred for 4 h, then quenched by acetone, diluted with DCM and washed with brine. The organic layer was dried over Na_2SO_4, filtered and concentrated under vacuum. The resulting residue was subsequently purified by column chromatography (10:1→4:1, hexanes–EtOAc) to afford **31** as a colorless syrup (88.3 mg, 99%): R_f 0.57 (2:1 hexanes–EtOAc); ^{1}H nMR (400 MHz, $CDCl_3$, δ_H) 7.69 (d, J = 16.0 Hz, 1H), 7.45–7.55 (m, 2H), 7.30–7.40 (m, 3H), 6.41 (d, J = 16.0 Hz, 1H), 5.12 (t, J = 9.0 Hz, 1H), 5.05 (br s, 1H), 4.60 (d, J = 7.7 Hz, 1H), 4.39 (d, J = 7.4 Hz, 1H), 4.26 (dd, J = 11.3 Hz, J = 3.7 Hz, 1H), 4.14 (br d, J = 12.0 Hz, 1H), 4.03 (t, J = 6.7 Hz, 2H), 3.85–3.94 (m, 2H), 3.55–3.84 (m, 5H), 3.44 (t, J = 8.0 Hz, 1H), 2.50–2.73 (m, 5H), 2.10 (s, 3H), 1.17–1.68 (m, 20H), 1.12 (d, J = 6.4 Hz, 3H), 0.80–0.95 (m, 18H), 0.17 (s, 3H), 0.13 (s, 3H); ^{13}C nMR (100 MHz, $CDCl_3$, δ_C) 172.3 (C=O), 172.2 (C=O), 170.8 (C=O), 166.1 (C=O), 146.4 (=CH), 134.2 (=C), 130.7 (=CH), 129.0 (2x=CH), 128.3 (2x=CH), 117.1 (=CH), 103.6 (O_2CH), 101.2 (O_2CH), 79.9 (OCH), 78.7 (OCH), 76.0 (OCH), 73.7 (OCH), 73.1 (OCH), 72.9 (OCH), 72.8 (OCH), 70.1 (OCH), 68.8 (OCH), 64.6 (OCH_2), 62.8 (OCH_2), 34.1 (CH_2), 33.4 (CH_2), 32.2 (CH_2), 32.0 (CH_2), 30.7 (CH_2), 29.1 (CH_2), 29.0 (CH_2), 26.0 ($C(\underline{C}H_3)_3$), 25.0 (CH_2), 24.4 (CH_2), 22.7 ($2xCH_2$), 20.9 (CH_3), 19.2 (CH_2), 18.0 ($Si\underline{C}(CH_3)_3$), 16.6 (CH_3), 14.2(0) (CH_3), 14.1(8) (CH_3), 13.8 (CH_3), −4.2 ($SiCH_3$), −4.5 ($SiCH_3$).

3.1.7. Synthesis of Compound **32**

To an ice-cold solution of **31** (96.0 mg, 0.104 mmol), tiglic acid (15.5 mg, 0.155 mmol, 1.5 eq.), CMPI (105.7 mg, 0.414 mmol, 4.0 eq.) and DMAP (12.7 mg, 0.104 mmol, 1.0 eq.) in DCM (2 mL), Et_3N was added (72.1 µL, 0.520 mmol, 5.0 eq.). The mixture was warmed to room temperature overnight. After evaporation, the resulting residue was then purified by column chromatography (10:1→4:1, hexanes–EtOAc) to afford **32** as a colorless syrup (91.0 mg, 87%): R_f 0.55 (2:1 hexanes–EtOAc); ^{1}H nMR (400 MHz, $CDCl_3$, δ_H) 7.59 (d, J = 16.0 Hz, 1H), 7.42–7.51 (m, 2H), 7.30–7.39 (m, 3H), 6.74–6.85 (m, 1H), 6.30 (d, J = 16.0 Hz, 1H), 5.20–5.34 (m, 2H), 5.04 (d, J = 3.3 Hz, 1H), 4.73 (d, J = 7.7 Hz, 1H), 4.39 (d, J = 7.6 Hz, 1H), 4.26 (dd, J = 11.3 Hz, J = 4.1 Hz, 1H), 4.14 (dd, J = 12.3 Hz, J = 2.3 Hz, 1H), 4.03 (t, J = 6.7 Hz, 2H), 3.92 (dd, J = 9.7 Hz, J = 7.6 Hz, 1H), 3.72–3.83 (m, 2H), 3.54–3.72 (m, 4H), 2.50–2.70 (m, 4H), 2.07 (s, 3H), 1.65–1.76 (m, 6H), 1.16–1.65 (m, 20H), 1.10 (d, J = 6.4 Hz, 3H), 0.80–0.95 (m, 18H), 0.17 (s, 3H), 0.12 (s, 3H); ^{13}C nMR (100 MHz, $CDCl_3$, δ_C) 172.2 (C=O), 172.1 (C=O), 170.9 (C=O), 167.6 (C=O), 165.5 (C=O), 146.3 (=CH), 138.3 (=CH), 134.1 (=C), 130.7 (=CH), 129.0 (2x=CH), 128.3 (2x=CH), 128.0 (=C), 116.8 (=CH), 103.9 (O_2CH), 101.1 (O_2CH), 79.6 (OCH), 78.7 (OCH), 74.4 (OCH), 73.2 (OCH), 72.9 (OCH), 72.7(7) (OCH), 72.7(6) (OCH), 68.8 (OCH), 68.4 (OCH), 64.6 (OCH_2), 62.7 (OCH_2), 34.0 (CH_2), 33.4 (CH_2), 32.2 (CH_2), 32.0 (CH_2), 30.7 (CH_2), 29.1 (CH_2), 29.0 (CH_2), 25.9 ($C(\underline{C}H_3)_3$), 25.0 (CH_2), 24.4 (CH_2), 22.8 (CH_2), 22.7 (CH_2), 21.0 (CH_3), 19.1 (CH_2), 17.9 ($Si\underline{C}(CH_3)_3$), 16.6 (CH_3), 14.5 (CH_3), 14.2(1) (CH_3), 14.1(9) (CH_3), 13.8 (CH_3), 12.1 (CH_3), −4.3 ($SiCH_3$), −4.5 ($SiCH_3$).

3.1.8. Synthesis of Analogue **3**

To a solution of **32** (83.4 mg, 0.083 mmol) and acetic acid (190 µL, 3.32 mmol, 40 eq.) in THF (3 mL), tetra-*n*-butylammonium fluoride (TBAF) solution was added in THF (1.0 M, 1.66 mL, 1.66 mmol, 20 eq.) at room temperature and the mixture was stirred overnight. The solution was diluted with CH_2Cl_2 and successively washed with 1N HCl, saturated aqueous $NaHCO_3$ and brine. The collected organic layer was then dried over Na_2SO_4 and filtered. The filtrate was concentrated under vacuum and the resulting residue purified by column chromatography (8:1→2:1, hexanes–EtOAc) to afford **3** as a colorless to pale yellow syrup (51.7 mg, 70%): R_f 0.31 (2:1 hexanes–EtOAc); ^{1}H nMR (400 MHz, $CDCl_3$, δ_H) 7.62 (d, J = 16.0 Hz, 1H), 7.43–7.54 (m, 2H), 7.30–7.42 (m, 3H), 6.79–6.91 (m, 1H), 6.31 (d, J = 16.0 Hz, 1H), 5.13–5.31 (m, 3H), 4.70 (d, J = 8.2 Hz, 1H), 4.38 (d, J = 7.3 Hz, 1H), 4.31 (br s, 1H), 4.15–4.26 (m, 2H), 4.04 (t, J = 6.7 Hz, 2H), 3.62–3.89 (m, 7H), 2.50–2.69 (m, 4H), 2.18 (s, 3H), 1.66–1.77 (m, 6H), 1.18–1.61 (m, 20H), 1.16 (d, J = 6.4 Hz, 3H), 0.80–0.95 (m, 9H); ^{13}C nMR (100 MHz, $CDCl_3$, δ_C) 172.3 (C=O), 172.1 (C=O), 171.4 (C=O), 168.0 (C=O), 165.6 (C=O), 146.6 (=CH), 139.1 (=CH), 134.0 (=C), 130.8 (=CH), 129s.0 (2x=CH), 128.4 (2x=CH), 127.8 (=C), 116.5 (=CH), 102.7 (O_2CH), 99.6 (O_2CH), 79.3 (OCH), 77.6 (OCH), 74.3 (OCH), 72.8 (OCH), 72.2 (OCH), 71.8 (OCH), 70.8 (OCH), 69.3 (OCH), 68.3 (OCH), 64.7 (OCH_2), 62.6 (OCH_2), 34.4 (CH_2), 33.4 (CH_2), 32.0 (CH_2), 31.9 (CH_2), 30.7 (CH_2), 29.1 (CH_2), 29.0 (CH_2), 24.8 ($2xCH_2$), 22.7 ($2xCH_2$), 21.1 (CH_3), 19.2 (CH_2), 16.3 (CH_3), 14.6 (CH_3), 14.1(9) (CH_3),

14.1(5) (CH_3), 13.8 (CH_3), 12.1 (CH_3). LRMS (ESI) m/z calcd for $C_{47}H_{70}NaO_{16}$ $[M+Na]^+$: 913. Found: 913. Purity: 97.1% ($MeCN/H_2O$ 83:17; 1.5 mL/min, t_R = 14.9 min, Figure S5).

3.2. Biological Analysis

All compounds from stock solutions in DMSO, or an equivalent volume of DMSO, were included at 5% (v/v) in membrane insertion assays or 20% (v/v) in cytotoxicity assays.

3.2.1. In Vitro Membrane Insertion Assay

Linear DNA of the short form of human HLA class II histocompatibility antigen gamma chain (Ii; P04232, isoform 2, residues 17–232) or human glycophorin C (GypC; P04921) were generated by PCR and transcribed into RNA using T7 RNA polymerase (Promega). Membrane insertion assays (20 μL, 1 h at 30 °C, containing 6.5% (v/v) nuclease-treated ER microsomes (from stock with OD280 = 44/mL)), endoglycosidase H_f (New England Biolabs) treatment, sample resolution by SDS-PAGE (16% polyacrylamide gels) and gel drying were performed as previously described [13,14,37,38,42]. Following exposure to a phosphorimaging plate for 24–72 h, radiolabeled products were visualized using a Typhoon FLA-7000 (GE Healthcare, Tokyo, Japan) and the ratio of the signal intensity for the N-glycosylated (XGly) and non-glycosylated (0Gly) forms obtained using AIDA v.5.0 (Raytest Isotopenmeβgeräte). This value was then expressed relative to the matched DMSO control (set to 100%) in order to estimate the mean relative insertion (± SEM) from insertion experiments performed in triplicate (n = 3, biologically independent experiments). IC_{50} value estimates were determined in Prism 8 (GraphPad, San Diego, CA, USA) using nonlinear regression to fit data to a curve of variable slope (four parameters) using the least-squares fitting method, with the top and bottom plateaus of the curve defined as 100% and 7.67% (the mean of all data at 500 μM across all compounds), respectively [13,14].

3.2.2. Cell Culture and Resazurin-Based Viability Assays

The human breast cancer cell line (MDA-MB-231) was maintained in a DMEM high glucose culture medium supplemented with 10% (v/v) fetal bovine serum (FBS) and 2 mM *L*-glutamine. Parental (Sec61α-WT) [13] HCT-116 (human colorectal cancer cells, ATCC, CCL-247) and mutant (Sec61α-G80W) [22] HCT-116 cells were maintained in McCoy's 5A (modified) medium ((ThermoFisher, Waltham, MA, USA, 16600-082)) supplemented with 10% (v/v) FBS (Gibco, 10500-064) and 100 units/mL penicillin and 100 mg/mL streptomycin (Gibco, cat: 15140-122). All cell lines were maintained in a 5% CO_2 humidified incubator at 37 °C. Cytotoxicity assays were performed in triplicate sets as previously described [13,22] and viable cells were counted using an automated cell counter (Bio-Rad TC20) immediately before each experiment. Compound stock solutions in DMSO (10 mM) were diluted with supplemented culture media (MDA-MB-231 cells: high glucose DMEM; HCT116 Sec61α-WT and Sec61α-G80W cells: McCoy's (modified) 5A) to make a series of gradient fresh working solutions at equal DMSO percentage immediately prior to each test. First, 100 mL of cells at a cell density of 5×10^4 cells/mL (MDA-MB-231 cells) or 2.5×10^4 cells/mL (Sec61α-WT or Sec61α-G80W HCT-116 cells) were seeded in black 96-well microtiter plates (Falcon, product 353219; 2500 cells/well) and incubated at 37° C for 24 h. Subsequently, the cells were treated with either 100 mL of the freshly made gradient working solution in a total volume of 200 mL/well (MDA-MB-231 cells) or 25 mL of freshly made gradient in a total volume of 125 mL/well (Sec61α-WT or Sec61α-G80W HCT-116 cells) at 37 °C for 72 h. For MDA-MB-231 cells, the media were discarded and 200 mL fresh medium containing 10% (v/v) alamarBlue HS cell viability reagent (resazurin stock solution) (ThermoFisher, Waltham, MA, USA, A50100) was added to each well. For Sec61α-WT or Sec61α-G80W HCT-116 cells, the media was not discarded and alamarBlue HS cell viability reagent (resazurin stock solution) (ThermoFisher, Waltham, MA, USA, A50100) was added to 10% (v/v). After that, all cells were incubated at 37 °C for a further 1–3 h and the emission of each well at 620 nM was detected using a Synergy H1 Hybrid multi-mode plate reader (BioTek, Agilent Technologies, Palo Alto, CA, USA) at excitation 580 nM. The percentage viability compared to the negative control (DMSO-treated cells) was determined and Prism 6 or 8 (GraphPad, San Diego, CA, USA) used to make a plot of viability (%) versus sample concentration and to calculate the concentration at which each compound exhibited 50% cytotoxicity (IC_{50}). IC_{50} value estimates were determined using nonlinear regression to fit data to a curve of variable slope (four parameters) using the least-squares fitting method. For HCT-116 cell IC_{50} curves, the top and bottom plateaus were defined as 100% and 23.70% (the mean of all data at 25 μM across all compounds), respectively.

3.2.3. Quantification and Statistical Analysis

Quantification procedures used in in vitro and in cellula experiments are described in Sections 3.2.1 and 3.2.2. For all in vitro data, quantifications are given as means ± SEM for independent membrane insertion experiments performed in triplicate ($n = 3$, biologically independent experiments) and statistical significance with respect to DMSO controls (set as 100%) was determined using Tukey's multiple comparison test (Figure 3C, two-way ANOVA) or unpaired t tests (Figure 6B, one-way ANOVA). For in cellula data using HCT116 Sec61α-WT cells, quantifications normalized to the DMSO control (set to 100%) are from one experiment (Figures 3G and 6D left, $n = 1$) or given as means ± SEM from two (Figures 3H and 6E, $n = 2$: Ipom-F, **22** and **3** (5–1 nM), **1**, **2** and **5** (1 nM)) or three (Figures 3H and 6E, $n = 3$: Ipom-F, **22** and **3** (25 μM–25 nM), **1**, **2** and **5** (25 μM–5 nM)) independent resazurin-based cytotoxicity experiments. For all in cellula data using HCT116 Sec61α-G80W cells, quantifications normalized to the DMSO control (set to 100%) are given as means ± SEM from three independent resazurin-based cytotoxicity screens ($n = 3$). Statistical significance comparing the viability of HCT116 Sec61α-WT and HCT116 Sec61α-G80W cells was determined by ordinary one-way ANOVA and Dunnett's multiple comparisons test (Figure 3G) or an unpaired t test (Figure 6E). In all cases, DF and F values are depicted in the appropriate figures and statistical significance is given as n.s., non-significant $p > 0.1$; *, $p < 0.05$; **, $p < 0.01$; ***, $p < 0.001$ and ****, $p < 0.0001$.

3.3. Homology Modeling and Docking Protocols

The 476 amino acid protein sequence of human Sec61α isoform 1 was retrieved from Uniprot (ID: P61619). The crystal structure of human Sec61α is not available; however, crystal structures of mammalian (canine) Sec61α have been reported [22]. Human Sec61α and Sec61α from *Canis lupus* (Uniprot ID: P38377) share 99.8% sequence identity. Hence, homology modeling was carried out to generate a three-dimensional conformational model of human Sec61α using a cryo-EM structure of the "inhibited state" of canine Sec61α as a template. Precisely, we used the cryo-EM structure reported for the inhibited state of canine Sec61α in the presence of mycolactone (6Z3T with resolution 2.6 Å) [8]. We added structural information for the missing part of 6Z3T by homology modeling based on 2WWB (EM structure with resolution 6.48 Å) [43]. The combined structure using 6Z3T and 2WWB was used as template for human Sec61α in homology modeling that was performed using MODELLER 9.21 [44]. After sequence alignment of target and template, MODELLER 9.21 was run locally with the automodel class to generate 50 different models. The model with the lowest DOPE score was selected as the final model and subjected to 1000 steps of energy minimization with the steepest descent algorithm, using the GROMACS (Available online: https://manual.gromacs.org/current/install-guide/index.html (accessed on 2 June 2022))package (version 5.0.7) [45] to relax side chain atoms. All compounds were modeled using structures drawn in ChemDraw Professional (CambridgeSoft, Waltham, MA, USA).

Docking of compounds was conducted using AutoDock4.2 [46] to predict energetically favorable binding poses of the compound inside or on the surface of human Sec61α. The docking calculations were performed in two consecutive steps. In the first docking step, we adopted a relatively large grid box (100 Å × 100 Å × 126 Å) covering the entire cavity of Sec61α, because the binding site(s) of these compounds are unknown. The Lamarckian genetic algorithm was employed with a population size of 150, 27×10^3 generations and

25×10^5 energy evaluations. All other docking parameters were set to the default values of AutoDock4.2. 1000 individual docking results were clustered according to a threshold for structural similarity of 2.0 Å RMSD. In each cluster, the representative conformation was set to the one with the lowest binding free energy for that cluster. Three independent sets of 1000 docking runs each were conducted in the first stage.

The first docking stage revealed that the compounds **22**, **2**, **3a** and **3b** dock favorably at two locations (the binding site of mycolactone and the lateral gate) within 1 kcal/mol. However, alternative poses for Ipom-F, **5** and **1** had predicted binding scores that are ~4, ~4 and ~2 kcal/mol less favorable than their best poses, respectively. Therefore, two small grid boxes are used at two different locations for compounds **22**, **2**, **3a** and **3b** in the second docking stage.

In the second docking stage, the size of the grid box was scaled down based on the population of the most stable binding positions of each compound. In this finer run, more stringent parameters were used; namely, cubic boxes of 86 Å × 86 Å × 70 Å, 0.5×10^6 generations and 100×10^6 energy evaluations. At this stage, we executed five independent fine docking runs yielding 50 docking results each. These 50 conformations were clustered similarly to those reported in a related docking study involving mycolactone [39]. The most favorable conformation of the Sec61α-analogue complex with lowest binding affinity score was selected for further analysis and considered as the final docking pose. Hydrogen bonding and contact residues ($\leq$ 4Å) were identified by LigPlot+ [47] using default parameters.

3.4. Data and Software

In vitro data were analyzed with AIDA v5.0 (Elysia-Raytest, Straubenhard, Germany). Homology modeling and docking analysis were performed with Modeller v9.24 and Autodock4.2, as described in the previous sections.

4. Conclusions

In summary, we conducted systematic studies on the first series of ring-opened analogues amongst all resin glycosides. We demonstrate that Ipom-F can be replaced with highly effective open-chain analogues that exert their cytotoxicity through the same molecular mechanism as their closed-chain counterparts; that is, by inhibiting Sec61-mediated protein translocation at the ER. The open-chain analogues **2** and **3** are defined as the most potent acyclic translocation inhibitors discovered to date. Thus, in contrast to coibamide A [35], opening of the Ipom-F macrocycle does not appear to result in a significant loss of either cytotoxicity or Sec61 inhibition (both in vitro and in cellula). We speculate that the disaccharide core provides the necessary [10] and sufficient conformational control so that the overall open-chain scaffold can still fit well into the binding pocket(s) of Sec61α. This is supported by our modeling studies, suggesting that these compounds can interact at one or more sites on the Sec61α subunit. Thus, we hypothesize that the flexibility of open-chain Ipom-F analogues may enhance their ability to perturb the normally stepwise binding of signal peptides to the Sec61 complex; a feature that may potentially be exploited towards the substrate-specific inhibition of Sec61-dependent protein clients in the future.

Synthetically, this acyclic structural framework allows us to bypass the two most challenging transition metal catalyzed reactions, namely ring-closing metathesis (RCM) and chemo-selective hydrogenation, which limit the production scale and flexibility of all current syntheses of Ipom-F [6–8]. Moreover, by incorporating the chemically advantageous features gleaned from our IC_{50} analyses of Ipom-F and analogues **1**, **2**, **5** and **22** (increased lipophilicity, expansion of the fatty acid portion, retention of the C-4 carbonyl group and bioisosteric replacement of the C-5 methylene with an oxygen atom; see also Table 1), we were able to synthesize open-chain analogues more efficiently by avoiding (i) a low-yielding Grignard reaction (10–30%) during the synthesis of the fatty acid fragment (4-oxo-8-nonenoic acid) at the 6″-OH-Glup position [8] and (ii) a three-step synthesis for the aglycone ((*S*)-4-hydroxy-1-nonene) at the C-1′-Fucp position through replacement of an expensive chiral starting reagent ((*S*)-(+)-epichlorohydrin) [8] with a commercially available, greener and non-chiral alternative (6-undecanol, cf. Scheme 1). Therefore, for the first time, we were able to remove the natural 11*S* configuration on the fatty acid chain, which has proven crucial for the biological activity of Ipom-F [8] and is a universal feature for all resin glycosides. Taken together, we have revolutionized the synthesis of a potent Sec61 inhibitor in a more scalable and flexible manner than the parent Ipom-F and present **3** as a new and the most synthetically accessible lead compound.

To conclude, the work presented here ensures future ipomoeassin research using the ring-opened scaffold, which will help our efforts to explore and exploit the complete function of the Sec61 translocon for drug discovery. More broadly, our findings may be extended to other resin glycosides that could inspire exploration of new ring-opened analogues derived from this unique category of macrolactone natural products.

Supplementary Materials: The following supporting information files can be downloaded at: https://www.mdpi.com/article/10.3390/molecules27144419/s1, Supporting information 1: NCI 60-cell line screening data of analogue **1**; Supporting information 2: Chemical syntheses and additional cytotoxicity and modeling data, including Figure S1: Structures of ring-closed analogues **13–21**; Figure S2: Correlation curves between ring-opened analogues **4–12** and ring-closed analogues **13–21**; Figure S3: Comparison of compound IC_{50} curves derived from in vitro and in cellula analyses; Figure S4: Contacts between Sec61α, Ipom-F and analogues; Figure S5: HPLC analysis of analogue **3**; Table S1: Cytotoxicity of open-chain analogues **4–12** and their corresponding closed-chain analogues **13–21** against MDA-MB-231 cells; Supporting information 3: 1D (^{1}H and ^{13}C) and/or 2D (COSY, HMQC, and/or HMBC) nMR spectra for compounds **S2**, **S3**, **S8**, **S9**, **3–12** and **24–32**.

Author Contributions: Conceptualization, S.H. and W.Q.S.; Methodology, S.O., P.B., G.Z., K.B.D. and B.S.H.; Formal analysis, S.O., P.B., K.B.D. and G.Z.; Investigation, S.O., P.B., K.B.D., G.Z., L.T., L.A., H.S., A.A., Z.H. and H.S.A.; Resources, B.S.H. and R.E.S.; Writing—Original draft preparation, S.O., P.B., K.B.D. and G.Z.; Writing—Review and editing, S.O., B.S.H., R.E.S., V.H., S.H. and W.Q.S.; Supervision, V.H., S.H. and W.Q.S.; Project administration, S.H. and W.Q.S.; Funding acquisition, R.E.S., V.H., S.H. and W.Q.S. All authors have read and agreed to the published version of the manuscript.

Funding: This work was funded by a Wellcome Trust Investigator Award in Science 202843/Z/16/Z (R.E.S.), a Deutsche Forschungsgemeinschaft (DFG) grant He 3875/15-1 (V.H.), a Wellcome Trust Investigator Award in Science 204957/Z/16/Z (S.H.), an AREA grant GM116032 from the National Institute of General Medical Sciences of the National Institutes of Health (NIH) and the start-up funds from Ball State University (W.Q.S.). The use of a ThermoFisher LTQ XL to obtain mass spectrometric data was supported by an NSF MRI grant under CHE-1531851.

Institutional Review Board Statement: Not applicable.

Informed Consent Statement: Not applicable.

Data Availability Statement: Data are contained within the article or Supplementary Material.

Acknowledgments: We are grateful to Sundeep Rayat and Tykhon Zubkov for their help with the Mass Spectrometry analyses and the HPLC purity tests of analogue **3**, respectively. We also thank the NCI for their NCI-60 cell line screen analysis of analogue **1**.

Conflicts of Interest: The authors declare no conflict of interest. The funders had no role in the design of the study; in the collection, analyses or interpretation of data; in the writing of the manuscript, or in the decision to publish the results.

Sample Availability: Samples of the final compounds (Ipom-F and **1–22**) are available from the authors.

Abbreviations

CL, cytosolic loop; cryo-EM, cryogenic electron microscopy; ER, endoplasmic reticulum; GypC, glycophorin C; HCT-116, human colorectal cancer cells; Ii, short form of HLA class II histocompatibility antigen gamma chain; Ipom-F, ipomoeassin F; MDA-MB-231, human epithelial breast cancer cells; NCI, National Cancer Institute; RCM, ring-closing metathesis; TM, transmembrane; Sec61α, alpha subunit of the Sec61 complex; WT, wild-type.

References

1. Ono, M. Resin glycosides from Convolvulaceae plants. *J. Nat. Med.* **2017**, *71*, 591–604. [CrossRef] [PubMed]
2. Pereda-Miranda, R.; Rosas-Ramirez, D.; Castaneda-Gomez, J. Resin glycosides from the morning glory family. *Prog. Chem. Org. Nat. Prod.* **2010**, *92*, 77–153. [CrossRef]
3. Cao, S.; Guza, R.C.; Wisse, J.H.; Miller, J.S.; Evans, R.; Kingston, D.G.I. Ipomoeassins A−E, Cytotoxic Macrocyclic Glycoresins from the Leaves of Ipomoea squamosa from the Suriname Rainforest1. *J. Nat. Prod.* **2005**, *68*, 487–492. [CrossRef]
4. Cao, S.; Norris, A.; Wisse, J.H.; Miller, J.S.; Evans, R.; Kingston, D.G.I. Ipomoeassin F, a new cytotoxic macrocyclic glycoresin from the leaves of Ipomoea squamosa from the Suriname rainforest. *Nat. Prod. Res.* **2007**, *21*, 872–876. [CrossRef] [PubMed]
5. Fürstner, A.; Nagano, T. Total Syntheses of Ipomoeassin B and E. *J. Am. Chem. Soc.* **2007**, *129*, 1906–1907. [CrossRef] [PubMed]
6. Nagano, T.; Pospíšil, J.; Chollet, G.; Schulthoff, S.; Hickmann, V.; Moulin, E.; Herrmann, J.; Müller, R.; Fürstner, A. Total Synthesis and Biological Evaluation of the Cytotoxic Resin Glycosides Ipomoeassin A–F and Analogues. *Chem. A Eur. J.* **2009**, *15*, 9697–9706. [CrossRef]
7. Postema, M.H.D.; TenDyke, K.; Cutter, J.; Kuznetsov, G.; Xu, Q. Total Synthesis of Ipomoeassin F. *Org. Lett.* **2009**, *11*, 1417–1420. [CrossRef]
8. Zong, G.; Barber, E.; Aljewari, H.; Zhou, J.; Hu, Z.; Du, Y.; Shi, W.Q. Total Synthesis and Biological Evaluation of Ipomoeassin F and Its Unnatural 11R-Epimer. *J. Org. Chem.* **2015**, *80*, 9279–9291. [CrossRef]
9. Zong, G.; Aljewari, H.; Hu, Z.; Shi, W.Q. Revealing the Pharmacophore of Ipomoeassin F through Molecular Editing. *Org. Lett.* **2016**, *18*, 1674–1677. [CrossRef]
10. Zong, G.; Hirsch, M.; Mondrik, C.; Hu, Z.; Shi, W.Q. Design, synthesis and biological evaluation of fucose-truncated monosaccharide analogues of ipomoeassin F. *Bioorg. Med. Chem. Lett.* **2017**, *27*, 2752–2756. [CrossRef]
11. Zong, G.; Sun, X.; Bhakta, R.; Whisenhunt, L.; Hu, Z.; Wang, F.; Shi, W.Q. New insights into structure–activity relationship of ipomoeassin F from its bioisosteric 5-oxa/aza analogues. *Eur. J. Med. Chem.* **2018**, *144*, 751–757. [CrossRef] [PubMed]
12. Zong, G.; Whisenhunt, L.; Hu, Z.; Shi, W.Q. Synergistic Contribution of Tiglate and Cinnamate to Cytotoxicity of Ipomoeassin F. *J. Org. Chem.* **2017**, *82*, 4977–4985. [CrossRef] [PubMed]
13. Zong, G.; Hu, Z.; O'Keefe, S.; Tranter, D.; Iannotti, M.J.; Baron, L.; Hall, B.; Corfield, K.; Paatero, A.O.; Henderson, M.J.; et al. Ipomoeassin F Binds Sec61α to Inhibit Protein Translocation. *J. Am. Chem. Soc.* **2019**, *141*, 8450–8461. [CrossRef]
14. Zong, G.; Hu, Z.; Duah, K.B.; Andrews, L.E.; Zhou, J.; O'Keefe, S.; Whisenhunt, L.; Shim, J.S.; Du, Y.; High, S.; et al. Ring Expansion Leads to a More Potent Analogue of Ipomoeassin F. *J. Org. Chem.* **2020**, *85*, 16226–16235. [CrossRef] [PubMed]
15. Aviram, N.; Schuldiner, M. Targeting and translocation of proteins to the endoplasmic reticulum at a glance. *J. Cell Sci.* **2017**, *130*, 4079–4085. [CrossRef]
16. Hegde, R.S.; Keenan, R.J. The mechanisms of integral membrane protein biogenesis. *Nat. Rev. Mol. Cell Biol.* **2022**, *23*, 107–124. [CrossRef]
17. O'Keefe, S.; Pool, M.R.; High, S. Membrane protein biogenesis at the ER: The highways and byways. *FEBS J.* **2021**. [CrossRef]
18. Gemmer, M.; Förster, F. A clearer picture of the ER translocon complex. *J. Cell Sci.* **2020**, *133*, jcs231340. [CrossRef]
19. Voorhees, R.M.; Fernández, I.S.; Scheres, S.H.; Hegde, R.S. Structure of the Mammalian Ribosome-Sec61 Complex to 3.4 Å Resolution. *Cell* **2014**, *157*, 1632–1643. [CrossRef]
20. Lang, S.; Pfeffer, S.; Lee, P.-H.; Cavalié, A.; Helms, V.; Förster, F.; Zimmermann, R. An Update on Sec61 Channel Functions, Mechanisms, and Related Diseases. *Front. Physiol.* **2017**, *8*, 887. [CrossRef]
21. Sicking, M.; Lang, S.; Bochen, F.; Roos, A.; Drenth, J.P.H.; Zakaria, M.; Zimmermann, R.; Linxweiler, M. Complexity and Specificity of Sec61-Channelopathies: Human Diseases Affecting Gating of the Sec61 Complex. *Cells* **2021**, *10*, 1036. [CrossRef] [PubMed]
22. Gérard, S.F.; Hall, B.S.; Zaki, A.M.; Corfield, K.A.; Mayerhofer, P.U.; Costa, C.; Whelligan, D.K.; Biggin, P.C.; Simmonds, R.E.; Higgins, M.K. Structure of the Inhibited State of the Sec Translocon. *Mol. Cell* **2020**, *79*, 406–415.e407. [CrossRef] [PubMed]
23. O'Keefe, S.; Zong, G.; Duah, K.B.; Andrews, L.E.; Shi, W.Q.; High, S. An alternative pathway for membrane protein biogenesis at the endoplasmic reticulum. *Commun. Biol.* **2021**, *4*, 828. [CrossRef] [PubMed]
24. Luesch, H.; Paavilainen, V.O. Natural products as modulators of eukaryotic protein secretion. *Nat. Prod. Rep.* **2020**, *37*, 717–736. [CrossRef] [PubMed]
25. Pauwels, E.; Schülein, R.; Vermeire, K. Inhibitors of the Sec61 Complex and Novel High Throughput Screening Strategies to Target the Protein Translocation Pathway. *Int. J. Mol. Sci.* **2021**, *22*, 12007. [CrossRef] [PubMed]
26. Linxweiler, M.; Schick, B.; Zimmermann, R. Let's talk about Secs: Sec61, Sec62 and Sec63 in signal transduction, oncology and personalized medicine. *Signal Transduct. Target Ther.* **2017**, *2*, 17002. [CrossRef]
27. Paatero, A.O.; Kellosalo, J.; Dunyak, B.M.; Almaliti, J.; Gestwicki, J.E.; Gerwick, W.H.; Taunton, J.; Paavilainen, V.O. Apratoxin Kills Cells by Direct Blockade of the Sec61 Protein Translocation Channel. *Cell Chem. Biol.* **2016**, *23*, 561–566. [CrossRef]
28. Tranter, D.; Paatero, A.O.; Kawaguchi, S.; Kazemi, S.; Serrill, J.D.; Kellosalo, J.; Vogel, W.K.; Richter, U.; Mattos, D.R.; Wan, X.; et al. Coibamide A Targets Sec61 to Prevent Biogenesis of Secretory and Membrane Proteins. *ACS Chem. Biol.* **2020**, *15*, 2125–2136. [CrossRef]
29. Garrison, J.L.; Kunkel, E.J.; Hegde, R.S.; Taunton, J. A substrate-specific inhibitor of protein translocation into the endoplasmic reticulum. *Nature* **2005**, *436*, 285–289. [CrossRef]

30. Junne, T.; Wong, J.; Studer, C.; Aust, T.; Bauer, B.W.; Beibel, M.; Bhullar, B.; Bruccoleri, R.; Eichenberger, J.; Estoppey, D.; et al. Decatransin, a new natural product inhibiting protein translocation at the Sec61/SecYEG translocon. *J. Cell Sci.* **2015**, *128*, 1217–1229. [CrossRef]
31. Hall, B.S.; Hill, K.; McKenna, M.; Ogbechi, J.; High, S.; Willis, A.E.; Simmonds, R.E. The Pathogenic Mechanism of the Mycobacterium ulcerans Virulence Factor, Mycolactone, Depends on Blockade of Protein Translocation into the ER. *PLoS Path.* **2014**, *10*, e1004061. [CrossRef] [PubMed]
32. McKenna, M.; Simmonds, R.E.; High, S. Mechanistic insights into the inhibition of Sec61-dependent co- and post-translational translocation by mycolactone. *J. Cell Sci.* **2016**, *129*, 1404–1415. [CrossRef] [PubMed]
33. Gamayun, I.; O'Keefe, S.; Pick, T.; Klein, M.-C.; Nguyen, D.; McKibbin, C.; Piacenti, M.; Williams, H.M.; Flitsch, S.L.; Whitehead, R.C.; et al. Eeyarestatin Compounds Selectively Enhance Sec61-Mediated Ca^{2+} Leakage from the Endoplasmic Reticulum. *Cell Chem. Biol.* **2019**, *26*, 571–583.e576. [CrossRef]
34. Klein, W.; Rutz, C.; Eckhard, J.; Provinciael, B.; Specker, E.; Neuenschwander, M.; Kleinau, G.; Scheerer, P.; von Kries, J.-P.; Nazaré, M.; et al. Use of a sequential high throughput screening assay to identify novel inhibitors of the eukaryotic SRP-Sec61 targeting/translocation pathway. *PLoS ONE* **2018**, *13*, e0208641. [CrossRef] [PubMed]
35. Hau, A.M.; Greenwood, J.A.; Löhr, C.V.; Serrill, J.D.; Proteau, P.J.; Ganley, I.G.; McPhail, K.L.; Ishmael, J.E. Coibamide A Induces mTOR-Independent Autophagy and Cell Death in Human Glioblastoma Cells. *PLoS ONE* **2013**, *8*, e65250. [CrossRef]
36. Tirincsi, A.; Sicking, M.; Hadzibeganovic, D.; Haßdenteufel, S.; Lang, S. The Molecular Biodiversity of Protein Targeting and Protein Transport Related to the Endoplasmic Reticulum. *Int. J. Mol. Sci.* **2022**, *23*, 143. [CrossRef]
37. O'Keefe, S.; High, S.; Demangel, C. Biochemical and Biological Assays of Mycolactone-Mediated Inhibition of Sec61. In *Mycobacterium Ulcerans: Methods and Protocols*; Pluschke, G., Röltgen, K., Eds.; Springer: New York, NY, USA, 2022; pp. 163–181.
38. O'Keefe, S.; Roboti, P.; Duah, K.B.; Zong, G.; Schneider, H.; Shi, W.Q.; High, S. Ipomoeassin-F inhibits the in vitro biogenesis of the SARS-CoV-2 spike protein and its host cell membrane receptor. *J. Cell Sci.* **2021**, *134*, jcs.257758. [CrossRef]
39. Bhadra, P.; Dos Santos, S.; Gamayun, I.; Pick, T.; Neumann, C.; Ogbechi, J.; Hall, B.S.; Zimmermann, R.; Helms, V.; Simmonds, R.E.; et al. Mycolactone enhances the Ca^{2+} leak from endoplasmic reticulum by trapping Sec61 translocons in a Ca^{2+} permeable state. *Biochem. J.* **2021**, *478*, 4005–4024. [CrossRef]
40. Bhadra, P.; Yadhanapudi, L.; Römisch, K.; Helms, V. How does Sec63 affect the conformation of Sec61 in yeast? *PLoS Comp. Biol.* **2021**, *17*, e1008855. [CrossRef]
41. Yang, Y.; Oishi, S.; Martin, C.E.; Seeberger, P.H. Diversity-oriented Synthesis of Inner Core Oligosaccharides of the Lipopolysaccharide of Pathogenic Gram-negative Bacteria. *J. Am. Chem. Soc.* **2013**, *135*, 6262–6271. [CrossRef]
42. Roboti, P.; O'Keefe, S.; Duah, K.B.; Shi, W.Q.; High, S. Ipomoeassin-F disrupts multiple aspects of secretory protein biogenesis. *Sci. Rep.* **2021**, *11*, 11562. [CrossRef] [PubMed]
43. Becker, T.; Bhushan, S.; Jarasch, A.; Armache, J.-P.; Funes, S.; Jossinet, F.; Gumbart, J.; Mielke, T.; Berninghausen, O.; Schulten, K.; et al. Structure of Monomeric Yeast and Mammalian Sec61 Complexes Interacting with the Translating Ribosome. *Science* **2009**, *326*, 1369–1373. [CrossRef] [PubMed]
44. Fiser, A.; Šali, A. Modeller: Generation and Refinement of Homology-Based Protein Structure Models. In *Methods in Enzymology*; Academic Press: Cambridge, MA, USA, 2003; Volume 374, pp. 461–491.
45. Van Der Spoel, D.; Lindahl, E.; Hess, B.; Groenhof, G.; Mark, A.E.; Berendsen, H.J.C. GROMACS: Fast, flexible, and free. *J. Comput. Chem.* **2005**, *26*, 1701–1718. [CrossRef] [PubMed]
46. Morris, G.M.; Huey, R.; Lindstrom, W.; Sanner, M.F.; Belew, R.K.; Goodsell, D.S.; Olson, A.J. AutoDock4 and AutoDockTools4: Automated docking with selective receptor flexibility. *J. Comput. Chem.* **2009**, *30*, 2785–2791. [CrossRef]
47. Laskowski, R.A.; Swindells, M.B. LigPlot+: Multiple Ligand–Protein Interaction Diagrams for Drug Discovery. *J. Chem. Inf. Model.* **2011**, *51*, 2778–2786. [CrossRef]

molecules

MDPI

Article

High Expression Level of α2-3-Linked Sialic Acids on Salivary Glycoproteins of Breastfeeding Women May Help to Protect Them from Avian Influenza Virus Infection

Li Ding [1], Yimin Cheng [2], Wei Guo [3], Siyue Sun [1], Xiangqin Chen [1], Tiantian Zhang [1], Hongwei Cheng [1], Jiayue Hao [1], Yunhua Lu [1], Xiurong Wang [4] and Zheng Li [1,*]

1 Laboratory for Functional Glycomics, College of Life Sciences, Northwest University, Xi'an 710069, China; liding@nwu.edu.cn (L.D.); sunsiyue@stumail.nwu.edu.cn (S.S.); chenxiangqin@stumail.nwu.edu.cn (X.C.); 202032666@stumail.nwu.edu.cn (T.Z.); 202133269@stumail.nwu.edu.cn (H.C.); haojiayue@stumail.nwu.edu.cn (J.H.); 201932005@stumail.nwu.edu.cn (Y.L.)

2 Department of Obstetrics and Gynecology, Xi'an Shiyou University, Xi'an 710065, China; ymcheng@xsyu.edu.cn

3 Department of Obstetrics and Gynecology, Shaanxi Provincial People's Hospital, Xi'an 710068, China; viking226@163.com

4 National Key Laboratory of Veterinary Biotechnology, Harbin Veterinary Research Institute, Chinese Academy of Agricultural Science, Harbin 150069, China; wangxiurong@caas.cn

* Correspondence: zhengli@nwu.edu.cn

check for updates

Citation: Ding, L.; Cheng, Y.; Guo, W.; Sun, S.; Chen, X.; Zhang, T.; Cheng, H.; Hao, J.; Lu, Y.; Wang, X.; et al. High Expression Level of α2-3-Linked Sialic Acids on Salivary Glycoproteins of Breastfeeding Women May Help to Protect Them from Avian Influenza Virus Infection. *Molecules* **2022**, *27*, 4285. https://doi.org/10.3390/molecules27134285

Academic Editors: Kyoko Nakagawa-Goto, Jean-Marc Sabatier and Jian Yin

Received: 27 March 2022
Accepted: 29 June 2022
Published: 3 July 2022

Abstract: Terminal sialic acids (Sia) on soluble glycoprotein of saliva play an important role in the clearance of influenza virus. The aim of this study is to investigate the alteration of sialylation on the salivary proteins of women during the lactation period and its effect on the saliva binding ability to virus. In total, 210 saliva samples from postpartum women with and without breastfeeding were collected, and the expression level of α2-3/6-linked Sia on the whole salivary proteins and specific glycoproteins of IgA and MUC5B from different groups were tested and verified using lectin microarray, blotting analysis and ELISA based method. The H1N1 vaccine and three strains of Avian influenza virus (AIV) were used for the saliva binding assay. Results showed that the variation in salivary expression level of α2-3-linked Sia was much more obvious than the α2-6-linked Sia, which was up-regulated significantly in the breastfeeding groups compared to the non-breastfeeding groups at the same postpartum stage. Furthermore, the binding abilities of salivary glycoproteins to AIV strains and H1N1 vaccine were increased in breastfeeding groups accordingly. This finding adds new evidence for the maternal benefit of breastfeeding and provides new thinking to protect postpartum women from AIV infection.

Keywords: breastfeeding women; saliva; sialic acids (Sia); avian influenza virus (AIV); glycoprotein

1. Introduction

Avian influenza virus (AIVs) can cause severe respiratory illness and has always been a threat to the life of people worldwide [1]. Furthermore, breastfeeding women are the most concerned population in addition to pregnant women in the past influenza pandemics due to their particular and important physiological statement [2]. During the lactation period, the mammary glands can be induced by the high expression level of prolactin to secrete sufficient maternal milk that is unusually rich in glycosylated proteins to provide the neonatal protection [3,4]. Besides which, salivary glands can also be regulated by hormones and secrete plenty of glycoproteins that are important in the maintenance of mucosal protection [5]. It has been reported that the glycosylation of salivary proteins can be changed with different physiological and pathological conditions [6–8]. However, little is known about the variation in salivary glycosylation during the special period of lactation and its effect on the salivary proteins binding ability to AIVs.

The oral cavity is an entry site where pathogens from outside can come into contact with the host cells, and human saliva is known as the most important of biological fluids, which can provide the first line of immune defense to resist virus attachment [9,10]. Now, up to more than a thousand proteins have been found in human saliva, but only a dozen of them are in abundance, such as Muc5B (mucin 5B), Muc7 (mucin 7), secretoryimmunoglobulinA (SIgA), cysteine-rich glycoprotein 340 (gp-340), prolactin inducible protein (PIP), and so on [11,12]. Most of these salivary proteins are highly glycosylated, and the carbohydrate part cantake up to 80% of the total protein weight (e.g., MUC5B). The glycans on salivary proteins are usually composed of carbohydrates such as glucose, galactose, mannose, fucose, amino sugars, et al. Among them, sialic acids (Sia) are the most typical ones, which are a diverse family of 9-carbon monosaccharides with N-acetylneuraminic acid (Neu5Ac) as their basic molecular structure and often added to the penultimate sugar—usually galactose (Gal) or its derivatives in α2-3 or α2-6 linkage [13]. This kind of Sia-containing structure usually appears on the outermost terminal of glycoconjugates and play an important role in the cell to cell signaling and microbe-host recognition, and immune regulation [14–16]. In particular, the Siaα2,3/6Gal-linked glycans can be used as receptors for the hemagglutinin (HA) and substrates for neuraminidase (NA) of influenza virus, by which the virus can bind to or release from the host epithelial cells and perform their infection process [17,18]. Different species of influenza viruses usually recognize different linkages of sialic acids, avian influenza viruses predominantly bind to α2,3-linked Sia, whereas human influenza viruses prefer to bind to α2,6-linked Sia on host cells [19]. The successful anti-influenza drugs, such as Oseltamivir, Zanamivir et al., are structural analogs that imitate sialosides and can be used as competitive substrate inhibitors of NA to resist the influenza viruses infection [20,21]. As the natural barrier, it has been proved that the soluble salivary proteins glycosylated with terminal α2-3/6-linked Sia can neutralize and inhibit influenza viruses in this way efficiently [22–24]. And the protein sialylation level in saliva has been expected to be used as an indicator to estimate the individual susceptibility to influenza virus [25].

As one of the most important glycosylation, the sialylation level on salivary proteins have been shown that can be varied according to different ages and sexes, and changed in elderly individuals with autoimmune diseases, chronic diseases, cancers and so on [26,27]. During female gestation, along with the fluctuation of hormone and the repression of the immune system, pregnancy-associated changes of sialylation have also been demonstrated both in serum and salivary proteins before [28]. After delivery, the hormones such as estrogens and progesterone start to resume, while the prolactin remains at a high level, especially in the early stage of postpartum and is quite different between postpartum women with and without breastfeeding. Therefore, the sialylation level on the glycoproteins of saliva during this period, which can directly involve in the anti-virus activity, is of great interest, to be explicit.

In this paper, the individual saliva samples from postpartum women with and without breastfeeding were collected and the alteration of terminal α2,3/6-linked Sia on salivary glycoproteins between these two groups were investigated and compared at different postpartum stages, and the binding affinities of salivary protein to the influenza virus were assessed by using the H1N1 vaccine and different AIV strains of H5N1-CK (A/Chicken/Guangxi/4/2009), H5N1-DK (A/Duck/Guangdong/17/2008) and H9N2-DK (A/Duck/Guangdong/S-7-134/2004). The results of this study will help us to further understand the influence of breastfeeding on the maternal risk of AIV infection, which will give a new thinking to improve the health statement of postpartum women.

2. Materials and Methods

2.1. StudyApprovaland Participants

The collection of human whole saliva for this study was approved by the Human Ethics Committees of Shaanxi Provincial People's Hospital, Xi'an Shiyou University and Northwest University, China. A total of 210 individual saliva samples from breastfeeding or non-breastfeeding postpartum women were collected, and healthy women in non-pregnant and non-breastfeeding statementwere used as control.

The participants that were recruited were in different periods of postpartum: first-period (FP, within six monthspostpartum); second-period (SP, from 7th to 12th month postpartum); and third-period (TP, from 13th to 24th month postpartum). Subjects were divided into seven groups (Supplementary Table S1), including the control group of healthy women without pregnancy and breastfeeding (HN) and the FP, SP, TP groups of breast-feeding women (FP-B, SP-B, TP-B) and non-breastfeeding women (FP-NB, SP-NB, TP-NB). Each group consisted of 30 samples.

2.2. Whole Saliva Sample Collection and Treatment

About 1mL of saliva sample was collected from each participant according to the protocol described previously [26–28]. The saliva sample was centrifuged at 12,000 rpm, 4 °C for 20 min to get rid of the insoluble materials. The supernatant was then collected and added with the cocktail inhibitor of protease (Sigma-Aldrich, St. Louis, MO, USA) to avoid protein degradation. Each of 200 μL saliva sample, from different individuals in the same group, were blended and the protein concentration of each blended group of samples was quantified by the BCA (bicinchoninicacid) assay (listed in Supplemental Table S1), and labeled with Cy3 (cyanine3fluorescent dye) (GE Healthcare, Buckinghamshire, UK).

2.3. Lectin Microarrays and Data Analysis

According to the previous protocol [26–28], the lectin microarray was produced by using 37 lectins with different binding preferences to N- and O-linked glycans. Each lectin was spotted in triplicate in one block, and with three blocks on one slide. The slide was blocked with 2% BSA in 1× PBS (0.01 mol/L phosphate buffer containing 0.15 mol/L NaCl, pH 7.4) (w/v, pH 7.4) and washed before use. Then, 5 μg of Cy3-labeled salivary proteins from each group diluted in 0.75 mL of hybridization buffer (2% BSA, 500 mM glycine and 0.1% Tween-20 in PBS, pH 7.4) was applied to each block of the slide, and incubated in the dark at 37 °C for 3h with gentle rotation. After being washed with 1× PBST (0.2% Tween 20 in 0.01 mol/L phosphate buffercontaining 0.15 mol/L NaCl, pH 7.4) and 1× PBS (0.01 mol/L phosphate buffer containing 0.15 mol/L NaCl, pH7.4), and dried by centrifugation, the slides were scanned using a Genepix 4000B confocal scanner (Axon Instruments Inc., Union City, CA, USA), and analyzed at 532 nm by the Genepix 3.0 software (version 6.0, Axon Instruments). For data analysis, the background signal was substracted, and the median value of the triplicate data for each lectin was selected and normalized to the sum median of all the lectins in one block. The average and standard deviation (SD) of the normalized medians for each lectin in different blocks tested for the same sample were calculated, and the difference between two groups was evaluated by the p value.

2.4. SDS-PAGE and Lectin Blotting Analysis

Equal amounts of salivary proteins from each group were separated by 10% sodium dodecylsulfate polyacrylamide gel electrophoresis (SDS-PAGE), and then stained either with alkaline silver or with Periodic Acid-Schiff staining (PAS) and Coomassie Brilliant Blue R250(CBB). For lectin blotting analysis, after SDS-PAGE, the proteins in each gel were transferred to a 0.45 μmpolyvinylidene fluoride (PVDF) membrane at a constant current of 300 mA for 45–80 min according to their molecular weight, blocked with Carbo-Free Solution (Vector, Burlingame, CA, USA) and then incubated with the cyanine5fluorescent dye (Cy5)-labeled maackia amurensis lectin II (MAL-II) and sambucus nigra lectin (SNA) at 4 °C in the dark overnight. After that, the membranes were washed with 1× TBST

(150 mM NaCl, 10 mM Tris-HCl, 0.05% *v*/*v* Tween 20, pH 7.5) and 1× TBS (150 mM NaCl, 10 mM Tris-HCl) respectively, and scanned by using the FluorImager (Storm 840, Molecular Dynamics Inc., Sunnyvale, CA, USA). The intensity of the binding band on the acquired image was further analyzed by the ImageJ software (NIH), and the relative fluorescence intensity (RFI) ofeachbandfor different groups was calculated and compared to the control group of HN.

2.5. Assessment of Salivary Proteins Binding Ability to AIVs

All the AIV strains used in this study were kindly presented from Harbin Veterinary Research Institute, China. The H1N1 influenza A vaccine (Split Virion, inactivated) was purchased from the company of Sinovac Biotech Ltd., Beijing. According to the method described in the previous literature [26,27], the viral proteins were extracted with a mixture of ethanol and ether by being thoroughly shaken. After static placement, the ether phase was discarded, and the proteins in the remaining liquid were dried and labeled with Cy5 dye. For the binding ability assay, an equal amount of 50μg proteins from each group were run on the 10% SDS-PAGE electrophoresis, transferred to 0.45 μm PVDF membrane, and then incubated with the Cy5-labeled viral proteins.The image scan and the analysis of band intensity were the same as that described above.

2.6. Western Blot Analysis

Equal amount of 20 μg salivary proteins from different groups were applied to the 10% SDS-PAGE and transferred from gels to the PVDF membrane as described in 2.4. After being blocked with 5% skimmed milk in TBST for 1 h, the membrane was incubated with primary antibody of mouse monoclonal anti-human MUC5B (1:1000 in TBST, Abcam, Ab105460) or rabbit monoclonal anti-human immunoglobulin A (IgA) (1:1000 in TBST, Abcam, Ab124716) at 4 °C overnight. The membrane was then washed three times with TBST and incubated with the horseradish peroxidase (HRP)-conjugated secondary antibody for 1 h at room temperature. The goat anti-mouse Immunoglobulin G heavy and light chains (IgG H&L) for MUC5B (1:5000 in TBST; Abcam, Ab205719) and the goat anti-rabbit IgG H&L for IgA (1:5000 in TBST, Abcam, Ab205718) were used, respectively. After being washed three times with TBST, the membranes were finally detected by using the chemiluminescent HRP substrate (Beyotime, Haimen, China) and scanned with the Tannon 5200 Imaging System (Tanon, Shanghai, China).

2.7. ELISA Based Lectin Binding Assay

For immobilization, 100 μL of 5 μg/mL MUC5B antibody (Abcam, Ab105460) or IgA antibody (Abcam, Ab124716) diluted in the coating buffer (0.1 M carbonate/bicarbonate, pH 9.4) was added to each well of the 96-well plate and incubated at 4 °C overnight. The wells used as the blank control were added with 100 μL of coating buffer without any antibody. The next day, the plate was washed three times with PBST and blocked with carbo-free solution at 37 °C for 1 h on the shaker. Then, the plate was washed again and 50 μg of salivary proteins (1 μg/μL) from each group was added to the well in triplicate and incubated at 37 °C for 1 h. After the plate being washed, 100 μL of the biotinylated SNA or MAL-II (Vector Laboratories Inc, Burlingame, CA, USA), which was 1:3000 or 1:200 diluted in carbon-free solution, was applied to each well and incubated at 37 °C for another 1 h. Next, the plate was washed again and 100 μL of streptavidin-HRP (Vector Laboratories) with 1:2000 dilution in carbon-free was added to the well and incubated at 37 °C for 50 min. The HRP signal was finally detected by 3,3′,5,5′-tetramethylbenzidine (TMB) substrate (Beyotime, Haimen, China) and the absorbance of 450 nm for each well was read by the microplate reader (Bio-Rad Laboratories, Hercules, CA, USA).

2.8. Statistical Analysis

Statistical analysis was performed by the SPSS software (version 20.0), and the *p* value between two different groups was acquired by Mann–Whitney test. The difference was considered statistically significant if the *p* value was <0.05.

3. Results

3.1. Relative Expression Levels of Terminal α2-3/6-Linked Siain Saliva Groups of Postpartum Women with and without Breastfeeding

In total, 210 saliva samples were collected from breastfeeding (B) and non-breastfeeding (NB) women in the first, second, third period (FP, SP,TP) of postpartum, and healthy women without pregnancy, and breastfeeding were used as the control group (Supplementary Table S1). The expression level of terminal α2-3 and α2-6-linked Sia on salivary proteins of different groups were detected by the specific lectin of SNA and MAL-II using lectin microarray (Figure 1A). The fluorescent images of Cy3-labeled salivary proteins binding to SNA or MAL-II on the lectin microarray were shown in Figure 1B. The normalized fluorescent intensities (NFIs) of SNA and MAL-II for different groups were listed in Supplementary Table S1 and further analyzed in Figure 1C.

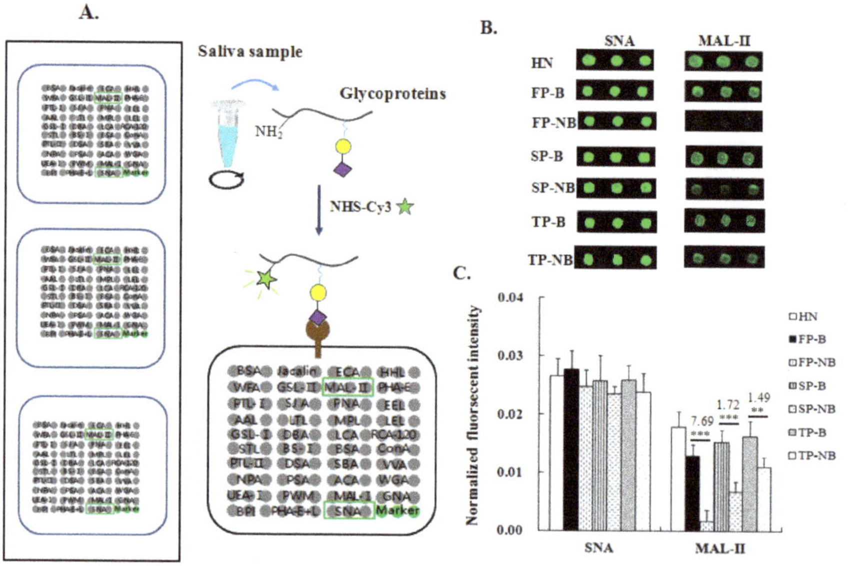

Figure 1. Comparison of the terminal α2-3/6-linked Sia expression level on salivary proteins of postpartum women with and without breastfeeding. (**A**) The label of saliva samples with Cy3 fluorescent dye and the layout of lectin microarray. (**B**) Fluorescent images of SNA detected forα2-6-linked Sia and MAL-II detected for α2-3-linked Sia from the lectin microarrays used for different groups. (**C**) Normalized fluorescent intensity (NFI) of SNA and MAL-II binding to salivary proteins of different groups. FP-B: first-period of postpartum women with breastfeeding; FP-NB: first-period of postpartum women without breastfeeding; SP-B: second-period of postpartum women with breastfeeding; SP-NB: second-period of postpartum women without breastfeeding; TP-B: third-period of postpartum women with breastfeeding; TP-NB: third-period of postpartum women without breastfeeding. Each sample was applied to three repeated slides, the standard deviation (SD) of NFIs for each lectin from all the repeated blocks was calculated. Data are shown as mean ± SD, Statistical significance between two different groups was analyzed by Student's *t*-test and indicated by the *p*-value. ** $p < 0.01$; *** $p < 0.001$.

From the result, it can be seen that the salivary expression level of α2-6-linked Sia was a little bit higher in the breastfeeding women than that of the non-breastfeeding ones at the same postpartum stage, but the discrepancy between these two groups was not too much (Figure 1C). While for the α2-3-linked Sia, its expression level was found to be significantly up-regulated in the breastfeeding groups of FP-B, SP-B, TP-B compared to non-breastfeeding groups of FP-NB, SP-NB, TP-NB, especially for the women within six months postpartum (fold change = 7.67, $p < 0.001$). The discrepancy was reduced along with the extension of postpartum time.

3.2. *Analysis for the Salivary Proteins and the Variation in Terminal α2-3/6-Linked Sia Expression Level on Glycoproteins of Different Groups*

Salivary proteins with and without glycosylation were shown by using different staining methods after SDS-PAGE (Figure 2A). Results showed that saliva samples from all the groups were composed of proteins with the similar molecular weight. PAS staining showed that there were two bands of proteins, which are heavily glycosylated, one of the band was shown in the stacking gel (band 1, b1) and the other one was around 150 kD (band 2, b2). Besides, there were another two glycosylated protein bands, which can be seen more apparently by the sensitive fluorescent detection of Cy5-labeled SNA (band 3, b3) and MAL-II (band 4, b4) below 100 kD.

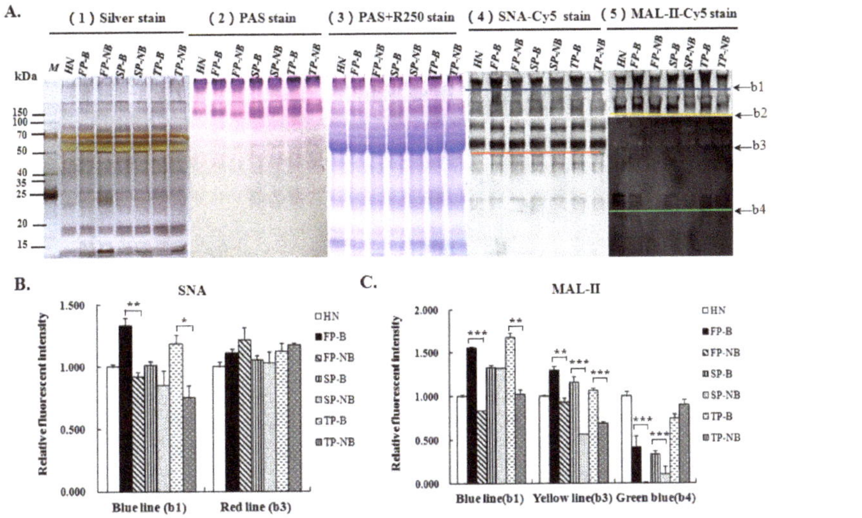

Figure 2. Analysis of the salivary proteins profiles and the variation in terminal α2-3/6-linked Sia expression level on glycoproteins of different groups. (**A**) Salivary proteins and glycoproteins from different groups were shown by the silver stain (1), PAS stain(1) and the combine of PAS and Coomassie Brilliant Blue R-250 stains (3), and the salivary proteins glycosylated with terminal α2-3/6-linked sialic acids were detected by the lectin blotting analysis of Cy5-labled SNA (4) and MAL-II (5). (**B**) Relative fluorescence intensities (RFIs) of the binding bands of salivary proteins to Cy5-labled SNA. (**C**) RFIs of the binding bands of salivary proteins to Cy5-labled MAL-II. Data shown are mean ± SD, * $p < 0.05$; ** $p < 0.01$; *** $p < 0.001$.

The expression level of terminal Siaα2-3/6Gal on salivary proteins of different groups were further investigated by the Lectin blotting analysis (Figure 2A(4,5)), and the samples treated with α2-3 or α2-3,6,8 sialidase were used as the blank control (Supplementary Figure S1). The Relative fluorescence intensities (RFIs) of the SNA and MAL-II binding bands between groups of women with and without breastfeeding were analyzed and shown in Figure 2B,C. The lectin binding image of SNA (Figure 2A(4)) showed that the salivary proteins glycosylated with terminal Siaα2-6Gal mainly appeared on the proteins of band 1 in the stacking gel and band 3 around 50 kD. The expression levels of Siaα2-6Gal on the proteins of band 3 showed no significant difference between women with and without breastfeeding. While for band 1, the expression level of Siaα2-6Gal in the saliva of the breastfeeding group was obviously increased compared with that of non-breastfeeding group at the same stage (Figure 2B). As to the lectin blotting of MAL-II, there were three distinct binding bands (band 1 in the stacking gel, band 2 around 150 kD, band 4 around 25 kD) to be observed. Apparent discrepancy of RFIs can be found for all of these three bands between breastfeeding groups and non-breastfeeding groups, the expression levels of Siaα2-3Gal on the salivary proteins of these bands were obviously higher in the former groups than that in the latter groups (Figure 2C).

3.3. Comparison of Salivary Proteins Binding Ability to AIVs between Postpartum Women with and without Breastfeeding

The binding abilities of salivary proteins to influenza virus were assessed using the inactivated H1N1 virus (H1N1 vaccine) and three AIV strains (H5N1-CK, H5N1-DK, H9N2). An equal amount of salivary proteins from different groups were separated by the SDS-PAGE and then blotted with the Cy5-labeled proteins of virus. From the binding images (Figure 3), there were two bands of proteins, which can bind strongly to all of the virus, one of the band presented in the stacking gel and the other band appeared at about 25 kD. The RFIs of the binding bands for different saliva groups were compared (Figure 3B,C). For the AIV strains, significant differences can be seen between the groups of women with and without breastfeeding at both of these two bands. Furthermore, the virus-binding intensities of salivary proteins from breastfeeding groups were much stronger than that from non-breastfeeding women, especially within six months postpartum. These results were consistent with the binding affinity of salivary proteins to MAL-II. As to H1N1 virus, the binding intensities of these two bands to different saliva groups were not obviously varied, as found in the AIVs, and that may be due to the HA of H1N1 preferring to bind to α2,6-linked SA receptor, so the change of α2,3-linked SA level in saliva could not affect its binding affinity too much. Besides, there was also a strong binding band at around 70 kD for H1N1 and H9N2 strains and a week binding band at around 150 kD for H5N1-DK, H1N1 and H9N2 strains. For the band around 150 kD, the binding intensity was weak and its variation in different groups was identical to the band in the stacking gel. As to the binding intensity around 70 kD, its variation in different groups was obviously not too much.

3.4. Expression Level of IgA and MUC5B in the Saliva of Postpartum Women with and without Breastfeeding

IgA and MUC5B are known as the primary glycoproteins in human saliva, which can exert high anti-influenza activity through their sialylated glycochains [23,29]. From the binding profile of lectin blotting analysis, it can be seen that the protein bands shown in the stacking gel and at around 50 kD, 25 kD were identical with the molecular weight of MUC5B and IgA heavy and light chains, respectively. Therefore, the protein expression level of MUC5B and IgA and the variation in Sia were further investigated in different groups.

After SDS-PAGE, the salivary proteins on each gel were incubated with the human anti-IgA or anti-MUC5B antibody, respectively. The binding band of salivary proteins corresponding to anti-IgA antibody and anti-MUC5B antibody appeared at around 50 kD and in the stacking gel as expected (Figure 4A). The RFIs of the corresponding bands between groups of women with and without breastfeeding were compared and shown in Figure 4B. The result showed that the protein expression level of IgA or MUC5B were not varied during the lactating period, and there was no significant difference between groups of breastfeeding and non-breastfeeding women.

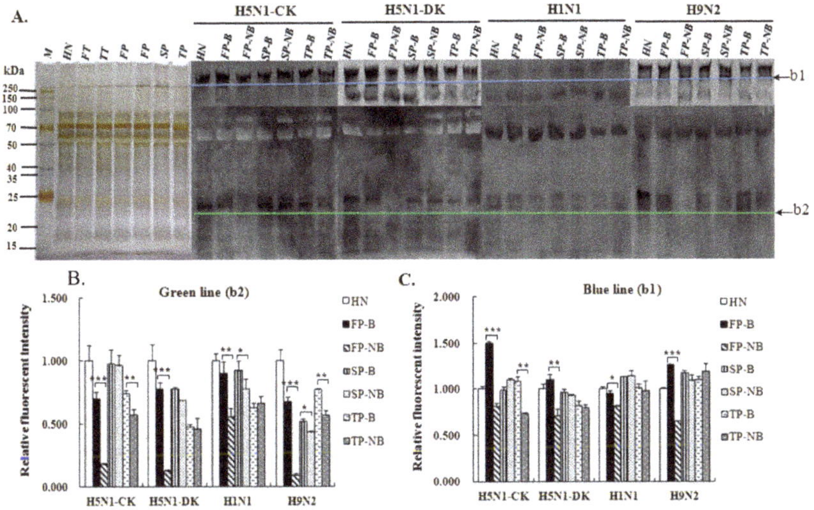

Figure 3. Assessment of the binding ability of salivary proteins from different groups to the AIV strains and the H1N1 influenza A vaccine. (**A**) The binding profiles of salivaryproteins to the H1N1 vaccine and the AIV strains of H5N1-CK (A/Chicken/Guangxi/4/2009), H5N1-DK (A/Duck/Guangdong/17/2008) and H9N2- DK (A/Duck/Guangdong/S-7-134/2004). (**B**) RFIs of the binding band shown at around 25 kD for different saliva groups. (**C**) RFIs of the binding band shown in the stacking gel for different saliva groups. Data shown are mean ± SD, * $p < 0.05$; ** $p < 0.01$; *** $p < 0.001$.

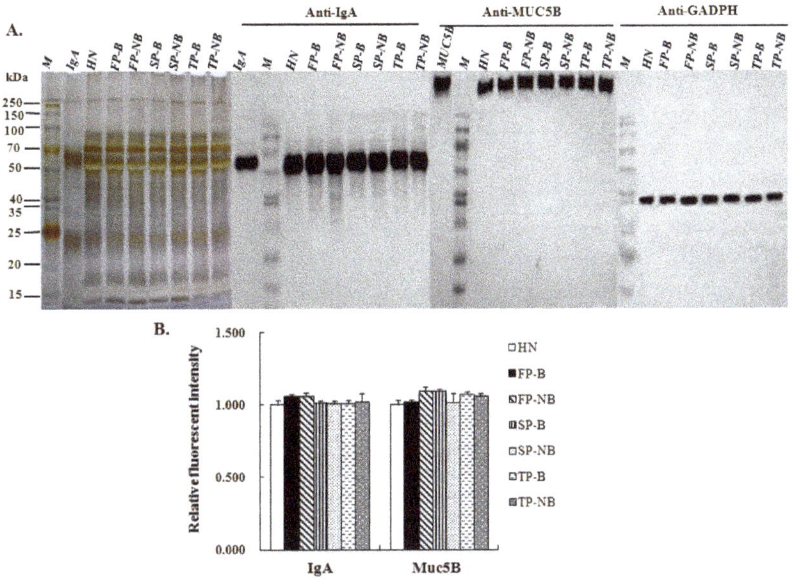

Figure 4. Analysis ofthe MUC5B and IgA expression level in different saliva groups. (**A**) The binding images of salivaryproteins corresponding to human anti-MUC5B or anti-IgA antibody, glyceraldehyde-3-phosphate dehydrogenase (GAPDH) was used as the loading control. (**B**) RFIs of the binding bands of salivary proteins corresponding to human anti-MUC5B or anti-IgA antibody. Data shown are mean ± SD, statistical notations and group abbreviations are same as shown in Figure 1.

3.5. Expression Level of Terminal α2-3/6-Linked Sia Expressed on IgA and MUC5B in Different Saliva Groups

Enzyme-linked immunosorbent assay (ELISA) based lectin binding assay was applied to evaluate the level of α2-3/6-linked sialic acids expressed on IgA and MUC5B of different groups. The IgA and MUC5B from different salivary groups were firstly captured by the human anti-IgA or anti-MUC5B antibodies that were immobilized on the 96-well plate. Then the Siaα2-3/6Gal structures on the specific protein of IgA and MUC5B were recognized by the biotinylated SNA or MAL-II and detected by the HPR labeled avidin. The results showed that both the expression levels of Siaα2-3/6Gal on MUC5B (Figure 5A) and IgA (Figure 5B) were up-regulated in the saliva of postpartum breastfeeding women compared with non-breastfeeding women. The increased expression level of Siaα2-3/6Gal on MUC5B of the breastfeeding women was consistent with the result of the lectin blotting analysis for the binding band that appeared in the stacking gel. As for IgA, the improved expression level of Siaα2-3Gal in breastfeeding women coincided with the lectin binding profiles analyzed for the band around 25 kD, which was identical with the molecular weight of IgA light chain. But for the expression of Siaα2-6Gal on IgA, its increased trend in the breastfeeding groups was not in agreement with the lectin blotting of SNA for the band at a little above 50 kD, where the heavy chain of IgA may present. This may be caused by the binding of other glycoproteins with the similar molecular weight in saliva.

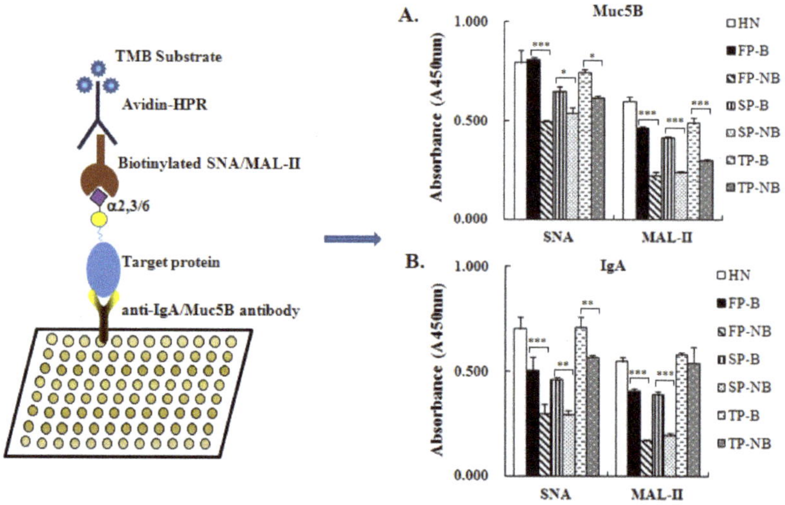

Figure 5. Comparison of theterminal α2-3/6-linked Sia expression level on MUC5B and IgA between saliva groups of women with and without breastfeeding. Salivary MUC5B or IgA was captured by the anti-human MUC5B or anti-human IgA antibody on 96-well plate, and the expression level of terminal α2-3/6-linked Sia, which was recognized by the biotinylated SNA/MAL-II and finally detected by the Avidin-HPR system. The absorbance at 450 nm was read for the analysis of SNA/MAL-II binding to IgA (**A**) and MUC5B (**B**) from different groups. Data shown are mean ± SD, statistical notations and group abbreviations are same as shown in Figure 1.

4. Discussion

Pregnancy and lactation are the particular physiological periods of women with dramatic fluctuation of hormones. It is reported that along with the hormone alteration, the innate immunity response to respiratory virus infection was affected [30,31]. The morbidity and mortality for pregnant women were obviously higher than the general population during the past influenza pandemics [2]. After delivery, the prolactin in breastfeeding women remains at a high level compared with non-breastfeeding women. However, their susceptibility to influenza virus has not yet been clear.

Considering the importance of α2-3/6-linked sialic acids involved in the anti-influenza virus activity, their expression level on the salivary proteins of women with and without breastfeeding were investigated in this study, and the associated salivary glycoprotein binding abilities to different strains of influenza virus were accessed. From the result of lectin microarray, the salivary expression level of α2-6-linked sialic acids between lactating women and non-lactating women was not varied too much on the whole level. While for the expression level of α2-3-linked sialic acids, the discrepancy between these two kinds of groups was much more significant, which is apparently higher in the breastfeeding groups than the non-breastfeeding groups at the same postpartum stage. This result was further verified by the lectin blotting analysis, the binding profile of salivary proteins to MAL-II showed that there are apparently three bands of proteins glycosylated with α2-3-linked sialic acids. From the RFI data, it can be seen that the expression level of α2-3-linked sialic acids on the proteins of all these three bands was much higher in the saliva of breastfeeding women and the binding abilities of salivary glycoproteins to AIV strains of H5N1-CK, H5N1-DK, H9N2 were found to be increased in breastfeeding groups accordingly. From the fluorescent binding images of salivary glycoproteins to different AIV strains, there are two binding bands with varied RFI between breastfeeding and non-breastfeeding groups. The one that appeared in the stacking gel was in accordance with the high molecular weight of salivary MUC5B, and the other band around 25 kD coincided with the molecular weight of IgA light chain. Given that both of the MUC5B and IgA are primary glycoproteins in human saliva with high anti-influenza activity, their protein and sialylation levels were further tested using Western blotting analysis and ELISA based MAL-II/SNA binding assay. The results showed that the protein level of MUC5B or IgA was almost the same among different saliva groups. As for the sialylation level, a significant discrepancy can be seen between the groups of women with and without breastfeeding, and the variation trend of α2-3 sialylation on the protein of MUC5B or IgA in different groups was consistent with that observed on the whole saliva level, which were significantly increased in the groups of breastfeeding women in comparison to the non-breastfeeding groups. These results suggested that the high expression of α2-3 sialylation in the whole saliva samples of lactating women mainly come from the glycosylation level rather than the protein level.

It is known that the terminal sialic acids on glycoproteins play an important role in the regulation of immune effects, and their expression can be varied according to different physiological and pathological conditions [14–16]. We have previously demonstrated that the expression of α2-3-linked sialic acids on salivary proteins was decreased in pregnant women, which was associated with their high susceptibility to AIV [28]. While in this study, the expression of α2-3-linked sialic acids was found to be increased significant in breastfeeding women under the high level of prolactin hormone. These results were accordant with the different influenza infection risk of women during pregnancy and postpartum periods that was reported in a recent study [2]. The different variation trend of α2-3 sialylation could be due to the physiological changes of hormones during pregnancy and lactation periods and their different influence on the immune system. It has been reported that the progesterone hormone has immunoinhibitory effects during pregnancy, whereas prolactin has the opposite effects of immunostimulatory [32,33]. Moreover, prolactin has also been proved to be involved in the regulation of glycosyltransferases for the synthesis of lactose in human milk [34]. The saliva glands and the glycosylation of their secretory products could also be modulated by hormones in this way [35], and the underling mechanism is of interest to be studied in the future.

It is well established that breastfeeding can provide numerous health benefits, not only for infants, but also for mothers [36]. The advantages of breastfeeding to mothers have been illustrated, including the improved glucose metabolism, the reduction in hypertension and hyperlipidemia, and the lower risks of breast and ovarian cancer, heart disease, type 2 diabetes et al. [37,38]. Here, our study further indicated that breastfeeding could be helpful to protect women from AIV infection by increasing the expression level of α2-3 linked sialic acids on their salivary proteins. This finding adds more evidence for the maternal benefit of breastfeeding, and provides a new support to encourage mothers to choose breastfeeding during postpartum.

In this study, the changes in sialylation level were detected and verified by using a different method, and investigated both on the whole saliva level and on the specific glycoproteins. However, there still some limitations. First, the sample size for each group is not too large. More samples need to be recruited to improve the accuracy. Second, the anti-influenza activity of saliva samples was not accessed by using the influenza virus directly, the saliva inhibition assays of hemagglutination and neutralization will make the result more meaningful. Finally, the mechanism of prolactin influence on the α2-3 sialylation of salivary proteins for breastfeeding women was not referred to in this study, and as such, requires further research.

Supplementary Materials: The following supporting information can be downloaded at: https://www.mdpi.com/article/10.3390/molecules27134285/s1, Figure S1: The binding profiles of SAN (A) and MAL-II (B) to salivary proteins treated with sialidase; Table S1: The demographic characteristics of the healthy control and postpartum women with and without breastfeeding.

Author Contributions: Conceptualization, Z.L.; methodology, L.D. and X.C.; validation, T.Z. and H.C.; formal analysis, J.H., S.S. and Y.L.; resources, W.G., Y.C. and X.W.; writing—original draft preparation, L.D.; writing—review and editing, X.C. and H.C.; project administration, Z.L. and L.D.; funding acquisition, L.D. All authors have read and agreed to the published version of the manuscript.

Funding: This work was supported by the National Natural Science Foundation of China (NSFC, Grant No. 81502820) and the Natural Science Foundation of Shaanxi Province, China (Grant No. 2016JQ8042).

Institutional Review Board Statement: The study was conducted in accordance with the Declaration of Helsinki, and approved by the Ethics Committee of Northwest University, China and Shaanxi Provincial People's Hospital, China.

Informed Consent Statement: Informed consent was obtained from all subjects involved in this study. Written informed consent has been obtained from the patients to publish this paper.

Data Availability Statement: Not applicable.

Conflicts of Interest: The authors declare no conflict of interest.

Sample Availability: Not applicable.

Abbreviations

AIV: Avian influenza virus; Sia, sialic acids; Gal, galactose; MUC5B, mucin5B; MUC7, mucin 7; HA, humagglutinin; NA, neuraminidase; IgG H&L, immunoglobulin G heavy and light chains; IgA, immunoglobulin A; SD, standard deviation; BCA, Bicinchoninicacid; Cy3/5, cyanine 3/5 fluorescent dye; SDS-PAGE, sodium dodecylsulfate polyacrylamide gel electrophoresis; PVDF, polyvinylidene fluoride; MAL-II, maackia amurensis lectin II; SNA, sambucus nigra lectin; RFI, relative fluorescence intensity; HRP, horseradish peroxidase; TMB, 3,3′,5,5′-tetramethylbenzidine; GAPDH, glyceraldehyde-3-phosphate dehydrogenase; ELISA, enzyme-linked immunosorbent assay; FP-B/NB, first-period of postpartum women with/without breastfeeding;SP-B/NB, second-period of postpartum women with/without breastfeeding; TP-B/NB, third-period of postpartum women with/without breastfeeding.

References

1. Li, Y.T.; Linster, M.; Mendenhall, I.H.; Su, Y.C.F.; Smith, G.J.D. Avian influenza viruses in humans: Lessons from past outbreaks. *Br. Med. Bull.* **2019**, *132*, 81–95. [CrossRef] [PubMed]
2. Prasad, N.; Huang, Q.S.; Wood, T.; Aminisani, N.; McArthur, C.; Baker, M.G.; Seeds, R.; Thompson, M.G.; Widdowson, M.A.; Newbern, E.C. Influenza-associated outcomes among pregnant, postpartum, and nonpregnant women of reproductive Age. *J. Infect. Dis.* **2019**, *219*, 1893–1903. [CrossRef] [PubMed]
3. Newburg, D.S. Glycobiology of human milk. *Biochemistry* **2013**, *78*, 771–785. [CrossRef] [PubMed]

4. Lis-Kuberka, J.; Królak-Olejnik, B.; Berghausen-Mazur, M.; Orczyk-Pawiłowicz, M. Lectin-based method for deciphering human milk IgG sialylation. *Molecules* **2019**, *24*, 3797. [CrossRef] [PubMed]
5. Pedersen, A.M.L.; Sørensen, C.E.; Proctor, G.B.; Carpenter, G.H.; Ekström, J. Salivary secretion in health and disease. *J. Oral Rehabil.* **2018**, *45*, 730–746. [CrossRef] [PubMed]
6. Fouani, M.; Basset, C.A.; Jurjus, A.R.; Leone, L.G.; Tomasello, G.; Leone, A. Salivary gland proteins alterations in the diabetic milieu. *J. Mol. Histol.* **2021**, *52*, 893–904. [CrossRef]
7. Fang, L.; Liu, Q.; He, P.; Wang, X.; Wang, Y.; Wei, M.; Chen, L. Alteration of salivary glycopatterns in oral lichen planus. *Biomarkers* **2018**, *23*, 188–195. [CrossRef]
8. Guruaribam, V.D.; Sarumathi, T. Relevance of serum and salivary sialic acid in oral cancer diagnostics. *J. Cancer Res. Ther.* **2020**, *16*, 401–404.
9. Tada, A.; Senpuku, H. The impact of oral health on respiratory viral infection. *Dent. J.* **2021**, *9*, 43. [CrossRef]
10. Dawes, C.; Pedersen, A.M.; Villa, A.; Ekström, J.; Proctor, G.B.; Vissink, A.; Aframian, D.; McGowan, R.; Aliko, A.; Narayana, N.; et al. The functions of human saliva: A review sponsored by the World Workshop on Oral Medicine VI. *Arch. Oral Biol.* **2015**, *60*, 863–874. [CrossRef]
11. Veerman, E.C.I.; van't Hof, W. Research on salivary proteins: From properties to functions. *Ned. Tijdschr. Voor Tandheelkd.* **2020**, *127*, 525–531. [CrossRef] [PubMed]
12. Jasim, H.; Olausson, P.; Hedenberg-Magnusson, B.; Ernberg, M.; Ghafouri, B. The proteomic profile of whole and glandular saliva in healthy pain-free subjects. *Sci. Rep.* **2016**, *6*, 39073. [CrossRef] [PubMed]
13. Schauer, R.; Kamerling, J.P. Exploration of the sialic acid world. *Adv. Carbohydr. Chem. Biochem.* **2018**, *75*, 1–213. [PubMed]
14. Pearce, O.M.; Läubli, H. Sialic acids in cancer biology and immunity. *Glycobiology* **2016**, *26*, 111–128. [CrossRef] [PubMed]
15. Adams, O.J.; Stanczak, M.A.; von Gunten, S.; Läubli, H. Targeting sialic acid-siglec interactions to reverse immune suppression in cancer. *Glycobiology* **2018**, *28*, 640–647. [CrossRef] [PubMed]
16. Gianchecchi, E.; Arena, A.; Fierabracci, A. Sialic acid-siglec axis in human immune regulation, involvement in autoimmunity and cancer and potential therapeutic treatments. *Int. J. Mol. Sci.* **2021**, *22*, 5774. [CrossRef]
17. Sieben, C.; Sezgin, E.; Eggeling, C.; Manley, S. Influenza A viruses use multivalent sialic acid clusters for cell binding and receptor activation. *PLoS Pathog.* **2020**, *16*, e1008656. [CrossRef]
18. Wen, F.; Wan, X.F. Influenza neuraminidase: Underrated role in receptor binding. *Trends Microbiol.* **2019**, *27*, 477–479. [CrossRef]
19. Byrd-Leotis, L.; Cummings, R.D.; Steinhauer, D.A. The interplay between the host receptor and influenza virus hemagglutinin and neuraminidase. *Int. J. Mol. Sci.* **2017**, *18*, 1541. [CrossRef]
20. Heida, R.; Bhide, Y.C.; Gasbarri, M.; Kocabiyik, Ö.; Stellacci, F.; Huckriede, A.L.W.; Hinrichs, W.L.J.; Frijlink, H.W. Advances in the development of entry inhibitors for sialic-acid-targeting viruses. *Drug Discov. Today* **2021**, *26*, 122–137. [CrossRef]
21. Han, J.; Perez, J.; Schafer, A.; Cheng, H.; Peet, N.; Rong, L.; Manicassamy, B. Influenza virus: Small molecule therapeutics and mechanisms of antiviral resistance. *Curr. Med. Chem.* **2018**, *25*, 5115–5127. [CrossRef] [PubMed]
22. Limsuwat, N.; Suptawiwat, O.; Boonarkart, C.; Puthavathana, P.; Wiriyarat, W.; Auewarakul, P. Sialic acid content in human saliva and anti-influenza activity against human and avian influenza viruses. *Arch. Virol.* **2016**, *161*, 649–656. [CrossRef] [PubMed]
23. Barnard, K.N.; Alford-Lawrence, B.K.; Buchholz, D.W.; Wasik, B.R.; LaClair, J.R.; Yu, H.; Honce, R.; Ruhl, S.; Pajic, P.; Daugherity, E.K.; et al. Modified sialic acids on mucus and erythrocytes inhibit influenza A virus hemagglutinin and neuraminidase functions. *J. Virol.* **2020**, *94*, e01567-19. [CrossRef] [PubMed]
24. Gilbertson, B.; Edenborough, K.; McVernon, J.; Brown, L.E. Inhibition of influenza A virus by human infant saliva. *Viruses* **2019**, *11*, 766. [CrossRef]
25. Tan, M.; Cui, L.; Huo, X.; Xia, M.; Shi, F.; Zeng, X.; Huang, P.; Zhong, W.; Li, W.; Xu, K.; et al. Saliva as a source of reagent to study human susceptibility to avian influenza H7N9 virus infection. *Emerg. Microbes Infect.* **2018**, *7*, 156. [CrossRef] [PubMed]
26. Qin, Y.; Zhong, Y.; Zhu, M.; Dang, L.; Yu, H.; Chen, Z.; Chen, W.; Wang, X.; Zhang, H.; Li, Z. Age- and sex-associated differences in the glycopatterns of human salivary glycoproteins and their roles against influenza A virus. *J. Proteome Res.* **2013**, *12*, 2742–2754. [CrossRef] [PubMed]
27. Zhong, Y.; Qin, Y.; Yu, H.; Yu, J.; Wu, H.; Chen, L.; Zhang, P.; Wang, X.; Jia, Z.; Guo, Y.; et al. Avian influenza virus infection risk in humans with chronic diseases. *Sci. Rep.* **2015**, *5*, 8971. [CrossRef]
28. Ding, L.; Fu, X.; Guo, W.; Cheng, Y.; Chen, X.; Zhang, K.; Zhu, G.; Yang, F.; Yu, H.; Chen, Z.; et al. Pregnancy-associated decrease of Siaα2-3Gal-linked glycans on salivary glycoproteins affects their binding ability to avian influenza virus. *Int. J. Biol. Macromol.* **2021**, *184*, 339–348. [CrossRef]
29. Maurer, M.A.; Meyer, L.; Bianchi, M.; Turner, H.L.; Le, N.P.L.; Steck, M.; Wyrzucki, A.; Orlowski, V.; Ward, A.B.; Crispin, M.; et al. Glycosylation of Human IgA Directly Inhibits Influenza A and Other Sialic-Acid-Binding Viruses. *Cell Rep.* **2018**, *23*, 90–99. [CrossRef]
30. Littauer, E.Q.; Skountzou, I. Hormonal Regulation of Physiology, Innate immunity and antibody response to H1N1 influenza virus infection during pregnancy. *Front. Immunol.* **2018**, *9*, 2455. [CrossRef]
31. Finch, C.L.; Zhang, A.; Kosikova, M.; Kawano, T.; Pasetti, M.F.; Ye, Z.; Ascher, J.R.; Xie, H. Pregnancy level of estradiol attenuated virus-specific humoral immune response in H5N1-infected female mice despite inducing anti-inflammatory protection. *Emerg. Microbes Infect.* **2019**, *8*, 1146–1156. [CrossRef] [PubMed]

32. Vieira Borba, V.; Shoenfeld, Y. Prolactin, autoimmunity, and motherhood: When should women avoid breastfeeding? *Clin. Rheumatol.* **2019**, *38*, 1263–1270. [CrossRef] [PubMed]
33. Peeva, E.; Zouali, M. Spotlight on the role of hormonal factors in the emergence of autoreactive B-lymphocytes. *Immunol. Lett.* **2005**, *101*, 123–143. [CrossRef] [PubMed]
34. Sadovnikova, A.; Garcia, S.C.; Hovey, R.C. A Comparative review of the extrinsic and intrinsic factors regulating lactose synthesis. *J. Mammary Gland. Biol. Neoplasia* **2021**, *26*, 197–215. [CrossRef] [PubMed]
35. Kurabuchi, S.; Matsuoka, T.; Hosoi, K. Hormone-induced granular convoluted tubule-like cells in mouse parotid gland. *J. Med. Investig.* **2009**, *56*, 290–295. [CrossRef] [PubMed]
36. Ramos-Roman, M.A.; Syed-Abdul, M.M.; Adams-Huet, B.; Casey, B.M.; Parks, E.J. Lactation versus formula feeding: Insulin, glucose, and fatty acid metabolism during the postpartum period. *Diabetes* **2020**, *69*, 1624–1635. [CrossRef]
37. Binns, C.; Lee, M.; Low, W.Y. The Long-term public health benefits of breastfeeding. *Asia Pac. J. Public Health* **2016**, *28*, 7–14. [CrossRef]
38. Sattari, M.; Serwint, J.R.; Levine, D.M. Maternal implications of breastfeeding: A review for the internist. *Am. J. Med.* **2019**, *132*, 912–920. [CrossRef]

molecules

MDPI

Article

Inhibition of Metastatic Hepatocarcinoma by Combined Chemotherapy with Silencing VEGF/VEGFR2 Genes through a GalNAc-Modified Integrated Therapeutic System

Xunan Li [1], Xiang Wang [2], Nian Liu [1], Qiuyu Wang [1] and Jing Hu [1,*]

[1] Wuxi School of Medicine, Jiangnan University, Lihu Avenue 1800, Wuxi 214122, China; lixunan3022@163.com (X.L.); lnxy713@163.com (N.L.); qyw9812@163.com (Q.W.)

[2] Key Laboratory of Carbohydrate Chemistry and Biotechnology, Ministry of Education, School of Biotechnology, Jiangnan University, Wuxi 214122, China; wx1111123456@163.com

* Correspondence: hujing@jiangnan.edu.cn; Tel.: +86-510-85197039

Abstract: Hepatocellular carcinoma (HCC) is a highly malignant tumor related to high mortality and is still lacking a satisfactory cure. Tumor metastasis is currently a major challenge of cancer treatment, which is highly related to angiogenesis. The vascular endothelial growth factor (VEGF)/VEGFR signaling pathway is thus becoming an attractive therapeutic target. Moreover, chemotherapy combined with gene therapy shows great synergistic potential in cancer treatment with the promise of nanomaterials. In this work, a formulation containing 5-FU and siRNA against the VEGF/VEGFR signaling pathway into N-acetyl-galactosamine (GalNAc)-modified nanocarriers is established. The targeting ability, biocompatibility and pH-responsive degradation capacity ensure the efficient transport of therapeutics by the formulation of 5-FU/siRNA@GalNAc-pDMA to HCC cells. The nano-construct integrated with gene/chemotherapy exhibits significant anti-metastatic HCC activity against C5WN1 liver cancer cells with tumorigenicity and pulmonary metastasis in the C5WN1-induced tumor-bearing mouse model with a tumor inhibition rate of 96%, which is promising for future metastatic HCC treatment.

Keywords: hepatocellular carcinoma; metastasis; ASGPR; gene therapy; VEGF/VEGFR2 signaling pathway

check for updates

Citation: Li, X.; Wang, X.; Liu, N.; Wang, Q.; Hu, J. Inhibition of Metastatic Hepatocarcinoma by Combined Chemotherapy with Silencing VEGF/VEGFR2 Genes through a GalNAc-Modified Integrated Therapeutic System. *Molecules* **2022**, *27*, 2082. https://doi.org/10.3390/molecules27072082

Academic Editor: Vito Ferro

Received: 25 February 2022
Accepted: 21 March 2022
Published: 24 March 2022

1. Introduction

Hepatocellular carcinoma (HCC) is a highly heterogeneous and malignant tumor with a poor prognosis. It remains the sixth most prevalent malignancy and the fourth leading cause of cancer-related mortality worldwide [1]. At early stages, HCC patients can usually receive liver transplantation or surgical resections as a curative treatment [2]. However, most patients are identified as advanced HCC at the time of diagnosis [3,4]. Despite remarkable progress in earlier diagnosis and novel therapeutic agents in the past decades, the 5 year survival rate of HCC patients remains remarkably poor due to serious extrahepatic metastases and a high recurrence rate [5,6]. To deal with this issue, nanomaterial-based delivery platforms attracted great attention to exploit multimodal therapies to achieve synergistic anti-tumor efficacies, such as liposomes [7], micelles [8], protein nanoparticles [9], carbon-based materials [10,11] and metal and covalent organic frameworks [12,13].

Tumor metastasis, the dissemination of primary tumor cells from the initial site to distant sites to form secondary tumors, is currently a major challenge in cancer treatment [14,15]. In many cases, metastases are already developed when primary cancers are diagnosed [16]. Cancer cells can escape to other tissues or organs through the blood vessels, and tumor vessels supply tumor cells with nutrients and oxygen to support their continued growth, invasion and metastasis [17–19]. The inhibition of tumor angiogenesis to block the tumor's blood supply has become an effective anti-tumor growth and metastasis strategy in cancer treatment [20]. The vascular endothelial growth factor (VEGF) is reported to be more expressed in liver cancer tissues than in normal liver tissues and para-neoplastic tissues, and the VEGF/VEGFR2 signal transduction pathway plays a major role in peritumoral angiogenesis, which is involved in the migration, proliferation and survival of perivascular new endothelial cells. It becomes a promising alternative to block the VEGF/VEGFR2 signaling pathway to inhibit the generation and migration of solid tumors [21].

Recently, the RNA interference (RNAi) system based on gene silencing mechanisms is utilized as a powerful tool in anti-tumor gene therapy [22,23]. Small interfering RNAs (siRNAs) have the capability to regulate the expression of RNA transcripts, induce mRNA degradation or repress protein translation [24]. To deliver the negatively charged naked siRNA, some challenges need to be faced during systemic circulation, including rapid metabolism, off-target effects and RNase degradation [25,26]. In view of this fact, to maximize the gene silencing efficiency of siRNA, it is highly needed to utilize a stable delivery system to make siRNAs effectively and accurately recognize and reach the targeted tumor sites after long-term circulation [26].

Herein, we synthesized an integrated delivery system based on two monomers [3-azido-2-hydroxypropyl methacrylate (AHPMA) and 2-(dimethylamino) ethyl methacrylate (DMAEMA)] to combine gene therapy against VEGF and VEGFR2 and chemotherapy to inhibit hepatocarcinogenesis and metastasis, which can be initiated with a cancer stem cell (CSC)-like cell line C5WN1 with tumorigenicity and pulmonary metastasis (Scheme 1) [27]. The monomer DMAEMA can form a stable complex with amino groups of negatively charged nucleic acids to prevent enzymatic degradation. The sponge effect triggered by acidic pH value can cause polymeric DMAEMA swelling and endosomal escape, thus achieving gene transfection [28,29]. These features make this platform suitable for nucleic acid concentration and transport. 5-Fluorouracil (5-FU), a commonly used anti-cancer reagent with an inhibitory effect on cell growth and migration, is encapsulated into the formulation for chemotherapy. Asialoglycoprotein receptor (ASGPR) is a mammalian lectin specifically expressed on the surface of hepatocytes and is an attractive hepatic delivery target [30,31]. GalNAc residues can mediate the specific cellular uptake of the nanoparticle by targeting ASGPR expressed liver cells as one of the specific ligands for ASGPR. The self-assembled nanoparticles are multiple-functionalized by modifying them with N-acetyl-galactosamine (GalNAc) for hepatic targeted delivery and rhodamine B (RhB) for fluorescein-based tracking. GalNAc residues can be linked to the azido group in AHPMA through click reaction.

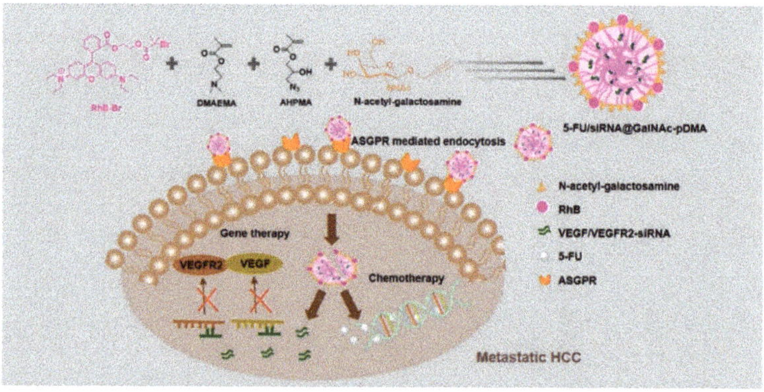

Scheme 1. Illustration of the preparation and the anti-metastatic HCC strategy of GalNAc-modified AHPMA and DMAEMA-based polymeric micelles carried with gene/chemotherapy through the VEGF/VEGFR signaling pathway.

2. Results and Discussion

2.1. Synthesis and Characterization of 5-FU/siRNA @GlaNAc-pDMA

The GalNAc-modified nanocarrier with fluorescence labeling was constructed. GalNAc [32] (Figures S1 and S2), RhB-based atom transfer radical polymerization (ATRP) initiator (RhB-Br) [33] (Figures S3 and S4) and monomeric AHPMA [34,35] (Figure S5) were first synthesized using previously reported methods. The fluorescently labeled and pH-responsive block copolymer p(RhB-DMAEMA-AHPMA-GalNAc) was prepared based on atom transfer radical polymerization (ATRP) with RhB-Br and AHPMA, which was verified by ^{1}H-NMR spectroscopy (Figure S6). Additionally, the copolymer p(RhB-DMAEMA-AHPMA-GalNAc) were obtained by a further "click" reaction with GalNAc. All proton signals of p(RhB-DMAEMA-AHPMA-GalNAc) were clearly shown in ^{1}H-NMR spectra (Figure S7), representing the successful synthesis of p(RhB-DMAEMA-AHPMA-GalNAc). The results of Fourier transform infrared (FT-IR) spectroscopy also confirmed the correct construction (Figure S8).

The glyco-copolymers p(RhB-DMAEMA-AHPMA-GalNAc) was then self-assembled into nanoparticles (GalNAc-pDMA) via a solvent exchange method. Transmission electron microscopy (TEM) micrographs showed that the assembled particles are roughly spherical in shape and with relatively homogeneous size (Figure 1A and Figure S9). Meanwhile, the size of GalNAc-pDMA particles was confirmed by dynamic light scattering (DLS) with an average diameter of 198.1 ± 0.4 nm and a polydispersity index (PDI) value of 0.24 ± 0.085 (Figure 1B). Hydrophobic 5-FU was trapped into the hydrophobic cores of GalNAc-pDMA by dialysis, which was confirmed with UV-spectroscopy (Figure S9). A simple dialysis method was applied to trap hydrophobic 5-FU into the hydrophobic cores of GalNAc-pDMA. The drug loading content (DLC) and drug entrapment efficiency (DEE) of GalNAc-pDMA were determined by measuring 5-FU absorption at 265 nm to be 12.6 ± 0.03% and 58.0 ± 3.4%, respectively (Figure S9).

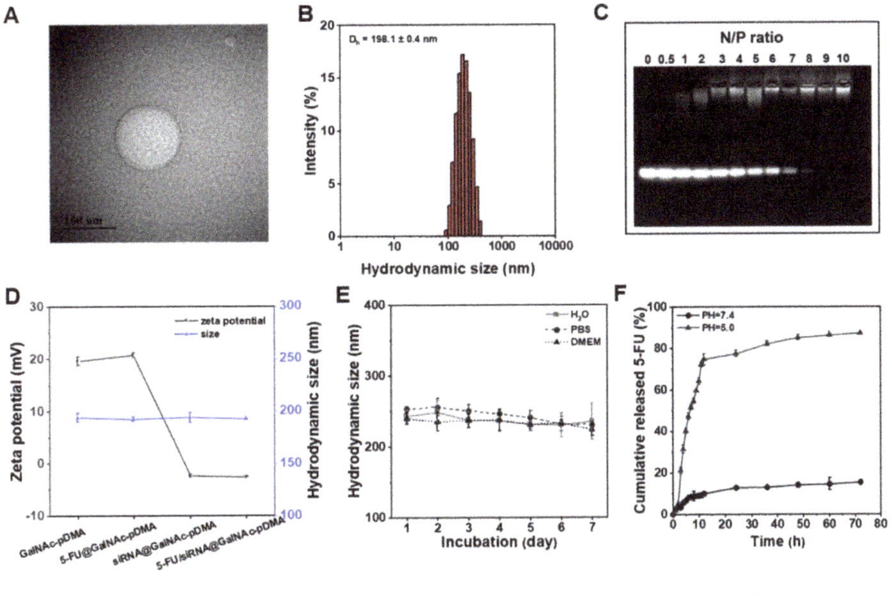

Figure 1. (**A**) TEM image of GalNAc-pDMA partilces. The scale bar:100nm. (**B**) Hydrodynamic size distribution of GalNAc-pDMA. (**C**) Gel electrophoresis assay for siRNA@GalNAc-pDMA condensing capacity at different N/P ratios. (**D**) The average particle size and zeta potential values of different particles. The error bars were obtained from three independent measurements. (**E**) Long-term hydrodynamic size change of 5-FU/siRNA@GalNAc-pDMA in different mediums at 37 °C. (**F**) Release of 5-FU from 5-FU/siRNA@GalNAc-pDMA at pH 5.0 and 7.4 Mean ± SD (n = 3).

The cationic DMAEMA unit can form complexes with negatively charged siRNA fragments. The siRNA condensing capacity of GalNAc-pDMA was determined by the nitrogen/phosphate ratio [N/P, nitrogen (N) of the PMAEMA moiety in GalNAc-pDMA and phosphate (P) in RNA] from 0.5 to 10, and naked siRNA was used as a negative control. The results of the agarose gel retardation assay showed that GalNAc-pDMA form complexes with a mixture of VEGF-siRNA andVEGFR2-siRNA with Cy5 labelling (sequence information see Table S1) at an N/P ratio of 9:1 (Figure 1F). After siRNA was loaded, the zeta potential reduced from 19.5 to −2.5 mV (Figure 1C), mostly due to the negative charge of siRNA, also confirming the siRNA condensing capability of GalNAc-pDMA.

To evaluate the stability of 5-FU/siRNA@GalNAc-pDMA, the hydrodynamic diameter and zeta potential of 5-FU/siRNA@GalNAc-pDMA in different solutions such as water, PBS (pH = 7.0) and cell culture medium (DMEM containing 10% fetal bovine serum (FBS) were measured for seven consecutive days (Figure 1D). The result showed that the particles remained relatively stable under physiological conditions.

The drug release of 5-FU/siRNA@GalNAc-pDMA was further measured under different pH conditions at 37 °C, mimicking the physiologic status (pH 7.4) and the periphery of the tumor (pH 5.0). The release profile of 5-FU showed that only 18% of 5-FU was released after 72 h in PBS at pH 7.4. Comparably, nearly 80% of 5-FU was released within 12 h under the acidic pH conditions (pH 5.0) (Figure 1E). It demonstrated that 5-FU/siRNA@GalNAc-pDMA has great pH-responsive drug release capacity, which thereby can reduce the occurrence of side reactions in normal tissues.

2.2. *In Vitro Targeted Synergistic Effect of 5-FU/siRNA@GalNAc-pDMA*

2.2.1. In Vitro Biosafety and Cytotoxicity Assessment

Good bioactivity and biocompatibility are essential characteristics for exogenous nanomaterials as drug delivery systems. The biocompatibility was assessed via MTT assays and hemolysis evaluation. C5WN1, HepG2 and Huh7 cells were tested, and HEK293 cells were selected as non-cancer cells incubating for 48 h with GalNAc-pDMA (0–600 $\mu g\ mL^{-1}$). The MTT assay results showed that viability of cells was greater than 90% in all cell lines even at the highest concentration of GalNAc-pDMA, indicating very low cytotoxicity of the nanomaterial (Figure S10). In addition, GalNAc-pDMA particles exhibited a very low hemolysis rate even at the highest concentration (<10%) compared to the serious hemolysis caused by water as a positive control with a hemolysis rate of 100% (Figure S10). It suggests that GalNAc-pDMA nanoparticles are suitable as delivery systems with high biocompatibility and low cytotoxicity.

2.2.2. ASGPR-Targeted Intracellular Uptake

The ASGPR-targeting capacity of GalNAc-pDMA particles as a co-delivery system was verified by observation with confocal laser scanning microscopy (CLSM) (Figure 2A) and flow cytometry analysis (Figure 2B). Hepatocarcinoma cells with high ASGPR expression on the cell membrane, such as C5WN1, HepG2 and Huh7, were used for the tests [36,37]. HEK293 cells with low ASGPR expression was used as a negative control. After 3 h incubation with 5-FU/siRNA@GalNAc-pDMA, the cellular uptake behavior of the nanoparticles was detected by CLSM and flow cytometry analysis. Compared with HEK293 cells, significant RhB fluorescence and cy5 fluorescence were both observed on the cell membrane and in the cytoplasm of C5WN1, HepG2 and Huh7 cells, demonstrating the active targeting of 5-FU/siRNA@GalNAc-pDMA towards high ASGPR expressing cells. In addition, there was no obvious difference in detected fluorescence when HEK293 cells were treated with a free galactose-supplemented medium and then incubated with 5-FU/siRNA@GalNAc-pDMA. In contrast, the fluorescence intensity in C5WN1, HepG2 and Huh7 cells became weaker after preincubation with free galactose, as free galactose competed with GalNAc-modified nanoparticles for binding to ASGPR on the cell membrane and decreased the subsequent endocytosis. All these results showed that 5-FU/siRNA@GalNAc-pDMA could effectively target HCC cell lines through the specific recognition of modified GalNAc residues with ASGPR.

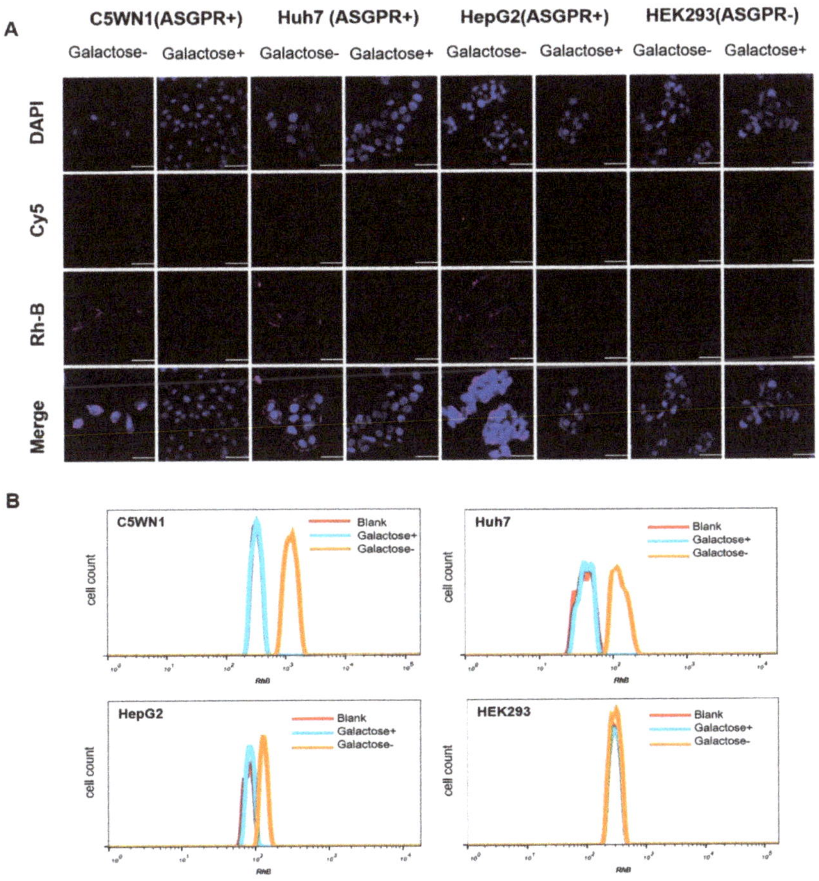

Figure 2. Targeted cellular delivery of 5-FU/siRNA@GalNAc-pDMA particles. (**A**) CLSM images of C5WN1, HepG2, Huh7 and HEK293 cells incubated with 5-FU/siRNA@GalNAc-pDMA with or without free galactose (1×10^{-3} M) competition. Scale bars: 50 μm. (**B**) Flow cytometry analysis of C5WN1, Huh7, HepG2 and HEK293 cells incubated with 5-FU/siRNA@GalNAc-pDMA with or without galactose (1×10^{-3} M) competition.

2.2.3. In Vitro Therapeutic Effect of Codelivery of 5-FU and siRNA

The in vitro therapeutic effect of 5-FU/siRNA@GalNAc-pDMA was investigated using an MTT assay (Figure 3). A series of concentrations of 5-FU, 5-FU@GalNAc-pDMA and 5-FU/siRNA@GalNAc-pDMA were cultured with C5WN1, Huh7 and HepG2 cells, respectively, in DMEM medium for 48 h. Comparably, the cell inhibitory effect of 5-FU/siRNA@GalNAc-pDMA was significantly more enhanced than those of free 5-FU and 5-FU/siRNA at the 5-FU concentration of 50 μg mL^{-1} (Figure S11), which was then used for the subsequent in vitro experiments. Compared with the control group treated with PBS, the cell viability of the free 5-FU treated group and the free siRNA treated groups was around 70%. The HepG2, Huh7 and C5WN1 cell growth of the encapsulated 5-FU group and encapsulated siRNA group, was greatly inhibited; however, there was no significant difference for HEK293 cells, suggesting the targeted delivery system efficiently increased the cytotoxicity of free 5-FU and siRNA towards the hepatocarcinoma cells. A glucose-modified pDMA (Glc-pDMA) was co-loaded with 5-FU and siRNA to be used as a control. The 5-FU/siRNA@Glc-pDMA treatment did not show apparent synergistic cytotoxicity on all cell lines. The 5-FU/siRNA@GalNAc-pDMA treated groups of HepG2, Huh7 and C5WN1 cells displayed significant cell death with cell viabilities of 32%, 36% and 24%, respectively, while there was no effect on HEK293 cells. These results indicate that 5-FU/siRNA@GalNAc-pDMA has significant superiority for targeted synergistic chemotherapy/gene therapy towards hepatocarcinoma cells.

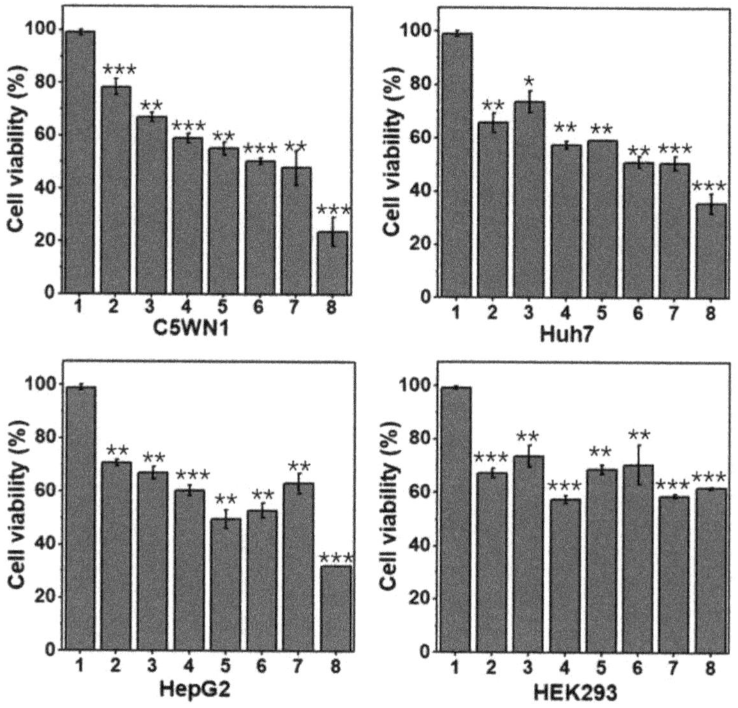

Figure 3. In vitro C5WN1, HepG2, Huh7 and HEK293 cells growth inhibition test. The cells were treated with different formulations for 48 h: (1) GalNAc-pDMA, (2) siRNA, (3) 5-FU, (4) 5-FU+siRNA, (5) siRNA@GalNAc-pDMA, (6) 5-FU@GalNAc-pDMA, (7) 5-FU/siRNA@Glc-pDMA, (8) 5-FU/siRNA@GalNAc-pDMA. (Each group was treated with the same dose of the following components: 5-FU at 50 μg mL^{-1}, GalNAc-pDMA at 300 μg mL^{-1} and siRNA at 3.5 μg mL^{-1}). Data are expressed as the mean ± SD. Compared with the control: * $p < 0.05$, ** $p < 0.01$ and *** $p < 0.001$.

2.2.4. Inhibition of C5WN1 Migration by Gene Therapy of 5-FU/siRNA@GalNAc-pDMA

It was shown that the VEGF/VEGFR2 signaling pathway plays a crucial role in regulating cell viability, proliferation and migration. The migration ability of HCC cells was also confirmed to be closely related to VEGF and VEGFR expression. In this work, co-loaded siVEGF and siVEGFR were utilized to inhibit the viability and migration of C5WN1 by interfering with the transcriptional process of both genes, thereby downregulating the expression of VEGF and VEGFR2.

To investigate the efficacy of gene silencing, the expression of VEGF and VEGFR2 protein expression levels were evaluated by Western blot. The protein expression levels of VEGF and VEGFR2 were downregulated to 68% and 70%, respectively, after incubation of C5WN1 cells with its respective siRNA alone (Figure 4A). After siVEGF and siVEGFR2 were, respectively, encapsulated with GalNAc-pDMA, the protein expression levels of VEGF and VEGFR2 were further downregulated to 38% and 44%, respectively. The expression of VEGF and VEGFR2 genes was further downregulated, respectively, after the simultaneous encapsulation of siRNA into GalNAc-pDMA particles. Surprisingly, the group treated with 5-FU/siRNA@GalNAc-pDMA showed the lowest VEGF and VEGFR2 protein levels of 28% and 27%, respectively, which suggested that the bimodal therapies of gene/chemotherapy by GalNAc-pDMA delivery system could effectively inhibit the VEGF/VEGFR signaling pathway, thus inhibiting HCC cells migration.

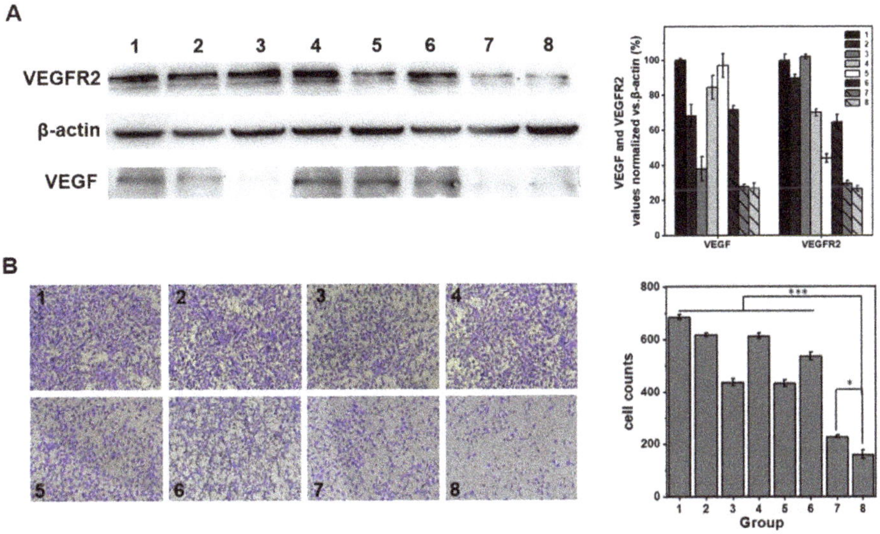

Figure 4. (**A**) Western blot analysis of the expression of VEGF and VEGFR2 after treatment with different formulations for 48h (left panel). Quantified results of Western blot analysis (right panel). (**B**) Transwell migration assay of C5WN1 cells with different formulations for 12 h. Photograph (magnification, ×100) (Left panel) and cell counts after treatment with different formulations (right panel). Group: (1) Blank group, (2) siVEGF, (3) siVEGF@GalNAc-pDMA, (4) siVEGFR2, (5) siVEGFR2@GalNAc-pDMA, (6) siRNA (siVEGF+siVEGFR2), (7) siRNA@GalNAc-pDMA, (8) 5-FU/siRNA@GalNAc-pDMA. Each group was treated with the same dose of the following components: 5-FU at 50 μg mL^{-1}, GalNAc-pDMA at 300 μg mL^{-1} and siRNA at 3.5 μg mL^{-1}. Data are expressed as the mean ± SD. Compared with the control: * $p < 0.05$ and *** $p < 0.001$.

The effect of 5-FU/siRNA@GalNAc-pDMA on the migration of C5WN1 cells was further studied. The cell migration of C5WN1 cells was detected by transwell assay after incubation with different formulations. The numbers of migrated cells of groups treated with encapsulated siRNA were much lower than those of groups treated with naked siRNA (Figure 4B), indicating the efficacy of targeted delivery. The group treated with 5-FU/siRNA@GalNAc-pDMA showed the highest inhibition on cell migration, suggesting the synergistic effect of this construct on the migration of C5WN1 cells.

2.3. *In Vivo Targeted Synergistic Effect of 5-FU/siRNA@GalNAc-pDMA*

In order to further evaluate the in vivo targeted synergistic effect, a C5WN1 subcutaneous tumor model was established [27]. Animal experiments were carried out with the approval of the experimental animal ethics committee of Jiangnan University (Approval Number: JN. No20210415c1350920). The in vivo biodistribution of GalNAc-pDMA was first evaluated. 5-FU@GalNAc-pDMA was injected into mice through the tail vein to observe the fluorescence signal of RhB in mice after 4, 8, 12, 24 and 48 h, and 5-FU@Glc-Micel was used as a non-targeted control group. After 4 h of injection, the nanoparticles were detected to be distributed throughout the whole body in both the targeted group and the non-targeted group (Figure 5A). After 8 h of injection, it was detected that 5-FU@GalNAc-pDMA started to accumulate in the tumor site. At 12 h post-injection, a significant accumulation of 5-FU@GalNAc-pDMA was detected in the tumor site. There was no obvious tumorous accumulation in the non-targeted group compared with the targeted group, indicating the excellent in vivo targeting capacity of 5-FU@GalNAc-pDMA mediated by specific ASGPR recognition. At 24 h after injection, a certain fluorescence intensity in the tumor site was still detected in the targeted group, suggesting prolonged drug retention and effectiveness achieved by the targeted delivery. After 48 h, the systemic fluorescence became very weak in both groups, showing that the nanoparticles could be systemically metabolized. Major organs of the heart, liver, spleen, lung, kidney and tumors were isolated from mice for imaging. Although there was also some accumulation in the tumor of the non-targeted group, mostly due to enhanced permeability and retention (EPR) effect, the fluorescence intensities in tumors of the targeted group were much stronger than those of the non-targeted group (Figure 5B). These results confirmed that 5-FU@GalNAc-pDMA can effectively accumulate in tumor tissues in the long term due to its good stability and targeting ability.

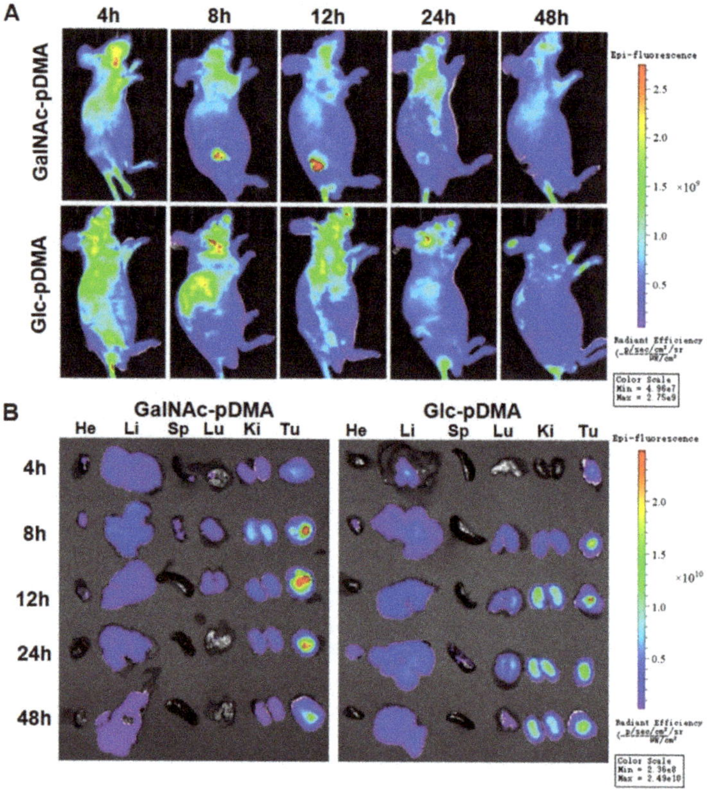

Figure 5. Biodistribution of 5-FU@GalNAc-pDMA and 5-FU@Glc-pDMA. (**A**) In vivo imaging of RhB fluorescence at different times after intravenous injection of 5-FU@GalNAc-pDMA and 5-FU@Glc-pDMA(30 mg kg^{-1}) via the tail vein. (**B**) Representative ex vivo images of RhB fluorescence of harvest organs (He: Heart, Li: Liver; Sp: Spleen; Lu: Lung; Ki: Kidney; Tu: Tumor) after different injection times. (n = 5 for each group).

The synergistic therapeutic effect of gene/chemotherapy carried by 5-FU/siRNA@GalNAc-pDMA was then further evaluated with the subcutaneous C5WN1 tumor-bearing mice, which is a CSC-like cell line with tumorigenicity and pulmonary metastasis. Nine groups of mice were randomly separated and intravenously injected with different formulations every two days during two weeks, including saline, GalNAc-pDMA, siRNA (siVEGF+siVEGFR2), Free 5-FU, 5-FU+siRNA, siRNA@GalNAc-pDMA, 5-FU@GalNAc-pDMA, 5-FU/siRNA@Glc-pDMA and 5-FU/siRNA@GalNAc-pDMA, respectively. The tumor volume and body-weight of mice were checked daily during the entire therapeutic period. On day 14, the tumor inhibition rates (TIR) were calculated based on the final tumor volume using saline as negative control (Figures 6A,B and S12). Compared with the saline control group, the blank micelles had almost no tumor inhibition effect. The free 5-FU and naked siRNA had weak tumor inhibition effects with the TIR of 32% and 33%, respectively. The TIR of encapsulated drug group and encapsulated gene group increased to 57% and 67%, respectively, showing the improved therapeutic effect of targeted delivery compared to individual chemotherapy or gene therapy. The 5-FU/siRNA@GalNAc-pDMA group reached the maximum tumor inhibition among all groups with the TIR of 96%. The tumors and major organs of all mice were dissected for a further check. The tumor mass of 5-FU/siRNA@GalNAc-pDMA group was found to have the lowest weight (Figure 6C), which was in good agreement with tumor volume measurement. In this study, a low intravenous injection dose of 5-FU at 5 mg kg^{-1} was selected in consideration of the low usage amount of nanoplatforms, compared with the reported 5-FU dose at 10–15 mg kg^{-1} in mice [38–40]. The significantly enhanced anti-tumor activity of this formulation with bimodal therapies at a low dose of 5-FU demonstrated the synergistic effect of combined gene therapy. The haematoxylin and eosin (H&E) sections of tumor tissues from different groups demonstrated that the most severe cell necrosis occurred in the 5-FU/siRNA@GalNAc-pDMA treated group, and different levels of cell necrosis were observed in other treatment groups, while the tumor cells in the saline and GalNAc-pMDA groups retained intact cell morphology (Figure S13), which supported the previous results. The above results indicated that the GalNAc-pDMA-based formulation carried synergetic gene/chemotherapy and elicited a potent therapeutic effect on metastatic HCC. The subcutaneous tumor model is not sufficient to study the effect of this formulation on extrahepatic metastasis; the inhibition efficacy on HCC metastasis still needs to be further studied in a tumor orthotopic transplantation model.

Moreover, the 5-FU/siRNA@GalNAc-pDMA-treated mice showed negligible body weight loss during the treatment, indicating particularly low systemic toxicity of the formulation (Figure 6D). The H&E sections of the main organs from each experimental group also showed no obvious lesions of the cell morphology, confirming the in vivo biosafety of 5-FU/siRNA@GalNAc-pDMA (Figure S14).

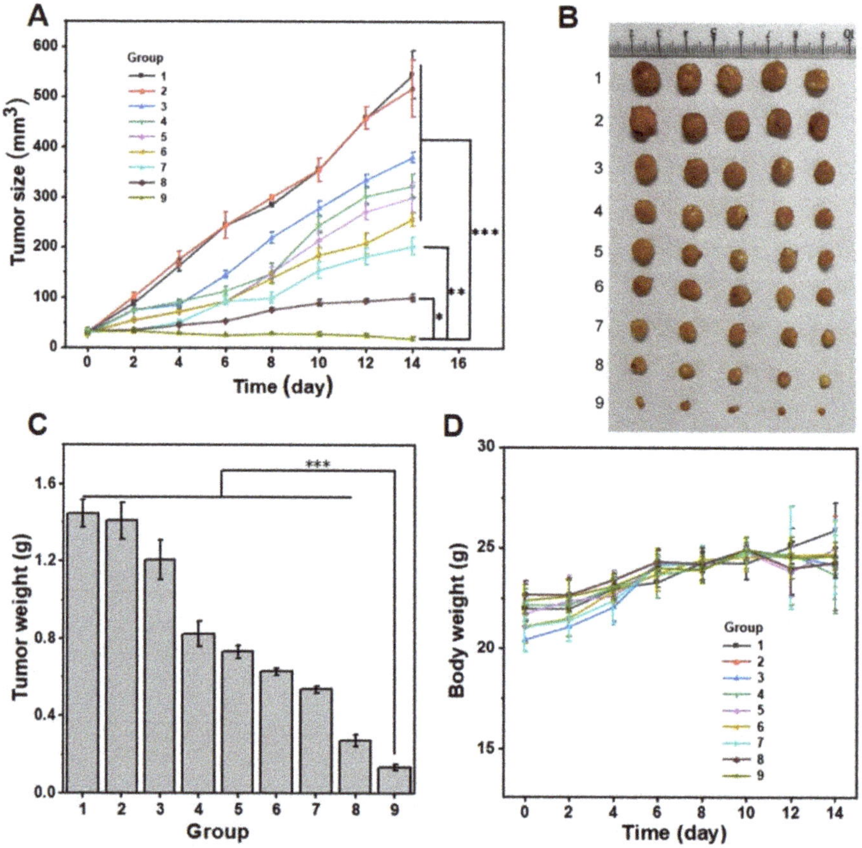

Figure 6. The synergistic anti-tumor effect of 5-FU/siRNA@GalNAc-pDMA in a subcutaneous HCC-bearing mouse model. (**A**) Growth curves of tumors in mice treated with different formulations. Data are shown as mean ± SD (n = 5). (**B**) Photographs of the dissected tumors after 14-day therapy in different groups. (**C**) Weight of the tumors on day 14. Data are shown as mean ± SD (n = 5). (**D**) Average body weight of tumor-bearing mice. Data are shown as mean ± SD (n = 5). * $p < 0.05$; ** $p < 0.01$; *** $p < 0.001$. Group: (1) saline, (2) GalNAc-pDMA, (3) siRNA, (4) 5-FU, (5) 5-FU+siRNA, (6) siRNA@GalNAc-pDMA, (7) 5-Fu@GalNAc-pDMA, (8) 5-FU/siRNA@Glc-pDMA and (9) 5-Fu/siRNA@GalNAc-pDMA. Each group was treated with the same dose of the following components: 5-FU at 5 mg kg^{-1}, GalNAc-pDMA at 30 mg kg^{-1} and siRNA at 0.35 mg kg^{-1}.

3. Materials and Methods

3.1. Materials

All chemicals and reagents were commercially available without the need for further purification, except 2-(Dimethylamino) ethyl methacrylate (DMAEMA, 98%, from Sigma Aldrich, Germany), which was distilled at reduced pressure just prior to use. VEGF siRNA and VEGFR2 siRNA were used as reported in the literature (the sequence and primer were listed in Table S1) and synthesized by Sangon Biotech, Shanghai, China. All of the aqueous solutions used in experiments were prepared using deionized water.

3.2. In Vitro Cytotoxicity Evaluation of GalNAc-pDMA

The HCC cell lines HepG2 and Huh7 were obtained from the Chinese Academy of Sciences Cell Bank. HEK293 cells were obtained from the Conservation Genetics CAS

Kunming Cell Bank. C5WN1 cells were self-established [27]. C5WN1, Huh7, HepG2 and HEK293 cells were maintained in DMEM medium supplemented with 10% FBS, 1% penicillin and 1% streptomycin (growth medium), in 5% CO_2 and 95% air at 37 °C. The MTT assay was carried out to determine in vitro cytotoxicity of GalNAc-pDMA. Cells were seeded at a density of 5×10^3 cells per well of 96-well plates in triplicate the day before GalNAc-pDMA treatment. To observe the cellular uptake of GalNAc-pDMA, the cells were incubated with 200 μL of fresh medium containing different formulations at a final GalNAc-pDMA concentration of 60–600 μg mL^{-1} and cultured for 48 h. The control group was treated with an equal volume of culture medium. Then, the medium was removed, and the cells were washed with PBS (pH 7.4) three times. In total, 100 μL of MTT solution (0.5 mg mL^{-1} PBS) were added to each well and kept in the dark for 4 h until purple formazan crystals formed. Finally, the MTT solution was aspirated gently, and 100 μL of DMSO were added. The absorbance of each well was measured at a wavelength of 470 nm with a microplate reader (Bio Tek, USA). Cell viability was calculated using the following equation:

$$\text{Cell Relative viability (\%)} = (OD_{Treated} - OD_{Blank})/(OD_{Control} - OD_{Blank}) \times 100\%.$$

3.3. Western Blotting Analysis

The VEGF and VEGFR2 expression of samples were tested by Western blotting analysis. C5WN1 cells were subject to different treatments for 48 h incubation. The total protein lysates were obtained after C5WN1 cells were lysed and centrifuged at 12,000× *g* at 4 °C for 15 min. Protein concentrations were measured by BCA protein assay (Biyotime). The same amounts of protein from various treatments were electrophoresed and resolved on 12% SDS-PAGE gels. After protein transferring onto PVDF membranes, the membranes were incubated with specific primary antibodies overnight including a rabbit polyclonal anti-VEGF antibody (Cat. ab46154, Abcam), a rabbit polyclonal anti-VEGFR2 antibody (Cat. ab256666, Abcam) and a rabbit polyclonal anti-β-actin anti-body (Cat. 20536-1-AP, Proteintech). All primary antibodies were incubated in PBS-Tween (PBS-T) and incubated overnight at 4 °C.

3.4. Transwell Assay

C5WN1 cells were incubated in the medium containing different formulations for 12 h. After trypsin digestion, the cells were adjusted to a concentration of 1×10^6 cells/mL and seeded on the upper chamber of Matrigel-coated transwells supplemented with 10% serum DMEM as a chemoattractant. Plates were harvested after 12 h incubation. The transwell membranes were fixed with methanol for 15 min and stained with 1% crystal violet for 10 min. Membranes were cleaned and then mounted on glass slides to record the number of invading cells under a microscope.

3.5. Intracellular Uptake Behavior

Intracellular uptake of 5-FU/siRNA@GalNAc-pDMA in hepatoma cells was observed by confocal laser scanning microscope (CLSM) and flow cytometry. For the CLSM assay, HepG2, Huh7 and C5WN1 cells were tested for their high ASGPR expression on the cell membrane; HEK293 cells are human embryonic kidney cells and were used as negative controls in this experiment due to the low expression of ASGPR. Cells (5×10^4 per dish) in DMEM medium supplemented with 10% FBS were seeded onto 35 mm glass-bottom Petri dishes and allowed to grow at 37 °C with 5% CO_2. After 24 h, the medium was removed, and cells were washed with PBS (pH 7.4) three times. 5-FU/siRNA@GalNAc-pDMA (10 μg mL^{-1}) in 1 mL of fresh medium were added and incubated for 4 h. Then, the fluorescence images of samples were acquired by CLSM (Wetzlar, Germany). The fluorescence of cell nuclei, RhB, and Cy5-siRNA was obtained using laser lines at 360, 560 and 650 nm, respectively.

For the flow cytometry assay, C5WN1, Huh7, HepG2 and HEK293 cells were seeded onto 24-well plates at a density of 1×10^4 cells per well. After 24 h incubation with 5% CO_2 at 37 °C, the cells were incubated in a medium containing siRNA@GalNAc-PDMA (10 or 20 μg mL^{-1}) for another 6 h. The cells were then measured by flow cytometry (Franklin, San Mateo, CA, USA). The experiment was repeated at least three times.

3.6. In Vivo Biodistribution

When the subcutaneous tumor reached 200 mm^3, 10 BALB/c nude mice were randomly divided into two groups and were i.v. injected with GalNAc-pDMAor Glc-pDMA (100 μL, 6 mg mL^{-1}) at 4, 8, 12, 24 and 48 h. The fluorescence signals of RhB in the anesthetized mice (using CO_2 asphyxiation) and organs were imagined by Bruker In Vivo Xtreme II (Woltham, MA, USA).

3.7. In Vivo Antitumor Effect

Tumor-bearing mice were randomly divided into nine groups (five mice per group) treated with: (1) saline, (2) GalNAc-pDMA, (3) siRNA(siVEGF+siVEGFR2), (4) 5-FU, (5) 5-FU+siRNA, (6) siRNA@GalNAc-pDMA, (7) 5-Fu@ GalNAc-pDMA, (8) 5-FU/siRNA@Glc-pDMA and (9) 5-Fu/siRNA@GalNAc-pDMA. In total, 200 μL of each formulation were injected via the tail vein on days 1, 3, 5, 7, 9, 11 and 13. Tumor sizes and mice body weights were recorded every day. On day 14, tumor tissues and major organs (liver, spleen, kidney, heart and lungs) were acquired for H&E staining. The stained specimens were examined by digital optical microscopy (Bruker, USA).

The tumor inhibition rates were defined as per the following formula:

$$TIR = 1 - (\text{average volume of tumors in experimental group/average volume of saline tumors}) \times 100\%.$$

3.8. Statistical Analysis

The experiments were repeated at least three times. The statistical significance of data was determined by one-way analysis of variance (ANOVA) with Origin Lab. Results were shown as mean ± standard deviation (mean ± SD) ($n \geq 3$; for animal tests $n > 6$).

4. Conclusions

In summary, a comprehensive therapeutic drug delivery system was designed and prepared. The platform was verified to be biocompatible and safe in vitro and in vivo. Modification of GalNAc residues enabled nanocarriers to efficiently and specifically target ASGPR expressed hepatocytes to increase the efficiency of therapeutics and reduce systemic toxicity in vivo. The fabricated nanoplatform achieved an elongated hepatic drug concentration by the stable transport of drug and therapeutic nucleic acids and pH-controlled drug release. The formulation of 5-FU and siRNA against the VEGF/VEGFR signaling pathway exhibited potent synergistic in vivo anti-tumor efficacy on metastatic HCC at a low dose of 5-FU. The strategy of the combined chemotherapy and gene therapy against cell migration has great potential to improve metastatic HCC treatment in future clinical applications.

Supplementary Materials: The following supporting information can be downloaded at: https://www.mdpi.com/article/10.3390/molecules27072082/s1. Scheme S1. Schematic illustration of the synthesis route of RhB-pDMAEMA-pAHPMA-GalNAc. Figure S1: ^1H NMR (CDCl3, 400 MHz) of compound 1. Figure S2. ^1H NMR ($CDCl_3$, 400 MHz) of compound 2. Figure S3. ^1H NMR ($CDCl_3$, 400 MHz) of compound 3. Figure S4. ^1H NMR (CDC13, 400 MHz) of compound 4. Figure S5. ^1H NMR (CDC13, 400 MHz) of compound 5. Figure S6. ^1H NMR (CDC13, 400 MHz) of RhB-pDMAEMA-pAHPMA. Figure S7. ^1H NMR (CDC13, 400 MHz) of RhB-pDMAEMA-pAHPMA-GalNAc. Figure S8. FT-IR spectrum of RhB-pDMAEMA-pAHPMA and RhB-pDMAEMA-pAHPMA-GalNAc. Figure S9. (A)TEM image of GalNAc-pDMA. (B) UV-vis absorption spectra of GalNAc-pDMA and 5-FU@GalNAc-pDMA. (C) Drug loading efficiency and encapsulation efficiency of 5-FU@GalNAc-pDMA. Figure S10. (A) Cell viability of HepG2, Huh7, C5WN1 and HEK293 cells incubated with GalNAc-pDMA at various concentrations for 48 h. (B)Hemolysis test results of GalNAc-pDMA, 5-FU@GalNAc-pDMA and 5-FU/siRNA@GalNAc-pDMA. Figure S11. Inhibitory effect of different concentrations of 5-FU, 5-FU@GalNAc-pDMA, and 5-FU/siRNA@GalNAc-pDMA (5-FU at 10–100 μg mL^{-1}, GalNAc-pDMA at 60–600 μg mL^{-1}, and siRNA at 0.7–7 μg mL^{-1}) were tested with C5WN1, Huh7 and HepG2 cells, respectively, in DMEM medium for 48 h. Figure S12. Morphological observation of subcutaneous HCC-bearing mice at the end of the treatment experiment. Mice were treated with (1) saline, (2) GalNAc-pDMA, (3) siRNA, (4) 5-FU, (5) 5-FU+siRNA, (6) siRNA@GalNAc-pDMA, (7) 5-Fu@GalNAc-pDMA, (8) 5-FU/siRNA@Glc-pDMA and (9) 5-Fu/siRNA@GalNAc-pDMA. Figure S13. H&E staining of tumor sections after various treatments with (1) saline, (2) GalNAc-pDMA, (3) siRNA, (4) 5-FU, (5) 5-FU+siRNA, (6) siRNA@GalNAc-pDMA, (7) 5-Fu@GalNAc-pDMA, (8) 5-FU/siRNA@Glc-pDMA and (9) 5-Fu/siRNA@GalNAc-pDMA. Figure S14. H&E staining of major organs in the subcutaneous HCC-bearing mice after the treatment experiment. (1) saline, (2) GalNAc-pDMA, (3) siRNA, (4) 5-FU, (5) 5-FU+siRNA, (6) siRNA@GalNAc-pDMA, (7) 5-Fu@GalNAc-pDMA, (8) 5-FU/siRNA@Glc-pDMA and (9) 5-Fu/siRNA@GalNAc-pDMA. Table S1: Sequences of VEGF-siRNA and VEGFR2-siRNA. Reference [41] are cited in the supplementary materials.

Author Contributions: Conceptualization, J.H.; methodology, X.L., X.W., N.L. and Q.W.; investigation, X.L. and X.W.; writing—original draft preparation, X.L.; writing—review and editing, J.H.; supervision and funding acquisition, J.H. All authors have read and agreed to the published version of the manuscript.

Funding: This research was funded by the National Natural Science Foundation of China, grant number 31700706; Natural Science Foundation of Jiangsu Province, grant number BK20202002, National Key R&D Program of China, grant number 2018YFA0901700.

Institutional Review Board Statement: The animal study protocol was approved by the experimental animal ethics committee of Jiangnan University (Approval Number: JN. No20210415c1350920 on 15 April 2021).

Conflicts of Interest: The authors declare no conflict of interest.

Sample Availability: Samples of the compounds are not available from the authors.

References

1. Llovet, J.M.; Kelley, R.K.; Villanueva, A.; Singal, A.G.; Finn, R.S. Hepatocellular carcinoma. *Nat. Rev. Dis. Primers* **2021**, *6*, 7. [CrossRef] [PubMed]
2. Siegel, R.L.; Miller, K.D.; Fuchs, H.E.; Jemal, A. Cancer Statistics, 2021. *CA Cancer J. Clin.* **2021**, *71*, 7–33. [CrossRef] [PubMed]
3. Llovet, J.M.; Zucman-Rossi, J.; Pikarsky, E.; Sangro, B.; Schwartz, M.; Sherman, M.; Gores, G. Hepatocellular carcinoma. *Nat. Rev. Dis. Primers* **2016**, *2*, 16018. [CrossRef]
4. Johnson, P.; Berhane, S.; Kagebayashi, C.; Satomura, S.; Kumada, T. Impact of disease stage and aetiology on survival in hepatocellular carcinoma: Implications for surveillance. *Br. J. Cancer* **2017**, *116*, 441–447. [CrossRef]
5. Jin Lan, H.; Shun Wang, C.; Qi Shui, O.; Bin, Y.; Shi Hao, Z.; Jing, T.; Jing, C.; Yan Wei, H.; Lei, Z.; Qian, W. The long non-coding RNA PTTG3P promotes cell growth and metastasis via up-regulating PTTG1 and activating PI3K/AKT signaling in hepatocellular carcinoma. *Mol. Cancer* **2018**, *17*, 93.
6. Roderburg, C.; Özdirik, B.; Wree, A.; Demir, M.; Tacke, F. Systemic treatment of hepatocellular carcinoma: From sorafenib to combination therapies. *Hepatic Oncol.* **2020**, *7*, 20. [CrossRef]
7. Wong, P.T.; Choi, S.K. Mechanisms of Drug Release in Nanotherapeutic Delivery Systems. *Chem. Rev.* **2015**, *115*, 3388–3432. [CrossRef]
8. Biswas, S. Polymeric micelles as drug delivery systems in cancer: Challenges and opportunities. *Nanomedicine* **2021**, *16*, 1541–1544. [CrossRef]
9. Simona, M.; Julien, N.; Patrick, C. Stimuli-responsive nanocarriers for drug delivery. *Nat. Mater.* **2013**, *12*, 991–1003.
10. Qiu, Z.; Hu, J.; Li, Z.; Yang, X.; Yin, J. Graphene oxide-based nanocomposite enabled highly efficient targeted synergistic therapy for colorectal cancer. *Colloids Surf. A Physicochem. Eng. Asp.* **2020**, *593*, 124585. [CrossRef]
11. Wang, Y.; Hu, J.; Xiang, D.; Peng, X.; Yin, J. Targeted nanosystem combined with chemo-photothermal therapy for hepatocellular carcinoma treatment. *Colloids Surf. A Physicochem. Eng. Asp.* **2020**, *596*, 124711. [CrossRef]
12. Hu, J.; Wu, W.; Qin, Y.; Liu, C.; Wei, P.; Hu, J.; Seeberger, P.H.; Yin, J. Fabrication of Glyco-Metal-Organic Frameworks for Targeted Interventional Photodynamic/Chemotherapy for Hepatocellular Carcinoma through Percutaneous Transperitoneal Puncture. *Adv. Funct. Mater.* **2020**, *30*, e1910084. [CrossRef]
13. Esrafili, A.; Wagner, A.; Inamdar, S.; Acharya, A.P. Covalent Organic Frameworks for Biomedical Applications. *Adv. Healthc. Mater.* **2021**, *10*, e2002090. [CrossRef] [PubMed]

14. Serviss, J.T.; Johnsson, P.; Grandér, D. An emerging role for long non-coding RNAs in cancer metastasis. *Front. Genet.* **2014**, *5*, 234. [CrossRef] [PubMed]
15. Marx, V. Tracking metastasis and tricking cancer. *Nature* **2013**, *494*, 133–138. [CrossRef] [PubMed]
16. Jian, Y.; Lu, C.; Wei, J.; Liu, W.; Xin, L. Abstract 1526: KPNA4 promotes prostate cancer metastasis through TNFα/β mediated cytokine crosstalk in tumor microenvironment. *Cancer Res.* **2016**, *76*, 1526.
17. Ehrenfeld, P.; Cordova, F.; Duran, W.N.; Sanchez, F.A. S-nitrosylation and its role in breast cancer angiogenesis and metastasis. *Nitric Oxide* **2019**, *87*, 52–59. [CrossRef]
18. Lane, A.N.; Higashi, R.M.; Fan, W.M. Metabolic reprogramming in tumors: Contributions of the tumor microenvironment. *Genes Dis.* **2020**, *7*, 185–198. [CrossRef]
19. Brown, M.; Assen, F.P.; Leithner, A.; Abe, J.; Schachner, H.; Asfour, G.; Bago-Horvath, Z.; Stein, J.V.; Uhrin, P.; Sixt, M. Lymph node blood vessels provide exit routes for metastatic tumor cell dissemination in mice. *Science* **2018**, *359*, 1408–1411. [CrossRef]
20. Strilic, B.; Offermanns, S. Intravascular Survival and Extravasation of Tumor Cells. *Cancer Cell* **2017**, *32*, 282–293. [CrossRef]
21. Bender, R.J.; Mac Gabhann, F. Dysregulation of the vascular endothelial growth factor and semaphorin ligand-receptor families in prostate cancer metastasis. *BMC Syst. Biol.* **2015**, *9*, 55. [CrossRef] [PubMed]
22. Setten, R.L.; Rossi, J.J.; Han, S.P. The current state and future directions of RNAi-based therapeutics. *Nat. Rev. Drug Discov.* **2019**, *18*, 421–446. [CrossRef]
23. Xin, Y.; Huang, M.; Guo, W.W.; Huang, Q.; Zhang, L.Z.; Jiang, G. Nano-based delivery of RNAi in cancer therapy. *Mol. Cancer* **2017**, *16*, 134. [CrossRef]
24. Zhang, M.M.; Bahal, R.; Rasmussen, T.P.; Manautou, J.; Zhong, X.B. The growth of siRNA-based therapeutics: Updated clinical studies. *Biochem. Pharmacol.* **2021**, *189*, 114432. [CrossRef] [PubMed]
25. Singh, A.; Trivedi, P.; Jain, N.K. Advances in siRNA delivery in cancer therapy. *Artif. Cells Nanomed. Biotechnol.* **2017**, *46*, 274–283. [CrossRef]
26. Dong, Y.; Siegwart, D.J.; Anderson, D.G. Strategies, design, and chemistry in siRNA delivery systems. *Adv. Drug. Deliv. Rev.* **2019**, *144*, 133–147. [CrossRef] [PubMed]
27. Wu, W.R.; Shi, X.D.; Zhang, F.P.; Zhu, K.; Zhang, R.; Yu, X.H.; Qin, Y.F.; He, S.P.; Fu, H.W.; Zhang, L.; et al. Activation of the Notch1-c-myc-VCAM1 signalling axis initiates liver progenitor cell-driven hepatocarcinogenesis and pulmonary metastasis. *Oncogene* **2022**, 1–17. [CrossRef]
28. Ye, Z.; Wu, W.R.; Qin, Y.F.; Hu, J.; Liu, C.; Seeberger, P.H.; Yin, J. An Integrated Therapeutic Delivery System for Enhanced Treatment of Hepatocellular Carcinoma. *Adv. Funct. Mater.* **2018**, *28*, 1706600. [CrossRef]
29. Cheng, Q.; Du, L.; Meng, L.; Han, S.; Wei, T.; Wang, X.; Wu, Y.; Song, X.; Zhou, J.; Zheng, S.; et al. The Promising Nanocarrier for Doxorubicin and siRNA Co-delivery by PDMAEMA-based Amphiphilic Nanomicelles. *ACS Appl. Mater. Interfaces* **2016**, *8*, 4347–4356. [CrossRef]
30. Stockert, R.J.; Morell, A.G.; Scheinberg, I.H. Hepatic binding protein: The protective role of its sialic acid residues. *Science* **1977**, *197*, 667–668. [CrossRef]
31. Hu, J.; Liu, J.; Yang, D.; Lu, M.; Yin, J. Physiological roles of asialoglycoprotein receptors (ASGPRs) variants and recent advances in hepatic-targeted delivery of therapeutic molecules via ASGPRs. *Protein Pept. Lett.* **2014**, *21*, 1025–1030. [CrossRef] [PubMed]
32. Thomas, B.; Pifferi, C.; Daskhan, G.C.; Fiore, M.; Renaudet, O. Divergent and convergent synthesis of GalNAc-conjugated dendrimers using dual orthogonal ligations. *Org. Biomol. Chem.* **2015**, *13*, 11529–11538. [CrossRef] [PubMed]
33. Cai, J.; Yue, Y.; Rui, D.; Zhang, Y.; Liu, S.; Wu, C. Effect of Chain Length on Cytotoxicity and Endocytosis of Cationic Polymers. *Macromolecules* **2011**, *44*, 2050–2057. [CrossRef]
34. Cai, X.; Zhong, L.; Su, Y.; Lin, S.; He, X. Novel pH-tunable thermoresponsive polymers displaying lower and upper critical solution temperatures. *Polym. Chem.* **2015**, *6*, 3875. [CrossRef]
35. Gao, C.; Zheng, X. Facile synthesis and self-assembly of multihetero-arm hyperbranched polymer brushes. *Soft Matter* **2009**, *5*, 4788–4796. [CrossRef]
36. Yik, J.H.; Saxena, A.; Weigel, P.H. The minor subunit splice variants, H2b and H2c, of the human asialoglycoprotein receptor are present with the major subunit H1 in different hetero-oligomeric receptor complexes. *J. Biol. Chem.* **2002**, *277*, 23076–23083. [CrossRef]
37. Treichel, U.; Meyer zum Büschenfelde, K.H.; Stockert, R.J.; Poralla, T.; Gerken, G. The asialoglycoprotein receptor mediates hepatic binding and uptake of natural hepatitis B virus particles derived from viraemic carriers. *J. Gen. Virol.* **1994**, *75*, 3021–3029. [CrossRef]
38. Cao, Z.; Liao, L.; Chen, X.; Lan, L.; Hu, H.; Liu, Z.; Chen, L.; Huang, S.; Du, J. Enhancement of Antitumor Activity of Low-Dose 5-Fluorouracil by Combination with Fuzheng-Yiliu Granules in Hepatoma 22 Tumor-Bearing Mice. *Integr. Cancer Ther.* **2013**, *12*, 174–181. [CrossRef]
39. Cao, Z.; Zhang, Z.; Huang, Z.; Wang, R.; Yang, A.; Liao, L.; Du, J. Antitumor and immunomodulatory effects of low-dose 5-FU on hepatoma 22 tumor-bearing mice. *Oncol. Lett.* **2014**, *7*, 1260–1264. [CrossRef]
40. Li, J.; Wang, X.; Hou, J.; Huang, Y.; Zhang, Y.; Xu, W. Enhanced anticancer activity of 5-FU in combination with Bestatin: Evidence in human tumor-derived cell lines and an H22 tumor-bearing mouse. *Drug Discov. Ther.* **2015**, *9*, 45–52. [CrossRef]
41. Karayianni, M.; Pispas, S. *Fluorescence Studies of Polymer Containing Systems*; Procházka, K., Ed.; Springer International Publishing: Cham, Switzerland, 2016; Volume 16.

MDPI
St. Alban-Anlage 66
4052 Basel
Switzerland
www.mdpi.com

Molecules Editorial Office
E-mail: molecules@mdpi.com
www.mdpi.com/journal/molecules

www.ingramcontent.com/pod-product-compliance
Lightning Source LLC
LaVergne TN
LVHW070050120526
838202LV00102B/1977
9783725804320